Nitrogen Fixation

Nitrogen Fixation: Achievements and Objectives

Proceedings of the 8th International Congress on Nitrogen Fixation
Knoxville, Tennessee, U.S.A.
May 20–26, 1990

Edited by

Peter M. Gresshoff, L. Evans Roth,
Gary Stacey, and William E. Newton

Chapman and Hall
New York and London

First published in 1990 by
Chapman and Hall
an imprint of
Routledge, Chapman & Hall, Inc.
29 West 35 Street
New York, NY 10001

Published in Great Britain by

Chapman and Hall
11 New Fetter Lane
London EC4P 4EE

Printed in the United States of America

Library of Congress Cataloging in Publication Data

International Congress on Nitrogen Fixation (8th : 1990 : Knoxville, Tenn.)
Nitrogen fixation : achievements and objectives : proceedings of the 8th International Congress on Nitrogen Fixation, Knoxville, Tennessee. U.S.A., May 20–26, 1990 / edited by Peter M. Gresshoff ...[et. al.].
p. cm.
Includes index.
ISBN 0-412-02591-4
1. Nitrogen—Fixation—Congresses. 2. Nitrogen-fixing microorganisms—Congresses. 3. Nitrogen-fixing plants—Congresses 4. Plant-microbe interrelationships—Molecular aspects—Congresses. I. Gresshoff, Peter M., 1948- . II. Title.
QR89.7.I565 1990 90-43106
589.9′0133—dc20

British Library cataloguing-in-publication data also available

List of sponsors

The University of Tennessee

The Center For Legume Research (UTK)

Tennessee Valley Authority

Oak Ridge Associated Universities

Organisation for Economic Co-operation and Development
(Project for Food, Agriculture and Fisheries)

U. S. Department of Energy (Bioscience Program)

U. S. National Science Foundation (Division of Molecular Biosciences)

U. S. National Institute of General Medical Sciences

U. S. Department of Agriculture
(Animal and Plant Health Inspecvtion Service)

United Nations Development Program

List of donors

Monsanto

LiphaTech

DuPont

Biotechnica International, Inc.

Asgrow Seed Company

International Society of Plant Molecular Biology

Federation of Tokachi Agricultural Cooperatives (Japan)

Enimont America, Inc.

Delta Airlines

Local Organizing Committee

Fred Allen
William Brown
Bob Conger
Nissa Dahlin-Brown
Robert Gibbs
Paul Giordano
John Graveel
Peter Gresshoff
Lea Anne Law
Owen Livingston
Beth Mullin
Dan Roberts
Evans Roth
Gary Stacey

International Program Advisory Committee

Maurice LaLonde (Canada)
Alva App (USA)
Peter Bottomley (USA)
Barry Rolfe (Australia)
Frans de Bruijn (Germany)
Joseph Thomas (India)
Hauke Hennecke (Switzerland)
Michael Dilworth (Australia)
San-Chuin Shen (P. R. China)
Adam Kondorosi (Hungary)
K. Harada (Japan)
Harold Woolhouse (UK)
Barbara Burgess (USA)
Ton Bisseling (The Netherlands)
Herman Bothe (Germany)
Julie Cullimore (UK)
Ben Lugtenberg (The Netherlands)
Claudine Elmerich (France)
Rafael Palacios (Mexico)
Klaus Lampe (Philippines)
Don Phillips (USA)
K. Mulongoy (Nigeria)
Sergey Shestakov (USSR)
Dietrich Werner (Germany)
Sharon Long (USA)
Yvon Dommerques (France)

International Steering Committee

Robert Burris (USA)
Ralph Hardy (USA)
Hauke Hennecke (Switzerland)
William Newton (USA)
Barry Smith (UK)
Cees Veeger (The Netherlands)

TABLE OF CONTENTS

OVERVIEW SECTION

SECTION A: Nitrogenase Biochemistry and Chemical Models

SECTION B: Plant-Microbe Interactions

SECTION C: Limiting Factors in Nitrogen Fixation

SECTION D: Nitrogen Fixation and the Environment

SECTION E: Bacterial Systems

SECTION F: Nitrogen Fixation in Agriculture and Industry

SECTION G: Plant Function

SECTION H: Nitrogen Fixation and Evolution

Preface

We are witnessing an increased awareness of the earth's environment. Examples are easily seen in the rise of 'Green Parties' across Europe, North America, Australasia, and lately Eastern Europe. The public outcry following industrial mishaps in Alaska, Chernobyl, Basel, and Bhopal, as well as the renewed legislative activity, such as the Clean Air Act in the USA and the European Community directive to member nation concerning the control of release of genetically engineered organisms are further examples of the general interest in the biosphere.

The 'Ozone hole', 'Greenhouse gases', and 'Genetically engineered Microorganisms' have gained public profiles, and are discussed widely in newspapers, magazines and the electronic media. A recent educational survey of nations, belonging to the Organisation for Economic Co-operation and Development (OECD) showed that school children are more literate with ecological terms (as listed above) than with 'pure' scientific terms, like 'phloem', 'mitosis', 'proton', or 'Jurasic period'.

Perhaps the increase in awareness is cyclical, being fed by non-scientific, sociological and economic advances. The late 1960s/early 1970s saw a major increase in environmental consciousness. Anti-pollution groups were founded, healthfood shops and naturopathy became acceptable as did recycling, the use of lead-free gasoline, and the reduced usage of environmental toxins, like DDT and PCB. For example, Monsanto Chemical Company instigated a self-imposed halt to the manufacture of PCB in the mid-seventies. Chemical companies started to look at biodegradable herbicides, slow release fertilizers, and specifically targeted pesticides. Slowly the concept of low input and sustainable agriculture developed. Renewable resources became an attractive concept beyond its use as an alliteration.

The 1973/74 oil price adjustment came right in the midst of this development and resulted in an upsurge in biological nitrogen fixation research. Significant momentum was gained by the realization of the large energy inputs required for industrially produced nitrogenous fertilizer. At the start of the last decade of this century, we can look at the results of this international focus on biological nitrogen fixation. What information has been provided so that we can approach the agronomic problems in a meaningful way? What have been our achievements and what are our objectives?

Not surprisingly, the 'young science' of nitrogen-fixation research has produced no major agronomic result. We do not have superior legumes or inocula. We do not have molecules that convert nitrogen gas directly into ammonia under the action of solar light. We do not have a hydrogen-based fuel economy, eliminating the demand for fossil fuel. What we have is a lot of valuable information and a recognition of the problems at hand. It becomes clear from the analysis of other technologies (electric, electronic, transportation, metallurgic, etc), that there is a distinct lag period between the development of the critical technology (e.g. the invention of the transistor or the airplane) and broad application (PCs and Boeing 747s, respectively). It would have been silly to ask the *Escherichia coli* geneticists of the late 1950s, having discovered the operon, Hfr strains and bacteriophages, to put their information to practical use, yet today we use these technologies to clone and express human insulin and pharmaceuticals.

The new decade is faced with a new set of problems, arising before the old ones are solved. Many nations are faced with trade deficits and internal economic problems. East European nations are changing to market-driven economies and democratic societies. Global food demand has exceeded the food supply for the first time in a decade. The time 'bought' by the Green Revolution is wearing thin. Population increases in the developing world continue to outstrip local productivity. The developed world progressively sees value in the quality of life as more people grow older and birth rates decline.

Thus biological nitrogen fixation research has renewed impetus and applicability. The old applications of our research are still to be fulfilled. The world needs better agricultural systems with plants, technologies, and microorganisms that are adapted to modern agricultural needs. The factors that influenced the evolution of the mechanisms underlying nitrogen fixation may not be critical factors in the broad acreage, high yield demand situation experienced today. We need to lower our demand for nitrogen fertilizer and make legume (and non-legume) crops, pastures and trees (!) more attractive to the grower. It seems essential to distinguish between our human role as manipulator versus participant in our biosphere. Difficult problems indeed!

Nitrogen fixation research does provide more than the above focus. The recent discoveries in all areas covered in this book permit one to recognize other benefits of our activity. Our primary product clearly is information, which in turn can help us to understand other biological processes. Only then will we, as a species, know how to interact, control, or be controlled by the biosphere of this planet.

The study of the signal exchange and regulatory circuits between the symbiotic partners and within the symbionts themselves relates to other plant-microbe interactions. The molecular mechanisms of nitrogenase function relate to electron transfer, metallo-enzymes, and cofactor chemistry. Evolutionary concepts dealing with divergence and redundancy of genetic systems show that gene duplication and/or convergent evolution led to functional duplication in all organisms involved in nitrogen fixation (e.g., alternative nitrogenases, multiple *nod*, *ala*, and *nif* genes, multiple nodulation controls, and multigene families involved in nodulin function). Our research into nodulation processes promises to accelerate plant biology in general through the elucidation of developmental cassettes and new signal compounds.

The 8th International Congress on Nitrogen Fixation took this global view as reflected by the stated theme, its program, and the title of this book. The cover design of the graphics on the first and second circulars and the abstract book images the sun, the green biosphere and water. Nitrogen fixation is linked to all of these. Coincidentally, the design may also be interpreted to represent the beautiful East Tennessean countryside experienced during the Congress by the over 600 registrants from 51 different countries

The large proportion of young and enthusiastic scientists, the high calibre of the oral and poster presentations, and the vigorous discussions, research planning, development of future collaborations, and exchange of materials during the Congress showed clearly that nitrogen-fixation research is significant and necessary and that synergism of researchers working at different levels of organizational complexity is fruitful.

In closing, we want to express our deepest thanks to our co-organizers Paul Giordano and Owen Livingston; the local and international committees that helped shaping the program, the UT Conference Department (especially Lea Anne Law, Nissa Dahlin-Brown and Robert Gibbs, and the rest of the 'Nitro' group); our sponsors around the world (see separate listing); our postdoctoral, graduate, and technical staff; Greg Payne; and the general staff of UT for making the Congress so successful. Special thanks must go to Janice Crockett for her untiring assistance in the execution of the planning of the Congress, Lorraine Coffey for excellent help with the abstract book, and Vicky Newton for the effort (a repeat performance (!)) of putting this volume together.

Finally, we want to thank our close families for putting up with us throughout this exciting period of bringing our colleagues and friends to Knoxville.

Knoxville May 1990

Peter M. Gresshoff
Gary Stacey
L. Evans Roth
William E. Newton

OVERVIEW SECTION

Gresshoff, Roth, Stacey & Newton (eds.) Nitrogen Fixation: Achievements and Objectives
Chapman and Hall New York • London 1990

Recent studies on the biochemistry and chemistry of nitrogenases

B.E. Smith

AFRC IPSR Nitrogen Fixation Laboratory, University of Sussex, Brighton, BN1 9RQ, U.K.

Introduction

In this short review I have assumed that the reader knows the background nitrogenase literature and have concentrated on the advances, over the past two years, in Mo nitrogenase biochemistry and related chemistry making suggestions on the chemical significance of the alternative enzymes. I have left the detail on the V and third nitrogenase systems to others (see R.R. Eady, P. Bishop, this volume).

There are two chemical approaches to understanding nitrogenase function. Modelling possible structures for the enzyme active site (see D. Coucouvanis, this volume) can help in interpreting spectroscopic data, however since isolated FeMoco cannot reduce N_2 this structural approach is limited in the mechanistic information that it can give.

The second approach attempts to model the substrate reduction chemistry carried out by the enzyme. Here metal complexes of potential intermediates formed during substrate reduction are isolated and characterised. The intermediate usually has to be stabilised by changing ligands to the metal. The result is a range of complexes, of a variety often bewildering to the biologist, but necessary to understand the possible substrate reduction pathways for comparison with the enzymic reactions.

Both chemical approaches are illustrated below.

The prosthetic groups of the MoFe protein

(a) Isolated FeMoco

FeMoco (55) is difficult to handle and papers on its isolation and purification continue to be published (12,42,68). It can be extracted from MoFe protein bound to DEAE cellulose by first denaturing the protein and then treating it with solvents containing tetraalkyl-ammonium or phosphonium salts, the chlorides being the most effective (42). Yields can be high and the technique is readily scaled up. Furthermore, FeMoco can be manipulated in a range of organic solvents (68). These advances should assist attempts at crystallization. FeMoco preparations are frequently contaminated with excess Fe and their activities of about 200±20 units/ng atom Mo, when combined with MoFe protein polypeptides from a nifB mutant, are usually lower than the optimal 275±25 units/ng atom Mo (18) (1 unit = 1nmole C_2H_4/min) implying that most preparations contain some inactive material. The Fe:Mo ratios have dropped slightly from an earlier consensus of 7±1:1 (36) to 6±1:1 (42,12).

The UV/visible absorption spectrum and the accessibility to o-phenanthroline of the iron of FeMoco varies from batch to batch and on handling and storage (12) indicating a lability in conformation or ligation which is not yet 'predictable, controllable or explicable in terms of structure'. A possible explanation is the identification, from electrochemical and epr measurements, of three distinct redox equilibria of FeMoco in NMF solution (48). The semi-reduced species of these equilibria had g values of 4.6, 3.4 and 2.0, 4.5, 3.6 and 2.0 and 4.9, 3.1 and 1.9 and were associated with differential pulse voltammetric reduction peaks at -0.37V, -0.32V and -0.43V respectively (vs NHE). The first of these FeMoco equilibria was observed in 'alkaline' and the other two in 'acidic' NMF, the ratio of forms present depending on sample history. The 'acid' in such solutions can be formed from dithionite by its spontaneous decomposition in FeMoco solutions. The biological significance of the above is unclear.

The Mössbauer isomer shifts for the oxidized and semi-reduced species of the 'alkaline' form of FeMoco have been reported (49) to differ by less than observed for characterized Fe/S cluster compounds, indicating that in FeMoco the added electron is delocalised less onto the Fe atoms than in the model complexes. Since X-ray absorption spectroscopy (20) indicates that the electron spends little time on the Mo atom it seems probable that there is considerable delocalisation onto the S atoms in FeMoco. The narrow linewidths of the Mössbauer spectra implied equivalent environments for each Fe atom, whereas for protein-bound FeMoco ENDOR spectra indicated at least five distinct environments (64).

(b) Biosynthesis of FeMoco

The nifQ, nifB, nifV, nifE, nifN and nifH gene products are involved in the biosynthesis of FeMoco. The nifV gene product is probably a homocitrate synthase and (R)-homocitrate is incorporated, apparently unchanged, into the FeMoco structure during biosynthesis in a 1:1 ratio with Mo (22; V.K. Shah, this volume). Several homocitrate analogues can replace it during the cofactor biosynthesis resulting in enzymes with altered substrate specificities (25). The 2-hydroxyl group, the 1- and 2-carboxyl groups and the R configuration of the chiral centre are required to generate catalytically competent cofactor.

Evidence has been presented (52) that FeMoco insertion into the MoFe protein polypeptides requires the Fe protein-MgATP complex. Faster activation but with no absolute dependence on the presence of Fe protein and MgATP was observed earlier (19). In the latter experiments the MoFe protein polypeptides contained significant quantities of Mo and the authors concluded that only protein stabilised by containing 1 Mo/mol could be activated by FeMoco.

The nifN and nifE gene products form an oxygen-labile $\alpha_2 \beta_2$ tetramer which has been purified from a nifB mutant strain of A. vinelandii to >95% purity (51). The protein was active in the in vitro FeMoco biosynthesis assay and probably contained an Fe/S cluster(s). The nifW and nifZ gene products are apparently required for the formation or accumulation of active MoFe protein in both K. pneumoniae (50) and A. vinelandii (26).

(c) The FeMoco centres within the MoFe proteins

The sharp g ≈ 6 line in the epr spectrum from FeMoco in Cp1 has been used (14) to quantify incompletely resolved hyperfine splittings observed with ^{95}Mo- and ^{57}Fe-enriched proteins. Simulation of the spectrum from the ^{57}Fe-enriched protein gave hyperfine coupling constants consistent with those observed for the 5 distinct Fe environments detected by ENDOR spectroscopy (64,65). One Fe atom had changed magnetic properties in NifV$^-$ Kp1 (65). However, the epr simulation (14) indicated hyperfine coupling to the ^{95}Mo nucleus of 2.9 MHz, whereas the ENDOR studies (65) suggested 7.5 MHz. The ENDOR studies (65) indicated that in NifV$^-$ Kp1 the Mo site is subject to the replacement, addition or subtraction of a non-sulphur ligand, weakly bound at or near to the Mo atom. In common with earlier ESE results (59) the ENDOR studies (65) detected an interaction (unchanged in NifV$^-$ Kp1) between the FeMoco cluster and a N atom. A number of ^{1}H interactions were also detected, some exchangeable, with NifV$^-$ Kp1 having more exchangeable protons than the wild-type (59). However, no strong interactions with ^{1}H or ^{17}O could be observed by epr spectroscopy when proteins in ^1H_2O and ^2H_2O or $H_2{}^{17}O$ and $H_2{}^{16}O$ were compared (14). It was therefore suggested that the exchangeable protons were not from directly bonded -OH or H_2O.

The FeMoco centres in Av1 and Cp1 have different epr temperature-dependencies and exhibit differing behaviour on reduction of the 6e-oxidized proteins, although their optical and C.D. spectra changed similarly on oxidation. It was suggested that the differences arose from the different environments of FeMoco in the two proteins (46).

The binding of metal clusters to the MoFe protein has been probed by site-directed mutagenesis. There is good evidence that Cys275 in the α-subunit provides a ligand to FeMoco. Mutating this residue to Ala in Kp1 made the FeMoco much more accessible with a broadened epr spectrum (28), and generated to a slow-moving electrophoretic form comparable in mobility to that found in nifB mutant extracts. This latter form, after incubation with FeMoco, became similar in mobility to the major component of active Kp1 (29; H. Kent et al., this volume). The slow-moving form was also detected in four randomly induced nifD mutants (16), viz. Gly186→Asp; Gly195→Gln; Ser443→Pro and Gly455→Asp. The Gly residues are unlikely ligands to FeMoco but could form part of its environment. In addition mutating His195→Asn or Gln191→Lys in the Av1 α-subunit rendered the organism Nif^- (54), changed the g values of the FeMoco epr spectrum and resulted in some C_2H_6, as well as C_2H_4, formation from the reduction of C_2H_2. Therefore it seems probable that (using K. pneumoniae residue numbers) FeMoco is bound by the side chains of Cys275, His196 and Gln192 of the α-subunit. Replacing Gly186 or Gly195 with larger, charged residues would probably perturb FeMoco binding but the involvement of Ser443 and Gly455 remains obscure.

(d) The 'P' clusters

Mössbauer spectroscopy of the 'P' clusters (56,57,69,41) has indicated that they include about 16 Fe atoms of three types: 10, designated D or M5; 4, designated Fe^{2+} or M4; and two designated S. These probably form two [3D + $1Fe^{2+}$] and two [2D + $1Fe^{2+}$ + 1S] unusual 4Fe-4S clusters (41). The Fe/S can be extruded from the protein as 4Fe-4S clusters (30), but its quantitation is still contentious since none of the MoFe protein preparations used for Mössbauer spectroscopy has contained the probable full complement of Fe atoms (36).

An S = 7/2 epr signal from the 'P' clusters which integrated to 1e/FeMoco centre was identified in solid thionine oxidized MoFe proteins (17) and interpreted in terms of either two types of 4Fe-4S cluster or one type of larger (6Fe or 8Fe) cluster. A Mössbauer comparison (35) of solid thionine with thionine solution oxidation showed that about half of the 'P' cluster iron was changed, notably by oxidation of half of the Fe^{2+} species in the former. These data support the view that there are two types of 'P' cluster but the suggestion that the S = 7/2 signal was induced by binding of the protein to the solid thionine is improbable since we have shown (15) that other, soluble oxidants can generate the S = 7/2 signal.

Fe/S clusters are usually bound to polypeptides by cysteine sulphur ligation of the iron atoms. If α-Cys275 binds FeMoco and α-Cys184 can be mutated to Ala (28) or Ser (1) and remain Nif^+ ,it seems likely that the 'P' clusters are bound by the remaining conserved Cys residues. These are α-Cys63, 89, 155 and β-Cys69, 94 and 152. The sequence similarity between the α- and β-subunits in the region of these residues (60) supports the view that they could be involved in binding similar clusters. However there are insufficient Cys residues to bind all the Fe atoms in the 'P' clusters and thus other amino-acid side-chains presumably also act as ligands. Work with model 4Fe-4S cluster compounds (10) has shown that this is possible with a variety of amino-acids. Such alternative ligands could account for some of the unusual properties of 'P' clusters. Additionally, electrostatic effects within the protein matrix have considerable influence on cluster redox potentials (47) and Mössbauer parameters (11). Mutation of any of these putative 'P' cluster Cys ligands to Ala resulted in total loss of diazotrophic growth, C_2H_2-reduction activity, normal FeMoco epr signal (28) and normal assembly of the subunits into tetramers (29). Mutation to Ser produced the same phenotype, except in the case of β-Cys152 where the mutant was Nif^+ (29,6). Furthermore, some substitutions for α-Cys89 are Nif^+ (7) as is the double mutation α-Cys89→Ala; β-Cys94→Ala (29). Thus none of the residues α-Cys89, α-Cys152 and β-Cys94 are absolutely required for 'P' cluster binding and apparently maintaining the symmetry between subunits is important for the assembly of active tetrameric protein.

Fe protein structure and interactions with nucleotides and the MoFe protein

The regulation of Fe protein activity in vivo by ADP-ribosylation of Arg101 has been demonstrated in Rhodospirillum rubrum (40) and Rhodobacter capsulatus (27; G. Roberts, this volume). The purified enzymes involved in the addition (40) and removal (38) of the ADP-ribosyl groups modify the same residue in Kp2, Cp2 and Av2 in vitro. Kp2 from K. pneumoniae strain UN1041 has His in place of Arg101 and has most properties of normal Kp2 including, with the MoFe protein, reductant-independent ATPase activity but, like ADP-ribosylated wild-type Kp2, will not support substrate reduction. Thus Arg101 is essential for effective protein-protein electron transfer (39). Carbodiimide cross-linking of the Av2/Av1 complex resulted in the formation of a 97K molecule consisting of one subunit of Av2 and one β- subunit of Av1 (67). Thus the latter must include at least part of the Av2 docking site. The cross-linked complex was inactive, thus supporting the earlier conclusion (61) that dissociation of the complex between each electron transfer is essential.

Mutations of the five conserved Cys residues in Av2 to Ser demonstrated that only Cys97 and Cys132 are absolutely required for activity (23) providing strong support for the hypothesis that these two Cys residues from each subunit ligand the iron atoms of the 4Fe-4S cluster with the cluster held between the subunits. Since Arg101 is a surface residue important for interactions with the MoFe protein (38) and this residue is very close to Cys97 it is likely that the 4Fe-4S cluster is close to the protein surface. The C97S and C132S mutants also lost Av1 activity which was recovered by incubation with FeMoco. Thus the 4Fe-4S cluster seems essential for FeMoco biosynthesis. Several nifM mutants produce normal MoFe protein so it seems unlikely that NifM is essential for Fe/S insertion.

One-electron oxidation of the 4Fe-4S cluster in the Fe protein by O_2 or H_2O_2 yielded active protein but further oxidation was damaging (62).

The 9 or 10 paramagnetically-shifted resonances in the ^1H NMR spectrum of Cp2 are similar in numbers, positions and temperature-dependencies to those of Bacillus stearothermophilus ferredoxin (43) indicating that at room temperature the 4Fe-4S cluster is in the S = 1/2 spin state. The magnetic susceptibility of the protein was relatively insensitive to MgATP, urea and ethylene glycol, which perturb the S = 1/2 / S = 3/2 spin equilibrium observed by epr which was therefore suggested to be observable only on freezing. MgATP and MgADP perturbed the ^{1}H spectra of Cp2 in different non-uniform ways indicating that only some of the cluster ligands were modified. MgATP induced a symmetry consistent with the 4Fe-4S cluster being held symmetrically between the subunits (23,43). However the kinetics of MgATP-dependent iron chelation from Av2 have been interpreted in terms of the existence of two, non-interconvertible, conformers of the MgATP complex (8).

Two molecules of MgADP bind tightly to each molecule of dye-oxidized and $S_2O_4^{2-}$-reduced MoFe protein. Dissociation of the complexes was slow and at different rates for the two MgADP molecules. MgATP binds to the oxidized protein very weakly (44). MgATP hydrolysis requires both proteins and these data indicate that MgATP bound initially to the Fe protein could be hydrolysed at a site formed at the protein-protein interface. Sequence analysis (53) identified a potential ATPase site on the MoFe protein β-subunit consistent with the Fe protein interaction site detected by cross-linking (67).

A stopped-flow spectrophotometric and calorimetric study of MgATP hydrolysis and Kp2→Kp1 electron transfer has revealed strong parallels with myosin ATPase (63). The hydrolysis reaction was <u>endothermic</u> resulting in the formation of an activated form of the enzyme complex with bound MgADP and P_i . At 6° the hydrolysis preceded electron-transfer, both reactions appearing reversible, but at 23° the reactions were essentially irreversible (see R.N.F. Thorneley, this volume).

Substrate reductions by nitrogenase

(a) <u>The reduction of alternative substrates</u>

The mechanisms of reduction of alternative substrates by nitrogenases can yield insights into the nature of the enzyme.

Nitrite is reduced by nitrogenase to NH_3 (66) but it also damages the Fe protein irreversibly. The process may involve a nitrosyl intermediate, although no free NO was detected. NO also irreversibly damages the Fe protein and apparently acts as a non-competitive inhibitor of N_2 , C_2H_2 and H^+ reduction and ATP hydrolysis. No products from NO reduction were reported (31).

Recent studies with both the wild-type (32) and <u>nifV</u> mutant enzyme (34) have been interpreted in terms of N_2O interacting with the same state of the enzyme as N_2 . However since N_2O can completely suppress H_2 evolution by the enzyme then it cannot displace H_2 on binding. Instead it was suggested (32) that the H atoms are used to form H_2O from the N_2O. However a membrane-leak mass spectrometry study of the time course of H_2 evolution by nitrogenase in the presence of N_2 , N_2O, C_2H_2, azide and cyanide showed a pre-steady-state burst of H_2 for each substrate, with one H_2 molecule being produced for each Mo atom in the MoFe protein (33).

The burst of H_2 evolution was noted earlier (61) for N_2 reduction but has not previously been reported for the other substrates. For cyanide reduction the result is in conflict with the data of Lowe et al. (37) who found a 3 s lag before H_2 evolution was inhibited and methane production began. This lag was equivalent to about 18 turnovers of the enzyme and was interpreted as a cyanide-induced once-only slow step, possibly ligand displacement. To reconcile these pre-steady-state data with earlier steady-state measurements, the authors proposed (37) that two cyanide-bound forms of the enzyme were in prototropic equilibrium ($pK_a = 6$) with only the protonated form being active in cyanide reduction. The discrepancy between the two studies of pre-steady-state H_2 evolution under cyanide may be associated with the methods employed; one group (33) used continuous membrane-leak mass spectrometry whilst the other (37) used the rapid-quenching technique.

Cyanide binds to isolated FeMoco (58) and a study of this interaction and that of CH_3NC with FeMoco, using ^{19}F nmr of $pCF_3C_6H_4S^-$ as a probe, showed that the binding of cyanide modified the FeMoco interaction with the thiol ligand (4). This modification may relate to that proposed (37) for the slow step in the reduction of cyanide by the enzyme.

The products of enzymic cyanide reduction are CH_4 and NH_3 (6 electrons) and CH_3NH_2 (4 electrons). A possible intermediate in the 6 electron reduction has been demonstrated electrochemically as $-Mo{\equiv}C{-}NH_2$ (24). Of further interest is the observed oxidative transformation of $-Mo{-}N{=}CH_2$ to $-Mo{-}C{\equiv}N$. The reverse of this reaction could provide a ready route to the formation of CH_3NH_2.

(b) Reduction of N_2

The most comprehensive description of N_2 reduction by nitrogenase is the Lowe-Thorneley mechanism (61) in which the MoFe protein is reduced by the Fe protein in a series of eight, one-electron steps with concomitant hydrolysis of MgATP. The rate-determining step, with dithionite as reductant, is the dissociation of the oxidized Fe protein (+ MgADP) from the reduced MoFe protein after each electron transfer. The following sequence for the formation of intermediates on sequential addition of electrons and protons has been proposed (61):

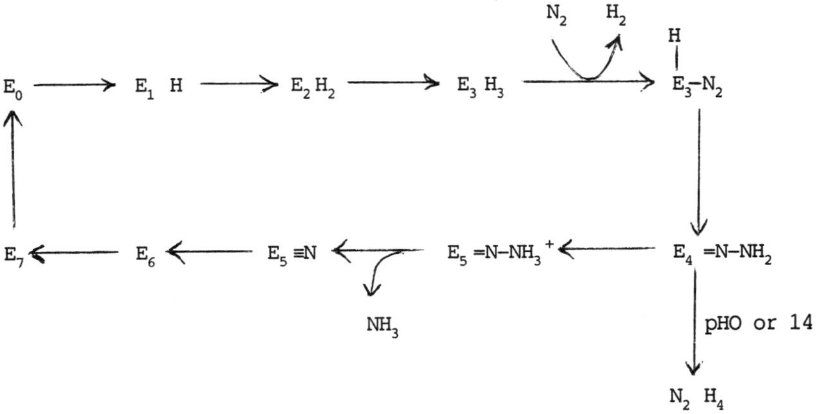

Possible intermediates of the form E=N-NH_3^+ have now been synthesized and structurally characterised (13) and a possible sequence of the reactions between E_5 and E_0 demonstrated. The nitride E_5 ≡N is probably reduced to an imide (E_6 =NH^+)which, under an atmosphere of N_2 ,yields a dinitrogen complex and NH_3 (45). The detailed chemistry of the processes involved is consistent with the formation of metallo-amides (E_7 -NH_2^+)as key intermediates.

The mechanism of the displacement by N_2 of H_2 from E_3H_3 probably proceeds through the formation of a dihydrogen complex from E_2 H_2 (<u>i.e</u>. $M-\overset{H}{\underset{H}{|}}$). The dihydrogen in such complexes is readily displaced by a number of ligands when the third proton is added (5).

Until recently phosphine complexes of Mo or W have been used in N_2 chemistry since they yield readily isolable intermediates. However the chemical environment of Mo in the enzyme includes sulphur and oxygen ligands and, to reproduce some of the model substrate chemistry on more biological sites, the hydride chemistry of Mo and W with bulky thiolate ligands is being developed (2). Significantly these hydride complexes will undergo $^1H/^2H$ exchange with $^2H_2$.

Additional chemical challenges are presented by the V and Fe-only nitrogenases. Fe-N_2 complexes have been known for some time (21,3) but their chemistry is relatively unexplored and very recently a characterised V-N_2 complex in which the N_2 forms a bridge between two V atoms has been reported (9). Do the alternative nitrogenases utilise similar chemistries to the Mo enzyme? Genetics indicates that all three enzymes have cofactors, <u>viz</u>. FeMoco, FeVaco and FeFeco. The structure of FeVaco seems very similar to that of FeMoco (see R. Eady, this volume). The common structural motif would seem to be the Fe and S atoms. What then is the function of the Mo or V atoms? One possibility is that they <u>are</u> directly involved in N_2 reduction but not alone, i.e. that the N_2 forms a bridge between a Mo or V atom and an Fe atom within the cluster. Reduction of such binuclear complexes can yield NH_3 (21,3). The bridges need not be present during the entire reduction process but may help to stabilise critical intermediates. Such rearrangements of end-on to side-on dinitrogen complexes during reduction are known (21).

Acknowledgements

I thank my colleagues at the Nitrogen Fixation Laboratory for many useful discussions and Brenda Hall for typing the manuscript.

References

1. Brigle, K.E., Setterquist, R.A., Dean, D.R., Cantwell, J.S., Weiss, M.C. & Newton, W.E. (1987) Proc. Natl. Acad. Sci. USA **84**, 7066–7069.
2. Burrow, T.E., Lazarowych, N.J., Morris, R.H., Lane, J. & Richards, R.L. (1989) Polyhedron **8**, 1701–1704.
3. Chatt, J., Dilworth, J.R. & Richards, R.L. (1978) Chem. Rev. **78**, 589–625.
4. Conradson, S.D., Burgess, B.K., Vaughn, S.A., Roe, A.L., Hedman, B., Hodgson, K.O. & Holm, R.H. (1989) J. Biol. Chem. **264**, 15967–15974.
5. Dawson, D., Henderson, R.A., Hills, A. & Hughes, D.L. (1989) Polyhedron **8**, 1870–1871.
6. Dean, D.R., Brigle, K.E., May, H.D. & Newton, W.E. (1988) in Nitrogen Fixation: Hundred Years After, eds. Bothe, H., de Bruijn, F.J. & Newton, W.E. (Gustav Fischer, Stuttgart, New York), pp. 107–113.
7. Dean, D.R., Setterquist, R.A., Brigle, K.E., Scott, D.J., Laird, N.F. & Newton, W.E. (1990) Molec. Microbiol., in the press.
8. Deits, T.L. & Howard, J.B. (1989) J. Biol. Chem. **264**, 6619–6628.
9. Edema, J.J.H., Meetsma, A. & Gambarotta, S. (1989) J. Am. Chem. Soc. **111**, 6878–6880.
10. Evans, D.J. & Leigh, G.J. (1988) J. Chem. Soc., Chem. Commun., 395–396.
11. Evans, D.J., Leigh, G.J., Houlton, A. & Silver, J. (1988) Inorg. Chim. Acta **146**, 5.
12. Frank, P., Gheller, S.F., Newton, W.E. & Hodgson, K.O. (1989) Biochem. Biophys. Res. Commun. **163**, 746–754.
13. Galindo, A., Hills, A., Hughes, D.L., Richards, R.L., Hughes, M. & Mason, J. (1990) J. Chem. Soc., Dalton Trans., 283–288.
14. George, G.N., Bare, R.E., Jin, H., Stiefel, E.I. & Prince, R.C. (1989) Biochem. J. **262**, 349–352.
15. Gormal, C.A., Lowe, D.J. & Smith, B.E., in preparation.
16. Govezensky, D. & Zamir, A. (1989) J. Bacteriol. **171**, 5729–5735.
17. Hagen, W.R., Wassink, H., Eady, R.R., Smith, B.E. & Haaker, H. (1987) Eur. J. Biochem. **169**, 457–465.

18. Hawkes, T.R., McLean, P.A. & Smith, B.E. (1984) Biochem. J. **217**, 317-321.
19. Hawkes, T.R. & Smith, B.E. (1984) Biochem. J. **223**, 783-792.
20. Hedman, B., Frank, P., Gheller, S.F., Roe, A.L., Newton, W.E. & Hodgson, K.O. (1988) J. Am. Chem. Soc. **110**, 3798-3805.
21. Henderson, R.A., Leigh, G.J. & Pickett, C.J. (1983) Adv. Inorg. Chem. Radiochem. **27**, 197-292.
22. Hoover, T.R., Imperial, J., Ludden, P.W. & Shah, V.K. (1989) Biochemistry **28**, 2768-2771 and references therein.
23. Howard, J.B., Davis, R., Moldenhauer, B., Cash, V.L. & Dean, D. (1989) J. Biol. Chem. **264**, 11270-11274.
24. Hughes, D.L., Mohammed, M.Y. & Pickett, C.J. (1989) J. Chem. Soc., Chem. Commun., 1399-1400.
25. Imperial, J., Hoover, T.R., Madden, M.S., Ludden, P.W. & Shah, V.K. (1989) Biochemistry **28**, 7796-7799.
26. Jacobson, M.R., Cash, V.L., Weiss, M.C., Laird, N.F., Newton, W.E. & Dean, D.R. (1989) Mol. Gen. Genet. **219**, 49-57.
27. Jouanneau, Y., Roby, C., Meyer, C.M. & Vignais, P.M. (1989) Biochemistry **28**, 6524-6530.
28. Kent, H.M., Ioannidis, I., Gormal, C., Smith, B.E. & Buck, M. (1989) Biochem. J. **264**, 257-264.
29. Kent, H.M., Baines, M., Gormal, C., Smith, B.E. & Buck, M. (1990) Molec. Microbiol., in the press.
30. Kurtz, Jr., D.M., McMillan, R.S., Burgess, B.K., Mortenson, L.E. & Holm, R.H. (1979) Proc. Natl. Acad. Sci. USA **76**, 4986-4989.
31. Liang, J. & Burris, R.H. (1988) Indian J. Biochem. Biophys. **25**, 636-641.
32. Liang, J. & Burris, R.H. (1988) Biochemistry **27**, 6726-6732.
33. Liang, J. & Burris, R.H. (1988) Proc. Natl. Acad. Sci. USA **85**, 9446-9450.
34. Liang, J. & Burris, R.H. (1989) J. Bacteriol. **171**, 3176-3180.
35. Lindahl, P.A., Papaefthymiou, V., Orme-Johnson, W.H. & Münck, E. (1988) Eur. J. Biochem. **263**, 19412-19418.
36. Lowe, D.J., Thorneley, R.N.F. & Smith, B.E. (1985) in Topics in Structural and Molecular Biology: Metalloproteins, Vol. 1, ed. Harrison, P.M. (Macmillan, London), pp. 207-249.
37. Lowe, D.J., Fisher, K., Thorneley, R.N.F., Vaughn, S.A. & Burgess, B.K. (1989) Biochemistry **28**, 8460-8466.
38. Lowery, R.G. & Ludden, P.W. (1988) J. Biol. Chem. **263**, 16714-16719.
39. Lowery, R.G., Chang, C.L., Davis, L.C., McKenna, M-C., Stephens, P.J. & Ludden, P.W. (1989) Biochemistry **28**, 1206-1212.
40. Ludden, P.W., Roberts, G.P., Lowery, R.G., Fitzmaurice, W.P., Saari, L.L., Lehman, L., Lies, D., Woehle, D., Wirt, H., Murrell, S.A., Pope, M.R. & Kanemoto, R.H. (1988) in Nitrogen Fixation: Hundred Years After, eds. Bothe, H., de Bruijn, F.J. & Newton, W.E. (Gustav Fischer, Stuttgart, New York), pp. 157-162.
41. McLean, P.A., Papaefthymiou, V., Orme-Johnson, W.H. & Münck, E. (1987) J. Biol. Chem. **262**, 12900-12903.
42. McLean, P.A., Wink, D.A., Chapman, S.K., Hickman, A.B., McKillop, D.M. & Orme-Johnson, W.H. (1989) Biochemistry **28**, 9402-9406.

43. Meyer, J., Gaillard, J. & Moulis, J-M. (1988) Biochemistry **27**, 6150-6156.
44. Miller, R.W. & Eady, R.R. (1989) Biochem. J. **263**, 725-729.
45. Mohammed, M.Y. & Pickett, C.J. (1988) J. Chem. Soc., Chem. Commun., 1119-1121.
46. Morgan, T.V., Mortenson, L.E., McDonald, J.W. & Watt, G.D. (1988) J. Inorg. Biochem. **33**, 111-120.
47. Moutet, J-C. & Pickett, C.J. (1989) J. Chem. Soc., Chem. Commun., 188-190.
48. Newton, W.E., Gheller, S.F., Feldman, B.J., Dunham, W.R. & Schultz, F.A. (1989) J. Biol. Chem. **264**, 1924-1927.
49. Newton, W.E., Gheller, S.F., Sands, R.H. & Dunham, W.R. (1989) Biochem. Biophys. Res. Commun. **162**, 882-891.
50. Paul, W. & Merrick, M. (1989) Eur. J. Biochem. **178**, 675-682.
51. Paustian, T.D., Shah, V.K. & Roberts, G.P. (1989) Proc. Natl. Acad. Sci. USA **86**, 6082-6086.
52. Robinson, A.C., Chun, T.W., Li, J-G. & Burgess, B.K. (1989) J. Biol. Chem. **264**, 10088-10095.
53. Robson, R.L. (1984) FEBS Lett. **173**, 394-398.
54. Scott, D.J., May, H.D., Newton, W.E., Brigle, K.E. & Dean, D.R. (1990) Nature **343**, 188-190.
55. Shah, V.K. & Brill, W.J. (1977) Proc. Natl. Acad. Sci. USA **74**, 3249-3253.
56. Smith, B.E. & Lang, G. (1974) Biochem. J. **137**, 169-180.
57. Smith, B.E., O'Donnell, M.J., Lang, G. & Spartalian, K. (1980) Biochem. J. **191**, 449-455.
58. Smith, B.E., Bishop, P.E., Dixon, R.A., Eady, R.R., Filler, W.A., Lowe, D.J., Richards, A.J.M., Thomson, A.J., Thorneley, R.N.F. & Postgate, J.R. (1985) in Nitrogen Fixation Research Progress, eds. Evans, H.J., Bottomley, P.J. & Newton, W.E. (Nijhoff, Dordrecht, Boston, Lancaster), pp. 597-603.
59. Thomann, H., Morgan, T.V., Jin, H., Burgmayer, S.J.N., Bare, R.E. & Stiefel, E.I. (1987) J. Am. Chem. Soc. **109**, 7913-7914.
60. Thöny, B., Kaluza, K. & Hennecke, H. (1985) Mol. Gen. Genet. **198**, 441-448.
61. Thorneley, R.N.F. & Lowe, D.J. (1985) in Molybdenum Enzymes, Ch. 5, ed. Spiro, T.G. (Wiley), pp. 221-284.
62. Thorneley, R.N.F. & Ashby, G.A. (1989) Biochem. J. **261**, 181-187.
63. Thorneley, R.N.F., Ashby, G., Howarth, J.V., Millar, N.C. & Gutfreund, H. (1989) Biochem. J. **264**, 657-661.
64. True, A.E., Nelson, M.J., Venters, R.A., Orme-Johnson, W.H. & Hoffman, B.M. (1988) J. Am. Chem. Soc. **110**, 1935-1943.
65. True, A.E., McLean, P., Nelson, M.J., Orme-Johnson, W.H. & Hoffman, B.M. (1990) J. Am. Chem. Soc. **112**, 651-657.
66. Vaughn, S.A. & Burgess, B.K. (1989) Biochemistry **28**, 419-424.
67. Willing, A.H., Georgiadis, M.M., Rees, D.C. & Howard, J.B. (1989) J. Biol. Chem. **264**, 8499-8503.
68. Wink, D.A., McLean, P.A., Hickman, A.B. & Orme-Johnson, W.H. (1989) Biochemistry **28**, 9407-9412.
69. Zimmerman, R., Münck, E., Brill, W.H., Shah, V.K., Henzl, M.T., Rawlings, J. & Orme-Johnson, W.H. (1978) Biochim. Biophys. Acta **537**, 185-207.

Gresshoff, Roth, Stacey & Newton (eds.) Nitrogen Fixation: Achievements and Objectives
Chapman and Hall New York • London 1990

Nodulation genetics: the plant-bacterial interface

Sharon R. Long

Dept. of Biological Sciences, Stanford University, Stanford, CA 94305, USA.

The interaction of *Rhizobium* bacteria with legume nodules proceeds in a series of stages controlled by both bacteria and plant. The success of the interaction is dependent, at each of many stages, on compatibility between the plant and the bacterium: thus the bacterial symbionts are said to display host specificity. The past few years have seen increasing use of bacterial genetics to explore the mechanism and regulation of this symbiosis. As seen elsewhere in this volume, the plant genetics and molecular biology of this process is also being studied increasingly. In early stages of most symbioses, bacteria cause the deformation of growth in plant epidermal root hairs, and stimulate host cortical cells to divide (see 12 for review). These early steps are controlled by nodulation (*nod*) genes in the bacteria (12). The plant expresses uniquely several early nodulins (ENOD sequences) (Franssen, Bisseling, *et al.* and other papers, this volume) at early stages. During subsequent invasion, bacterial genes for a number of possibly interchangeable surface components are important (references 9 and 17 provide examples in one *Rhizobium* for equivalent action by more than one surface polysaccharide). Other bacterial functions are also needed for infection and release. Plants respond by expression of more nodule-specific genes. These are the late nodulins, including leghemoglobin (3, 4; see papers by Sengupta-Gopalan and others, this volume). Nitrogen fixation occurs as bacteria and plant differentiate and establish metabolic cooperation. As discussed extensively in this volume, lowered oxygen is a major controlling factor for differentiation at this stage (see papers by Layzell and by de Bruijn, and others, this volume). Also, the barrier provided by the peribacteroid membrane turns out to be complex in its structure and biogenesis, as discussed in papers by D. Day and by N. Brewin among others, in this volume. Another interesting feature of *Rhizobium* behavior at this stage is the eventual cessation of the bacteria's cell division: it is possible that the bacterial replication and division functions come under control of the host. In our lab, we have recently cloned the critical cell division gene, *ftsZ*, from *Rhizobium meliloti* as a prelude to study of cell division regulation (J. Corbo, W. Margolin, and S. R. Long, manuscript in unpublished data.

For all symbiotic stages, research has been greatly advanced in the past two years by genetic analysis. This is particularly apparent in the study of nodulation genetics. Advances have occurred in several major areas: the identification of new nodulation genes; the use of different genetically modified *Rhizobium* to study mechanism; and the analysis of *nod* gene regulation.

Identification of new *nod* genes: themes of recent progress.
The details of new *nod* gene identification are reported in a number of papers in this volume, and space precludes detailed presentation here. However, it is important to point out several major themes. First, as *nod* genes are identified in each *Rhizobium* it is apparent the phenotype of a mutation in a particular gene may differ on one host plant as compared to another host plant. The *Bradyrhizobium japonicum nodVW* genes provide an example of this, being required for siratro but not for soybean nodulation (M. Göttfert, P. Grob and H. Hennecke, paper in press).

A second observation is that a *nod* gene may be present in more than one copy. Several groups have established this for *nodD*, as discussed later. It is also true for *nodPQ*: there are two very highly conserved copies of these two linked genes (14). We have mapped the second copy of *nodPQ* to the second megaplasmid (pRmSU47b) of *R. meliloti*, distinct from the location of the first *nodPQ* copy in the major *nod-nif* gene cluster on pRmSU47a (J. Schwedock and S.R. Long, unpublished observations). Thus, study of *nod* genes should include tests for gene redundancy.

A third interesting new development relating to the identification of *nod* genes is presented in a paper by Downie in this volume. In *R. leguminosarum* bv. *viciae*, the *nodFE* genes and *nodO* genes appear to be able to complement each other's functions, at least in part. This is intriguing since the sequence homologies and probable functions for *nodFE* and *nodO* are completely different (6; Downie, this volume).

Genetics used for recent studies of *nod* gene mechanism.
A number of phenotypes have been used to characterize *nod* gene mutants or transconjugants. These include microscopic observation of root hair deformation, or of nodule cell divisions, in host or non-host roots inoculated with genetically altered *Rhizobium* cells. Our laboratory is now investigating the potential usefulness of direct electrophysiological measurements in single root hairs during infection (D. W. Ehrhardt and S. R. Long, unpublished observations).

These phenotypic studies have recently been extended by preparation of cell-free supernatants or filtrates, collected from cultures of wild-type or mutant *Rhizobium*cells. Among the studies establishing the combination of genetics and cell-free bioassay were those of Van Brussel *et al.* (16) and Faucher *et al.* (7). It has been found that, according to the genetic constitution of a strain, the supernatant contains a factor or factors that can cause host-specific root hair branching or deformation. This has been pursued by using bioassay to purify single molecules, whose structure can then be defined. Studies reporting such combinations of genetics and chemistry are described by Dénarié and by Philip-Hollingsworth in this volume. The definition of chemical structures such as NodRm-1, a modified tetra-N-acetylglucosamine, should be the starting point for elucidating mechanism of action of individual *nod* genes in a number of *Rhizobium* species.

Regulation of *nod* genes: new complexities.
In the previous Nif congress, several groups described their work showing the flavonoid inducers from plants, acting in some manner that required the bacterial *nodD* gene, induce expression of *nod* genes in *Rhizobium* and *Bradyrhizobium*. Further advances in these systems and in *Azorhizobium* are described in this volume. Several new complexities are now being seen: first, it is documented in several systems that mutating *nodD* or changing the species source of *nodD* can affect the *nod*-induction response to one set of flavonoids compared to another. This is seen in a number of papers (10; 15; 2), and many others that cannot be detailed for reasons of space. Another interesting complication found in recent work is the existence of circuits that control or interact with *nodD* genes in regulation of inducible *nod* promoters. Work in our lab showed that *nodD3* protein acts as a constitutive activator of *nodA*, *nodH* and *nodF* promoters in *R. meliloti* SU47; in turn, *nodD3* appears to be controlled by a regulatory gene, *syrM* (13). We have recently sequenced the *syrM* gene and have found it to be homologous to the *nodD-lysR* family of proteins (1). This suggests the possibility that a hierarchy of regulatory proteins has evolved for symbiotic control, each based on a similar structural plan.

More complex circuitry is found in a different *R. meliloti* strain, Rm41, as reported by Kondorosi and colleagues (reference 11, and in this volume). They observed that Rm41 has a repressor that binds to a sequence between the *nodD* and *nodA* promoters. Because the gene for this repressor is found in Rm41 but not SU47, this research advance illustrates the value of studying diverse systems both for regulation and for mechanism.

It is anticipated that the circuitry for *nod* gene regulation should connect at some level with other regulatory genes. Evidence of this has been published by Dusha *et al.* (5) (see also Kondorosi *et al.*, this volume), who showed that a chromosomal locus, mutated in *R. meliloti* SU47 derivative 399, mediates a response of *nod* gene regulation to combined nitrogen provided as NH_3. Recent work in our lab has revealed a distinct regulatory element, mutated in strain B4: the gene this locus appears to be required for NodD proteins to be active, and possibly for *nodD3* to be expressed (J. Ogawa and S. R. Long, unpublished observations). More regulatory genes should be sought, to enlarge and complete our understanding of how *nod* gene expression is connected to both symbiotic and housekeeping genes, and to general growth and physiology of the bacteria. A third major area in which regulation looks complicated is that of promoter function. The footprint of Nod-D containing extracts (11) and of purified NodD1 and NodD3 proteins (8) have been established for *nod* box promoters in *R. meliloti*. The protected areas are very large, corresponding well to the conserved *nod* box region. This result, predicted from previous genetic studies, calls now for detailed molecular explanation. How does the 35 kDa NodD protein bind to the *nod*-box? What are the essential bases in the *nod* box? In results reported in this volume, Hong and Stacey have sequenced a *B. japonicum nod* box and Goethals *et al.* have shown function in and sequenced an *Azorhizobium caulinodaus nod* box. Because the sequences of these are very divergent from previously studied *nod* boxes, their sequences may permit us to deduce the minimal conserved structure for *nod* box function. It will be interesting to test the binding of these sequences to homologous and heterologous NodD proteins.

The structure of the *nod* box is complex; also, we have found that the high level transcriptional activator *R. meliloti* NodD3 causes hypersensitive cleavage of *nod* boxes by Dnase I (8), we wondered whether there exists bending of the *nod* box DNA, and/or cooperativity between halves of the *nod* box. Our recent results indicate that spacing between the 3' and 5' halves of the *nod* box affects NodD binding, in a manner suggesting cooperation of NodD protein bound at each half of the box (R. F. Fisher and S. R. Long, unpublished observations).

Final comments.
Throughout studies of mechanism and of regulation, several themes are apparent. First, it is very clear that genetic studies continue to be a driving force for identifying phenomena and for controlling of background effects. The study of individual genes and their roles will doubtless be a major feature of work in the next few year.
A second important theme emerging from current research is that *nod* genes may differ from one *Rhizobium* to another, and that even where *nod* genes may be universal, their ease of identification may vary in different systems. This reinforces the lesson that study of diverse systems is crucial for our advancing understanding of *nod* genes and nodulation mechanism.

Finally, it should be a goal for all *nod* genetic studies to find ways in which results can be extended to providing universal new principles for microbiology. The NodD-*nod* box interaction may be one example of this, in its possible applications to understanding the more general behavior of LysR-NodD type proteins. The synthetic activities of enzymes encoded by *nod* genes may have some novel features interest to biochemists studying reactions in diverse bacteria. Thus with genetic analysis as fuel, and with attention both to diversity and to universality, the *nod* gene voyage will progress vigorously towards our next congress!

Acknowledgments.
Work in the author's laboratory is supported by NIH grant GM30962 (gene regulation), DOE contract AT03-82ER12084 (*nod* genes and mechanism), USDA grant 88-37262-3978 (*Rhizobium* cell division), and by an NSF-PYI award.

References.

1. Barnett, M. & Long, S. R. (1990) *J. Bacteriol.* In press.
2. Burn, J., Rossen, L. & Johnston, A. W. B. (1987) *Genes Dev.* **1**, 456-464.
3. Delauney, A. & Verma, D. P. S. (1988) *Plant Mol. Biol. Rep.* **6**, 279-285.
4. Dickstein, R., Bisseling, T., Reinhold, U. N. & Ausubel, F. (1988) *Genes and Devel.* **2**, 677-687. 5. Dusha, I., Bakos, A., Kondorosi, A., De Bruijn, F. J. & Schell, J. (1989) *Mol. Gen. Genet.* **219**, 89-96.
6. Economou, A., Hamilton, W. D. O., Johnston, A. W. B. & Downie, J. A. (1990) *EMBO J.* **9**, 349-354.
7. Faucher, C., Maillet, F., Vasse, J., Rosenberg, C., Van Brussel, A. A. N., Truchet, G. & Denarie, J. (1988) *J. Bacteriol.* **170**, 5489-5499.
8. Fisher, R. & Long, S. R. (1989) *J. Bacteriol.* **171**, 5492-5502.
9. Glazebrook, J. & Walker, G. C. (1989) *Cell* **56**, 661-672.
10. Horvath, B., Bachem, C. W. B., Schell, J. & Kondorosi, A. (1987) *EMBO J.* **6**, 841-848.
11. Kondorosi, E., Gyuris, J., Schmidt, J., John, M., Duda, E., Hoffmann, B., Schell, J. & Kondorosi, A. (1989) *EMBO J.* **8**, 1331-1340.
12. Long, S. R. (1989) *Cell* **56**, 203-214.
13. Mulligan, J. T. & Long, S. R. (1989) *Genetics* **122**, 7-18.
14. Schwedock, J. & Long, S. R. (1989) *Mol. Plant-Microbe Interactions* **2**, 181-194.
15. Spaink, G. P., Wijffelman, C. A., Okker, R. J. H. & Lugtenberg, B. E. J. (1989) *Plant Mol. Biol.* In press.
16. Van Brussel, A. A. N., Zaat, S. A. J., Canter Cremers, H. C. J., Wijffelman, C. A., Pees, E., Tak, T. & Lugtenberg, B. J. J. (1986) *J. Bacteriol.* **165**, 517-522.
17. Williams, M. N. V., Hollingsworth, R. I., Klein, S. & Signer, E. R. (1990) *J. Bacteriol.* **172**, 2622-2632.

Gresshoff, Roth, Stacey & Newton (eds.) Nitrogen Fixation: Achievements and Objectives
Chapman and Hall New York • London 1990

Physiological, metabolic and developmental implications of O_2 regulation in legume nodules

D.B. Layzell, S. Hunt, A.H.M. Moloney, S.M. Fernando & L. Diaz del Castillo.
Department of Biology, Queen's University, Kingston, Ontario, Canada K7L 3N6

INTRODUCTION

The fixation of atmospheric N_2 by the bacterial enzyme nitrogenase is energetically expensive, requiring at least 16 ATP per N_2 fixed. The ATP is derived from oxidative phosphorylation, so a high flux of O_2 is required to the bacteria. However, nitrogenase is irreversibly inhibited by O_2, so the O_2 concentration at the site of the enzyme must be regulated stringently (3). In legume nodules, the infected cells housing the nitrogenase-containing bacteroids occupy a distinct central zone surrounded by inner and outer cortical layers. The O_2 concentration in this central zone is maintained at a low level by the high rate of respiration in support of nitrogenase activity and by a barrier to O_2 diffusion in the inner cortex (27, 49 58). Recent studies have shown that the cortical diffusion barrier has a variable resistance and that it maintains the O_2 concentration in the infected cells at a level which limits nitrogenase activity at all times (20, 28). The barrier also responds to changes in physiological and environmental conditions which affect nodule metabolism and therefore the ability of the nodule to consume the O_2 that diffuses into the central zone from the rhizosphere (27, 58). The purpose of this review is to summarise the evidence for a variable diffusion barrier in legume nodules, to propose a possible mechanism for diffusion barrier regulation and to discuss the implications of O_2 regulation on nodule physiology, biochemistry and development. Although the majority of the work in this field has been done with various soybean X Bradyrhizobium symbioses, considerable evidence indicates that these basic regulatory mechanisms are likely to occur in other legume / Rhizobium symbioses such as those found in clover (58), pea (58, 9), and lupin (9).

[1] **Abbreviations:** EAC; Electron allocation coefficient of nitrogenase. O_e; Rhizosphere pO_2. O_i; concentration of O_2 in the infected cells. TNA; Total nitrogenase activity.

THE COMPONENTS OF O_2 REGULATION.

Under steady state conditions O_2 enters the nodule by diffusion according to Fick's first law:

$$F = \frac{O_e - O_i}{R} \qquad \text{(Eqn. 1)}$$

where O_e and O_i are the rhizosphere and infected cell O_2 concentrations (units of moles / m^3 in the aqueous phase), respectively, and F is the flux of O_2 to the infected cells (units of moles/m^2/s) which, under steady state conditions, is equivalent to the rate of respiratory O_2 consumption. R is the resistance to the diffusion of O_2 through the nodule cortex from O_e to O_i (units of sec/m). At normal atmospheric O_2 pressure (20.9 kPa) and 20°, O_e is ca. 260 mmol/m^3, whereas measured values for O_i are usually between 3 to 30 umol/m^3 (22, 28): i.e. 10^{-5} to 10^{-4} times the rhizosphere O_2 concentration. In nodules, the low O_i (and hence large O_e-O_i) is maintainted by both a high F (in support of nitrogenase activity) and a high R in the nodule cortex (27).

The existence of a cortical diffusion barrier was first noted by Tjepkema and Yocum (49) who inserted an O_2 micro-electrode into a soybean nodule and showed that the O_2 concentration declined sharply in the region of the inner cortex. The O_2 concentration in the central zone was below the level of detection of the O_2 electrode. Similar results have been reported elsewhere (59) and anatomical studies of soybean nodules (3, 5) have shown that the diffusion barrier is in the region of a layer of densely packed cells with few intercellular spaces. Mathematical models of gas diffusion indicate that if most or all of the intercellular spaces of this layer are plugged with water to a depth of ca. 25-50 um (equivalent to ca. 1-5 cell layers), the barrier will provide sufficient resistance to O_2 diffusion to maintain O_i at measured values (18, 45, 47). Varying the depth of the water in the intercellular spaces, or varying the number of spaces that are plugged with water, would allow the barrier to vary its resistance and regulate O_i in response to changing physiological or environmental conditions (18, 44, 45).

Due to the relatively large number and volume (ca. 1-5% of tissue) of intercellular spaces in the central zone (3), there is likely to be little resistance to the diffusion of O_2 from the inside of the inner cortex to each infected and uninfected cell. However, the low O_2 concentration in this region of the nodule results in a shallow gradient of O_2 between the surface of the infected cells and the bacteroids which they contain. Therefore, to provide the flux of O_2 required to maintain high rates of bacteroid respiration, facilitated diffusion of O_2 is required (18, 43). This is achieved by the reversible binding of O_2 to leghemoglobin (Lb), a heme protein, which occurs in high concentrations (ca. 3 mol/m^3 in soybean nodules, 3) in the infected cell cytosol. The oxyleghemoglobin (LbO_2) thus formed diffuses down its own concentration gradient to the surface of the bacteroid where O_2 dissociates from LbO_2 as it equilibrates with the pool of free O_2 being consumed in respiration. The Lb then diffuses back to the cell surface, where the free O_2 concentration is slightly higher, and associates with another O_2 molecule. Lb has a high affinity for O_2: at a free O_2 concentration of ca. 37 umoles/m^3, 50% of the total Lb would be in the form of LbO_2 (3).

THE REGULATION OF O_i AND EVIDENCE FOR DIFFUSION BARRIER CONTROL.

In 1975, Pankhurst and Sprent (36) found that nitrogenase activity in drought-stressed legume nodules was O_2 limited. They suggested that the drought treatment inhibited nodule metabolism and N_2 fixation by causing a collapse in lenticels, thereby increasing the resistance to O_2 diffusion into the nodule. Eight years later, Minchin et al. (33) reported an C_2H_2-induced decline in nitrogenase activity in various legume symbioses. Witty and coworkers (57) showed that O_2 affected the nature and magnitude of this decline and suggested that legume nodules had physiological control over the resistance to O_2 diffusion to the infected cell. In subsequent papers, they and others (1, 5, 6, 9, 11, 13, 16, 18-22, 26-28, 31-34, 38, 41, 44, 45, 47, 52-59) have provided strong evidence that nodules of various legume symbioses are able to regulate diffusive resistance in response to a wide range of environmental and physiological treatments. From these studies, it now seems likely that legume nodules regulate diffusion resistance in response to "sensors" for central zone O_2 concentration, phloem sap supply and N assimilation rate.

Evidence for an O_2 sensor was provided by experiments in which nitrogenase activity and respiration were monitored as O_e was rapidly increased from 20 to 30% (v/v) of the gas phase (19). This treatment caused an inhibition of nitrogenase activity and respiration, followed by a recovery to initial rates within 30 min. Spectroscopic measurements of O_i showed that O_i increased with the increase in O_e but returned to its initial value within 10 min (22). Therefore, since O_2 consumption rate and O_i were similar at 20 and 30% O_2, according to Fick's Law (Eqn 1), the resistance to O_2 diffusion must have increased. Direct measurements of resistance from analyses of the time course of C_2H_4 efflux from control and high O_2-adapted nodules following exposure to 10% C_2H_2 have confirmed that resistance increases in response to increases in O_e (56).

The inhibition of nitrogenase activity and respiration that occurs when phloem sap supply to the nodule is restricted (e.g. by treatments such as stem girdling, nodule excision, leaf removal or continuous dark) was originally thought to be due to carbohydrate deprivation and consequent limitation of reductant and ATP supply in support of nitrogenase activity. However, this inhibition may be largely overcome by increasing O_e, indicating that O_2 rather than carbohydrate availability limits nitrogenase activity under these conditions (16, 52, 54). Nitrogenase inhibition by NO_3^- fertilization may also be overcome by increasing O_e (31, 52). Since NO_3^- is known to reduce phloem sap supply to nodules (53), it is possible that NO_3^- inhibits nodule metabolism in the same way as stem girdling or a continuous dark treatment. The restriction of nitrogenase activity by O_i under adverse conditions may be a protective mechanism resulting in a reduced rate of carbohydrate depletion and an increase in the time that the nodule can maintain respiratory O_2 consumption to maintain a noninhibitory level of O_i (27).

Evidence for a N assimilation sensor is found in treatments which specifically inhibit nitrogen fixation such as exposing nodules to 10% C_2H_2 (33, 57) or to atmospheres lacking N_2 gas (e.g. $Ar:O_2$; 119, 58), or by inhibiting the pathway of NH_4^+ assimilation with, for example, allopurinol, an inhibitor of ureide biosynthesis (1). In all cases total nitrogenase activity declines, but the inhibition can be overcome by increasing O_e, indicating that the legume nodules increase their diffusion resistance and restrict O_2 supply to the infected cells. This mechanism may prevent the bacteroids from becoming parasitic by coupling their O_2 supply to the supply of fixed nitrogen to the plant.

Further evidence for the role of N assimilation in the regulation of nitrogenase activity may be found in a recent study in which Alfalfa nodules were shown to fix N_2 at higher rates when infected with Pseudomonas syringae cv. tabaci (24). This soil microorganism produces tabtoximine-β-lactam which inhibits the glutamine synthetase of the plant cells but not that of the bacteroids. Such differential inhibition causes a change in the glutamate and glutamine pool sizes in the nodules and has an apparent stimulatory effect on nitrogenase activity. Whether or not this stimulation is related to changes in diffusion barrier resistance is, as yet, not clear.

DIFFUSION BARRIER MECHANISM

The mechanism by which the nodule regulates diffusion barrier resistance in response to changes in O_e, phloem sap supply or the production of fixed nitrogen is not known. Neither is it known whether each perturbation is detected by a separate, or by the same, sensing mechanism, nor how the sensor(s) mediate the changes in diffusion barrier resistance. However, if the barrier consists of an aqueous layer, as discussed above, its resistance may be varied by an osmotic mechanism which controls the proportion, or path length, of the O_2 that diffuses into the nodule through the aqueous rather than the gaseous phase (18). The supply, synthesis, or degradation of an osmoticant may be controlled by the nodule sensor(s), causing a movement of water into or out of the intercellular spaces of the diffusion barrier with a resulting change in diffusion resistance.

Since interrupting phloem supply by stem girdling causes a decline in the concentration of nodule sucrose (the largest carbohydrate pool in the nodule), sucrose is an obvious candidate for a regulatory osmoticant. Although one study (14) reported that sucrose concentration increases in drought-streased nodules (which have high resistances), recent studies (21) have shown that the sucrose pool in soybean nodules declines as diffusion resistance increases when nodules are exposed to an $Ar:O_2$ atmosphere or to an increase in O_e. Even though the mesasured decline in sucrose causes only a small (ca. 5%) reduction in the total osmolarity of nodule cells, a mathematical model of nodule water relations (21) predicts that this reduction may be sufficient to move enough water into the intercellular spaces to produce the observed increase in nodule diffusion resistance. The model predicts that the small intercellular spaces characteristic of the nodule inner cortex would be the first to fill with water, creating a localised barrier to diffusion in this region.

A recent study (37) has provided evidence that the conversion of soluble carbohydrate to polysaccharide in plant tissues is inhibited by low O_2 concentration. If this occurs in root nodules, it may represent a mechanism by which the O_2 concentration regulates diffusion barrier resistance, since at low O_2 concentration, sucrose would accumulate and draw water from the intercellular spaces causing a relaxation of diffusion resistance to O_2 entry. The biochemical basis for the O_2 effect on sucrose - starch conversion is not known, but it may be related to O_2 effects on the adenylate energy charge in the plant fraction and the requirement for ATP in the synthesis of starch from sucrose (21). It is also interesting to note that the majority of starch grains in the nodule are found in the inner cortex (29), where their interconversion with sucrose may regulate diffusion barrier resistance.

The effects of treatments which inhibit nitrogen assimilation may also be related to this putative mechanism for diffusion barrier control. Nitrogen assimilation has a net ATP cost (26), and a reduction of the nitrogen assimilation rate by allopurinol, C_2H_2 or $Ar:O_2$ may cause ATP to accumulate in the plant fraction of the nodule, favoring starch formation (21). The consequent reduction of the sucrose pool would result in a higher diffusion barrier resistance localized in the inner cortex. Also, the cessation of ureide production in the inner cortex (35) would further reduce the concentration of osmotically active metabolites in the nodule, perhaps contributing to the increase in diffusion barrier resistance.

NITROGENASE LIMITATION BY O_i

A spectroscopic technique for measuring O_i _in vivo_ has shown that when nitrogenase in intact nodules is inhibited by stem girdling, continuous darkness or NO_3 treatment, nitrogenase activity is correlated with the magnitude of O_i (28). This indicates that O_i and, presumably, ATP supply to nitrogenase, is the basis of each inhibition (28). This conclusion is consistent with the fact that exposure of NO_3^--inhibited or stem-girdled nodules to high O_e (>20%) results in a 3 to 5-fold increase in nitrogenase activity to ca. 70% or more of that in control plants (52). Even in control plants, gradual increases in O_e result in a significant stimulation of nitrogenase activity (20) indicating that nodules are O_2-limited at all times, and that this limitation is exacerbated under adverse environmental or physiological conditions. A similar conclusion has also been made in recent studies with lupins and peas (9). The implications of this limitation, and other potential effects of high nodule resistance to gas diffusion, are discussed in the remainder of this review.

REGULATION OF NODULE METABOLISM BY O_i

Values of O_i in soybean nodules have been estimated by a technique involving the in vivo spectroscopic measurement of leghemoglobin oxygenation (22, 28). In nodules having high nitrogenase activities, O_i values were found to be much higher (17 - 30 nM, or 30-48% Lb oxygenation) than O_i values in nodules with nitrogenase activities inhibited by stem girdling, continuous dark or NO_3^- treatment (4.5 - 7.5 nM, or 11-17% Lb oxygenation)(28). In soybean nodules, the bacteroids contain various terminal oxidases, the one responsible for ATP synthesis in support of nitrogenase activity having the highest affinity for O_2 with a $Km(O_2)$ of about 5nM (3, 4). In contrast, the terminal oxidase in the mitochondria of cowpea nodules has a $Km(O_2)$ of about 100 nM (40). Assuming that soybean mitochondria are similar to those from cowpea, it seems likely that mitochondrial respiration would be severely O_2 limited. However, bacterial respiration may be more or less O_2 limited, depending on the magnitude of the ATP requirements of nitrogenase and the gradient in O_2 concentration across the peribacteroid space between the infected cell cytosol and the bacteroid terminal oxidase. Certainly the low O_2 concentration in the central zone of soybean nodules is a very important regulator of many aspects of C metabolism (e.g. PHB synthesis in bacteroids, leghemoglobin, ureide and organic acid synthesis in the plant fraction; 30, 58), but at this time we do not know whether it is the plant or the bacterial terminal oxidase that is the reason for the O_2 limitation of nodule metabolism and nitrogenase activity in vivo.

The simplest explanation for the physiological data is that bacterial oxidative phosphorylation limits the rate of ATP synthesis in support of nitrogenase activity. The reducing power required for both nitrogenase and oxidative phosphorylation would be provided by carbon skeletons from the plant and, due to O_2 limitation, the supply of carbon skeletons may exceed the metabolic capacity of the bactria. In this case, the low O_2 concentration in the plant fraction would favour bacteroid metabolism by preventing the operation of a reductive Tricarboxylic Acid (TCA) cycle in favour of an oxidative TCA cycle (51, 58). In the oxidative cycle, C would flow from sucrose, through the glycolyis or pentose-phosphate pathway (2), to malate (and/or succinate) via phosphoenolpyrvate carboxylase (PEPC) and malate dehydrogenase (MDH) (23, 42). MDH would reoxidize the NADH formed by glyceraldehyde 3-phosphate dehydrogenase in the glycolytic breakdown of phloem-supplied sucrose and therefore malate (or succinate) would be both an anaerobic (or microaerobic) end product of plant metabolism and a substrate for bacteroid metabolism. In soybeans, full TCA cycle activity may also be inhibited by malonate, a succinate analog and succinate dehydrogenase inhibitor found in high concentrations in the plant fraction (48, 17).

It is also possible that the primary O_2 limitation is not in the bacteria, but in the plant fraction of the nodule and that bacteroid metabolism and nitrogenase activity are dependant upon an O_2-sensitive supply of carbon skeletons from the plant fraction. In this case, bacteroids would be operating aerobically (although at a low O_2 concentration) while ATP synthesis in the mitochondria would be very sensitive to O_i. This would account for the observation that in the central zone, the mitochondria are found clustered around the air spaces that separate the cells. Within the infected and uninfected cells, there is a significant ATP requirement for processes such as growth and maintenance respiration, and ureide or amide biosynthesis (26). In addition, Day and coworkers (8) isolated intact peribacteroid membranes from soybean nodules and showed that they contained specific transporters for malate and succinate and an ATPase which seems to be required for these transporters to operate. If O_i decreases in the infected cells due, for example, to a reduction in phloem sap supply or rhizosphere pO_2, the corresponding reduction in ATP synthesis in the plant fraction may reduce organic acid transport across the peribacteroid membrane and the bacteroids would become more carbohydrate limited. Therefore, an O_2 limitation in the plant fraction may limit C metabolism and nitrogenase activity in the bacteroid fraction. If malate transport is sensitive to O_i the resultant accumulation of malate in the plant cytosol would likely reduce PEPC activity (42).

At present very litle evidence exists to distinguish between the plant and bacteroid fractions as the primary site of O_2 limitation in legume nodules and further studies are needed in this area. One approach to this question has been to expose soybean nodules to step changes in pO_2 and to measure nodule metabolite pool sizes at different times after the treatment (13).

THE DIFFUSION BARRIER AND H_2 INHIBITION OF N_2 FIXATION

The fixation of N_2 by nitrogenase involves the obligate production of H_2 gas, and at least one molecule of H_2 is produced for each molecule of N_2 fixed. The proportion of total electron flow through nitrogenase which is allocated to N_2 fixation rather than H_2 evolution is defined as the electron allocation coefficient (EAC), and since a minimum of 25% of the electrons are used for proton reduction, the EAC has a maximum value of 0.75. In most legumes, measured EAC values vary between 0.5 and 0.70 with environmental and physiological conditions, indicating suboptimal conditions for N_2 fixation (34, 27).

Little is known about how EAC is regulated in legume nodules. *In vitro* studies have shown that increases in EAC are positively correlated with ATP/ADP ratio, the adenylate energy charge (AEC) and the rate of electron flow through nitrogenase (15). However, in *in vivo* studies, these parameters seem to be either not correlated or negatively correlated with EAC (12, 53), suggesting that they are either not involved, or are not the only factor involved in EAC regulation in legume nodules.

Dixon and coworkers (10) suggested that the H_2 gas produced by nitrogenase may be involved in EAC regulation through its role as a competitive inhibitor of N_2 fixation and a stimulator of electron flow to H_2 production. Recently, Rasche and Arp (39) measured the $K_i(H_2)$ in bacteroids from soybean and found it to be equivalent to an atmospheric concentration of approximately 3% H_2. Within the nodule, the H_2 concentration in the infected cells (and therefore the degree to which N_2 fixation is inhibited) would depend on the rate of H_2 production from nitrogenase and the rate of H_2 diffusion out of the nodule. If the cortical diffusion barrier is an aqueous barrier, as current evidence suggests (18, 59), then it should restrict the diffusion of H_2 out of the central zone of the nodule and the diffusion of N_2 into the nodule. If this is true, the diffusion barrier may play a major role in the regulation of EAC. To test this hypothesis, Moloney et al. (34) have constructed a mathematical model to predict EAC from information on the nature of the cortical O_2 diffusion barrier and on the effect that the central zone H_2 and N_2 concentrations should have on the competitive inhibition of N_2 fixationby H_2. The good fit between EAC values predictaed by the model and measurements of EAC at various external pH_2, pO_2 and pN_2 strongly support the suggestion that the resistance of the diffusion barrier plays a important role in the regulation of EAC in soybean nodules.

IMPLICATIONS OF O_2 REGULATION DURING NODULE DEVELOPMENT

In Rhizobium and Bradyrhizobium species, the proteins required for nitrogenase activity are coded for by the nif and fix genes. Recent studies with R. meliloti (7) and B. japonicum (25) have shown that these genes are controlled by the regulatory gene, nifA, the expression of which is either directly (25) or indirectly (7) dependent upon the bacteria being in a low O_2 environment. Therefore, a low O_2 concentration in the infected cells is required both to initiate nitrogenase activity in young, developing nodules and to maintain nitrogenase activity in mature nodules.

On theoretical grounds, it would be very difficult for a small, young nodule to provide the low O_2 concentration needed to initiate nifA gene expression (18, 46). This is because small nodules have a large surface area for O_2 diffusion, relative to the small volume of plant tissue available to consume the O_2 that diffuses into the nodule. In addition, very young nodules would lack the large respiratory demand that is associated with nitrogenase activity in mature nodules, and would thus have difficulty in consuming the O_2 diffusing in from the rhizosphere. Assuming that young nodule do not possess intercellular spaces, or that these spaces are water-filled, and that the central zone respiration is similar to that which occurs to support nitrogenase activity in mature nodules, it has been estimated that a minimum nodule radius of 0.55 mm would be required to provide the low O_2 concentration needed for nifA expression (46).

Since 3-day-old soybean nodules of ca. 0.4 mm radius have active nitrogenase (3), it has been suggested that developing nodules must have high rates of O_2 consumption that are not associated with nitrogenase activity, and which may be coupled to alternative chain respiration (46).

As nodules become active and grow in size, their surface area to volume ratio drops sharply together with the magnitude of the diffusion barrier needed to maintain a low O_i (18). This trend is reflected in an increase in the size and volume of intercellular spaces occupying the central zone of soybean nodules during development (SM Fernando, DB Layzell and Wm Newcomb, unpublished data). In contrast, the cells of the inner cortex remain tightly packed as the nodule ages (3), but even the development of this tissue layer is sensitive to the O_2 requirements of the developing nodule. This has been shown in the recent work of Dakora (5) who studied the morphology and physiology of soybean and cowpea roots that were allowed to form nodules in the presence of various pO_2 (1 to 80%). It was shown that in the nodules grown at the lowest O_2 concentrations, the cells of the inner cortex were larger, more spherical and had a greater volume of intercellular space than the inner cortex of nodules grown at normal (20% O_2) or high pO_2.

Very little is known about the specific plant genes (i.e. nodulin genes) that provide legume nodules with the ability to regulate the supply and concentration of O_2 in the infected cells under such a wide range of environmental and physiological conditions. Recently, van de Wiel et al. (50) have made a significant contribution to this area by showing that the nodulin, ENOD2, is specifically expressed in the inner cortex of pea and soybean nodules from the nodule primoridia stage (6-8d) to mature nodules (20-21d). The amino acid sequence derived from the cloned ENOD2 cDNA led them to suggest that the gene product was a cell wall protein and to propose that it may have a role in contributing to the unique structure of the cells that make up the inner cortex (50).

CONCLUSIONS

The majority of experimental evidence available to date indicates that the supply of O_2 to the infected cells of legume nodules limits their nitrogenase activity under virtually all environmental conditions. It is therefore tempting to speculate that manipulation of the factors controlling O_i may lead to increased rates of N_2 fixation in crop legumes, and thereby increases in yield. The diffusion barrier presents an obvious target for such manipulation since gradual increases in O_e, which are apparently sufficient to partially overcome diffusional restraints to O_2 influx, may increase nitrogenase activity by 25% or more (9, 20). However, a relaxation of diffusion resistance in the early stages of nodule development may prevent the expression of nif genes and the establishment of the entire N_2 fixing system.

The question also arises about the possible benefit to the N_2 fixing symbiosis of maintaining an O_i which limits nitrogenase activity. One possibility is that this provides a buffer against sudden increases in O_i which may occur when nodule respiration drops in response to environmental perturbations such as a decline in soil temperature. It has also been suggested, on a theoretical basis, that limitation of nodule respiration by O_i is an essential component of O_2 regulation, since a change in O_i between the limits normally measured in nodules would require extremely precise control of diffusion barrier resistance in the absence of any change in respiratory O_2 consumption (27). A permanent relaxation of diffusion barrier resistance may not, therefore, be a feasible way to increase N_2 fixation in legumes.

Factors other than the diffusion barrier may also contribute to the regulation of O_i in nodules. Evidence has been provided that isolated bacteroids are able to regulate their metabolism to maintain an O_2 concentration in their aqueous environment suitable for nitrogenase activity (4). The mechanism for this apparently involves the use of components in the pathway of O_2 consumption having different affinities for O_2, and with different efficiencies for coupling respiratory energy to nitrogenase activity. Whether such a mechanism for O_2 protection of nitrogenase occurs in symbiotic bacteroids is not known, and these studies await the development of a sensitive method to measure O_2 consumption in intact nodule tissue (11).

In conclusion, it can be appreciated that a much fuller understanding of the physical and physiological mechanisms involved in interactions between O_2 concentration and nitrogenase activity must be attained before the goals of reducing limitations to N_2 fixation in legumes, and transferring N_2 fixation capability to other crop species, can be realized.

Acknowledgments:

Supported by an NSERC (Canada) Operating Grant to DBL and by partial travel support from Queen's University to DBL, AHM, SMF and LDDC.

LITERATURE CITED

1. Atkins, C.A., Sanford, P.J., Storer,P.J. & Pate, J.S. (1988) Plant Physiol. **88**, 1229-1234.
2. Anthon, G.E. & Emerich, D.W. (1990) Plant Physiol. **92**, 346-351.
3. Bergersen, F.J. (1982) Root nodules of legumes: Structure and functions. (Research Studies Press, Chichester, U.K.).
4. Bergersen, F.J. & Turner, G.L. (1990) Proc. R. Soc. Lond. B **238**, 295-320.
5. Dakora, F.D. (1989) Ph.D. Thesis, University of W. Australia.
6. Davey, A.G. & Simpson, R.J. (1989) J. Exp. Bot. **40**, 149-158.

7. David,M., Daveran,M., Batut,J., Dedieu,A., Domergue,O., Ghai, J., Hertig, C., Boistard, P.& Kahn, D. (1988) Cell **54:** 671-683
8. Day, D.A. Price, G.D. & Udvardi, M.K. (1989) Aust. J. Plant Physiol. **16:** 69-84.
9. Diaz del Castillo, L., Hunt, S. & Layzell, D.B. (1990) Abstract in this volume.
10. Dixon, R.O.D., Blunden, E.A.G. & Searl, J.W. (1981) Plant Sci. Lett. **23**, 109-116.
11. Denison, R.F. & Layzell, D.B. (1990) Abstract in this volume.
12. Edie, S.A. & Phillips, D.A. (1983) Plant Physiol. **72**, 156-160.
13. Fernando, S.M., Hunt, S., Smith, R. Turpin, D.H., Atkins, C.A. & Layzell, D.B. (1990) Abstract in this volume.
14. Fellows, R.J., Patterson, R.P., Raper, C.D., Harris, D. (1987) Plant Physiol **84:** 456-460.
15. Hageman,R.V. & Burris,R.H. (1980) Biochim. Biophys. Acta **541**, 63-75.
16. Hartwig, U., Boller, B.C., Baur-Hoch, B. and Nosberger, J. (1990) Ann. Bot. **65**, 97-105.
17. Hunbeck, C. & Werner, D. (1987) Endocyt. C. Res. **4**, 185-196.
18. Hunt,S., Gaito,S.T. & Layzell,D.B. (1988) Planta **173**, 128-141.
19. Hunt, S., King, B.J., Canvin, D.T. & Layzell, D.B. (1987) Plant Physiol. **84**, 164-172.
20. Hunt, S., King, B.J. & Layzell, D.B. (1989) Plant Physiol. **91**, 315-321.
21. Hunt, S., King, B.J., Denison, R.F., Kouchi, H. Tajima, S. & Layzell, D.B. (1990) Abstract in this volume.
22. King, B.J., Hunt, S., Weagle, G.E., Walsh, K.B., Pottier, R.H., Canvin, D.T. & Layzell, D.B. (1988) Plant Physiol. **87**, 296-299.
23. King, B.J., Layzell, D.B. & Canvin, D.T. (1986) Plant Physiol. **81**, 200-205.
24. Knight,T.J. & Langston-Unkefer,P.J. (1988) Science 241: 951-954
25. Kullik, I., Hennecke, H. & Fischer, H-M. (1989) Arch. Microbiol. **151**, 191-197.
26. Layzell,D.B., Gaito,S.T. & Hunt,S. (1988) Planta **173**, 117-127.
27. Layzell, D.B. & Hunt, S. (1990) Physiol. Plant. (in press)
28. Layzell, D.B., Hunt, S. & Palmer, G.R. (1990) Plant Physiol. **92**, 1101-1107.
29. Lin, J., Walsh, K.B., Canvin, D.T. & Layzell, D.B. (1988) Can. J. Bot. **66**, 526-534.
30. McDermott, T.R., Griffith, S.M., Vance, C.P. & Graham, P.H. (1989) FEMS Microbiol. Reviews **63**, 327-340.
31. Minchin, F.R. Minguez, M.I., Sheehy, J.E., Witty, J.F. & Skot, L. (1986) J. Exp. Bot. **37**, 1103-1113.
32. Minchin, F.R., Sheehy, J.E. & Witty, J.F. (1986) J. Exp. Bot. **37**, 1581-1591.
33. Minchin, F.R., Witty, J.F., Sheehy, J.E. & Muller, M. (1983) J. Exp. Bot. **34**, 641-649.
34. Moloney, A.H., Guy, R., & Layzell, D.B. (1990) Abstract in this volume.
35. Newcomb, E.H. Kaneko, Y. VandenBosch, K.A. (1989) Protoplasma **150**, 150-159.

36. Pankhurst, C.E. & Sprent, J.I. (1975) Protoplasma **85**, 85-98.
37. Priestley, C.A., Treharne, K.J. & Lenz, F. (1988) Ann. Bot. **61**, 159-167.
38. Ralston, E.J., & Imsande, J. (1982) J. Exp. Bot. **33:** 208-214.
39. Rasche, M.E. & Arp, D.J. (1989) Plant Physiol. **91**, 663-668.
40. Rawsthorne, S. & LaRue, T.A. (1986) Plant Physiol. 81, 1092-1096.
41. Schuller, K.A., Minchin, F.R. & Gresshoff, P.M. (1988) J. Exp. Bot. **39**, 865-877.
42. Schuller, K.A., Turpin, D.H. and Plaxton, W.C. (1990) Abstract in this volume.
43. Sheehy, J.E. & Bergersen, F.J. (1986) Ann. Bot. **58**, 121-136.
44. Sheehy, J.E., Bergersen, F.J. Minchin, F.R. & Witty, J. (1987) Ann. Bot. **60**, 345-351.
45. Sheehy, J.E., Minchin, F.R. & Witty, J. (1985) Ann. Bot. **55**, 549-562.
46. Sheehy, J.E. & Thorneley,J.H.M. (1988) Ann. Bot. **61**, 605-609.
47. Sinclair,T.R. & Goudriaan,J. (1981) Plant Physiol. **67**, 143-145.
48. Streeter, J. (1987) Plant Physiol. **85**, 768-773.
49. Tjepkema, J.D. & Yocum, C.S. (1974) Planta **119**, 351-360.
50. van de Wiel,C., Scheres,B., Franssen,H., van Lierop,M-J., van Lammeren,A., van Kammen,A., Bisseling,T. (1990) EMBO Journal **9**, 1-7.
51. Vanleberghe, G.C., Horsey, A.K., Weger, H.G. & Turpin, D.H. (1989) Plant Physiol. 91: 1551-1557
52. Vessey, J.K., Walsh, K.B. & Layzell, D.B. (1988) Physiol. Plant. **73**, 113-121.
53. Vessey, J.K., Walsh, K.B. & Layzell, D.B. (1988) Physiol. Plant. **74**, 137-146.
54. Walsh, K.B., Vessey, J.K., Layzell, D.B. 1987 Plant Physiol. **85**: 137-144.
55. Weisz, P.R., Denison, R.F. & Sinclair, T.R. (1985) Plant Physiol. **78**, 525-530.
56. Weisz, P.R. & Sinclair, T.R. (1988) Plant Physiol. **84**, 906-910.
57. Witty, J.F., Minchin, F.R., Sheehy, J.E. & Minguez, M.I. (1984) Ann. Bot. **53**, 13-20.
58. Witty, J.F., Minchin, F.R., Skot, L. & Sheehy, J.E. (1986) Oxford Surveys of Plant Molecular and Cell Biology **3**, 275-314.
59. Witty, J.F., Skot, L. & Revsbech, N.P. (1987) J. Exp. Bot. **38**, 1129-1140.

Gresshoff, Roth, Stacey & Newton (eds.) Nitrogen Fixation: Achievements and Objectives
Chapman and Hall New York • London 1990

Regulation of nitrogen fixation and assimilation genes in the free-living versus symbiotic state

Frans J. deBruijn*, Uwe Hilgert, John Stigter*, Maria Schneider*, Heiner Meyer z.A., Ulrike Klosse and Katharina Pawlowski

Max-Planck-Institut fur Züchtungsforschung, Carl-von-Linné-Weg 10, D-5000 Köln 30, FRG. *Present address: MSU-DOE Plant Res. Lab & Dept. of Microbiology & Public Health, Michigan State Univ.,E. Lansing, MI 48824, USA.

Biological nitrogen fixation is an extremely energy intensive process. Under ideal conditions the reduction of one molecule of dinitrogen (N_2), catalysed by the nitrogenase enzyme complex, requires 16 molecules of ATP. However, in vivo ATP requirements of up to 42 per N_2 reduced have been measured (80; see 66). It is therefore not surprising that free-living nitrogen-fixing bacteria, such as Klebsiella pneumoniae, only reduce N_2 when nitrogen (N-) starved and tightly regulate their nitrogen-fixation (nif) genes in response to the intracellular concentration of combined nitrogen (N-regulation). In the case of symbiotic nitrogen fixing bacteria such as rhizobia, the situation is different. As docile endosymbionts in nodules, having differentiated into a clearly different physiological state (bacteroids), they are provided by the plant host with (ample) carbon (C-) source(s) for the production of ATP and reducing equivalents, to facilitate nitrogen fixation for the plant's benefit (see 66). Therefore, the nif/fix genes of these bacteria may not need to be subject to N-regulation.

Oxygen (O_2) supports the production of ATP necessary for nitrogenase activity in aerobic nitrogen fixing bacteria. However, in obligate aerobes, as well as facultative or obligate anaerobes, O_2 is capable of severely inhibiting nitrogenase and repressing the synthesis of the Nif/Fix polypeptides (see 43). Therefore, the intracellular O_2 concentration represents a very important second "signal" for free-living and symbiotic nitrogen-fixing bacteria, which have evolved a variety of mechanisms to regulate their nif/fix (and associated) genes accordingly (O_2-regulation; see 43).

In K. pneumoniae, the NH_3 produced by the nitrogen fixation process, is (primarily) assimilated via the bacterial glutamine synthetase (GS) - glutamate synthase (GOGAT) pathway (see 39). In fact, the expression of the GS gene (glnA) is coordinately regulated with the nif genes, in response to the cellular N-status (see 39, 59), to facilitate a tight coupling of nitrogen fixation and assimilation processes. This coupling does not appear to occur in rhizobia. In the free-living state most rhizobia do utilize the GS-GOGAT pathway to assimilate NH_3 for growth (see 62). However, free-living diazotrophy is not generally found in rhizobia and thus a coordinate N-regulation (activation) of nif/fix and gln/glt (GOGAT) genes is not required. In the nodule the bacteroids excrete the NH_3 produced by nitrogen fixation into the plant cell cytoplasm, where it is assimilated by plant glutamine synthetase(s) (see 62; 35). Rhizobial GS and GOGAT activities are generally very low in the nodule (see 62). Therefore, an opposite mode of nif/fix and gln/glt gene regulation appears to exist in the symbiotic state. Thus, switches in regulatory circuitries must occur in the free-living bacterium to bacteroid differentiation process, making it interesting to compare nif/fix and gln/glt gene regulation in these two different states.

Here we will summarize the data available on N- and O_2-regulation of nif/fix and gln/glt genes in the free-living versus symbiotic state. We will focus on the regulatory circuitries found in K. pneumoniae, Rhizobium meliloti and Bradyrhizobium japonicum, since these bacteria have served as model systems. In addition, we will relate these circuitries with those found in Azorhizobium caulinodans (31). A. caulinodans is unique amongst the Rhizobiaceae since it is capable of nitrogen fixation and N_2-dependent growth in the free-living state at an unusually high O_2 concentration and temperature. Moreover, A. caulinodans induces highly efficient nitrogen fixing nodules on both the roots and stems of its host, the tropical legume Sesbania rostrata (see 5). These characteristics have made A. caulinodans very interesting for (molecular) genetic studies on the processes of free-living versus symbiotic nitrogen fixation and assimilation (see 5).

In this review merely data relating to the transcriptional control of nif/fix and gln/glt gene expression will be discussed and selected examples presented. The reader is referred to the following references for comprehensive reviews on the general topic of N- and O_2-control (27, 39, 43, 52, 56, 60; and contributions by Hennecke et al., Buck et al., Klipp et al., Ditta et al. and Elmerich et al., this volume).

N- and O_2-regulation in the free-living diazotroph K. pneumoniae

K. pneumoniae has been a model system to study the structure, function and regulation of nitrogen fixation and assimilation genes (see 60; 27; Buck et al., this volume). The 21 K. pneumoniae nif genes are clustered on the chromosome and are organized in 8 different transcription units (see 59). The regulation of nif gene expression takes place on a nif-specific and a general nitrogen regulation level, in response to the cellular N- and O_2-status. Two of the nif genes (nifA; nifL) are responsible for controlling the expression of all the nif operons, including the nifLA unit itself. The nifA gene product (NifA) activates the nif promoters by binding to a specific DNA sequence (TGT N_{10} ACA; Upstream Activator Sequence, UAS; 12), which is located usually around 100 bp upstream of the transcriptional start site. The UAS can function when placed further upstream or in the opposite orientation (14, 63, 64). NifA is capable of nif gene activation only under N-limiting, anaerobic conditions and is otherwise counteracted by the gene product of the nifL gene (NifL; 2, 11). It has been suggested that NifL acts by destabilizing nif mRNA's in the presence of O_2 (18; Fig.1) and its activity has been shown to be dependent on metal ions (Fe^{2+} or Mn^{2+}; 40). In fact, homology of NifL to an Escherichia coli exonuclease has been found, which could account for its mRNA destabilization (degradation) capacity (32). It is not known how NifL senses the cellular N- and O_2 status, but NifL action appears to be independent of the glnB/glnD nitrogen sensing system (44; see below). In addition to the nifLA genes, a third locus (nifX) has recently been identified, which plays a role in negative regulation of nif gene expression in response to the cellular N- and O_2-status (37).

The expression of the nifLA operon, in addition to being autoregulated, is under control of a general cellular nitrogen regulation (ntr) system, which responds to fluctuating concentrations of combined nitrogen in the cell (see 27, 27, 52, 59, 60). The ntr system of enteric bacteria such as K. pneumoniae consists of (at least) 6 different loci, named ntrA, ntrB, ntrC, glnB, glnD and glnE, whose gene products interact in a complex fashion (see 39, 52, 55, 56). In addition to regulating the nif gene cluster, the ntr system also controls the expression of a number of other genes (operons) involved in N-metabolism, such as glnA (GS), and the catabolism/transport of aminoacids such as histidine (hut), proline (put) and arginine (aut) (see 39, 55). The ntrC gene product (NtrC; NRI) is a DNA binding protein which specifically interacts with a DNA motif (GCAC-N_5-TGGTGC) found in the promoter region of N- (ntr) regulated genes such as nifLA and glnA (see 52, 56). NtrC binding is mediated by a characteristic helix-turn-helix motif in its C-terminal region (21), although its central domain is also required (52). NtrC activity is modulated by phosphorylation, in response to the cellular N-status. Under N-starvation conditions, NtrC is phosphorylated to NtrC-P by the gene product of the ntrB gene (NtrB; NRII), binds to its target site(s) and activates ntr-regulated genes. Under N-excess conditions, NtrC-P is dephosphorylated by NtrB and although it is still capable of binding ntr regulated promoters, it is no longer capable of activating transcription (52, 56; see Fig.1). The ntrB and ntrC genes form one transcription unit (ntrBC), which, in turn, is part of an operon with glnA (glnAntrBC). The glnAntrBC operon is transcribed from three promoters, one of which is NtrC-P activated (glnAp2) and two of which are NtrC repressible (glnAp1; ntrBp; 27, 52, 56; Fig.1). The ntrBC genes belong to a family of regulatory protein pairs involved in prokaryotic signal transduction. The pair usually consists of a sensor (in this case NtrB; NRII) and regulatory (NtrC; NRI) protein, whereby the C-terminal part of sensor modifies (phosphorylates; dephosphorylates) the N-domain of its regulator partner in response to the environmental signal (N-status), which it senses with its N-domain (see 1). The sensor proteins of the superfamily of so-called bicomponent regulatory systems share significant homology in their C-terminal domains, while the regulatory proteins share homology in their N-terminal domains (see 1).

The mechanism by which NtrB senses the actual N-status of the cell is complex and depends on the activity of two additional proteins, the products of the glnB (GlnB; PII) and glnD (GlnD; see 56). When the intercellular N (NH_3) concentration is low, this is reflected in a low glutamine (Gln) to 2-ketoglutarate (2-KG) ratio, which in turn leads to the uridylylation of GlnB to GlnB-UMP by GlnD. GlnB-UMP does not affect the transcription of ntr-regulated genes. At a high NH_3 concentration and correspondingly elevated Gln/2-KG ratio, GlnB-UMP is deuridylylated by GlnD and the GlnB protein now stimulates the dephosphorylation of NtrC-P to NtrC, thereby abolishing its capacity to activate ntr regulated promoters (see 56). However, recent studies using glnB mutations have revealed that the GlnB protein is not absolutely required for NtrC(-P) mediated regulation in response to the cellular N-status (44).

Activation of ntr-regulated promoters by NtrC-P is strictly dependent on the presence of the ntrA gene product (NtrA; RpoN) which constitutes an alternative sigma factor (sigma-54; see 52, 56, 59). Interestingly, NifA-mediated activation of the nif promoters is dependent on the same NtrA protein. The nif- and ntr-regulated promoters share a highly conserved DNA sequence in their -24 to -12 region (GGYAYR-N_4-TTGCA; 39), of which the GG-N_{10}-GC motif is absolutely essential (12). This motif (ntr-box) constitutes the recognition site for the RNA-polymerase-sigma 54 complex and its interaction with the NtrC-P regulator protein results in the transcriptional activation of these promoters by stimulating the isomerization of the closed- to open RNA polymerase-DNA complex (see 56, 63; Buck et al., this volume). The ntrA gene itself does not appear to be N-regulated, since it is expressed independent of the cellular N-status (6) and therefore probably serves to set aside a fraction of the RNA polymerase molecules for transcription of ntr regulated genes (56). Recent results suggest that mutations in genes downstream of the ntrA gene affect expression of ntr controlled promoters, but this effect does not appear to be related to ntrA expression itself (61). The NtrA (sigma-54) product its not only required for the activation of nif and ntr-regulated promoters (via NifA and NtrC-P), but also for the regulation of unrelated metabolic functions in multiple bacterial species. This important fuction of NtrA in global regulatory networks has led to the proposal to call it RpoN and its target sequence simply the -24/-12 consensus promoter (see 52, 83).

Thus, in the free-living diazotroph K. pneumoniae, the nitrogen-fixation (nif) and assimilation (glnA), as well as aminoacid catabolism/transport genes (aut, hut, put), all of which are involved in generating ammonium and glutamine to the cell, are coordinately regulated by a complex ntr system, which is capable of sensing fluctuations in the cellular N-status (Fig.1). The glnAntrBC operon constitutes the center of the ntr system, since the level of its expression determines the concentration (activity) of GS (essential in determining the Gln/2-KG ratio in the cell), as well as of NtrB and NtrC (required to activate the nif genes; see 56).

Fig.1: Regulation of nif genes in K. pneumoniae.

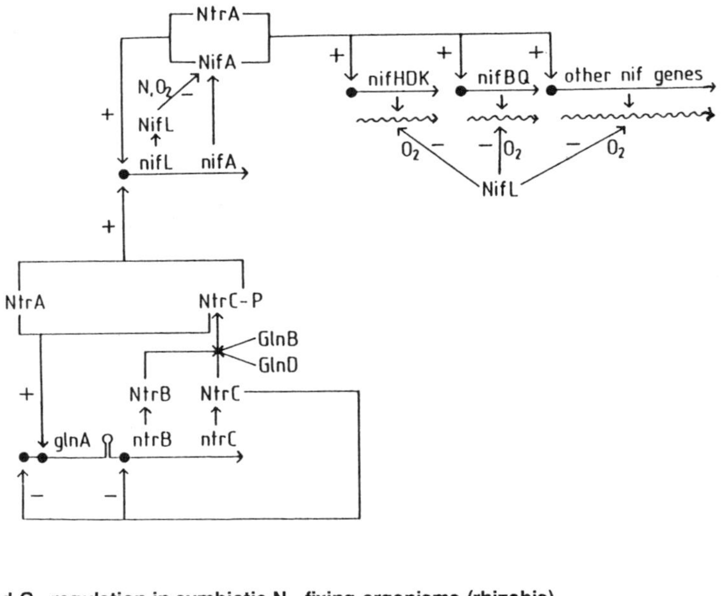

N- and O_2 regulation in symbiotic N_2-fixing organisms (rhizobia).

A number of features of the K. pneumoniae system outlined above are directly applicable to rhizobial systems. The expression of the nif/fix genes is strictly controlled by NifA in concert with NtrA and the corresponding UAS and -24/-12 (ntr) motifs have been found in their promoter regions (see 9, 39, 52, 71, 83). The nifA genes of R. meliloti, R. leguminosarum, B. japonicum and A. caulinodans have been shown to have a high degree of overall homology and to share conserved regions, such as a DNA binding (helix-turn-helix) and a nucleotide binding motif (see 71 and references contained therein). NtrA genes of R. meliloti (73), Rhizobium sp. NGR234 (81), B. japonicum (Fischer et al., this volume) and A. caulinodans (M. Schneider, J. Stigter and F.J. de Bruijn, unpublished) have been identified and ntrA mutants constructed. R. meliloti and Rhizobium sp. NGR234 ntrA mutants are strictly Fix^- (unable to fix nitrogen in symbiosis) and unable to grow on nitrate (Nar^-; 73, 81). A. caulinodans ntrA mutants are not only Fix^- and Nar^-, but also Nif^- in the free-living state (M. Schneider, J. Stigter and F.J. de Bruijn, unpublished). In B. japonicum two ntrA genes have been found and only the double mutant has a Fix^- phenotype (Fischer et al., this volume). Thus NifA and NtrA are essential components of the nif/fix regulatory circuitry in rhizobia.

The presence of ntrC-like genes has been reported for R. meliloti (82), B. japonicum (57) and A. caulinodans (8, 70). R. meliloti and B. japonicum ntrC mutants are not disturbed in symbiotic nitrogen fixation, although in the case of R. meliloti, ntrC appears to be required for transcriptinal activation of nif promoters under N-starvation conditions (Fig.2; 82). However, A. caulinodans $ntrC^-$ mutants are severely affected in free living ($Nif^{-/+}$) as well as symbiotic nitrogen fixation ($Fix^{-/+}$ delayed; 70). Thus, a R. meliloti and B. japonicum N-regulation via ntrC does not play a significant role in symbiotic nitrogen fixation but in A. caulinodans it appears to be important for N_2-fixation in the free-living and symbiotic state.

O_2-regulation of R. meliloti nif/fix gene expression.

R. meliloti nif/fix expression has been shown to be O_2-regulated, at the level of nifA transcription (26, 87). In addition, in vivo activation of the nif/fix genes by NifA has been shown to be oxygen sensitive (4). A cluster of fix genes, mapping approximately 200 kb away from the nifHDK genes on the pSym megaplasmid (3, 72) has been shown to be responsible for regulation of nifA expression (24). The nifA promoter is positively regulated by the fixJ gene product (FixJ), which in turn is modified ($FixJ^*$; phosphorylated?) by the product of the fixL gene, in response to the cellular O_2 status (Fig.2; 24, 41). $FixJ^*$ also activates another fix gene in the cluster (fixK), which shares homology with two regulatory proteins in enteric bacteria (Fnr; Crp). Fnr is involved in regulating genes which facilitate the use of alternative electron acceptors, such as NO_3, NO_2 or fumarate, under anaerobic conditions (38, 78). Crp is the cAMP receptor protein which shares homology with Fnr (75). FixK is involved in negative regulation of nifA and its structural gene (fixK), but positively regulates the nifA-independent fix gene fixN (Fig.2; 3, 41). Fnr-regulated genes in E. coli share a distinct DNA motif which has been postulated to be the Fnr binding site (TTGAT-N_4-ATCAA; 46, 79). A similar motif was found in the R. meliloti fixN, but not the nifA promoter (3), although it has been found in the A. caulinodans nifA promoter (see below; 9, 71). An alternative consensus sequence ("J-box") which maybe involved in NifJ mediated control has been postulated in the 5' upstream region of the nifA, fixK and fixL genes (G. Ditta, personal communication; Ditta et al., this volume). An fnr-like gene, with structural and functional homology to fixK has also been identified in R. leguminosarum, but its role is not entirely clear since mutations in this locus only slightly affect symbiotic N_2-fixation (20). Thus a two tier, cascade-type regulation of nif/fix expression in R. meliloti exists, which does not involve N-control or the NtrC protein but rather

O_2-control via the fixLJ/fixK system (Fig.2; 24). Sequencing studies have shown that the fixLJ products belong to the family of two-component regulatory systems (see 1), whereby FixL is the (O_2) sensor which transduces the signal to the FixJ regulator (24). Both FixL and FixJ appear to be phosphoproteins and have been purified (Ditta et al., this volume). However, no concrete information about the oxygen sensing by FixL exists.

Fig.2: Regulation of nif/fix gene expression in R. meliloti.

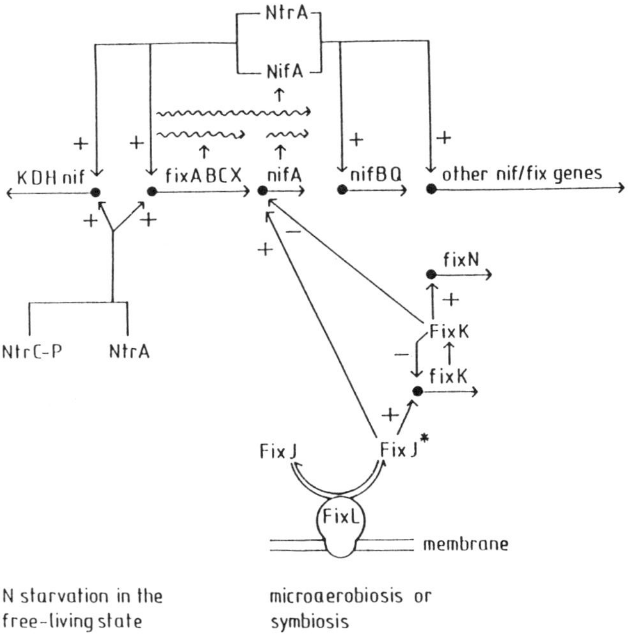

O_2-regulation of B. japonicum nif/fix gene expression.

In B. japonicum nif/fix gene expression has also been shown to be O_2-regulated, but by a slightly different mechanism. The B. japonicum nifA gene is part of a fixRnifA operon, which is transcribed from a single promoter (fixRp; 85). Under aerobic conditions, the fixRnifA operon is transcribed at a basal level from the fixRp, which contains a -24 to -12 (ntr) consensus sequence and therefore ought to be NtrA (sigma -54) dependent (84, 85). An UAS is also present in the fixRp but activation is via an unknown activator protein (not NtrC; not NifA; 85). However, the NifA protein produced under these conditions is inactive, due to direct O_2 inactivation (49). Under microaerobic conditions, the NifA produced is active and capable of stimulating fixRnifA expression 5-fold, which in turn results in enough NifA production to activate the other NifA-dependent nif/fix genes (85). Thus both transcriptional and posttranslational levels of O_2 regulation exist in B. japonicum. An fixLJ operon has recently been identified and a fixL mutation results in a 90% reduction of Fix activity (Anthamatten and Hennecke, this volume). However fixLJ control of fixRnifA expression has not yet been demonstrated. The O_2 inactivation of the NifA protein may involve a cluster of cysteines, shown to be essential for NifA activity (33). These cysteines may form a metal binding domain and a switch from anaerobiosis to aerobiosis could result in a change oxidation state of this metal (or of the cysteines) which may induce a confirmational change in the NifA protein from an active to inactive state (33, 49).

Regulation of A. caulinodans nif/fix gene expression.

It appears as if in A. caulinodans both N- and O_2 regulation of nif/fix gene expression proceeds via the nifA promoter. The expression of a nifA-lacZ fusion is repressed by NH_3 and/or O_2 and is subject to NtrC control. The nifA promoter contains the -24 to -12 consensus motif (GG-N_{10}-GC; probably responsible for NtrC+NtrA control), a NifA UAS motif (GGT-N_{10}-ACA; may be involved in autoregulation by NifA), and a FNR binding consensus sequence (TTGAT-N_4-ATCAA; see above; may be involved in microaerobic induction by an Fnr-like protein; 9, 65, 71). Using oligonucleotide mutagenesis, all three of these motifs have been found to be important for nifA expression (Stigter et al., this volume). The presence of the Fnr-box in the nifA promoter has prompted a search for a corresponding fnr gene. Using a heterologous probe from E. coli, an A. caulinodans fnr-like gene has been cloned and an fnr-like mutant constructed. This mutant strain is affected in symbiotic, but not free-living N_2-fixation and in dissimilatory, but not assimilatory nitrate reductase activity (K. Pawlowski, G. Felix and F.J. de Bruijn, manuscript submitted). NifA expression is only marginally affected in this strain, suggesting that another fnr- or fixK-like gene is responsible for O_2-regulation of the nifA promoter. Using DNA gyrase inhibitors it has been shown that supercoiling of the nifA promoter may also play a role in O_2 regulation (9, 71), as has been found in Rhodopseudomonas capsulata and K. pneumoniae (25, 28, 51).

Ntr-mediated control of nif (nifA) gene expression in A. caulinodans involves another locus (ntrYX), belonging to the family of bicomponent regulatory systems (9; K. Pawlowski, U. Klosse and F.J. de Bruijn, manuscript in preparation). NtrY shares homology with other sensor proteins of bicomponent pairs and with transmembrane proteins, suggersting it is an integral membrane signal transducer. It may be responding to both N- and O_2-signals, and act by modifying (activating; phosphorylating?) NtrX, which shares a high degree of homology with NtrC proteins from various bacteria. NtrYX mutations have varying effects on free-living and symbiotic N_2 fixation (nifA expression), nitrate utilization, amino acid catabolism (K. Pawlowski, U. Klosse and F.J. de Bruijn, manuscript in preparation). Thus, A. caulinodans nif/fix gene expression is N- (ntr) regulated in the free-living and symbiotic state. The regulatory circuitry involved in N- and O_2 regulation is complex, involving in addition to the NifA and NtrA proteins at least two (bicomponent) systems (NtrB?/NtrC; NtrY/NtrX) and shares elements with circuitries in strictly free-living and symbiotic N_2 fixing organisms.

N- and O_2 regulation of nitrogen assimilation genes in R. meliloti and B. japonicum.

As pointed out in the Introduction, members of Rhizobiaceae generally use the GS-GOGAT pathway for NH_3 assimilation (see 62). Rhizobiaceae are unusual in having at least two different GS forms (GSI; GSII; 23; 36). GSI resembles the single GS found in Enterobacteriaceae, such as K. pneumoniae, in its structure (12 subunits), molecular weight (59,000 dalton/subunit), heat stability and reversible modification (activation/inactivation) via adenylylation. GSII consists of 8 subunits of 36,000 dalton, is heat labile and not subject to adenylylation (23). It was postulated that one of the GS forms (GSII) could be playing a special role in plant-bacterial interactions, especially since GSII was found to share homology with the GS of higher plants (15). The latter observation also suggested horizontal gene transfer, but this hypothesis has been questioned recently (74). Moreover, the creation of well defined insertion mutations in the glnA (GSI) and glnII (GSII) genes of R. meliloti and B. japonicum have revealed that neither gene plays a special role in symbiotic N_2 fixation (10, 16, 76, 77). In fact, a R. meliloti glnAglnII double mutant remains Fix^+ on alfalfa, even though it is a glutamine auxotroph (10, 77), while a B. japonicum glnAglnII mutant is Nod^-, preventing the determination of its Fix phenotype (16). A third R. meliloti gln locus (glnT) has been described recently, which is capable of complementing an E. coli glnA mutant strain, but appears to be 'cryptic' in free-living Rhizobium (10). GlnT mutants are also unaffected in symbiotic N_2 fixation. It is not known whether glnT is expressed under special phyiological conditions, or if the glnT gene product(s) correspond to a third form of GS identified in rhizobia (50, 86).

These three gln loci have been shown to be differentially regulated. Both in R. meliloti and B. japonicum glnA expression appears to be independent of N-regulation via the ntr system, while glnII expression is strictly NtrC controlled (10, 57). A -24 to -12 consensus promoter, as well as two NtrC (NRI) binding sites have been found in the B. japonicum glnII promoter (57). The R. meliloti glnT locus is also ntrC dependent (10). Interestingly, under microaerobic conditions and in symbiosis, the B. japonicum glnII gene is not ntrC controlled but rather O_2 regulated via nifA (57). Thus nitrogen assimilation in B. japonicum is both N- and O_2-regulated and subject to separate regulatory networks in the free-living versus symbiotic state (57).

A glnB gene has been identified in B. japonicum (58), as well as in R. leguminosarum (19, 45) and found to be linked to glnA. The B. japonicum glnB gene is regulated by NtrC at tandem promoters, both in the free-living as well as symbiotic state (58). The R. leguminosarum glnB gene expression has also been shown to be ntrC dependent, at least in K. pneumoniae (M. Chiurazzi and M. Iaccarino, personal communication). NtrC mediated regulation of a nitrogen regulation gene (glnB) in the nodule raises some interesting questions about N-control in the symbiotic state. If, as commonly assumed, the intracellular NH_3 concentration in bacteroids is high, NtrC must be active independent of the GlnB/GlnD/NtrB nitrogen sensing system. Assuming it is low, possibly due to the rapid export of NH_3 from the bacteroids, the ntr system is functional as usual and could regulate glnB as well as other nitrogen fixation/assimilation genes (58). In the latter case, it may not be so surprising to observe ntrC mediated nif/fix gene regulation in planta in the case of the A. caulinodans - S. rostrata symbiosis (see above).

Much less is known about the structure and regulation of (brady)rhizobial glt (GOGAT) genes. R. meliloti $GOGAT^-$ mutants are unable to assimilate NH_3, but are Nod^+ Fix^+ on alfalfa (48, 53, 68), just like their glnAglnII counterpart (10), suggesting that the capacity to assimilate NH_3 is not required for the establisment of an effective symbiosis. This, in turn, supports the hypothesis that the host plant supplies the bacteroids with a source of combined nitrogen such as glutamate (47). B. japonicum $GOGAT^-$ mutants, in contrast, are Fix^- on soybean (67).

In enteric bacteria, such as Klebsiella, no simple correlation between GOGAT activity levels and the cellular nitrogen status has been found (see 60), unlike in the case of GS. The slight N-response described for Klebsiella, may be due to an indirect effect of the ntr system on glt expression and has been postulated to occur via ntr control of an unlinked (repressor) gene (nac; 54). In E. coli gltBC expression may be affected by the tightly linked gltF gene product, which plays a (general) role in N-regulation in the cell (17). No evidence for the existence of nac or gltF genes in (brady)rhizobia has been reported.

N-regulation of gln/glt genes in A. caulinodans.

A. caulinodans is unusual again in lacking thge GSII enzyme and its corresponding gene (29). The glnA gene has been cloned (7) and glnA::Tn5 mutants are strict glutamine auxotrophs (Gln^-), Nif^- and Nod^+Fix^- on S. rostrata (9; M. Schneider, U. Hilgert and F.J. de Bruijn, unpublished). The glt locus has also been cloned and glt::Tn5 mutants are Asm^-, Nif^- and Fix^- as well (30, 42). No ntrC regulation of glnA or glt expression has been found (42, 70). The Nif^- phenotype of GS^- or $GOGAT^-$ mutants in culture appears to be due to feedback inhibition of nitrogenase activity or synthesis by NH_3 or glutamine (42). Studies with glnA-lacZ and glt-lacZ fusions have revealed that glnA expression seems to be hardly affected by increasing NH_3 concentrations, whereas glt is induced in the presence of NH_3 or glutamine (U. Hilgert and F.J. de Bruijn, unpublished observations). Thus, although A. caulinodans in the free living state resembles the diazotroph K. pneumoniae in its capacity to grow at the expense of (fixed) N_2, surprisingly no evidence for a coordinate regulation of nif/fix and gln/glt gene expression by the ntr system could be documented thus far. In the symbiotic state, A. caulinodans GS and GOGAT activities are clearly essential for the establishment of a Fix^+ nodule. This, however, may be due to the inherent disadvantage of free-living Asm^- bacteria during the infection process, especially on the S. rostrata stem, which may constitute an extremely N-poor environment.

Conclusions.

It is clear from the studies summarized above that major differences exist in the N- and O_2 regulation pathways of nitrogen fixing bacteria in the free-living versus symbiotic state. While in K. pneumoniae a coordinate regulation of nitrogen fixation (nif/fix) and assimilation (gln/glt) by the ntr system exists, in (brady)rhizobia the regulatory circuitries are clearly distinct in culture versus in planta. The A. caulinodans N- and O_2 regulatory pathways have elements in common both with K. pneumoniae and (brady)rhizobia, but this bacterium has in addition novel components lacking in other system. Much remains to be learned, but it is hoped that the comparative analysis of these different N_2 fixing systems will ultimately yield a clearer insight into the processes of N- and O_2 control, plant-bacterium signalling and the bacterium-bacteroid transition.

Acknowledgements.

We would like to thank all the collegues who sent (p)reprint materials at a short notice and apologize for not being able to include all the information provided in this review. We also thank Jeff Schell for support and encouragement and Frau Gisela Kobert for help with preparing the manuscript.

References.

1. Albright, L.M., Huala, E. & Ausubel, F.M. (1989) Ann. Rev. Genet. 23, 311-336.
2. Austin, S., Henderson, N. & Dixon, R. (1990) Eur. J. Biochem. 187, 353-360.
3. Batut, J., Daveran-Mingot, M.-L., David, M., Jacobs, J., Garnerone, A.M. & Kahn, D. (1989) EMBO J. 8, 1279-1286.
4. Beynon, J.L., Williams, M.K. & Cannon, F.C. (1988) EMBO J. 7, 7-14.
5. de Bruijn, F.J. (1989) in Plant-Microbe Interactions, eds. Kosuge, T. & Nester, E.W. (Macmillan, New York), pp. 457-493.
6. de Bruijn, F.J. & Ausubel, F.M. (1983) Mol. Gen. Genet. 192, 342-353.
7. de Bruijn, F.J., S. Rossbach & Schell, J. (1985) in Nitrogen Fixation Research Progress, eds. Evans, H.J., Bottomley, P.J. & Newton, W.E. (Martinus Nijhoff, The Hague), p. 218.
8. de Bruijn, F.J., Pawlowski, K., Ratet, P., Hilgert, U. & Schell, J. (1987) in Molecular Genetics of Plant-Microbe Interactions, eds. Verma, D.P.S. & Brisson, N. (Martinus Nijhoff, The Hague), pp. 266-271.
9. de Bruijn, F.J., Pawlowski, K., Ratet, P., Hilgert, U., Wong, C.H., Meyer z.A., H. & Schell, J. (1988) in Nitrogen Fixation: Hundred Years After, eds. Bothe, H. & de Bruijn, F.J. (Fischer, Stuttgart), pp. 351-355.
10. de Bruijn, F.J., Rossbach, S., Schneider, M., Ratet, P., Messmer, S., Szeto, W.W., Ausubel, F.M. & Schell, J. (1989) J. Bacteriol. 171, 1673-1682.
11. Buchanan-Wollaston, V., Cannon, M.C. & Cannon, F.C. (1981) Mol. Gen. Genet. 184, 102-106.
12. Buck, M. (1986) J. Bacteriol. 166, 545-551.
13. Buck, M., Miller, S., Drummond, M. & Dixon R. (1986) Nature 320, 374-378.
14. Buck, M., Woodcock, J., Cannon, W., Michenall, L. & Drummond, M. (1987) Mol. Gen. Genet. 210, 140-144.
15. Carlson, T.A. & Chelm, B.K. (1986) Nature 322, 568-570.
16. Carlson, T.A., Martin, G.B. & Chelm, B.K. (1987) J. Bacteriol. 169, 5861-5886.
17. Castano, I., Bastarrachea, F. & Covarrubias, A. (1988) J. Bacteriol. 170, 821-827.
18. Collins, J.J., Roberts, G.P. & Brill, W.J. (1986) J. Bacteriol. 168, 173-178.
19. Colonna-Romano, S., Riccio, A., Guida, M., Defez, R., Lamberti, A., Iaccarino, M., Arnold, W., Priefer, U. & Pühler, A. (1987) Nucl. Acids Res. 15, 1951-1964.
20. Colonna-Romano, S., Arnold, W., Schlüter, A., Boistard, P., Pühler, A. & Priefer, U. (1990) Mol. Gen. Genet., in press.
21. Contreras, A. & Drummond, M. (1988) Nucl. Acids Res. 16, 4025-4039.
22. Darrow, R.A. & Knotts, R.R. (1977) Biochem. Biophys. Res. Commun. 78, 554-559.
23. Darrow, R.A., Christ, D., Evans. W.R., Jones, B.L., Keister, D.L. & Knotts, R.R. (1981) in Current Perspectives in Nitrogen Fixation, eds. Gibson, A.H. & Newton, W.E. (Australian Academy of Sciences, Canberra), pp. 182-185.
24. David, M., Daveran, M.-L., Batut, J., Dedieu, A., Domergue, O., Ghai, J., Hertig, C., Boistard, P. & Kahn, D. (1988) Cell 54, 671-683.
25. Dimri, G.P. & Das, K.H. (1988) Mol. Gen. Genet. 212, 360-363.
26. Ditta, G., Stanfield, S., Corbin, D. & Helinski, D.R. (1980) Proc. Natl. Acad. Sci. USA 77, 7347-7351.
27. Dixon, R.A. (1988) in The Nitrogen and Sulphur Cycles, eds. Cole, J.A. & Ferguson, S.J. (Cambridge University Press, Cambridge).
28. Dixon, R.A., Henderson, N.C. & Austin, S. (1988) Nucl. Acids Res. 16, 9933-9946.
29. Donald, R.G.K. & Ludwig, R.A. (1984) J. Bacteriol. 158, 1144-1151.

30. Donald, R.G.K., Lapointe, J. & Ludwig, R.A. (1988) J. Bacteriol. 170, 1197-1204.
31. Dreyfus, B., Garcia, J.L. & Gillis, M. (1988) Int. J. System. Bacteriol. 38, 89-98.
32. Drummond, M., Whitty, P. & Wootton, J. (1986) EMBO J. 5, 441-447.
33. Fischer, H.-M., Bruderer, T. & Hennecke, H. (1988) Nucl. Acids Res. 16, 2207-2224.
34. Fischer, H.-M., Fritsche, S., Herzog, B. & Hennecke, H. (1989) FEBS Lett. 255, 167-171.
35. Forde, B.G. & Cullimore, J.V. (1989) Oxford Surveys on Plant Molecular and Cell Biology 6, 247-296.
36. Fuchs, R.L. & Keister, D.L. (1980) J. Bacteriol. 144, 641-648.
37. Gosink, M.M., Franklin, N.M. & Roberts, G.P. (1990) J. Bacteriol. 172, 1441-1447.
38. Griffiths, L. & Cole, J.A. (1987) Arch. Microbiol. 147, 364-369.
39. Gussin, G.N., Ronson, C.W. & Ausubel, F.M. (1986) Ann. Rev. Genet. 20, 567-591.
40. Henderson, N., Austin, S. & Dixon, R. (1989) Mol. Gen. Genet. 216, 484-491.
41. Hertig, C., Li, R.Y., Louarn, A.-M., Garnerone, A.-M., David, M., Batut, J., Kahn, D. & Boistard, P. (1989) J. Bacteriol. 171, 1736-1738.
42. Hilgert, U., Schell, J. & de Bruijn, F.J. (1987) Mol. Gen. Genet. 210, 195-202.
43. Hill, S. (1988) FEMS Microbiol. Rev. 54, 111-130.
44. Holtel, A. & Merrick, M. (1989) Mol. Gen. Genet. 217, 474-480.
45. Holtel, A., Colonna-Romano, S., Guida, M., Riccio, A., Merrick, M.J. & Iaccarino, M. (1989) FEMS Microbiol. Lett. 58, 203-208.
46. Jayaraman, P.S., Peakman, T.C., Busby, S.J.W., Quincey, R.V. & Cole, J.A. (1987) J. Mol. Biol. 196, 781-788.
47. Kahn, M.L., Kraus, J. & Somerville, J.E. (1985) in Nitrogen Fixation Research Progress, eds. Evans, H.J., Bottomley, P.J. & Newton, W.E. (Martinus Nijhoff, Boston), pp. 193-195.
48. Kondorosi, A., Svab, Z., Kiss, G.B. & Dixon, R.A. (1977) Mol. Gen. Genet. 151, 221-226.
49. Kullik, I., Hennecke, H. and Fischer, H.-M. (1989) Arch. Microbiol. 151, 191-197.
50. Kumar, P.S. & Rao, S.L.N. (1986) Curr. Microbiol. 14, 221-226.
51. Kranz, R.G. & Haselkorn, R. (1986) Proc. Natl. Acad. Sci. USA 83, 6805-6809.
52. Kustu, S., Santero, E., Keener, J., Popham, D. & Weiss, D. (1988) Microbiol. Rev. 53, 367-376.
53. Lewis, T.A., Gonzalez, R. & Botsford, J.L. (1990) J. Bacteriol. 172, 2413-2420.
54. Macaluso, A. & Bender, R.A. (1986) in Abstracts of the American Society for Microbiology, (American Society for Microbiology, Washington), p. 154.
55. Magasanik, B. (1982) Ann. Rev. Genet. 16, 135-168.
56. Magasanik, B. (1988) Trends Biochem. Sci. 13, 475-479.
57. Martin, G.B., Chapman, K.A. & Chelm, B.K. (1988) J. Bacteriol. 170, 5452-5459.
58. Martin, G.B., Thomashow, M.F. & Chelm, B.K. (1989) J. Bacteriol. 171, 5638-5645.
59. Merrick, M. (1988) in Nitrogen Fixation: Hundred Years After, eds. Bothe, H. & de Bruijn, F.J. (Fischer, Stuttgart), pp. 293-302.
60. Merrick, M. (1988) in The Nitrogen and Sulphur Cycles, eds. Cole, J.A. & Ferguson, S.J. (Cambridge University Press, Cambridge).
61. Merrick, M. & Coppard, J.R. (1989) Mol. Microbiol. 3, 1765-1775.

62. Miflin, B.J. & Cullimore, J.V. (1984) in Genes Involved in Microbe-Plant Interaction, eds. Verma, D.P.S. & Hohn, R. (Springer, Wien), pp. 129-178.
63. Morett, E. & Buck, M. (1988) Proc. Natl. Acad. Sci. USA 85, 9401-9405.
64. Morett, E., Cannon, W. & Buck, M. (1988) Nucl Acids Res. 16, 11469-11488.
65. Nees, D.W., Stein, P.A. & Ludwig, R.A. (1988) Nucl. Acids Res. 16, 9839-9853.
66. O'Brian, M.R. & Maier, R.J. (1989) Biochim. Biophys. Acta 974, 229-246.
67. O'Gara, F., Sundaram, M. & Meade, J. (1984) FEMS Microbiol. Lett. 24, 241-245.
68. Osburne, M.S. & Signer, E.R. (1980) J. Bacteriol. 143, 1234-1240.
69. Pawlowski, K. (1989) PhD thesis (Cologne University, Cologne).
70. Pawlowski, K., Ratet, P., Schell, J. & de Bruijn, F.J. (1987) Mol. Gen. Genet. 206, 207-219.
71. Ratet, P., Pawlowski, K., Schell, J. & de Bruijn, F.J. (1989) Mol. Microbiol. 3, 825-838.
72. Renalier, M.H., Batut, J., Ghai, J., Terzaghi, B., Gherardi, M., David, M., Garnerone, A.-M., Vasse, J., Truchet, G., Huguet, T. & Boistard, P. (1987) J. Bacteriol. 169, 2231-2238.
73. Ronson, C.W., Nixon, B.T., Albright, L.M. & Ausubel, F.M. (1987) J. Bacteriol. 169, 2424-2431.
74. Shatters, R.G. & Kahn, M.L. (1989) J. Mol. Evol. 29, 422-428.
75. Shaw, D.J. and Guest, J.R. (1982) Nucl. Acids Res. 10, 6119-6130.
76. Somerville, J.E. & Kahn, M.L. (1983) J. Bacteriol. 156, 168-176.
77. Somerville, J.E., Shatters, R.G. & Kahn, M.L. (1989) J. Bacteriol. 171; 5079-5086.
78. Spiro, S. & Guest, J.R. (1987a) Mol. Microbiol. 1, 53-58.
79. Spiro, S. & Guest, J.R. (1987b) J. Gen. Microbiol. 133, 3279-3288.
80. Stam, H., van Verseveld, H.W. & Stouthamer, A.H. (1986) FEMS Microbiol. Lett. 35, 215-220.
81. Stanley, J., van Slooten, J., Dowling, D.N., Finan, T. & Broughton, W.J. (1989) Mol. Gen. Genet. 217, 528-532.
82. Szeto, W.W., Nixon, B.T., Ronson, C.W. & Ausubel, F.M. (1987) J. Bacteriol. 169, 1423-1432.
83. Thöny, B. & Hennecke, H. (1989) FEMS Microbiol. Rev. 63, 341-358.
84. Thöny, B., Fischer, H.-M., Anthamatten, D., Bruderer, T. & Hennecke, H. (1987) Nucl. Acids Res. 15, 8479-8499.
85. Thöny, B., Anthamatten, D. & Hennecke, H. (1989) J. Bacteriol. 171, 4162-4169.
86. Tsuprun, V.L., Zograf, O.N., Orlova, E.V., Kiselev, N.A., Pushkin, A.V., Shiffelova, G.E., Scolovieva, N.A., Evstigneeva, Z.G. & Kretovich, W.L. (1987) Biochim. Biophys. Acta 913, 368-376.
87. Virts, E.L., Stanfield, S.W.,, Helinski, D.R. & Ditta, GB.S. (1988) Proc. Natl. Acad. Sci. USA 85, 3062-3065.

Gresshoff, Roth, Stacey & Newton (eds.) Nitrogen Fixation: Achievements and Objectives
Chapman and Hall New York • London 1990

Evolution, structure and function of nitrogen-fixing root nodules: Confessions of ignorance

Janet I.Sprent

Department of Biological Sciences, University of Dundee, Dundee, DD1 4HN, Scotland

INTRODUCTION

In being provocative as requested, I have selected a few aspects from recent developments in nodule science and extended the current state-of-play into some speculations which I hope may be used as testable hypotheses – to be upheld, demolished or modified by the 9th Conference in this series.

I shall begin with some evolutionary considerations, continue with some comments on structure and functioning of well-loved symbioses such as those of soybean, then speculate as to how lesser-known legume symbioses may operate, taking a leaf out of the book on the *Frankia* – nodulated plants. Finally I shall make some comments on the problems of extending nodulation to new plants, both legume and non-legume.

EVOLUTION OF NITROGEN FIXATION

Three major bursts of microbial evolution have been suggested[17], (a) about 4.0 Ga BP when organisms first evolved, in an essentially anaerobic environment; (b) 1.8 to 2.0 Ga BP when aerobes proliferated following development of oxygenic photosynthesis; (c) 0.5 Ga BP when macroscopic plants and animals provided new niches. The latter opened up possibilities for symbiotic N_2 fixation and may have included 3 periods when peaks of atmospheric O_2 (hence, arguably, N-demand) coincided with peaks of species diversity[32] (Fig. 1).

Estimates of when the angiosperms first evolved vary from fossil evidence (about 150 Ma BP) to molecular clock data (200 and 300 Ma BP). A concensus is emerging that the dicot and monocot lineages separated very early and this may relate to the absence of known true symbiotic N_2-fixing monocots.

Evolution of Rhizobia. Extant rhizobia vary considerably both in the free living and symbiotic state. The ancestral types appear most likely to have been free living phototrophs, which (like many other soil organisms) produced substances which could modify plant growth[32]. During the evolution of symbiotic forms a number of gains and losses occurred (Table 1).

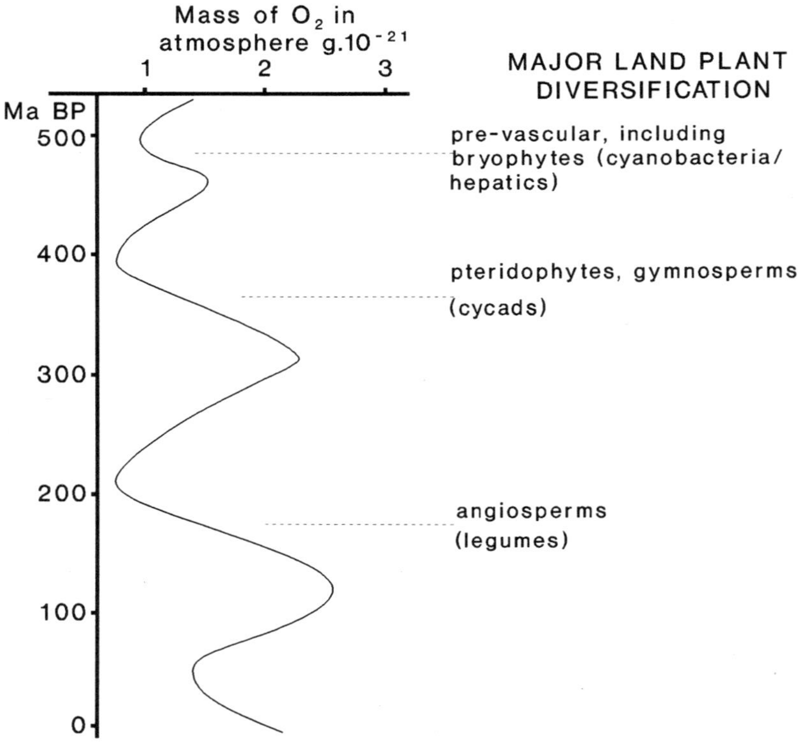

Fig.1. Hypothetical relationships between possible fluctuations in atmospheric oxygen levels and rapid species diversification in groups of plants containing nitrogen-fixing symbionts (as suggested in brackets)[32].

Table 1. Features of extant rhizobia seen in relation to a hypothetical photosynthetic, free living, nitrogen fixing ancestor.

LOSSES	Light-harvesting pigments (except *Photorhizobium*)
	Ability to grow autotrophically (except some strains of *B. japonicum* and *R. meliloti* which contain RUBISCO
	Control of N_2-fixation by ammonia and oxygen (some aspects retained by *Azorhizobium*)
GAINS	Ability to infect certain plants and induce nodule formation
	Adaptations to existence in plant cystosolic environment, for example control by high $[K^+]$ low pO_2
NO CHANGE?	Ability to produce phytohormones

Evolution of legume nodules. We have previously suggested that early nodules were formed following wound infections, probably associated with the formation of lateral roots[31]. *Photorhizobium* and *Azorhizobium* both inhabit stem nodules: the latter are associated with adventitious root initials. Since these genera retain characters which may be 'primitive'(Table 1) we have recently proposed that the ancestral nodule may have been formed on stems of legumes growing in marshy areas[32]. In due course nodules in axils of lateral roots were formed: phototrophy was no longer advantageous but a subterranean existence was (ready access to combined C and soil nutrients via roots, protection from desiccation) and thus the (arguably) more advanced rhizobia (*Rhizobium*, *Bradyrhizobium*) evolved.

Perhaps the major weakness in the above suggestion is that no stem nodules are known on either of the sub-families Caesalpiniodeae and Mimosoideae. The former is generally considered to be the more primitive although when the Mimosoideae and Papilionoideae (where wound infections are probably common) diverged from the Caesalpiniodeae is far from clear.

Extant caesalpinioid nodules are confined to a very few (8 fully confirmed[32]) genera. Their most characteristic feature is the retention of bacteria within infection threads during their nitrogen fixing state (as in *Parasponia*)[10]. This feature and the frequent lignification might be advantageous for fossilization, and a renewed search for fossil legume nodules is warranted. A scheme for legume nodule evolution based on known features is given in Fig. 2.

THE VERSATILE LEGUME NODULE

Legumes nodulate with at least four genera of rhizobia, using at least three modes of infection, producing a wide range of nodule shapes, sizes, and structures. However, within any one host-rhizobial combination, events from infection to nodulation follow an ordered sequence and most people would agree that the interaction between host and symbiont is "tightly regulated"[36], Is this true, or is it a consequence of trying to grow uniform plants for experimental purposes? Are the symbionts balanced on a knife edge which can pivot within certain limits before exceeding those required for nitrogen fixation? Work from various sources suggests that this may be so. First, the structure of nodules formed on a single host genotype by a single rhizobial genotype may vary considerably both within and between nodules. This is clear from work on *R. loti* which nodulates *Lotus pedunculatus*[36]. Also in unpublished work from our laboratory on tree rhizobia which infect *Phaseolus vulgaris*, nodules from a single strain and a single plant varied from pink to white and with considerable variation in internal structure. Further, nodule developmental processes may be altered by stress factors such as NaCl[13]. In the field, nodules on a single plant may vary in effectiveness possibly as a result of their containing different rhizobia. Could it equally result from either the same rhizobia but with nodulation modified by (a) host factors such as

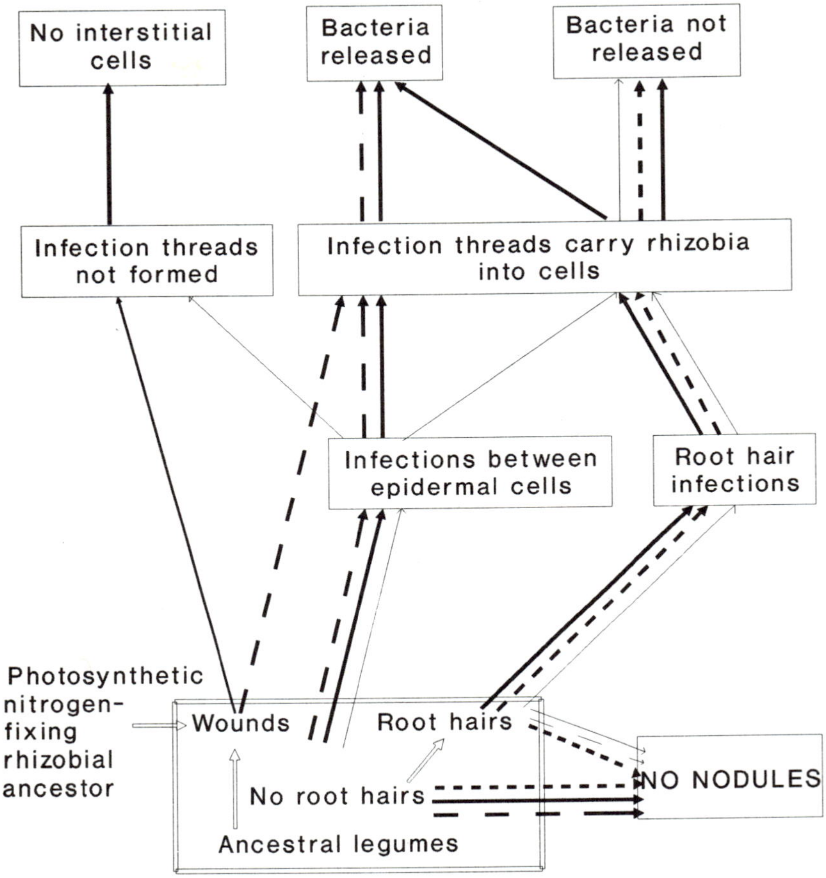

Fig.2. Possible sequences of events leading to extant nodule forms. All pathways indicated by bold lines have been demonstrated: faint lines remain to be confirmed. Pathways indicated by continuous lines are for papilionoid, long dashes for mimosoid and short dashes for caesalpinioid species. The papilionoid pathway leading from an intercellular infection *via* infection threads to released bacteria is based on unpublished data of Cordeiro and Sprent for *Lonchocarpus muelhbergianus*. References to other pathways may be found in [32].

different hormone levels on different parts of the root system, or (b) changes in the soil pH, salinity, etc?

GLYCOPROTEINS – FLAVOUR OF THE DECADE

Glycoproteins and proteoglycans are compounds containing both protein and carbohydrate (Table 2). The distinction is frequently unclear and the term 'glycoprotein' is often used in a general sense. In this sense, glycoproteins appear to be very important in nodulation.

Table 2. Some known features of 'glycoproteins'.

GLYCOPROTEINS -------->	PROTEOGLYCANS
Mostly protein	Mostly carbohydrate
Less water soluble	More water soluble
Generally located within cell wall, often rich in hydroxyproline, often highly basic	Generally extracellular, often very polyanionic
May be involved in cell wall growth	May be involved in recognition processes

A glycoprotein with similarities to gum arabic has been found in the matrix of infection threads in pea nodules[15] and subsequently in intercellular spaces[5]. This led to the suggestion that infection threads are extensions of the intercellular space system[35]. In parallel with these studies, some nodulin genes (e.g. ENOD2[37] and ENOD12) were found to code for hydroxyproline rich proteins, substances characteristic of plant cell walls[8] where they are normally glycosylated. Specific molecules may be recognised at the tissue level by various techniques, including *in situ* hybridization[37] and use of monoclonal antibodies[5]. Nodule host cells thus have the machinery to synthesise a suite of substances ranging from classical glycoproteins to classical proteoglycans (Table 2). All of these may have roles in non-nodule cells, but their production may be amplified by the presence of symbionts, pathogens or damage.

At one end of the spectrum we have the hydroxyproline-rich glycoproteins (HRGP) such as those coded for by ENOD2 and ENOD12[26] and which are also specifically expressed in lateral root initiation [16]. They may be essential for growth of new cells and/or mechanical strengthening. The interesting observation that the ENOD2 product may be co-located with a hypothetical oxygen diffusion barrier in nodules[37] will be examined later.

At the other end of the spectrum are the predominantly carbohydrate substances (such as gum arabic) which are usually extracellular and very water soluble. Their polyanionic polysaccharide parts have gel properties which vary with esterification, pH, Ca^{2+} (or Mg^{2+})[8]. They may partially or completely occlude intercellular spaces in nodules to an extent that may vary with the pO_2 under which plants were grown (James *et al*, this volume).

Glycoproteins and oxygen diffusion barriers Advanced legume nodules appear to operate both fixed and variable resistances to oxygen diffusion in order that they may function properly in environments of varied oxygen supply (flooded v aerated soils) and under different levels of oxygen demand (e.g. lower when photosynthate is in short supply). Both cell wall and intercellular glycoproteins have been implicated in the oxygen diffusion resistance mechanisms (see above). Some speculations on how such a system may operate follow. The underlying assumption will be that a diffusion resistance can be achieved by a fixed or variable water-filled pathway of suitable length[25]. Effectively this could mean that nodules must have an internal boundary layer which lacks intercellular spaces[21].

The synthesis of HRGP is oxygen dependent in that a mixed function oxidase is required to hydroxylate proline[6]. Such an oxygen-requiring reaction could act as sensor, leading to wall modifications altering cell geometry in a fashion which could modify oxygen diffusion resistance. Circumstantial supporting evidence for an effect on a fixed resistance could be obtained from experiments similar to those reported from ENOD2 studies[37], but on nodules grown under different pO_2. The fixed resistance also responds to non-ambient pO_2 by adjusting the volume of airspaces[9], a rather standard plant response, which at least in the case of sub-ambient pO_2 may be effected by ethylene accumulation.

In terms of variable resistance to O_2, the kinetics suggest that a biophysical/biochemical response is involved. The sensor would then be lack of oxygen for some essential reaction (e.g. a terminal oxidase) triggering anaerobiosis or other metabolic change (see also[3]). Variations in gas conductance in nodules have been compared with stomatal functions[25] and ways in which changes in cell geometry could be effected suggested elsewhere[20]. Driving forces for such alterations could be turgor changes driven by ion pumps and/or alterations in wall elasticity or yield turgor. pH changes may also play a role [6].

Intercellular glycoproteins could be involved in both fixed and variable diffusion resistances by altering in amount or molecular configuration. Ionic and pH environments are known to affect gel properties of glycoproteins and we have preliminary NMR evidence that water distribution within the molecule may change with cation environment. Variations in viscosity and molecular configuration could affect the volume of intercellular space occupied by glycoprotein and this effectively would change the length of any water-filled pathway. A cation pump which might affect cell turgor, could equally be invoked as a possible way of altering intercellular-space glycoprotein configuration.

Glycoproteins similar to those found in intercellular spaces of nodules are not exclusively a property of potentially nodulated legumes[35] and could be put to use if rhizobia were introduced into other species. Whether rhizobia transit roots within infection threads or in intercellular spaces, it is likely that they are surrounded by glycoprotein[5, 14], a potential nutrient source. We are currently testing whether rhizobia can grow on such substrates.

OXYGEN DIFFUSION IN OTHER NODULES

The possibilities outlined above have been based on experiments with advanced crop species such as soybean and pea and usually on laboratory grown plants. Bacteroids from such plants have been shown rapidly to adjust their metabolisms to altered supplies of O_2 and other metabolites[3, 4]. How does such plasticity impinge upon the need for, or operation of oxygen diffusion resistances? How do laboratory results relate to field situations which may have other over-riding constraints such as water-shortage? Do all legume nodules have the same type of response? In view of the wide spectrum of nodule structure and the variations known in non-legumes, probably not.

Frankia-induced nodules occur in a range of dicot families[33]. Their anatomy has the root-like feature of a central vascular system, with the infected tissue being in the cortex, surrounded on each side by a few uninfected layers. There is no analogue to the uninfected 'cortex' of legume root nodules. Perhaps therefore it is not surprising that there is no evidence for a cortically or otherwise located variable gas diffusion resistance[27]. Instead there are various ways in which O_2 transport to the N-fixing regions may be enhanced or retarded depending on species and conditions. Both host and symbiont are involved, in the latter by presence of nitrogenase-containing vesicles (except in *Casuarina* nodules) whose walls vary in thickness in direct relation to external pO_2[22]. Host cells may respond to insufficient O_2 by producing haemoglobin to facilitate O_2 diffusion (as in most legume nodules which have been studied), by producing negatively gravitropic aerating roots (*Myrica*) or by assymetric production of cork (*Coriaria*)[27]. Actinorhizal nodules indicate that there are many different ways of balancing oxygen supply and demand, without invoking a variable oxygen diffusion resistance.

In nodules of *Parasponia*, which have root-like anatomy and retain rhizobia within host-cell-wall bounded threads, an added possibility is that this wall material contributes to a diffusion resistance. Such a situation may also obtain in those primitive legume nodules which have 'persistent infection threads'[10]. Further, some woody legumes may produce corky cells as in *Coriaria*, though not asymmetrically – there is no information as to how these corky layers relate to pO_2. However it has been suggested[28] that cork is produced as protection against desiccation and may, concomitantly, impede O_2 transport. Although haemoglobin has been found in some of these primitive legume nodules[11], the amount may be variable and its function remains to be demonstrated. Thus, with the exception of structures analogous to vesicle walls, but with the possible addition of infection thread walls, legume nodules may display as wide a variety of adaptations to oxygen supply as actinorhizal nodules. Studies on more types of legume nodules are planned: these have not only intrinsic interest, but also implications in understanding the constraints on attempts to construct new symbioses. In this connection we would be unwise to ignore the stem-like nature of legume nodules and the root-like nature of nodules outside the Leguminoseae.

WHAT IS AN N_2-FIXING ROOT NODULE?

Taking both legume and non-legume nodules, which criteria are essential? Possibly: (a) a differentiated outgrowth, most commonly from a root, with functional vasculature (b) the ability of some host cells to accomodate N_2-fixing eubacteria. There are numerous optional extras such as haemoglobin to transport oxygen. On present evidence (b) is harder to achieve than (a). We will discuss these criteria with a view to predicting ways in which functional novel symbioses may be designed.

Formation of outgrowths. The induction of nodule-like (but uninfected) outgrowths on legumes by growth substances has been known for many years[1]. Recent work has shown that auxin inhibitors can induce genes such as ENOD2[12] and that spontaneous nodulation can occur[34]. All these outgrowths have the typical legume nodule feature of a peripheral vasculature.
Thus it appears that the machinery for making nodules is inherent in legumes and does not necessarily require the presence of rhizobia. The corresponding experiments in non-legumes are effectively precluded because nodules in these are root-like structures and thus relate to normal plant morphology. Production of roots may be stimulated by auxins[19]. In addition to these structured outgrowths, relatively unorganised ones may form naturally, for example in the axils of lateral roots[29]. Cell division may be induced by microorganisms including rhizobia and the resulting cells may burst through the surface – in the case of *Parasponia* to be followed by infection and nodulation[18].
Rhizobium meliloti produces substance(s) which can stimulate cell division in protoplasts from a variety of species, including monocots such as barley and wheat[24]. Coupled with other treatments such as osmotica and wall-degrading enzymes, cell divisions induced by rhizobia may lead to formation of outgrowths such as recently observed on rice[2]. Here rhizobia occurred in intercellular spaces and inside dead or dying cells. Nitrogenase activity was not detected and the 'nodule structures' lacked an organised anatomy. Given the right mix of cell division and growth factors, of microbial and other origin, many plants, both monocots and dicots, can form outgrowths of various types, sometimes with an organised anatomy.

Getting the bacteria into living host cells. This has so far proved virtually impossible. Clearly the invading organisms must produce suitable wall-degrading enzymes. Reproducible evidence of these has proved hard to obtain. This may be because the search has generally been for enzymes such as cellulase or pectinases. Penetration of a root hair wall or intercellular passage of bacteria may require the breakdown or loosening of only one or a few of the many cell wall components. This may demand production of only small amounts of highly specific enzymes, not readily detected in culture.

Also, very precise conditions may be necessary for the production and/or operation of such enzymes, as has been shown recently for polysaccharide-hydrolysing enzymes of *Frankia*[23]. Penetration into individual host cells poses similar problems, together with the need for bacterial cell division without invoking a defence reaction from the plant.

Some aspects of this problem have been discussed elsewhere[30]. If a nodule is to fix nitrogen on a scale appropriate for the growth of the host plant, intracellular proliferation of symbiont cells is almost certainly essential. This division must be at the expense of host metabolites, so the situation remains pathogenic and does not become mutualistic until nitrogen fixation begins.

A prognosis for effective novel symbioses. The rather pragmatic approach to date, which has yielded structures such as that reported on rice[2] has certainly raised hopes that we are not too far from making new symbioses. The time is now ripe for asking highly specific questions. Each laboratory is likely to have its own favourite host plant and possibly its favourite symbiont – from *Frankia* to various types of rhizobia. Many of the steps once regarded as essential (e.g. release of bacteria from infection threads) in the light of recent observations on natural symbioses[30], are not. Some of the keys will lie in the nature of the host cell wall and the symbiont's ability to breach it. This and other structural features may render monocots generally more intractable than dicots. As mentioned earlier, the 2 groups probably diverged early in evolution; this is reflected in differences in structure of many of the major cell wall components[6]. Further, there are gross differences in anatomy, including root structure. If we understand the differences essential to the conversion of (say) a monocot wall to a form amenable to symbiont penetration, we can then, in theory transfer the necessary gene(s) for the required substance(s). There is now evidence that this may not be easy, because monocots have some different sets of preferred codons to dicots[7]. For this reason it has been suggested that rather than transfer a dicot gene to a monocot, in some cases it would be better to make an artifical gene using the monocot preferred codons and insert that into the monocot[7]. For a number of reasons therefore, I think it may prove more difficult to get functional nodules on monocots than dicots. Even the latter are going to require exercise of considerable ingenuity.

REFERENCES

1. Allen, E.K., Allen, O.N. & Newman, A.S. (1953) Am. J. Bot. **40**, 429–435.
2. Al-Mallah, M.K., Davey, M.R. & Cocking, E.C. (1989) J. Exp. Bot., **40**, 473–478.
3. Appels, S. (1989) *The Symbiosis Between* Rhizobium leguminosarum *and* Pisum sativum: *regulation of the nitrogenase activity.* (Ph.D. thesis, Landbouwuniversiteit te Wageningen).
4. Bergersen, F.J. & Turner, G.L. (1990) Proc. Roy. Soc. Lond. B **238**, 295–320.
5. Bradley, D.J., Wood, E.A., Larkins, A.P., Galfre, G., Butcher, G.W. & Brewin, N.J. (1988) Planta **173**, 149–160.
6. Brett, C. & Waldron, K. (1990) *Physiology and Biochemistry of Plant Cell Walls* (Unwin Hyman, London).

7. Campbell, W.H. & Gowri, G. (1990) Plant. Physiol. **92**, 1–11.
8. Cassab, G.I. & Varner, J.E. (1988) Ann. Rev. Plant Physiol. Plant. Mol. Biol. **39**, 321–353.
9. Dakora, F.D. & Atkins, C.A. (1988) in *Microbiology in Action*, eds. Murrell, W.G. & Kennedy, I.R. (Research Studies Press, Letchworth), pp.133–142.
10. Faria, S.M. de, McInroy, S.G. & Sprent, J.I. (1987) Can. J. Bot. **65**, 533–538.
11. Faria, S.M. de, McInroy, S.G., Rowell, P. & Sprent, J.I. (1988) in *Nitrogen Fixation: Hundred Years After* eds. Bothe, H., Bruijn, F.G. de & Newton, W.E. (Gustav Fischer, Stuttgart). p. 524.
12. Hirsch, A.M., Bhuvaneswari, T.V., Torrey, J.G. and Bisseling, T. (1989) Proc. Natl. Acad., Sci. USA **86**, 1244–1248.
13. James, E.K. & Sprent, J.I. (1988) in *Nitrogen Fixation: Hundred Years After* eds. Bothe, H., Bruijn, F.G. de & Newton, W.E. (Gustav Fischer, Stuttgart). p. 525.
14. James, E.K., Brewin, N.J., Chudek, J.A., Minchin, F.R. & Sprent,J.I. (1989) in *Proceedings of Meeting on Plant and Soil Nitrogen Metabolism* (AFRC, Lancaster).
15. Johnston, A.W.B., Downie, J.A., Rossen, L., Shearman, C.A., Firmin, J.L., Borthakur, D., Wood, E.A., Bradley, D. & Brewin, N.J. (1987) Phil. Trans. R. Soc. Lond. B **317**, 193–207.
16. Keller, B. & Lamb, C.J. (1989) Genes & Development **3**, 1639–1646.
17. Knoll, A.H. & Bauld, J. (1989) Trans. Roy. Soc. Edinb., Earth Sciences **80**, 209–223.
18. Lancelle, S.A. & Torrey, J.G. (1984) Protoplasma **123**, 26–37.
19. Middleton, W., Jarvis, B.C. & Booth, A. (1980) New Phytol. **84**, 251–259.
20. Parsons, R. (1989). *Physiological Regulation of Nitrogen Fixation in Soybean Root Nodules.* (Ph. D. Thesis, Australian National University).
21. Parsons, R. & Day, D.A. (1990) Plant, Cell Environ. (in press).
22. Parsons, R., Silvester, W.B., Harris, S., Gruijters, W.T.M. & Bullivant, S. (1987) Plant Physiol. **83**, 728–731.
23. Safo-Sampah, S. & Torrey, J.G. (1988) Plant and Soil **112**, 89–97.
24. Schmidt, J., Wingender, R., John, M., Wieneke, U. & Schell, J. (1988) Proc. Natl. Acad. Sci. USA **85**, 8578–8582.
25. Sheehy, J.E., Minchin, F.R. & Witty, J.F. (1985) Ann. Bot. **55**, 549–562.
26. Sheres, B., Wiel, C van de, Zalensky, A., Horvath, B., Spaink, H, Eck, H. van, Zwartkruis, F., Wolters, A-M., Gloudemans, T., Kammen, A. van & Bisseling, T. (1990) Cell **60**, 280–294.
27. Silvester, W.B. & Winship, L.J. (1990) Plant Physiol. **92**, 480–486.
28. Sprent, J.I. (1985) in *Plants for Arid Lands* eds. Wickens, G.E., Goodin, J.R. & Field, D.V. (Allen & Unwin, London), pp. 215–229.
29. Sprent, J.I. (1986) International Tree Crops J.**4**, 47–54.
30. Sprent, J.I. (1989) New Phytol. **111**, 129–153.
31. Sprent, J.I. & Raven, J.A. (1985) Proc. Roy. Soc. Edinb., B **85**, 215–237.
32. Sprent, J.I. & Raven, J.A. (1990) in *Biological Nitrogen Fixation* eds. Stacey, G., Burris, R.H. & Evans, H.J. (Chapman & Hall, New York), in press.
33. Sprent, J.I. & Sprent, P. (1990). *Nitrogen Fixing Organism – Pure and Applied Aspects.* (Chapman & Hall, London).
34. Truchet, G., Barker, D.G., Camut, S., Billy, F. de, Vasse, J. & Huguet, T. (1989) Mol. Gen. Genet. **219**, 65–68.
35. VandenBosch, K.A., Bradley, D.J., Know, J.P., Perotto, S., Butcher, G.W. & Brewin, N.J. (1989) EMBO J. **8**, 335–342.
36. Ward. L.J.H., Rockman, E.S., Ball, P., Jarvis, B.D.W. & Scott, D.B. (1989) Mol. Plant-Microbe Interactions **2**, 224–232.
37. Wiel, C van de, Scheres, B., Franssen, H., Lierop, M-J. van, Lammeren, A van, Kammen, A. van & Bisseling, T. (1990) EMBO J. **9**, 1–7.

Gresshoff, Roth, Stacey & Newton (eds.) Nitrogen Fixation: Achievements and Objectives
Chapman and Hall New York • London 1990

Nitrogen cycling and human intervention

A.J.M. Schoot Uiterkamp

TNO-MT, Department of Biology,
PO Box 217, 2600 AE Delft, the Netherlands

1. INTRODUCTION

Ever since Antoine Lavoisier, over 200 years ago, called the element nitrogen "azote" meaning "without life", subsequent history has proved nitrogen and its compounds such as food, poisons, fertilizers and explosives to be anything but lifeless. Moreover if the bacterium *Azotobacter* were to "live up" to the literal meaning of its name, it would not exist and neither would perhaps conferences on nitrogen fixation!

In the following I will briefly review the various, often detrimental and ever increasing human interventions with the biogeochemical cycling of nitrogen, and indicate some approaches to improve the present situation and aim at sustainable environmental compatibility of human actions.

2. BRIEF HISTORIC REMARKS

While we are now concerned about mankind fixing too much nitrogen - or rather too much, too fast and in the wrong places - we should not forget that only a century ago people were concerned about the opposite. The major sources of fixed nitrogen at that time were Chile salpeter (i.e. sodium nitrate) and Peru guano (droppings from birds, a kind of cormorant called the guanay). These sources were decreasing while the demand for fixed nitrogen, for both fertilizers and explosives, was increasing. There was some concern among scientists that the newly discovered denitrifying bacteria would be able to deplete fixed nitrogen stocks in the soil, leading to disastrous results for food production. Sir William Crookes in an address before the Royal Society of London in 1898 pointed to the dire consequences if no synthetic way of producing nitrogen fertilizer was to be developed. In the end it was mainly the demand for explosives in World War I, which led Haber and Bosch in 1913 to the large scale industrial use of their ammonia synthesis. In the second half of this century concern about excess production of nitrogen oxides from large scale industrial combustion processes, car engines and the like, gave rise to a variety of control technologies from catalytic denitrification to three-way catalysts. Throughout this period farmers have remained concerned about sources of nitrogen for plant production - while governments are worried about the consequences of excess fertilizers - and scientists are very active at finding novel nitrogen fixing systems for plant growing purposes.

3. BIOGEOCHEMICAL CYCLING OF NITROGEN

If it were not for the existence of biogeochemical cycling, atmospheric nitrogen should be largely oxidized to nitrate and dissolved in the oceans or locked up in sediments. Earth without life would have an atmosphere with a chemical composition somewhere between those of Venus and Mars. Combined action of photosynthetic and non-photosynthetic organisms, chemical and geological processes and transport by wind and water, gave rise to the pre-industrial nitrogen rich atmosphere. In this atmosphere the amount of nitrogen removed by abiotic and biotic fixation, was closely balanced by the amount of nitrogen returned to the atmosphere by organisms releasing nitrogen from other nitrogen compounds. Since the advent of the industrial era human intervention has upset this balance. Mainly due to ever increasing energy demands fertilizer use and cultivation of leguminous plants, fixed nitrogen is now showing excess concentrations on a local, regional and global scale. Natural denitrifying processes cannot keep up with these nitrifying processes. This gives rise to phenomena like acid rain, eutrophication and ground-water contamination with nitrates, and contributes to global climate changes. Biogeochemical cycles of elements like C, O, N, P and S are coupled. All cycles are now being disturbed individually. These disturbances can therefore be felt in other cycles as well. For general aspects of the nitrogen cycle I refer to the literature [Refs. 3, 7, 8, 18]. Table 1 shows a recent compilation of estimated sources, sinks and accumulation of atmosferic NO_x, NH_3 and N_2O [Ref. 13].

Table I. *Sources, sinks and accumulation of atmospheric NO_x, NH_3 and N_2O in 10^{12} g N/year.*

Compound	Sources			Sink	Accumulation
	Anthropogenic	Natural	Total		
NO_x	30 (16-51)	26 (10-51)	56 (26-102)	50 (24-85)	–
NH_3	98 (86-110)	31 (29-33)	129 (115-143)	169 (113-225)	–
N_2O	8 (2-13)	12 (5-22)	20 (7-35)	9 (6-13)	5 (3-7)

The large ranges in estimates are indicative of the large gaps in our knowledge concerning nitrogen cycling.

4. ANTHROPOGENIC NITROGEN COMPOUNDS

Figure 1 schematically indicates the main environmental compartments and anthropogenic nitrogen compounds that will be briefly reviewed below.

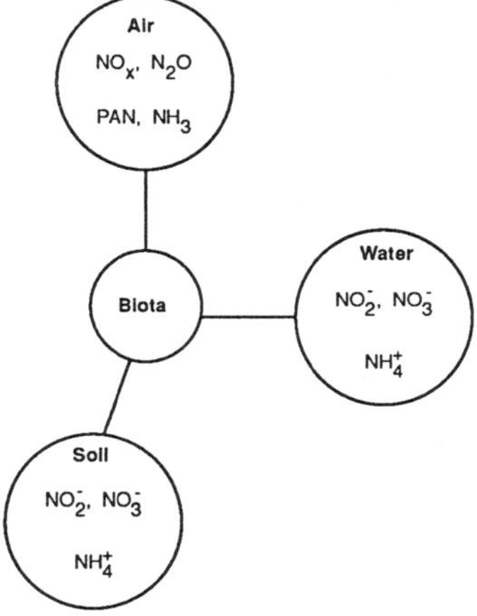

Figure 1. Environmental compartments and anthropogenic nitrogen compounds.

These compounds are NO_x, N_2O, NH_3 and NO_3^-. I will per compound give some examples of type, scale and consequences of the problems caused by the compound. My examples will often be drawn upon the European situation. Europe and more specifically N.W. and Central Europe is an environmental laboratory in which the prolonged combination of a high density of population, traffic, industrial and agricultural activities, has led to severe environmental consequences.

4.1 NO_x : NO and NO_2

The major reactive nitrogen oxides NO and NO_2 are collectively referred to as NO_x. In nature NO_x is produced by the abiotic reduction of nitrite in acid soils, by biotic processes related to nitrification and denitrification, by lightning and by biomass burning. The latter processes are dominated by "fire" the other classic environmental element, besides water, air and soil. Fire and lightning, or more generally high temperatures, play an important role in the nitrogen cycle. Main sources of anthropogenic NO_x are fossil fuel combustion for industrial purpose and transportation. The gases themselves may cause health problems (especially pulmorary disorders) and damage crops and materials. On a local and regional scale the most obvious and direct consequences of NO_x emissions is the formation of tropospheric photochemical smog and more specifically ozone. The troposphere is the lower 15 km of the earth' atmosphere. Estimates of the future growth in NO_x emissions - by 2020 over 65% of these emissions could originate in developing countries - show that concentrations of tropospheric ozone will continue to rise at a rate faster than the doubling shown during the past 100 years [Ref. 14]. The present background ozone concentrations on the median latitude in the northern hemisphere are around 35 ppbv. This means that plant life is threatened already (Dutch advisory levels during the growing season are 25 ppbv). The Dutch advisory health protection level is 80 ppbv. This means that ozone levels even at moderate latitudes have a high likelihood of approaching these advisory levels. Ozone is also a major greenhouse gas. On a molar basis its greenhouse warming potential is 2000 times higher than that of CO_2. Tropospheric NO_x is converted within days to nitric acid that represents a substantial contribution to the acidity of precipitation within and up to 2000 km horizontally and downwind from industrial and population concentrations. Acid precipitation has well-known detrimental consequences for flora, fauna, crops and materials. Transported vertically into the stratosphere, the atmospheric compartment extending roughly from 15 to 50 km, NO_x probably contributes to the breakdown of the ozone layer and the formation of the antarctic and arctic ozone holes with the concomitant UV radiation increase potential [Ref. 11].

4.2 PAN

While sources, sinks, fluxes and effects of NO_x are well documented, much less attention is given to organic nitrates such as peroxyacetylnitrate (PAN) $CH_3C(O)OONO_2$ [Ref. 22]. PAN is formed from oxidized hydrocarbons and NO_x. Atmospheric measurements show that PAN is often present in concentrations much greater than NOx in the global atmosphere. While higher concentrations are occurring close to anthropogenic source regions (i.e. urban areas), PAN is also observed in pristine tropospheric ocean air indicative of the existence of natural sources and/or long-range transport. PAN and other organic nitrates represent a major area of interaction between the N and C cycles. PAN has been observed to act as a gas-phase plant mutagen and as a potent Ames-active mutagen [Ref. 10].

4.3 N_2O

Nitrous oxide (N_2O) is a relatively stable gas - tropospheric lifetime over 100 years - that is present in the atmosphere at concentrations around 300 ppbv. Its concentrations are presently increasing at ~ 0.25% per year. On a molar basis the greenhouse warming potential of N_2O is 200 times higher than that of CO_2. This

means that N_2O is a greenhouse gas of increasing importance. N_2O is also oxidized to nitric oxide in the stratosphere. This nitric oxide reacts with ozone. N_2O is therefore implicated as well in the breakdown of the stratospheric ozone layer.
Although N_2O is clearly important for the understanding of global atmosphere and climate, relatively little work has been done to quantify potential sources of N_2O. It is generally agreed that the observed increase in atmospheric N_2O is anthropogenic, but there is little consensus as to the relative importance of different sources. The idea that fossil fuel combustion was the major source, has recently been disputed. Sampling artefacts are mentioned as the culprit [Ref. 15]. This means that other sources will increase in relative importance. The major source remaining is chemically fertilized soil. This is corroborated by the observation that the concentration of N_2O in phreatic aerobic aquifers, contaminated by anthropogenic activities, are up to three orders of magnitude higher than the concentration expected as a result of atmospheric equilibrium. N_2O would originate from nitrification and denitrification of nitrogen compounds [Ref. 24].

4.4 NH_3

Ammonia is an important atmospheric pollutant. Over large areas in Europe acid precipitation falls down, in which up to 70% of the original acid is neutralized by NH_3 [Ref. 4]. Furthermore in some places serious acidifying effects have been observed which could be attributed to the wet deposition of ammonium ions and the dry deposition of NH_3 and ammonium aerosols. Van Breemen et al. [Ref. 28] suggest that nitrification and acidification in soil occurs according to the overall reaction:

$$NH_4^+ + 2O_2 \rightarrow 2H^+ + NO_3^- + H_2O$$

Animal wastes and fertilizers are thought to be responsible for 90% or more of the anthropogenic NH_3 emission. In Western European countries like the Netherlands, Belgium, and Denmark intensive animal husbandry occurs alongside industrial activities. It was recently shown that in animal husbandry concentration areas in the Netherlands the quantity of ammonium (NH_4^+) and sulphate (SO_4^{2-}) in atmospheric precipitation was 1.5-4 times higher than in relatively remote areas, whereas 70-90% of the deposition of nitrogen consisted of NH_x compounds [Ref. 26]. This is an example of the coupling between N and S cycles. The oxidation of SO_2 in wet deposition can apparently be enhanced by NH_3. The enhancing effect of NH_3 is especially pronounced if the oxidation occurs with an oxidant such as O_3, which also originates from anthropogenic activities (see above) [Ref. 2].

4.5 NO_3^-

Nitrate is the anionic nitrogen species normally found in soil and groundwater. Under certain local conditions its naturally occurring concentrations can be quite high (e.g. in South America). Excess nitrate in soils and groundwater can also result from acid precipitation of oxidized conversion products of NO_x.
In areas of intensive animal husbandry ammonia released from manure is partly released into the atmosphere and partly converted by nitrifying microorganisms to soluble nitrates in soil (cf. 4.4). Another source of nitrates in soil and groundwater is the direct application of nitrate containing fertilizers. Since nitrate has a high water solubility and poorly binds to soil particles, it has become one of the main pollutants of groundwater, often reaching concentration levels exceeding the World Health Organization drinking water quidelines. Nitrate pollution of groundwater has become widespread and of ever increasing importance in Western and Central Europe. Nitrate pollution (and N_2O emissions) parallels increasing usage of N and P fertilizers.
Table 2 shows the use of nitrogen and phosphorus containing fertilizers in some OECD countries [Ref. 5].

Table 2. *Use of N and P containing fertilizers in some OECD countries per ha cultivated land.*

	N-containing fertilizer			P-containing fertilizer		
	1975	1980	1986	1975	1980	1986
		kg N			kg P_2O_5	
the Netherlands	22.5	24.0	25.0	4.0	4.1	4.5
Belgium	12.1	12.9	13.3	8.0	6.8	6.0
FR of Germany	10.2	12.9	13.2	6.5	7.0	5.7
Denmark	12.0	13.2	13.5	4.6	3.9	3.8
Italy	4.4	6.0	6.2	2.9	4.4	4.5
Spain	2.5	2.9	3.5	1.4	1.5	1.6
Japan	12.1	11.5	12.9	11.6	12.9	13.8
USA	2.2	2.5	2.2	1.1	1.1	0.8

Table 3 shows main emission sources of nitrogen in the Netherlands in 1986 [Ref. 5]. The table shows quite graphically the large contribution from fertilizers.

Table 3. *Emission sources of nitrogen in the Netherlands in 1986.*

		10^6 kg N			10^6 kg N
Air	NO_x	172	**Soil**	Animal manure	554
	NH_3	215		Chemical fertilizer	502
				wastes	34
Water	leaching	160		Wet and dry deposition	142
	riverine import	540			
	waste water	61			
	Wet and dry deposition	15			

Table 4 shows the effect of application of high loads of N-containing fertilizer to agricultural land in the Netherlands [Ref. 5].

Table 4. *Average groundwater concentration of some substances at a depth of 9 m in the Netherlands.*

	unit	underneath agricultural sandy soil			underneath non cultivated sandy soil		
		1985	1986	1987	1985	1986	1987
pH	pH	5.9	5.9	5.8	5.6	5.5	5.6
nitrate	mg N/l	23.0	24.0	23.0	1.8	2.0	2.2
ammonium	mg N/l	1.1	1.1	1.3	0.2	0.2	0.2
phosphorus	mg P/l	0.1	0.1	0.1	0.1	0.1	0.1
potassium	mg K/l	13.0	13.0	13.0	2.3	2.3	2.4
sulphate	mg SO_4/l	97	101	89	39	41	40

N-containing fertilizers and other N-containing substances in soils are subjected to various nitrogen conversion processes. In general agricultural practice pesticides are often applied. Table 5 shows that commonly used pesticides may have an inhibitive effect on N-cycle processes in soil. The criterion used in Table 5 is an inhibitive effect of at least 25% at the highest common application dose [Ref. 30].
It is important that the side effect of pesticides on the soil effect is assessed. At present a number of test methods are available [Ref. 30].

Table 5. *Effect of pesticides on N-cycle processes in soil.*

Process	Number of pesticides investigated	% with effects	Types
Nitrification	140	14	Dithiocarbamates, fumigants
Denitrification	29	10	Phenylmercury, fumigants
Nitrogen fixation (symbiotic)	25	12	Fumigants
Nitrogen fixation (asymbiotic)	33	9	Dalapon, dinoseb
Nitrogen mineralization	63	6	Phenylmercury urea herbicides

Nitrate in groundwater may react further with sulphides like pyrite to yield sulphates and reduced nitrogen [Ref. 27].
Nitrate contamination is not restricted to groundwater. In can affect rivers, lakes, seas and oceans as well. It reaches surface waters through riverine and atmospheric input and can, in combination with phosphorus, give rise to a variety of eutrophication phenomena such as algal blooms, which occur regularly in the North Sea [Ref. 1]. It is estimated that in 1985 18% (i.e. 243,000 tons N) of the total N input into the North Sea (i.e. 1,343,000 tons N) was of atmospheric origin [Ref. 20]. For the Mediterranean Sea it is estimated that atmospheric nitrogen corresponds to about 10% of the average "new" phytoplankton production and may reach values as high as 50% during oligotrophic periods, which occur during periods of stratified water [Ref. 17]. Similar phenomena are observed in open oceans as well. Wherever high levels of pollutant atmospheric nitrogen occur downwind from populated shores, oceanic P becomes growth rate limiting and there is an increase in the downwind incidence of undetectable phosphate on the surface of the neighboring ocean. In the absence of high anthropogenic nitrogen over the continents, oceanic nutrient nitrogen depletion seems to predominate as in the central Indian Ocean [Ref. 9].

5. SOME (PARTIAL) SOLUTIONS

Space does not allow a description of all possible solutions towards the control and abatement of emissions of the substances indicated under 4. I will only briefly mention a few kinds of approaches to combustion (emissions to air), waste water treatment (emissions to water) and fertilization (emissions to soil) that have been implemented and some other that are subject of research and development.

5.1 Combustion

In industrial fuel combustion, oxides of nitrogen are formed either from atmospheric air (as in the case of natural gas) or from the N present in fuel itself. Therefore control technologies can be divided in technologies preventing the formation of NO_x and technologies removing NO_x from flue gases. The former case centers on flue gas circulation and combustion with excess air. Another approach is stepwise combustion by first blowing off nitrogen from fuel at relatively low temperatures followed by a high temperature burn. Another possibility is catalytic combustion. Processes that remove NO_x after formation can be separated in dry and wet processes. In the case of wet processes one distinguishes between thermal NO_x and catalytic reduction. In both cases ammonia is added with the objective to remove oxidized with reduced nitrogen. The use of the three way catalyst in combating vehicle emissions is well established. New processes centre on the addition of cyananuric acid $(HNCO)_3$ to remove NO_x from diesel exhaust [Ref. 31] and on the use of aqueous emulsions containing yellow phosphorus and an alkali for the combined removal of SO_2 and NO_x from flue gas [Ref. 6]. The use of fuel cells allows the combustion of fuels at extremely low NO_x emission levels [Ref. 29]. It is interesting that atmospheric flames either need a carbon source (to make CN intermediates) or O radicals to generate NO_x from atmospheric nitrogen [Ref. 12]. It is therefore worthwile to study fossil combustion processes with pure or enriched oxygen as the oxidant. Presently commercially available gas membrane modules offer the possibility to separate N_2 from O_2 and to enrich oxygen in air at relatively low costs. The potential benefits from moving towards hydrogen as an energy source should be obvious as well..

5.2 Water treatment

In the case of nitrates in groundwater, biological denitrification is attempted together with physico-chemical techniques. Once nitrates are in the groundwater, it may take decades for them to come out of it again. In the Netherlands 25% of all groundwater used for drinking water preparation will have to be denitrified in the future despite measures to control emissions from animal manure application. In classical waste water treatment emphasis was on carbon removal and not so much on nitrogen removal. Given the effects of acidification, eutrophication and groundwater contamination described above, much more attention will have to be given to nitrogen removal. Interesting possibilities are offered e.g. sludge on carrier systems, i.e. systems allowing a better survival potential under reactor conditions, for the slow-growing microorganisms involved in nitrogen removal. Many of these systems are still in the development stage, but emphasis was and is still mainly on reactor design. Microbiological routes will have to be explored as well. Bacterial species such as *Thiosphaera pantotropha* show aerobic denitrification and heterotrophic nitrification, i.e. they are able to simultaneously reduce nitrate and convert ammonium to nitrogen under aerobic conditions [Ref. 23]. In the "Anamnox" process ammonium and nitrate are converted simultaneously to nitrogen under anaerobic conditions without the need for additional carbon sources [Ref. 19]:

$$5NH_4^+ + 3\,NO_3^- \rightarrow 4N_2 + 9N_2O + 2H^+$$

The "Anamnox" process is in fact the equivalent of the wet gas treatment processes described in 5.1 and the classical lab scale preparation of nitrogen by heating NH_4NO_2. Both the application of *Thiosphaera pantotropa* and the "Anamnox" process are actively investigated, as are various other microbiological systems. Besides these more or less classical microbiological applications, it is to be expected that the use of genetically modified microorganisms will be explored as well.

5.3 Fertilizers

Use of fertilizers has many side effects, as indicated above. Without going to genetic engineering, a great deal can be improved by returning to traditional agricultural methods such as composting, crop rotation, nitrogen fixing manures such as clover, and extending the use of *Azolla* or cyanobacteria as green manure. The latter two methods have traditionally been applied in many Asian countries such as China and India. Controlled releases of fertilizers and better ways to recycle nitrogen are but a few other approaches beside all the efforts directed towards plant and microbiological nitrogen fixation. One thing seems certain, despite the (return to) greater exploitation of biological nitrogen fixing systems, the use of synthetic nitrogen containing fertilizers will increase due to the ever increasing demand for food.

6. SUSTAINABLE DEVELOPMENT

Human intervention in nitrogen cycling is full of paradoxes. On the one hand we unwillingly fix too much nitrogen in unwanted forms at the wrong times and in the wrong places with consequences which we are subsequently forced to face. On the other hand we deliberately fix large quantities of nitrogen for agricultural and industrial purposes. The present nitrogen balance sheet of the Netherlands shows that of 4350 x 10^6 kg of fixed nitrogen entering the country, 92% is exported again through air and water or as durable products. This means that every year some 8% (360 x 10^6 kg) of fixed nitrogen remains, representing a considerable burden on the

local and regional environment. On a global scale the picture seems different. The global nitrogen cycle shows that soil nitrogen lost through farming and general wastage, has to be suppleted with about 12% of new nitrogen [Ref. 21]. Even when we succeed in engineering more plants and microbiological systems with nitrogen fixing capabilities, limits would remain. Some emphasize the possibilities of biotechnology [Ref. 25] and others the limits [Ref. 16]. Given the world is finite, one thing is certain: when the nitrogen supply is not limiting any longer, something else will be. In nature this can be among others, the availability of other nutrients like phosphate and water, the supply of trace elements like iron, molybdenum and vanadium and/or the energy supply. In industrial practice it is bound to be the energy supply and environmental constraints. Industrial and biological de novo fixing of nitrogen have a high energy cost. The Haber-Bosch process requires two tons of fuel for every ton of ammonia produced. The biological conversion of one molecule of dinitrogen into two moles of ammonia requires from 12 to 24 moles of ATP. Given the diagnostic elements in sections 3 and 4, some (partial) treatments and cures presented in section 5 and some basic constraints mentioned in this section, it is only appropriate to conclude by pointing to the need for prevention.
In 1987 the Brundtland Commission published the report "Our common future". It concludes that environmental crises, development crises and energy crises are strongly linked. None of these crises can be solved without simultaneously solving the other. This brings the Commission to the concept of sustainable development: a development that provides the present needs of society without endangering the requirements of future generations. In other words: the Commission states that a development is sustainable if it in the long run, enhances environmental quality and the resource base on which human endeavour depends, provides for basic human needs, is economically viable and enhances the quality of life for individuals and society. In practice this implies a strong emphasis on closing elemental cycles, on energy conservation, on reusing rather than on disposing of materials and on product quality over product quantity. Last but not least it implies human population control and appropriate environmental management ranging from individual to planetary level (without having the necessary manuals and flow sheets: they did not come with the planet!). This is quite a challenge given the prevailing political systems and opinions, legislative and regulatory approaches, different economic systems and different levels of education and public awareness.

7. REFERENCES

1. Barth, H. & Nielsen, A. (1989) EC Water Pollution Research Reports **19**, 1-3.
2. Behra, P., Sigg, L. & Stumm, W. (1989) Atmos. Envir. **23**, 2691-2707.
3. Brill, W.J. (1977) Sci. Amer. **236** (3), 60-61.

4. Buysman, E., Maas, H.F.M. & Asman, W.A.M. (1987) Atmos. Envir. **21**, 1009-1022.
5. CBS (1990) Statistisch Jaarboek (SDU, the Hague).
6. Chang, S.G. & Lin, D.K. (1990) Nature **343**, 151-153.
7. Cook, R. (1984) Episodes **7**, 3-7.
8. Delwiche, C.C. (1970) Sci. Amer. **223** (3), 136-146.
9. Fanning, K.A. (1989) Nature **339**, 460-463.
10. Gaffney, J.S., Streit, G.E., Spall, D.E. & Hall, J.H. (1987) Envir. Sci. Technol. **21**, 519-524.
11. Graedel, T.E. & Crutzen, P.J. (1989) Sci. Amer. **261** (3), 28-36.
12. Hayhurst, A.N. & Vince, I.M. (1977) Nature **266**, 524-525.
13. Heederik, P.J. (1987) SCMO-TNO Report, pp. 45.
14. Hough, A.M. & Derwent, R.G. (1990) Nature **344**, 645-648.
15. Lyon, R.K., Kramlich, J.C. & Cole, J.E. (1989) Envir. Sci. Technol. **23**, 392-393.
16. MacKenzie, D. (1990) New Scientist (Mar. 31) 28-29.
17. Martin, J.M., Elbaz-Poulichet, F., Guieu, C., Loije-Pilot, M.D. & Han, G.C. (1989) Marine Chem. **28**, 159-182.
18. Mooney, H.A., Vitousek, P.M. & Matson, P.A. (1987) Science **238**, 926-932.
19. Mulder, A. (1989) EP O 327 184 A1.
20. North Sea Conference (1990) Interim Report on the Quality States of the North Sea (Netherlands Min. of Transport & Public Works, the Hague).
21. Postgate, J. (1990) New Scientist (Feb. 3), 57-61.
22. Roberts, J.R. (1990) Atmos. Envir. **24A**, 243-287.
23. Robertson, L.A., Van Niel, E.W.J., Torremans, R.A.M. & Kuenen, J.G. (1988) Appl. Envir. Microbiol. **54**, 2812-2818.
24. Ronen, D., Magaritz, M. & Almon, E. (1988) Nature **335**, 57-59.
25. Schneiderman, H.A. & Carpenter, W.D. (1990) Envir. Sci. Technol. **24**, 466-473.
26. Schuurkes, J.A.A.R., Maenen, M.M.J. & Roelofs, J.G.M. (1988) Atmos. Envir. **22**, 1689-1698.
27. Stumm, W. & Morgan, J.J. (1981) Aquatic Chemistry 2nd Ed. (Wiley-Intersience, New York) pp. 450.
28. Van Breemen, N., Burrough, P.A., Velthorst, E.J., Van Dobben, H.F., De Wit, T., Ridder, T.B. & Reijnders, H.F.R. (1982) Nature **299**, 548-550.
29. Van der Klein, C.A.M. (1990) Energiespectrum (Mar), 77-82.
30. Vonk, J.W. (1990) Toxicol. Envir. Chem. (in press).
31. Wicke, B.G., Grady, K.A. & Ratcliffe, J.W. (1989) Nature **338**, 492-493.

Gresshoff, Roth, Stacey & Newton (eds.) Nitrogen Fixation: Achievements and Objectives
Chapman and Hall New York • London 1990

H_2 cycling in N_2 fixation: Past, present, and future outlook

Daniel J. Arp

Laboratory for Nitrogen Fixation Research, Oregon State University, Corvallis, Oregon 97331, USA.

I. Introduction

H_2 production by nitrogenase is a major factor limiting the efficiency of biological N_2 fixation. This production appears to be an integral part of the reaction mechanism leading to the reduction of N_2 (i.e., N_2 reduction by nitrogenase cannot occur in the absence of H_2 production) and it would appear unlikely that this inefficiency can be improved at the source. The H_2 produced by nitrogenase can either be lost to the environment or oxidized by hydrogenase. H_2 oxidation typically benefits the system and results in an increased efficiency of N_2 fixation. This process of H_2 production by nitrogenase followed by H_2 oxidation via hydrogenase is referred to as "hydrogen cycling". The purpose of this chapter is to provide an historical account of the major accomplishments which have led to our current view of H_2 cycling in N_2 fixing microorganisms. Such an account will also serve to focus attention on the contributions of Dr. Harold Evans and his colleagues during the last 15 years of rapid development in this field.

II. Background

A relationship between H_2 and N_2 fixation was first recognized in 1937 by P.W. Wilson and his colleagues (44) while attempting to measure the apparent Km for N_2 in intact clover plants. When H_2 was used as the diluent gas, an inhibition of N_2 fixation was observed. This inhibition was later shown to be competitive vs. N_2. In further efforts to understand this inhibition, Wilson and colleagues (33) looked for and found hydrogenase activity in bacteroids isolated from pea root nodules. This followed the original observation in 1931 by Stephenson and Stickland (41) of hydrogenase activity in microorganisms. Although Wilson found hydrogenase activity in bacteroids and in *Azotobacter vinelandii*, the relationship between H_2 and N_2 fixation remained obscure. Sixteen years later, Burris and colleagues (25) demonstrated that soybean nodules were capable of H_2 evolution. They suggested that H_2 evolution was an activity of nitrogenase, but proof of this awaited purified preparations of nitrogenase.

In 1967, R.O.D. Dixon (10) demonstrated that both H_2 evolution and H_2 oxidation activities could occur simultaneously in pea root nodules. Thus, it became apparent that two enzyme activities were required to explain the H_2 metabolism of at least some N_2-fixing systems. Dixon also made a number of seminal contributions in subsequent works (11, 12). Among them was the demonstration that the host could influence the hydrogenase activity expressed by the bacteroids and that not all bacteroids expressed hydrogenase activity. Dixon suggested that the hydrogenase might increase the efficiency of N_2 fixation and proposed three potential mechanisms (Dixon's postulates) to account for this increased efficiency. Thus, the potential for an important--though not essential--role for hydrogenases in N_2 fixation was established.

Efforts with nitrogenase continued and progressed rapidly with the establishment of methods to purify the enzyme. Not only could the purified enzyme evolve H_2, but N_2 reduction by nitrogenase was specifically and competitively inhibited by H_2. Furthermore, H_2 inhibition and HD formation (8) by nitrogenase share the same mechanistic basis and H_2 inhibition actually results in the generation of additional H_2. Nitrogenase was also shown to be capable of diverting all the available electrons to H_2 evolution either in the absence of any other reducible substrates or in the presence of the inhibitor CO (35). Although it had been shown that N_2 inhibited H_2 formation (35), a nagging concern was whether at high enough N_2 concentrations, H_2 evolution would be completely prevented. Extrapolations had suggested that even at infinite pN_2, about 25% of the electrons would go to H_2 formation. The mechanistic implications of this extrapolation were profound. This concern was finally addressed in 1984 when Simpson and Burris (40) carried out inhibition experiments at a pN_2 of 50 atm and still observed that about 25% of the electron flux through nitrogenase went to H_2 formation. H_2 formation is now considered an obligate step in N_2 reduction and the reduction of one N_2 will be accompanied by the formation of a minimum of one H_2.

In summary, N_2 fixing systems interact with H_2 in a number of ways:

1). Under appropriate conditions, all or a portion of the electron flux through nitrogenase can be diverted to the generation of H_2.

2). Under optimal conditions, N_2 reduction by nitrogenase is accompanied by a stoichiometric production of H_2 (for each N_2 reduced, one H_2 is produced).

3). N_2 reduction by nitrogenase is specifically and competitively inhibited by H_2. The mechanism of this inhibition is consistent with that of HD formation by nitrogenase and results in the formation of additional H_2.

4). The H_2 produced by nitrogenase can be consumed by an H_2-oxidizing hydrogenase.

III. H_2 Losses and Distribution of Hydrogenases

By 1976, the groundwork for an understanding of the dihydrogen metabolism of N_2 fixing systems had been laid. However, it was not apparent how significant H_2 evolution was as a factor which limited the efficiency of N_2 fixation. In that year, Schubert and Evans (38) reported on the extent of H_2 losses in a number of legume and nonlegume symbioses. The study included alfalfa, clovers, peas, soybeans and alder and the inoculants included commercial mixtures, native strains and pure cultures. The results revealed that with most symbionts only 40-60% of the electron flux through nitrogenase was transferred to N_2, the remainder was used to generate H_2. Recognizing that H_2 evolution was a major factor affecting the efficiency of nitrogen fixation in nodulated symbionts, Schubert and Evans (38) defined the term "relative efficiency" to provide an index of the efficiency of a N_2-fixing system relative to its H_2 metabolism. The rate of H_2 evolution from nodules was compared to the rate of electron flow through nitrogenase. A value of 1.0 indicated that no H_2 was evolved from the system under investigation. While most relative efficiencies ranged from 0.4 to 0.6, a few symbioses gave values approaching 1.0. These were later shown to express an "uptake" or H_2-oxidizing hydrogenase which consumed the H_2 produced by nitrogenase.

It then became of interest to determine how common hydrogenase expression was among various symbionts. For the next several years, a number of surveys were carried out in various laboratories to assess the distribution of hydrogenase. For example, Ruiz-Argueso et al. (36) demonstrated that only 6 of 15 strains of *Rhizobium leguminosarum* which were tested expressed H_2-oxidizing activity in pea nodules. Furthermore, none of the 6 H_2-oxidizing strains expressed sufficient activity to recover all of the H_2 produced by nitrogenase. In general, the results revealed that H_2-oxidizing activity was not observed in the majority of *Rhizobium*/legume symbioses. While hydrogenase is more common in the *Bradyrhizobium*/legume symbioses, the majority still do not express hydrogenase. In contrast to *Rhizobium* and *Bradyrhizobium*, the majority of *Frankia*/actinorhizal symbioses tested did not evolve H_2.

IV. Benefits of H_2 Cycling

If the loss of H_2 from nodules lowers the efficiency of N_2 fixation, then those symbioses with no H_2 evolution were expected to have an increased efficiency. Would this increased efficiency be manifested at the whole plant level? To answer this question, Albrecht et al. (1) compared biomass accumulation and nitrogen content of soybean plants inoculated with either H_2-uptake positive or H_2-uptake negative strains of *B. japonicum*. The plants inoculated with H_2-uptake positive strains accumulated higher biomass and N contents than the plants inoculated with H_2-uptake negative strains. The study included a comparison of different *B. japonicum* strains as well as a comparison of H_2-uptake negative mutants and their H_2-uptake positive parental strains. However, this study did not compare isogenic strains (the mutants turned out not to be revertible), so the increased productivity could not be unambiguously correlated with the presence of H_2-oxidizing activity. Similar experiments were repeated with ever increasing sophistication, not only with respect to the isogenecity of the strains, but also with regard to the experimental design. For example, in one experiment a HUP^- *B. japonicum* mutant was compared with a stable HUP^+ derivative obtained by recombinant techniques and the soybean plants were grown to maturity in large concrete cylinders (16). Again, increases in total biomass (9%) and N content (11%) were correlated with the presence of the H_2-oxidizing activity. The experiment included 10 replicates of each HUP phenotype, and the differences observed were statistically significant.

The question of increased efficiency of N_2 fixation correlated with H_2-oxidizing activity has also been considered with other systems. Indeed, the most conclusive evidence to indicate a beneficial role for hydrogenase came from experiments with *Azotobacter chroococcum* which were carried out by Geoff Yates and coworkers (45). Continuous cultures *A. chroococcum* under N_2-fixing conditions were initiated with mixtures of an H_2-oxidizing recombinant strain and a non-H_2-oxidizing mutant parent. The dominance pattern was then assessed when various nutrients were used to limit the growth of the cultures. Under energy-limited conditions (primarily sucrose limitation), the H_2-oxidizing phenotype rapidly dominated the culture. The competitive benefit demonstrated by the H_2-oxidizing strain was attributed to the only apparent difference between the two strains, namely, the presence or absence of hydrogenase. Interestingly, when cultures were limited for O_2, the non-H_2-oxidizing phenotype slowly dominated the cultures. This result suggested that H_2 oxidation could actually be detrimental to a culture under certain conditions.

V. The Mechanism of Increased Efficiency

As mentioned above, Dixon (12) suggested three mechanisms by which an H_2-oxidizing hydrogenase might improve the efficiency of N_2 fixation in nodules, and each of these has now received some attention.

1) The cycling of H_2 generated by nitrogenase via a hydrogenase would provide an additional source of reductant to the system. Theoretically, this reductant could be used by the cell for any purpose, including reduction of N_2 or generation of ATP. The use of H_2-linked reductant to power nitrogenase was demonstrated by Emerich et al. (14) with bacteroids isolated from soybean nodules. Additional experiments demonstrated that the level of ATP in bacteroids was increased by the addition of H_2 to H_2-oxidizing bacteroids, while no increase was observed by the addition of H_2 to non-H_2-oxidizing bacteroids. The use of H_2 as a source of reductant might also be expected to lead to a conservation of organic carbon substrates. Dixon (11) demonstrated that the utilization of carbon substrates by pea root nodule bacteroids was inhibited when H_2 was included.

2) Because H_2 oxidation is coupled to O_2 consumption, the presence of an H_2-oxidizing hydrogenase could lead to a decreased pO_2 within the nodule, thereby providing an additional mechanism to protect nitrogenase from O_2 damage. Using isolated soybean nodule bacteroids, Emerich et al. (14) demonstrated that the optimal pO_2 for nitrogenase activity could be shifted to higher levels in the presence of H_2 if the bacteroids expressed an H_2-oxidizing hydrogenase. The ability of nitrogenase to function at a higher applied pO_2 was taken as evidence that the hydrogenase had lowered the *in situ* pO_2.

3) H_2 is a specific and competitive inhibitor of N_2 reduction by nitrogenase. Hydrogenase could serve to lower the *in situ* pH_2, thereby

decreasing or eliminating any H_2 inhibition of N_2 reduction. At first, this mechanism was not considered to be viable. It was assumed that H_2 would diffuse rapidly from a nodule and would not reach inhibitory concentrations, especially given that the inhibition constants for H_2 inhibition of N_2 reduction which had been measured were quite high (0.11-0.55 atm H_2) (34). However, in a sophisticated model of the gas concentrations within a nodule, Layzell and coworkers (26) estimated that the H_2 concentration within the infected tissue could reach 0.3 atm. Then, Rasche and Arp (34) measured the Ki for H_2 inhibition of N_2 reduction by isolated bacteroids and found a surprisingly low Ki of 0.03 atm. Rasche and Arp also demonstrated that detectable levels of H_2 were released from non-H_2-oxidizing nodules when the nodules were crushed in an ice-cold solution of NaOH, while no H_2 was released from H_2-oxidizing nodules. These results call for a re-examination of the importance of hydrogenase in decreasing the amount of H_2 inhibition of N_2 reduction.

VI. Physiology and Biochemistry of H_2 Metabolism

In the last 15 years, there have been a number of advances in our understanding of the physiology and biochemistry of H_2 metabolism in bacteroids and free-living microorganisms. The apparent Km for H_2 in intact bacteroids was determined to be 50 nM, a remarkably low value (15). A number of studies indicated that electron flow from hydrogenase to O_2 proceeded via the electron transport chain (see 17 for review). Eisbrenner et al. (13) identified a spectral component in *B. japonicum* that is typical of b-type cytochromes and which appeared to be closely linked to the H_2 metabolism of the cells. The possibility was raised that this component accepts electrons from hydrogenase at a low redox potential.

Evans and coworkers had long been interested in the enzymes which fix CO_2 in bacteroids and nodules. This interest eventually led to the characterization of ribulose-bisphosphate carboxylase activity in bacteroids and the establishment of methods to express this activity in free-living cultures (39). The conditions for expression of carboxylase activity were consistent with the conditions required for expression of hydrogenase activity in free-living cultures (see below). Given that *B. japonicum* could express the two critical enzymes for lithoautotrophic growth with H_2 as the energy source (hydrogenase and RuBP carboxylase), the question arose as to whether *B. japonicum* could also grow lithoautotrophically. Hanus et al. (21) demonstrated that several strains of *B. japonicum* which expressed an H_2-oxidizing hydrogenase were capable of lithoautotrophic growth. Although previously viewed as strict chemoorganotrophs, these results revealed that at least some strains of *Bradyrhizobium* were also lithoautotrophs.

In developments with other H_2-oxidizing lithoautotrophic bacteria, Friedrich et al.(18) demonstrated in 1981 that the dependency of the lithoautotrophic growth of *Alcaligenes eutrophus* on nickel was related to hydrogenase. Shortly thereafter, Graf and Thauer (20) demonstrated that Ni was associated with purified hydrogenase from *Methanobacterium thermoautotrophicum*. In 1982, Friedrich et al. (19) demonstrated that radioactive Ni copurified with both the soluble and particulate hydrogenases from *A. eutrophus*. This presented the question as to whether the hydrogenase from *B. japonicum* was also a Ni-containing enzyme. The first experiments to address this problem were carried out by Klucas, Hanus and Russell in Evans' laboratory. They showed that *B. japonicum* required Ni for expression of hydrogenase activity and for growth under lithoautotrophic conditions (27). Harker et al. (22) and Stults et al. (43) demonstrated that radioactive Ni copurified with hydrogenase isolated from free-living *B. japonicum* cultures. Arp (2) demonstrated that hydrogenase isolated from bacteroids of *B. japonicum* contained 0.6 Ni/molecule of hydrogenase. Hydrogenases from a number of N_2-fixing and other microorganisms are now known to contain Ni (see 24 for review).

In addition to Ni, the purified hydrogenase contains Fe, and the ratio of Ni to Fe is about 1:10 (2). The iron is present in iron sulfur centers as indicated by the absorption spectra (2) and EPR spectra (Arp, unpublished). The molecular weight of the native enzyme is 104,000 (2) and consists of two subunits of ca. 65,000 and 35,000 molecular weight (2; 22). This metal content and subunit composition is typical of a large number of hydrogenases from diverse physiological groups including phototrophs, desulfovibrians, hydrogen bacteria and N_2-fixing microorganisms.

For the purified enzyme from *B. japonicum* bacteroids, the K_m for H_2 was determined to be near 1 uM. While considerably higher than the appK_m measured in intact bacteroids, this value is still much lower than the values reported for other hydrogenases. Purified hydrogenase from *B. japonicum* was shown to catalyze both isotope exchange and H_2 evolution in the purified state (3). However, both reactions were slow relative to the rate of H_2 oxidation. The catalytic properties of this enzyme are well-suited to its physiological role, namely, to scavenge the H_2 generated by nitrogenase. Active *B. japonicum* hydrogenase is sensitive to O_2. For this reason, the hydrogenase is typically purified under anaerobic conditions (2,22), although Stults et al. (43) were able to purify *B. japonicum* hydrogenase aerobically.

Recently, Boursier et al. (7) demonstrated that selenium increases hydrogenase expression in *B. japonicum*. Furthermore, radioactive selenium co-purified with the hydrogenase, and radioactivity was associated with both subunits following SDS-PAGE. There is no internal TGA codon in the DNA sequence of *B. japonicum* which would support the existence of a selenocysteine residue in this hydrogenase (37). The role of selenium in *B. japonicum* hydrogenase remains to be elucidated.

VII. Regulation and Genetics of H_2 Oxidation

In the early experiments of Wilson and colleagues (33), H_2 oxidation was observed in bacteroids isolated from pea nodules, but no hydrogenase activity was detected in free-living cultures. Subsequent studies have confirmed the absence of hydrogenase activity in typical laboratory cultures of *Bradyrhizobium* and *Rhizobium*. However, Maier et al. (30) described conditions for expressing hydrogenase activity in free-living cultures of *B. japonicum*. Limited organic carbon, the presence of H_2 and, most importantly, low pO_2 were all factors which contributed to the expression of hydrogenase. This development facilitated physiological studies and provided a mechanism for selection of colonies based on H_2-oxidizing phenotype. Coupled with a plate assay for hydrogenase activity (31), it was now possible to screen large numbers of bacteria for the presence or absence of hydrogenase. Accordingly, it was only a short time before Maier, Postgate and Evans (31) reported on the isolation of the first mutants which lacked the H_2-oxidizing phenotype. A number of regulatory mutants have subsequently been isolated by Maier's group (32), including strains with increased and decreased sensitivity to O_2 as well as a constitutive mutant.

The question of how the host controls the expression of hydrogenase is still largely unanswered, despite the importance of this aspect of the regulation of hydrogenase. Bedmar et al. (5) showed that even within a single species of peas, different cultivars produced nodules with significantly different levels of hydrogenase activity from a single strain of *R. leguminosarum*. Shoot/root grafting studies demonstrated that a transmissible shoot factor was involved in this regulation (6), though the factor has not been identified. Lambert et al. (29) showed that a region of pHU52 not present on pHU1 was required for the expression of hydrogenase in free-living cultures, but was not required for expression in nodules. This further emphasized that the factors which regulate hydrogenase expression in free-living cultures are not identical with those which regulate expression in the nodule.

Evans and coworkers were the first to apply recombinant DNA techniques to the study of a hydrogenase. A gene bank was constructed from *B. japonicum* and used to isolate the hydrogenase genes (9). Two recombinant cosmids, pHU1 and pHU52, from this library proved most useful in analyzing the *hup*-specific genes. Analysis of Tn5 insertions into the *B. japonicum* genome by complementation with pHU1 indicated that *hup*-specific sequences occurred over a region spanning about 15 kilobases. pHU1 was later shown not to contain all of the *hup* genes. Therefore, additional cosmids were isolated. One, pHU52, complemented 5 of 6 HUP^- mutants (PJ-18, which is thought to be a dominant mutation,was not complemented) (28). When pHU52 was transferred to HUP^- *R. meliloti* and *R. leguminosarum*, transconjugants were HUP^+. The transconjugants were also HUP^+ in their respective symbioses, although the activities were low (apparently due to the instability of the cosmid insert into the recipients). The structural genes for hydrogenase were identified by expression of subclones of pHU52 in *E. coli* and immunological detection of the hydrogenase polypeptides (46). This led to the determination of the nucleotide sequence which encodes the amino acid sequence of the hydrogenase structural genes (37). Comparison of the sequences with those derived for other NiFe hydrogenases revealed considerable similarities. In particular, the positions of a number of cysteines, the putative metal-binding ligands, were conserved through a number of sequences.

VIII. Conclusions

The majority of the experimental evidence, both with symbioses and with free-living bacteria, indicates that the presence of a hydrogenase can provide a measureable benefit to a N_2-fixing system. Therefore, it remains a worthwhile goal to select or engineer agronomically important strains of *Rhizobium* and *Bradyrhizobium* to express a hydrogenase. However, the molecular basis for this benefit is not yet fully understood. Clearly, research efforts must continue to be directed towards elucidating the mechanism of the hydrogenase benefit. Although *hup*-specific genes span several kilobases, only two *hup*-specific genes have been identified, namely, those coding for the subunits of hydrogenase. The number and roles of the remaining genes remain to be established. Experiments are already underway in several laboratories to sequence and characterize additional *hup*-specific genes. Further efforts are needed to stably integrate the required *hup* genes into the chromosome of a HUP^- recipient, especially where the donor and recipient are different species. Another area which requires additional attention is the mechanism of regulation of hydrogenase expression by the host. Clearly, if *hup*-specific genes are not expressed in the symbiosis, then no H_2-cycling benefit can be realized.

In summary, much progress has been made in the last 15 years in our understanding of H_2 cycling in N_2-fixing systems. Major contributions were made during that time by Dr. Harold Evans and his associates. For example, 1) the magnitude of H_2 losses from nodulated plants was quantified, 2) the distribution of hydrogenase was determined, 3) the "whole-plant" benefit associated with a hydrogenase was demonstrated, 4) the physiological basis and result of H_2 cycling was examined, 5) hydrogenase was characterized at the molecular level and 6) the *hup*-specific region of the genome was identified and characterized, including sequencing of two *hup*-specific genes. Through Dr. Evans' work, and the work of others, H_2 cycling is now recognized as an important mechanism to improve the efficiency of biological N_2 fixation.

Acknowledgements

I would like to dedicate this contribution to Harold Evans on the occasion of his retirement as Director of the Laboratory for Nitrogen Fixation Research and advancement to the position of Distinguished Professor Emeritus at Oregon State University. His thorough investigations of H_2 cycling in legume nodules provide an excellent example of how to approach a scientific problem.

References

1. Albrecht, S.L., Maier, R.J., Hanus, F.J., Russell, S.A., Emerich, D.W. & Evans, H.J. (1979) Science **203**, 1255-1257.
2. Arp, D.J. (1985) Arch. Biochem. Biophys. **237**,504-512.
3. Arp, D.J. & Burris, R.H. (1982) Biochim. Biophys. Acta **700**, 7-15.
4. Arp, D.J. & Burris, R.H. (1979) Biochim. Biophys. Acta **570**, 221-230.
5. Bedmar, E.J., Edie, S.A. & Phillips, D.A. (1983) Plant Physiol. **72**, 1011-1015.
6. Bedmar, E.J. & Phillips, D.A. (1984) Plant Physiol. **75**, 629-633.
7. Boursier, P., Hanus, F.J., Papen, H., Becker, M.M., Russell, S.A. & Evans, H.J. (1988) J. Bacteriol. **170**, 5594-5600.
8. Burgess, B.K., Wherland, S., Stiefel, E.L. & Newton, W.E. (1980) in Molydenum Chem. Biol. Significance, ed. Newton, W.E. & Otsuka, S. (Plenum, New York) pp. 37-84.
9. Cantrell, M.A., Haugland, R.A. & Evans, H.J. (1983) Proc. Natl. Acad. Sci. USA **80**, 181-185.
10. Dixon, R.O.D. (1967) Annals Botany, N.S. **31**, 179-188.
11. Dixon, R.O.D. (1968) Arch. Mikrobiol. **62**, 272-283.
12. Dixon, R.O.D. (1972) Arch. Mikrobiol. **85**, 193-201.
13. Eisbrenner, G. & Evans, H.J. (1982) Plant Physiol. **70**, 1667-1672.
14. Emerich, D.W., Ruiz-Arqueso, T., Ching, T.M. & Evans, H.J. (1979) J. Bacteriol. **137**, 153-160.
15. Emerich, D.W., Ruiz-Argueso, T., Russell, S.A. & Evans, H.J. (1980) Plant Physiol. **66**, 1061-1066.
16. Evans, H.J., Hanus, F.J., Haugland, R.A., Cantrell, M.A., Xu, L.-S., Russell, S.A., Lambert, G.R. & Harker, A.R. (1985) in World Soybean Research Conference III: Proceedings, ed. Shibles, R. (Westview Press, London) pp. 935-942.
17. Evans, H.J., Harker, A.R., Papen, H., Russell, S.A., Hanus, F.J. & Zuber, M. (1987) Ann. Rev. Microbiol. **41**, 335-361.
18. Friedrich, B., Heine, E., Finck, A., & Friedrich, C.G. (1981) J. Bacteriol. **145**, 1144-1149.
19. Friedrich, C.G., Schneider, K. & Friedrich, B. (1982) J. Bacteriol. **152**, 42-48.
20. Graf, E.-G. & Thauer, R.K. (1981) FEBS Lett. **136**, 165-169.
21. Hanus, F.J., Maier, R.J. & Evans, H.J. (1979) Proc. Natl. Acad. Sci. USA **76**, 1788-1792.
22. Harker, A. R., Xu, L.-S., Hanus, F.J. & Evans, H.J. (1984) J. Bacteriol. **159**, 850-856.
23. Haugland, R.A., Cantrell, M.A., Beaty, J.S., Hanus, F.J., Russell, S.A. & Evans, H.J. (1984) J. Bacteriol. **159**, 1006-1012.
24. Hausinger, R.P. (1987) Microbiol. Reviews **51**, 22-42.
25. Hoch, G.E., Schneider, K.C. & Burris, R.H. (1960) Biochim. Biophys. Acta **37**, 273-279.
26. Hunt, S., Gaito, S.T. & Layzell, D.B. (1988) Planta **173**, 128-141.
27. Klucas, R.V., Hanus, F.J., Russell, S.A. & Evans, H.J. (1983) Proc. Natl. Acad. Sci. USA **80**, 378-382.
28. Lambert, G.R., Cantrell, M.A., Hanus, F.J., Russell, S.A., Haddad, K.R. & Evans, H.J. (1985) Proc. Natl. Acad. Sci. USA **82**, 3232-3236.
29. Lambert, G.R., Harker, A.R., Cantrell, M.A., Hanus, F.J.,

Russell, S.A., Haugland, R.A. & Evans, H.J. (1987) Appl. Environ. Microbiol. **53**, 422-428.
30. Maier, R.J., Campbell, N.E.R., Hanus, F.J., Simpson, F.B., Russell, S.A. & Evans, H.J. (1978) Proc. Natl. Acad. Sci. USA **75**, 3258-3262.
31. Maier, R.J., Postgate, J.R. & Evans, H.J. (1978) Nature **276**, 494-495.
32. O'Brian, M.R. & Maier, R.J. (1989) Biochim. Biophys. Acta **974**, 229-246.
33. Phelps, A.S. & Wilson, P.W. (1941) Soc. Exp. Biol. Med. **47**, 473-476.
34. Rasche, M.E. & Arp, D.J. (1989) Plant Physiol. **91**,663-668.
35. Rivera-Ortiz, J.M. & Burris, R.H. (1975) J. Bacteriol. **123**, 537-545.
36. Ruiz-Argueso, T., Hanus, J. & Evans, H.J. (1978) Arch. Microbiol. **116**, 113-118.
37. Sayavedra-Soto, L.A., Powell, G.K., Evans, H.J. & Morris, R.O. (1988) Proc. Natl. Acad. Sci. USA **85**, 8395-8399.
38. Schubert, K.R. & Evans, H.J. (1976) Proc. Nat. Acad. Sci. USA **73**, 1207-1211.
39. Simpson, F.B., Maier, R.J. & Evans, H.J. (1979) Arch. Microbiol. **123**, 1-8.
40. Simpson, F.B. & Burris, R.H. (1984) Science **224**, 1095-1097.
41. Stephenson, M. & Stickland, L.H. (1931) Biochem. J. **25**, 205-215.
42. Stults, L.W., Moshiri, F. & Maier, R.J. (1986) J. Bacteriol. **166**,795-800.
43. Stults, L.W., O'Hara, E.B., & Maier, R.J. (1984) J. Bacteriol. **159**,153-158.
44. Wilson, P.W. & Umbreit, W.W. (1937) Arch. Mikrobiol. **8**, 440-457.
45. Yates, M.G.& Campbell, F.O. (1989) J. Gen. Microbiol. **135**, 221-226.
46. Zuber, M., Harker, A.R., Sultana, M.A. & Evans, H.J. (1986) Proc. Natl. Acad. Sci. USA **83**, 7668-7672.

SECTION A:

NITROGENASE BIOCHEMISTRY AND CHEMICAL MODELS

Invited papers
Workshop summary
Poster summaries

Gresshoff, Roth, Stacey & Newton (eds.) Nitrogen Fixation: Achievements and Objectives
Chapman and Hall New York • London 1990

Analysis of *Azotobacter vinelandii* strains containing defined deletions in *nif* genes required for FeMo-co biosynthesis

Sara Tal, Jia-ge Li, Teresa Chun, Amy Robinson and Barbara Burgess
Department of Molecular Biology and Biochemistry, University of California at Irvine, Irvine, CA 92717 USA.

The substrate-reducing site of the MoFe protein of nitrogenase is believed to be a cluster called FeMo-co, which has a stoichiometry Mo:Fe_{6-8}:S_{8-9}:homocitrate (2, 8, 17, 18). It has been known for some time that *in vivo*, FeMo-co biosynthesis requires the participation of at least the *nif*Q,B,N,E and V gene products (7). In 1984, Ugalde *et. al.* demonstrated that the biosynthesis of FeMo-co did not require the participation of the *nif*DK (MoFe protein) polypeptides (19). Subsequent reports, which demonstrated that the Fe protein (*nif*H) polypeptide was required for FeMo-co biosynthesis in both *Klebsiella pneumoniae* (6) and *A. vinelandii* (13), were therefore quite surprising. In the *A. vinelandii* study we discovered that although FeMo-co was present in a *nif*DK deletion strain, it was not present in a *nif*HDK deletion strain. This led us to construct a new *A. vinelandii* strain, called DJ54, which was deleted only for the *nif*H structural gene. When derepressed for nitrogenase synthesis this strain contained the MoFe protein polypeptides but did not contain FeMo-co (15). Here we report the further characterization of the FeMo-co-deficient MoFe protein synthesized by DJ54.

METHODS

The following procedures were carried out using published methods: (a) *A. vinelandii* MoFe protein and Fe protein purification and FeMo-co isolation (3); (b) extract preparations (13,15); (c) FeMo-co reconstitution assays with and without the addition of Fe protein/MgATP (14); (d) anaerobic native gel electrophoresis (14); and (e) Western blotting (14). *A. vinelandii* strains used were: DJ54 (ΔnifH) (15); DJ75 (Δ*nif*K) (13); DJ141 (Δ*nif*B) (kindly provided by Dennis Dean); and DJ35 (Δ*nif*E) (1).

RESULTS AND DISCUSSION

Table 1 compares the phenotype of the A. vinelandii ΔnifH strain, DJ54, to those of strains AvUW45 and DJ35 which are unable to synthesize FeMo-co due to defects in the nifB and nifE genes, respectively. Unlike these strains our DJ54 strain does not make the Fe protein of nitrogenase. With respect to the MoFe protein, however, all three strains have a very similar phenotype. All contain the MoFe protein polypeptides but have only extremely low levels of MoFe protein activity. If isolated FeMo-co is added to extracts from any of the three strains, an active MoFe holoprotein is assembled. Our interpretation of these data is that the Fe protein is required for some step in FeMo-co biosynthesis and that if the Fe protein is missing an inactive, FeMo-co-deficient MoFe protein is synthesized.

Table 1. Specific Activity[a] of A. vinelandii Strains.

STRAIN	Fe protein	MoFe Protein	MoFe protein with added FeMo-co
wild type	30	43.00	40
DJ54 (ΔnifH)	0	0.20	25
UW45 ($NifB^-$)	32	0.25	17
DJ35 (ΔnifE)	35	0.29	17

a) nmoles C_2H_4/min/mg crude extract protein

Purification of FeMo-co-deficient MoFe protein from DJ54. Figure 1 is an SDS coomassie-stained gel of fractions of the DJ54 MoFe protein from various steps in the purification. The protein shown in lane 6 is estimated to be ca 80% pure. Our characterization of that protein reveals that: a) it is an $\alpha_2\beta_2$ tetramer; b) it does not contain Mo; c) it does not give rise to an FeMo-co EPR signal and ; d) it contains ca 13 Fe atoms per $\alpha_2\beta_2$ tetramer. As shown in Figure 2 these Fe atoms can be "extruded" or removed from the protein to give a species with the spectroscopic properties of synthetic Fe_4S_4 clusters.

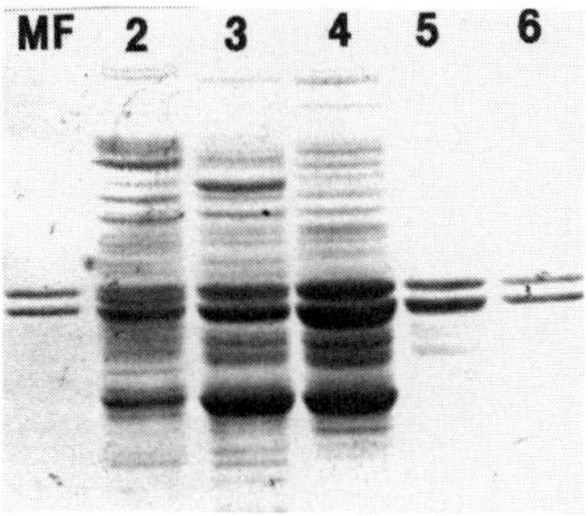

Figure 1.
Purification of the DJ54 MoFe protein. Fractions from each purification step (14 μg total protein) were run on an SDS polyacrylamide gel and the proteins were visualized by coomassie staining: (MF) purified wild type MoFe protein; (2) cell-free extracts; (3) first DEAE cellulose column; (4) second DEAE cellulose column; (5) sephacryl S-200 column; (6) Q-sepharose column.

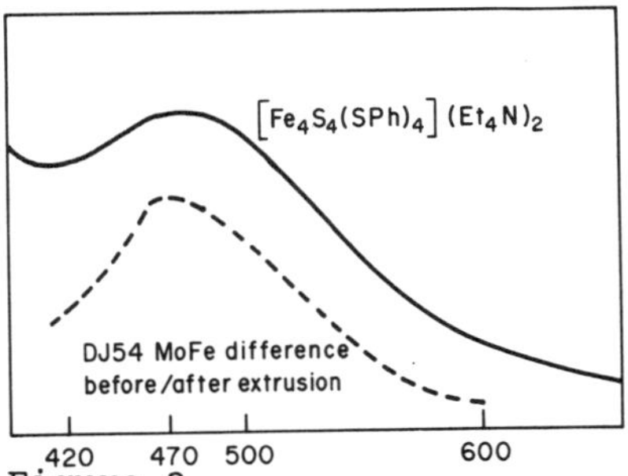

Figure 2.
Extrusion of [Fe-S] clusters from the DJ54 MoFe protein. A 0.7 ml Q-sepharose fraction was mixed anaerobically with an 0.21 ml solution containing 25% Triton, 72.5% DMF, 2.5% benzene thiol and 0.4% SDS and centrifuged at 2500 rpm for 10 min. Visible spectra were recorded before and after extrusion and compared to synthethic $[Fe_4S_4(SPh)_4](Et_4N)_2$.

<u>Co-purification and Characterization of a contaminating protein</u>. During the purification of the DJ54 MoFe protein we noticed that in our coomassie-stained gels (Figure 1, lane 5) the β subunit band appeared to be much darker than the α subunit band. This phenomenon is also found in literature reports which show gels of purified <u>A. vinelandii</u> wild type and mutant MoFe proteins (e.g. 12). The increased intensity of the β subunit band is due to the presence of a contaminating protein. As shown in Figure 3, we have purified that protein from an <u>A. vinelandii</u> strain Δ<u>nif</u>K DJ75, that does not synthesize the β subunit of the MoFe protein. The contaminating protein runs in an SDS gel in the identical position to the β subunit of the MoFe protein. The native protein also appears to be a tetramer. Although we do not have evidence that this protein is related to nitrogenase, it is important to recognize its presence because it contains <u>ca</u> 1 Fe atom per subunit which, as shown in Figure 4, appears to be in the form of a heme.

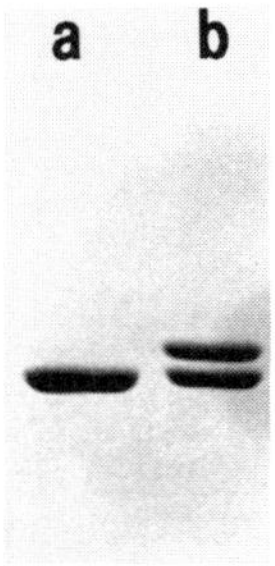

Figure 3.
Coomassie-stained SDS gel electrophoresis of (a) contaminating protein as purified from A. vinelandii ΔnifK strain, DJ75, and (b) purified MoFe protein.

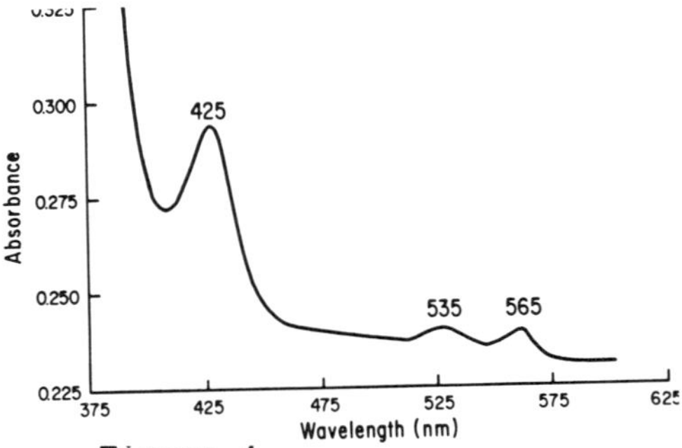

Figure 4.
Visible spectra of the protein shown in Figure 3(a) in the presence of excess $Na_2S_2O_4$.

Activation of the DJ54 protein with FeMo-co. The purification data given above are consistent with the idea that we have a FeMo-co-deficient MoFe protein which contains [Fe-S] clusters and is simply missing FeMo-co. As shown in Table 1 the DJ54 protein could also be activated with isolated FeMo-co in cell free extracts to give similar activity to that of the wild type or other mutant strains. However, unlike the situation in cell-free extracts, as we began to purify the DJ54 MoFe protein we found that only low levels of activity were obtained (14). This observation led us to investigate and optimize the activation assay conditions and to conclude that: a) the activation of the FeMo-co-deficient MoFe protein synthesized by the ΔnifH strain DJ54 is highly sensitive to protein concentration, salt concentration, pH and temperature; b) some active MoFe holoprotein (ca 25% of wild type) can be assembled by addition of isolated FeMo-co alone; and c) the addition of Fe protein together with MgATP is required to obtain 100% of wild type activity in a cell-free extract and to optimize activity after purification (14).

Comparison of FeMo-co-deficient MoFe proteins from ΔnifH, ΔnifB and ΔnifE strains. FeMo-co-deficient forms of the MoFe protein have also been purified from Nif B^- strains of K. pneumoniae and (6) A. vinelandii (4,12). The protein purified from a K. pneumoniae Nif B^- point mutant was reported to be an $\alpha_2\beta_2$ tetramer, to copurify with an unknown 21kD protein, to contain normal MoFe protein P-clusters, and ca 0.4-0.9 Mo atoms per tetramer. It could be reconstituted with FeMo-co, but only to a maximum of 40-50% wild type activity (6,16). The protein purified from _A. vinelandii_ UW45, which also has a point mutation in the _nif_B gene (10), was recently reported to be an $\alpha_2\beta_2\gamma_2$ protein with the γ subunit being an unknown 20KD protein (12). It does not contain Mo, does contain _ca_ 13 Fe atoms per molecule and can be easily reconstituted with FeMo-co to 100% wild type activity (4,12). Importantly, the latter protein does not appear to require the addition of Fe protein/MgATP, with FeMo-co, in order to obtain maximum activity (12).

These data indicate that there are significant differences between the FeMo-co-deficient proteins purified from different Nif B^- strains and differences between $NifB^-$ proteins and our Δ_nif_H protein. To investigate this further we have compared the properties of FeMo-co-deficient MoFe proteins synthesized by three specifically-constructed _Av_ Δ_nif_B, Δ_nif_E and Δ_nif_H strains. Table 2 compares the activation of these three proteins with FeMo-co in cell-free extracts. We are able to easily reconstitute Δ_nif_H, DJ54 to wild type levels of activity but are unable to reconstitute the Δ_nif_B or Δ_nif_E extracts to greater than 50% of wild type activity.

Table 2. Maximum Activity is Different.

	WT	Δ_nif_H DJ54	Δ_nif_B DJ141	Δ_nif_E DJ35
Sp.Act.	77.3 ± 1.8	74.5 ± 1.6[a]	39.2 ± 0.6[a]	38.5 ±0.9[a]
% Activity	100	100	50	50

a) nmoles C_2H_4/min/mg after reconstitution of extracts with FeMo-co

These three strains were also compared for their stability toward heat treatment by heating extracts to 56^oC for 5 minutes, followed by immediate cooling and centrifugation at 148,000 x g for one hour (Table 3). The wild type MoFe protein activity increases two fold by this purification step. Both the Δ_nif_E and Δ_nif_B strains are more heat sensitive and their specific activities do not increase after this treatment. The Δ_nif_H FeMo-co-deficient MoFe protein is extremely unstable leaving virtually no reconstitution activity in the supernatant. Thus, the maximum activation levels (Table 2) and heat sensitivity (Table 3) of the Δ_nif_E and Δ_nif_B proteins are similar to each other but are different from the Δ_nif_H protein.

Table 3. Heat Stability is Different.

	WT	Δ*nif*H	Δ*nif*B	Δ*nif*E
Before Heat	77.9 ± 0.5	72.4 ± 1.8	31.2 ± 1.0	30.0 ± 0.6
After Heat Supernatant	160.8 ± 3.8	6.9 ± 0.1	31.7 ± 0.6	29.8 ± 0.9
% In-activation	0	95	50	50

Figure 5 shows a native anaerobic gel electrophoresis separation of cell free extracts from Δ*nif*H, Δ*nif*B and Δ*nif*E strains. The wild type extracts give a single tight band that runs in the same position as purified MoFe protein. The MoFe protein synthesized by Δ*nif*H strain runs above the wild type band. After activation to 100% wild type activity with the Fe protein/MgATP and FeMo-co the Δ*nif*H protein band tightens up, moves down and looks just like the native protein (14). In contrast the FeMo-co-deficient MoFe proteins synthesized by the Δ*nif*B and Δ*nif*H strains always run as doublets on native gels with the top band running in an identical position to the wild type MoFe protein. This pattern does not change after reconstitution with FeMo-co to 50% wild type activity, suggesting that the bottom band cannot be reconstituted. The data in Figure 5 clearly show that the activatable MoFe proteins synthesized by DJ141 (Δ*nif*B) and DJ35 (Δ*nif*E) strains run in the same place on native gels but at a very different position from the DJ54 (Δ*nif*H) protein. Taken together, these data show that all FeMo-co-deficient forms of the MoFe proteins are not alike. The protein present in Δ*nif*B and Δ*nif*E strains are like each other but are unlike the protein synthesized by the Δ*nif*H strain.

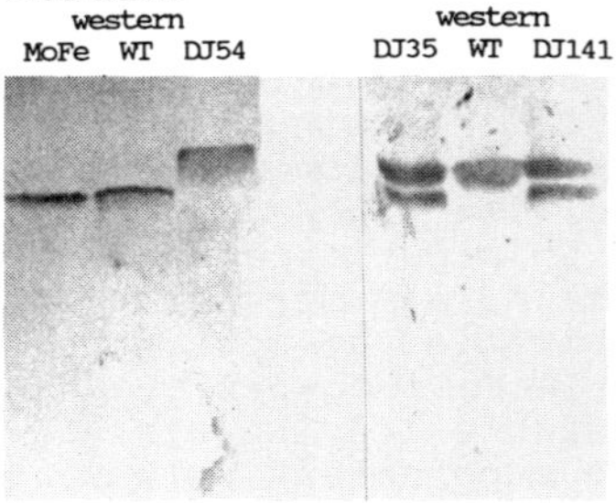

Figure 5. Native anaerobic gel electrophoresis of purified MoFe protein and cell free extracts of wild type *A. vinelandii*, Δ*nif*H (DJ54), Δ*nif*B (DJ141) and Δ*nif*E (DJ35) after blotting and reacting with anti-MoFe protein antibodies.

CONCLUSIONS

FeMo-co is biosynthesized in a process that requires the participation of the *nif*Q,B,N,E,V and H gene products (7). If the *nif*B,N,E and H gene products are defective *in vivo* due to mutation, then their function can be circumvented *in vitro* by addition of isolated FeMo-co (15, 17). We suggest that if the *nif*B,N and E only participate in FeMo-co biosynthesis then the FeMo-co-deficient MoFe proteins synthesized by $NifB^-$, N^- or E^- stains should be the same. Indeed we have shown they appear to be for the *A. vinelandii* Δ*nif*E (DJ35) and Δ*nif*B (DJ141) strains.

In a separate step of MoFe protein maturation, the *nif*D,K gene products are assembled with P-clusters, into a "FeMo-co-deficient" MoFe protein. This process appears to require at least the *nif*Z and *nif*W gene products (9), and possibly the 20 KD protein(s) which copurifies with the Nif B^- MoFe proteins (6,12). Our data can be explained if the Fe protein and MgATP are also involved in the assembly of the FeMo-co-deficient MoFe protein and we suspect that active Fe protein is required. This requirement can be circumvented *in vitro* if necessary (e.g. for Δ*nif*H, DJ54), by addition of the Fe protein with MgATP (14). This interpretation explains a published report showing that MgATP is not needed to insert FeMo-co into the purified Nif B^- *Av*UW45 protein(12). Thus, if the Fe protein and MgATP are present *in vivo* (e.g. for Nif B^- mutants) they fulfill their role in the synthesis of the FeMo-co-deficient MoFe protein, and do not have to be added *in vitro*.

Many of the gene products involved in synthesis of the FeMo-co-deficient MoFe protein seem to "fine tune" the MoFe protein. Thus, if they are missing, an active MoFe protein is still assembled, it just has less activity than the wild type protein. For example, *A. vinelandii* Δ*nif*Z and Δ*nif*W strains have about 50% of wild type MoFe protein activity (9). Strains that do not contain active Fe protein (e.g. Nif M^-, point mutations in *nif*H) also have lowered MoFe protein activity. For example, an *A. vinelandii* Δ*nif*M strain has 30% of wild type MoFe protein activity. This amount of activity is comparable to the 25% we see for our Δ*nif*H (DJ54) if we circumvent the requirement for *nif*H in FeMo-co biosynthesis by adding FeMo-co *in vitro*, but do not circumvent the requirement for Fe protein in MoFe protein maturation by adding Fe protein and MgATP (14). If this view of MoFe protein maturation is correct then the FeMo-co-deficient MoFe protein synthesized by a Δ*nif*H strain should be different from the MoFe proteins synthesized by Δ*nif*B and Δ*nif*E strains, as we have shown they are. Purification of the Δ*nif*H, Δ*nif*B and Δ*nif*E proteins will expose their differences and thus, the function of the Fe protein/MgATP in MoFe protein maturation.

ACKNOWLEDGEMENTS

We thank Dennis Dean for Av strains DJ141 and DJ35 and NIH grant #R01GM43144.

1. Brigle, K.E., Weiss, M.C., Newton, W.E. & Dean, D.R. (1987) J.Bacteriol **169**, 1547.
2. Burgess, B.K. (1990) Chemical Reviews in press, and references therein.
3. Burgess, B.K., Jacobs, D.B. & Stiefel, E.I. (1980) Biochim. Biophys. Acta. **614**, 196.
4. Burgess, B.K. Yang, S.-S., You, C.-B., Li, J.-G., Friesen, D., Pan, W., Steifel, E.I., Newton, W.E., Conradson, S.D. & Hodgson, K.O. (1981) in Current Perspectives in Nitrogen Fixation, eds. Gibson, A.H. & Newton, W.E. (Australian Academy of Sciences, Canberra), pp.71.
5. Dean, D.R. & Brigle, K.E. (1985) Proc. Natl. Acad. Sci. USA **82**, 5720.
6. Filler, W.A., Kemp, R.M., Ng, J.C., Hawkes, T.R., Dixon, R.A. & Smith, B.E. (1986) Eur. J. Biochem. **160**, 371.
7. Hoover, T.R., Imperial, J., Ludden, P.W. & Shah, V.K. (1988) BioFactors **1**, 199, and references therein.
8. Hoover, T.R., Imperial, J., Ludden, P.W. & Shah, V.K. (1989) Biochemistry **28**, 2768.
9. Jacobson, M.R., Cash, V.L., Weiss, M.C., Laird, N.F., Newton, W.E. & Dean, D.R. (1989) Mol. Gen. Genet. **219**, 49.
10. Joerger, R.D., Premakumar, R. & Bishop, P.E. (1986) J. Bacteriol. **168**, 673.
11. Paustian, T.D., Shah, V.K. & Roberts, G.P. (1989) Proc. Natl. Acad. Sci. USA **86**, 6082.
12. Paustian, T.D., Shah, V.K. & Roberts, G.P. (1990) Biochemistry **29**, 3515.
13. Robinson, A.C., Burgess, B.K. & Dean, D.R. (1986) J. Bacteriol. **166**, 180.
14. Robinson, A.C., Chun, T.W., Li, J.-G., & Burgess, B.K. (1989) J. Bacteriol. **264**, 10088.
15. Robinson, A.C., Dean, D.R. & Burgess, B.K. (1987) J. Biol. Chem. **262**, 14327.
16. Robinson, A.E., Richard, A.J.M., Thomson, A.J., Hawkes, T.R. & Smith, B.E. (1984) Biochem. J. **219**, 495.
17. Shah, V.K. & Brill, W.J. (1977) Proc. Natl Acad. Sci USA **74**, 3249.
18. Stiefel, E.I. & Cramer, S.P. (1985) in Molybdenum Enzymes, ed. Spiro, T.G. (John Wiley & Sons, New York), p. 411 and references therein.
19. Ugalde, R.A., Imperial, J., Shah, V.K. & Brill, W.J. (1984) J. Bacteriol. **159**, 888.

Gresshoff, Roth, Stacey & Newton (eds.) Nitrogen Fixation: Achievements and Objectives
Chapman and Hall New York • London 1990

In vitro synthesis of the iron-molybdenum cofactor and its analogs: Requirement of a non-*nif* gene product for the synthesis, and altered properties of dinitrogenase

Vinod K. Shah, Mark S. Madden, and Paul W. Ludden.

Department of Biochemistry and Center for the Study of Nitrogen Fixation, College of Agricultural and Life Sciences, University of Wisconsin-Madison, Madison, Wisconsin 53706, U.S.A.

I. Introduction

Nitrogenase catalyzes the ATP- and reductant-dependent reduction of N_2 to ammonia and the reduction of other substrates (i.e. acetylene, cyanide). In the absence of any reducible substrate, nitrogenase catalyzes the reduction of protons to H_2. Nitrogenase consists of two proteins, dinitrogenase (MoFe protein, component I) and dinitrogenase reductase (Fe protein, component II) (1,5). Dinitrogenase reductase donates electrons one at a time to dinitrogenase with the concomitant hydrolysis of ATP. Electrons passed to dinitrogenase are channelled to a unique prosthetic group called the iron-molybdenum cofactor (FeMo-co) which is composed of Fe, Mo, S, (27) and homocitrate (10). FeMo-co has been proposed as the site for substrate reduction (6, 28). In *Klebsiella pneumoniae,* at least six *nif* (nitrogen fixation) gene products (NIF Q, NIF B, NIF V, NIF N, NIF E, and dinitrogenase reductase) are involved in the synthesis of active FeMo-co. The NIF Q is required for early steps in the processing of Mo for FeMo-co biosynthesis (11). Mutations in *nif* B, *nif* N, or *nif* E result in formation of an apo-dinitrogenase (lacking FeMo-co), that can be activated *in vitro* with purified FeMo-co (24,27,31). Certain strains of *K. pneumoniae* and *Azotobacter vinelandii* with mutations in *nif* H (the structural gene for dinitrogenase reductase) fail to synthesize FeMo-co (4,25).

Dinitrogenase from Nif V^- mutants effectively reduces acetylene and protons, but not N_2; and proton reduction is inhibited by carbon monoxide (18,19). When apo-dinitrogenase is activated with FeMo-co from dinitrogenase of a Nif V^- mutant, the resulting holo-dinitrogenase has substrate reducing properties of the enzyme from the Nif V^- mutant, showing that the defect in Nif V^- mutants is in the FeMo-co (6). The dinitrogenase from Nif V^- mutants has been extensively studied with EPR, EXAFS, Mössbauer, magnetic circular dichroism, and ENDOR spectroscopy. No differences have been detected between the wild-type and NifV^- dinitrogenases, except for minor differences in the ENDOR spectra (3,20,21,26).

The structure and biosynthesis of FeMo-co has remained undefined and elusive since its discovery in 1977. To address this problem, a system for *in vitro* synthesis of the FeMo-co has been developed that requires molybdate, ATP, at least the Nif B, Nif N, Nif E, (29); dinitrogenase reductase (12,25,30); and homocitrate (R-2-hydroxy-1,2,4 butanetricarboxylic acid) (7,8). *In vitro* FeMo-co synthesis system has allowed: a) a sensitive biological assay of homocitrate and its subsequent quantitation in the FeMo-co (7,10), b) an enzymatic assay and purification of the Nif NE protein (22), c) an enzymatic assay and purification of the Nif B protein, d) determination of the structural requirements of organic acids for incorporation of ^{99}Mo into the FeMo-co (9), e) incorporation of homocitrate and its analogs (i.e. citrate, isocitrate, 1-OH citrate, homoisocitrate, etc.) into FeMo-co which results in altered substrate specificities and inhibitor susceptibilities of dinitrogenase (9,13).

II. Substrate Specificities and Inhibitor Susceptibilities of Dinitrogenase Activated with Homocitrate Analogs :

Use of citrate in place of homocitrate in the *in vitro* FeMo-co synthesis system resulted in the formation of holo-dinitrogenase with effective proton and acetylene reduction activities, but with poor N_2 reduction and no HD formation (9). In addition, proton reduction is inhibited by carbon monoxide (CO). These results are analogous to dinitrogenase obtained from Nif V^- mutants (6,18,19), and recently citrate was isolated from dinitrogenase of the Nif V^- mutant of *K. pneumoniae* (15). In this chapter, apo-dinitrogenase activated by FeMo-co synthesized *in vitro* using e.g. citrate will be referred to as citrate-dinitrogenase or dinitrogenase containing citrate or dinitrogenase activated with citrate.

Imperial *et al.* (13) tested several organic acids for their ability to replace homocitrate in the FeMo-co synthesis system and suggested a model for the general structural requirements of organic acid for incorporation of molybdenum into FeMo-co : 1) stereochemistry at the C-2 chiral carbon analogous to that of R-homocitrate; 2) a hydroxy group on the chiral carbon atom; 3) a carboxy group on the chiral carbon atom; 4) a terminal carboxy group beta to the chiral center; 5) a 4-6 carbon chain length containing two terminal carboxy groups.

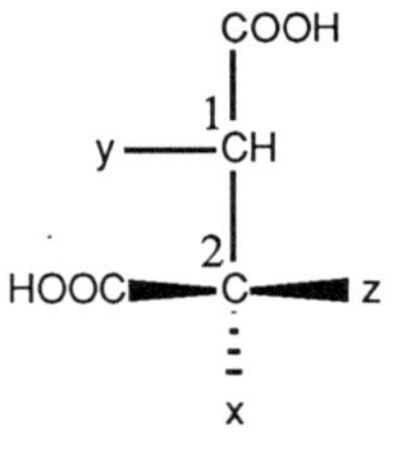

y = -H, z = -OH
x = $-CH_2CH_2COOH$ ('R'-homocitrate)
$-CH_2COOH$ (citrate)

y = -OH, z = -H
x = $-CH_2CH_2COOH$ (homoisocitrate)
x = $-CH_2COOH$ (isocitrate)

y = -OH, z = -OH
x = $-CH_2COOH$ (1-OH citrate)

Figure 1. Structural features of organic acids employed to study the role of homocitrate in FeMo-co.

To further study the structural requirements of homocitrate and its analogs, Madden *et al.* (17) prepared two diastereomers of fluorohomocitrate and expanded studies to other substrates of nitrogenase (i.e. cyanide) as well as other inhibitors (i.e. COS). The substrate reduction properties of dinitrogenase activated by the two diastereomers of fluorohomocitrate are markedly different. These differing substrate reduction properties and inhibitor susceptibilities of dinitrogenase activated by the diastereomers and other homocitrate analogs to inhibition by CO and COS, provide further insight into the functional groups on homocitrate required for active FeMo-co.

Dinitrogenase activated with *erythro*-fluorohomocitrate is capable of N_2 reduction, while dinitrogenase activated with *threo*-fluorohomocitrate is unable to catalyze N_2 reduction. On the other hand, dinitrogenases activated with either diastereomer of fluorohomocitrate reduce cyanide and protons as effectively as homocitrate-dinitrogenase, and acetylene 50-60% as effectively as homocitrate-dinitrogenase. Dinitrogenase activated with racemic mixtures of 1-OH acids (homoisocitrate, isocitrate, and 1-OH citrate) reduce cyanide 50-80% as effectively as homocitrate-dinitrogenase. These data clearly indicate that dinitrogenases containing homocitrate analogs with a fluorine atom or a hydroxy group at the C-1 carbon can still catalyze effective cyanide reduction. However, cyanide reduction of citrate-dinitrogenase was only 19% that of homocitrate-dinitrogenase. Because isocitrate-dinitrogenase can effectively reduce cyanide but citrate-dinitrogenase cannot, the position of the hydroxy group appears to be critical in the formation of dinitrogenase competent for cyanide reduction. In addition, dinitrogenase activated with 1-OH citrate shows cyanide reduction levels between that of isocitrate- and citrate-dinitrogenase, suggesting that the addition of a C-1 hydroxy group to citrate may counteract the effect associated with the C-2 hydroxy group of citrate.

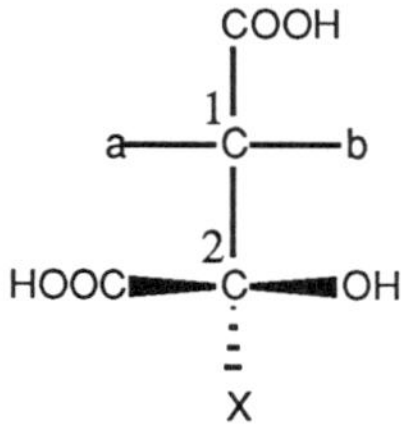

Figure 2. Model for the minimum organic moiety required for the synthesis of FeMo-co. Fluorine substitution at 'a' (as in *erythro*-fluorohomocitrate) shows proton, cyanide, C_2H_2, and N_2 reduction. Fluorine substitution at 'b' (*threo*-fluorohomocitrate) results in proton, cyanide, C_2H_2 reduction, but no N_2 reduction. Interchanging the hydrogen atom at 'a' with the hydroxy group at the C-2 carbon results in proton and cyanide reduction without CO (COS) inhibition.

Dinitrogenases containing homocitrate or *erythro*-fluorohomocitrate were not prone to CO inhibition of proton reduction. However, proton reduction of dinitrogenase containing *threo*-fluorohomocitrate or citrate were inhibited 59% by CO. These results suggest that CO inhibition of proton reduction may be associated with electronic alterations around the hydroxy or carboxy groups attached to the C-2 carbon of homocitrate analogs. The fluorine atom of *threo*-fluorohomocitrate is 'cis' with respect to the C-2 hydroxy group and 'trans' with respect to the C-2 carboxy group, while the fluorine of *erythro*-fluorohomocitrate is 'trans' with respect to the C-2 hydroxy group and 'cis' with respect to the C-2 carboxy group. The size of the fluorine atom is very similar to that of a hydrogen atom attached to carbon, therefore, steric effects from the fluorine atom of fluorohomocitrate are probably not causing the dramatic differences in the properties of these two analog-activated dinitrogenases. However, the three electron pairs of the fluorine atom may be perturbing the metal ligating properties of the C-2 hydroxy or carboxy oxygen atoms and contribute toward altered substrate specificities of dinitrogenases activated with fluorinated analogs of homocitrate. Imperial *et al.* (13) postulated that the carboxy and hydroxy groups of homocitrate may be ligating metal atoms when incorporated into FeMo-co. The hydroxy group and the carboxy group attached to the C-2 carbon of homocitrate analogs are essential to activate dinitrogenase, since tricarballylate (1,2,3-propanetricarboxylic acid), 1,2,4-butanetricarboxylic acid or 2-hydroxy-2-methyl-1,3-propanedicarboxylic acid (β-hydroxy-β-methylglutarate) did not facilitate the incorporation of molybdenum into dinitrogenase or generate active FeMo-co (9,13). Dinitrogenase could not be activated with tri-t-butyl fluorohomocitrate, providing further evidence that the carboxy groups are essential for enzyme activation.

Dinitrogenase containing homoisocitrate, isocitrate or 1-OH citrate effectively reduced protons without CO inhibition. However, dinitrogenase containing *threo*-isocitrate shows CO inhibition of proton reduction similar to dinitrogenase containing *threo*-fluorohomocitrate or citrate. Isocitrate does not contain a C-2 hydroxy group, yet the *threo*-diastereomer of isocitrate contains a hydroxy group 'trans' to the C-2 carboxy group and shows CO inhibition. This suggests that dinitrogenase containing homocitrate analogs with atoms having electron pairs attached to the C-1 carbon will show CO inhibition of proton and cyanide reduction if these atoms are 'trans' to the C-2 carboxy group (i.e. 'cis' to the C-2 hydroxy group or in the *threo*-conformation). The relationship between CO inhibition and the atoms attached to the C-1 carbon relative to the C-2 chiral center in homocitrate analogs was further investigated by analyzing CO inhibition of cyanide reduction with analog-activated dinitrogenases. CO effectively inhibited cyanide reduction by dinitrogenase containing homocitrate, *erythro*-fluorohomocitrate, *threo*-fluorohomocitrate, *threo*-isocitrate, or citrate. On the other hand, very little CO inhibition was observed for dinitrogenases activated with FeMo-co synthesized using the racemic mixtures of homoisocitrate, isocitrate or 1-OH citrate. These results show that the presence of a hydroxy group 'cis' to the C-2 carboxy group of homocitrate analogs prevents CO inhibition.

CO (COS) inhibition of substrate reduction is dependent upon the homocitrate analog incorporated into FeMo-co. CO and COS inhibit proton reduction by analog-activated dinitrogenase to similar extents. However, dinitrogenase containing homocitrate, citrate, or *threo*-isocitrate show high CO inhibition of cyanide reduction, while COS inhibition of cyanide reduction is low. These dramatic differences in CO and COS inhibition suggest that COS may act at a different site or by a different mechanism than CO. Lowe *et al.* (16) and Liang & Smith (14) proposed that cyanide binds to a more oxidized form of FeMo-co than that required for proton or N_2 reduction. Since CO inhibition was dependent upon the substrate being reduced (protons < cyanide

< acetylene), it would seem that the more reduced forms of dinitrogenase are more susceptible to CO inhibition.

These results further refine the model proposed by Imperial *et al.* (13) and establish that the stereochemistry at C-1 of homocitrate is crucial to the function of FeMo-co in dinitrogenase. Substrate reduction properties of altered forms of FeMo-co are dependent upon atoms attached to the C-1 carbon relative to the hydroxy and carboxy groups attached to the C-2 carbon. Fluorine substitution of hydrogen atoms on the C-1 carbon of fluorohomocitrate dramatically alters N_2 reduction as well as CO (COS) inhibition properties of dinitrogenase.

III. Preferential Incorporation of Erythro-diastereomers into the FeMo-co :

Proton and cyanide reduction activities of dinitrogenase containing the *threo*-diastereomer of isocitrate were 33% and 37% that of the dinitrogenase activated by FeMo-co synthesized using a racemic mixture of isocitrate. Dinitrogenase containing *threo*-isocitrate is susceptible to CO inhibition of proton and cyanide reduction. However, use of the racemic mixture of isocitrate for FeMo-co synthesis yields dinitrogenase that effectively reduces protons and cyanide without CO or COS inhibition. This suggests that the *erythro*-diastereomer may be preferred for incorporation into FeMo-co. In addition, CO inhibition of proton reduction by dinitrogenase activated with FeMo-co synthesized using a racemic mixture of fluorohomocitrate was 5%, characteristic of the *erythro*-fluorohomocitrate-containing dinitrogenase. If the FeMo-co synthesis system was unable to discriminate between the two fluorinated diastereomers, then CO inhibition of proton reduction by dinitrogenase activated using the racemic mixture should be at levels intermediate between *erythro*- and *threo*-fluorohomocitrate, (about 30%). Since identical CO inhibition was observed using either *erythro*-fluorohomocitrate or the racemic mixture of fluorohomocitrate, it seems that *erythro*-fluorohomocitrate is preferentially incorporated into dinitrogenase.

IV. Requirement of a Non-nif Gene Product :

Our primary goal is to synthesize FeMo-co *in vitro* under defined conditions and to understand the role of each of the gene product involved in its synthesis and insertion into apo-dinitrogenase. This information will be very valuable in understanding the mechanism of biosynthesis and the structure of FeMo-co. With this in mind, we have purified to homogeneity NIF NE (22) and apo-dinitrogenase (23) from *A. vinelandii* mutant UW45, and partially purified NIF B from *A. vinelandii* mutant CA12 and *K. pneumoniae* mutant UN1217 (30, V. Shah, unpublished data). In *K. pneumoniae,* >75% of the NIF B activity was found associated with the membranes. During purification of NIF B from *K. pneumoniae* membrane preparations, we observed inhibition of *in vitro* FeMo-co synthesis by Triton X-100 elutes of the membranes. A low molecular weight compound was isolated and purified from the membrane which specifically inhibits NIF B activity (2).

Most of the commercially available detergents at 0.5-1.0% concentrations failed to elute significant quantities of NIF B activity from *K.*

pneumoniae membrane suspensions (30). N-tetradecylsulfobetaine (SB-14) at a concentration of 1% eluted 60-90% of the NIF B activity from the membrane suspensions in 0.3 M NaCl (NaBr or LiBr) in 0.1 M Tris-HCl pH 7.4. NIF B was further purified from the SB-14 extracts by chromatography in the presence of SB-14.

In vitro FeMo-co synthesis using partially purified apo-dinitrogenase, dinitrogenase reductase, and NIF NE from *A. vinelandii* and NIF B from *K. pneumoniae* membranes produced approximately 10-15% of expected activity. Addition of 0.05% cholate, taurocholate, or N-laurylsarcosine during FeMo-co synthesis increased the activity 4 to 8 fold. Addition of lipid preparations of *K. pneumoniae* or *Escherichia coli* also resulted in a 2-3 fold increase in activity. These data suggest that lipid is required during some phase of FeMo-co synthesis or for FeMo-co stability. FeMo-co is known to be unstable in aqueous media (27), hence it is possible that *in vivo,* FeMo-co is protected by a lipid bilayer during its synthesis and is transported in the lipid bilayer to apo-dinitrogenase or to some other protein. Other possible roles of lipids may be that: 1) specific lipids may be involved *in vivo* for optimal enzyme activity, and some detergents can fulfill the lipid requirement *in vitro,* 2) the FeMo-co synthesis or some aspect of the synthesis may occur *in vivo* in the lipid bilayer, 3) the lipids act as allosteric effectors in modulating enzyme activity. It is not possible to differentiate between these possibilities at this time, because of the complexity of the FeMo-co synthesis system. If an independent assay is developed for NIF NE and NIF B, it may be possible to pinpoint the role of lipids. It will be interesting to know if lipids (detergents) stimulate *in vitro* FeMo-co synthesis using soluble NIF B from *A. vinelandii* or *K. pneumoniae.*

References

1. Bulen, W.A. & LeComte, J.R. (1966) *Proc. Natl. Acad. Sci. U.S.A.* ***56,*** 979-986.
2. Downs, D.M., Ludden, P.W., & Shah, V.K. (1990). *J. Bacteriol.* (submitted).
3. Eidness, M.K., Flank, A.M., Smith, B.E., Flood, A.C., Garner, C.D., & Cramer, S.P. (1986). *J. Am. Chem. Soc.* ***108,*** 2746-2747.
4. Filler, W.A., Kemp, R.M., Ng, J.C., Hawkes, T.R., Dixon, R.A., & Smith, B.E. (1986) *Eur. J. Biochem.,* **160,** 371-377.
5. Hageman, R.V. & Burris, R.H. (1978) *Proc. Natl. Acad. Sci. U.S.A.* ***75,*** 2699-2702.
6. Hawkes, T.R., McLean, P.A., & Smith, B.E. (1984) *Biochem. J.* **217,** 317-321.
7. Hoover, T.R., Shah, V.K., Roberts, G.P., & Ludden, P.W. (1986) *J. Bacteriol.* ***167,*** 999-1003.
8. Hoover, T.R., Robertson, A.D., Cerny, R.L., Hayes, R.N., Imperial, J., Shah, V.K. & Ludden, P.W. (1987) *Nature (London)* **329,** 855-857.
9. Hoover, T.R., Imperial, J., Liang, J., Ludden, P.W. & Shah, V.K. (1988) *Biochemistry* **27,** 3647-3652.
10. Hoover, T.R., Imperial, J., Ludden, P.W. & Shah, V.K. (1989) *Biochemistry,* **28,** 2768-2771.

11. Imperial, J., Ugalde, R.A., Shah, V.K., & Brill, W.J. (1984) *J. Bacteriol.,* **158,** 187-194.
12. Imperial, J., Shah, V.K., Hoover, T.R., & Ludden, P.W. (1988) in *Nitrogen Fixation: Hundred Years After,* (Bothe, H., de Bruijn, F.J., & Newton, W.E., Eds.) p. 128, Gustav Fischer Verlag, Stuttgart, New York.
13. Imperial, J., Hoover, T.R., Madden, M.S., Ludden, P.W., & Shah, V.K. (1989) *Biochemistry,* **28,** 7796-7799.
14. Liang, J., & Smith, B.E. (1984) in *Advances in Nitrogen Fixation Research,* (Veeger, C., & Newton, W.E., Eds.) p. 155, Nijhoff/Junk, Pudoc.
15. Liang, J., Madden, M.S., Shah, V.K., & Burris, R.H. (1990). *Biochemistry* (submitted).
16. Lowe, D.J., Fisher, K., Thorneley, R.N.F., Vaughn, S.A., & Burgess, B.K. (1989) *Biochemistry* **28,** 8460-8466.
17. Madden, M.S., Kindon, N.D., Ludden, P.W., & Shah, V.K. (1990). *Proc. Natl. Acad. Sci.* U.S.A. (in press).
18. McLean, P.A., & Dixon, R.A. (1981) *Nature* **292,** 655-656.
19. McLean, P.A., Smith, B.E., & Dixon, R.A. (1982). *Biochem J.* ***211,*** 589-597.
20. McLean, P.A., True, A., Chapman, S., Nelson, M., Teo, B.-K., Munck, E., Hoffman, B., & Orme-Johnson, W.H., (1985). in *Nitrogen Fixation Research Progress* (Evans, H.J., Bottomley, P.J., & Newton, W.E., Eds.) p.620, Martinus Nihoff Publishers, Boston.
21. McLean.P.A., True, A.E., Nelson, M.J., Chapman, S., Godfrey, M.R., Teo, B.-K., Orme-Johnson, W.H., & Hoffman, B.M. (1987). *J. Am. Chem. Soc.* ***109,*** 943-945.
22. Paustian, T.D., Shah, V.K., & Roberts, G.P. (1989). *Proc. Natl. Acad. Sci. U.S.A.* ***86,*** 6082-6086.
23. Paustian, T.D., Shah, V.K., & Roberts, G.P. (1990) *Biochemistry* , **29,** 3515-3522.
24. Roberts, G.P., MacNeil, D., & Brill, W.J. (1978) *J. Bacteriol.* ***136,*** 267-279.
25. Robinson, A.C., Dean, D.R. & Burgess, B.K. (1987) *J. Biol. Chem.* ***262,*** 14327-14332.
26. Smith, B.E., Bishop, P.E., Dixon, R.A., Eady, R.R., Filler, W.A., Lowe, D.J., Richards, A.J.M., Thomson, A.J., Thorneley, R.N.F., & Postgate, J.R. (1985). in *Nitrogen Fixation Research Progress* (Evans, H.J., Bottomley, P.J., & Newton, W.E., eds.) pp. 597-603. Martinus Nijhoff Publishers, Boston. 27. Shah, V.K. & Brill, W.J. (1977) *Proc. Natl. Acad. Sci. U.S.A.* ***74,*** 3249-3253.
28. Shah, V.K., Chisnell, J.R., & Brill, W.J. (1978). *Biochem. Biophys. Res. Commun.**81,*** 232-236.
29. Shah, V.K., Imperial, J., Ugalde, R.A., Ludden, P.W. & Brill, W.J. (1986) *Proc. Natl. Acad. Sci. U.S.A.* **83,** 1636-1640.
30. Shah, V.K., Hoover, T.R., Imperial, J., Paustian, T.D., Roberts, G.P., Ludden, P.W. (1988) in *Nitrogen Fixation: Hundred Years After,* (Bothe, H., de Bruijn, F.J., & Newton, W.E., Eds.) pp. 115-120, Gustav Fischer Verlag, Stuttgart, New York.
31. Ugalde, R.A., Imperial, J., Shah, V.K., & Brill, W.J. (1984). *J. Bacteriol.* ***159,*** 888-893.

Gresshoff, Roth, Stacey & Newton (eds.) Nitrogen Fixation: Achievements and Objectives
Chapman and Hall New York • London 1990

Identification of FeMoco domains within the nitrogenase MoFe protein

Dennis R. Dean, Deborah J. Scott, and William E. Newton

Departments of Anaerobic Microbiology and Biochemistry, The Virginia Polytechnic Institute and State University, Blacksburg VA, 24061

Introduction

During nitrogenase turnover, the Fe protein (a homodimer, Mr ≈ 60,000) sequentially delivers single electrons to the MoFe protein (an $\alpha_2\beta_2$ tetramer, Mr ≈ 220,000) upon which is located the substrate-reduction site (for a review, see Smith et al., this volume). Because the Fe protein is a single electron donor yet multiple electrons are required for substrate reduction, the MoFe protein must be able to accept and store multiple electrons. Very little is known about where and how these electrons are accumulated within the MoFe protein nor is anything known about their intramolecular delivery to the substrate-reduction site. The MoFe protein is known to contain redox centers of two distinct types; two iron-molybdenum cofactors, called FeMoco, and four proposed [4Fe-4S] centers, called P clusters. There is compelling biochemical and genetic evidence that FeMoco is the redox center which is, contains or is part of the substrate-reduction site. In its semi-reduced state, FeMoco exhibits a characteristic S = 3/2 electron paramagnetic resonance (EPR) signal that is unique to metallobiomolecules. This EPR signal can be recognized in whole cells, purified MoFe protein, and in FeMoco extracted from its polypeptide matrix. It is generally believed that P clusters are redox centers involved in the accumulation and intramolecular delivery of electrons to the substrate-reduction site.

Combined genetic and biochemical studies have been ongoing in our laboratories to examine the distribution and catalytic function of MoFe protein's metallocluster types. This study has as its basis the detection and analysis of altered catalytic and spectroscopic features that arise from substitution of specific amino acids located within metallocluster-binding domains. Studies of this nature require two considerations. First, individual residues within the polypeptide sequence must be targeted for substitution. Second, the appropriate substituting amino acids must be selected. To satisfy the first requirement, we previously developed a working model where major polypeptide environments for each of the metallocluster types were assigned. This model (Fig. 1) was based largely on primary sequence comparisons and metallocluster extrusion requirements. We propose that regions labeled domains I, II, and III within the MoFe protein α- and β-subunits provide nearly equivalent P-cluster environments and domains labeled IV and V within the α-subunit provide FeMoco environments. Various aspects of the development of this working model have been reported previously (2, 3, 4, 13). Here, we describe how primary sequence comparisons have been further used to provide a successful rationale for placing particular substitutions within the MoFe protein α-subunit which result in alteration of the functional properties of the FeMoco environment.

Figure 1 Alignment of interspecifically conserved Cys residues from the MoFe protein α- and β-subunits from A. vinelandii (Av), K. pneumoniae (Kp) and C. pasteurianum (Cp). Residues which are conserved in both subunits are indicated by asterisks. Numbers refer to the A. vinelandii sequence. The A. vinelandii MoFe protein α- and β-subunits have 492 and 523 residues, respectively. Domains I, II and III are targeted as potential P-cluster environments and domains IV and V are targeted as FeMoco environments.

```
       Domain I       Domain II       Domain III      Domain IV          Domain V
       62              83   88        154           183                      275
       ▼               ▼    ▼         ▼             ▼                        ▼
Av-α RGCAYAGSKGVV-8-HISHGPVGCGQYSR-60-CPIGLIGDDI-19-CEGFRGVSQSLGHHIAN-74-NLVHCYRSMNY
Kp-α RGCAYAGSKGVV-8-HISHGPVGCGQYSR-60-CPVGLIGDDI-19-CEGFRGVSQSLGHHIAN-73-NLVHCYRSMNY
Cp-α RGCAYAGCKGVV-8-HITHGPIGCSFYTW-60-CPVGLIGDDI-19-CEGYKGVSQSAGHHIAN-70-NLVQCHRSINY
       *   *           **  **  *      *     ***
Cp-β KTCQPVGAMYAA-7-PHSHGSQGCCSYHR-52-CLSETLGDDL
Kp-β KACQPLGAVLCS-7-PYVHGSQGCVAYFR-52-CMAEVIGDDL
Av-β KACQPLGAVLCA-7-PYVHGSQGCVAYFR-52-CMAEVIGDDL
       ▲               ▲    ▲         ▲
       70              90   95        153
       Domain I       Domain II       Domain III
```

General considerations

All site-specific amino-acid substitution studies suffer from the caveat that the effect of any particular substituting residue upon the global structure of a polypeptide, and thus long range effects, cannot be accurately predicted. This problem, evident even when a three-dimensional structure of the polypeptide is available, is further complicated in the case of nitrogenase because the role of the metalloclusters in controlling its protein folding patterns is not known. There are two ways this problem can be approached. One way involves "saturation substitution". Namely, many different amino acids can be individually substituted for a particular position and the observed differential effects upon catalysis interpreted in terms of the potential functionality of the substituting residues. Another approach, "domain switching", requires comparison of structurally similar domains from related proteins. Such comparisons can provide a strategy whereby amino-acid substitutions can be placed within a domain of one protein such that the primary sequence of a similar domain from a different protein having structural similarity but functional inequivalence is duplicated. Both of the above approaches have intrinsic merit. Domain switching provides a reasonable means to identify important residues while saturation substitution is particularly useful once the important residues are recognized.

Analysis of domain V within the MoFe protein α-subunit

The primary sequences of the MoFe-protein subunits, deduced from gene sequence data, are known for a wide variety of diazotrophic organisms. Such information is of particular value because exhaustive biochemical and biophysical data are available for MoFe proteins isolated from Azotobacter vinelandii, Klebsiella pneumoniae, and Clostridium pasteurianum. The metalloclusters contained within the MoFe protein from each of these organisms are believed to be identical. Thus, it should be possible to correlate differences in the catalytic and spectroscopic properties of MoFe proteins isolated from different species with constraints placed upon the individual metalloclusters by their respective polypeptide environments. One example of this was reported by Morgan et al. (11). They observed differences in the temperature relaxation patterns for the S = 3/2 EPR signal from MoFe protein isolated from C. pasteurianum when compared to MoFe protein isolated from A. vinelandii. If it is assumed that such spectroscopic differences arise from differences within the FeMoco polypeptide environments in these proteins, this feature should also be reflected in differences in primary sequence within their respective FeMoco-binding domains. Comparison of domain V from the C. pasteurianum and A. vinelandii MoFe proteins does indeed reveal significant differences in primary sequence. Of particular interest are residues that immediately flank Cys-275, which has been proposed to provide a direct mercaptide ligand to FeMoco. The A. vinelandii sequence within this region, His-Cys-Tyr, is replaced with the sequence Gln-Cys-His in C. pasteurianum. Thus, we reasoned that, if Cys-275 is a direct ligand to FeMoco (and there is now other evidence supporting this hypothesis, see 8, 9), the location of a His residue on opposite sides of the Cys ligand in MoFe protein from the different organisms is very likely to place different constraints upon FeMoco's environments. To examine this possibility, we replaced this portion of the A. vinelandii domain V sequence with the corresponding C. pasteurianum sequence.

Substitution by Gln for His-274 and His for Tyr-276 does not have any significant effect upon the diazotrophic growth capability of the mutant strain (the mutant strain is designated DJ386). Thus, FeMoco's functional environment has not been compromised in DJ386. This result is consistent with the hypothesis that domain V from both A. vinelandii and C. pasteurianum are functionally equivalent with respect to catalysis. However, the FeMoco environment within the MoFe protein from DJ386 is clearly altered because a significant change in the EPR spectrum of MoFe protein from DJ386 is observed. For example, the MoFe protein from the wild type exhibits g values of 4.32 and 3.65 in its S = 3/2 EPR spectrum while the MoFe protein from DJ386 exhibits g values of 4.47 and 3.52. Although these data do not necessarily indicate a direct interaction of the two altered residues with FeMoco, our results, and related studies performed in Sussex (8, 9), demonstrate that domain V within the MoFe protein α-subunit does provide a FeMoco environment.

Analysis of domain IV within the MoFe protein α-subunit

Comparison of the domain IV regions from A. vinelandii and C. pasteurianum reveal few differences in primary sequence and those differences that are observed can be considered as conservative replacements (see Fig. 1). Thus, we concluded that the same strategy used to examine domain V was unlikely to be as profitable when applied to domain IV. A different domain switching approach was therefore used to examine the functional properties of domain IV.

We have previously suggested that the products of the nifE and nifN genes provide a scaffold upon which FeMoco is assembled prior to its donation to the apo-MoFe protein (2). There are several experimental observations which support this idea: (i) FeMoco can be synthesized in the complete absence of the MoFe protein subunits (14); (ii) Both the nifE and the nifN products are absolutely required for biosynthesis of FeMoco; (iii) There is considerable primary sequence conservation between the nifE product and the MoFe α-subunit and between the nifN product and the MoFe protein β-subunit (see, for example, 2); and (iv) The nifEN products form an Fe-S containing $\alpha_2\beta_2$ tetramer analogous to the MoFe protein (12). Because FeMoco must escape from the nifEN- products complex during maturation of the MoFe protein, the respective FeMoco-binding sites within the MoFe protein and within the nifEN-products complex are likely to be structurally similar but functionally inequivalent. For example, similar folding patterns within domain IV of the MoFe protein α-subunit and within the corresponding region from the nifE gene product are indicated by the spatial conservation of the three Gly residues found within their respective sequences (see Fig. 2).

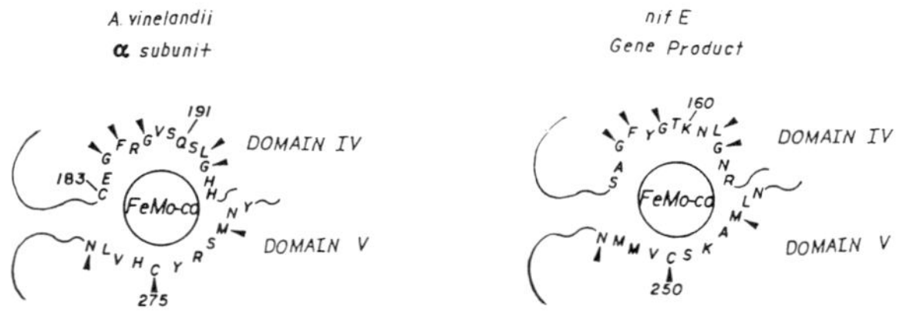

Figure 2 Schematic representation of residues conserved within the A. vinelandii MoFe protein α-subunit and the A. vinelandii nifE product proposed to provide FeMoco-binding or FeMoco intermediate-binding domains. Arrowheads indicate residues conserved within both the MoFe protein α-subunit and the nifE product.

Certain residues, however, that are found within the MoFe protein α-subunit domain IV region, which could potentially serve as direct ligands to FeMoco (e.g. Cys-183, His-195, His-196), are not found at the corresponding positions within the _nifE_ product. We have, therefore, systematically substituted residues within the MoFe protein α-subunit domain IV region by the corresponding residues from the _nifE_ product. In this way, we expected to alter the functional FeMoco-binding properties of domain IV without severely altering the global polypeptide structure. Results concerning the analysis of one of these mutants are described below.

Substitution of the MoFe protein α-subunit residue Gln-191 by the corresponding residue from the _nifE_ product, Lys, eliminated diazotrophic growth in the resultant mutant strain (the mutant strain is designated DJ255). The global structure of the MoFe protein present in DJ255 was not, however, severely altered by this substitution because it exhibits a fully developed EPR spectrum. Nevertheless, alterations in the g values ($g = 4.25$ and $g = 3.76$) and line shapes for the $S = 3/2$ EPR spectrum of MoFe protein from DJ255 clearly indicates a perturbation in its FeMoco environment. Catalytic properties of MoFe protein isolated from DJ255 also indicate an alteration in its FeMoco environment. In an atmosphere of 10% acetylene, nitrogenase isolated from wild type catalyzes the reduction of acetylene to ethylene with no trace of ethane production. By contrast, nitrogenase isolated from DJ255 produced ethane as well as ethylene (see Table 1). These changes in the spectroscopic and catalytic features of MoFe protein isolated from DJ255, as well as for MoFe protein from mutant strains having other substitutions within domain IV (our unpublished data), demonstrate that domain IV within the MoFe protein α-subunit provides a FeMoco environment.

Table 1 Acetylene Reduction Activities of Purified Proteins

MoFe protein	Ethylene	Ethane
wild type	1844 nm/min/mg	none
DJ255	54 nm/min/mg	7.8 nm/min/mg

Because nitrogenase from DJ255 is capable of reducing acetylene to ethane, a property also demonstrated for the V-dependent nitrogenase (5), it was important to establish that we have not observed an artifact. Namely, the possibility that ethane production in DJ255 results indirectly from accumulation of an alternative nitrogenase or a hybrid species containing conventional and alternative nitrogenase components must be considered. The following experiments firmly established that the spectroscopic and catalytic properties observed for MoFe protein from DJ255 is, in fact, the consequence of changes within the Mo-dependent nitrogenase: (i) The respective one-dimensional electrophoretic mobilities of the α- and β-subunits from both the altered and the native MoFe protein are identical, while those of the α- and β-subunits from alternative nitrogenases are different; (ii) Acetylene reduction assays using MoFe protein isolated from DJ255 result in ethylene and ethane formation, while mutant strains derived from DJ255 for which nifDK, nifB, nifE, or nifN are deleted, are incapable of reducing acetylene to either ethylene or ethane.

Alterations within domain IV can elicit a nifV⁻-like phenotype

Homocitrate is an organic constituent of FeMoco and is present within FeMoco in stoichiometric amounts with Mo (7). Biosynthesis of homocitrate requires the nifV gene product. In K. pneumoniae, inactivation of the nifV gene results in the accumulation of an altered form of FeMoco, one which apparently contains citrate rather than homocitrate. Nitrogenase produced from K. pneumoniae nifV mutants is capable of reducing acetylene but reduces nitrogen only at very low rates. Also, hydrogen evolution catalyzed by the nifV⁻ nitrogenase is sensitive to inhibition by CO, a property not exhibited by the native enzyme (10). These alterations in the properties of the nifV⁻ nitrogenase could arise either directly from changes in the interaction of the metal-sulfur core with the different organic component or indirectly from a change in the orientation of FeMoco within its polypeptide matrix. Because nitrogenase isolated from DJ255 is capable of reducing acetylene but not N_2 we wondered if it would also exhibit the other characteristic property of the nifV⁻ nitrogenase, H_2 evolution sensitive to CO. As shown in Table 2, H_2 evolution catalyzed by nitrogenase isolated from DJ255 is indeed sensitive to CO.

Table 2 Sensitivity of Hydrogen Evolution to CO (3.0 kPa)

MoFe Protein	H_2 Evolution -CO	H_2 Evolution +CO
wild type	2494 nm/min/mg	2841 nm/min/mg
DJ255	544 nm/min/mg	254 nm/min/mg

Such results suggest that the nifV⁻ phenotype could arise from a change in protein-FeMoco interactions. Thus, homocitrate might have an important function in directing the proper orientation of FeMoco within the polypeptide matrix. These results also indicate the possibility that the homocitrate portion of FeMoco might directly interact with domain IV within the MoFe protein α-subunit. Any of homocitrate's carboxylate groups could theoretically form strong electrostatic interactions with positively charged residues found within domain IV or domain V.

The nifV$^-$-like phenotype of nitrogenase isolated from DJ255 also raises the possibility that the structure of FeMoco is altered once inserted into the apo-DJ255-MoFe protein. To test this possibility, we performed reconstitution experiments analogous to those reported by Hawkes et al. (6) in their analysis of the nifV$^-$ nitrogenase. In one experiment, we isolated a derivative of DJ255 for which nifB is deleted. Apo-MoFe protein produced in this mutant was then reconstituted with FeMoco isolated from the wild type. The reconstituted protein exhibited substrate-reduction properties and a S = 3/2 EPR spectrum identical to the original DJ255 strain. This result shows that the DJ255 phenotype arises from an alteration in the polypeptide and not from a defect in FeMoco biosynthesis. In a separate experiment, apo-MoFe protein from a nifB-deletion mutant, which otherwise produced a wild type MoFe protein, was reconstituted using FeMoco extracted from the DJ255 MoFe protein. The reconstituted MoFe protein prepared in this way exhibited catalytic and spectroscopic properties characteristic of the wild type protein. These experiments show that the DJ255 phenotype results from an alteration in the MoFe protein α-subunit polypeptide and that FeMoco does not rearrange once it is inserted into the apo-DJ255-MoFe protein.

Summary

Results described in this brief review demonstrate the feasibility of altering the polypeptide environment of FeMoco such that changes in the catalytic and spectroscopic properties of nitrogenase are elicited. The success of our approach also illustrates the utility of domain switching as a rational method to analyze the functional properties of metallocluster-binding regions without necessarily compromising their structural integrity. With the identification of two FeMoco binding domains within the MoFe protein α-subunit in hand, we have now initiated saturation substitution studies on these domains. The further biophysical characterization of such altered MoFe proteins should aid in the development of a functional description of FeMoco's polypeptide environments and the contribution of those environments to nitrogenase catalysis.

Acknowledgements

We thank our colleagues H. May, K. Brigle, R. Setterquist, M. Weiss, J. Cantwell, N. Laird, V. Cash, and J. Windle for their contributions to this work and M. Buck and B. Smith for communicating results prior to publication. This work was supported by grant DMB-8510029 from the National Science Foundation.

References

1. Arnold, W., Rump, A., Klipp, W., Priefer, U. B. & Puhler, A. (1988) J. Mol. Biol. **203**, 715-738.

2. Brigle, K. E., Weiss, M. C., Newton, W. E. & Dean, D. R. (1987) J. Bacteriol. **169**, 1547-1553

3. Dean, D. R., Brigle, K. E., May, H. D. & Newton, W. E. (1988) in Nitrogen Fixation: Hundred Years After, eds. Bothe, de Bruijn & Newton (Gustav Fischer, Stuttgart, New York) pp. 107-113.

4. Dean, D. R., Setterquist, R. A., Brigle, K. E., Scott, D. J., Laird, N. F. & Newton, W. E. (1990) Molec. Microbiol. in Press.

5. Dilworth, M. J., Eady, R. R., Robson, R. L. & Miller, R. W. (1987) Nature **327**, 167-168.

6. Hawkes, T. R., McLean, P. A. & Smith, B. E. (1984) Biochem. J. **217**, 317-321.

7. Hoover, T. R., Imperial, J., Ludden, P. W. & Shah, V. K. (1989) Biochemistry **28**, 2768-2771.

8. Kent, H. M., Baines, M., Gormal, C., Smith, B. E. & Buck, M. (1990) Molec. Microbiol. in Press.

9. Kent, H. M., Ioannidis, I., Gormal, C., Smith, B. E. & Buck, M. (1989) Biochem. J. **264**, 257-264.

10. McLean, P. A., Smith, B. E. & Dixon, R. A. (1983) Biochem. J. **211**, 589-597.

11. Morgan, T. V., Mortenson, L. E., McDonald, J. W. & Watt, G. D. (1988) J. Inorg. Biochem. **33**, 111-120.

12. Paustian, T. D., Shah, V. K. & Roberts, G. P. (1989) Proc. Natl. Acad. Sci. USA **86**, 6082-6086.

13. Scott, D. J., May, H. D., Newton, W. E., Brigle, K. E. & Dean, D. R. (1990) Nature **343**, 188-190.

14. Ugalde, R. A., Imperial, J., Shah, V. K. & Brill, W. J. (1984) J. Bacteriol. **159**, 888-893.

Gresshoff, Roth, Stacey & Newton (eds.) Nitrogen Fixation: Achievements and Objectives
Chapman and Hall New York • London 1990

Kinetics and mechanisms of ATP hydrolysis, electron transfers and proton release by *Klebsiella pneumoniae* nitrogenase

Roger N.F. Thorneley.

AFRC Institute of Plant Science Research, Nitrogen Fixation Laboratory, University of Sussex, Brighton BN1 9RQ, U.K.

Eady et al. (3) and Hageman et al. (4) used the rapid acid quench technique to demonstrate that the transfer of one electron from the 4Fe-4S centre in the Fe protein to the MoFe protein is coupled to the hydrolysis of two equivalents of MgATP; a stoichiometry also given by the steady state rates of substrate reduction and P_i formation under optimum conditions.

MgATP hydrolysis is associated with the Fe protein-MoFe protein complex, since neither protein hydrolyses MgATP on its own (5). However, electron transfer from Fe protein to MoFe is not required for MgATP hydrolysis, since in the absence of $Na_2S_2O_4$, when the Fe protein becomes oxidised, hydrolysis of MgATP continues (2,5,6). This reductant-independent ATPase activity can explain why the limiting stoichiometry of 2MgATP hydrolysed for each electron transferred to substrate is not observed under all conditions.

These results can be explained by the sequential and competing reactions shown in scheme 1. This shows two competing cycles which represent reductant-dependent (cycle 1) and reductant-independent (cycle 2) ATPase activities. Cycle 1 is an expanded version of the Fe-cycle of Lowe & Thorneley (7).

Step 1 in this cycle is the rapid reversible formation of a protein complex involving reduced $Kp2(MgATP)_2$ and $Kp1^{\dagger}$. Stopped-flow calorimetry (Fig.1) and stopped-flow spectrophotometry have allowed the next two steps in cycle 1 to be resolved into an MgATP-cleavage step occurring on the protein complex prior to the transfer of a single electron from Kp2 to Kp1 (9). This resolution was only possible at 5°C, since at the temperature we usually use for our kinetic studies, 23°C, the MgATP cleavage reaction is too fast

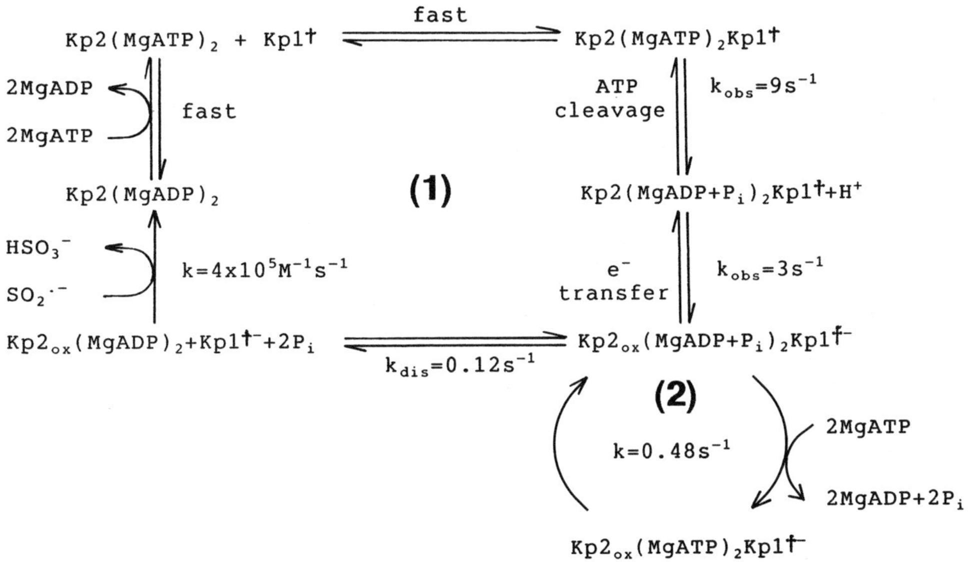

Scheme 1. Kp2 signifies the Fe-protein with two MgATP or MgADP binding sites. Kp1† represents one of two independently functioning halves of the tetrameric Kp1, MoFe-protein. Each Kp1† is assumed to contain one Mo substrate-binding site and one Kp2-binding site. The rate constants have been determined by stopped-flow spectrophotometry, stopped-flow calorimetry and steady-state H_2 and P_i assays at 6°C, pH 7.0. The ATP:2e ratio under these conditions was *ca*. 20:1.

($k>140s^{-1}$) to be monitored by the current version of the stopped-flow calorimeter.

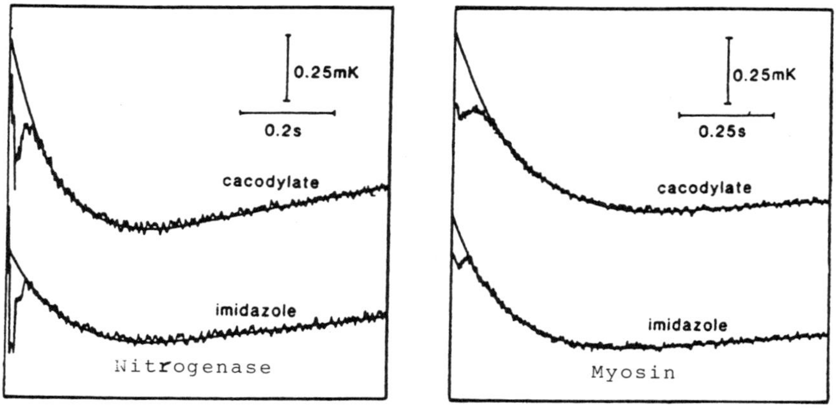

Fig.1. **ATP splitting by Myosin and Nitrogenase is Endothermic**

The temperature changes on mixing either Myosin-S1 (100 μM) or Nitrogenase (Kp1 = 76 μM, 110 μM Mo, Kp2 = 250 μM) with MgATP (10 mM) at 5°C in cacodylate ($\Delta H_{ionisation}$ = 0) and imidazole buffers ($\Delta H_{ionisation}$ = 36 kJ mol^{-1}) are shown. The different amplitudes in the two buffers allow the number of protons released concomitant with ATP cleavage to be calculated.

Therefore the first important conclusion is :-

1) **MgATP cleavage to yield bound MgADP + P_i precedes electron transfer from Kp2 to Kp1 within the protein complex.**

By studying the cleavage reaction in two different buffer systems, it is possible by comparing the amplitudes of the stopped-flow calorimeter traces to monitor proton uptake and release (Fig.1). This allows the second important conclusion :-

2) **MgATP cleavage to yield MgADP + P_i is accompanied by the release of a proton into solution.**

It would therefore appear that the proton that is necessarily liberated when MgATP is cleaved to MgADP +

P_i at pH7 is not retained within the protein complex and is not therefore the proton that acts as a substrate at the FeMo-cofactor to generate H_2, $2NH_3$ from N_2, or C_2H_4 from C_2H_2. This is consistent with the model developed below which suggests that the MgATP cleavage sites are on the surface of the protein complex (at the interface of Kp2 and Kp1†) while the electron transfer and substrate reduction sites are buried at the protein interface.

The third conclusion that comes from the stopped-flow calorimetry data is that :-

3) On Enzyme MgATP cleavage to yield MgADP + P_i is endothemic and reversible.

This leads to the concept of an energised complex ($Kp2(MgADP+P_i)_2Kp1$†) formed prior to electron transfer from the Fe- to MoFe-protein. A similar situation is thought to occur within the catalytic cycle of rabbit muscle myosin. Fig. 1 and Table 1 shows a comparison of data and the key parameters.

Table 1. **ATP cleavage on Nitrogenase and Myosin at pH7;5 and 23°C**

$$\text{E.MgATP} \underset{k_{obs}}{\overset{K}{\rightleftharpoons}} \text{E(MgADP+P}_i\text{)} + H^+$$

$$\Delta H_{obs}$$

	Nitrogenase	Myosin	T°C
k_{obs} (s^{-1})	9	7	5
k_{obs} (s^{-1})	140	160	23
ΔH_{obs}(kJ mol.ATP^{-1})	+36(72)*	+32(64)	5
H^+ release (observed)	0.5(1.0)	0.2(0.4)	5
Equilibrium constant K	~1	1	5
	>10	9	23

* Numbers in parentheses corrected for K represent complete cleavage. Data taken from reference 9.

In the myosin system, P_i release is exothermic, rate limiting and associated with a protein dissociation reaction (myosin-actin) which is the 'work step', in muscle contraction.

The similarities with the nitrogenase system are obvious and suggest that the P_i release step, which so far we have not been able to monitor without the perturbations associated with quenching methods, is the key to understanding 'energy transduction' by nitrogenase. We suggest that some of the free energy available after MgATP cleavage to yield bound (MgADP + P_i) but before or concomitant with P_i release, is used to lower the potential of the reduced substrate-reduction site on the MoFe-protein (i.e. super-reduced, epr-silent, FeMo-cofactor). This would explain why $Kp2(MgATP)_2$ is such a specific reductant of Kp1 and why $Kp2(MgADP)_2$ in the presence of added P_i will not reduce Kp1. In the latter case any complex that forms between $Kp2(MgADP)_2$, Kp1 and phosphate ion is not 'energised' by the cleavage of the β-γ phosphate bond of ATP.

It is interesting to note that the MgADP forms of Kp2 and Av2 have lower redox potentials, E_m = -350 mV and -450 mV, than the corresponding MgATP forms, E_m = -320 mV and 350 mV, respectively (8). This is consistent with the active form of Kp2 in the complex with Kp1, prior to electron transfer but after MgATP cleavage, resembling the MgADP-bound conformation of the free Kp2 protein (i.e. MgATP cleavage generates the more strongly reducing conformations of Kp2 and Av2).

The cleavage of MgATP and subsequent electron transfer from Kp2 to Kp1 is shown as being reversible in scheme 1. Evidence for this is the dependence of the absorbance change at 430 nm used to monitor oxidation of Kp2 on temperature (9). At 5°C, the overall equilibrium constant must be close to unity but at 23°C in excess of ten. Further evidence for reversible cleavage of MgATP comes from ^{18}O-exchange experiments when ^{18}O-labelled P_i was incubated with MgADP and $Kp2_{ox}$ and Kp1 (see Poster contribution by Thorneley et al. elsewhere in this book).

At 23°C, the rate of dissociation of the protein complex ($k = 6.4 s^{-1}$), measured by mixing $Kp2_{ox}(MgADP)_2$ with Kp1 in the absence of P_i (Thorneley & Lowe, 1983), predicts the measured substrate reduction rate (i.e. protein dissociation is rate limiting in the catalytic cycle). This means that P_i release after MgATP cleavage does not directly limit the rate of protein dissociation in a system that is hydrolysing MgATP. Indeed, data obtained at 5°C, when the ATP:2e ratio was 20:1 show that protein complex dissociation, although obligate for electron transfer to $Kp2_{ox}(MgADP)_2$ from $SO_2^{\cdot -}$, is not necessary for MgADP + P_i replacement by MgATP. This is shown in cycle 2 of scheme 1.

Thus we reach the fourth important conclusion, namely:-

4) **Dissociation of the protein complex, $Kp2_{ox}(MgADP)_2$ Kp1 is obligatory for reduction of oxidised Fe-protein by $SO_2^{\cdot -}$ but not for the replacement of MgADP by MgATP.**

This causes us to 'bury' the electron-transfer site at the interface of the Fe-protein with the MoFe protein, but to place the two MgATP/MgADP-binding sites on the surface, at the interface. It is important to remember that reductant dependent and independent ATPase activities are a function of the protein complex not of either of the isolated component proteins.

Thus the protein complex $Kp2_{ox}(MgADP + P_i)_2Kp1^{\dagger}$ is a common intermediate in cycles 1 and 2 of scheme 1. In the absence of $Na_2S_2O_4$, it is this protein complex that we propose hydrolyses MgATP and accounts for 'reductant-independent' ATPase activity. How far the ATP:$2e^-$ ratio deviates from a limiting value of 4 (two turns of cycle 1) depends on the steady state level of the intermediate. This in turn depends on the pH, temperature, relative and absolute concentrations of the Fe- and MoFe-proteins and the concentration of reductant ($Na_2S_2O_4$).

The last step of the Fe-cycle of Lowe & Thorneley (7), the reduction of $Kp2_{ox}(MgADP)_2$ by $SO_2^{\cdot -}$ and replacement of 2MgADP by 2MgATP, has been resolved into two reactions in scheme 1. Kinetic studies by Ashby & Thorneley (1) showed that $Kp2_{ox}$ bound 2MgADP and was reduced by one electron equivalent. They also showed that during turnover reduction preceded MgADP replacement by MgATP.

The importance of the protein complex formed from the Fe- and MoFe-proteins in hydrolysing MgATP, affecting energy transductions and substrate reduction causes us once again to criticise those who refer to the MoFe-protein as nitrogenase and the Fe-protein as nitrogenase reductase. **Nitrogen fixation is a property of the protein complex not just of the MoFe-protein** and the Fe-protein has a role significantly greater than that of an electron donor.

References

1. Ashby, G. and Thorneley, R.N.F. (1987), Biochem J., 246, 455-465.
2. Cordewener, J., Asbroek, A., Wassink, H., Eady, R.R., Haaker,H. and Veeger, C. (1987), Eur.J.Biochem., 162, 265-270.
3. Eady, R.R., Lowe, D.J. and Thorneley, R.N.F. (1978), FEBS LETT., 95, 211-213.
4. Hageman, R.V., Orme-Johnson, W.H. and Burris, R.H. (1980), Biochemistry, 19, 2333-2341.
5. Imam, S. and Eady, R.R. (1980), FEBS LETT., 110, 35-38.
6. Jeng, D.Y., Morris, J.A. and Mortenson, L.E. (1970), J.Biol.Chem., 245, 2809-2814.
7. Lowe, D.J. and Thorneley, R.N.F. (1983), Biochem.J., 215, 393-403.
8. Thorneley, R.N.F. and Ashby, G. (1989), Biochem.J., 261, 181-187.
9. Thorneley, R.N.F., Ashby, G., Howarth, J.V., Millar, N.C. and Gutfreund, H. (1989), Biochem.J., 264, 657-661.

Gresshoff, Roth, Stacey & Newton (eds.) Nitrogen Fixation: Achievements and Objectives
Chapman and Hall New York • London 1990

Crystal structure of the nitrogenase iron protein from *Azotobacter vinelandii*

Millie M. Georgiadis, Pinak Chakrabarti, and Douglas C. Rees

Division of Chemistry 147-75 CH, California Institute of Technology
Pasadena, CA 91125 USA

Introduction

The nitrogenase enzyme system consists of two component proteins, an iron (Fe-) protein and a molybdenum-iron (MoFe)-protein. Prior to substrate reduction, electrons are transferred from Fe-protein to MoFe-protein in an ATP-dependent process. The coupling of ATP hydrolysis to electron transfer is mediated by Fe-protein, which is the only known reductant of MoFe-protein supporting catalytic activity (8). Fe-protein is a dimer of two identical 32kDa subunits, the sequences of which are highly conserved in both conventional and alternate nitrogenases of a diverse group of prokaryotes. The Fe-protein dimer symmetrically binds one 4Fe-4S cluster to Cys 97 and Cys 132 from each subunit (sequence numbering of the *Azotobacter vinelandii* Fe-protein) (2,4). Unusual aspects of Fe-protein include the ATP-electron transfer coupling and the unique cluster environment, which is unlike any observed in ferredoxin structures. To address the structural basis of these properties, we have determined the crystal structure of the *A. vinelandii* Fe-protein at 3.0Å resolution. At present, 247 of the 289 residues have been modeled as a poly-alanine chain, and detailed fitting of the actual side chains is in progress. The current model is sufficient to begin addressing questions concerning the overall shape and fold of the protein, the environment of the cluster, the probable nucleotide binding site, the site(s) of interaction with MoFe-protein, and the interactions at the dimer interface.

Methods

Crystals of Fe-protein from *A. vinelandii* were grown in space group $P2_1$ with cell constants a=56.8Å, b=92.9Å, c=63.6Å, β=100° and contain one dimer per asymmetric unit. These crystals diffract to at least 2.5 Å resolution. The structure determination has involved multiple methods of phase determination and refinement. Initial phases were determined using multiple isomorphous replacement and anomalous scattering information. The non-crystallographic symmetry axis relating the two identical subunits was refined and subsequently used to average the electron density maps. A molecular envelope was determined using a modified Wang algorithm (10) and was then used in an iterative density averaging and solvent flattening procedure. A partial model consisting of isolated secondary structure elements was built and refined using the restrained least squares refinement program PROLSQ (3). Phases calculated from this partial model were combined with the original phases, followed by phase refinement with the iterative density averaging and solvent flattening procedure. The resulting electron density map was found to be interpretable, and the majority of the polypeptide backbone was modeled as poly-alanine. Currently the actual sequence is being fit to a new electron density map calculated using combined model and experimental phases, followed by phase refinement with the iterative density averaging and solvent flattening procedure described above.

Results and Discussion

Overall shape and fold of the protein

The overall shape of the Fe-protein dimer might be described as resembling a butterfly (Fig. 1). Each subunit of the Fe-protein is a single domain with an α/β type fold. The cluster is located near the "top" of the dimer molecule. A cleft exists in the middle of the dimer at the interface between the two subunits. At the core of each subunit is a 6 stranded parallel β sheet, which contains a nucleotide binding fold. An α helix extending from the cluster (the "cluster helix") includes residues 98 to 110. This helix contains Arg 100 that is reversibly ADP-ribosylated in photosynthetic bacteria. Additionally, there are 5 other α helices per subunit that consist of 7 or more residues.

The dimensions of each monomer are approximately 40Å across in the direction parallel to the two fold axis, with a cross section of about 35Å by 45Å in the perpendicular directions. The dimer is likewise 40Å across in the direction parallel to the two fold axis, with a perpendicular cross section of about 45Å by 75Å.

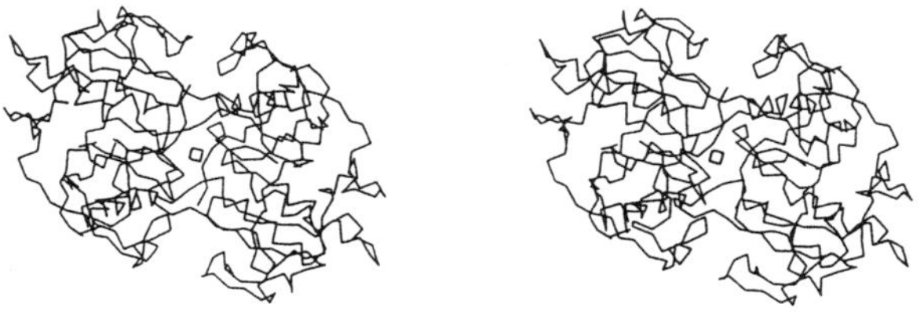

Fig. 1 Stereo pair of the Fe-protein dimer (Cα trace) looking down the two-fold axis relating the two subunits. The "top" of the molecule is closest to the viewer.

The cluster environment

The cluster environment may be described as a "gem setting" similar to a diamond setting in a ring. The 4Fe-4S cluster can be envisioned as a "gem" in a protein "setting" created by the interaction of the two identical subunits in the Fe-protein dimer. The "gem setting" may be visualized as follows. Residues 126-131 from each of the two-fold related dimers support the "gem setting" from below the cluster. Residue 132 from each subunit ligates the cluster. On the other sides of the cluster, residues 91-96 from each subunit form a strand/loop that wraps around the sides of the cluster and packs against residues 130-133 of the two-fold related subunits. Cys 97 provides the second ligand to the cluster from each subunit. Residues C-terminal from the cluster ligands of each subunit complete the "setting" for the cluster as seen in Fig. 2. The two-fold symmetric ligation of the 4Fe-4S cluster is consistent with the results of chemical modification (2), genetic data (4), and NMR studies (5).

Unlike the ferredoxin 4Fe-4S clusters which are completely surrounded by protein and hence buried from exposure to the solvent (8), the Fe-protein cluster is exposed to solvent along the "top face". The solvent accessibility was observed in the pulsed EPR studies of Morgan (6). The exposed nature of the cluster may help explain the extreme oxygen sensitivity of the Fe-protein. The surface accessibility of the cluster is presumably important for electron transfer interactions with the MoFe-protein.

Fig. 2 Stereo pair of a poly-alanine model showing a view of the protein environment surrounding the cluster as viewed from the bottom of the cluster.

The nucleotide binding fold

The nucleotide binding profile defined by Wierenga *et al.* was designed to identify βαβ ADP-binding folds in proteins (11). This profile consists of a consensus Gly-X-Gly-X-X-Gly sequence pattern and an acidic residue in the second β strand. Fe-protein has two sequences containing the Gly-X-Gly-X-X-Gly pattern beginning at Gly 9 and Gly 94. The βα part of the fold has been identified in the Fe-protein structure starting at Gly 9. The terminal phosphate of the nucleotide moiety in many mononucleotide binding proteins binds near the consensus sequence. We have been able to occupy this site with a molybdate anion, which we believe is acting as a phosphate analog (Fig. 3). We have so far been unable to obtain crystals with bound nucleotide that diffract to high resolution, however, and thus do not have any direct information about the binding of the adenine portion of the nucleotide. The molybdate site is located approximately 20Å from the cluster. The distance between the molybdate sites related by the two-fold symmetry of the dimer is also about 20Å. If we assume that the molybdate site is indeed the terminal phosphate site, it is impossible for the phosphate groups of ATP to interact directly with the cluster. Once the sequence modelling has been completed, it may be possible to model in an adenine nucleotide based on the conformations observed in other ADP binding proteins.

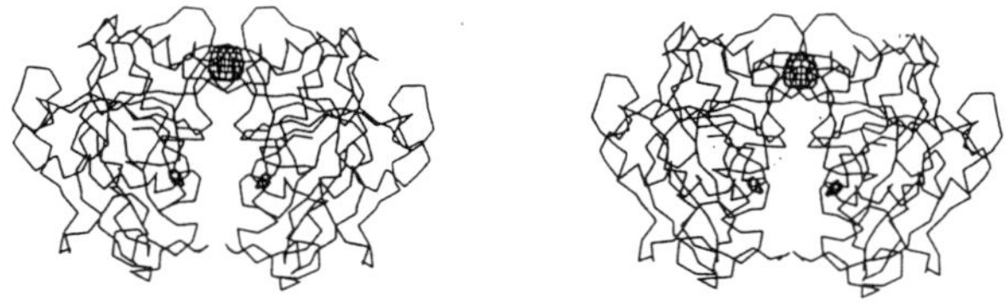

Fig. 3 Stereo pair of the anomalous difference Fourier map superimposed on the Cα trace of the Fe-protein dimer. The large peak in the center is the 4Fe-4S cluster, and the two small peaks are the molybdate sites of the two subunits. The molybdate sites are believed to be acting as analogs of the nucleotide terminal phosphate groups.

Interaction with the MoFe-protein

One of the more interesting questions to be answered in nitrogenase studies is the nature of the site(s) of interaction between the Fe-protein and the MoFe-protein and the proximity of these sites to the cluster and nucleotide binding sites of the Fe-protein. Willing *et al.*(12,13) have isolated a cross-linked species from *A. vinelandii* consisting of the β subunit of MoFe-protein and one Fe-protein subunit. The residues involved in the cross-link have been identified as Glu 112 from Fe-protein and Lys 399 from the MoFe-protein. Glu 112 is located in a loop that is directly C-terminal to the "cluster helix" and is located at the "top" of the molecule. This portion of the molecule is exposed and thus available for binding to the MoFe-protein. The cross-linking site is approximately 25Å away from the cluster and is some 30Å away from the molybdate site. The distance between Glu 112 and the two-fold related residue is approximately 55Å.

Another residue involved in the interaction between the Fe-protein and the MoFe-protein is Arg 100 which is located in the "cluster helix". ADP ribosylation of Arg 100 in photosynthetic bacterial nitrogenases blocks binding of the Fe-protein to the MoFe-protein in physiological response to the presence of fixed nitrogen (7,9). The location of Glu 112 and Arg 100, along with the 4Fe-4S cluster, at the "top" of the molecule suggests that this region of Fe-protein contacts the MoFe-protein. It also is possible that the Fe-protein subunits in the dimer interact asymmetrically with the MoFe-protein, as indicated by an analysis of the cross-linking reaction between the two proteins (12,13).

Interactions at the dimer interface

The majority of the packing contacts between the two Fe-protein subunits occur in the vicinity of the cluster. In addition to the close packing of the strands and loops containing the cluster ligands described above, there are several other packing interactions that can be identified. The cluster-ligand (Cys 97) containing loop interacts with residues in the middle of a long α helix that includes residues 150-178 of the two-fold related subunit. The C-terminal strand interacts with helix residues of the two fold related subunit in the vicinity of residue 250. Additional contacts involve the two-fold related C-terminal strands that pack together across the two-fold axis between the two subunits (Fig. 4). In addition to these protein-protein contacts, the location of the putative nucleotide phosphate site near the cleft between the subunits suggests that this interface region might also be involved in nucleotide binding.

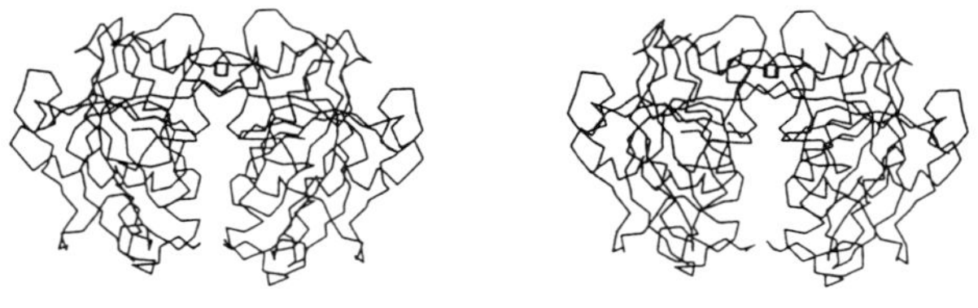

Fig. 4 Stereo pair of the Cα trace of the Fe-protein dimer viewed perpendicular to the two-fold axis relating the two subunits. This view shows the cleft between the subunits.

Although our preliminary structure analysis does not provide any specific details about the nature of the conformational change that is believed to occur upon the binding of nucleotide by Fe-protein, the structure of the dimer suggests a possible mechanism for generating a conformational change. Cluster chelation experiments have shown that the complex kinetics observed in the presence of ATP are consistent with two conformers of the Fe-protein (1). An alternate Cys ligation model for the cluster can be identified that might represent a second conformational state of Fe-protein. The isolated cluster has 3 two-fold axes relating the 4 tetrahedrally arranged Fe sites. If the ligands to the cluster from one subunit are fixed, the ligands from the second subunit may be placed in two distinct arrangements, corresponding to application of alternate two fold operations through the cluster in positioning the second subunit (Fig. 5).

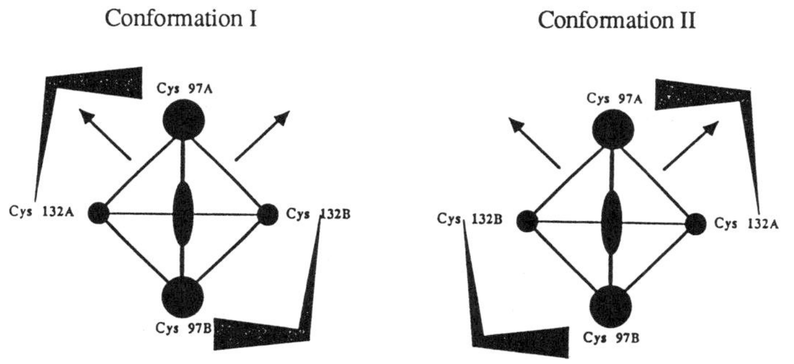

Fig. 5 Illustration of the two alternate Cys ligation patterns possible for the two subunits of Fe-protein, designated as conformations I and II.

These two conformations (if, in fact, they both exist) could possibly interchange through some type of cluster-ligand exchange reaction. Although the immediate environment of the cluster will be similar in these two conformations (since the same ligands coordinate the cluster in each case), and although each conformation will have a molecular two-fold axis, the relative orientations of the subunits in the two conformations will differ significantly. If one form has the two subunits arranged side-by-side (as observed in the present structure), the second form will have the two subunits oriented in an approximately antiparallel, head-to-head manner. In the second form, the cleft region between the subunits would be opened up to the solvent. If the cleft region does represent the site of ATP binding, the dynamic formation of this second conformation during the electron transfer reaction could provide a mechanism for the MoFe-protein to directly interact with ATP and participate in the ATPase activity. While this proposal is highly speculative, it serves to illustrate the possible structural, dynamic and mechanistic surprises that await us in the nitrogenase system.

References

(1) Deits, T. L. and Howard, J. B. (1989) J. Biol. Chem. **264**, 6619-6628.

(2) Hausinger, R. P. and Howard, J. B. (1983) J. Biol. Chem. **257**, 13468-13492.

(3) Hendrickson, W. A. (1985) Methods Enzymol. **115**, 252-270.

(4) Howard, J. B., Davis, R., Noldenhauer, B., Cash, V. L., and Dean, D. (1989) J. Biol. Chem. **264**, 11270-11274.

(5) Meyer, J., Gaillard, J., and Moulis, J.-M.(1988) Biochem. **27**, 6150-6156.

(6) Morgan, T. V., McCracken, J., Orme-Johnson, W. H., Mims, W. B., Mortenson, L. E., and Peisach, J. (1990) Biochem. **29**, 3077-3082.

(7) Murrell, S. A., Lowery, R. G., and Ludden, P. W. (1988) Biochem. J. **251**, 609-612.

(8) Orme-Johnson, W. H. (1985) Annu. Rev. Biophys. Biophys. Chem. **14**, 419-459.

(9) Pope, M. P., Murell, S. A. and Ludden, P. W. (1985) Proc. Natl. Acad. Sci. USA **82**, 3173-3145.

(10) Wang, B. C. (1985) Methods Enzymol. **115**, 90-112.

(11) Wierenga, R. K., Terpstra, P., and Hol, W. G. J. (1986) J. Mol. Biol. **187**, 101-107.

(12) Willing, A., Georgiadis, M. M., Rees, D. C., and Howard, J. B. (1989) J. Biol. Chem. **264**, 8499-8503.

(13) Willing, A. and Howard, J. B. (1990) J. Biol. Chem. **265**, 6596-6599.

Gresshoff, Roth, Stacey & Newton (eds.) Nitrogen Fixation: Achievements and Objectives
Chapman and Hall New York • London 1990

Structure of the nitrogenase MoFe protein: Spatial distribution of the intrinsic metal atoms determined by X-ray anomalous scattering

J.T. Bolin[1], A.E. Ronco[1], L.E. Mortenson[2], T.V. Morgan[2], M. Williamson[3] and N.-h. Xuong[3].

[1]Biological Sciences, Purdue Univ., W. Lafayette, IN 47907 USA.
[2]Center for Metalloenzyme Studies, Univ. Georgia, Athens, GA 30602 USA.
[3]Biology Dept., Univ. California-San Diego, LaJolla, CA 92093 USA.

Introduction

A focal point for discussions of the organization of the metal atoms bound to nitrogenase MoFe protein is the model proposed by Zimmerman *et al.* on the basis of a series of EPR and Mössbauer experiments [1]. This model defined two types of metal-sulfur clusters, M-centers and P-clusters. The original and current versions of the model both identify the M-centers as FeMo-cofactors. Two are bound per $\alpha_2\beta_2$ tetramer; each comprises 1 Mo and 6-7 Fe atoms. P-clusters account for an additional 16 Fe atoms organized into four Fe_4S_4 clusters. The current model recognizes two pairs of two distinct types of P-clusters. Thus MoFe protein should bind one FeMo-cofactor and one P-cluster of each type per $\alpha\beta$ unit [2].

In the experiments leading to the above model, MoFe protein was titrated to states more oxidized than the native form. Alternative models offer differing views of the course of oxidation, and also differ with respect to the organization of the non-cofactor Fe atoms. On the basis of oxidative and reductive titrations monitored by a variety of techniques, Watt and co-workers [3] have argued for the presence of three centers possessing the properties of P-clusters and one center of a different type. Recently, two models were proposed by Hagen *et al.* [4]. One of these collects 8 Fe atoms into each of two clusters, while the other holds to the dominant view of four Fe_4S_4 P-clusters, but suggests that oxidation of P-clusters by more than 1 e^- from the native state splits the set into two pairs with different properties.

As part of our effort to determine the three-dimensional crystal structure of the MoFe protein from *Clostridium pasteurianum* (hereafter Cp1), we have obtained superb X-ray diffraction data that allow measurement of the X-ray anomalous scattering that originates from the intrinsic metal atoms of Cp1. These data define the organization of the metal atoms into clusters as well as the spatial distribution of the clusters with respect to each other.

Methods

Cp1 was purified from *Clostridium pasteurianum* strain W5 by a procedure derived from the original protocol [5]. Single crystals were grown at 4° C in a nitrogen atmosphere glove box by a microdialysis method based on the previously reported batch crystallization [6]. The space group of the crystals is $P2_1$, with unit cell parameters a = 69.9 Å, b = 151.2 Å, c = 121.8 Å, and β =110.2°, and one molecule per asymmetric unit. For X-ray diffraction studies, crystals were transferred to a freshly prepared stabilizing solution containing 8% polyethyleneglycol-8000, 0.4M $MgCl_2$, 50 mM tris-Cl pH 7.5, and 1 mM $Na_2S_2O_4$. After several hours they were mounted and sealed, within the glove box, in quartz glass X-ray capillaries. A wet mount was used, and thus the crystals were always completely immersed in the stabilizing solution during X-ray experiments. Diffraction data were measured on the Mark II and Mark III area-detector diffractometers at the U.C.S.D. National Research Resource for Protein Crystallography using a strategy designed to optimize measurement of Bijvoet differences. Details of the diffraction experiments and data processing will be reported elsewhere.

Technical Background

For the benefit of readers unfamiliar with crystallographic terminology, we present a brief introduction to the theory which underlies the experiments described below.

In a typical single-crystal X-ray diffraction experiment, one measures a set of reflection intensities, I(h), where h is a triplet of integers that identifies the reflection and specifies its direction. I(h) is proportional to the square of the amplitude of a complex number called the structure factor, F(h), which can be considered a vector sum of the scattering contributions from all atoms in the structure. When the wavelength of the radiation used is far removed from wavelengths corresponding to the X-ray absorption frequencies of the atoms in the crystal, the contribution of each atom can be modeled by a real-valued atomic scattering factor, f°. In this case $|F(h)|$ and $|F(-h)|$ are equal.

When the wavelength used, λ, is near an X-ray absorption edge of the atoms in the crystal, this equality no longer holds and anomalous scattering can be observed. According to standard treatments, the atomic scattering factor becomes highly λ-dependent, and the net atomic scattering factor, f, is given by the complex-valued sum $f = f^{\circ} + f' + if''$, where f' and f'' incorporate the λ-dependent anomalous scattering. Each term in the sum is measured in electrons and depends on atomic number. Thus, Fe exhibits significant anomalous scattering for Cu-Kα radiation, with f°, f', and f'' equal to 26, -1.1, and 3.4 e^-, respectively, at zero scattering angle. In contrast, S shows weaker anomalous scattering with f°, f', and f'' equal to 16, 0.3, and 0.6 e^-, whereas C, N, and O are normal scatterers and have negligible values for f' and f''.

The presence of anomalously scattering atoms allows measurement of difference signals useful for locating the anomalous scatterers and solving the crystallographic phase problem. Bijvoet differences, $(|F(h)| - |F(-h)|)$, arise from f'' and are obtained from measurements of both I(h) and I(-h) at a single wavelength. A common method for locating anomalous scatterers involves calculation of a Bijvoet difference Patterson map [7]. This three-dimensional function should have peaks at positions corresponding to the endpoints of vectors between anomalously scatterering atoms, and the amplitude of each peak should be proportional to the product of the f'' values of the atoms at the ends of the vector. Peaks arising from atoms whose positions are related by crystallographic symmetry occur only in certain locations and are called Harker peaks, whereas general peaks arise from atoms not related by crystallographic symmetry, and can occur anywhere in the map.

In the case of MoFe proteins, the intrinsic metal atoms give rise to significant anomalous scattering at wavelengths near the Fe-K absorption edge, 1.74 Å. We have measured diffraction data from crystals of Cp1 using Cu-Kα and Co-Kα radiation and have calculated Bijvoet difference Pattersons maps from the data. On the basis of these Patterson maps, we can deduce the spatial distribution of the metal atoms in Cp1, but only to a resolution where a typical metal-sulfur cluster should scatter as a unit. That is, the results reported below give information about the nature and position of clusters of metal atoms, not individual metal atoms.

Results

Roughly 450,000 reflections to 2.6 Å Bragg spacing were measured using Cu-Kα radiation, λ = 1.54 Å. We are concerned here with a subset including data to 5 Å which specifies an Rsym value of 3% when the data are merged in Laue group 2. For this subset 97% of the possible reflections in Laue group 2 were measured; the average redundancy was 5 measurements per reflection.

Using Co-Kα radiation, λ= 1.79 Å, 651,000 measurements were made to a limiting spacing of 3 Å. The subset to 5 Å is virtually complete and specifies an Rsym value of 5.1% in Laue group 2, with an average redundancy of more than 10 measurements per reflection.

Before reporting the actual results from analysis of Bijvoet difference Patterson maps, we wish to describe the expected result of the Cu-Kα experiment if one assumes two FeMo-cofactors of stoichiometry $Mo_1Fe_7S_8$, and 4 separate Fe_4S_4 P-clusters per tetramer. For data with Bragg spacings greater than 5 Å, each cluster should scatter as a unit and one can add the f'' values for the atoms in the cluster to estimate the anomalous scattering power. Thus, based on standard values [8] for f'', $Mo_1Fe_7S_8$ clusters have a total f'' of 30 e^-, whereas Fe_4S_4 clusters have a total of 15 e^-. The Bijvoet difference Patterson map should have two Harker peaks of relative magnitude 4 and four peaks of magnitude 1. Two general peaks of magnitude 4, twelve of magnitude 1, and sixteen of intermediate magnitude are also predicted.

The Harker section from a Patterson map based on data between 20 and 5 Å is shown in Figure 1. Clearly, there are four major peaks of roughly equivalent magnitude, rather than the result expected from the four P-cluster model. The relative maximum values of peaks 1, 2, 3, and 4 are 1050, 860, 910, and 610, respectively, where the standard deviation of the entire map is roughly 100. Furthermore, there are exactly 12 general peaks elsewhere in the map whose peak maxima range from 560 to 990. These 4 Harker and 12 general peaks strongly suggest that Cp1 binds four large metal-sulfur clusters, all containing roughly the same number of metals. Therefore, we propose that the Fe atoms assigned to P-clusters in the standard model are found in two 8 Fe groups that are approximately as well integrated, in a spatial sense, as are the metal atoms in the FeMo-cofactor.

By further analysis of this Patterson map we determined the positions of the four clusters in the crystallographic asymmetric unit. These positions were expanded by crystallographic symmetry into nine contiguous unit cells and the resulting 72 positions were examined for groups of four consistent with the known molecular symmetry and approximate molecular size of Cp1. Thus, we sought two pairs of positions such that the line segments between the paired positions are both perpendicular to the known direction of the molecular two-fold symmetry axis and not inconsistent with a molecular diameter of roughly 80 Å. A clearly superior solution was found that places cluster sites derived from peaks 1 and 3 in one pair related by the molecular two-fold, and sites derived from peaks 2 and 4 in another. In the interest of clarity, we will describe the geometry of this solution after presenting the results obtained with Co-Kα radiation.

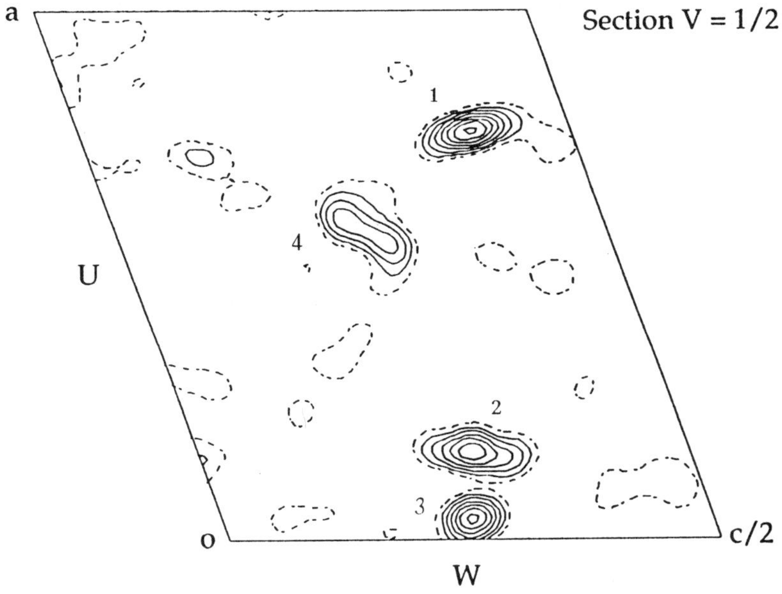

Figure 1. The Harker section (V=1/2) of a Cu-Kα Bijvoet difference Patterson map based on 20 - 5 Å data. The first contour level (dashed) and the interval between levels are both set to one standard deviation of the map. The numbers 1 - 4 indicate the four significant Harker peaks described in the text.

The use of Co-Kα radiation allows us to distinguish between $Mo_1Fe_7S_8$ and Fe_8S_8 clusters because the anomalous scattering power of Fe is only 16% of its value at Cu-Kα whereas that of Mo is 30% larger. Thus, one expects a total f'' of 13 e^- for a $Mo_1Fe_7S_8$ cluster and 9 e^- for an Fe_8S_8 cluster. Harker peaks arising from vectors between $Mo_1Fe_7S_8$ clusters should be twice the magnitude of peaks arising from vectors between Fe_8S_8 clusters. The Harker section of a Bijvoet difference Patterson map based on data between 20 and 5 Å is shown in Figure 2. As expected, two of the Harker peaks found in the Cu map, peaks 1 and 3, are enhanced relative to the others. The peak positions have shifted slightly relative to the Cu map. However, the maxima in the areas assigned to peaks 1 - 4 in the Cu map agree fairly well with the expected 2:1 ratio: peaks 1 and 3 have maxima of 520 and 740; peaks 1 and 4 have maxima of 374 and 350. We assign peaks 1 and 3 to the FeMo-cofactors, a result consistent with the pairing made on the basis of the Cu maps.

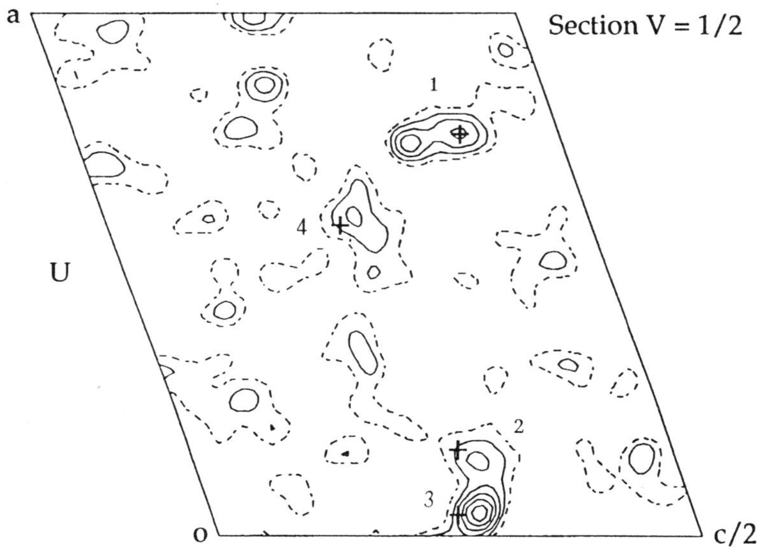

Figure 2. The Harker section (V=1/2) of a Co-Kα Bijvoet difference Patterson map based on 20 - 5 Å data, as described in the text. Contour levels and peak numbers are as described in the caption for Figure 1. A '+' symbol marks the center of the analogous peak in the Cu-Kα map.

Taken together, the two Patterson studies lead to the following model for the spatial distribution of the metal-sulfur clusters in Cp1. The two FeMo-cofactors are bound at positions related by the molecular two-fold symmetry axis and are separated by roughly 70 Å. Similarly, a second pair of large metal-sulfur clusters, which we may reasonably assume contain roughly 8 Fe atoms each, are also found at positions related by molecular symmetry axis and separated by 70 Å. The distance between either FeMo-cofactor site and the closest '8-Fe' cluster site is 19 Å, and thus these nearby sites probably belong to the same half-molecule. It should be noted that the two 70 Å inter-site vectors do not lie in the same plane, but are separated by roughly 9 Å at the point of closest approach (along the molecular symmetry axis). A schematic drawing of the cluster arrangement is given in Figure 3.

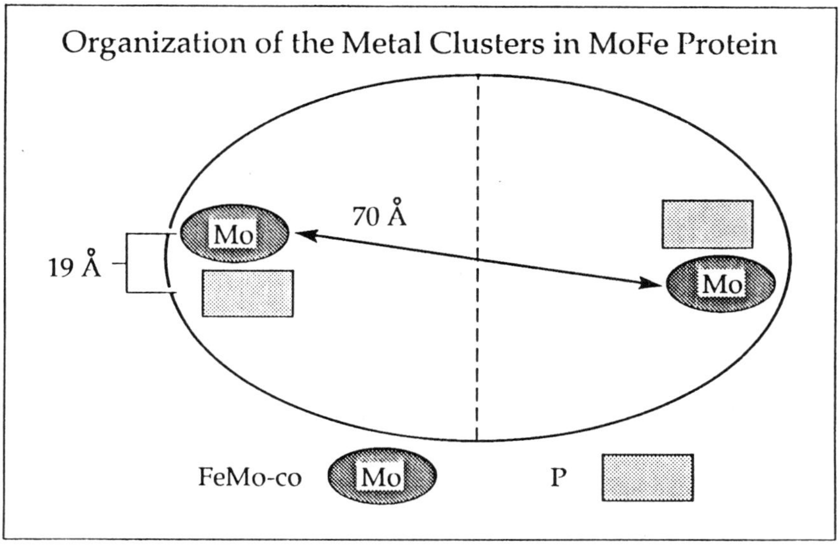

Figure 3. A schematic representation of the spatial arrangement of the metal-sulfur clusters bound to Cp1 as determined by the X-ray anomalous scattering studies described in the text. The representation of the large, '8-Fe' cluster with a P symbol indicates only that it must contain the Fe atoms normally assigned to P-clusters.

Discussion

With regard to the implications of the above model, we draw attention to several considerations. First, we note that our results do not distinguish among the several schemes for the organization of the non-cofactor Fe atoms outlined in the introduction. We argue for the presence of an 8-Fe center, and this group might be a new type of protein-bound metal-sulfur cluster. But, given the resolution of these studies, the 8-Fe center also might be a pair of closely associated Fe_4S_4 clusters, as found, for example, in the structure recently reported by Challen *et al.* [9].

The distances between pairs of clusters also raise interesting possibilities. A 70 Å separation between similar clusters makes it possible that the clusters are located near the surface of Cp1; if spherical, Cp1 should have a diameter of 80 Å. We also note that the 19 Å distance between FeMo-cofactor and the nearest 8-Fe group is well within the range of distances compatible with long range electron transfer in other systems [10].

Our model strongly disagrees with the cluster arrangement proposed by Sosfenov *et al.* on the basis of an attempt to phase diffraction data from crystals of *Azotobacter vinelandii* MoFe protein by direct methods [11]. We feel their approach was fragile at best, whereas our data directly probe the cluster arrangement. Our data are also incompatible with models for the chemistry of N_2 reduction that depend on participation of both Mo atoms in the binding and reduction of a single substrate molecule [12].

References

1. Zimmerman, R., Münck, E., Brill, W. J., Shah, V. K., Henzl, M. T., Rawlings, J. & Orme-Johnson, W. H. (1978) Biochim. Biophys. Acta, **537**, 185- 207.
2. McLean, P. A., Papaefthymiou, V., Orme-Johnson, W. H. & Münck, E. (1987) J. Biol. Chem., **262**, 12900-12903.
3. Watt, G. D. & Wang, Z.-.C. (1986) Biochemistry, **25**, 5196-5202.
4. Hagen, W. R., Wassink, H., Eady, R. R., Smith, B. E. & Haaker, H. (1987) Eur. J. Biochem., **169**, 457-465.
5. Zumft, W. G. & Mortenson, L. E. (1973) Eur. J. Biochem., **35**, 401-409.
6. Weininger, M. S. & Mortenson, L. E. (1982) Proc. Natl. Acad. Sci. USA, **79**, 378-380.
7. Rossmann, M.G. (1961) Acta Crystallogr., **14**, 383-388.
8. *International Tables for X-ray Crystallography. Volume III.* (1962) Macgillavry, C. H. & Rieck, G. D., Editors, Kynoch Press, Birmingham, England, p. 214.
9. Challen, P.R., Koo, S.-H., Dunham, W. R. & Coucouvanis, D. (1990) J. Am. Chem.Soc., **112**, 2445-2456.
10. May, S. L., Ellis, W. R., Jr, Crutchley, R. J. & Gray, H. B. (1986) Science, **233**, 948-952.
11. Sosfenov, N. I., Andrianov, V. I., Vagin, A. A., Strokopytov, B. V., Vainstein, B. K., Shilov, A. E., Gvozdev, r. I., Likhtenstein, G. I., Mitsova, I. Z. & Blazhchuk, I. S. (1986) Dokl. Akad. Nauk. SSSR, **291**, 1123-1127.
12. Shilov, A. (1988) in Nitrogen fixation: Hundred Years After, Bothe, H, de Bruijn, F. J. & Newton, W. E., Editors, Gustav Fischer, Stuttgart, pp. 57-61.

Gresshoff, Roth, Stacey & Newton (eds.) Nitrogen Fixation: Achievements and Objectives
Chapman and Hall New York • London 1990

Vanadium nitrogenase of *Azotobacter*

R.R. Eady, R. Pau, D.J. Lowe and F.J. Luque

AFRC Institute of Plant Science Research, Nitrogen Fixation Laboratory, University of Sussex, Brighton, BN1 9RQ, U.K.

Vanadium-containing nitrogenases have been purified and characterised from Azotobacter chroococcum and Azotobacter vinelandii, the structional genes cloned, sequenced and assigned to the polypeptides of the enzyme. In this synthesis paper on V-nitrogenase we review material published since the International Nitrogen Fixation Congress in Cologne in 1988 and also discuss some of our unpublished data.

Vanadium is present in V-nitrogenase in a vanadium and iron-containing protein (the VFe protein), analogous in function to the MoFe protein of Mo-nitrogenase (11,7). V-nitrogenases have specific Fe proteins associated with them, these proteins ($Ac2^{v}$ from A. chroococcum and $Av2^{v}$ from A. vinelandii) have very similar properties to the Fe proteins associated with Mo-nitrogenase (8,10).

The structural genes for the V-nitrogenase of A. chroococcum (23,24) and A. vinelandii (25,16) have been cloned and sequenced. The derived amino acid sequences are very similar, the Fe proteins (encoded by vnfH) have 12 and VFe protein subunits (encoded by vnfDK) have 5 and 26 differences respectively. In A. chroococcum a gene (vnfG located between vnfDK encodes a small subunit of the VFe protein (24). A. vinelandii has a gene homologous with vnfG located downstream from vnfD (16). It is likely that both $Av1^{v}$ as well as $Ac1^{v}$ has an $\alpha_2\beta_2\delta_2$ subunit structure, the additional subunit being a major difference from the MoFe proteins.

Spectroscopic and cofactor extrusion experiments indicate that $Ac1^{v}$ (the VFe protein of A. chroococcum) and $Av1^{v}$ (the VFe protein of A. vinelandii) have redox centres analogous to FeMoco (1,3,7,11,19,28) and the

'P' clusters of the MoFe proteins (20). Sequence comparison with MoFe proteins enables potential metal cluster binding domains in these enzymes, with distinct but similar cofactors to be identified. NifD and vnfD show 33%, nifK and vnfK show 31% amino acid sequence identity. In particular the conserved cysteinyl residues that are conserved among MoFe proteins, (18a) one of which has been implicated in binding FeMoco, are also conserved in the VFe protein. (See ref.21 for discussion).

SUBSTRATE REDUCTION BY V-NITROGENASE

In addition to N_2 both V- and Mo-nitrogenases will reduce H^+ or C_2H_2. A characteristic of substrate reduction by V-nitrogenase when assayed at 30°C is the decreased effectiveness with which N_2 or C_2H_2 compete with H^+ as reducible substrate when compared with Mo-nitrogenase. Consequently, even when functioning under optimum conditions of electron flux and saturating N_2, approximately half of the electron flux through V-nitrogenase results in H_2 formation. This apparently occurs in vivo since when uptake hydrogenase activity is inhibited, cultures of Azotobacter evolve considerably more H_2 when growth is dependent on V-nitrogenase compared with Mo-nitrogenase.

A pre-steady state study of the MgATP-dependent electron transfer reaction between $Ac2^v$ and $Ac1^v$ by stopped-flow spectrophotometry (31) showed that the rate of association of the two proteins and the rate of electron transfer between them are very similar to those of Mo-nitrogenase and are too fast to be rate-limiting in turnover. The dependence of the rate of electron transfer on MgATP concentration and its competitive inhibition by MgADP gave K_D of 230 μM for binding of MgATP and a K_i of 30 μM for MgADP, similar to those obtained for Mo-nitrogenase. The kinetics of these reactions which constitute the first two steps in the Lowe-Thorneley catalytic cycle of nitrogenase function (30), are so similar to those of Mo-nitrogenase that they suggest a common mechanism for N_2 reduction by both types of nitrogenase.

The effect of assay temperature on the reduction of N_2 by V and Mo-nitrogenase has been shown to be different (18). As the temperature is decreased from 30°C, (the usual assay temperature in vivo and in vitro) to 5°C, N_2 remained an effective substrate for V but not Mo-nitrogenase. This behaviour was conferred by the Fe protein of V-nitrogenase when assayed with MoFe protein. This property of V-nitrogenase has yet to be demonstrated in vivo, but if it occurs it may account for the persistance in Nature of an apparently less efficient nitrogenase system.

A characteristic of V-nitrogenase is the reduction of C_2H_2 to form some C_2H_6 in addition to C_2H_4. This property is conferred by the VFe protein (5). Although both V- and Mo-nitrogenase will slowly reduce C_2H_4 to C_2H_6, Mo-nitrogenase does not produce C_2H_6 from C_2H_2 (6,4). A study of the properties of mutant MoFe proteins constructed by site-directed mutagenesis, has however shown C_2H_6 as a product, as well as the separation of C_2H_2 reductase activity from nitrogenase activity (20). This separation of substrate specificity of Mo-nitrogenase also occurs at 10°C, or suboptimal ratios of MoFe to Feprotein (29) and in the heterologous nitrogenase formed between Cp2 and Kp1 (27). There appears to be subtle interactions between polypeptide and cofactor which result in altered substrate specificity, either achieved by kinetic or structural changes in the cofactor or polypeptide. (See below).

EPR SPECTROSCOPY OF VFe PROTEINS

The epr spectra of $S_2O_4^{2-}$-reduced VFe proteins are more complex than those of MoFe proteins since several paramagnetic centres are present. In the case of MoFe proteins, epr spectra are characterised by an S=3/2 signal arising from the FeMoco centres with g values near 4.3, 3.7 and 2.01 and which integrates to 0.94 spin/Mo atom.

$Av1^V$ shows a broad poorly-resolved signal at g=5.5 which has been assigned to the low field inflexion of the transition from the ground state to the first excited state of the Kramer's doublets of an S=3/2 spin system and integrates to 0.89 spin/V atom (19). In this region of the spectrum $Ac1^V$ has signals at g=5.6, 4.35 and 3.77 assignable to an S=3/2 spin system but which appear to arise from a mixture of species reflecting different environments of the cofactor centre, or the presence of contaminating FeMoco (7, D.J. Lowe unpublished).

Both $Ac1^V$ and $Av1^V$ show an axial signal with g values near 2.04 and 1.93 assigned to an S=1/2 spin state system. This signal which is not present in spectra of MoFe proteins, integrates to 0.2-0.3 spin/mol and becomes oxidized before the S=3/2 species suggestive of it having a lower redox potential.

A third signal, also not present in MoFe proteins, has recently been detected in preparations of $Av1^V$ (12). This signal most readily observed at 2K in the dispersion mode and high power is a broad structureless absorption envelope which dominates the spectrum, and is characteristic of a paramagnetic centre coupled to a metal centre.

Improved purification simplifies the epr spectrum of $Ac1^V$

These signals described above arise from preparations of VFe proteins which are homogeneous by the criteria of SDS electrophoresis and, in the case of $Ac1^V$, by NH_2-terminal amino acid sequence analysis. However, the generally lower metal content of VFe proteins compared with MoFe proteins may indicate that current preparations of the VFe proteins are not homogeneous with respect to their metal content. Some evidence for this was reported in the initial description of V-nitrogenase (7,22).

We have recently separated some of these species of $Ac1^V$ by gradient elution from DEAE Sephacel. Fractions were obtained, which although essentially homogeneous on SDS electrophoresis, ranged in specific activity from 550 to 2050 nmol H_2 evolved min^{-1} mg $protein^{-1}$. The g=1.94 epr feature was associated only with the early fractions which had a low Fe content and specific activity. This lack of correlation of the intensity of the S=1/2 species with specific activity make it very unlikely that it is associated with active VFe protein.

In addition, fractions which lacked the S=1/2 feature also exhibited a simpler, low field spectrum and are very similar to those of $Av1^V$ in this region. Detailed analysis of the earlier data had proved intractable since it appeared to arise from a mixture of species with S=3/2 spin states. The resolution of these species should simplify the analysis of the types of redox centre associated with active VFe protein.

X-RAY SPECTROSCOPIC STUDIES OF VFe PROTEIN

The only direct information on the environment of V in VFe proteins has been provided by vanadium K-edge X-ray spectroscopy of $Ac1^V$ (1,13) and $Av1^V$ (9). The edge structure and XANES (X-ray absorption near edge structure) of such spectra contain information about coordination geometry and probable oxidation level, and the extended X-ray absorption fine structure (EXAFS) the distance, type and number of the nearest neighbour atoms of V. The edge and XANES of dithionite-reduced $Ac1^{v}$, and $Av1^{v}$ and the model compound $\{VFe_3S_4Cl_3[HCON(CH_3)_2]_3\}^-$(A) are very similar. The relatively low intensity of the pre-edge feature precludes the presence of V=O bonds or tetrahedral symmetry at the vanadium atom as in $[VS_4FeCl_2)_2][Et_4N]_3$ where this feature is more pronounced. The spectra are consistent with a distorted octahedral coordination of the vanadium atom as has been established by X-ray structure analysis for complex (A). Little change in the spectrum of $Ac1^{v}$ was observed in the dye-oxidized or 'super reduced' species generated during enzyme turnover. The small differences were consistent with angular rather than radial changes (3).

Analysis of the EXAFS region of the vanadium K-edge X-ray absorption spectrum confirmed V in $Ac1^{v}$ and $Av1^{v}$ to be in an environment similar to compound(A). The spectra could be simulated with back-scattering contributions to three components V-(O,N), V-(S) and V-(Fe) shells (see Table 1). These data show the chemical environment of V in VFe proteins to be very similar to that of Mo in MoFe proteins.

Table 1. The vanadium nitrogenase of Azotobacter The environment of molybdenum in the MoFe protein and vanadium in the VFe protein, deduced from X-ray absorption spectroscopy

	Distance Å	Number of atoms
VFe protein from *A. chroococcum*		
V-S	2.31	3±1
V-Fe	2.75	3±1
V-O(N or C)	2.15	3±1
VFe protein from *A. vinelandii*		
V-S	2.34	3-4
V-Fe	2.74	2-4
V-O(N or C)	2.15	2-3

MoFe protein from *Klebsiella pneumoniae*		
Mo-S	2.37	4.5±1
Mo-Fe	2.67	3.5±1
Mo-O(N or C)	2.12	2±1

THE IRON AND VANADIUM CONTAINING COFACTOR

The iron and vanadium containing cofactor (FeVaco) has only been extracted from $Ac1^v$ (28). The N-methylformamide extract contains V:Fe:S in the ratio of approximately 1:6:5 and is capable of activating purified $NifB^-Kp1$. The chimeric protein formed by FeVaco and components of Mo-nitrogenase does not reduce N_2 but reduces C_2H_2 to form C_2H_6 and C_2H_4, a characteristic of VFe proteins.

Iron K-edge EXAFS data have recently been collected and are consistent with the V atom being in a V-Fe-S cluster confirming that the N-methylformamide extract of $Ac1^v$ does indeed contain a polynuclear metal-sulphur cluster (13). The EXAFS data are very similar to those of FeMoco (2) and show clear evidence for the presence of FeS_2M rhombs (M = Mo, V, Fe) with FeFe separations of *ca* 2.61Å plus a longer Fe--Fe separation at *ca* 3.65Å. These cofactor clusters must possess long range order for the long Fe--Fe back-scattering contribution to remain coherent. Although the synthetic cluster (A) provides a reasonable model for both V and Fe K-edge EXAFS data, no cluster has been synthesised that mimics both the mixed second shell and long range structure now shown to be present in FeVaco and FeMoco.

SPECIFICITY OF COFACTOR BIOSYNTHESIS

Of the assigned *nif* genes, *nifQBENV* and *nifH* have a role in FeMoco biosynthesis (see ref. 14), NifB (17), is required for V-nitrogenase activity, and by inference FeVaco biosynthesis. Some of the genes in this functional grouping are reiterated in *Azotobacter*, i.e. the *vnfEN* (15) and the structural genes for the Fe proteins of Mo-independent nitrogenases, *vnfH* and *anfH*.

We have initiated a programme to determine the function of EN-like genes by constructing strains of *A. vinelandii* carrying several combinations of deletions in the structural genes of different nitrogenases and also of *vnfEN* and *nifE*. These deletions were examined for diazotrophic growth and their ability to derepress nitrogenase activity.

A strain carrying the double deletion of nifE and vnfEN was unable to grow under any conditions tested, consistent with the involvement of vnfEN in cofactor biosynthesis i.e. FeVaco and the cofactor of nitrogenase-3. As expected, a strain carrying only a deletion of nifE grew in medium lacking Mo or V (nil medium) or in the presence of V. No growth occurred in the presence of Mo as a consequence of the repression of Mo-independent nitrogenase structural genes.

To test the ability of vnfEN to form a functional cofactor for Mo-nitrogenase a strain carrying a double deletion of Mo-independent nitrogenase structural genes (ΔvnfDGK, ΔanfHD) and also a deletion of nifEN was constructed. No growth occurred in nil medium or when supplemented with Mo, V, or Mo+V. When a Mo-resistant derivative of this strain was constructed growth occurred in the presence of Mo; the growth was inhibited in medium containing both Mo and V and to a greater extent in medium containing V alone. These data indicate that vnfEN can function in the biosynthesis of a cofactor, competent in the reduction of N_2, with Mo-nitrogenase components. Mo-resistance is required to prevent Mo repression of vnfEN and V inhibits growth since the chimeric protein formed between FeVaco and MoFe protein is unable to reduce N_2, as shown in *in vitro* experiments (28). Derepression of this strain shows that nitrogenase (measured as acetylene reductase) is present in nil medium or in medium supplemented with V or Mo+V; in all cases some C_2H_6 is formed as a product of C_2H_2 reduction.

Similarly in a Mo-resistant strain deleted for nifD, anfHD and vnfEN, nifE allows the synthesis of active acetylene reductase activity in the presence of Mo or V, i.e. a cofactor which functions in V-nitrogenase. C_2H_6 is also formed from C_2H_2 by these chimeric proteins but in this case absence of growth indicates that N_2 is not reduced.

These cofactors synthesised by the 'wrong' processing genes exhibit substrate reduction patterns similar to those obtained by site directed mutagenesis of MoFe proteins. The separation of nitrogenase activity from acetylene reductase activity associated with C_2H_6 formation potentially provides a powerful tool for probing cofactor polypeptide interactions specifically required for the reduction of N_2 but not other substrates.

ACKNOWLEDGEMENTS. We thank Rosemary Foote for efficient typing of the manuscript, Marie Eldridge and Lesley Mitchenall for expert technical assistance.

REFERENCES

1. Arber, J.M., Dobson, B.R., Eady, R.R., Stevens, P., Hasnain, S.S., Garner, C.D. and Smith, B.E. (1987) Nature (London), 325, 372-374.
2. Arber, J.M., Flood, A.C., Garner, C.D., Gormal,C.A., Hasnain, S.S. and Smith, B.E. (1988) Biochem J., 252, 421-425.
3. Arber, J.M., Dobson, B.R., Eady, R.R., Hasnain, S.S., Garner, C.D., Matsushita, T., Nomura, M. and Smith, B.E. (1989) Biochem J., 258, 733-737.
4. Ashby, G., Dilworth, M.J. and Thorneley, R.N.F. (1987) Biochem J., 247, 547-554.
5. Dilworth, M.J., Eady, R.R., Robson, R.L. and Miller, R.W. (1987) Nature (London), 327, 167-168.
6. Dilworth, M.J., Eady, R.R. and Eldridge, M.E. (1988) Biochem J., 248, 745-751.
7. Eady, R.R., Robson, R.L., Richardson, T.J., Miller, R.W. and Hawkins, M. (1987) Biochem J., 244, 197-207.
8. Eady, R.R., Richardson, T.H., Miller, R.W., Hawkins, M. and Lowe, D.J. (1988) Biochem J., 256, 189-196.
9. George, G.N., Coyle, C.L., Hales, B.J. and Cramer, S. (1988) J. Am. Chem. Soc., 110, 4057-4059.
10. Hales, B.J., Langosch, D.J. and Case, E.E. (1986) J. Biol. Chem., 261, 15301-15306.
11. Hales, B.J., Case, E.E., Morningstar, J.E., Ozeda, M.F. and Mauterer (1986) Biochemistry, 25, 7251-7255.
12. Hales, B.J., True, A.E. and Hoffman, (1989) J. Am. Chem. Soc., 111, 8519-8520.
13. Harvey, I., Arber, J.M., Eady, R.R., Smith, B.E., Garner, C.D. and Hasnain, S.S. (1990) Biochem J., 266, 929-931.
14. Hoover, T.R., Imperial, J., Ludden, P.W. and Shah, V.K. (1988) Biofactors, 1, 199-206.
15. Joerger, R.D., Jacobson, M.R. and Bishop, P.E. (1989) J. Bacteriol., 171, 3258-3267.
16. Joerger, R.D., Loveless, T., Pau, R.N., Mitchenall, L. and Bishop, P.E. (1990) J. Bacteriol., in press.
17. Kennedy, C.K., Gamal, R., Humphrey,R., Ramos, J., Brigle, K. and Dean, D. (1986) Mol. Gen. Genet., 205, 318-325.
18a. Kent, H., Ionnidis, I., Gorman, C., Smith, B.E. and Buck, M. Biochem. J., (1989), 264, 257-264.

18. Miller, R.W. and Eady, R.R. (1988) Biochem J., 256, 429-432.
19. Morningstar, J.E. and Hales, B.J. (1987) J. Am. Chem. Soc., 109, 6854-6855.
20. Morningstar, J.E., Johnson, M.K., Case, E.E. and Hales, B.J. (1987) Biochemistry, 26, 1795-1800.
21. Pau, R.N. (1990).
22. Robson, R.L., Eady, R.R., Richardson, T.J., Miller, R.W., Hawkins, M. and Postgate, J.R. (1986) Nature (London), 322, 388-390.
23. Robson, R.L., Woodley, P.R. and Jones, R. (1986) EMBO J., 5, 1159-1163.
24. Robson, R.L., Woodley, P.R., Pau, R.N. and Eady, R.R. (1989) EMBO J., 8, 1217-1224.
25. Raina, R., Reddy, M.A., Ghosal, D. and Das, H.K. (1988) Mol. Gen. Genet., 214, 121-127.
26. Scott, D.J., May, H.D., Newton, W.E., Brigle, K.E. and Dean, D.R. (1990) Nature (London), 343, 188-190.
27. Smith, B.E., Thorneley, R.N.F., Eady, R.R. and Mortenson, L.E. (1976) Biochem J., 157, 439-447.
28. Smith, B.E., Eady, R.R., Lowe, D.J. and Gormal, C. (1988) Biochem J., 250, 299-302.
29. Thorneley, R.N.F. and Eady, R.R. (1977) Biochem J., 167, 457-461.
30. Thorneley, R.N.F. and Lowe, D.J. (1985) in Molybdenum Enzymes (Spiro, T.G. Ed.) (Wiley & Son: New York), pp.221-284.
31. Thorneley, R.N.F., Bergström, N.H.J., Eady,R.R. and Lowe, D.J. (1989) Biochem J., 257, 789-794.

Gresshoff, Roth, Stacey & Newton (eds.) Nitrogen Fixation: Achievements and Objectives
Chapman and Hall New York • London 1990

Possible structural analogs for the Fe/M/S sites in the nitrogenases. The$[Fe_6S_6(L)_6(M(CO)_3)_2]^{n-}$ clusters and the $[(MFe_3S_4)_2(\mu_2\text{-}S)(\mu_2\text{-}L)]^{n-}$ singly- and doubly-bridged double cubanes

D. Coucouvanis

Dept. of Chemistry, University of Michigan, Ann Arbor, MI 48106, USA.

Introduction

The MoFe-protein components in the nitrogenases (31) from various nitrogen-fixing organisms contain a common Fe/Mo/S structural unit that can be isolated as an extractable cofactor (FeMoco) (36). The intimate relation of this cofactor to N_2 reduction is provided by the observation that inactive extracts of derepressed cells of mutant strains, which do not contain FeMoco, are activated upon its addition . The cofactor appears to be a small anionic cluster that does not contain amino acids and may contain homocitrate (26). The determination of its structure has not been as simple a task as it might seem. Indeed, since its isolation nearly 13 years ago, FeMoco still has not been obtained in crystalline form.

Extensive studies on either the MoFe protein or isolated FeMoco have made available a large amount of data concerning the properties of the Fe/Mo/S center in nitrogenase (N_2ase). Particularly important in the design of synthetic analog clusters is: a) numerous elemental analyses (9,30,36,) with the latest of these (30) suggesting a Fe:Mo:S ratio of 7±1:1:8±1; b) Mo extended X-ray absorption fine structure (EXAFS) analyses (18) that indicate a Mo coordination sphere that contains 3-4 sulfur atoms at 2.35Å, 2-3 iron atoms at 2.7Å and 2-3 light atoms (O,N) at 2.1Å; c) Fe-EXAFS analyses (4,5) that are consistent with an average Fe coordination environment that consists of 3.4 ± 1.6 S (Cl) atoms at 2.25 Å, 2.3±0.9 Fe atoms at 2.66 Å, 0.4±0.1 Mo atoms at 2.76 Å, 1.2±1.0 O (N) atoms at 1.81 (7) Å, and perhaps a second shell of Fe atoms (5) at a distance of 3.75Å; d) Mossbauer studies that have shown an aggregate of six iron atoms each in a distinct magnetic environment coupled to an overall S=3/2 spin system (28,35); e) ^{57}Fe electron nuclear double resonance (ENDOR) studies (24) that show a multitude of ^{57}Fe signals (five or six doublets) and indicate a rather asymmmetric structure in which the iron atoms roughly can be grouped into two sets of trios, each set having very similar hyperfine parameters; and f) ^{95}Mo ENDOR studies (24) that identify the presence of one molybdenum per spin system in a possible Mo oxidation state of (+4). Collectively, these data define formidable constraints that must be adhered to in the design of synthetic Fe/Mo/S clusters as analogs for the Fe/Mo/S aggregate in nitrogenase and the nitrogenase cofactor.

Numerous Fe-Mo-S complexes have been synthesized in search for synthetic analogs for the N_2ase cofactor. Without exceptions, these complexes all contain FeS_2Mo units and have been obtained by: a) the use of the MoS_4^{2-}anion as a ligand in $MoS_4^{2-}/Fe(L)_n$ ligand exchange reactions (14); b) by spontaneous self assembly reactions that employ MoS_4^{2-} as one of the reagents (12,25); c) the use of preformed Fe/S clusters ($[Fe_6S_6(L)_6]^{3-}$ (15), $[Fe_2S_2(CO)_6]^{2-}$ (8) and Mo reagents other than MoS_4^{2-}; and d) the coupling of

$[Fe_3MoS_4(Cl_4cat)(Cl)_4]^{2-}$ clusters and the synthesis of doubly-bridged double-cubanes (16).

Oligonuclear Fe-M-S Complexes

The syntheses, structures and electronic properties of oligonuclear Fe-M-S (M=Mo,W) complexes have been described and reviewed in detail (14,25). These dinuclear or trinuclear complexes that contain at least one $M(\mu_2\text{-}S)_2Fe$ rhombic unit, originally were conceived as potential "building blocks" for the synthesis of higher nuclearity aggregates. Thus far, there has been little success in using them to obtain larger Fe/Mo/S clusters that could quality as synthetic analogs on the basis of stoichiometry.

Polynuclear Fe-M-S Complexes from "spontaneous self assembly" reactions

A great number of polynuclear Fe-M-S clusters have been obtained (25) by "spontaneous self-assembly" procedures, that employ simple reactants and the $[MS_4]^{2-}$ anions as a soluble source of Mo or W sulfides. This approach is based on the assumption that the cores of the Fe/M/S centers in N_2ase are thermodynamically stable units that should be accessible from appropriate reagents even in the absence of a protein environment. Clusters obtained thus far contain the common structural feature of $MFe_3(S)_4$ cubane type units (M-Mo,W), and can be classified into two general structural types; the double-cubanes that contain two bridged $MFe_3(S)_4$ cubane units and the single cubanes that contain the $MFe_3(S)_4$ cores. The former include $[M_2Fe_6S_9(SR)_8]^{3-}$ (1,37), $[M_2Fe_6S_8(SR)_9]^{3-,4-}$ (39), $[M_2Fe_6S_8(OMe)_3(SR)_6]^{3-}$ (1), $M_2Fe_7S_8(SR)_{12}]^{3,4-}$ (39) and $[M_2Fe_6S_8(SR)_6(3,6\text{-}R^{2'}cat)_2]^{4-}$ (6). The reactivity of some of the double-cubanes in bridge cleavage reactions has led to the synthesis of single cubane clusters that include $[MFe_3(S)_4(SR)_3(3,6\text{-}R_2'cat)L]^{2-,3-}$ (32), (L=p-$ClC_6H_4S^-$, THF, N_3^-, and CN^-) and $[MoFe_3(S)_4(SEt)_3Fe(cat)_3]^{3-}$(40). The coordination environment around the Mo atom in various of these clusters resembles that suggested by EXAFS for the Fe/Mo/S cluster in N_2ase. However, none of the synthetic clusters possesses the desirable 7±1:1:8±1 Fe/Mo/S atomic ration.

Polynuclear Fe-M-S Complexes from preassembled Fe/S units

This approach has been followed in the synthesis of such clusters as $[MoOFe_5S_6(CO)_{12}]^{2-}$ (8) and $[MoFe_6S_6(CO)_{16}]^{2-}$ (20), from $[Fe_2S_2(CO)_6]^{2-}$ and low-valent Mo reagents. These clusters approximate quiet well the N_2ase cofactor Mo:Fe ratio and should be considered acceptable synthetic analogs provided that they retain their stoichiometry and core structure with terminal ligands other than CO. In our attempts to obtain synthetic analog clusters, we have used preassembled Fe/S and Fe/Mo/S clusters in either Lewis acid addition reactions or controlled cluster coupling reactions. The results of our studies are summarized in this paper.

Heteronuclear clusters with M_8S_6 Pentlandite type cores.

Outstanding among the various proposed structural models (25) for the N_2ase active site are certain clusters that can be envisioned as derivatives of known Fe/S compounds. One of these models (Fig. 1A) contains the cubic $MoFe_7S_6$ core structurally analogous to that in the mineral pentlandites (34) and the known $[Co_8S_6(SPh)_8]^{4-}$ (13) and $[Fe_8S_6(I)_8]^{3-}$ (33) clusters. An entry to the synthesis of clusters, related in core structure to the M_8S_6 pentlandite unit (Fig. 1A), became available following the successful isolation of the

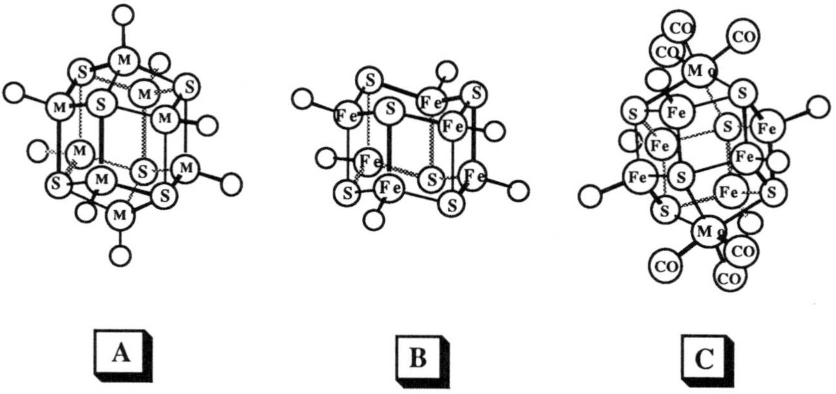

Fig.1. A, the M_8S_6 "pentlandite" type core. B, general structure of the $[Fe_6S_6(L)_6]^{n-}$ prismanes. C, general structure of the $-M(CO)_3$ prismane adducts.

$[Fe_6S_6L_6]^{2-,3-}$ prismane clusters (27). The triangular array of sulfur donors in the Fe_3S_3 "face" of the $[Fe_6S_6L_6]^{n-}$ prismanes (Fig. 1B), with S-S distances of ~3.8Å, suggested a possible tridendate ligand well suited for bonding interactions with coordinatively unsaturated metal complexes. The synthesis of the $[Fe_6S_6(L)_6(M(CO)_3)_2]^{n-}$ adducts (n=3,4) is accomplished (2,15) readily by the reaction of the $M(CO)_3(CH_3CN)_3$ complexes with the $[Fe_6S_6L_6]^{n-}$ prismanes (eq.1).

$$[Fe_6S_6L_6]^{3-} + 2M(CO)_3(CH_3CN)_3 \rightarrow [Fe_6S_6(L)_6(M(CO)_3)_2]^{3-} + 6CH_3CN \qquad (1)$$

Pure products are obtained only when an approximate two-fold molar excess of the $M(CO)_3(CH_3CN)_3$ reagent is used. When stoichiometric amounts of $M(CO)_3(CH_3CN)_3$ are used, mixtures of the 1:1 and 2:1 adducts are obtained for L= halide or p-R-PhO- (p-R - an electron releasing group). The final oxidation level (n=3 or 4) of the isolated $[Fe_6S_6(L)_6(M(CO)_3)_2]^{n-}$ adducts depends on the potential of their 3-/4- redox couple. $M(CO)_3(CH_3CN)_3$ will reduce those adducts with a 3-/4 redox couple more positive than -0.30V. For the $[Fe_6S_6(p\text{-}RPhO)_6(M(CO)_3)_2]^{n-}$ clusters, with electron releasing R groups such as $-CH_3$, $-OCH_3$ or $-N(CH_3)_2$, the trianionic adducts are obtained, while with the electron withdrawing $-CO(CH_3)$ group and with terminal halides, only the tetraanionic adduct can be obtained (eq.2).

$$[Fe_6S_6(X)_6(M(CO)_3)_2]^{3-} + M(CO)_3(CH_3CN)_3 \rightarrow$$
$$[Fe_6S_6(X)_6(M(CO)_3)_2]^{4-} + [M(CO)_3(CN)_3]^{+} \qquad (2)$$

Chemical oxidation by $Fe(Cp)_2^{+}$ or reduction by BH_4^- can be used for the synthesis of the trianionic and tetraanionic clusters from their one-electron redox counterparts. $[(Fe_6S_6Cl_6)(M(CO)_3)_2)^{3-}$ undergoes oxidative decarbonylation with substituted o-quinones to give the known (25), $(MoFe_3(S)_4(Cl)_4(cat)]^{3-}$ single cubane clusters that contain Mo(III) coordinated by the catecholate dianion (3). In the $[Fe_6S_6(X)_6(M(CO)_3)_2]^{n-}$

adducts (X=Cl,Br,I; n=3), the C=O vibrations are found between 1948cm^{-1} and 1905cm^{-1}, while for n=4, the C=O frequencies are found between 1912cm^{-1} and 1847cm^{-1}. For the $[Fe_6S_6(OR)_6(Mo(CO)_3)_2]^{n-}$ adducts, with n=3, the C=O frequencies are observed between 1912 cm^{-1} and 1864cm^{-1} while in the tetraanions, they are found between 1895cm^{-1} and 1830cm^{-1}. $[Fe_6S_6(p\text{-}CH_3C_6H_4O)_6(Mo(CO)_3)_2]^{3-}$ shows two sets of isotopically shifted resonances for the o-, m- and p-CH_3 protons in the ^{1}H NMR. One set is due to the $[Fe_6S_6(p\text{-}CH_3C_6H_4O)_6(Mo(CO)_3)_2]^{3-}$ anion. The other two are tentatively assigned to the monocapped 1:1 adduct $[Fe_6S_6(p\text{-}CH_3C_6H_4O)_6(Mo(CO)_3]^{3-}$ (16a). The monocapped $[Fe_6S_6(OAr)_6(Mo(CO)_3)]^{3-}$ clusters may be appropriate models for the Fe/Mo/S center in the fully reduced form of N_2ase.

The electronic spectra of the $[Fe_6S_6(ArO)_6(Mo(CO)_3)_2]^{n-}$ adducts are dominated by an intense ($\varepsilon \sim 1\times10^4$) absorption between 400nm and 500nm that corresponds to a L→Fe charge-transfer process. This absorption is bathochromically shifted (by c.a. 30-50nm) when compared to the same absorption in the "parent" prismanes. The shift to lower energies is expected for the adducts where the $M(CO)_3$ units deplete electron density from the core sulfido ligands and indirectly from the iron atoms. Within the $[Fe_6S_6(ArO)_6(M(CO)_3)_2]^{n-}$ clusters, the L→ Fe transition occurs at higher energies (by ca. 20-40nm) for n=4 than for n=3 and shifts to lower energies with increasing electron-releasing strength of the para substituents.

The $[Fe_6S_6X_6)(M(CO)_3)_2]^{3-}$ trianions are characterized by S=1/2 ground states and their EPR spectra, obtained only at temperatures <15K, resemble the spectra of the $[(Fe_6S_6X_6)]^{3-}$ "parent" prismanes (27). The corresponding tetraanions are EPR silent and possess diagmagnetic ground states. The six, high-spin Fe sites in the $[(Fe_6S_6X_6)(M(CO)_3)_2]^{n-}$ anions are antiferromagnetically coupled as evidenced by the reduced ambient temperature magnetic moments and the temperature dependence of the isotopically shifted resonances in the aryloxide derivatives.

The crystal structures of the centrosymmetric $[Fe_6S_6(L)_6(M(CO)_3)_2]^{n-}$ anions (L=Cl$^-$, M=Mo,n=4, (15b); L=p-CH_3(CO)-$C_6H_4O^-$,M=Mo, n=4, (2); L=Cl$^-$, Br$^-$, M=Mo, n=3, (15d); L=PhO$^-$, M=W, n=3, (15c) have been determined. All show very similar $[Fe_6M_2S_6]^{2+,3+}$ cores (Fig. 1C). These cores are best described as M_2Fe_6 rhombohedral units of nearly exact D_3d symmetry, with the two M atoms located in opposite corners on the idealized 6 axis of the rhombohedrom. A quadruply bridging S^{2-} ligand is located on each of the six faces of the rhombohedron (Fig. 1C). The Fe-Mo and Mo-S distances (ranges: Fe-Mo; 2.929Å-3.005Å and Mo-S; 2.572Å-2.649Å.) are significantly longer than those in the Fe/Mo/S center in N_2ase as determined by EXAFS. The trianions and tetraanions shows no significant differences within the Fe_6S_6 subunits. However, a significant shortening of the Mo-Fe and the Mo-S distances is found in the trianions. The crystallographic and Mossbauer data (vide infra) suggest that the highest occupied M.O in the tetraanions consists mainly of Mo and S atomic functions and is antibonding in character.

The ^{57}Fe Mossbauer spectra of the $[Fe_6S_6(L)_6(Mo(CO)_3)_2]^{n-}$ clusters show similar isomer shift (I.S.) and quadrupole splitting (ΔEq) parameters for the trianions (L=Cl$^-$, 0.56 mm/sec, 1.00 mm/sec; L=Br$^-$, 0.59 mm/sec, 1.06 mm/sec; vs. Fe, T=92K) and tetraanions (0.63 mm/sec, 1.00 mm/sec and 0.62 mm/sec, 1.02 mm/sec, respectively, for L=Cl$^-$ and Br$^-$). These I.S. values are significantly larger than those in the "parent" $[Fe_6S_6(L)_6]^{3-}$ clusters, (I.S. ~0.49mm/sec, Δ Eq~1.10 vs. Fe, T=125K). The data suggest that, in the $[Fe_6S_6(L)_6(Mo(CO)_3)_2]^{n-}$ adducts, the Fe_6S_6 cores assume an oxidation level lower than the one in the $[Fe_6S_6(L)_6]^{3-}$ clusters. In a formal sense, both adduct oxidation levels can be described as containing the $[Fe_6S_6]^{2+}$ core. In the 3- adducts, this core is bound to Mo in a formal +0.5 oxidation state, while in the 4- adducts, it is bound to Mo(0). It appears that a reduction of the Fe_6S_6 core is important for the stability of either the 3- or the 4- levels of the adducts.

The Coupling of Homo- and Hetero- Metallic Cubane Clusters

We initiated out studies in this area of analog cluster design and synthesis in an attempt to obtain a molecular framework suitable for the bimetallic activation of N_2ase substrates. Furthermore, recent discoveries of new N_2ase systems, that contain vanadium (19) and perhaps iron (22) in place of molybdenum, and the structural similarity of the iron-vanadium cofactor to the FeMoco prompted us to search for a synthetic analog with a common Fe/S structural framework capable of accomodating different metals at one site. On this basis, we have proposed (10) yet another structure for the Fe/M/S sites in the various N_2ase (Fig. 2A). The proposed model is a composite cluster that contains the known Fe_4S_4 (7) and $MoFe_3S_4$ (25) or VFe_3S_4 (29) structural units coupled by a single Fe-μ_2-S-Fe bridge.

As for the single $MoFe_3S_4$ cubanes, comparative V EXAFS studies between single cubanes containing the VFe_3S_4 core and the V/Fe protein of the vanadium nitrogenase from *A. chroococcum* (21) suggest very similar coordination environments for the vanadium atoms. The coupling of a Fe4S4 cubane to a MFe_3S_4 cluster (M=Mo,V) will result in a new cluster that retains the undisputedly attractive structural features of the MFe_3S_4 clusters and, in addition, will: a) possess an acceptable Fe/M/S stoichiometry of 7:1:9; b) contain a site for the bimetallic activation of molecular nitrogen as proposed previously (23); c) contain at least four distinct iron sites; and d) have a flexible (M) site capable of accomodating Mo, V (or Fe).

The synthesis of the singly-bridged double-cubane $[(Fe_4S_4Cl_3)_2S]^{4-}$, (**I**), is accomplished readily in reactions of the $[Fe_4S_4Cl_4]^{2-}$ anion with Na_2S in 1 : 1 molar ratio in CH_3CN solution and the crystalline Et_4N^+ and n-Bu_4N^+, Ph_4P^+ salts of the tetraanion have been structurally characterized (10) (Fig. 2B). A similar coupling reaction between $(Et_4N)_2[MoFe_3S_4Cl_3(Cl_4cat)(CH_3CN)]$, (**II**), and $(Et_4N)_2S$ in a 1:1 molar ratio in CH_3CN solution gives a good yield the $[[MoFe_3S_4Cl_2(Cl_4cat)]_2(\mu_2\text{-}S)_2]^{6-}$ doubly-bridged, double-cubane cluster, (**III**), that is obtained in crystalline form as the $(Et_4N)^+$ salt (16). A variation of the same reaction in which the cluster $(Et_4N)_2[MoFe_3S_4Cl_3(Cl_4cat)(CH_3CN)]$ first is reacted with 0.5 equiv. of Et_4NOH followed by 0.5 equiv of Li_2S gives a good yield the $[[MoFe_3S_4Cl_2(Cl_4cat)]_2(\mu_2\text{-}S)(\mu_2\text{-}OH)]^{5-}$ cluster, (**IV**). that can be isolated as a crystalline Et_4N^+ salt (16). The reaction of Na_2S, $(Et_4N)_2[Fe_4S_4Cl_4]$ and (**II**) in a 1:1:1 molar ratio in CH_3CN affords a microcrystalline solid with an EPR spectrum different than that of (**II**) but indicative of a S=3/2 ground state (16). Analytical data for this compound are consistent with the $(Et_4N)_5\{[Fe_4S_4Cl_2](\mu_2\text{-}S)_2[MoFe_3S_4Cl_2(Cl_4cat)]\}$ formulation, (**V**). This cluster, which most likely has a core structure similar to the one proposed in Fig. 2A, possesses an acceptable stoichiometry and many of the distinct features associated with the N_2ase cofactor. Attempts to obtain single crystals of (**V**) from a CH_3CN/benzene mixture gave only small crystals of (**III**) apparently by disproprotionation of (**V**). The reactions of (**IV**) with $(R_3Si)_2S$ in CH_3CN solution gives (**III**) in excellent yield. Similar reactions with $(R_3Si)X$ (X=CN,N_3) also occur readily and crystalline products, analogous to (**IV**) are obtained. The structure of the $[[MoFe_3S_4Cl_2(Cl_4cat)]_2(\mu_2\text{-}CN)]^{5-}$ cluster (**VI**) has been determined (17) (Fig. 2C). In a more recent development, we have obtained

Gresshoff, Roth, Stacey & Newton (eds.) Nitrogen Fixation: Achievements and Objectives
Chapman and Hall New York • London 1990

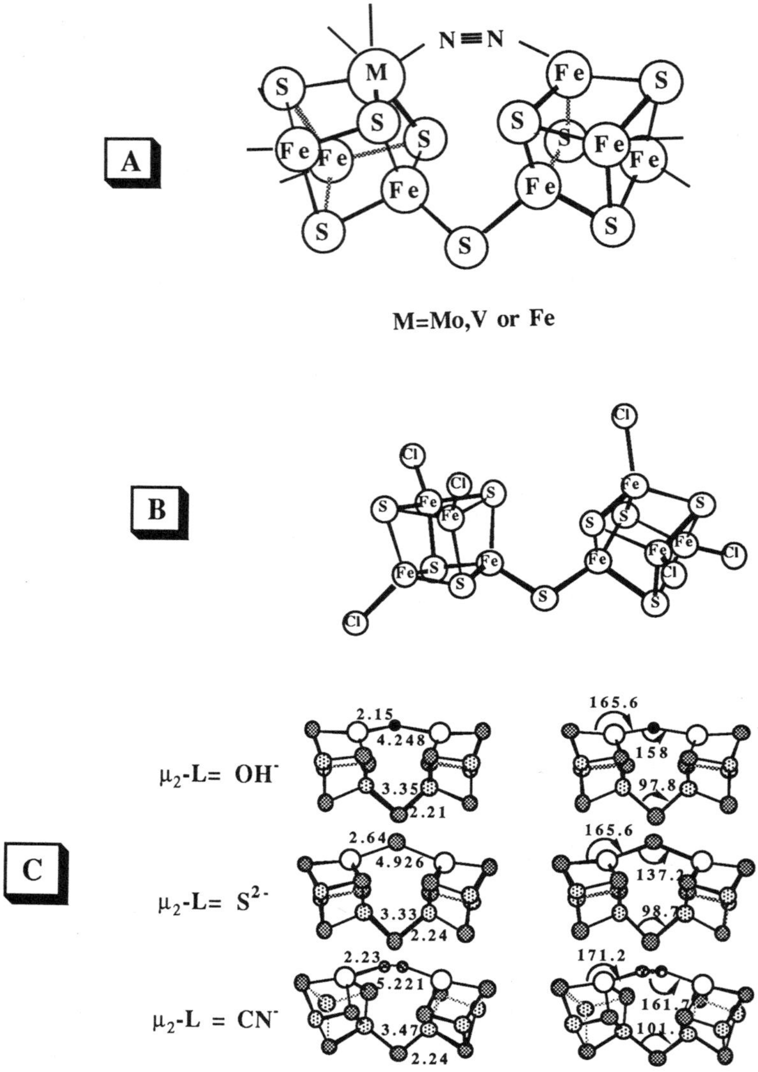

Fig. 2. **A**) A proposed model[10] for the active site in nitrogenase. **B**) The structure of the $[(Fe_4S_4Cl_3)_2S]^{4-}$ cluster. **C**) The $[Fe_6Mo_2S_9(\mu_2\text{-}L)]$ cores in the doubly bridged double Cubanes

$[[MoFe_3S_4Cl_2(Cl_4cat)]_2(\mu_2\text{-}S)(\mu_2\text{-}N_2H_4)]^{4-}$ in the reaction of (**IV**) with $N_2H_5^+Cl^-$. The reactivity characteristics of this new cluster are currently under investigation.

Characteristic band patterns in the far infrared spectra of the singly-bridged and doubly-bridged double cubanes between 400 and 300 cm^{-1} (an area free from cation vibrations) have been found very useful as diagnostic indicators for these molecules and make it possible to distinguish these compounds from the corresponding single cubanes. Specifically, the far I.R. spectrum of (**I**) with vibrational absorption bands at 394, 376, 361, 346, and 331cm^{-1} is quite distinct from that of $[Fe_4S_4Cl_4]^{2-}$ which shows bands at 383 and 352 cm^{-1}. Similarly, distinct "fingerprints" are observed in the far I.R. spectra of (**III**), (**IV**), (**V**) and (**VI**) that differentiate them from (**II**). The electronic spectrum of (**I**) shows absorption at 670nm (ε=10,700), at 480nm (sh, ε=17,700) and 240nm (ε=57,500). The spectrum indeed is very similar to that of $[Fe_4S_4Cl_4]^{2-}$ (7). The electronic spectra of (**III**), (**IV**), (**V**) and (**VI**) are essentially featureless and show a steadily increasing absorption from 750nm to the UV region of the spectrum.

The cyclic voltammetry of (**I**), in CH_3CN solution vs SCE, shows two quasireversible waves at -0.80V and -1.10V. The two waves presumably arise as a result of the 1e- reduction of each of the two subclusters. The 300mV separation between these waves indicate that intercluster interactions are substantial . The cuclic voltammetry of the double-bridged double-cubanes shows features common to all these clusters; a quasireversible reduction wave between -1.0V and -1.3V, (in CH_3CN solution vs Ag/AgCl), and two oxidation waves between -0.10 and +0.04V. By comparison, the single cube (**II**) shows a quasireversible reduction at -0.8V and irreversible oxidation at ~+0.4V.

Mossbauer spectrum of (**I**) consists of a broad asymmetric quadrupole doublet with average I.S. and ΔEq values of 0.48 and 0.98 mm/sec, respectively. Each of the Mossbauer spectra for (**III**), (**IV**), and (**V**) can be fitted by assuming two doublets in a 2:1 intensity ratio. The major components show I.S. values that range from 0.50 to 0.53 mm/sec and ΔEq values that range from 1.10 to 1.19 mm/sec. The minor components have I.S. values in the range from 0.30 to 0.34 mm/sec and ΔEq values between 1.05 and 1.13 mm/sec. The Mossbauer spectra are consistent with the structures of these clusters, which show two distinct Fe sites. Furthermore, it appears that the two Fe atoms bound to the μ_2-S bridging ligands are at a higher formal oxidation level than the remaining four Fe atoms.

A crystal structure determination (10) of (**I**) shows the tetraanion (Fig. 2B) as a μ_2-S^{2-} singly-bridged double-cubane. The anion is located on a crystallographic two-fold axis that passes through the μ_2-S^{2-} ligand. Inter-cluster S-S repulsions in (**I**), are relieved by a twist of one cubane subunit relative to the other by 38.3°. The Fe-Fe, Fe-S and Fe-Cl bond lengths at 2.787 (11) Å, 2.293 (5) Å and 2.227(5) Å, respectively, are unexceptional and very similar to those reported (7) for the $[Fe_4S_4Cl_4]^{2-}$ cluster at 2.766(5)Å, 2.295(3)Å, and 2.216(2)Å. The Fe-S bridge bonds in (**I**) at 2.206(4)Å are similar to those found (7) in $[Fe_2S_2Cl_4]^{2-}$(2.200(1)Å). The Fe-Fe distance across the Fe-μ_2-S-Fe bridge in (**I**), at 3.433(4)Å, is a direct consequence of the oblique Fe-μ_2-S-Fe angle of 102.2°.

The structure of (**III**) shows the hexaanion, (Fig. 2C), as a di-μ_2-sulfido double-bridged double-cubane with two homometallic M-μ_2-S-M bridges (M=Mo, Fe). The pentaanions in (**IV**) and (**V**) have a similar structure (17,40) with Mo-μ_2-OH-Mo and Mo-μ_2-CN-Mo bridges in place of the Mo-μ_2-S-Mo in (**III**). The basic Fe/Mo/S framework in these clusters and selected structural features are shown in Fig. 2C. The three anions show nearly exact, non-crystallographically required,C_{2v} symmetry. Among the outstanding structural features in **(III)** and **(IV)** and **(V)** are included short Fe-S bonds in the Fe-μ_2-S-Fe bridge, and oblique Mo-μ_2-E-Mo angles of 137.2(7)° (E=S^{2-}), 158(2)° (E=OH^-) and 161° (E=CN^-). The structures of **(III)**, **(IV)**, and **(V)** (Fig. 2) show a remarkable steric flexibility in accepting widely different Mo-Mo distances which is accomplished by pivoting the two cubane subunits around each of the Fe atoms within the Fe-μ_2-S-Fe bridges. A decrease in the Mo-Mo distance is observed from 5.221(8)Å in **(V)** to 4.926(8) in **(III)** to 4.248(9)Å in **(IV)** as the steric demands of the bridging ligand change. The Fe-μ_2-S-Fe angles and the Fe-Fe distances within the inter-cubane bridges do not show as pronounced a variation and are found at 101.1(9)° and 3.47(1)Å in **(V)**, 98.7(8)° and 3.33(1)Å in **(III)** and 97.8(8)° and 3.35(2)Å in **(IV)**.

Acknowledgements

The author wishes to acknowledge the invaluable contributions of his associates: P. R. Challen, Sang-Man Koo, S. Al Ahmad, M. G. Kanatzidis, A. Salifoglou, W. R. Dunham and W. R. Hagen who have been the prime contributors in this research. Most of the Mossbauer spectroscopy results were obtained by A. Simopoulos, A. Kostikas and V. Papaefthymiou of the Solid State Physics Group in the Nuclear Research Center Demokritos, Aghia Paraskevi, Attiki, Greece. Generous support of this work by a grant from the National Institute of Health (GM-26671) also is gratefully acknowledged.

References

1. a) Acott, S.R., Christou, G., Garner, C.D., King, T.J., Mabbs, F.E., Miller, R.E., (1979) Inorg. Chim. Acta **35**, L337. b) Christou, G., Garner, C.D., Mabbs, F.E., (1978) Inorg. Chim. Acta. **28**, L189. c) Christou, G., Garner, C.D., Mabbs, F.E., King, T.J. (1978) J. Chem. Soc. Chem Commun., 740.
2. Al-Ahmad, S.A., Salifoglou, A., Kanatzidis, M.G., Dunham, W.R. & Coucouvanis, D. (1990) Inorg. Chem **29**, 927.
3. Al-Ahmad, S.A. (1990) Ph.D. Thesis, Univ. of Michigan.
4. Antonio, M.R., Teo, B.K., Orme-Johnson, W.H., Nelson, M.J., Groh, S.E., Lindahl, P.A., Kauzlarich, S.M. & Averill, B.A. (1982) J. Am. Chem. Soc. **104**, 4703.
5. Arber, J.M., Flood, A.C., Garner, C.D., Hasnain, S.S. & Smith, B.E. (1986) J. de Physique **47**, C8-1159.
6. a) Armstrong, W.H.& Holm, R.H. (1981) J. Am. Chem. Soc. **103**, 6246. b) Armstrong, W.H., Mascharak, P.K. & Holm, R.H. (1982) J. Am. Chem. Soc. **104**, 4373.
7. a) Bobrik, M.A., Hodgson, K.O.& Holm, R.H. (1977) Inog. Chem. **16**, 1851. b) Wong, G.B., Bobrik, M. A.& Holm, R.H. (1978) Inorg. Chem. **17**, 578.
8. Bose, K.S., Lamberty, P.E., Kovacs, J.E., Sinn, E. & Averill, B.A. (1986) Polyhedron **5**, 393-398.
9. a) Burgess, B. K., Jacobs, D.J. & Stiefel, E.I. (1980) Biochim, Biophys. Acta. **614**, 196. b) Newton, W.E., Burgess, B.K. & Stiefel E.I. (1980) in *Molybdenum Chemistry of Biological Significance*, eds. Newton, W.E. and Otsuka, S. (Plenum Press, NY) p. 191. c) Burgess, B.K. & Newton, W.E. (1983) in *Nitrogen Fixation*, eds. Mueller, A. & Newton, W.E. (Plenum Press, NY) p. 83.
10. Challen, P.R., Koo, S. M., Dunham, W.R.& Coucouvanis, D. (1990) J. Am. Chem. Soc. **112**, 2455.
11. Christou, G. Garner, C.D., Mabbs, F.E. & King, T.J. (1978) J. Chem Soc. Chem Commun., 740.

12. Christou, G., Mascharak, P.K., Armstrong, W.H., Papaefthymiou, G.C., Frankel, R.B. & Holm, R.H. (1982) J. Amer. Chem. Soc. **104**, 2820.
13. Christou, G., Hagen, K.S. & Holm, R.H. (1982) J. Am. Chem. Soc. **104**, 1744.
14. Coucouvanis, D. (1981) Acc. Chem. Res. **14**, 201-209.
15. a) Coucouvanis, D.& Kanatzidis, M.G. (1985) J. Amer. Chem. Soc. **107**, 5005. b) Kanatzidis, M.G. & Coucouvanis, D. (1986) J. Amer. Chem. Soc. **108**, 337. c) Salifoglou, A., Kanatzidis, M.G. & Coucouvanis, D. (1986) J. Chem. Soc. Chem. Commun. 559., d) Coucouvanis, D., Salifoglou, A., Kanatzidis, M.G.,Simopoulos, A. & Kostikas, A. (1987) J. Amer. Chem. Soc. **109**, 3807.
16. Coucouvanis, D. Challem, P.R., Koo, S.M., Davis, W.M., Butler, W.& Dunham, W.R. (1989) Inorg. Chem. **28**, 4181.
17. Coucouvanis, D. & Challen, P.R., unpub. data.
18. a) Cramer, S.P., Gillum, W.D., Hodgson, K. O., Mortenson, L.E., Stiefel, E. I., Chisnell, J. R., Brill, W. J. & Shah, V. K. (1978) J. Amer. Chem. Soc. **100**, 4630. b) Cramer, S.P., Hodgson, K. O., Gillum, W. O. & Mortenson, L.E. (1978) J. Am. Chem. Soc. **100**, 3398. c) Conradson, S.D., Burgess, B.K., Newton, W.E., Hodgson, K.O., McDonald, J.W., Rubinson, J.F., Gheller, S. F., Mortenson, L.E., Adams, M.W.W., Mascharak, P.K., Armstrong, W.H. & Holm, R.H. (1985) J. Am. Chem. Soc. **107**, 7935. d) Conradson, S.D., Burgess, B. K., Newton, W. E., Mortenson, L. E. & Hodgson, K. O. (1987) J. Am. Chem. Soc. **109**, 7507.
19. a) Eady, R., Robson, R. & Postgate, J. (1987) New. Sci. **18**, 59. b) Hales, B.J., Case, E., Morningstar, J., Djeda, M. & Mauterer, L.A. (1987) Biochem. **25**, 7251.
20. Eldredge, P.A., Bryan, R.F., Sinn, E. & Averill, B.A. (1988) J. Am. Chem. Soc. **110**, 5573.
21. a) George, G.N., Coyle, C.L., Hales, B.J. & Cramer, S.P (1988) J. Am. Chem. Soc. **110**, 4057. b) Arber, J.M., Dobson, B.R., Eady, R.R., Stevens, P., Hasnain, S.S., Garner, C.D. & Smith, B.E. (1987) Nature **325**, 273.
22. Hales, B.J., Scorsone, K.A. & Moore, V. (1989) Abstracts for the Fourth Conference on Bioinorganic Chem., Inorg. Biochem. **36**, 71.
23. Hardy, R.W.F., Burns, R.C. & Parshall, G. W. (1973) in *Inorganic Biochemistry,* ed. Eichhorn, G.L. (Elsevier, Amsterdam) pp. 745-793.
24. a) Hoffman, B.M., Roberts, J.E. & Orme-Johnson, W.H. (1982) J. Am. Chem. Soc. **104**, 860. b) Hoffman, B.M., Venters, R.A., Roberts, J.E., Nelson, M. & Orme-Johnson, W.H. (1982) J. Am. Chem. Soc. **104**, 4711. c) Venters, R., Nelson, M.J., McLean, P.A., True, A.E., Levy, M.A., Hoffman, B.M., & Orme-Johnson, W.H. (1986) J. Am. Chem. Soc. **108**, 3487.
25. Holm, R.H. & Simhon, E.D. (1985) in *Molybdenum Enzymes*, ed. Spiro, T.G. (Wiley-Interscience, NY) Chapter 1 and references therein.
26. Hoover, T., Imperial, J., Ludden, P. & Shah V.K. (1989) Biochem. **28**, 2768.
27. a) Kanatzidis, M.G., Dunham, W.R., Hagen, W.R. & Coucouvanis, D. (1984) J. Chem. Soc. Chem. Comm., 365. b) Coucouvanis, D., Kanatzidis, M.G., Dunham, W. R. & Hagen, W.R. (1984) J. Am. Chem. Soc. **106**, 2978. c) Kanatzidis, M.G., Hagen, W.R., Dunham, W.R., Lester, R.K. & Coucouvanis, D. (1985) J. Am. Chem. Soc. **107**, 953. d) Kanatzidis, M.G., Salifoglou, A. & Coucouvanis, D. (1985) J. Am. Chem. Soc. **107**, 3358. e) Kanatzidis, M.G., Salifoglou, A., Coucouvanis, D. (1986) Inorg. Chem. **25**, 2460.
28. Huynh, B.H., Munck, E. & Orme-Johnson, W.H. (1979) Biochim. Biophys. Acta. **527**,192.
29. Kovacs, J.A. & Holm, R.H. (1986) J. Am. Chem. Soc. **108**, 340.
30. a) Nelson, M.J., Levy, M.A. & Orme-Johnson, W.H. (1983) Proc. Natl. Acad. Sci. USA **80**, 147. b) Orme-Johnson, W.H., Wink, D.A., Mclean, P.A., Harris,

G.S., True, A.E., Hoffman, B., Munck, E. & Papaefthymiou, V. (1987) Proc. 3rd I.C.B.I.C., Rec. Trav. Chim. Pays-Bas. **106**, 299.
31. Orme-Johnson, W.H. (1985) Ann. Rev. Biophys. Biophys. Chem. **14**, 419-459.
32. Palermo, R.E., Singh, R., Bashkin, J.K. & Holm, R.H. (1984) J. Am. Chem. Soc. **106**, 2600.
33. Pohl, S. & Saak, W. (1984) Angew. Chem. Int. Ed. Engl. **23**, 907.
34. Rajamani, V. & Prewitt, C.T. (1973) Can. Mineral. **12**, 178.
35. Rawlings, J., Shah, V.K., Chisnell, J.R., Brill, W.J., Zimmerman, R., Munck, E. & Orme-Johnson, W.H. (1978) J. Biol. Chem. **253**, 1001.
36. Shah, V.K. & Brill, W.J. (1977) Proc. Natl. Acad. Sci. USA **74**, 3249.
37. Wolff, T.E., Berg, J.M., Warrick, C., Hodgson, K.O., Holm, R.H. & Frankel, R.B. (1978) J. Am. Chem. Soc. **100**, 4630.
38. a) Wolff, T.E., Berg, J., Hodgson, K.O., Frankel, R.B. & Holm, R.H. (1979) J. Am. Chem. Soc. **101**, 4140. b) Wolff, T.E., Berg, J., Power, P.P., Hodgson, K. O., Holm, R. H. & Frankel, R.B.(1979) J. Am. Chem. Soc. **101**, 5454.
39. Wolff, T.E., Power, P.P., Frankel, R.B. & Holm, R.H. (1980) J. Am. Chem. Soc. **102**, 4694.
40. Wolff, T.E., Berg, J.M. & Holm, R.H. (1981) Inorg. Chem. **20**, 174.

Gresshoff, Roth, Stacey & Newton (eds.) Nitrogen Fixation: Achievements and Objectives
Chapman and Hall New York • London 1990

Workshop summary: How is the MoFe protein organized to fix nitrogen?

William E. Newton

Department of Biochemistry and Nutrition, Virginia Tech, Blacksburg, VA 24061, USA.

The complexities of the structural organization and biochemical-biophysical features of the nitrogenase MoFe protein and its complement of metalloclusters has intrigued and confounded researchers for the past several decades. Although there has been recent dramatic progress, certain fundamental questions concerning nitrogenase catalysis and the role of the MoFe protein in that process remain unanswered. For example, how does the MoFe protein manage its interactions with such a wide variety of molecules, including the Fe protein and certain non-physiological electron donors, the large range of small-molecule electron acceptors (e.g., N_2, H^+, C_2H_2, HCN, N_2O) and inhibitors (e.g., CO, H_2, CN^-, NO)? And how are all these inputs correlated intra-molecularly such that ambient temperature and pressure nitrogen fixation ensues! Also, what is the molecular structure of the metalloclusters, how are they distributed within the MoFe protein and what role in catalysis does each cluster type have? In this Workshop, progress made in research aimed at addressing these and related questions was discussed. These discussions are summarized here and placed within the context of relevant data and ideas reported prior to and during the Congress.

From the middle 1970's through the early 1980's, research on the MoFe protein had a biophysical focus and was concerned with the protein's internal organization. Major efforts in this area used labeling studies, with [57]-Fe and [95]-Mo, of the metal-containing prosthetic groups. These labeled MoFe proteins, when subjected to a combination of electron paramagnetic resonance (EPR) and Mossbauer spectroscopies, allowed two types of prosthetic groups to be identified [Smith & Lang (1974) Biochem. J. **137**, 169; Zimmermann et al. (1978) Biochim. Biophys. Acta **537**, 185]. These were called the "M" centers and the "P" clusters. Using metallocluster extrusion techniques and oxidation-reduction experiments, the "P" clusters were later reported as comprising a total of four [4Fe-4S] ferredoxin-like clusters per MoFe protein, while the "M" centers were correlated with two iron-molybdenum cofactors (FeMoco; of composition Mo:7+/-1Fe:9+/-1S) per protein [Shah & Brill (1977) Proc. Natl. Acad. Sci. USA **74**, 3249]. More recent work has led to the suggestion that the "P" clusters may be of higher nuclearity than four and that they may comprise two [8Fe-8S] centers [Hagen et al. (1987) Eur. J. Biochem. **169**, 457]. Although much has been gained from such studies, including the likely binding of FeMoco to the protein by a histidinyl nitrogen [Thomann et al. (1987) J. Am. Chem. Soc. **109**, 7913] and a cysteinyl sulfur [Burgess et al. (1980) J. Biol. Chem. 255, 353], no definitive information was produced concerning either the modes of intra-molecular communication among these two types of metal-containing prosthetic groups or their distribution within the polypeptide matrix and their individual responsibilities in catalysis.

More recently, the research has assumed a genetic bias. Through the planned introduction of mutations into selected nif genes, insight into the biosynthesis, and therefore the organization, of the holo-MoFe protein has been forthcoming. It was found that: (i) the FeMoco-deficient apo-MoFe protein (containing a full complement of "P" clusters) and FeMoco are biosynthesized independently and that FeMoco is inserted into the apoprotein in a separate process in vivo [Ugalde et al. (1984) J. Bacteriol. **159**, 888]; (ii) the apo-MoFe protein can be activated in vitro by adding isolated FeMoco to an extract of an appropriate mutant; and (iii) an altered FeMoco is responsible for the distinctive phenotype of the $nifV^-$ mutant [Hawkes et al. (1984) Biochem. J. **217**, 317], which implicates FeMoco as the N2-fixing site.

Very recently, oligonucleotide-directed mutagenesis has been used to identify two FeMoco-binding domains within the MoFe protein a-subunit. The first FeMoco-related domain surrounds 195-histidine [Scott et al. (1990) Nature **343**, 188], while the second domain involves 275-cysteinyl [Kent et al. (1989) Biochem. J. **264**, 257; Brigle et al. (1985) Gene **37**, 37]. All substitutions incorporated at either residue resulted in a Nif^- phenotype. Substitution of 275-cys by alanine resulted in an extract with an EPR signal reminiscent of that of isolated rather than protein-bound FeMoco, which suggests that FeMoco is very loosely bound. This was confirmed by the amount of (and ease with which) FeMoco transferred from the "275-cys-to-ala" MoFe protein to reconstitute the apo-MoFe protein in another extract. Substitution of 195-his by asparagine produced a mutant whose MoFe protein had both an altered EPR signal and changed catalytic properties, indicating that the EPR chromophore and the substrate-reducing site are one and the same. Similar effects were observed on substituting 191-glutamine by lysine.

The application of genetic techniques, including site-directed mutagenesis, the discovery of the two alternative nitrogenase systems [reported in depth by Eady et al., Bishop et al., Pau et al. and Oliver et al. in these Proceedings] and the information now beginning to flow from the x-ray crystallographic and absorption studies have breathed new life into the biochemical-biophysical studies of the nitrogenase MoFe protein. Work reported in these Proceedings, both formally and through poster summaries, clearly describes this new upsurge.

In the area of biosynthesis and maturation of the MoFe protein, a binary plasmid system was used to produce Klebsiella pneumoniae MoFe protein in Escherichia coli. It was suggested that a minimum set of genes (nifHDKTYUSWZM plus nifA) was required for maximal accumulation of FeMoco-reactivatable apo-MoFe protein. Also, "P" cluster assembly was suggested to be dependent on nifHS plus either nifM or nifTYUWZ, and that correct FeMoco binding requires "P" clusters in the apo-MoFe protein [White et al.]. This work (done anaerobically) assigns a different role for nifS, which has been implicated, along with nifU, in O_2 protection of nitrogenase [Walmsley et al.]. It also assigns another role to nifH, which already is involved in production of the Fe-protein polypeptide and in the biosynthesis of FeMoco, plus the sometime requirement for Fe protein in FeMoco insertion into the apo-MoFe protein. This last question was apparently satisfied by the report that not all apo-MoFe proteins are created equal! Apo-MoFe protein produced in a nifH-deletion mutant strain requires the Fe protein for FeMoco insertion, while those from nifB$^-$, nifE$^-$ or nifN$^-$ strains apparently do not [Tal et al.]. A role for molecular chaperones, specifically GroEL, in correct assembly of the MoFe protein was reported both in strains of E. coli transformed with the K. pneumoniae nif cluster and in K. pneumoniae overexpressing the groE transcriptional unit [Govezensky et al.].

Through x-ray crystallography, we now have firmer structural information, including suggestions of how the Fe protein might interact with ATP (using molybdate as a phosphate analog) and with the MoFe protein [Georgiadis et al.]. The proposed ATP-binding site of the Fe protein has been subjected to site-directed mutagenesis [Seefeldt et al.]. Stopped-flow calorimetry [Thorneley] and oxygen-exchange studies [Thorneley et al.] indicated that ATP hydrolysis precedes electron transfer from Fe protein to MoFe protein during catalysis and that the ATP-cleavage site resides at the interface of the proteins, implicating the two-protein complex as the catalytic entity.

It is now clear that the MoFe protein is composed of two independent operational units. Each unit comprises one "M" center (bound FeMoco) and, 19A away, half of the "P" clusters, which are in the form of either one large cluster or two very closely grouped (bonded?) [4Fe-4S] clusters. The two units are separated by 70A [Bolin et al.], which must place them close to the protein surface. A near-surface location of the "M" centers was also inferred from its observed interaction with exogenous spin probes [Oliver & Hales]. Modulation of the electron spin echo envelope of the paramagnetic center of the VFe protein was shown to be nearly identical [Thomann et al.] to that observed for the "M" center of the MoFe protein [Thomann et al., J. Am. Chem. Soc. **109**, 7913], thus indicating the same ligation to the polypeptide by histidine. This histidine ligand binding FeMoco to the protein is almost certainly 195-his in the a-subunit because, when this residue was substituted by asparagine, the major N-based modulation disappearred [reported orally by Newton in collaboration with Thomann and Dean]. Surprisingly, using x-ray absorption spectro-electrochemistry at the Mo L_2- and L_3-edges, the Mo atom in isolated

FeMoco did not participate in redox when cycled among its oxidized, semi-reduced and fully reduced states [Newton et al.]. If this carries over into the holoprotein, then Fe and S are implicated as the major redox components of bound FeMoco.

Because the exact nature of the "P" clusters is unclear and the MoFe-protein x-ray data allows the possibility of a nuclearity exceeding four, the 6-Fe cluster identified recently by W.R. Hagen was used to demonstrate the increased redox "flexibility" (and thus potential advantage) that a nuclearity higher than four in a cluster can bring to its parent protein [presented orally by Haaker for Hagen].

Homocitrate has been used to cure the $NifV^-$ growth phenotype of *Klebsiella pneumoniae* cells [Hoover et al. (1988) J. Bacteriol. **170**, 1978], but both positive [Madden et al.] and negative [Collet et al.] results were reported with the same $NifV^-$ mutant of *Azotobacter vinelandii*. Incorporation of homocitrate and its analogs into the apo-MoFe protein of *K. pneumoniae* can, however, be effected *via* the *in vitro* synthesis of FeMoco. Depending on the analog, differing patterns of substrate reduction and inhibition result, with citrate producing identical results to the $NifV^-$ phenotype. A model for the minimum organic moiety for FeMoco synthesis was determined [Shah et al.]. The sensitivity to CO of $NifV^-$ nitrogenase-catalyzed H_2 evolution was also found in an altered "a-191-gln-to-lys" MoFe protein. Thus, this phenotype might arise from a changed FeMoco-protein interaction, which is influenced by the organic moiety [Dean et al., Scott et al.]. Neither *K. pneumoniae* cells nor *A. vinelandii* cells would take up 1-fluorohomocitrate [Madden et al.], but *in vitro* incorporation of either of its two diastereoisomers gave very different results. The *erythro-*isomer (with the -F and the $-CO_2H$ on the same side) allows nitrogenase-catalyzed reduction of N_2 and its H_2 evolution is insensitive to CO, while nitrogenase containing the *threo-* isomer cannot catalyze N_2 reduction and its H_2 evolution is sensitive to CO [Shah et al., Madden et al.]. ^{19}F nuclear magnetic resonance spectra of these MoFe proteins showed single ^{19}F resonances with different chemical shifts. The presence of ^{13}CO over the sample split the ^{19}F resonance, indicating close proximity of CO to the 1-fluorohomocitrate and so to FeMoco [Madden et al.]. Because CO and C_2H_2 appear to bind independently to nitrogenase [Lowe et al.], this result suggests that C_2H_2 may bind elsewhere (as well as ?) to FeMoco.

It was very rewarding to see the substantial progress made in this area of research since the last Congress in Cologne. But what progress can we expect in the near future? We shall almost certainly see the MoFe protein crystal structure. Also we can expect greater insight into how the individual metalloclusters contribute to the dynamic catalytic process; vital information that will complement the static information obtained from the crystal structure. And many other contributions, that we cannot even predict now, will surely surface! It is exciting indeed to anticipate what awaits us at the next Congress in Mexico.

Gresshoff, Roth, Stacey & Newton (eds.) Nitrogen Fixation: Achievements and Objectives
Chapman and Hall New York • London 1990

Physical characterization of the molybdenum-iron cofactor of nitrogenase

Thomas A. Collet, Patricia D. Christie, Alison B. Hickman, Anne-Frances Miller, Thanos Salifoglou, David W. Wright and William H. Orme-Johnson

Dept. Chemistry, Massachusetts Inst. Tech., Cambridge, MA 02139, USA.

We are using a number of physical and biological probes of the physical characteristics of the molybdenum-iron cofactor of nitrogenase, including: (i) X-ray photoelectron spectroscopy (XPS) to investigate the elemental composition, overall charge, and Mo oxidation state; (ii) stability studies and EPR spectroscopy in various solvents; and (iii) experiments to determine the catalytic and structural role of homocitrate.

XPS identifies all elements in the top 10-50Å of a sample and, by comparison with known compounds, gives information on the relative amounts of each and their oxidation states. A large data set on Mo/Fe/S model compounds and various FeMoco samples indicates a limiting number of counterions (approx. 4) and, thus, a likely -4 charge on FeMoco in DMF. Changes in the Mo signal upon oxygen exposure suggest a change from MoIV to MoVI. Efforts to deduce the elemental composition are hampered by excess dithionite and/or its decomposition products.

The stability of FeMoco (initially extracted in DMF) was followed both in DMF and CH_3CN over time, during reconstitution experiments employing <u>nifB</u>$^-$ mutant extracts and saturating Fe protein, and showed that FeMoco is stable for at least five days.

When FeMoco is dissolved in DMF, its EPR spectrum is dramatically altered from g values close to 4 to g values close to 2 indicating a spin change from S=3/2 to S=1/2 and a different electronic structure to that in NMF.

Complementation experiments with a <u>nifV</u>$^-$ mutant of <u>Azotobacter vinelandii</u> (DJ71; a gift from D. Dean), using up to 50 mM R,S-homocitric acid lactone, failed to restore wild-type growth as did various derivatives. Upon repeating the experiments of Hoover (1) with <u>Klebsiella pneumoniae</u> (UN and UN1990), homocitric acid did cure the growth phenotype. Subsequent data base searches of GenBank and EMBL revealed a Na-ion translocation oxalacetate decarboxylase in <u>K. pneumoniae</u> with significant amino acid similarity to the <u>nifV</u> gene product.

1. Hoover, T.R. et al (1987) Nature **329**, 855-857.

Gresshoff, Roth, Stacey & Newton (eds.) Nitrogen Fixation: Achievements and Objectives
Chapman and Hall New York • London 1990

Relevance of conserved amino acid residues for activity and stability of iron protein of *K. pneumoniae* expressed in *E. coli*

T.A. Collet, T. White, K. Howard*, M.P. Reeve, and W.H. Orme-Johnson

Department of Chemistry, Massachusetts Institute of Technology, Cambridge, MA 02139, USA, *Proctor and Gamble, Cincinnati, OH 45241 USA.

The iron protein (Fe protein) is the specific reductant for the molybdenum-iron protein of nitrogenase (MoFe protein), the actual site of nitrogen reduction in biological nitrogen fixation. Electron transfer is characterized by concomitant hydrolysis of two MgATP for each electron passed from the Fe protein to the MoFe protein. The iron protein exists as an α_2 dimer holding its Fe_4S_4 cluster at the interface between the identical subunits. Our goal is to define the role of five conserved cysteine residues during catalytic turnover and apoMoFe protein synthesis.

A binary plasmid system is used to produce the iron protein of _K. pneumoniae_ in _E. coli_: _nifA_ on the activator plasmid is under control of the tac promoter; _nifH_ and _nifM_ are under control of their native promoters on the expression plasmid. The amino acid sequence of the iron protein shows a high degree of evolutionary conservation. The five conserved cysteine residues (#38,#85,#97,#132,#184 in _K. pneumoniae_) were changed one at a time to serine residues using site-directed mutagenesis. Mutant proteins were studied in their catalytic behavior and by native, anaerobic gel electrophoresis followed by Western immunoblotting. Extracts prepared from Fe^{55}-labeled cultures were also probed by autoradiography.

Single amino acid changes from cysteine to serine result in mutant proteins that are not active in catalysis. Strong thiol reductants will restore native conformation as seen on native gels, however, these proteins are present at levels significantly lower than wild type protein and they do not contain any iron under the conditions studied. Experiments to determine whether the mutant proteins will function in the synthesis of apoMoFe protein are in progress.

Gresshoff, Roth, Stacey & Newton (eds.) Nitrogen Fixation: Achievements and Objectives
Chapman and Hall New York • London 1990

Modification of dinitrogenase reductase in the cyanobacterium *Anabaena variabilis*

A. Ernst and P. Böger

Lehrstuhl für Physiologie und Biochemie der Pflanzen, Universität Konstanz, D-7750 Konstanz, Germany

In the heterocystous cyanobacterium *A. variabilis* ATCC 29413 a change in nitrogenase activity and concomitant change in the electrophoretic mobility of the dinitrogenase reductase (Fe-protein) was detected with an antibody produced against this protein (4). Modification was induced under aerobic conditions either by NH_4Cl at pH 10 or by cessation of C-supply from photosynthesis, stored glycogen or exogenously added fructose. This modification was reversed under anaerobic conditions (in presence of H_2, light, and DCMU) or upon reestablishing diazotrophic growth conditions in presence of air-level oxygen (2). Under diurnal growth conditions nitrogenase exhibited lowest activity at the end of the dark period and maximum activity after 6 h of illumination. The Fe-protein became completely modified during the dark period. Demodification and activation of nitrogenase activity occurred within 1 h illumination and was followed a by further increase of activity due to nitrogenase synthesis. Addition of NH_4Cl (pH 7) represses synthesis of nitrogenase but not the activation response toward light of existing nitrogenase (3).

All results are consistent with the hypothesis that photosynthate limitation modulates nitrogenase activity and modification. The experiments suggest that modification renders the Fe-protein more O_2-tolerant (5) or allow the cell to control activity other then by reductant supply in order to regulate the C:N balance (2,3). Regulation by an intermediate of N-assimilation seems unlikely because the Fe-protein is not modified during NH_3 inundation in absence of O_2. So either the decrease of a C-compound or the increase of O_2 or an intermediate of O_2 reduction in heterocysts are possible effectors of modification. The latter hypothesis is supported by the fact that methyl viologen when causing H_2O_2 release of filaments inhibits nitrogenase and induces Fe-protein modification (1).

(1) S.N. Bagchi, A. Ernst, P. Böger, manuscript in preparation.
(2) A. Ernst, S. Reich, P. Böger, 1990, J. Bacteriol., **172**, 748-755.
(3) A. Ernst, Y.-D.- Liu, S. Reich, P. Böger, 1990, Botanica Acta, in press.
(4) S. Reich, P. Böger, 1989, FEMS Microbiol. Lett. **58**, 81-85.
(5) Tabita, F.R., Van Baalen, C., Smith, R.L., Kumar, D. (1988) in *Nitrogen Fixation: Hundred Years After*, eds. Bothe, H., de Bruijn, F.L., Newton, W.E. (G. Fischer, Stuttgart, NY), pp. 189-194.

Gresshoff, Roth, Stacey & Newton (eds.) Nitrogen Fixation: Achievements and Objectives
Chapman and Hall New York • London 1990

Inorganic models of nitrogenase attached to a polymer support. Control of redox reactions in ammonia formation

T. Adrian George* and Bharat B. Kaul

Department of Chemistry, University of Nebraska, Lincoln, Nebraska 68588-0304, USA.

Understanding the mechanism of biological nitrogen fixation is a contemporary problem of academic and practical interest. Molybdenum and tungsten complexes of molecular nitrogen (N_2) of the type trans-[$Mo(N_2)_2(PMePh_2)_4$] (**1**) have been investigated as "models" of the N_2-reduction site of nitrogenase because in reactions with acid ammonia is formed. The initial phase of this reaction produces an intermediate containing the hydrazido(2-) (NNH_2) ligand; examples of which have been identified in solution and isolated and characterized. In the case of the molybdenum NNH_2 complexes, it has been proposed that ammonia is formed by disproportionation of the ligand. In order to investigate this hypothesis and further study the electron transfer behavior of these complexes, we have site-isolated an N_2 complex of molybdenum on a microreticular, phosphine (-$P(Ph)CH_2CH_2PPh_2$)-substituted polystyrene-2% divinylbenzene resin (PS-diphos). Two independent methods were used to attach the complex; (i) Phosphine exchange between **1** and PS-diphos and (ii) N_2 exchange with H_2 in [MoH_4(PS-diphos)$(PMePh_2)_2$ that had been prepared by the reduction of $MoCl_4(PMePh_2)_2$ with $LiBHEt_3$ in the presence of PS-diphos. The solvent-swollen trans-[$Mo(N_2)_2$(PS-diphos)$(PMePh_2)_2$] (**2**) reacted with acid in THF to yield the NNH_2 intermediate but *no ammonia*. Thus, preventing molybdenum complexes from interacting prevents completely ammonia formation which supports the disproportionation hypothesis. In CH_2Cl_2 hydrazine was produced but *no ammonia*. The four-electron reduction of N_2 to hydrazine occurs at a single metal center. In order to "reactivate" the NNH_2 intermediate to yield ammonia the hypothesis would suggest that an external reducing agent is necessary. Indeed, an equimolar mixture of **1** and **2** reacted with HBr in THF to afford ammonia in yields 33% higher per mol of **1** than when **1** was reacted alone. To demonstrate that at least some of the extra ammonia came from the polymer-bound complex, $^{15}N_2$-labelled **2** was employed. Both $^{15}NH_3$ and $^{14}NH_3$ were formed (verified by hypochlorite oxidation of ammonium ion to N_2 and subsequent analysis by mass spectrometry). The tungsten analogue of **1** also "activated" **2** to produce ammonia. Attempts were made to attach tungsten-N_2 complexes to the polymer but the extent of loading was low.

Kaul, B. B., Hayes, R. K. & George, T. A. (1990) J. Am. Chem. Soc. **112**, 2002-2003.

Gresshoff, Roth, Stacey & Newton (eds.) Nitrogen Fixation: Achievements and Objectives
Chapman and Hall New York • London 1990

Metal complexes as models for nitrogenase substrate-reducing sites

K. Fisher[1], D.L. Hughes[1], D.J. Lowe[1], R.H. Morris[2], A.J.L. Pombeiro[3] and R.L. Richards[1]

[1] AFRC-IPSR Nitrogen Fixation Laboratory, University of Sussex, Brighton BN1 9RQ, U.K.
[2] Chemistry Department, University of Toronto, Ontario M5S 1A1, Canada.
[3] Centro de Quimica Estrutural, Instituto Superior Tecnico, 1096 Lisboa, Portugal.

Molybdenum complexes having phosphorus or sulfur ligation are able to bind and reduce nitrogenase substrates such as N_2, H^+, RC≡CH and $H_2NC{\equiv}N$ and thus indicate likely mechanisms for nitrogenase function.
Although the stepwise reduction of N_2 to NH_3, and the binding of H_2 and its displacement by incoming N_2 are well established at molybdenum-phosphorus centres, only recently has such chemistry been developed at molybdenum-sulfur centres. In particular, polydentate thioether ligands allow the binding of N_2 at molybdenum(1) and thiolate ligands allow the binding of the proton and H_2 e.g. in $[MoH(SR)_3L_n]$ (R=bulky group, L=ligand, n=1 or 2) for the latter case. The binding of alternative substrates of nitrogenase at metal centres has been thoroughly studied for RNC and RC≡CH(2) but cyanide ligands have received less attention. $H_2NC{\equiv}N$ is reduced by nitrogenase and binds at N_2-binding metal centres in the form M=N-C≡N. However, although organic cyanides (RC≡N) were reported to be reductively cleaved to RCH_3 by crude nitrogenase, with the exception of $CH_2{=}CH_2CN$, they show no formation of hydrocarbon reduction products with pure nitrogenase preparations. When bound at N_2-binding metal sites, RC≡N protonate to give RCH=N- ligands but are not cleaved.

References

1. Adachi, T., Kaminaka, M. and Ueda, T., and Yoshida, T. (1988) J. Amer. Chem. Soc., 110, 4872-4873.
2. Hughes, D.L., Pombeiro, A.J.L.P. and Richards, R.L. (1988). in Nitrogen Fixation:Hundred Years After, eds. Bothe, H., de Bruijn, F.J. and Newton, W.E., (G. Fischer, Stuttgart) p66.

Gresshoff, Roth, Stacey & Newton (eds.) Nitrogen Fixation: Achievements and Objectives
Chapman and Hall New York • London 1990

Role of molecular chaperones in *nif* expression and nitrogenase assembly

D. Govezensky, T. Greener, G. Segal and A. Zamir

Biochemistry Department, Weizmann Institute, Rehovot 76100 Israel

Introduction. The emerging significance of molecular chaperones in assisting the correct folding of other proteins and/or their assembly into functional supramolecular structures has led us to examine the involvement of GroEL, a member of the chaperonin subclass of molecular chaperones, in nitrogenase formation.

Results. An *E. coli* mutant in *groEL* and an isogenic wild type strain were transformed with the *K. pneumoniae nif* cluster and analyzed for native or denatured Kp1 by gel-electrophoresis and immunoblotting with anti Kp1 antibodies. The level of both assembled Kp1 and NifDK polypeptides was much lower in the mutant compared with the wild type. Furthermore, when pulse-labeled, the mutant exhibited a slower rate of synthesis of Nif polypeptides but not of proteins synthesized in repressed cells. Beta-galactosidase activity expressed from a *nifH'-'lacZ* fusion in the presence of a constitutively expressed *nifA*, but not expression from the endogenous *lac* operon, was considerably lower in the *groEL* mutant than in the wild type. In derepressed wild type *K. pneumoniae,* overexpression of the *E. coli groE* operon accelerated the appearance of Kp1 and its constituent polypeptides. To examine whether chaperonins were also involved in nitrogenase assembly, the kinetics of mature Kp1 and Kp2 formation was followed by ^{35}S-pulse-chase analyses. The results indicated that the Fe protein assembled without a noticeable lag while assembly of the MoFe protein was incomplete even after 1 h of chase. The immediate precursors of Kp1 remained invisible in this analysis. Immunoprecipitation of pulse-chased cell extracts with anti-GroEL antibodies indicated that immediately following their synthesis, NifH, NifDK and several other, presumably Nif polypeptides were transiently associated with GroEL. The analysis with anti-Kp1 antibodies indicated an association of NifDK with several other Nif products in complexes whose decay followed kinetics approaching those of Kp1 assembly.

Conclusions. (**i**) GroEL is required for maximal expression from *nif* promoters, probably because interaction with the chaperonin is necessary for the correct folding of NifA. (**ii**) MoFe protein assembles very slowly via intermediates which remain undetected in the electrophoretic analysis. (**iii**) Fe protein assembles without a measurable lag period. (**iv**) Newly synthesized NifH, NifDK and several other, mostly Nif polypeptides are transiently associated with GroEL. Evidence for the existence of additional intermediate complexes in MoFe protein formation, containing additional Nif proteins has been obtained.

Gresshoff, Roth, Stacey & Newton (eds.) Nitrogen Fixation: Achievements and Objectives
Chapman and Hall New York • London 1990

Characterization of a large scale fermentation of *Azotobacter vinelandii*

J.-F. P. Hamel; T. Salifoglou; R. J. Murray; A.-F. Miller; D. W. Wright; D. Hopkins; A. B. Hickman; C. L. Cooney and W. H. Orme-Johnson

Depts. of Chemistry and Chem. Engineering, Biotechnology Process Engineering Center, Massachusetts Institute of Technology, Cambridge, MA 02139 USA.

Nitrogenase consists of two proteins: the MoFe protein and the Iron (Fe) protein. Present in the MoFe protein is a unique heterometallic cluster of composition $MoFe_{6-8}S_{4-9}$(hc), called FeMoco, which is implicated as the site of N_2 binding and reduction. FeMoco has been the subject of vigorous investigation due to its pivotal role in N_2 reduction. Our group is actively involved in research on N_2 fixation, most particularly in the study of the FeMoco cluster of nitrogenase. Our efforts have focused on the extraction and isolation of active FeMoco from the MoFe protein of nitrogenase, purification and subsequent study of it using analytical, spectroscopic and structural techniques. From the classic work of Shah and Brill[1] we have evolved a FeMoco purification scheme adaptable to large scale operations[2]. Here we report studies on the efficient production of optimally derepressed cells from which to isolate substantial quantities of FeMoco. Fermentations at the 1000-l scale were employed. Fermentation as well as downstream processing conditions were investigated. The organism employed in these fermentations as well as in the subsequent extractions of the MoFe protein and its FeMoco cluster is *Azotobacter vinelandii* (*A.v.*). A fed-batch procedure was implemented on a 1500-l Chemap bioreactor for the production of *A.v.* The 1000-l batch included food grade sucrose as the carbon source. The air supplied to the culture provided both oxygen and nitrogen. Foaming was controlled using a mechanical foam breaker and polypropylene glycol. The pH was maintained at 7.2 with potassium hydroxide. Temperature and dissolved oxygen were set at 30°C and 10% (of air saturation), respectively. In vivo and in vitro activities were measured by the C_2H_4 reduction assay. During the exponential phase of a typical *A.v.* fermentation, the specific growth rate was $0.3h^{-1}$. Both in vivo and in vitro activities displayed similar profiles and reached 32 nmol C_2H_4/min.ml and 60 nmol C_2H_4/min.mg, respectively. The final cell mass reached about 1.6 g/l (dry cell weight). A yield of 0.08 gcell/gsugar was obtained. During cell harvesting by centrifugation, more than 99.1% cells were recovered at a feed rate of 160 kg/h producing about 12 kg of cell slurry. Processing[3] of 5 kg of cell paste routinely yields 1.8g of MoFe protein/kg cell paste after the second DEAE anion exchange column[4].

REFERENCES

1. Shah, V. K., Brill, W. J. (1977) Proc. Natl. Acad. Sci. USA **74**, 3249-3253.
2. Wink, D. A., McLean, P. A., Hickman, A. B., Orme-Johnson, W. H. (1989) Biochemistry **28,** 9407-9412.
3. Burgess, B. K., Jacobs, D. B., Stiefel, E. I. (1980) Biochim. Biophys. Acta **614,** 196-209.
4. Operations were conducted in a glove box with $[O_2]$< 2 ppm.

Gresshoff, Roth, Stacey & Newton (eds.) Nitrogen Fixation: Achievements and Objectives
Chapman and Hall New York • London 1990

Site-directed mutagenesis of the *Azotobacter vinelandii* ferredoxin I gene

Siiri E. Iismaa, Ana E. Martin, Barbara K. Burgess, C.Dave Stout, Gerard M. Jensen, Philip J. Stephens, Julea N. Butt & Fraser A. Armstrong

Dept. Molecular Biology and Biochemistry, U.C.-Irvine, CA 92717 USA.

INTRODUCTION. Azotobacter vinelandii Ferredoxin I (AvFdI) contains one [4Fe-4S] cluster coordinated by cysteine residues at positions 20, 39, 42 and 45, one [3Fe-4S] cluster coordinated by Cys residues 8, 16 and 49 and two free Cys residues, 11 and 24. Cys 24 is in van der Waal's contact with the 4Fe cluster. Here we describe two site-directed mutants (C20AFdI and C24AFdI) in which Cys 20 and Cys 24, respectively, has been changed to Ala.

METHODS. Mutants were constructed and characterized as described (2).

RESULTS AND DISCUSSION. In C20AFdI, removal of the Cys 20 ligand to the 4Fe cluster causes considerable protein rearrangement to retain the cluster, with Cys 24 becoming the fourth cluster ligand (2). This results in a 100 mV decrease in the reduction potential of the 4Fe cluster.
In contrast, the C24AFdI shows very little change in structure except at position 24. The 4Fe cluster in C24AFdI, however, has increased thermal motion or disorder relative to the native because there are fewer van der Waal's contacts between residue 24 and the 4Fe cluster in C24AFdI than in native FdI. This 4Fe cluster has a reduction potential 30 mV more positive than the native.
The [3Fe-4S] cluster was not affected by either mutation.
To investigate what type of cluster forms in the absence of a fourth ligand to the 4Fe cluster, C20A/C24AFdI has also been constructed and its purification is in progress.

Although the specific in vivo function of AvFdI is unknown, an assay that tests whether our mutant FdIs are functional in vivo has been developed (1). Preliminary results indicate that C20AFdI and C24AFdI are functional, while C20A/C24AFdI is not.

REFERENCES
1. Martin, A.E., Burgess, B.K., Iismaa, S.E., Smartt, C.T., Jacobson, M.R. & Dean, D.R. (1989) J. Bacteriol. **171**, 3162-3167
2. Martin, A.E., Burgess, B.K., Stout, C.D., Cash, V.L., Dean, D.R., Jensen, G.M. & Stephens, P.J. (1990) Proc. Natl. Acad. Sci. USA **87**, 598-602.

Gresshoff, Roth, Stacey & Newton (eds.) Nitrogen Fixation: Achievements and Objectives
Chapman and Hall New York • London 1990

A hybrid *Azotobacter vinelandii - Clostridium pasteurianum* Fe protein that has both *in vivo* and *in vitro* catalytic activity

Marty R. Jacobson[1], John S. Cantwell[2] and Dennis R. Dean[1]

The Department of Anaerobic Microbiology[1] and the Department of Biochemistry and Nutrition[2], Virginia Polytechnic Institute and State University, Blacksburg, VA 24061, USA.

Site-directed mutagenesis and gene replacement procedures were used to construct a mutant strain of Azotobacter vinelandii which expresses a hybrid nitrogenase Fe protein. This hybrid Fe protein has its carboxy-terminal eighteen residues replaced with the five analogous residues from the Clostridium pasteurianum Fe protein sequence. Thus, the hybrid Fe protein is 13 amino acids smaller than wild-type Fe protein and has a net loss of 4 negatively charged residues, resulting in a change in the size and charge as confirmed by SDS-PAGE and two-dimensional gel electrophoresis. Although the strain which produces this hybrid Fe protein (DJ319) was capable of diazotrophic growth, its growth rate was lower than that observed for the wild type strain. Interestingly, strain DJ319 was observed to accumulate substantially more (~3x) MoFe protein and Fe protein polypeptides when compared to the wild-type. The basis for this result is not known. In vitro biochemical analyses of the wild-type Fe protein and the hybrid Fe protein revealed that at low electron flux (low MoFe:Fe protein molar ratios) the two proteins are catalytically indistinguishable. These results demonstrate that the tight, inactive complex which is formed when A. vinelandii MoFe protein and C. pasteurianum Fe protein are mixed in heterologous reconstitution experiments cannot be accounted for only by differences in the A. vinelandii and C. pasteurianum Fe protein primary sequences located at their respective carboxy-termini. Under conditions of higher electron flux the native Av2 is considerably more active. Although the basis for this result is not known, one possible explanation is, recycling of oxidized hybrid Av2 protein is less efficient resulting in less reduced, ATP bound Fe protein available at high MoFe:Fe protein ratios.

These results do not eliminate the possibility that the carboxyl-end of the Fe protein plays any role in nitrogenase component protein association or dissociation, however, they do show that the terminal 18 amino acids in the A. vinelandii Fe protein are not essential to that process. Also, neither the truncation of the C. pasteurianum Fe protein when compared to the A. vinelandii Fe protein, nor the absence of a concentration of charged residues present at the carboxyl-end of the A. vinelandii Fe protein but not present in the C. pasteurianum Fe protein can account for the formation of the tight, ineffective complex in heterologous A. vinelandii MoFe protein - C. pasteurianum Fe protein component protein reconstitution experiments. Nevertheless, it is a reasonable hypothesis that localized charge differences between the A. vinelandii and C. pasteurianum Fe proteins are likely to be involved in formation of the catalytically inactive complex formed in heterologous in vitro reconstitution experiments.

Gresshoff, Roth, Stacey & Newton (eds.) Nitrogen Fixation: Achievements and Objectives
Chapman and Hall New York • London 1990

Properties of four ferredoxins from *Rhodobacter capsulatus*: biochemical and genetic studies of their role in N_2 fixation

Y. Jouanneau, C. Grabau, E. Schatt, C. Duport, C. Meyer & P.M. Vignais

Biochimie Microbienne / LBIO (CNRS URA 11300) - Centre d'Etudes Nucléaires de Grenoble - 85 X - 38041 Grenoble cedex - France

Two ferredoxins (FdI & FdII) have previously been isolated from the phototrophic bacterium *Rhodobacter capsulatus* (1, 2). We have recently purified two new ferredoxins, called FdIII and FdIV, and characterized those proteins. In this report, we present a comparison of the four ferredoxins so far isolated from *R. capsulatus* and provide biochemical and genetic evidence of their possible involvement in N_2 fixation. FdI, a 2[4Fe-4S] ferredoxin of 64 amino acids was previously shown to catalyze *in vitro* electron transfer to nitrogenase (1). Its structural gene, *fdxN*, was recently cloned, sequenced and mapped inside a main *nif* gene cluster of the *R. capsulatus* chromosome; as with *nif* genes, *fdxN* transcription was repressed by excess ammonium ions (3). FdII was about 5 times less abundant than FdI in cells grown under conditions of N-limitation but its synthesis was not repressed in cells supplied with excess ammonium. FdII appeared to be as effective an electron carrier to nitrogenase as FdI in *in vitro* tests. FdII exhibited physical and chemical properties similar to the well-characterized 7Fe ferredoxin I from *Azotobacter vinelandii*. The gene coding for FdII was cloned and sequenced. The FdII polypeptide contained 111 amino acids and shared 65% identical amino acids with *A. vinelandii* FdI. Based on these similarities, we propose to call the gene encoding *R. capsulatus* FdII, *fdxA*, as is the structural gene for *A. vinelandii* FdI.

A third ferredoxin, FdIII, was purified from N-limited cells. FdIII consisted of a polypeptide of 100 residues including 9 Cys, and contained two [4Fe-4S] clusters. FdIII had no detectable activity as an electron carrier to nitrogenase. The gene encoding FdIII has previously been identified as an open reading frame (ORF5) being part of a *nif* operon including *nifE*, *nifN*, *nifX* and *nifQ* (4). We now propose to call it *fdxB*. Considering the location of *fdxB* on the chromosome, it is anticipated that FdIII is somehow involved in the processing of nitrogenase component 1. FdIV is a [2Fe-2S] ferredoxin composed of 95 amino acids. Its sequence displays no more that 25% amino acid identity with the sequences of plant ferredoxins. The gene encoding FdIV, that we called *fdxC*, was found immediately upstream of *fdxN* (5). mRNA transcript analyses showed that *fdxC* is not transcribed in NH_4^+ grown cells. Recently, FdI and FdIV have been successfully co-expressed in *Escherichia coli* from a cloned DNA fragment encompassing *fdxC* and *fdxN*.

1. P.C. Hallenbeck *et al.* (1982) Biochim. Biophys. Acta *681*:168-176.
2. A.F. Yakunin & I.N. Gogotov (1983) Biochim. Biophys. Acta *725*:298-308.
3. E. Schatt *et al.* (1989) J. Bacteriol. *171*:6218-6226.
4. C. Moreno-Vivian *et al.* (1989) J. Bacteriol. *171*:2591-2598.
5. E. Schatt (1989) Ph. D. Thesis, Grenoble University I.

Gresshoff, Roth, Stacey & Newton (eds.) Nitrogen Fixation: Achievements and Objectives
Chapman and Hall New York • London 1990

Site-directed mutagenesis of the MoFe protein (Kp1) of *Klebsiella pneumoniae* nitrogenase

H.M. Kent, I. Ioannidis, C. Gormal, M. Buck and B.E. Smith

AFRC Institute of Plant Science Research Nitrogen Fixation Laboratory, University of Sussex, Brighton, BN1 9RQ, U.K.

Using directed mutagenesis amino acid substitutions have been made in the α- and β-subunits of Kp1 at positions normally occupied by conserved cysteine[1] or tyrosine residues. The mutants have been analysed in terms of their growth on N_2, C_2H_2 reduction, the accessibility of FeMoco, their e.p.r. spectroscopy, and by non-denaturing gel electrophoresis followed by probing with polyclonal antibodies. Four electrophoretic forms of Kp1 were identified in wild-type extracts: form a, the major form, probably associated with fully active Kp1; forms b and c, which are probably biosynthetic precursors of form a; and form d, which is probably from oxygen damaged Kp1. Nif^+, Nif^- and intermediate phenotypes have been obtained and have provided evidence that Cys275 in the α-subunit and nearby amino acids are involved in binding FeMoco. Extracts of all Nif^+ mutants exhibited a normal e.p.r. signal and a distribution of electrophoretic forms comparable to the wild-type, the abundance of the major form a correlating with the specific activity of the extract. Nif^- mutants exhibit an altered or no e.p.r. signal and either a major form of low electrophoretic mobility form c or a fast-moving form e. Form c is the major form in a nifB mutant which lacks cofactor. It is converted to form a upon incubation with FeMoco. Two single mutations, C89A in the α-subunit and C94A in the β-subunit, which are tightly Nif^- can be combined to produce a weakly active nitrogenase indicating important interactions between subunits.

1 H.M. Kent et al. (1989) Biochemical Journal 264, 257-264.

Gresshoff, Roth, Stacey & Newton (eds.) Nitrogen Fixation: Achievements and Objectives
Chapman and Hall New York • London 1990

Kinetic and ENDOR studies on the interaction between nitrogenase and acetylene

David J. Lowe, Karl Fisher, Barry D. Howes and Roger N.F. Thorneley
AFRC-IPSR Nitrogen Fixation Laboratory, The University of Sussex, Falmer, Brighton, Sussex, BN1 9RQ, U.K.

This work seeks to extend our kinetic scheme for nitrogenase action (3) to C_2H_2 reduction to C_2H_4. The steady-state decrease in the ability of Klebsiella pneumoniae nitrogenase to reduce C_2H_2 as [MoFe protein]:[Fe protein] increases (2), as well as the pre-steady-state kinetics of C_2H_4 formation, can be simulated with the following additional assumption: 1) C_2H_2 binds reversibly to Kp1 reduced by 1 or 2 electrons/Mo; 2) C_2H_4 is not released (even on acid quench) until the Kp1 has been reduced by 3 or 4 electrons/Mo. At 30µM Kp1, C_2H_2 reduction is inhibited by 80% whereas H^+ reduction under Ar (3) is only inhibited by 20%; this must be because C_2H_2 increases the tightness of a Kp2:Kp1 complex. Since CO, in the presence of C_2H_2, inhibits C_2H_4 evolution although the C_2H_2 induced inhibition of product formation remains, CO and C_2H_2 must bind independently to nitrogenase. Proton ENDOR of Kp1 shows no effect of C_2H_2 at pH7.4 or 9.5 and hence it is unlikely that it binds close to the FeMoco centre of Kp1.

This evidence supports a suggestion (1) that sideways bound C_2H_2 is reduced to sideways bound C_2H_4 (a) in equilibrium with $-CH_2-CH_3$ (b) and that C_2H_4 is released from (a) whereas the further slow reduction to C_2H_6 takes place via (b).

BDH is supported by an AFRC Link Grant to Dr. R.C. Bray.

References

1. Ashby, G.A., Dilworth, M.J. and Thorneley, R.N.F. (1987) Biochem. J., 247, 547-554.
2. Hageman, R.V. and Burris,R.H. (1980) Biochim. Biophys. Acta. 591, 63-75.
3. Thorneley, R.N.F. and Lowe D.J. (1985) in Metal Ions in Biology Vol 7: Molybdenum Enzymes (Spiro T.G. Ed.) pp.221-284, Wiley Interscience, New York.

Gresshoff, Roth, Stacey & Newton (eds.) Nitrogen Fixation: Achievements and Objectives
Chapman and Hall New York • London 1990

Diastereomer-dependent substrate reduction properties of dinitrogenase containing 1-fluorohomocitrate in the iron-molybdenum cofactor

M. S. Madden, P. W. Ludden and V. K. Shah

Department of Biochemistry and Center for the Study of Nitrogen Fixation, University of Wisconsin-Madison, Madison, WI 53706 USA.

In vitro synthesis of the iron-molybdenum cofactor (FeMo-co) using homocitrate and its analogs produces modified forms of FeMo-co which alter the substrate specificities (N_2, acetylene, cyanide, or proton reduction) of dinitrogenase. The (1R, 2S) *threo-* and (1S, 2S) *erythro-* fluorinated diastereomers of homocitrate have now been incorporated *in vitro* into dinitrogenase. Dinitrogenase activated with FeMo-co containing *threo-*fluorohomocitrate reduces protons, cyanide, and acetylene, but not N_2. In addition, proton reduction is inhibited by carbon monoxide (CO), a characteristic of $NifV^-$ mutants. Dinitrogenase activated with FeMo-co containing *erythro*-fluorohomocitrate reduces protons, cyanide, acetylene, and N_2. In this case, proton reduction is not inhibited by CO, a wild-type characteristic. No CO inhibition of proton reduction was observed when dinitrogenase was activated with FeMo-co synthesized using the racemic mixture of flurohomocitrate. These results strongly suggest that the *erythro-* diastereomer is preferentially incorporated. NMR spectra of partially-purified dinitrogenase containing *threo*-fluorohomocitrate have been recorded.

Cyanide reduction properties of dinitrogenase activated with FeMo-co containing either fluorohomocitrate diastereomer are similar, and CO strongly inhibits cyanide reduction. Dinitrogenases activated with FeMo-co containing homocitrate analogs with a hydroxy group on the C-1 carbon are much less susceptible to CO inhibition of cyanide reduction. However, proton and cyanide reduction by dinitrogenase activated with FeMo-co containing (1R, 2S) *threo*-isocitrate is only one-third that of dinitrogenase activated with the racemic mixture of isocitrate and shows strong CO inhibition of substrate reduction. These results suggest that CO inhibition of proton and cyanide reduction occurs when the hydroxy group on the C-1 carbon of analogs is 'trans' to the C-2 carboxy group (i.e. in the *threo-* conformation) and that the *erythro-* form is preferentially incorporated.

Carbonyl sulfide (COS) inhibition of substrate reduction by dinitrogenase was dependent upon the homocitrate analog incorporated. Dinitrogenase containing homocitrate, citrate, or *threo*-isocitrate show high CO inhibition of cyanide reduction while COS inhibition is quite low, suggesting that COS may act at a different site or by a different mechanism than CO.

Hoover *et al* (1988) (1) have shown that homocitrate cures the *nifV*$^-$ phenotype of *Klebsiella pneumoniae*. We find this occurs with *Azotobacter vinelandii*. However, neither *A. vinelandii* nor *K. pneumoniae* would transport fluorohomocitrate into cells and, therefore, do not incorporate fluorohomocitrate *in vivo*.

1. Hoover, T.R. *et al.*, (1988) J. Bacteriol. **170,** 1978-1979.

Gresshoff, Roth, Stacey & Newton (eds.) Nitrogen Fixation: Achievements and Objectives
Chapman and Hall New York • London 1990

Evidence for cysteine ligation to a metal at site 153 in the ß-subunit of the nitrogenase MoFe protein of *Azotobacter vinelandii*

H. D. May[1], W. E. Newton[1,2], K. E. Brigle[3] and D. R. Dean[3]

[1]USDA-ARS-WRRC, Albany, CA 94710, USA. Depts. of Biochemistry[2] and Anaerobic Microbiology[3], Virginia Tech, Blacksburg, VA 24061, USA.

The α and β subunits of the nitrogenase MoFe protein are encoded by *nifD* and *nifK* respectively. Interspecies comparison of MoFe protein sequences reveals that 5 conserved cysteines exist in the α subunit (residues 62, 88, 154, 183 and 275) and that 3 are in the β subunit (residues 70, 95 and 153). Site-directed mutagenesis (1-3) was used to substitute the cysteine at site 153 in the ß-subunit by alanine. Although this substitution results in a Nif^- phenotype, derepressed whole cells (K153CA) exhibit a $S=^3/_2$ electron paramagnetic resonance spectrum identical to that of wild type cells but reduced in intensity. Acetylene reduction activity of crude extracts from K153CA cells was about 3% of wild-type activity levels. Substitution of the K153 cysteine with serine or deletion of the cysteine resulted in a Nif^+ phenotype but a slowed growth rate. Electron paramagnetic resonance analysis of whole cells of these mutants (K153CS and K153ΔC) showed no change in either *g* value or intensity of the $S=^3/_2$ signal, and the acetylene reduction activity of MoFe protein in extracts of these mutants was approximately 50% of wild-type extracts.

After purification (ion exchange, hydrophobic interaction or crystallization) electron distribution to the reduction of dinitrogen or protons was the same for the altered (K153CS) enzyme as it was for the native enzyme. Maximum ammonia, hydrogen and ethylene production by purified K153CS MoFe protein was approximately half that of native protein enzyme regardless of purification. The K153CS enzyme's activity equalled that of native enzyme when electron flux was diminished by lowering the Fe protein to MoFe protein ratio. The change in the activity of the altered MoFe protein versus native protein activity as electron flux increases suggests that the K153 ligand may be involved with an intermediate electron donor/acceptor within the MoFe protein, possibly a "P"-type Fe-S cluster.

1. Brigle, K.E., Setterquist, R.A., Dean, D.R., Cantwell, J.S., Weiss, M.C. & Newton, W.E. (1987) Proc. Natl. Acad. Sci. USA **84**,7066-7069.
2. Dean, D.R., Brigle, K.E., May, H.D. & Newton, W.E. (1988) in *Nitrogen Fixation: Hundred Years After,* eds. Bothe, H., de Bruijn, F. J. & Newton, W.E. (Gustav Fischer, Stuttgart & NY), pp. 107-113.
3. Kent, H.M., Ioannidis, I., Gormal, C., Smith, B.E. & Buck, M. (1989) Biochem. J. **264**, 257-264.

Gresshoff, Roth, Stacey & Newton (eds.) Nitrogen Fixation: Achievements and Objectives
Chapman and Hall New York • London 1990

Electron flux studies of cyclopropene and 3,3-difluorocyclopropene reductions by *Azotobacter vinelandii* (Av) and *Klebsiella pneumoniae* (Kp) nitrogenases

C.E. McKenna[a,b], M. Bravo-Leerabhandh[a], J.P. Gemoets[a], and C.A. Gormal[c]

[a]Department of Chemistry, the University of Southern California, Los Angeles, CA 90089-0744. [b]Corresponding author. [c]Agricultural and Food Research Council Institute of Plant Science Research, Nitrogen Fixation Laboratory, University of Sussex, Brighton BN1 9RQ, UK.

Cyclopropene (Cpe) is reduced by wild-type (wt) Av and Kp nitrogenases, and also by the Kp-*nifV*$^-$ nitrogenase, to two isomeric 2e$^-$ products, propene (P) and cyclopropane (C) [1]. The P:C product ratio is ~ 2:1 when Fe protein (Av2 or Kp2) is in excess over complementary wt MoFe protein (Av1 or Kp1), and 1.4:1 under similar conditions with Kp1-*nifV*$^-$ [1]. In contrast, 3,3-difluorocyclopropene (DFCpe) is reduced by Av nitrogenase to 2-fluoropropene (a 4 e$^-$ product) and propene (a 6 e$^-$ product) [2]. The structural relationship and unique product chemistries of these non-ionic, strained-ring reduction substrates have prompted us to initiate a collateral study of electron flux effects on their interactions with nitrogenases.

At an Av2/Av1 ratio of 15 (high electron flux), Cpe (0-0.2 atm in Ar) was less effective in competing with H_2 evolution than at a component ratio of 0.5 (lower electron flux). However, extrapolation of the data indicates that Cpe at sufficiently high pressure can eliminate H_2 evolution at either component ratio. The greater effectiveness of Cpe at lower electron flux corresponds at least in part to a decreased Km value. As electron flux was lowered significantly (Fe/MoFe < 2), the P/C product ratio increased in reductions of Cpe by Av and also by Kp nitrogenase; a smaller change was seen with Kp-*nifV*$^-$ nitrogenase. Extrapolation of the Av data suggests a limit of exclusive propene formation as the Av2/Av1 ratio approaches zero.

Fluoride ion (F^-) was identified by ion-specific potentiometry as a product of DFCpe reduction by nitrogenase (Av). On a 2e$^-$ basis, total electron flux to C_3 products (13%) + H_2 (79%) in reduction of DFCpe (0.06 atm) was 8% less than H_2 formed (100%) in the absence of DFCpe. Net F^- produced in the DFCpe reduction was approximately consistent with C_3 product stoichiometry. DFCpe was also shown to be a substrate for Kp and Kp-*nifV*$^-$ nitrogenases, which reduce it to the same C_3 products as formed by the Av enzyme. Product formation required ATP, dithionite, and both Kp nitrogenase components. DFCpe total C_3 product Km values for Kp and Kp-*nifV*$^-$ (Fe/MoFe = 10) were 0.41 $\pm$ 0.019 and 0.0074 $\pm$ 0.0014 atm, respectively, reflecting a 9-fold greater relative allocation of 2e$^-$ to DFCpe (0.06 atm) by the mutant vs. wt enzyme (20.0% vs. 2.4%; balance to H_2). This selectivity is not observed with Cpe [1].

References

1. Gemoets, J.P., Bravo, M., McKenna, C.E., Leigh, G.J. & Smith, B.E. (1989) *Biochem. J.* **258**, 487-491; and references therein.

2. McKenna, C.E., Nakajima, T., Jones, J.B., Huang, C., McKenna, M.C., Eran, H. & Osumi, A. (1980) in *Molybdenum Chemistry of Biological Significance*, Newton, W.E. & Otsuka, S., eds. Plenum Press, New York, pp. 39-57; Bravo-Leerabhandh, M., Ph. D. Thesis, USC, 1990; McKenna, C.E., *et al.*, in preparation.

Gresshoff, Roth, Stacey & Newton (eds.) Nitrogen Fixation: Achievements and Objectives
Chapman and Hall New York • London 1990

Interrelation of Fe-protein and MoFe-protein production from plasmid-borne *Klebsiella pneumoniae nif* genes in *Escherichia coli*

Anne-Frances Miller and William H. Orme-Johnson

Department of Chemistry, Massachusetts Institute of Technology, Cambridge, MA 02139, USA.

Plasmids bearing nitrogen-fixation (nif) genes support production by E. coli of both the nitrogenase Fe protein (2) and the apo-MoFe protein lacking only FeMoco (1). These systems have been exploited to determine the roles of specific nif genes in Fe protein and apo-MoFe protein production (1,2). We are using the same approach to learn about the production of FeMoco and holoMoFe protein.

Table 1 shows that nifBQ are required for FeMoco production in E. coli. With excess Fe citrate and in the absence of nifBQ, higher Fe-protein activities may be obtained. By contrast, in the presence of nifBQ, neither MoFe-protein activity nor Fe-protein activity increases. Therefore, nifB may be involved in Fe metabolism and allocation in nitrogenase synthesis. NifQ is involved in Mo metabolism (3). Alternatively, MoFe-protein synthesis could be competing for Fe with Fe-protein synthesis. However, when nifM was deleted, preventing Fe-protein synthesis, the level of MoFe-protein activity was drastically decreased (Table 1), indicating that diversion of Fe to Fe protein is not limiting MoFe-protein production. The decrease in MoFe-protein activity is much greater than the 50% decrease in the nifHDK peptides observed by antibodies and SDS PAGE. Thus, the multiple steps in MoFe-protein production may be affected by deletion of nifM and depletion of nifHDK.

Table 1. Fe-protein and MoFe-protein activities of extracts of E. coli bearing nif gene-encoding plasmids (cultures grown in minimal medium with only 3 uM Fe nitrilotriacetate).

nif genes present in entirety	Fe Protein activity*	MoFe protein activity#
AHDKTYENXUSVWZM	65	0.1
BQAHDKTYENXUSVWZM	58	5.7
AHDKTYENXUSVWZ	0	0.1
BQAHDKTYENXUSVWZ	0	0.45

(Measured in the presence of excess MoFe protein* or excess Fe protein#; in units of nmoles ethylene/mg protein/min.)

1. Harris, et al. (1990) J. Biol. Chem. (in press).
2. Howard et al. (1986) J. Biol. Chem. **261**, 772-778.
3. Ugalde et al. (1985) J. Bacteriol. **164**, 1081-1087.

Gresshoff, Roth, Stacey & Newton (eds.) Nitrogen Fixation: Achievements and Objectives
Chapman and Hall New York • London 1990

X-ray absorption spectro-electrochemistry of the nitrogenase iron-molybdenum cofactor

W.E. Newton[1,2], J.S. Cantwell[1,2], B.J. Feldman[2], S.F. Gheller[2], F.A. Schultz[3], P. Frank[4], B. Hedman[4] and K.O. Hodgson[4]

[1]Dept. of Biochemistry, Virginia Tech, Blacksburg, VA 24061 USA.
[2]USDA-ARS Western Regional Res. Center, Albany, CA 94710 USA.
[3]Dept. of Chemistry, Indiana-Purdue Univ., Indianapolis, IN 46223 USA. [4]Dept. of Chemistry, Stanford Univ., Stanford, CA 94305 USA.

Enzyme-bound iron-molybdenum cofactor (FeMoco) is the catalytic center of nitrogenase. After extrusion, it undergoes two redox processes (1) as occurs when protein-bound. These interconvert the semi-reduced (sr), EPR-active state with either the oxidized (ox) state or the fully reduced (red) state. To study FeMoco with stringent redox-state control and without dithionite, we have built an anaerobic electrochemical cell compatible with XAS measurements at the S K- and Mo L_2- and L_3-edges. The high sensitivity of these edges to changes in structural and electronic environment (2) allows insight to how FeMoco accommodates conversion among its three redox states.

Dithionite, which obscures the S K-edge spectrum, was removed by a anaerobic G-25 column. The two-compartment cell has a polypropylene window with the sample, a reticulated vitreous carbon (RVC) working electrode and a miniature Ag/AgCl reference electrode separated by an ion-conducting Nafion membrane from a Pt auxillary electrode. RVC's high surface area ensures rapid electrolysis without stirring.

Coulometric data collected for FeMoco(ox)/(sr)/(red) <u>during the XAS runs</u> gave an average n value of 0.77. The Mo L_2- and L_3-edges did not change in either energy or intensity for all three FeMoco redox states. However, in the S K-edge spectra, the "white-line" intensity at ca. 2470 eV decreased/increased incrementally as reduction/oxidation proceeded and the two transitions at ca. 2480 eV, assigned to bound thiosulfate (2), were lost in the (red) state and were not recovered on re-oxidation. We conclude that: (i) both FeMoco's redox processes involve one electron; (ii) its Mo atom is redox-inactive; (iii) electrons involved in redox utilize an orbital with a S-derived component; and (iv) ligation by thiosulfate (and its loss) may be related to FeMoco's multiplicity of forms and its decreased reactivity.

Supported by PHS (DK-37255), NSF (CHE-88-17702) and SSRL.

1. Schultz, F.A. et al. (1985) J. Am. Chem. Soc. **107**, 5364-5368; (1988) Biochem. Biophys. Res. Comm. **152**, 629-635.
2. Hedman, B. et al. (1988) J. Am. Chem. Soc. **110**, 3798-3805.

Gresshoff, Roth, Stacey & Newton (eds.) Nitrogen Fixation: Achievements and Objectives
Chapman and Hall New York • London 1990

Investigation of the nitrogenase metal clusters using spin probes

Melinda E. Oliver and Brian J. Hales

Department of Chemistry, Louisiana State University, Baton Rouge, LA 70803 USA.

Extrinsic paramagnetic probes have been used to enhance electron paramagnetic resonance signals by relieving power saturation in biological systems. Dysprosium ion (J = 15/2) has been used extensively as a spin probe because of its short spin-lattice relaxation time. The interaction between the biological paramagnet and the added lanthanide is dipolar in nature and results in a relief from saturation of electron paramagnetic resonance signals. By observing changes in power saturation when the dysprosium is added, information can be obtained related to the distance between the perturbing ion and the radical or metal cluster being investigated i.e., the distance of closest approach of the ion to the center of the metal cluster. This is indeed a powerful tool that can be used to estimate the distance a reactive center may be from the surface of a protein or a membrane.

The effect of dysprosium on the power saturation of cytochrome c (as a control model) as well as the MoFe protein and the Fe protein of conventional nitrogenase was studied. Using this technique, the calculated distance of the heme in cytochrome c from the surface of the protein was 3.7 Å in excellent agreement with 4 Å as determined from previous studies (1). The MoFe protein shows a significant effect with added dysprosium yielding a distance of 6 Å. This distance indicates that the M center is relatively close to the surface of the protein. This value can be compared with the diameter of the MoFe protein, which would be around 80 Å, assuming a globular protein.

It was found that, in addition to Dy^{3+}, the saturation of the S = 1/2 EPR signal of component 2 is affected by the presence of MgATP and salt. The salt effect may be due to the fact that high salt hinders the ability of the dysprosium complex to interact with the Fe protein. Another possibility is that low salt promotes aggregation of the Av2 into small clusters. Extrapolating the saturation data to the high salt extreme gives a value of 18 Å for the distance from the Fe-S cluster to the surface. The effect of adding MgATP to component 2 caused this distance to decrease to 14 Å suggesting a more open protein in the presence of MgATP.

1. Blum, H.,Leigh, J. S. & Ohnishi, T. (1980) Biochim. Biophys. Acta *626*, 31-40,

Gresshoff, Roth, Stacey & Newton (eds.) Nitrogen Fixation: Achievements and Objectives
Chapman and Hall New York • London 1990

Investigation of the paramagnetism of the VFe protein of V-nitrogenase from *Azotobacter vinelandii*

Melinda E. Oliver, Roland Tittsworth, Carol Blanchard, P. Scott Barrilleaux and Brian J. Hales

Department of Chemistry, Louisiana State University, Baton Rouge, LA 70803 USA

In the as-isolated state, component 1 of V-nitrogenase from *Azotobacter vinelandii* exhibits an EPR spectrum containing two different signals in the derivative absorption mode (1). One signal is axial with $g_{av} < 2$ and most likely arises from a reduced Fe-S cluster in the S = 1/2 state. The other signal occurs in the g = 5.5 region and has been equated with two of the inflections of an S = 3/2 state with high rhombic symmetry. In order to better understand the origin of these two paramagnetic centers, the effects of the following conditions on them were studied: 1) changes in pH, 2) addition of glycerol, 3) enzymatic turnover, and 4) changes in temperature and incident microwave power.

Changes in both the S = 3/2 and S = 1/2 signal were observed as a function of pH. The S = 3/2 signal decreased in rhombicity with increases in pH. On the other hand, the amplitude for the S = 1/2 signal increased significantly with increasing pH without any detectable shifts in g-factor. The changes for the S = 3/2 signal may arise from conformational changes in the protein while the changes in the S = 1/2 signal may arise from a combination of pH-induced conformational changes and/or changes in the environmental potential of the protein since raising the pH lowers the effective midpoint potential of dithionite. When 50% glycerol was added to the protein solution, the S = 1/2 signal remained unaffected while the S = 3/2 signal sharpened, implying an increase in rhombicity such that the ground and excited state signals of the S = 3/2 occur at virtually the same field position. This alignment of the g-factors of the ground and excited state enhances the detectability of the high-field inflections at g = 2.0 and 1.5. It was also observed that changes in temperature and incident microwave power greatly affected the shape of the S = 1/2 signal. Finally, under the condition of enzymatic turnover, the intensities of all the signals of both proteins decreased with the S = 3/2 signal being further augmented in the g = 4.3 region.

These results suggest that the S = 1/2 signal arises from an atypical Fe-S cluster possessing a very low mid-point potential. Although this signal is affected by enzymatic turnover, it is not necessarily associated with a cluster that's directly involved in the catalytic process. Furthermore, like the cofactor cluster in the conventional MoFe protein, the symmetry of the S = 3/2 signal of the VFe protein is influenced by pH and solvent changes in the medium.

(1) Morningstar, J. E. & Hales, B. J. (1987) J. Am. Chem. Soc. *109*, 6854-6855,

Gresshoff, Roth, Stacey & Newton (eds.) Nitrogen Fixation: Achievements and Objectives
Chapman and Hall New York • London 1990

Metal regulation of nitrogenase gene expression in *Azotobacter vinelandii*

R. N. Pau, L. A. Mitchenall and F. J. Luque

AFRC Institute of Plant Science Research, Nitrogen Fixation Laboratory, University of Sussex, Brighton BN1 9RQ, UK

A. vinelandii has two alternative nitrogenases, a vanadium nitrogenase, and nitrogenase-3, which apparently lacks both molybdenum (Mo) and vanadium (V). The structural genes for nitrogenase-3, *anfHDGK*, are located downstream from *anfA*, whch is required for their expression (1). We have established that the structural genes, vnfHDGK, for the V nitrogenase (2) are located downstream from *vnfA* and *vnfEN* (1), as shown below (boxes with continuous outlines show gene sequences in GenBank).

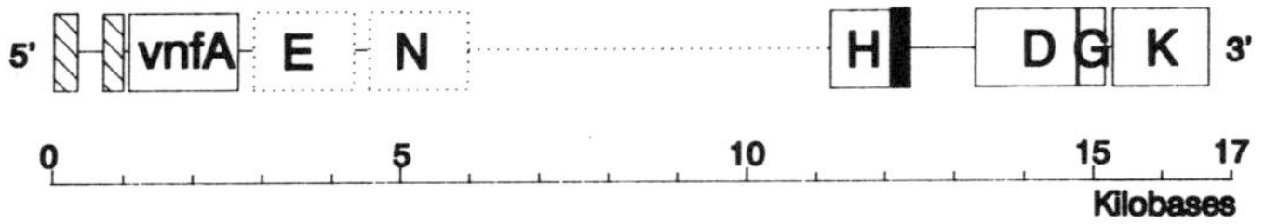

V represses synthesis of the third nitrogenase in the wild type strain UW136. However, growth (3) and northern analysis of mutant strains showed that a deletion in *vnfDGK* enables *anfHDGK* to be transcribed in the presence of V. The effect of the deletion in *vnfDGK* on regulation of *anfHDGK* transcription by V is not due to inability of the deletion strains to synthesize an active V nitrogenase; for a strain that does not synthesize active nitrogenases (mutated in *nifB*), showed normal metal regulation of both *vnfHDGK* transcription.

Mo represses expression of both alternative nitrogenases. Spontaneous mutants that permit their expression in the presence of Mo occur with relatively high frequency, and may arise from mutations in a number of genes. We have characterized two independent Tn*5* mutants which confer the ability to express the V nitrogenase in the presence of Mo.

1. Joerger, R.D., Jacobson, M.R. & Bishop, P.E. (1989) J. Bacteriol., **171**, 3258-3267
2. Joerger, R.D., Loveless, T.M., Pau, R.N., Mitchenall, L.A., Simon, B.H. & Bishop, P.E. (1990) . Bacteriol. In press.
3. Pau, R.N., Mitchenall, L.A. & Robson, R.L. (1989) J. Bacteriol. **171**, 124-129.

Gresshoff, Roth, Stacey & Newton (eds.) Nitrogen Fixation: Achievements and Objectives
Chapman and Hall New York • London 1990

Alterations in the α-subunit of the nitrogenase MoFe protein at amino acid residues 191 and 195 define a domain that changes the catalytic properties of FeMoco

Deborah J. Scott[1], Dennis R. Dean[2], and William E. Newton[1,3]

1 USDA, ARS-WRRC, 800 Buchanan St., Albany, CA 94710, USA
2 Dept. of Anaerobic Microbiology, VPI, Blacksburg, VA 24061, USA
3 Dept. of Biochemistry, VPI, Blacksburg, VA 24061, USA.

The two subunits of MoFe protein from a variety of bacterial sources have a high degree of sequence identity. Interspecies and intersubunit comparisons were used to identify cysteines and histidines proposed as a ligand for FeMoco in the MoFe protein (1). Recently, two sites in the a-subunit were shown to affect the FeMoco because substitution at each site uniquely altered the substrate reduction pattern and EPR signal. The sites and the amino acid substituted are a-191 Q->K (strain DJ255) and a-195 H->N (strain DJ178). The amino acid substituted was chosen based on sequence comparison with nifE gene product (3).

Did the amino acid substitutions alter the FeMoco? To this end, we constructed cofactor-deficient strains from DJ178 and DJ255 by deletion of the nifE gene to give strains DJ379 and DJ470, respectively. A wild-type strain containing a deletion in nifEN, DJ42, was used as control. In an in vitro complementation assay, we added native FeMoco to crude extracts of DJ379, DJ470 and DJ42; we measured acetylene reduction activity and, after concentration of the complemented extract, we recorded EPR signal. We repeated these experiments with the FeMoco purified from the MoFe of DJ255 using column method described by McLean et al (2). Both cofactors complemented the apoprotein equally to yield acetylene-reduction patterns and EPR signals equivalent to their respective Nif E^+ strain. We conclude that the FeMoco is identical to native FeMoco and so must contain homocitrate. Additionally we conclude that protein constraints are responsible for inducing the altered EPR signal and for producing ethane from acetylene.

After purifying the MoFe proteins from DJ255, DJ178 and AvOP, they were assayed at 40:1 ratio Av2/Av1± CO for substrate reduction patterns under acetylene, Ar, and N_2 atmosphere. The DJ178 MoFe protein produces neither H_2 nor NH_3. The DJ255 MoFe protein evolves H_2 under acetylene, Ar and N_2 in a reaction that is sensitive to CO. Because the evolution of H_2 is sensitive to CO, but its FeMoco contains homocitrate, we suggest that the glutamine at amino acid residue 191 in the a-subunit of the MoFe protein may be involved in an interaction with homocitrate.

The work was supported by NSF DMB 85-10029.

1. Dean, D.R. et al in Nitrogen Fixation: Hundred Years After, eds. Bothe, H., de Bruijn, F.J. & Newton, W.E. (Gustav Fischer, Stuttgart, 1988), pp. 107-113.
2. McLean, P.A., et al (1989) Biochemistry **28**, 9402-9406.
3. Scott, D.J., et al (1990) Nature **343**, 188-190.

Gresshoff, Roth, Stacey & Newton (eds.) Nitrogen Fixation: Achievements and Objectives
Chapman and Hall New York • London 1990

Site-directed mutagenesis of the proposed MgATP binding region of nitrogenase iron protein from *Azotobacter chroococcum*

L.C. Seefeldt, R. Robson and L.E. Mortenson

Center for Metalloenzyme Studies, University of Georgia, Athens, GA, 30606, USA

Nitrogenase, the enzyme complex responsible for reduction of N_2 to ammonia, consists of two component proteins. One component, the MoFe protein, is thought to be the site of N_2 reduction. The other component, Fe protein, contains two identical subunits (γ_2) with one Fe_4-S_4 center and is responsible for donation of low potential electrons to the MoFe protein. For each electron transferred to the MoFe protein, the Fe protein must bind and hydrolyze 2 MgATP to 2MgADP + 2 Pi. The site of MgATP binding to Fe protein and the role of MgATP hydrolysis in the mechanism of electron donation is at present unknown. The *A. chroococcum* Fe protein sequence $G^9K^{10}G^{11}G^{12}I^{13}G^{14}K^{15}$ has the same conserved motif (GXXXXGK) found in the loop structure of mononucleotide binding proteins. Thus, this region represents a possible site of interaction with ATP in the nitrogenase Fe protein. We have undertaken site-directed mutagenesis as one method to identify amino acids that are involved in ATP binding and hydrolysis. The Fe protein gene (*nif* H) and *nif* M, which appears to be necessary for expression of active Fe protein, were cloned into plasmids for expression in *E. coli.* The wild-type Fe protein was expressed in *E. coli* as determined by Western immunoblots of cell-free extracts and was active in acetylene reduction activity when complimented with purified MoFe protein. Mutation of G^{11} in the proposed ATP binding loop to a proline was accomplished by site-directed mutagenesis. The mutant Fe protein ($G^{11}P$) was expressed in *E. coli* at the same levels as the wild-type, but showed no acetylene reduction activity when complimented with MoFe protein. These results suggest that G^{11} in the proposed ATP binding loop plays an important role in Fe protein activity.

Gresshoff, Roth, Stacey & Newton (eds.) Nitrogen Fixation: Achievements and Objectives
Chapman and Hall New York • London 1990

FeMo-cofactor-deficient MoFe proteins synthesized by Δ*nifB* and Δ*nifH* strains are different

Sara Tal, Jia-ge Li, Teresa Chun, Amy Robinson and Barbara Burgess

Department of Molecular Biology and Biochemistry, UC Irvine, Irvine, CA 92717, USA.

INTRODUCTION. The biosynthesis of FeMo-cofactor _in vivo_ requires at least six genes: Q,B,N,E,V and H. Mutations in _nif_ B,N,E or H give rise to inactive FeMo-co-deficient forms of the MoFe protein that can be activated _in vitro_ by adding isolated FeMo-co.

RESULTS. We have compared the inactive FeMo-co-deficient MoFe proteins in cell-free extracts from Δ_nif_H (DJ54) (1), Δ_nif_B (DJ141) and Δ_nif_E (DJ35). It was found that: a. We are able to reconstitute Δ_nif_H extracts to wild type (WT) levels of activity by adding FeMo-co but are unable to reconstitute the Δ_nif_B or Δ_nif_E extracts to greater than 50% of WT activity; b. Δ_nif_H extracts require addition of Fe protein/MgATP to optimize the activation (2). Preliminary results indicate Δ_nif_E and Δ_nif_B do not; c. When extracts were heated to 56°C for 5 minutes the WT MoFe protein activity increases two-fold. The activity for Δ_nif_E and Δ_nif_B MoFe protein remains the same. The Δ_nif_H FeMo-co-deficient MoFe protein is extremely unstable and about 95% is denatured; d. The WT extracts give a single tight band on western blot of anaerobic native gel electrophoresis, that runs in the same position as purified MoFe protein. The Δ_nif_H MoFe protein runs above the WT band. Δ_nif_B and Δ_nif_H MoFe protein run as doublets with the top band running in the position of the WT MoFe protein.

DISCUSSION. We conclude that the FeMo-co-deficient MoFe proteins synthesized by _av_ Δ_nif_B and Δ_nif_E are like each other but unlike Δ_nif_H MoFe protein. If the _nif_B and _nif_E genes are only involved in FeMo-co biosynthesis then their FeMo-co-deficient proteins should be the same. The Fe protein polypeptide appears to be required both for FeMo-co biosynthesis and for the maturation of FeMo-co-deficient MoFe protein. Therefore deletion of _nif_H gives a FeMo-co-deficient MoFe protein different from the Δ_nif_E or Δ_nif_B protein.

REFERENCES.

1. Robinson, A.C., Dean, D.R. & Burgess, B.K. (1987) J. Biol. Chem. 262, 14327.
2. Robinson, A.C., Chun, T.W., Li, J.-G., & Burgess, B.K. (1989) J. Bacteriol. 264, 10088.

Gresshoff, Roth, Stacey & Newton (eds.) Nitrogen Fixation: Achievements and Objectives
Chapman and Hall New York • London 1990

Nitrogen coordination to VFe protein of alternative nitrogenase from *Azotobacter vinelandii*

Hans Thomann[1], Marcelino Bernardo[1] and Brian J. Hales[2]

(1) Corporate Research Laboratories, Exxon Research and Engineering Co. Annandale, NJ 08801 USA.
(2) Department of Chemistry, Louisiana State University, Baton Rouge, LA 70803 USA.

Electron spin echo (ESE) spectroscopy is a technique that allows us to gain information concerning the general structure in the neighborhood of a paramagnetic center. Often, neighboring nuclei next to the center produce a regular time-dependent magnetic field on the unpaired electron, causing a modulation of the echo amplitude. Typically, this modulation is due to weak hyperfine interactions from these neighboring nuclei and, therefore, can be used to obtain structural information.

The EPR spectrum of the VFe protein of alternative nitrogenase from *Azotobacter vinelandii* exhibits two signals (one, labeled S1, an S = 1/2 system at $g_{av} < 2$ and the other, labeled S2, is an S = 3/2 system at $g = 5.5$) in the derivative absorption mode (1) and a recently discovered third signal (S3) in the dispersion mode (2). By analogy with the MoFe protein of conventional nitrogenase, S2 is felt to originate at the VFe cofactor.

Using a three pulse stimulated echo pulse sequence at 1.7 K and $g = 1.685$, 1.95 and 2.075, modulated ESE waveforms were obtained. Phase corrected Fourier transforms of these waveforms yielded frequencies characteristic of nitrogen atoms covalently coupled to transition metal ions. Pure ^{14}N quadrupole transitions were observed in ESE modulation spectra when the local magnetic field due to the paramagnetic electron was comparable in magnitude to the external applied (Zeeman) magnetic field. In this case, the local magnetic field canceled the external magnetic field for one of the electron spin manifolds.

These results allow the assignment of this spectrum to a weakly coupled nitrogen that is covalently coordinated to FeVco with a coupling constant of A = 1.8 MHz. Furthermore, the quadrupole parameters of this nitrogen identify it as N-donor ligand in the range observed for the imino nitrogen of imidazole complexes. These parameters are also nearly identical to those previously observed for the histidine ligated to FeMoco in the MoFe protein from *Clostridium pasteurianum* (3).

1. Morningstar, J. E. & Hales, B. J. (1987) J. Am. Chem. Soc. *109*, 6854-6855.
2. Hales, B. J., True, A. E. & Hoffman, B. M. (1989) J. Am. Chem. Soc. *111*, 8519-8520.
3. Thomann, H., Morgan, T. V., Jin, H., Burgmayer, S. J. N., Bare, R. E. & Stiefel, E. I. (1987) J. Am. Chem. Soc. *109*, 7913-7914

Gresshoff, Roth, Stacey & Newton (eds.) Nitrogen Fixation: Achievements and Objectives
Chapman and Hall New York • London 1990

Kinetic studies on the mechanism of ATP-hydrolysis by *Klebsiella pneumoniae* nitrogenase

R.N.F. Thorneley*[1], G.A. Ashby[2], C. Julius[1], J. Hunter[2] and M.R. Webb[2].
[1]AFRC Institute of Plant Science Research, Nitrogen Fixation Laboratory, University of Sussex, Falmer, Brighton, Sussex, BN1 9RQ, U.K.
[2]MRC National Institute for Medical Research, The Ridgeway, Mill Hill, London, NW7 1AA, U.K.

Reductant-independent ATPase activity is a property of the complex formed between oxidised Fe-protein and reduced MoFe-protein i.e. $Kp2_{ox}Kp1$ (see also article by R.N.F. Thorneley elsewhere in this book). Titration of Kp1 (15μM) with $Kp2_{ox}$ (0 to 120 μM) showed tight binding (one Kp2 to each Mo site on Kp1) with a limiting activity of 0.60 ± 0.05 P_i released $Mo^{-1}s^{-1}$. This represents about 5% of the rate of reductant-dependent ATPase activity. Reductant-independent ATPase activity showed a hyperbolic dependence on MgATP concentration with K_m = 400 μM and competitive inhibition by MgADP with K_i = 145 μM. Evidence for reversible cleavage of MgATP to MgADP + P_i was obtained by incubating the component proteins alone and together with MgADP and ^{18}O-labelled P_i followed by conversion of P_i to triethylphosphate and analysis by mass spectrometry (Table 1).

Table I. **Oxygen-exchange catalyzed by nitrogenase in presence and absence of ADP at 23°C. pH7.4**

The assays contained (^{18}O)P (5 mM), Ap_5A (20μM), Hepes (25 mM), $MgCl_2$ (10mM) and ADP (250 μM). The protein concentrations were [Kp1]=33 μM and [Kp2]=100 μM. Replicate determinations are given.

Sample	$^{18}O_{ex}$/time (mM/min)
$Kp2_{ox}$	0.80, 0.82
$Kp2_{ox}$+ADP (250 μM)	0.76, 0.76
Kp1	0.30, 0.32
Kp1+ADP (250 μM)	0.39, 0.38
$Kp2_{ox}$+Kp1	1.52, 1.42
$Kp2_{ox}$+Kp1+ADP (250μM)	2.89, 2.93

These results clearly show that MgADP only significantly catalyses ^{18}O-loss from ^{18}-P_i in the presence of the $Kp2_{ox}$ Kp1 protein complex, providing evidence for the reversible cleavage of MgATP to MgADP + P_i and for the MgATP hydrolysis site being formed at the interface of the two component proteins.

Gresshoff, Roth, Stacey & Newton (eds.) Nitrogen Fixation: Achievements and Objectives
Chapman and Hall New York • London 1990

Genetic and biochemical analysis of the roles of *K. pneumoniae nif* genes in the production of the apoMoFe protein of nitrogenase

Theresa C. White, Georgianna S. Harris, Janice E. Flory, and William H. Orme-Johnson

from the Department of Chemistry, Massachusetts Institute of Technology
Cambridge, MA 02139 USA.

INTRODUCTION: Using a binary plasmid system to produce *K. pneumoniae* nitrogenase components in *E. coli*, we have determined that the minimal set of *nif* genes required for the maximal production of FeMoco-reactivatable apoMoFe protein is *nifHDKTYUSWZM+A* (pVL222+pVL15)[1]. In the present study, we attempt to elucidate, using ^{55}Fe-radiolabel and immunodetection techniques, the roles of *nifH, T, Y, U, S, W, Z, M* in apoMoFe assembly.
MATERIALS AND METHODS: *E. coli* strain W3110 and *K. pneumoniae* strains UN (*nif*$^{+}$) and UN106 (*nifB*$^{-}$) were used. For ^{55}Fe labeling, 0.5 mCi of [^{55}Fe] ferric citrate were added to 250ml MM9CA cultures. Crude cell extracts, both FeMoco reactivated and unactivated, were resolved on anaerobic non-denaturing polyacrylamide gels. For immunoblot analysis, proteins were transferred to Immobilon™which was probed with anti-Kp1 antiserum. For ^{55}Fe autoradiography, the gels were dried immediately and exposed to X-ray film.
RESULTS AND DISCUSSION: FeMoco titrations of UN106 and pVL222+pVL15 crude extracts generate the same curves. Immunoblot analysis of the titration samples resolved on non-denaturing gels shows that addition of FeMoco to both extracts increases the electrophoretic mobility of the apoMoFe protein in a two-step process (fig.1). We propose that the band comigrating with purified MoFe protein (Kp1) represents the species containing 2 mol FeMoco/mol tetramer while the intermediate band has only 1 mol FeMoco/mol tetramer. This FeMoco-induced two-step mobility shift was used to further characterize the mutant apoMoFe proteins produced in *E. coli* by *nif* deletion plasmids. Individually deleting *nifY, U, W, Z, or M* from the set of *nif* genes on pVL222, which results in only partial loss (25-75%) of apoMoFe activity, still allows for production of a species that can be FeMoco-reactivated in a two-step process. However, deletion of *nifH, nifS*, or the combined deletion of *nifTYUM* or *nifWZM* from pVL222 results in production of an inactive species which comigrates with apoMoFe protein but whose mobility increases, upon FeMoco addition, to that of MoFe protein in a single step! ^{55}Fe labeling of these extracts lacking apoMoFe activity, followed by autoradiography of non-denaturing gels, showed that immunoreactive species exhibiting this altered FeMoco-induced mobility shift contain no detectable iron. These results suggest that assembly of the P-clusters requires the genes *nifH, nifS*, and either *nifM* or the combined presence of *nifTYUWZ*, and that proper FeMoco binding to apoMoFe protein is dependent on the presence of P-clusters.

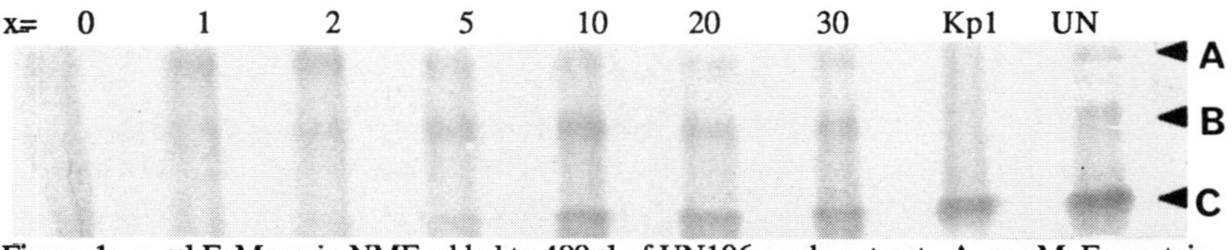

Figure 1: x=µl FeMoco in NMF added to 400µl of UN106 crude extract. A=apoMoFe protein. B=intermediate with 1 FeMoco/tetramer?. C=MoFe protein (Kp1).

[1] Harris, *et al.*(1990) J. Biol. Chem. (in press).

SECTION B:

PLANT-MICROBE INTERACTIONS

Invited papers
Workshop summary
Poster summaries

Gresshoff, Roth, Stacey & Newton (eds.) Nitrogen Fixation: Achievements and Objectives
Chapman and Hall New York • London 1990

Rhizobium meliloti nodulation genes specify the production of an alfalfa-specific sulfated lipo-oligosaccharide signal

Patrice Lerouge, Philippe Roche, Jean-Claude Promé, *Catherine Faucher, *Jacques Vasse, *Fabienne Maillet, *Sylvie Camut, *Francoise de Billy, *David G. Barker, *Jean Dénarié and *George Truchet

Centre de Recherche de Biochimie et de Genetique Cellulaires, CNRS-UPS, 118 route de Narbonne, 31062 Toulouse Cedex, France and *Laboratoire de Biologie Moleculaire des Relations Plantes-Microorganismes, CNRS-INRA, BP27, Castanet-Tolosan Cedex, France.

INTRODUCTION

The infection of leguminous plants by Rhizobium and the subsequent co-differentiation of both organisms leads ultimately to the formation of the unique nitrogen-fixing plant organ known as the root nodule (16). Within the nodule, the Rhizobium are furnished with photosynthate-derived energy and, in exchange, the endosymbiotic bacterium reduces atmospheric nitrogen to a form which can be assimilated by the host plant. This remarkable symbiosis has been the subject of extensive research for many years, but it is only recently that it has become possible to study, at the molecular level, the role of diffusible factors in plant-bacterial signalling and recognition, with the exciting prospect that such molecules might also be responsible for triggering plant morphogenesis.

In the case of temperate legumes, bacterial infection proceeds via the root hair, and involves curling of the hair tip, followed by the formation of an infection thread within the root hair. While these early stages of infection are proceeding, nodule meristematic activity is induced in the inner cortical cell layers of the root. By means of the infection thread, the Rhizobium penetrate the root cortex, invade the newly divided cells and subsequently differentiate into the nitrogen-fixing bacteroid form within the central tissue of the nodule. The use of various types of Rhizobium mutants has revealed that the micro-symbiont is capable of eliciting nodule organogenesis at a distance (10,20). Recent studies have even demonstrated that, for certain alfalfa clones, nodulation can occur in the total absence of Rhizobium (21, also see Joshi et al. these Proceedings). This suggests that the plant possesses the entire genetic programme for nodule organogenesis and that, under normal circumstances, the Rhizobium provides only a trigger to switch on this programme.

Fast-growing rhizobia, such as R. meliloti and R. leguminosarum, have a narrow host range, with the former able to nodulate Medicago, Melilotus and Trigonella species and the latter Pisum and Vicia. Figure 1 shows the genes of R. meliloti which are necessary for nodulation (nod genes) and which can be classified as either "common" (i.e., genes such as nodABC, which can complement mutations in analogous genes of another Rhizobium), "regulator" (i.e., nodD) or "host-specific" (such as nodH and nodQ in R. meliloti). For example, a mutation in the R. meliloti nodH or nodQ genes causes a shift in host-range from Medicago to Vicia (3,6).

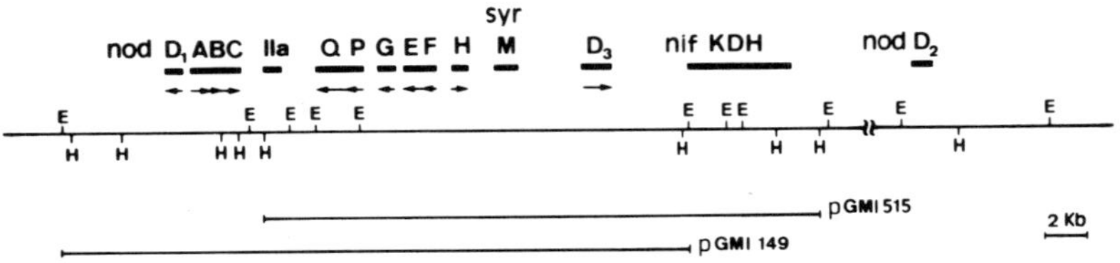

Figure 1. Genetic and physical map of the nodulation region of R. meliloti 2011. The horizontal lines represent the restriction map (E, EcoR1; H, HindIII). The plasmids pGM1149 and pGM1515 are shown below the map, and the arrows indicate the direction of transcription of the nod genes (16).

It is now clear that part of the early interaction between the plant host and Rhizobium involves the transcriptional activation of bacterial nodulation genes by the combined action of specific flavonoid compounds present in plant exudates and the regulator nodD gene products (16). However, it is also clear that this very early stage of the interaction cannot account for the specificity of infection and nodulation (9). Recently, it has been shown that when R. meliloti nod genes are induced in batch culture by the flavone luteolin, soluble factors can be recovered in the supernatant which are able to provoke plant reactions such as root hair deformations (1,8,9). Most significantly, the specificity of this response mirrors the specificity of infection and nodulation by the Rhizobium itself.

In this article, we would like to present the approach that we have undertaken in order to identify Rhizobium extracellular signals which elicit a host-specific response in legume root hairs. Two of these factors have been purified and their structures determined and, in the light of this new information, we will discuss the possible role of such molecules in bacterial-plant recognition and nodule organogenesis.

RESULTS

Bioassays for Rhizobium extracellular signals

An important advance in our understanding of how Rhizobium nod genes operate came with the discovery that sterile supernatants of R. leguminosarum cultures are able to elicit a thick and short root (Tsr) reaction on seedlings of the R. leguminosarum host, common vetch (Vicia sativa subsp. nigra) (22,23). The authors showed that the nodD and nodABC genes are required for the Tsr response and, furthermore, that the flavonoid nod gene inducer, naringenin, must also be present in the bacterial growth medium.

Unfortunately, the Tsr reaction could not be detected on any of a range of R. meliloti hosts tested (8), but it was observed that treating plants with the sterile supernatants of flavonoid-activated cultures can induce a generalized deformation of root hairs (Had). This bioassay can be used not only for common vetch (23), but also for alfalfa (8) and white clover (2). Furthermore, it is more rapid than the Tsr assay and significantly more sensitive (9). As with Tsr, the Had reaction is dependent on functional nodDABC genes. The role of host-specific genes in the production of hair deformation factors will now be discussed.

R. meliloti nodH and nodQ genes determine the host-specific modification of extracellular signals

The observation that supernatants of R. meliloti were able to induce root hair deformations on alfalfa but not on common vetch and vice versa for the supernatants of R. leguminosarum cultures (8) demonstrated clearly that there is a correlation between host specificity and extracellular factor activity. To investigate the possible role of host specificity genes of R. meliloti in the production of this symbiotic signal, we examined the Had activities of various R. meliloti mutants on both homologous and heterologous hosts. Mutations in nodH lead to a shift in signal activity [Had$^-$ on alfalfa (HadA$^-$) and Had$^+$ on vetch (HadV$^+$)], and mutations in nodQ led to an extended activity (HadA$^+$ HadV$^+$). These modifications precisely mirror the changes in host specific nodulation of the corresponding mutated strains (3,6). It should be pointed out that, in all the above experiments, the same flavonoid inducer, luteolin, was used to induce nod activity.

The introduction of the R. meliloti host range genes into R. leguminosarum (plasmid pGM1515 containing nodPQ, nodFEG, nodH, nodD3, and syrM, Fig. 1) results in the production of an extracellular factor that is now able to deform alfalfa hairs, and at the same time has decreased Tsr activity on vetch (9). Most significantly, mutations in either nodH or nodQ restore the original R. leguminosarum phenotype (Had$^-$ on alfalfa and Tsr$^+$ on vetch). This clearly shows that the introduction of R. meliloti nodH and nodQ into R. leguminosarum results in a modification of the specificity of the extracellular Had factors. The additional observation that a sterile filtrate of a R. meliloti strain carrying only nodD1 and the common nodABC genes is able to deform root hairs on vetch but not on alfalfa (8), has led us to propose the model shown in Fig. 2. Both R. leguminosarum and R. meliloti common nod genes lead to the synthesis of a factor which is Had$^+$ on vetch, but Had$^-$ on alfalfa. The nodH and nodQ genes of R. meliloti then convert this common factor to an alfalfa-specific signal, which is no longer recognized by vetch root hairs. The reason why the nodQ gene is apparently required for the production of the alfalfa signal in R. leguminosarum, but not in R. meliloti itself, could be due to the presence of more than one nodQ gene in R. meliloti (19), and that this second gene has weak activity.

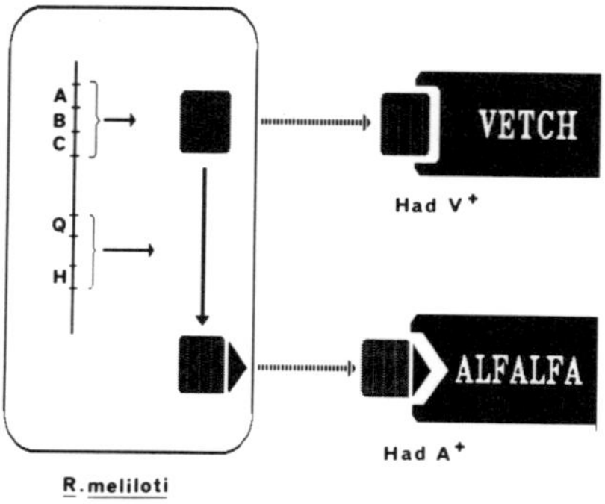

Figure 2. A model for the role of the R. meliloti common (nodABC) and host specific (nodH and nodQ) nod genes in the production of symbiotic extracellular signals which are able to deform root hairs on either alfalfa (HadA$^+$) or vetch (HadV$^+$). See text for explanation.

Purification and structure of NodRm-1, an alfalfa-specific signal

Preliminary analysis of the hair deformation factors present in filtrates of luteolin-induced R. meliloti cultures had shown that most of the activity could be extracted in a butanol-soluble fraction. Unfortunately, the concentration of the active molecule(s) proved to be insufficient for further structural analysis. In order to amplify factor production by R. meliloti, we introduced into the strain Rm2011 an Inc-P1 plasmid derivative, pGMI149 (Fig. 1), carrying all the common and host specific nod genes as well as the three regulatory genes, nodD1, nodD3 and syrM. The presence of this plasmid led to an increase of more than a hundred-fold in the Had activity on alfalfa. As a result of this amplification, it was now possible to observe two far-UV (220 nm) absorbing peaks when the butanol soluble fraction was further fractionated by HPLC on a C_{18} reverse-phase column (13). As had been previously observed for both nodulation and Had signal production

(8), Tn5 insertions into either the nodA or nodC genes resulted in the simultaneous disappearance of these two UV-absorbing peaks.

Subsequently, processing of the supernatant of a large-scale culture was carried out with the expolysaccharide-deficient R. meliloti strain EJ355 (containing plasmid pGMI149) because of its non-mucoid characteristics (10). The signals produced by Rm 2011 (pGMI149) and Rm EJ355 (pGMI149) are identical as assessed by both chemical criteria (HPLC, NMR and mass spectrometry) and the biological Had assays. Filtrates were extracted and extracellular factors purified by a combination of reverse-phase C_{18} HPLC, gel permeation on a Sephadex LH20 column and by ion-exchange chromatography on a DEAE column (13). Ten liters of a luteolin-induced R. meliloti culture yielded about 4 mg of purified factors, which showed Had activity on alfalfa (at concentrations between 10^{-8} and 10^{-11} M), but not on common vetch.

Mass spectrometry, chemical modification and NMR spectroscopy were used to establish the structure of the factors as N-acyl tri N-acetyl β-1,4-D-glucosamine tetrasaccharides, bearing a sulfate group on C-6 of the reducing sugar. The aliphatic chain carried by the non-reducing terminal sugar residue is a 2,9-hexadecadienoic N-acyl group (13; and Fig. 3). The two factors correspond to the α- and β-anomers of the same molecule at the C-1 position of the reducing end sugar, and we propose that this molecule be called NodRm-1. The choice of this nomenclature is based on the nod genes responsible for its production rather than the effects (such as hair deformation) that this factor can elicit on the plant host (see below). The fact that we have always observed a strict correlation both qualitatively and quantitatively between this molecule and the specific Had activity (Had^+ on alfalfa and Had^- on vetch) is strong evidence that NodRm-1 is indeed the plant-specific symbiotic signal previously characterized in the hair deformation assays.

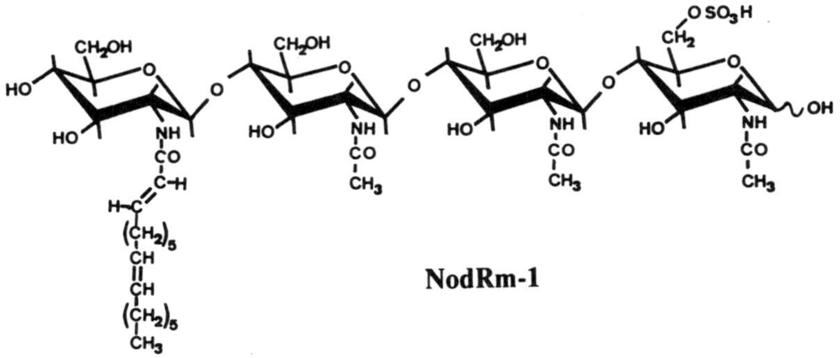

Figure 3. Structure of the sulfated and acylated glucosamine oligosaccharide symbiotic signal (NodRm-1) purified from supernatants of flavonoid-induced cultures of R. meliloti. This structure has been established from a combination of NMR spectroscopic, mass spectrometric, ^{35}S-labeling and methylation analyses (14).

Purification and structure of NodRm-2

To study the role of the nodH host-range gene in the synthesis of NodRm-1, we constructed a derivative of the Rm EJ355(pGMI149) strain carrying a copy of a nodH::Tn5 insertion in both the pSym and the pGMI149 plasmids. No NodRm-1 peaks could be detected in the butanol extract of the nodH mutant supernatant, after reverse-phase HPLC analysis. Instead, peaks corresponding to more hydrophobic compounds were observed and these fractions were Had^- on alfalfa and Had^+ on vetch. Mass spectrometry, NMR spectroscopy and methylation studies revealed that the new peaks correspond to the α- and β-anomers of a lipo-oligosaccharide which we propose to call NodRm-2 and which differs from NodRm-1 only by the absence of the sulfate group.

A Rm EJ355 (pGMI149) $nodQ^-$ derivative was constructed in the same way and the HPLC profile of the butanol extract of its supernatant showed two clusters of peaks corresponding to the NodRm-1 and NodRm-2 compounds. Further purification and chemical characterization confirmed that the $nodQ^-$ mutant produced both NodRm-1 and NodRm-2 signals.

Thus, the R. meliloti wild-type strain, which infects and nodulates alfalfa, produces NodRm-1; a $nodH^-$ mutant, which infects and nodulates common vetch, produces NodRm-2; and a $nodQ^-$ mutant, which has an extended host-range (alfalfa and vetch), produces both signals. A $nodA^-$ mutant, which does not infect either host, does not produce either of these signals. There is, therefore, a striking correlation in R. meliloti between the production of NodRm factors and the ability to specifically infect and nodulate a particular legume. These results indicate that the nodH and nodQ genes are responsible for the transfer of a sulfate group onto the lipo-oligosaccharide NodRm-2 factor and, by this means, determine host-specificity.

Biological activity of NodRm-1 in the alfalfa root cortex

Since R. meliloti is able to elicit cortical cell divisions and nodule organogenesis "at a distance", it was proposed that diffusible nodule-inducing signals could be involved in these processes (10,12,20). We, therefore, studied the effects of adding purified NodRm-1 to the roots of aseptically grown alfalfa seedlings.

We found that NodRm-1 was able to elicit cortical cell division. At concentrations of 10^{-9} - 10^{-10}M, discrete foci of meristematic cells were observved in the inner cortex of secondary roots. At higher concentrations of NodRm-1, the mitotic acivitity extended over the entire cortex and even reached the epidermis over large areas of the secondary root. This mitotic activity in the cortex gave rise to various root deformations, varying from root bumps to elongate or multilobate structures having the appearance of nodules. Cytological observations revealed that most of these structures could not be classified as secondary roots but rather exhibited the following ontological, anatomical and phsiological features of genuine alfalfa nodules: (i) they originated from the inner cortex, (ii) endodermis and vascular bundles were peripheral, (iii) their formation was totally repressed by the addition of combined nitrogen (15 mM potassium nitrate).

In conclusion, the nodABC, nodH and nodQ genes of R. meliloti are required for both the bacterial induction of root hair-curling and nodule formation of alfalfa and for the production of NodRm-1. Moreover, the additioin of NodRm-1 to aseptically-grown alfalfa seedlings elicits root hair deformation, cortical cell division and nodule formation. The simplest hypothesis for the mechanism by which the nodABC, nodH and nodQ genes operate in determining host-range, infection and nodulation is by coding for proteins which contribute to the synthesis of the NodRm-1 signal.

DISCUSSION AND PERSPECTIVES

Current research into the earliest stages of the interaction between Rhizobium and its legume host is now giving us a clearer view of how these two organisms exchange information at the molecular level even before they come into contact with each other. Firstly, plant signals in the form of flavonoids present in the root exudates activate bacterial nodulation genes in conjunction with an appropriate regulator nodD gene. Efficient nod gene activation does require a correct matching between the nodD gene product and the flavonoid content of the plant exudate (16).

The second, and highly specific, stage of the bacteria-plant interaction occurs with the production of extracellular signals by a combination of common and host specific Rhizobium nod gene activities. Using the hair deformation assay as a means of identifying and purifying these signals, we have always observed a strict correlation between the specificity of the bioassay and the symbiotic behavior of the intact Rhizobium meliloti. The structure of the signal molecule for R. meliloti (NodRm-1) has been determined to be a sulfated lipo-oligosaccharide (Fig. 3), which is active in the root hair deformation assay at extremely low concentrations (down to 10^{-11} M). It is worth noting that most phytohormones are only active in the 10^{-7} M range when added exogenously (24), and only certain fungal oligosaccharide elicitors which trigger the plant defense response, are known to be active in the nanomolar range (4).

The results immediately pose a certain number of questions concerning symbiotic signalling between Rhizobium and its legume host which we are now in a position to investigate. (i) What are the modifications in the structure of NodRm-1 which are caused by nodH or nodQ mutations, and which cause a shift or extension of host range? (ii) How do these modified signals compare in structure with the wild-type signal produced by R. leguminosarum, both of which are able to deform root hairs of common vetch? (iii) Does the production of this extracellular molecule involve de novo synthesis from N-acetyl D-glucosamine or rather the degradation and/or modification of pre-existing cell wall macromolecules such as peptidoglycan? (iv) Does NodRm-1 also have the potential to induce mitotic activity in cortical root cells of alfalfa? If so, it will be important to establish to what extent NodRm-1 resembles the R. meliloti factors identified by Schmidt et al. (18), which stimulate mitosis in plant protoplasts. The structure of these molecules is likely to be different since the production of these latter factors does not require either nodC or the host specificity genes. (v) Finally, could this molecule be responsible for the triggering of nodule organogenesis? An elegant series of experiments carried out with combinations of nodulation and infection mutants R. meliloti have recently demonstrated that a diffusible bacterial signal can trigger nodulation even when the bacteria and the plant are separated by a filter (12). Furthermore, only mutations in the nodABC and nodH genes of R. meliloti give rise to a Nod$^-$ phenotype on alfalfa.

The identification and purification of a bacteria-derived molecule, which can initiate hair branching on legume hosts in a specific manner, is also extremely important for future research on the plant host itself. It is intriguing that the NodRm-1 molecule has structural features which are typical of lectin ligands such as hexosamine oligosaccharides (15). The proposition that plant lectins play a role in the specificity of the plant-Rhizobium interaction has long been the subject of debate (for review, see ref. 5). Further direct evidence in favor of such an idea has come from recent experiments in which the introduction of a pea lectin gene into white clover led to successful infection and nodulation by a pea-specific Rhizobium (7). The same authors (13) go on to suggest that a plant root lectin located in the cytoplasmic membrane of susceptible root hairs might be a part of a Nod signal receptor. With purified signal molecules such as NodRm-1 now available, it should be possible to examine the role of lectin or other receptors present at the root tip. Radioactive labeling of NodRm-1 can easily be carried out via the sulfate moiety, and this should also facilitate the subsequent purification of such receptors.

Finally, Scheres et al. (17) have reported that supernatants of flavonoid-induced R. leguminosarum cultures are able to elicit expression of an infection-specific plant gene Enod12 in pea root hairs. The transcription of this gene is normally detected in root hairs only 24 h after infection with Rhizobium (17) and the fact that its expression is triggered by hair deformation factors suggests that it is one of the earliest plant genes to be activated in the interaction with Rhizobium. It will be very interesting to examine whether this gene responds to Had factors in a host-specific manner, and to use transgenic plant technology to monitor the expression of this gene after factor induction. By this means, we should gain access to the signal transduction pathway within the root hair and perhaps also to the pathway which leads ultimately to the triggering of nodule organogenesis.

Thus, new approaches towards studying the molecular basis of specific plant-microbe interactions are opening up, and the availability of molecules such as NodRm-1 will surely prove invaluable in our search to understand some of the mechanisms by which signalling can take place between a Rhizobium and its host legume.

References

1. Banfalvi, Z. & Kondorosi, A. (1989) Plant Mol. Biol. **13**, 1-12.
2. Bhuvaneswari, T.V. & Solheim, B. (1985) Physiol. Plant. **63**, 25-34.
3. Cervantes, E., Sharma, S.B., Maillet, F., Vasse, J., Truchet, G. & Rosenberg, C. (1989) Molec. Microbiol. **3**, 745-755.
4. Darvill, A.G. & Albersheim, P. (1984) Ann. Rev. Plant. Physiol. **35**, 243-298.
5. Dazzo, F.B. & Gardiol, A.E. (1984) in Genes Involved in Microbe-Plant Interactions, eds., Verma, D.P.S. and Hohn, T.H. (Springer-Verlag, Wien/New York), pp. 3-31.
6. Debelle, F., Rosenberg, C., Vasse, J., Maillet, F., Martinez, E., Denarie, J. & Truchet, G. (1986) J. Bacteriol. **168**, 1075-1086.
7. Diaz, C.L., Melchers, L.S., Hooykaas, P.J.J., Lugtenberg, B.J.J. & Kijne, J.W. (1989) Nature **338**, 579-581.
8. Faucher, C., Maillet, F., Vasse, J., Rosenberg, C., van Brussel, A.A.N., Truchet, G. & Denarie, J. (1988) J. Bacteriol. **170**, 5489-5499.
9. Faucher, C., Camut, S., Denarie, J. & Truchet, G. (1989) Mol. Plant-Microbe Interact. **2**, 291-300.
10. Finan, T.M., Hirsch, A.M., Leigh, J.A., Johansen, E., Kuldau, G.A., Deegan, S., Walker, G.C. & Signer, E.R. (1985) Cell **40**, 869-877.
11. Hollingsworth, R., Squartini, A., Philip-Hollingsworth, S. & Dazzo, F. (1989) in Signal Molecules in Plant and Plant-Microbe Interactions, ed. Lugtenberg, B.J.J. (Springer-Verlag, Berlin/Heidelberg), pp. 387-393.
12. Kapp, D., Niehaus, K., Quandt, J., Muller, P. & Puhler, A. (1990) The Plant Cell **2**, 139-151.
13. Kijne, J.W., Diaz, C.L. & Lugtenberg, B.J.J. (1989) in Signal Molecules in Plant and Plant-Microbe Interactions, ed. Lugtenberg, B.J.J. (Springer-Verlag, Berlin/Heidelberg), pp. 351-358.
14. Lerouge, P., Roche, P., Faucher, C., Maillet, F., Truchet, G., Prome, J.C. & Denarie, J. (1990) Nature **344**, 781-784.
15. Lis, H. & Sharon, N. (1986) Ann. Rev. Biochem. **55**, 35-67.
16. Long, S.R. (1989) Cell **56**, 203-214.
17. Scheres, B., Van De Wiel, A., Zalensky, A., Harvath, B., Spaink, H., Van Eck, H., Zwartkruis, F., Wolters, A.-M., Gloudemans, T., Van Kammen, A. & Bisseling, T. (1990) Cell **60**, 281-294.
18. Schmidt, J., Wingender, R., John, M., Wieneke, U. & Schell, J. (1988) Proc. Nat. Acad. Sci. USA **85**, 8578-8582.
19. Schwedock, J. & Long, S.R. (1989) Molec. Plant-Microbe Interact. **2**, 181-194.
20. Truchet, G., Michel, M. & Denarie, J. (1980) Differentiation **16**, 163-173.

21. Truchet, G., Barker, D.G., Camut, S., deBilly, F., Vasse, J. & Huguet, T. (1989) Mol. Gen. Genet. **219**, 65-68.
22. van Brussel, A.A.N., Zaat, S.A.J., Canter Cremers, A.C.J., Wijffelman, C.A., Pees, E., Tak, T. & Lugtenberg, B.J.J. (1987) J. Bcteriol. **161**, 517-522.
23. Zaat, S.A.J., van Brussel, A.A.N., Tak, T., Pees, E.E. & Lugtenberg, B.J.J. (1987) J. Bacteriol. **169**, 3388-3391.
24. Zeroni, M. & Hall, M.A. (1980) in Hormonal Regulation of Development. 1. Molecular Aspects, ed., MacMillan, J. (Springer-Verlag, Berlin/Heidelberg), pp. 511-586.

Gresshoff, Roth, Stacey & Newton (eds.) Nitrogen Fixation: Achievements and Objectives
Chapman and Hall New York • London 1990

Surface interactions between rhizobia and legume root hairs

J.W. Kijne[1,2], C.L. Diaz[2], B.S. dePater[2], G. Smit[1], R. Bakhuizen[2] and B.J.J. Lugtenberg[1,2]

[1]Department of Plant Molecular Biology, [2]Center for Phytotechnology RUL/TNO, Leiden University, Nonnensteeg 3, 2311 VJ Leiden, The Netherlands.

Root hairs show apical growth

Epidermal cells of plant roots can locally change the direction of cell wall growth resulting in root hair formation (6). Root hairs grow by apical extension (apical growth or tip growth), and function in the absorption of water and nutrients. Epidermal cells which form a root hair are designated as trichoblasts. Also, pollen tubes, rhizoids (19) and fungal hyphae (26) are characterized by tip extension in their cell wall during growth.

For apical wall growth in fungi, a model has been proposed which may also apply to root hair growth. According to this model (26), cell wall constituents are secreted at the extreme cell apex contributing to formation of a hydrated visco-elastic wall that expands under internal pressure. During growth, an apical wall segment is displaced in a subapical direction moving to a more external location because of continuous apposition of new elastic wall material. At the same time, this wall segment undergoes gradual rigidification by a cross-linking process. At the base of the apical dome of the cell, rigidification has advanced to such an extent that the wall no longer yields to the internal pressure and the maximal diameter of the cell is attained. The rate of extrusion of wall constituents and the rate of cross-linking are supposed to be independent. Thus, cessation of cell growth will result in rigidification of the wall over the whole apex. An electron-microscopy study of growing pea root hairs demonstrated that the cell wall at the hair tip shows an amorphous sponge-like structure, whereas in the subapical area, a fibrillar arrangement starts from the outside of the wall (12). This observation is consistent with the model for apical growth in fungal hyphae. Similar observations have been made in root hairs of the non-legume radish (Kijne et al., unpublished results). Therefore, rigidification of the primary cell wall of a root hair appears to be a process which is essentially different from the apposition of the so-called fibrillar beta-layer at the inside of the cell wall of the root hair shaft (2). In contrast to the situation with fungal hyphae, little is known of the molecular architecture of a growing root hair wall. Moreover, neither in fungi nor in root hairs have the particular properties of the plasma membrane at the cell tip, which enables polar transport, secretion and assembly of polymers, been properly characterized.

Such growing root hair tips with a spongy, extending, rigidifying cell wall and a polarized growth-supporting plasma membrane are the meeting points for many rhizobia and legume host plants at the start of the their symbiotic association.

Rhizobia induce apical growth

Under appropriate conditions, rhizobia release signal molecules which may trigger formation of tip-growing branches in root hairs (4,17). Furthermore, rhizobia induce formation of infection threads in root hairs as well as in adjacent cortical cells. Infection threads are tubular structures showing apical growth for which continuous presence of infective rhizobia is required. These observations suggest that rhizobia can induce apical growth in legume root cells.

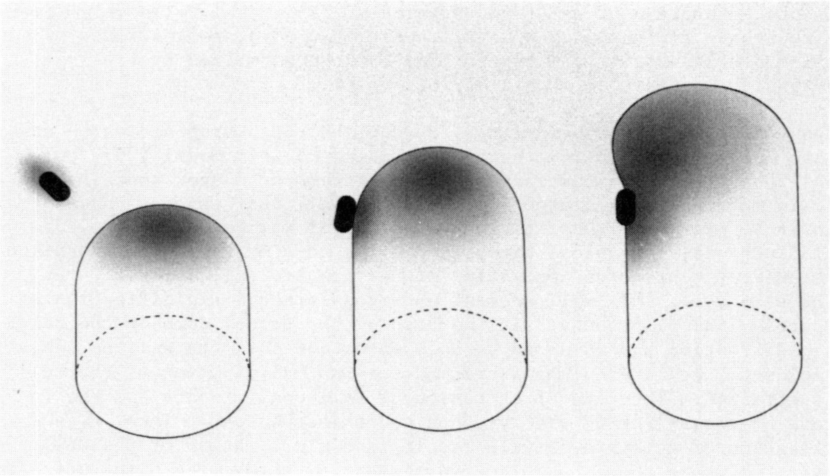

Figure 1. Redirection of apical growth by a root hair-bound Rhizobium cell (see ref. 23).

Usually, tight curling of growing root hairs is a prerequisite for induction of infection-thread formation. A computer simulation model of root hair curling (23) predicts that tight hair curling results from ultimate dominance of Rhizobium-induced apical growth over normal tip-growth of the root hair (Figure 1). Attachment of rhizobia to the surface of the apical dome, followed by displacement in subapical direction, is an essential element in this model. Indeed, Yao and Vincent (27) showed that the presence of living rhizobia at the root hair surface is a prerequisite for induction of tight ("marked") curling. Thus, attachment of rhizobia may play a dual role in root hair curling by (i) enabling delivery of rhizobial growth signals at the target site, and (ii) directing the curling process by subapical displacement of the growth-inducing cell(s) over the extending apical dome. In this respect, it is significant that, in pea, tips of growing root hairs are the preferred sites of attachment by rhizobia (20). Since, however, several rhizobial cells may attach to a root hair tip, competitive growth induction may hamper formation of a tight curl. Not surprisingly, only one out of four/five growing pea root hairs with bound infective rhizobia curls (combined data from 13 and 16). Heterogeneity in both populations of rhizobia and root hairs with regard to signal production and signal reception, respectively, may also influence the ultimate amount of tightly curled root hairs.

Interestingly, low-inoculum spot-inoculation experiments with pea and a non-motile nodulating mutant of Rhizobium leguminosarum biovar viciae (R.l.viciae) (9) demonstrated that the majority of pea trichoblasts, which were susceptible to rhizobial induction of curling and infection thread formation, had not yet formed root hairs at the time of inoculation, whereas only a small number of infections resulted from spot-inoculation among growing root hairs. Apparently, trichoblasts are more responsive to rhizobial growth signals shortly after establishment of their own apical growth area. As judged from nodulation experiments with motile rhizobia, a similar situation exists in other legumes (3,5,18).

Attachment is a two-step process

Attachment of R.l.viciae to pea root hair tips is a two-step process (22). In the first step, single rhizobial cells adhere to the surface of the root hair. In the second step, other rhizobia adhere to the hair-bound bacterial cells, resulting in accumulation of rhizobia at the attachment site. A rhizobial surface protein, rhicadhesin, is probably involved in binding rhizobia to the surface of plant root hairs (22). Rhicadhesin is a calcium-binding protein of about 14 kDa, and appears to be common among Rhizobiaceae, including Agrobacterium. Sym plasmid-borne nodulation genes are not required for synthesis of rhicadhesin. Since rhicadhesin appeared to mediate rhizobial attachment to the root hairs of various legumes and non-legumes, (a) common plant cell wall component(s) must be responsible for the interaction with the rhizobial surface.

Rhizobial cellulose fibrils are involved in accumulation of R.l. viciae bacteria at pea root hair tips (21). Under appropriate culture conditions, this accumulation is accelerated by pea lectin (Psl) molecules secreted by the roots (9,16). Psl is a 49 kDa protein with two high-affinity binding sites for oligosaccharides containing a trimannoside core with a neighboring fucosyl-alpha-1,6-N-acetylglucosamine group (8,15). In pea roots, Psl is present on the outer surface of the plasma membrane in the growth area of trichoblasts and in the apical dome of growing root hairs. Furthermore, Psl is secreted into the rhizosphere (9). Sym plasmid-borne nodulation genes are not required for Psl-mediated accumulation of R.l.viciae, which is consistent with the observations that several Rhizobium species are able to bind Psl (24) and that Psl can precipitate extracellular polysaccharides of various fast-growing Rhizobium species (14). Mutants of R.l.viciae unable to bind Psl are not yet available, and the role of Psl-enhanced accumulation in nodulation is still unclear. On the other hand, cellulose fibril-negative R.l.viciae mutants have been isolated (21). Since these mutants show normal nodulation properties, fibril-mediated accumulation is apparently not a prerequisite for successful root hair infection.

Interestingly, carbon-limited (potentially infective) R.l.viciae bacteria combine good attachment properties with the inability to induce hair curling and infection thread formation in pea and the inability to be agglutinated by Psl (16). This observation suggests that these bacteria do not produce growth signals, and points at a direct or indirect relationship between signal production and cell surface composition.

Results from attachment studies with pea and R.l.viciae differ from those with clover and R.l.trifolii. In contrast to the situation with pea, the clover lectin trifoliin A is present on the outer surface of the cell wall of growing clover root hairs. Infective R.l.trifolii cells can be agglutinated by trifoliin A and show lectin-mediated attachment to the root hair surface (attachment step 1) (for a review, see 7). It is unknown if this type of attachment is sufficient for induction of tight root hair curling or if rhicadhesin, also produced by R.l.trifolii cells, additionally must be involved. Evidently, a crucial element in the study of rhicadhesin in induction of root hair curling is the isolation of rhicadhesin-negative rhizobial mutants, which, unfortunately, are not yet available.

Lectin may be a signal reseptor

Induction of tight root hair curling by rhizobia has a high degree of host-plant-specificity (e.g., 13,27). R. meliloti is unable to induce significant curling in pea, whereas alfalfa root hairs do not tightly curl in presence of R.l.viciae. Because rhicadhesin-mediated attachment of rhizobia is not host-plant-specific, the growth-inducing signals may be determinants of this specificity. Recently, an alfalfa-specific signal of R. meliloti has been identified as being a sulfated beta-1,4-tetrasaccharide of D-glucosamine in which three amino groups are acetylated and one is acylated with a C_{16} bis-unsaturated fatty acid (Mw 1102) (17). Common as well as host-range nodulation genes are involved in the production of this signal designated as NodRm-1. The heterologous host plant Vicia is unresponsive to NodRm-1. If an (acylated) oligosaccharide structure turns out to be a general characteristic of rhizobial growth signals, legume lectins, which are characterized by specific binding sites for oligosaccharides and by conserved hydrophobic binding sites, may be good candidates for representing signal receptor molecules. Available data indicate that lectins from host plants in different rhizobial cross-inoculation groups differ in sugar-binding specificity (25).

In pea, the pattern of Psl location at the root surface corresponds completely to the susceptibility of root epidermal cells to infection by R.l.viciae (9). Whereas the surface of the pea root tip is Psl-negative, lectin appears at the surface of epidermal cells at the time of their differentiation into trichoblasts. Trichoblasts with emerging root hair tips represent the majority of Psl-positive pea root epidermal cells, whereas the surface of the mature root hair zone is characterized by absence of Psl and by an almost total resistance to rhizobial infection (11). Furthermore, Psl on the membrane surface of trichoblasts is capable of binding specific saccharides (Diaz et al., these Proceedings). Genetic evidence for a role of Psl as a determinant of host-plant-specificity in legume root hair infection has recently been presented (10). Transformation of white clover (hairy) roots with a functional psl gene confers upon these roots the ability to form infection threads in response to R.l.viciae, the pea symbiont. Interesting, R.l.viciae is able to induce root hair curling in wild type clover roots, in contrast to other rhizobia. However, this type of curling is different from normal tight curling, and appears to be similar to the type of curling induced by R.l.trifolii strains carrying mutations in certain genes encoding the host range of nodulation (Wijffelman, personal communication). According to the computer simulation model of root hair curling (see above), signals of such mutants are unable to attract all apical growth activity to the site of attachment of growth-inducing rhizobia. In turn, R.l.trifolii is able to induce root hair curling in pea. These observations suggest that both rhizobial biovars release closely related growth signals.

It is tempting to combine the above mentioned data into a working hypothesis in which rhizobial oligosaccharide signals interact with functional oligosaccharide-binding lectin molecules at the outside of the target membrane to produce redirection of apical growth and triggering of subsequent root hair infection. Experimental results of Abe et al. (1984), which show that clover root hair infection is stimulation by lectin-binding oligosaccharides from the capsular and extracellular polysaccharides of R.l.trifolii, are consistent with this hypothesis. However, before the hypothesis can be tested, additional information is required. Growth signals of R.l.viciae and R.l.trifolii have not yet been fully characterized, no data are available on alfalfa root lectin, clover root lectin is poorly characterized and its location on the clover root surface has to be defined more precisely. Part of the missing information may, however, be found on other pages of these Proceedings.

REFERENCES

1. Abe, M., Sherwood, J.E., Hollingsworth, R.I. & Dazzo, F.B. (1984) J. Bacteriol **160**, 517-520.
2. Belford, D.S. & Preston, R.D. (1961) J. Exp. Bot. **12**, 157-168.
3. Bhuvaneswari, T.V., Bhagwat, A.A. & Bauer, W.D. (1981) Plant Physiol. **68**, 1144-1149.

4. Bhuvaneswari, T.V & Solheim, B. (1985) Physiol. Plant **68,** 1144-1149.
5. Bhuvaneswari, T.V., Turgeon, B.G. & Bauer, W.D. (1980) Plant Physiol. **66**, 1027-1031.
6. Cormack, R.G.H. (1962) Bot. Rev. **28**, 446-464.
7. Dazzo, F.B., Hollingsworth, R.I., Sherwood, J.E., Abe, M., Hrabak, E.M., Gardiol, A.E., Pankratz, H.S., Smith, K.B. & Yang, H. (1985) in Nitrogen Fixation Research Progress, eds. Evans, H.J., Bottomley, P.J. & Newton, W.E. (Martinus Nijhoff, Dordrecht), pp. 239-245.
8. Debray, H., Decout, D., Strecker, G., Spik, G. & Montreuil, J. (1981) Eur. J. Biochem. **117**, 41-55.
9. Diaz, C.L. (1989) Root lectin as a determinant of host-plant specificity in the Rhizobium-legume symbiosis, PhD Thesis, Leiden University, The Netherlands.
10. Diaz, C.L., Melchers, L.S., Hooykaas, P.J.J., Lugtenberg, B.J.J. & Kijne, J.W. (1989) Nature (London) **338**, 579-581.
11. Diaz, C.L., Van Spronsen, P.C., Bakhuizen, R., Logman, G.J.J., Lugtenberg, B.J.J. & Kijne, J.W. (1986) Planta **168**, 350-359.
12. Goosen-de Roo, L., Mommaas-Kienhuis, A.M. & Kijne, J.W. (1984) in Advances in Nitrogen Fixation Research, eds. Veeger, C. & Newton, W.E. (Nijhoff/Junk/Pudoc, Den Haag, Wageningen), pp. 417.
13. Haack, A. (1964) Zentr. Bakteriol. Parasitenk. Abt. II **117**, 343-366.
14. Kamberger, W. (1979) Arch. Microbiol. **121**, 83-90.
15. Kornfeld, K., Reitman, M.L. & Kornfeld, R. (1981) J. Biol. Chem. **256**, 6633-6640.
16. Kijne, J.W., Smit, G., Diaz, C.L. & Lugtenberg, B.J.J. (1988) J. Bacteriol. **170**, 2994-3000.
17. Lerouge, P., Roche, P. Faucher, C., Maillet, F., Truchet, G., Prome, J.C. & Denarie, J. (1990) Nature **344**, 781-784.
18. Pueppke, S.G. (1982) Can. J. Microbiol. **29**, 69-76.
19. Schnepf, E. (1986) Ann. Rev. Plant Physiol. **37**, 23-47.
20. Smit, G., Kijne, J.W. & Lugtenberg, B.J.J. (1986) J. Bacteriol. **168** 821-827.
21. Smit, G., Kijne, J.W. & Lugtenberg, B.J.J. (1987) J. Bacteriol. **169**, 4294-4301.
22. Smit, G., Logman, T.J.J., Boerrigter, M.E.T.I., Kijne, J.W. & Lugtenberg, B.J.J. (1989) J. Bacteriol. **171**, 4054-4062.
23. Van Batenburg, F.H.D., Jonker, R. & Kijne, J.W. (1986) Physiol. Plant **66**, 476-480.
24. Van der Schaal, I.A.M., Kijne, J.W., Diaz, C.L. & Van Iren, F. (1983) in Lectins, Biology, Biochemistry, Clinical Biochemistry, Vol. 3, eds. Bog-Hansen, T.C. & Spengler, G.A. (Walter de Gruyter & Co, Berlin), pp. 531-538.
25. Van Driessche, E. (1988) in Advances in Lectin Research, Vol. 1, ed. Franz, H. (Springer Verlag, Berlin, Heidelberg, NY), pp. 73-134.
26. Wessels, J.G.H. (1988) Acta Bot. Neerl. **37**, 3-16.
27. Yao, P.Y. & Vincent, J.M. (1969) Aust. J. Biol. Sci. **22**, 413-423.

Gresshoff, Roth, Stacey & Newton (eds.) Nitrogen Fixation: Achievements and Objectives
Chapman and Hall New York • London 1990

Isolation, characterization, and structural elucidation of a "*nod* signal" excreted by *Rhizobium trifolii* ANU843 which induces root hair branching and nodule-like primordia in axenic white clover seedlings

Rawle I. Hollingsworth[1], Saleela Philip-Hollingsworth[2], and Frank B. Dazzo[2]

[1]Depts. of Biochemistry & Chemistry, and [2]Dept. of Microbiology, Michigan State University, East Lansing, MI 48824, USA.

Introduction

Rhizobium excretes various metabolites which are biologically active on its leguminous host. For example, a substituted tetrasaccharide excreted by *R. meliloti* has been reported to induce alfalfa-specific root hair branching and control host specificity in this symbiosis (3). We study the excreted metabolites from the clover symbiont, *R. trifolii*, which are responsible for inducing the morphological changes that occur during infection and nodulation by this bacterial symbiont. We are particularly interested in the metabolites whose production/export require pSym *nod* genes; hence, these are called "nod signals". Our general strategy has been to grow wild-type *R. trifolii* ANU843 on defined BIII medium with or without flavones to enhance *nod* gene expression, fractionate the culture supernatant, and perform axenic seedling bioassays on the purified excreted metabolites. Our results to date have shown that wild-type *R. trifolii* excretes a diversity of bacterial factors capable of affecting development of the root nodule symbiosis (1,2this paper, Fig. 1).

Figure 1

DIVERSITY OF SIGNALS FROM *R. trifolii* INDUCE MORPHOLOGICAL CHANGES IN CLOVER

1. FLAVONE—INDEPENDENT ACCUMULATION
   - BF—1 Root hair proliferation
   - BF—2 Root hair curvature
2. FLAVONE—DEPENDENT ACCUMULATION
   - BF—3 Root hair deformation, Noi
   - BF—4 Enhances BF—3 Noi activity
   - BF—5 Root hair branching, hair tip swelling, weak Noi

Here we report on BF-5, an extracellular *nod* signal from *R. trifolii* which induces root hair branching and enhances nodule formation on the legume host, clover.

Methods

Our axenic seedling bioassay (1) requires small volumes of filter-sterilized test sample and can be examined microscopically without disturbing the root system or introducing microbial contamination. Root hair deformation is assigned a score of 0-4 depending upon the percentage of hair cells affected by the introduced compound (0 = 0%; 1 = 1-25%; 2 = 26-50%; 3 = 51-75%; and 4 = >75%). Then, roots are cleared with Clorox, stained with methylene blue, and examined to enumerate the foci of cortical cell divisions (4). This procedure distinguishes the rounded meristem of nodule primordia from cone-shaped lateral root primordia. To assay for infection-related biological activity, seedling roots are preincubated with test samples before inoculation with *R. trifolii*. Later, infected root hairs and root nodules are counted. Sterile -N Fahraeus medium was evaluated as a control in all bioassays.

BF-5 was purified by a combination of ethanol precipitation, reverse phase HPLC, gel permeation HPLC, and gel filtration chromatography. Purity was established by TLC, HPLC, proton N M R, and G C / M S.

Results

TLC analysis was performed on purified BF-5 and culture filtrates from *R. trifolii* ANU843 grown with and without 4'-7'-dihydroxyflavone (DHF), and from a *nodD::Tn5* mutant derivative (*R. trifolii* ANU851 from B. Rolfe, ANU) grown with DHF. The results indicate that DHF and *nodD* are required for accumulation of BF-5 in the culture supernatant.

BF-5 induces at least 3 morphological responses in axenic white clover seedlings: branching of young root hairs (Fig. 2A), tip swelling followed by branching and resumed elongation in older root hairs (Fig. 2B), and foci of cortical cell divisions leading to nodule-like meristems within the root. These responses occurred in white clover (*Trifolium repens*) and strawberry clover (*T. fragiferum*), but not alfalfa (*Medicago sativa*) or birdsfoot trefoil (*Lotus corniculatus*), suggesting that BF-5 responses are restricted to the clover cross-inoculation group. Pretreatment of white clover roots with BF-5 increased nodule formation by *R. trifolii* ANU843.

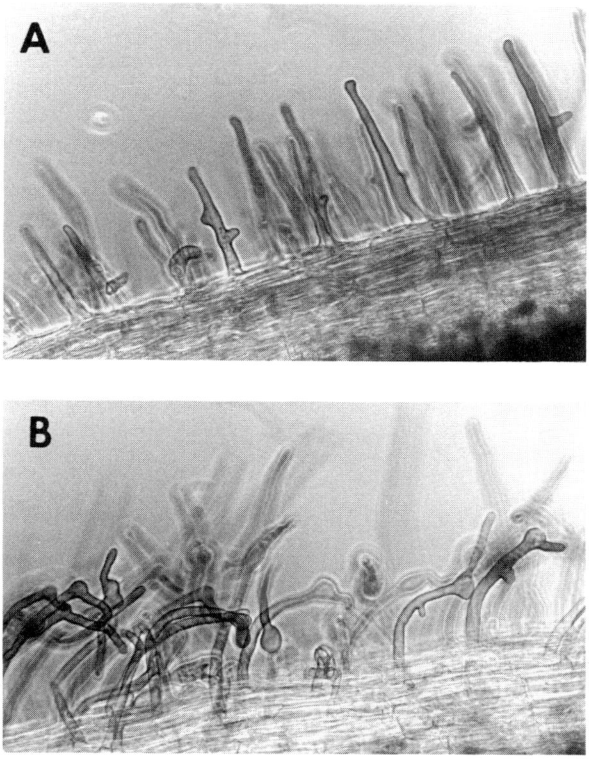

Figure 2. BF-5 from R. *trifolii* induces branching of young root hairs (A) and tip swellings followed by branching on older root hairs (B) of white clover.

The structure of BF-5 was established as N-acetylglutamic acid by proton NMR, Fourier transform IR, GC/MS of the methylated derivative, and chemical synthesis (Fig. 3). The biological activity of BF-5 on clover roots was confirmed with authentic N-acetylglutamic acid obtained from chemical synthesis and a commercial product.

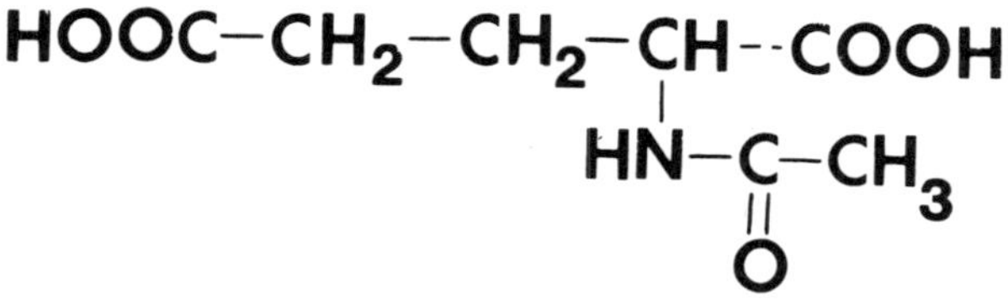

BF-5 = N-acetylglutamic acid

- Proton NMR
- FTIR
- GC / MS
- Organic Synthesis
- Commercial Product

Figure 3.

In summary, extracellular accumulation of N-acetylglutamic acid by *R. trifolii* requires flavones to induce pSym *nod* genes, the positive regulatory gene *nodD*, and can elicit branching and tip swellings on clover root hairs. Under normal conditions, the level of BF-5 exposed to clover roots is a limiting factor to nodule formation by *R. trifolii*, and BF-5 supplements enhance this symbiotic step. We are currently evaluating the requirement of other pSym *nod* genes in production/export of BF-5, and its possible role in determining host specificity. N-acetylglutamic acid is one of several extracellular signals from *R. trifolii* which induces morphological responses leading to development of the nitrogen-fixing, *R. trifolii*-clover symbiosis.

Acknowledgements

We thank Andrea Squartini, David Baker, Ann Aggarwal, Terry Mapes, and Jane McCoy for technical assistance, Barry Rolfe for the bacterial strains, and Georges Truchet for his critical comments. This work was supported by the MSU-NSF Science and Technology Center for Microbial Ecology, DOE grant DEFG-0289ER14029 and an affirmative-action post-doctoral fellowship from the College of Natural Sciences, MSU.

Literature Cited

1. Hollingsworth, R. I., Squartini, A., Philip-Hollingsworth, S., and Dazzo, F. (1989) in Signal Molecules in Plants and Plant-Microbe Interactions, ed. Lugtenberg, B. (Springer-Verlag, Berlin), pp. 387-393.

2. Hollingsworth, R. I., Squartini, A., Philip-Hollingsworth, S., and Dazzo, F. (1990) in Nitrogen Fixation: Achievements and Objectives, eds. Gresshoff, P., Roth, I., Stacey, G., and Newton, W. (Chapman and Hall, New York), in press, this book.

3. Lerouge, P., Roche, P., Faucher, C., Maillet, F., Truchet, G., Promo, J., and Denarie, J. (1990) Nature 344, 781-784.

4. Truchet, G., Camu, S., DeBilly, F., Odorico, R., and Vasse, J. (1989) Protoplasma 149, 82-88.

Gresshoff, Roth, Stacey & Newton (eds.) Nitrogen Fixation: Achievements and Objectives
Chapman and Hall New York • London 1990

Recent studies on the *Rhizobium*-legume symbiosis

Frank Dazzo[1], Rawle Hollingsworth[2], Saleela Philip-Hollingsworth[1], Andrea Squartini[1], Kenneth Chapman[1], Lucette Cargill[1], Janet Salzwedel[1], Maureen Petersen[1], Stuart Pankratz[1], Guy Orgambide[1], Paul De Troch[1], Thomas Olen[1], David Baker[1], Jaime Maya-Flores[1], Annshu Aggarwal[1], and Gary Hollander[1]

Departments of Microbiology[1] and Biochemistry[2], Michigan State University, East Lansing, MI 48824, U.S.A.

This brief paper summarizes our research progress since the 7th International Congress on Nitrogen Fixation at Köln, West Germany. We published studies reported at that meeting on the structural similarities and differences between the capsular acidic heteropolysaccharide (CPS) of Rhizobium trifolii and R. leguminosarum (4), and the effect of interspecies transfer of Rhizobium host-specific nodulation genes on CPS structure and in situ binding by trifoliin A lectin (5). We also published on the application of computer-aided image analysis for studies of the Rhizobium-legume symbiosis (2). Interesting measurements reported in that paper include: (i) R. leguminosarum bacteroids are 7.2 times the volume of vegetative bacteria within pea nodules, (ii) the average vertex angle of root hair branching induced by R. trifolii on white clover is 67°, (iii) R. trifolii rotates its flagella at 330 rpm and swims at 53 µm/sec in the external root environment of agarless Fahraeus slide cultures of white clover, (iv) inoculation of white clover roots with R. trifolii leads to a 35% increase in cytoplasmic streaming in root hairs, and (v) white clover root hairs elongate during active growth at an average rate of 38 µm/hr.

We have submitted two papers on R. trifolii glycoconjugates: (i) "pSym nodI Function Downregulates 3-Hydroxybutyrate Substitution in the Acidic Heteropolysaccharide of Rhizobium trifolii"; and (ii) "Rhizobium Lipopolysaccharide Modulates Infection Thread Development in White Clover Root Hairs". We are preparing a third manuscript entitled "Chemical Analysis of the Lipopolysaccharide from Rhizobium trifolii at Mid-exponential and Early Stationary Phases of Growth".

At plant-microbe conferences in Acapulco, Mexico (1) and Biddinghuizen, the Netherlands (3), we reported on extracellular signals from R. trifolii which are active on clover roots. During this Nitrogen Fixation Congress, we are reporting our most recent results on 3 bacterial factors purified from culture supernatants of flavone-induced R. trifolii: (i) synergistic effects of BF-3 and BF-4 which are had and noi signals, and (ii) BF-5 which is a hab signal requiring nodD. These extracellular metabolites from R. trifolii add to our growing list of diverse signal molecules requiring flavones for production/export and which modulate specific morphological responses on white clover roots during infection and nodulation.

Studies which we reported during the 12th North American Symbiotic Nitrogen Fixation Conference at Iowa State University include: (i) "Toxicity of Unbalanced pSym nod Gene Expression in R. trifolii" (collaboration with M. Djordjevic & B. Rolfe, Australian National University); (ii) "Pit Erosion of Root Epidermal Cell Walls in the Rhizobium-White Clover Symbiosis" (collaboration with D. Hubbell, University of Florida) and (iii) "Effects of Rhizobium phaseoli Inoculants on Bean Yield, Root Growth, and Early Nodule Development in Mexico" (collaboration with M. Valdes, National Polytechnic Institute).

We are presenting two posters on R. trifolii glycoconjugates at this meeting: (i) alterations in CPS due to nodI function, and (ii) the use of monoclonal antibodies to detect subtle alterations in LPS of R. trifolii Tn5 insertion mutants and their symbiotic phenotypes (collaboration with S. Wright, USDA-ARS). We are presenting another poster on induction of root hair peroxidases by flavone-dependent, excreted elicitors from Rhizobium. We plan to present our results of collaborative studies with B. Rolfe and M. Djordjevic (ANU) on impaired encapsulation in an R. trifolii nodE::Tn5 mutant at the conference on Molecular Genetics of Plant-Microbe Interactions in Interlochen, Switzerland. We are also collaborating with Professor J. Vanderleyden and P. De Troch of KULeuven, Belgium, on exopolysaccharide chemistry in Azospirillum brasilense.

Acknowledgements

These studies were supported by NIH GM34331 and the MSU Center for Microbial Ecology. We thank Jane McCoy, Ines Toro, Terry Mapes, Samuel Pogats, and Jacquelin Kowalski for technical assistance.

Literature Cited

1. Dazzo, F., Hollingsworth, R., and 17 others. (1988) in Molecular Genetics of Plant-Microbe Interactions 1988, eds. Palacios, R. & Verma, D.P. (American Phytopathological Society, St. Paul), pp. 35-40.

2. Dazzo, F. & Petersen, M. (1989) Symbiosis 7, 193-210.

3. Hollingsworth, R., Squartini, A., Philip-Hollingsworth, S. & Dazzo, F. (1989) in Signal Molecules in Plants and Plant-Microbe Interactions, ed. Lugtenberg, B. (Springer-Verlag, Berlin), pp. 387-393.

4. Philip-Hollingsworth, S., Hollingsworth, R. & Dazzo, F. (1989) J. Biol. Chem. 264, 1461-1466.

5. Philip-Hollingsworth, S., Hollingsworth, R., Dazzo, F., Djordjevic, M. & Rolfe, B. (1989) J. Biol. Chem. 264, 5710-5714.

Gresshoff, Roth, Stacey & Newton (eds.) Nitrogen Fixation: Achievements and Objectives
Chapman and Hall New York • London 1990

The *Rhizobium leguminosarum* bv.*viciae* NodO protein compensates for the exported signal made by the host-specific nodulation genes

J. A. Downie, A. Economou, A. K. Scheu, A. W. B. Johnson*, J. L. Firmin, K. E. Wilson, M. T. Cubo, A. Mavridou, C. Marie, A. Davies and B. P. Surin**

John Innes Institute, John Innes Center for Plant Science Research, Colney Lane, Norwich NR4 7UH, UK. Present Address: *School of Biological Sciences, University of East Anglia, Norwich, NR4 7TJ, UK, **Research School of Chemistry, Australian National University, ACT 2601, Australia.

Introduction

The nodulation of peas and vetches by *R. leguminosarum* biovar *viciae* requires the expression of several bacterial nodulation (*nod*) genes (1). One of these (*nodD*) is expressed constitutively and is involved in the positive regulation of the other *nod* genes (2). Flavonoids such as naringenin and eriodictyol are secreted from pea roots (3,4) and appear to be recognized by the NodD protein (5,6) which is then involved in the activation of transcription of the other *nod* genes. This regulation is mediated by the NodD protein binding to the promoters of the other *nod* genes (7) at a highly conserved DNA sequence (*nod*-box). In *R. meliloti* there also appears to be a repressor of *nod* gene expression (8) but we have not found such a repressor in b.v. *viciae*. However, we have identified mutations that lower the level of induction of nod gene expression. Recombinant plasmids which suppress these mutations have been identified and DNA hybridization using these clones has shown that they are from outside the *nod* gene region.

The *nod* gene region from one strain of bv. *viciae* is shown in Fig. 1. The *nod* operons *nodABCIJ*, *nodFEL*, *nodMNT* and *nodO* are all preceded by a *nod*-box and DNA hybridization experiments using oligonucleotide probes homologous to the nod-box have identified only four hybridizing regions, indicating that all of the nod operons in that strain are represented in Fig. 1. The DNA sequences of all of the *nod* genes have been determined. Fig. 1 shows the order and physical organization of the *nod* genes, the arrows corresponding to the relative size of the open reading frames identified from the DNA sequence. Mutations in *nodA*, *nodB*, *nodC* or *nodD* block nodulation (9,10), but although mutations in the other nod genes may reduce nodulation, they do not abolish it (9,16). This paper focuses on the roles of genes other than *nodABCD* in nodulation.

Results

Deletion of the nodFELMNTO region blocks nodulation

Plasmid pIJ1089, which carries all of the nodulation genes shown in Fig. 1, allows strains of *R. leguminosarum* lacking a symbiotic plasmid to nodulate peas and vetch. A fragment of DNA (about 20 kb) carrying the *nodF-nifH* region was deleted from pIJ1089 and replaced with a kanamycin-resistance gene. This deletion mutation (*nod*Δ99::*kan*) was then transferred to the symbiotic plasmid pRL1JI. Whereas individual mutations within the *nodFELMNTO* region did not block nodulation, no nodules were formed on peas or vetch inoculated with the deletion mutant. The deletion mutant did retain root hair deformation activity but no infection threads were found, indicating that nodulation is blocked at the stage of infection thread development. Further work described here is aimed at determining the genetic and biochemical roles of those genes deleted from the strain carrying the *nod*Δ99::*kan* allele.

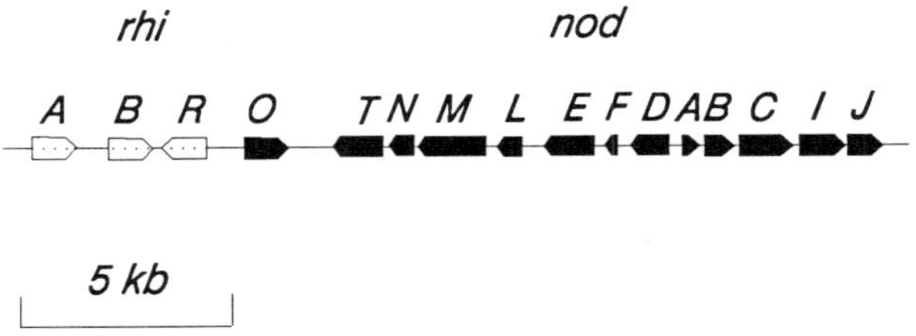

Fig. 1. Map of the nodulation region of *Rhizobium leguminosarum* by *viciae* strain 248. Black arrows represent the nodulation (*nod*) genes and open arrows the rhizosphere-expressed (*rhi*) genes, present on the symbiotic plasmid pRL1JI (11,12,13,14,23,30).

Coinoculation experiments indicate that the exopolysaccharide is not modified by the host-specific *nod* gene products

Three classes of *R. leguminosarum* mutants have been defined that are unable to nodulate peas or vetch. These include mutants affected in the *nodDABC* region (9,10), exopolysaccharide-deficient mutants (17) and the deletion mutant described above. It has been suggested that the host-specific *nod* genes of *R. leguminosarum* may be involved in modification of the acidic exopolysaccharide (18). It is possible to test this hypothesis genetically by testing the nodulation ability of mixed cultures containing the exopolysaccharide-deficient mutant and the deletion mutant lacking the host-specific nod gene region. It would be predicted that if the host-specific *nod* genes were involved in modification of the exopolysaccharide, then the mixed inoculant would not form nodules, since the strain that has the host-specific nodulation genes lacks the exopolysaccharide substrate that is expected to be modified. however, the mixture of the exopolysaccharide-deficient mutant and the deletion (*nod*Δ*99::kan)* mutant formed normal numbers of nodules on peas and vetch. Therefore, we conclude that the host-specific *nod* genes are unlikely to be involved in a modification of the exopolysaccharide that is necessary for nodulation. Interestingly, the proportion of bacteria recovered from nodules was very consistent, with only 5% of the bacteria being the exopolysaccharide-deficient mutant. This suggests that these bacteria may have to be enveloped by the exopolysaccharide from the deletion mutant before infection can occur.

Either of two different *nod* regions can complement the *nod* deletion *nod*Δ*99::kan*

Plasmids pIM164 (*nodFE*) and pIM186 (*nodFEL*) were transferred to strain A69 (which carries the *nod*Δ*99::kan* deletion) and the transconjugants were tested for their ability to nodulate vetch (*Viciae hirsuta*). In each case nodules were formed; with the strain carrying pIM164 (*nodFE*) about 30% of the normal number of nodules was formed and with pIM186 (*nodFEL*) normal numbers of nodules were found. These results confirm earlier observations (12,19,20) about the crucial role of the *nodFEL* region in nodulation.

Significantly, when plasmid pIJ788 (*nodO*) was transferred to the deletion mutant A69, the transconjugant was found to nodulate vetch (at 30% of normal levels) even though the *nodFELMNT* genes were absent from that strain. Because the NodO protein is unrelated to NodF, NodE or NodL (see below) we conclude that efficient nodulation can be mediated by at least two separate pathways, one encoded by *nodFEL* and the other by *nodO*.

On the basis of these observations we have attempted to identify the biochemical roles of these nodulation genes and also tried to evaluate the functional significance of the other genes deleted in strain A69 (*nod*Δ*99::kan*).

The *nodFEL* genes determine host specificity by modifying the root hair curling factor.

It was established previously that NodF is homologous to acyl-carrier protein (11) and that NodE is homologous to the *E. Coli* enzyme (encoded by *fabB*) ß-ketoacyl synthase, the condensing enzyme activity of fatty acid synthase (21) (J.A.D. unpublished observations). The *nodL* gene product is homologous to acetyl transerases (22). On the basis of these and earlier genetic observations it is very likely that these genes are involved in the biosynthesis of a signal molecule that determines host specificity. Furthermore, it would be predicted that NodL would be involved in an acetylation reaction.

To test this, bv. *viciae* was grown in minimal medium plus flavonoids (to induce *nod* gene expression) and pulse labelled with C^{14}-acetate. Reverse phase hydrophobic chromatography of the growth medium supernatant revealed a peak of radioactivity that was present in a wild-type strain but absent from a mutant strain lacking all of the *nod* genes. Previously (23) it was found that wild-type *R. leguminosarum* secretes into the growth medium a root hair deformation factor that is necessary for nodulation and whose formation requires the *nodABC* genes. When the fractions containing the peak of radioactivity (described above) were assayed for root hair deformation on *V. hirsuta*, they were found to be active. Since that peak of C^{14}-labelled material was absent from a *nodC* mutant strain (which lacks the root hair deformation factor) it is concluded that the C^{14}-acetate was labelling the root hair deformation factor. Moreover, the radioactivity labelled peak was absent from the deletion mutant A69 (*nod*Δ*99::kan*), showing that this locus is involved in the synthesis of the root hair deformation factor. It is to be recalled that the deletion mutant A69 (*nod*Δ*99::kan*) retains root hair deformation activity. However, unlike the activity from the wild-type strain the root hair deformation activity from the deletion mutant did not bind to the reverse phase column. A similar observation was made with a NodE mutant strain. It was reported recently (24) that the root hair deformation factor from *R. meliloti* consists of a sulphated tetraglucosamine which is acylated on one of the N-groups and acetylated on the other amino groups. On the basis of the results described above, it is likely that NodL is involved in the acetylation of the glucosamine molecules while nodFE are most probably involved in determining the type of fatty acid that is present on one of the glucosamine molecules. Such a biosynthetic pathway shows marked similarity to the pathway of lipopolysaccharide synthesis in Gram-negative bacteria (25).

From these results, it is concluded that the *nodABC* gene products are involved in the formation of a compound that induces root hair deformation, that host specificity is determined in part by a NodL-dependent acetylation of the root hair deformation factor and that the *nodFE* gene products confer host specificity by a modification of the compound made by the nodABC proteins such that the modified root hair deformation factor can bind to the reverse phase column. The nature of these modification must determine one aspect of the host specificity in nodulation.

The *nodMNT* genes

Relatively little is known about the role of these genes in nodulation. NodM is strongly homologous to the *E. coli* enzyme (encoded by *glmS*) involved in the synthesis of glucosamine phosphate, an intermediate on the pathway to N-acetyl glucosamine. We have biochemical evidence to suggest that NodM is involved in the formation of glucosamine phosphate which is likely to be an intermediate in the biosynthesis of root hair curling factor. Since mutants lacking *nodM* grow normally it is likely that these bacteria have an alternate route of synthesising glucosamine. If this is the case, then it could explain why mutation of *nodM* has relatively little effect on nodulation. The roles of *nodN* and *nodT* remain obscure although it has been suggested that nodT may encode an outer membrane lipoprotein (13).

The *nodO* gene

The nodO gene encodes a Ca^{2+}-binding (26) secreted protein that shows homology to the haemolysin family of proteins (26,27) which are secreted by Gram-negative bacteria. It is thought that the NodO protein may interact directly with plant root cell membranes in such a way that it facilitate normal infection.

Since the export of the NodO protein is crucial to its role in nodulation we have analyzed the mechanism of its export. In view of its homology to haemolysin and other secreted proteins (26) it seemed possible that it could share a common mechanism of export. The haemolysin protein requires the *hlyBD* gene products for its export from *E. coli* (28) and similar genes are found to be required for the export of related proteins such as the protease secreted from *Erwinia chrysanthemi* (29). When the cloned *nodO* gene was transferred to a non-haemolytic *E. coli* strain, the NodO protein was secreted but only if the *hlyBD* genes were present. Therefore, it is expected that similar genes will be required for NodO export in *R. leguminosarum*.

Five secreted proteins showing homology to NodO have been identified (26) and for each of these, export genes homologous to *hlyBD* have been identified closely linked to the genes encoding the secreted protein. However, when the *nodO* gene, cloned behind a constitutive vector promoter was transferred to a strain of *R. leguminosarum* lacking its symbiotic plasmid, the exported NodO protein was found in large quantities in the growth medium supernatant. This demonstrates that the genes required for NodO secretion are not on the symbiotic plasmid and therefore are unlinked to *nodO*.

Furthermore, since the NodO protein was secreted from this strain even in the absence of flavonoids, it is concluded that the genes required for NodO export are expressed constitutively unlike nodulation genes which require flavonoids for their induction.

Using DNA hybridization with a *nodO*-specific probe we have been unable to identify *nodO* homologues in *R. meliloti* or *B. japonicum,* suggesting that NodO is a host-specific nodulation protein. Presumable it plays some role in the initiation of infection thread information.

The rhiABR genes

The *rhi* (*rhi*zosphere) genes are adjacent to the *nod* genes and three *rhi* genes (*rhiABR*) have been identified. The *rhiR* gene product positively regulates the *rhiAB* genes which appear to be in one operon. Mutations in the *rhi* genes have no effect on nodulation and the subcloned *rhi* gene region does not confer on strain A69 (*nod*Δ*99::kan)* any nodulation characteristics. These genes however are influenced in their expression by flavonoids. Whereas flavonoids induce *nod* gene expression they lower the level of *rhi* gene expression (14). This effect appears to be mediated via the *rhiR* gene since flavonoids repress its transcription. Significantly, this flavonoid-dependent repression of *rhiR* depends upon *nod* gene expression; in a strain carrying a mutation in *nodD* no repression of *rhiR* by flavonoids is seen. Therefore, it is possible that these genes influence normal nodulation and their role is currently being investigated.

Conclusion

The host specificity encoded by the nod gene region of *R. leguminosarum* bv. *viciae* functions via at least two routes. One of these is dependent on a modification of the root hair curling factor specific to *nodFEL* while the other depends on the secreted calcium binding protein, NodO. The *nodO* gene product can compensate for the absence of the *nodFEL* genes but the mechanism for this has yet to be established.

References

1. Downie, J.A. & Johnston, A.W.B. (1988) Plant Cell Environ. **11**, 403-412.
2. Rossen, L., Shearman, C.A., Johnston, A.W.B. & Downie, J.A. (1985) EMBO **4**, 3369-3373.
3. Zaat, S.A.J., Wijffelman, C.A., Mulders, I.H.M., van Brussel, A.A.N. & Lugtenberg, B.J.J. (1988) Plant Physiol **86**, 1298-1303.
4. Firmin, J.L., Wilson, K.E., Rossen, L. & Johnston, A. (1986) Nature **324**, 90-92.
5. Burn, J.E., Hamilton, W.D., Wootton, J.C. & Johnston, A.W.B. (1989) Mol. Microbiol. 3, 1567-1577.
6. Spaink, H.P., Wijffelman, C.A., Pees, E., Okker, R.J.H. & Lugtenberg, B.J.J. (1987) Nautre **328**, 337-339.
7. Hong, G-F., Burn, J. & Johnston, A.. (1987) Nucl. Acids. Res. **15**, 9677-9690.
8. Kondorosi, E., Gyuris, J., Schmidt, J., John, M., Duda, E., Hoffmann, B., Schell, J. & Kondorosi, A. (1989) EMBO J. **8**, 1331-1340.
9. Downie, J.A., Knight, C.D., Johnston, A.W.B. & Rossen, L. (1985) Mol. Gen. Genet. **198**, 255-262.
10. Wijffelman, C.A., Pees, E., can Brussel, A.A.N., Okker, R.J.H. & Lugtenberg, B.J.J. (1985) Arch. Microbiol. **143**, 225-232.
11. Shearman, C.A., Rossen, L., Johnston, A.W.B. & Downie, J.A. (1986) EMBO **5**, 647-652.

12. Surin, B.P. & Downie, J.A. (1988) Mol. Microbiol. **2**, 173-183.
13. Surin, B.P., Watson, J.M., Hamilton, W.D.O., Economou, A., Downie, J.A. (1990) Mol. Microbiol. **4**, 245-252.
14. Economou, A., Hawkins, F.K.L., Downie, A. J. & Johnston, A.W.B. (1989) Mol. Microbiol. 3 (1), 87-93.
15. De Maagd, R.A., Spaink, H.P., Pees, E., Mulders, I. H. M., Wijfjes, A., Wijffelman, C.A., Okker, R.J.H. & Lugtenberg, B.J.J. (1989a) J. Bacteriol. **171**, 1151-1157.
16. Canter-Cremers, H.C.J., Wijffelman, C.A., Pees, E., Rolfe, B.G., Djordjevic, M.A. & Lugtenberg, B.J.J. (1988) J. Plant Physiol. **132**, 398-404.
17. Borthakur, D., Barber, C.E., Lamb, J.W., Daniels, M.J., Downie, J.A. & Johnston, A.W.B. (1986) Mol. Gen. Genet. **200**, 278-282.
18. Phillip-Hollingsworth, S., Hollingsworth, R.I. & Dazzo, F. (1989) J. Biol. Chem. **264**, 1461-1466.
19. Surin, B.P. & Downie, A.J. (1989) Plant Mol. Biol. **12**, 19-29.
20. Spaink, H.P., Wiffelman, C.A., Pees, E., Okker, R.J.H. & Lugtenberg, B.J.J. (1987) Nature **328**, 337-339.
21. Bibb, M.J., Biro, S., Motamedi, H., Collins, J.F. & Hutchinson, C.R. (1989) EMBO J. **8**, 2727-2736.
22. Downie, J.A. (1989) Mol. Microbiol. **3**, 1649-1651.
23. Zaat, S.A.J., Van Brussel, A.A.N., Tak, T., Pees, E. & Lugtenberg, B. J. (1987) J. Bacteriol. **169**, 3388-3391.
24. Lerouge, P., Roche, P., Faucher, C., Maillet, F., Truchet, G., Promé J.E. & Dénarié, J. (1990) Nature **344**, 781-784.
25. Raetz, C.R.H. (1986) Ann. Rev. Genet. **20**, 253-295.
26. Economou, A., Hamilton, W.D.O., Johnston, A.W.B. & Downie, J.A. (1990) EMBO J. **9**, 349-354.
27. De Maagd, R.A., Wijfjes, A.H.M., Spaink, H.P., Ruiz-Sainz, J.E., Wijffelman, C.A., Okker, R.J.H. & Lugtenberg, B.J.J. (1986b) J. Bacteriol. **171**, 6746-6770.
28. Holland, I.B., Wang, R., Seror, S.J. & Blight, M. (1989) Symp. Soc. Gen. Micro. **44**, 219-254.
29. Létoffé, S., Delepelaire, P. & Wandersman, C. (1990) EMBO J. **9**, 1375-1382.
30. Rossen, L., Johnston, A. & Downie, J.A. (1984) Nucl. Acids Res. **12**, 9497-9508.

Gresshoff, Roth, Stacey & Newton (eds.) Nitrogen Fixation: Achievements and Objectives
Chapman and Hall New York • London 1990

Rhizobium meliloti nodulation genes and their regulation

E. Kondorosi[1], Z. Györgypal[2], I. Dusha[2], N. Baev[2], M. Pierre[1], B. Hoffmann[1], A. Himmelbach[1], Z. Banfalvi[2] and A. Kondorosi[1,2]

[1]Institut des Sciences Végétales, CNRS, 91198 Gif-sur-Yvette Cedex, France
[2]Institute of Genetics, Biological Research Center, Hungarian Academy of Sciences, H-6701, Hungary.

INTRODUCTION

Nodulation of legume roots by rhizobia leads to the establishment of nitrogen fixing symbiosis. Genes determining nodule induction have been identified in several *Rhizobium*, *Bradyrhizobium* and *Azorhizobium* species. One set of nodulation genes, the common *nod* genes (*nodABC*) are essential and conserved in all rhizobia both functionally and at the DNA sequence level, while other sets of genes determine the host-specificity of nodulation (*hsn*) and/or contribute to a more efficient nodulation ability (19, for review). These nodulation genes are barely expressed in free-living cells. When rhizobia are exposed to exudates of the plant hosts, the nodulation genes are transcriptionally activated which requires also the regulatory NodD protein. The plant signals activating *nod* gene expression are various products of the phenylpropanoid pathway, primarily flavonoids or isoflavones (22, for review). The two classes of induced nodulation genes provide functions leading to root hair curling and induction of meristematic cell division. These plant responses are evoked by a family of signal molecules synthetized by different sets of *nod* gene products (2, 3, 8, 28). A host-specific signal has been recently identified by Lerouge et al. (21) as sulphated and acylated glucosamine oligosaccharide.

We have been studying the genetic determinants controlling nodulation in *R. meliloti*, the microsymbiont of legumes belonging to the genera *Medicago*, *Melilotus* and *Trigonella*. Recently, we have identified and characterized several new nodulation genes. Moreover, we found that fine-tuning of *nod* gene expression is controlled by several different trans-acting factors including products of the 3 *nodD* alleles and *syrM* which mediate regulation by the plant signals or by the combined nitrogen status of the environment. Finally, the possible common ancestry of the signal molecule (ligand) - binding domains of the NodD and of the human hormone receptors is discussed.

NEW *NOD* GENES IN *R. MELILOTI*

In *R. meliloti* strain AK631 we have identified nodulation genes organised into at least eight transcriptional units (Fig. 1) forming the *nod* regulon. Previously we have shown that inducible *nod* transcriptional units are preceded by 5'-flanking regions containing the highly conserved *nod* box sequences. In *R. meliloti* we have demonstrated 6 *nod* box sequences (*n*1-*n*6 ; 26) which are likely *cis*-acting control elements binding the regulatory NodD

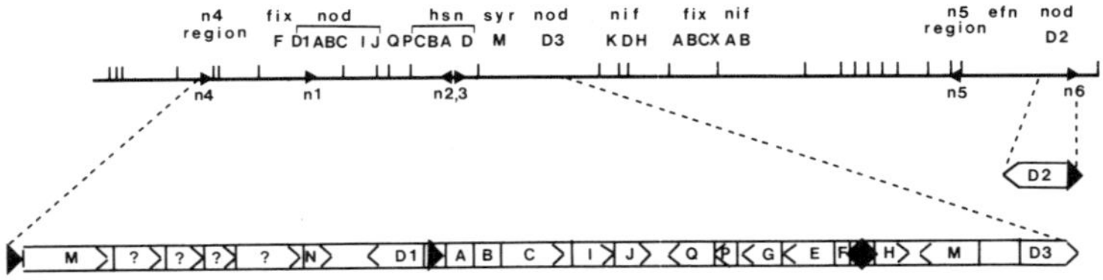

Fig. 1. Organization of *nod-nif* region in *R. meliloti* strain AK631.

protein (9 , 20). The regulatory region of the common *nodABC*(*IJ*) genes contains *n*1, while *n*2 and *n*3 are in front of the *hsnABC* (*nodFEG*) and *hsnD* (*nodH*) genes. Phenotypes of mutations and analysis of *lacZ* fusions downstream of the *n*4 and *n*5 boxes indicated that these transcriptional units are functional in nodule induction on a host-specific manner (10). Subsequent Tn5 mutagenesis of the region downstream of *n*4 until the common *nod* genes delimited a DNA region where mutations slightly delayed nodulation and decreased nodulation competitiveness. The nucleotide sequence of this region revealed six open reading frames. Homology comparisons with *nod* genes from other *Rhizobium* species indicated that this DNA region contains the *nodM* and *nodN* genes. The *nodMN* genes have been described before in *R. leguminosarum* biovars *viciae* and *trifolii* where *nodN* is next to *nodM* in the transcription unit (30), while in *R. meliloti* these two genes are separated by several new *nod* genes which have not been found in other *Rhizobium* species so far. Mutational and/or DNA homology studies indicate that *R. meliloti* strain AK631 also contains *the nodI,J,P* and *Q* genes. The phenotype of these mutations are hardly distinguishable from the wild type strain, except one *nodP* mutant.

Genetic analysis of the *n*5 and *n*6 regions is likely to result in the identification of further new *nod* genes in *R. meliloti* strain 41, indicating that the number of nodulation genes (including the regulatory genes, the three nodD alleles (14) and *syrM* is well over twenty. Fig.1 shows that the megaplasmid region from the *n*4 box extending until the *nif* structural genes contains a large cluster of *nod* genęs almost contiguously. Between *nodN* and *nodD1* a *fix* gene is located (31) recently described as *nifN*, 1) and the remaining DNA regions between the other transcriptional units are rather small. Nevertheless, further *nod* genes may be revealed on these segments by appropriate methods.

BOTH POSITIVE AND NEGATIVE CONTROLS OF *NOD* GENE EXPRESSION ARE MEDIATED BY MULTIPLE GENES.

Previously, we showed that the expression of common nodulation genes in *R. meliloti* is under both positive and negative control (20). Each *nodD* allele can activate *nod* gene expression, if appropriate flavonoids are provided. The nucleotide sequences of the three *nodD* alleles in *R. meliloti* AK631 are highly homologous but not identical (14 , 18, unpublished) . Moreover, the *NodD* proteins have distinct ability to interact with the plant signals from the seed or root exudates which results in host-dependent activat*ion of nod* genes (11, 16). The three different NodD proteins have the ability to bind to the *nod* boxes even in the absence of the plant signal (9, 20). The "non-activated" NodD competes with the plant signal-activated NodD for binding to the *nod* box, leading to a decrease of *nod* gene expression (13, 25, submitted). Since plant exudates contain a range of flavonoids or related inducers (or inhibitors) and different plant signals interact differently with the NodD1, D2 and D3 proteins, respectively, the level of *nod* gene expression is modulated by all these interactions. The range of plant signals may vary within the same plant genus. By analyzing different host plants a correlation between the level of *nod* gene expression and nodule induction ability was demonstrated (13, submitted).

The *nodD* genes generally express constitutively in the absence of plant signals. Mulligan and Long (25) reported for *R. meliloti* strain 1021 that another gene, *syrM* may be involved in controlling the production of significant amounts of the NodD3 protein and the NodD3 so produced does not require a plant signal for its *nod* gene activation function. The activation of *nodD3* expression by *syrM* was observed only when both *syrM* and *nodD3* were on a plasmid present in the cells at least in 6 copies.

Our results indicate that the NodD3 protein of strain AK631 exhibits some distinct properties to that of strain 1021. We observed that *nod* gene induction ability of the NodD3 protein can be increased by specific plant signals. A strain carrying a plasmid with intact *nodD3* and Tn5 mutation in *syrM* (this mutation slightly affects the nodulation ability of AK631) allowed to detect at least tenfold induction of the *nodC-lacZ* fusion by several plant signal molecules (13, submitted). At the same time, the presence of *syrM* on plasmid allowed constitutive *nodC-lacZ* expression which was not abolished by a *nodD3* mutation.

Further studies also indicate that the introduction of *syrM* on a plasmid in several copies into *R. meliloti* is essential for the constitutive expression of the *nodC-lacZ* fusion. We found that the presence of *syrM* and *nodD3* in extra copies on plasmid pNID6 (11), extended the host range of strain AK631. *R. meliloti* strains including AK631 normally do not nodulate siratro (*Macroptilium atropurpureum*). Strain AK631(pNID6) however, could induce nodules on siratro. Mutation in *syrM*, but not in *nodD3*, abolished this host range extension. Previously, we showed that AK631 carrying the *nodD* gene of *Rhizobium* sp. NGR234 (a microsymbiont of siratro) gained the ability to nodulate siratro, due to its new NodD protein which could interact with plant signals from siratro exudates (17). Our data suggest that *syrM* not only activates *nodD3* expression but has an additional role in plant-signal independent induction of other nodulation genes.

We determined the nucleotide sequence of *syrM* from strain AK631. Interestingly, the putative amino acid sequence of SyrM shows extensive homology with the NodD proteins and other members of the LysR regulatory gene family (15). The same observation was made for the SyrM of strain 1021 (S. Long, personal communication). The *syrM*, therefore, may be considered as the fourth *nodD* allele (nodD4). At the same time, *syrM* controls also the expression of exopolysaccharide genes (25) involved at a later stage of nodule development. One can speculate that the *syrM* may coordinate these two symbiotic functions during nodule differentiation or may mediate a yet unknown external signal influencing both exopolysaccharide synthesis and *nod* gene regulation.

In the majority of *R. meliloti* strains including AK631 a repressor protein binds to the overlapping *nodD1* and *nodA* as well as to the *nodD2* promoters at the RNA polymerase binding site. The repressor was shown to bind to 33 a bp sequence in which 21 bp are conserved in the binding sites. The repressor keeps the expression of *nodD* and *nodABC* low. The *nod* repressor-containing strains exhibit more efficient nodulation ability compared to the Rep^- derivatives, suggesting that fine-tuning of *nod* gene induction by both positive and negative *trans*-acting factors provides more optimal nodulation ability for the microsymbiont (20).

The DNA region responsible for the repressor production has been cloned and sequenced. The determined open reading frames of a 1.8 kb fragment encoding repressor production together with their mutational analysis indicate that the production of an active repressor requires at least two ORFs.

Recently we showed that high level of combined nitrogen (NH_4^+) exerts an additional negative control of *nod* gene expression (6). Mutations in genes encoding for the activators of the nitrogen regulation system (*ntrA* and *ntrC*) decreased the *ex-planta* expression of *nodC-lacZ* fusions as well as the nodulation efficiency of *Medicago sativa* . Moreover, it was found that under low combined nitrogen conditions the expression of *nodD3*, but not *nodD1*, was elevated in the presence of luteolin.. Our studies indicate that nitrogen control mediated by the *ntr* genes involves control via *nodD3*. Since *nodD3* is controlled by *syrM*, perhaps the SyrM protein mediates nitrogen control to *nodD3*, although *ntr* regulation of the *nodD3* promoter is also a possibility which can be experimentally verified. One can speculate that the *syrM-nodD3* genes were evolved to mediate *nod* gene control by environmental factors (such as combined nitrogen) in the soil.

Our studies on the nitrogen control of *nod* gene expression resulted in the identification of a novel negative control element of nitrogen regulation in *Rhizobium*. By mutational analysis a *R. meliloti* derivative was obtained where the *nod* gene expression escaped repression by NH_4^+. These mutants exhibited an increased level of *nodABC* expression at high combined nitrogen concentrations. This repressor controls at least another *ntr* regulated gene gln B encoding glutamine synthetase II (6). The mutant was more efficient in nodulating the plant host at elevated NH_4^+ concentrations. Moreover, plant nodulation assays with mixed inoculations showed that the mutant is more competitive in nodulation than the wild type strain.

The involvement of several regulatory genes in positive as well as in negative control of nodulation by *R. meliloti* indicates a fairly complex mechanism for fine-tuning of *nod* gene regulation. In addition, the suspected further interactions among the regulatory proteins and the plant signals may further modulate of nodule induction.

RELATIONSHIP BETWEEN NODD AND HUMAN HORMONE-BINDING PROTEINS

Transduction of extracellular molecular signals by mammalian steroid/thyroid hormone receptors (7) show resamblance to plant signal transduction by the NodD protein. The hormonal ligands of these receptors are hydroxyl-substituted polycyclic compounds similar to flavonoids or chalcones. For instance, flavonoid-type phytoestrogens bind to mammalian estrogen receptors (24), or flavonoids are ligands of the uterine type II receptor (23). We found that several hormone-analogous compounds (p. ex. aurons, estradiol, coumestrol) were able to induce the expression of a *nodC-lacZ* fusion in conjunction with specific NodD proteins. A related compound, catechin showed synergism with luteolin suggesting cooperative binding of luteolin and catechin to NodD (12).

Protein sequence comparisons of hormone receptors and NodD revealed two polypeptide modules (M1 and M2) in NodD. The M1 region shared homology with a conserved section of the human steroid/thyroid hormone receptors, suggesting that the M1 modules in NodD and in the receptors had a common ancestor. The module M2 showed similarity to the ligand-binding region of the human thyroxine-binding prealbumin and of the human estrogen receptor. Mutations affecting the interaction of NodD with flavonoids within the M1 and M2 regions have been described which lend further support for the suggested role of both modules in ligand binding (4, 5, 17, 29). Based on the stronger conservation of the M1 module compared to the M2 regions, we propose that M1 establishes a general ligand-binding matrix by forming a hydrophobic cavity, whereas M2 contributes to the specificity of the protein towards ligands. The two regions might be in physical interaction by protein folding to form a ligand-binding pocket (12).

Recently Schlaman et al. (27) reported that NodD is located in the cytoplasmic membrane. We suggest that the hypothetised ligand-binding pocket (M1 and M2) is in that region of the cytoplasmic membrane which is accessible to the flavonoids present in the periplasmic space, while the N-terminal DNA-binding domain is in the cytosol. Binding of the flavonoids to the pocket results in a conformational change of the NodD which is then capable of activating transcription.

REFERENCES

1. Aguilar, O.M., Reiländer, H., Arnold, W. & Pühler, A. (1987) J. Bacteriol. **169**, 5393-5400.
2. Banfalvi, Z. & Kondorosi, A. (1989) Plant Mol. Biol. **13**, 1-12.
3. Bhuvaneswari, T.V. & Solheim, B. (1985) Physiol. Plant. **63**, 25-34.
4. Burn, J., Rossen, L. & Johnson, A.W.B. (1987) Genes Dev. **1**, 456-464.
5. Burn, J.E., Hamilton, W.D., Wootton, J.C. & Johnston, A.W.B. (1989) Molec. Microbiol. **3**, 1567-1577.
6. Dusha, I., Bakos, A., Kondorosi, A., de Bruijn, F. & Schell, J. (1989) Mol. Gen. Genetic **19**, 89-96.
7. Evans, R.M. (1988) Science **240**, 889-895.
8. Faucher, S., Maillet, F., Vasse, J., Rosenberg, C., van Brussel, A.A.N., Truchet, G. & Denarie, J. (1988) J. Bacteriol. **170**, 5489-5499.
9. Fisher, R.F., Egelhoff, T. T., Mulligan, J.T. & Long, S. R. (1988) Genes Dev. **2**, 282-293.
10. Gerhold, D., Stacey, G. & Kondorosi, A. (1989) Plant Mol. Biol. **12**, 181-188.
11. Györgypal, Z., Iyer, N. & Kondorosi, A. (1988) Mol. Gen. Genet. **212**, 85-92.
12. Györgypal, Z.& Kondorosi, A. (1990) submitted.
13. Györgypal, Z., Kondorosi, E. & Kondorosi, A.(1990) submitted.
14. Gottfert, M.,Horvath, B.,Kondosori, E., Putnoky, P., Rodriquez-Quinones, F., & Kondorosi, A. (1986) J. Mol. Biol. **191**, 411-420.
15. Henikoff, J.A., Haughn, G.W., Calvo, J.M. & Wallace, J.C. (1988) Proc. Natl. Acad. Sci. USA **85**, 6602-6606.
16. Honma, M.A., Asomaning, M. & Ausubel, F.M. (1989) J. Bacteriol. **172**, 901-911.
17. Horvath, B., Bachem, C.W.B., Schell, J. & Kondorosi, A. (1987) EMBO J. **6** 841-848.
18. Iyer, N. & Kondorosi, A. (1990) submitted.
19. Kondorosi, E. & Kondorosi, A. (1986) Trends in Biochem. Sci. **11**, 296-299.
20. Kondorosi, E., Gyuris, J., Schmidt, J., John, M., Duda, E., Hoffman, B., Schell, J. & Kondosori, A. (1989) EMBO J. **8**, 1331-1340.

21. Lerouge, R., Roche, P., Faucher, C., Maillet, F., Truchet, G., Promé, J.C. & Dénarié, J. (1990) Nature **344**, 781-784.
22. Long. S.R. (1989) Cell **56**, 203-214.
23. Markaverich, B.M., Roberts, R.R., Alejandro, M.A., Johnson, G.A., Middleitch, B.S. & Clark, J.H. (1988) Steroid Biochem. **30**, 71-78.
24. Martin, P.M., Horwitz, K. B., Ryan, D. S. & McGuire, W.L. (1978) Endocrinology, **103**, 1860-1867.
25. Mulligan , J.T. & Long, S.R. (1989) Genetics **122**, 7-18.
26. Rostas, K., Kondosori, E., Horvath, D., Simoncsits, A. & Kondorosi, A. (1986) Proc. Natl. Acad. Sci. US **83**, 1757-1761.
27. Schlaman, H.R.M., Spaink, H.P., Okker, R.J.H. & Lugtenberg, B.J.J. (1989) J. Bacteriol. **171**, 4686-4693.
28. Schmidt, J., Wingender-Drissen, R., John, M., Wieneke, U. & Schell, J. (1988) Proc. Natl. Acad. Sci. USA **85**, 8578-8582.
29. Spaink, H.P., Wijffelman, C.A., Okker, R.J.H. & Lugtenberg, B.J.J. (1989) Plant Mol. Biol. **12**, 59-73.
30. Surin, B.P., Watson, J.M., Hamilton, W.D.O., Economou, A. & Downie, J.A. (1990) Mol. Microbiol. **4** (2),245-252.
31. Török, I., Kondorosi, E., Stepkowski, T., Posfai, J. & Kondorosi, A. (1984) Nucleic Acids Res. **12**, 9509-9524.

Gresshoff, Roth, Stacey & Newton (eds.) Nitrogen Fixation: Achievements and Objectives
Chapman and Hall New York • London 1990

Regulatory steps in nodulation by *Rhizobium leguminosarum* bv *viciae*

Ben Lugtenberg, Rund de Maagd, Ton van Brussel, Hayo Canter Cremers, Lenie Goosen-de Roo, Rob Okker, Kees Recourt, Helmi Schlaman, Herman Spaink, Carel Wijffelman and Bas Zaat

Leiden University, Department of Plant Molecular Biology, Nonnensteeg 3, 2311 VJ Leiden, The Netherlands.

Expression of *nod* genes during symbiosis: The expression of inducible *nod* genes of *Rhizobium* requires three components: (i) an inducible *nod* gene promoter, (ii) a functional *nodD* gene, whose product acts as a positive regulator when activated, and (iii) flavonoid compounds of the plant, which activate the *nodD* gene product. The chemical nature of the flavonoid inducers, either present in roots or seeds or exuded from them, have been determined for host plants of several species of *Rhizobium* and *Bradyrhizobium*. In general, there are numerous inducers, which differ between plant species, and can contribute to host specificity (13,17). Also, there appear to be differences between different organs of the same plant species, as the seeds of alfalfa exude different inducers to the roots (10,11).

We recently observed that the inducer profile of one specific organ of one plant species can even be changed. The resence of *R. leguminosarum* bv. *viciae* bacteria increased exudation, resulting in a dramatic alteration of the profile of exuded inducers of *Vicia sativa*. HPLC analysis showed that the number of major *nod* gene inducers increased from 2 to 8 and ^{1}H-NMR analysis revealed 2 chalcones and 6 flavanones. Only one of these flavanones is released as a major *nod* gene inducer by both non-inoculated and inoculated roots (see Recourt et al. these Proceedings). All released flavonoids can be linked to known biosynthetic pathways. Initial studies on *V. sativa* roots show that inoculation increases transcription of key enzymes (Recourt et al., unpublished).

***nodD* Hybrid Insensitive to Inducers and Anti-Inducers:** By constructing hybrids of the *nodD1* gene of *R. meliloti* and the *nodD* gene of *R. leguminosarum* bv. *trifolii*, we obtained a class of hybrid genes with surprising properties. (i) The product of these hybrids was completely independent of flavonoids for activation of inducible *nod* genes and was therefore designated FITA (flavonoid-independent transcription activation) (15). (ii) It is insensitive to all known anti-inducers tested (e.g., umbelliferone) as well as to methanol extracts containing possible inducers and anti-inducers of *V. sativa, Trifolium repens* and *T. pratense* seedlings (14). *Rhizobium* strains containing such a hybrid *nodD* gene, e.g., *nodD*604, as the only functional *nodD* gene had a largely extended host range and (iii) showed increased nitrogen fixation while the number of root nodules was the same as in the control. These properties are very promising for applications since it may be that a *Rhizobium* strain containing a FITA *nodD* gene in the presence of anti-inducers, which should inhibit indigenous rhizobia, would be highly competitive.

Repression of *nod* Gene Expression in Bacteroids: A Novel Mechanism? Recent studies by Schlaman et al. have indicated that the level of expression of inducible *nod* genes is very low in bacteroids whereas *nodD* expression is normal. The most likely explanations, either absence of inducing flavonoids or inhibition of NodD protein activation by anti-inducers, are probably incorrect since the same results were found for a strain containing the *nodD*604 hybrid gene. These results indicate that (i) the improved nitrogen fixation of strains with a FITA type *nodD* is not due to a higher level of *nod* gene products in bacteroids and (ii) the transcription of *nod* genes is repressed during symbiosis. This repression is not due to the absence of inducing flavonoids and probably also not by the presence of anti-inducers. It may be due to a negative regulatory mechanism which becomes active between the entrance of the bacterium into the plant root and its differentiation into a bacteroid.

A Novel Bacterial Activity Secreted upon Inducible *nod* Gene Activation: Co-cultivation of *V. sativa* ssp. *nigra* with the homologous symbiont *R. leguminosarum* bv. *viciae* results in a strongly increased *nod* gene inducing activity in root exudates. Heterologous *Rhizobium* strains do not cause this effect. Increased activity is the result of the action of a biovar-specific signal molecule, the Ini-factor, whose synthesis requires the *nodDABCEL* genes. The genes *nodF,I* and *J* are involved in the timely appearance and in reaching the maximal level of Ini-factor (see van Brussel et al., these Proceedings). The apparent molecular weight of the Ini-factor is between 1000 and 10,000 Da. The factor is being purified and may be related to the one recently described by the Toulouse group (8).

Hair Curling and Exo$^-$ Mutants: Exo$^-$ mutants of *R. leguminosarum* are deficient in Hac (hair curling). Recently, Canter Cremers et al. (manuscript submitted) biochemically identified a class of these mutants as defective in the synthesis of UDP-glucose-4'-epimerase. The mutation can be functionally complemented by *exoB* of *R. meliloti*. The mutants are defective in the synthesis of lipopolysaccharide, capsular polysaccharide and exopolysaccharide but form apparently normal 1,2-glucan. Since it is not clear why UDP-glucose-4'-epimerase should be involved in the synthesis of the secreted bacterial Nod-related factors and since attachment of *exoB* mutants is normal, it may be that, in addition to attachment and the Hac factor, at least one other factor is required for successful hair curling.

Exchangeable Roles of *nodE* and *nodO* Genes: We have recently identified a new *nod* gene of *R. leguminosarum* biovar *viciae*, *nodO*, which encodes a secreted protein with a predicted molecular weight of 30 kDa. The *nodO* gene of the Sym plasmid pRL1JI is located at the left of the earlier identified *nod* genes, and is preceded by a poorly conserved *nod* box (1). The *nodO* protein is identical to the secreted, flavonoid-inducible, sym plasmid-dependent protein with a relative molecular weight of 50 kDa [as observed with SDS-PAGE in our earlier studies (2)]. The discrepancy between predicted and observed molecular weights is so far unexplained.

Our studies of *nodO* mutants showed only a minor effect of these mutations on nodulation in one host plant/bacterium combination, whereas no effect was found in other combinations (2). However, recent studies by Johnston et al. (6), using cloned *nod* DNA fragments, have shown that the effect of a *nodO* mutation is much more dramatic in combination with a *nodE* mutation. These observations imply that the functions of *nodE* or *(nodFEL)* and *nodO* are (at least partially) interchangeable, and that there may be at least two pathways in the early steps of nodulation. Although normally both pathways may be operating, the plant/bacterium combination-dependent effects of *nodO* mutations observed by us, suggest that the relative importance of the two pathways may vary from one plant/bacterium combination to the other. Since NodO protein is secreted, we argued that it probably acts on the plant. Indeed, preliminary results of immunofluoresence studies suggest binding of NodO protein to the roots of *Vicia hirsuta* and *V. sativa* (de Maagd et al., unpublished).

Consistent with and supporting these observations is that the NodO protein binds Ca^{2+} ions (4) and that it shows homology with the domain of *Escherichia coli* hemolysin, which is responsible for the initial Ca^{2+}-dependent binding to a receptor on the surface of erythrocytes (9). This strongly suggests that the functioning of NodO protein involves binding to the plant cell surface.

Given these data and assuming that the role of *nodE* is only in the synthesis of a secreted low molecular weight factor, which somehow elicits the expression of nodulin ENOD12 (12), the data suggest that at least two pathways can result in nodulation. In the first pathway, a *nodE*-requiring secreted product elicits the activation of ENOD12 expression, possibly in a lectin-mediated way. In the second pathway, NodO protein binds to the plant cell surface. This is presumably followed by the generation or modification of a signal molecule that is involved in nodulation, perhaps in combination with the non-*nodE*-modified form of the signal mentioned earlier.

Regulation of Molecular Changes During Development from Bacteria to Bacteroids: During the development of free-living bacteria to bacteroids, a number of cell surface changes occur in outer membrane proteins and LPS (3). More extensive studies of different plant/bacterium combinations have shown that most of these changes are common to all studied combinations (Roest et al., unpublished), and therefore may be essential for effective nodulation. The stage in which at least an important part of these changes takes place has been narrowed down to the release from the infection thread, possibly presence in the infection droplet (see Goosen-de Roo et al., these Proceedings). One of the main aims of our present research is the identification of the plant factor(s) or signal(s) responsible for the molecular changes during bacteroid development described above. To this end, we have cloned a bacterial gene presumably encoding an outer membrane protein that is hardly or not at all present in bacteroids (see de Maagd et al., these Proceedings). By replacing its promoter by one that is not repressed in bacteroids, we will test whether synthesis of this protein interferes with normal bacteroid development. Moreover, the promoter of the outer membrane protein will be used in an attempt to identify the signal responsible for this symbiotic regulation.

It remains to be seen whether repression of *nod* gene expression and of the synthesis of several outer membrane proteins and the possible repression of the synthesis of part of the LPS molecule are regulated by the same signal, e.g., a low oxygen concentration (7,16).

1. de Maagd, R.A., Wijfjes, A.H.M., Spaink, H.P., Ruiz-Sainz, J.E., Wijffelman, C.A., Okker, R.J.H. & Lugtenberg, B.J.J. (1989) J. Bacteriol. **171**, 6764-6770.
2. de Maagd, R.A., Spaink, H.P., Pees, E., Mulders, I.H.M., Wijfjes, A.H.M., Wijffelman, C.A., Okker, R.J.H. & Lugtenberg, B.J.J. (1989) J. Bacteriol. **171**, 1151-1157.
3. de Maagd, R.A., de Rijk, R., Mulders, I.H.M. & Lugtenberg, B.J.J. (1989) J. Bacteriol. **171**, 1136-1142.
4. Economou, A., Hamilton, W.D.O., Johnston, A.W.B. & Downie, J.A. (1990) EMBO J. **9**, 349-354.
5. Hartwig, U.A., Maxwell, C.A., Joseph, C.M. & Phillips, D.A. (1990) Plant Physiol. **92**, 116-122.
6. Johnston, A.W.B., Latchford, J.W., Davis, E.O., Economou, A. & Downie, J.A. (1989) in Signal Molecules in Plants and Plant-Microbe Interaction, ed. Lugtenberg, B.J.J. (Springer-Verlag, Heidelberg), pp. 311-318.
7. Kannenberg, E. & Brewin, N. (1989) J. Bacteriol. **171**, 4543-4548.
8. Lerouge, P., Roche, P., Faucher, C., Maillet, F., Truchet, G., Prome, J.C. & Denarie, J. (1990) Nature (London) **344**, 781-784.
9. Ludwig, A., Jarchau, T., Benz, R. & Goebel, W. (1988) Mol. Gen. Genet, **214**, 553-561.
10. Maxwell, C.A., Hartwig, U.A., Joseph, C.M. & Phillips, D.A. (1989) Plant Physiol. **91**, 842-847.
11. Peters, N., Frost, J. & Long, S. (1986) Science **233**, 977-980.
12. Scheres, B., Van der Wiel, C. Zalensky, A., Horvath, B., Spaink, H., Van Eck, H., Zwartkruis, F., Wolters, A.-M., Gloudemans, T., Van Kammen, A. and Bisseling, T. (1990) Cell **60**, 281-294.
13. Spaink, H.P., Wijffelman, C.A., Pees, E., Okker, R.J.H. & Lugtenberg, B.J.J. (1987) Nature (London) **328**, 337-340.
14. Spaink, H.P., Okker, R.J.H., Wijffelman, C.A., Tak, T., Goosen-de Roo, L., Pees, E., Van Brussel, A.A.N. & Lugtenberg, B.J.J. (1989) J. Bacteriol. **171**, 4045-4053.
15. Spaink, H.P., Okker, R.J.H., Wijffelman, C.A. & Lugtenberg, B.J.J. (1989) Plant Molec. Biol. **12**, 59-73.
16. Vandenbosch, K.A., Brewin, N.J. & Kannenberg, E.L. (1989) J. Bacteriol. **171**, 4537-4542.
17. Zaat, S.A.J., Wijffelman, C.A., Mulder, I.H.M., Van Brussel, A.A.N. & Lugtenberg, B.J.J. (1988) Plant Physiol. **86**, 1298-1303.

Gresshoff, Roth, Stacey & Newton (eds.) Nitrogen Fixation: Achievements and Objectives
Chapman and Hall New York • London 1990

Nutrient exchange across the peribacteroid membrane of isolated symbiosomes

David A. Day, Li-Jun Ou Yang and Michael K. Udvardi

Botany Department, Australian National University, Canberra A.C.T. 2601, Australia.

INTRODUCTION

Nitrogen fixation in legume nodules involves a complicated exchange of regulatory signals and nutrients between the plant cytoplasm and the bacteroid, and this exchange involves transport across two membranes of selective permeability: the bacteroid membrane and the plant-derived envelope surrounding the bacteroid, the peribacteroid membrane (PBM). Following a recently suggested nomenclature (16), we will use the term "symbiosome" to refer to the unit comprised of the PBM and the enclosed bacteroids (formerly called peribacteroid units). However, the term "peribacteroid" remains, in our opinion, an accurate description of the membrane limiting the symbiosome and we will therefore continue to use that term.

There have been many studies of metabolite transport in free-living rhizobia and isolated bacteroids, which have yielded valuable information about the interactions between the host cell and the endophyte. In particular, investigations of the *Rhizobium* Dct system (15) and associated metabolism (5, 17) have highlighted the essential role of dicarboxylates in the symbiosis. However, these studies must be supplemented by similar ones involving the PBM before they can be confidently placed in an *in vivo* context, since it is the PBM which determines which metabolites the bacteroid will encounter in the nodule. With this in mind, we have developed techniques for the isolation of intact symbiosomes from soybean and siratro nodules, and have used them to investigate the permeability of the PBM to a range of metabolites likely to be present in the nodule. A major advantage of using intact symbiosomes over PBM vesicles is that the correct orientation of the membrane is maintained, as is the important bacteroid sink within the symbiosome. The results of these investigations are summarized in this paper.

METHODS

Intact symbiosomes were isolated from soybean nodules using Percoll step-gradients as described by Day *et al* (4). The essential features of this technique are very gentle grinding of nodules, layering the homogenate directly onto the gradient and centrifuging at low speeds for a short time; this ensures intact symbiosomes free from contamination by plant organelles (as judged by microscopy and enzyme assays) and only slightly contaminated by free bacteroids.

Uptake of radioactively labelled compounds by symbiosomes was measured by the silicon-oil centrifugal-filtration method (4), using short (10-30 s) incubation times to avoid confusion between transport and metabolism whenever possible. The silicon-oil technique has the advantage of allowing accurate internal volume measurements to be made, in comparison with the filtration technique often used with bacteria; the latter method is also likely to damage the fragile symbiosomes when reactions are terminated by suction. ATPase activity and PBM energisation were measured according to (20).

RESULTS AND DISCUSSION

Carbon supply to the bacteroid

The PBM of soybean and siratro symbiosomes is essentially impermeable to a range of sugars, including mannitol, sucrose, glucose and fructose: uptake rates with these compounds are very slow (see Fig. 1 for glucose uptake) and increase linearly with concentration (23), indicating transport by passive diffusion. Free bacteroids are also impermeable to sucrose and glucose but possess transport systems for active fructose uptake (23). *In vivo*, the latter system will be severely limited by the PBM.

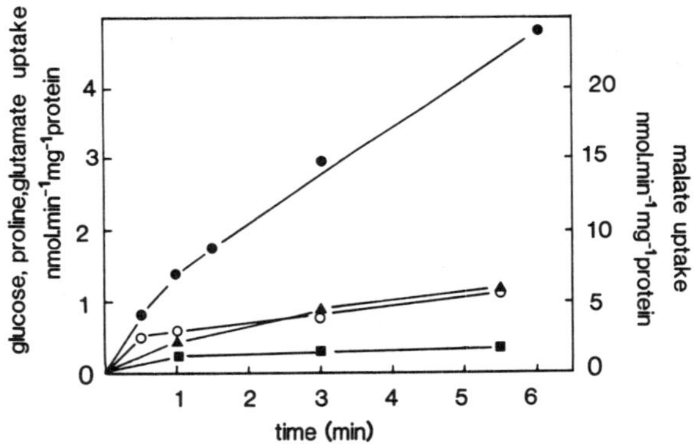

Figure 1. Uptake of ^{14}C-labelled compounds by isolated soybean symbiosomes. The following (final) concentrations were used: glucose (○), 1 mM; proline (■) and glutamate (▲), 0.1 mM; malate (●), 0.3 mM. Note that the scale for malate uptake is different from that for the other compounds.

A similar situation exists for a number of aminoacids. Bacteroids possess carriers for some of these compounds, such as glutamate (14, 22), and can rapidly accumulate them from the external medium. Aminoacid transport across the PBM, however, is much slower and does not lead to accumulation against a concentration gradient (23). Rates of glutamate and proline uptake by symbiosomes are shown in Fig. 1 but similar results have been obtained with several other aminoacids. Addition of ATP to the reaction medium, to energize the PBM, has no effect on glutamate and proline uptake (22, 23). Proline uptake was studied in more detail because it has recently been implicated in the supply of reducing power to bacteroids (12). Fig. 2 shows the response of the rate of proline uptake by symbiosomes to increasing proline concentration: no evidence of saturation kinetics was seen between 0.01 and 2 mM, using short incubation times. Inclusion of glycine (Fig. 2) or D-proline and hydroxyproline (not shown), inhibitors of proline carriers on other membranes, had no effect on the rate of uptake. We therefore conclude that the PBM lacks a carrier for proline and that the observed slow uptake is via passive diffusion; in a previous study (22) we came to the same conclusion for glutamate and it is therefore doubtful whether these compounds can play a major role in providing reducing power to the bacteroid during nitrogen fixation. Nonetheless it should be noted that diffusion of neutral aminoacids such as proline is relatively rapid and at high concentrations (say, 10 mM) the rate of uptake could become signficant, especially if metabolism within the symbiosome was rapid.

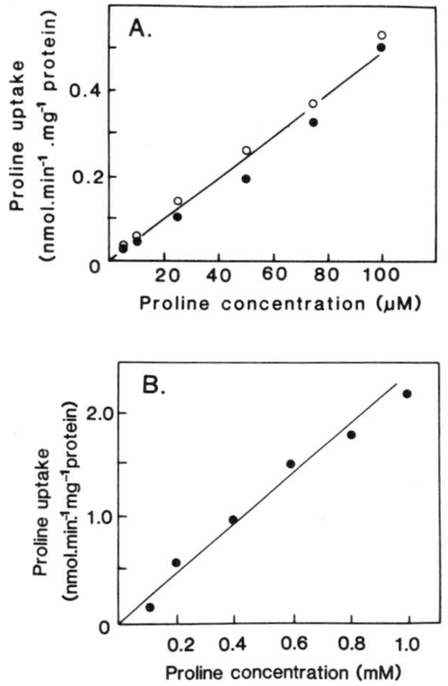

Figure 2. Uptake of [^{14}C]-L-proline by soybean symbiosomes. The reaction was terminated after 30 s. (●): proline alone; (○): plus 1 mM glycine in the reaction medium.

In contrast to the sugars and aminoacids, uptake of dicarboxylic acids across the PBM is rapid (Fig. 1) and catalysed by a distinct carrier (21). The PBM carrier is distinguished from the bacteroid Dct system and dicarboxylate carriers on other plant membranes by its sensitivity to inhibitors, by the range of organic acids which it recognizes, and by its kinetics (Table 1). Interestingly, two of the most potent inhibitors of the PBM carrier (at least in soybean and siratro symbiosomes), phthalonate and hydroxycyanocinnanic acid (21), are inhibitors of some mitochondrial and plastid transport systems but not their dicarboxylate carriers (8).

Table 1. Properties of dicarboxylate transport systems

Membrane	Affinity (Km, mM)	Substrates	Inhibitors (selective for system)
Peribacteroid membrane	0.1 - 0.2	malate, succinate, OAA, fumarate	phthalonate, cyanocinnamic acid derivates
Mitochondria	0.2 - 0.5	mal, succ, malon, Pi, sulphate	n-butlymalonate, benzylmalonate
tonoplast	4 - 5	large range of organic acids	1,2,3-benzene tricarboxylic acid, pyridoxal phosphate
bacteroid	0.01 - 0.02	mal, succ, OAA, fumarate	none of the above
plastid	0.1 - 0.5	mal, OAA, glut, asp, oxoglut	none of the above

The PBM dicarboxylate carrier recognizes malate, oxaloacetate, fumarate, succinate and oxoglutarate as substrates but has only a very weak affinity for malonate (L-J. Ou Yang and D.A. Day, unpublished results). The latter result agrees with those of Humbeck and Werner (10). Of these substrates, the first four stimulate oxygen uptake by soybean symbiosomes significantly and are all potential substrates *in vivo*. The high concentration of malate in the nodule cytoplasm (18), however, suggests that malate would be the favoured candidate for carbon substrate supplied to the bacteroid. Rates of malate uptake at sub-millimolar concentrations are greater than ten-fold those of other substrates tested (Fig. 1) and are rapid enough to meet the ATP-generation rates required by nitrogenase (4). From our results, it is unlikely that any sugar can contribute to this process and any contribution by aminoacids is likely to be minor. Nonetheless, uptake of ^{35}S-methionine over 60 min by isolated symbiosomes is rapid enough to label bacteroid proteins (11; S. Young and D.A. Day, unpublished results) and it is possible that *in vivo* the plant supplies aminoacids to support bacteroid protein synthesis.

Ammonia transport

In return for carbon and other nutrients from the plant, the bacteroid provides the plant with ammonia. It is now well documented that free-living rhizobia synthesize an ammonium transporter under nitrogen-limited conditions and that in nitrogen-fixing bacteroids the synthesis of that transporter is repressed so that ammonia leaves the bacteroid as NH_3 via passive diffusion (3, 7). We have investigated ammonia transport across the PBM, using the analogue ^{14}C-methylamine (MA) with intact symbiosomes. From both uptake and counterflux experiments using ammonia and ^{14}C-MA (M.K. Udvardi and D.A. Day, unpublished results), we have concluded that there is no ammonium (methylammonium) transport system on the PBM and that ammonia crosses both the PBM and the bacteroid membrane as NH_3. Since diffusion of NH_3 is rapid, ammonia efflux from the symbiosome could be very fast *in vivo* provided that nodule glutamine synthase maintains a large concentration gradient.

PBM energization and anion transport

Several studies (1, 2, 6, 20) have shown that the PBM from different symbioses possess ATPase activity. In soybean, this ATPase activity can lead to energization of the PBM via inward proton-pumping which creates a membrane-potential with the inside positive (2, 20). This can be conveniently followed as oxonol fluorescence quenching (Fig. 3). The properties of PBM ATP-hydrolysis are summarized in Table 2. Although most of these properties, particularly inhibitor sensitivity, are consistent with a plasmalemma-type ATPase (2, 6, 20), the presence of a second, tonoplast-type ATPase has been suggested on the basis of nitrate sensitivity and pH profiles (1). This question has yet to be resolved but PBM energization seems to be catalysed primarily by the former (20).

Table 2. Properties of the PBM ATPase

Km	80 μm
pH optimum	6.5
KCl	stimulates
vanadate DCCD calcium	inhibit severely (50-80%)
nitrate	inhibits slightly (20-30 %)
oligomycin azide oubain molydbate NEM	do not inhibit significantly

ATP-dependent oxonol fluorescence quenching is rapidly reversed by the anions NO_3^- and Cl^- (20), indicating rapid transport of these ions across the PBM. When low (< Km) ATP concentrations are used to drive H^+-pumping, organic anions, including malate and succinate, can also cause partial collapse of the $\Delta\varphi$ (Fig. 3). The pH optimum of malate uptake suggests that it is the univalent mal^- species which is transported by the dicarboxylate carrier (21, hereafter referred to as the dicarb carrier) and it seems that this is driven in part by the PBM $\Delta\varphi$.

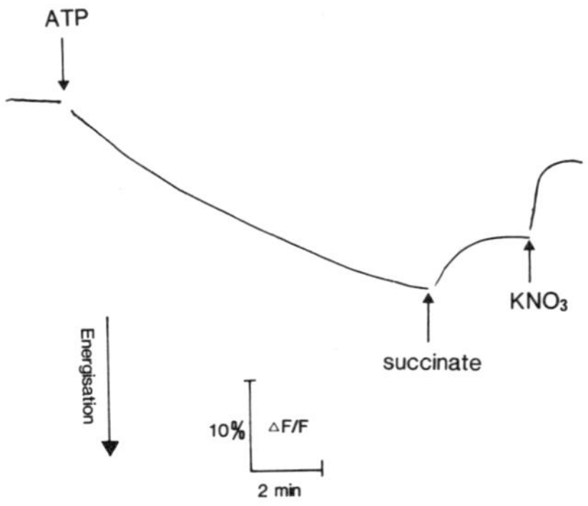

Figure 3. Effect of anions on PBM membrane potential. The Dj formation was measured as oxonol fluorescence quenching by intact soybean symbiosomes: a downward deflection of the fluorimeter indicates energization of the PBM by H^+ pumping during ATP hydrolysis. An upward deflection indicates collapse of Dj by anion uptake into the symbiosomes. ATP was added at 10 mM, malate at 10 mM and nitrate at 1 mM.

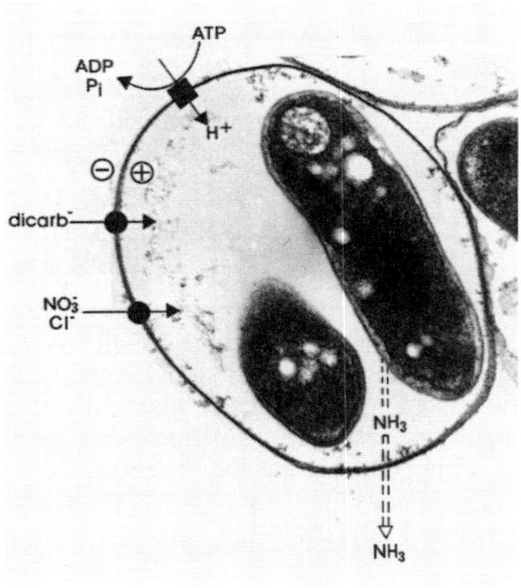

Figure 4. Summary of transport across the PBM. ■: H^+ pumping ATPase; ●: anion carriers or channels. The broken arrow indicates passive efflux of NH_3 from the symbiosome.

In this context, it is noteworthy that ATP can stimulate malate uptake, albeit only slightly (20%, on average), and NO_3^- prevents this stimulation (L-J. Ou Yang and D.A. Day, unpublished results). However, NO_3^- inhibits in the absence of ATP and this inhibition appears to be competitive in nature, implying that NO_3^- can use the dicarb carrier. Since the inhibition is only slight (approx. 20%), and since NO_3^--induced depolarization of the PBM is much greater than that by malate (Fig. 3; 20), it is probable that there is also a separate anion channel which is responsible for most of the transport of NO_3^- (and Cl^-). These ion movements are summarized in Fig. 4.

ATP stimulation of malate uptake suggests a role for PBM energization in the regulation of the dicarb carrier and cytosolic ATP/ADP ratios in infected cells may be important here. However, malate uptake is severely inhibited by the protonophore CCCP and by the inhibitor of bacteroid respiration, KCN (L-J. Ou Yang and D.A. Day, unpublished results), showing that energization of the bacteroid membrane is also important to maintain rapid dicarboxylate transport into symbiosomes. *In vivo,* O_2 levels in the infected zone will be important in this context.

Inhibition of the rates of ATP-induced oxonol fluorescence quenching provides a simple screen for anion permeation across the PBM; on this basis the PBM is permeable to the following compounds (in order of ease of penetration): thiocyanate, nitrate, chloride, ascorbate, malate/succinate, fumarate, oxoglutarate and malonate (M.K. Udvardi and D.A. Day, unpublished result). Glutamate and isocitrate have no effect on oxonol quenching. It is important to note that these results are completely in agreement with our direct measurements of metabolite uptake (see above), lending credibility to the techniques employed.

SUMMARY AND FUTURE WORK

Our studies with symbiosomes isolated from soybean nodules have indicated the presence of a dicarboxylate carrier, a general anion channel and a H^+-translocating ATPase on the PBM (Fig. 4). Our preliminary results with siratro and those of Herrada et al (9) with french-beans suggest that the dicarb carrier may be a common feature of hosts of *Bradyrhizobium* species. It is important now that these studies are extended to other *Rhizobium* symbioses. It is also important that the dicarb carrier and ATPase(s) are identified at the molecular level and the regulation of their synthesis is determined: it may well be that some of the previously identified PBM nodulins (24) have transport functions. Such molecular studies will also be of help in deciding on the origin of the PBM (13). Most of our experiments have used short reaction times, carefully controlled ionic environments and aerobic conditions, and long term experiments with nitrogen-fixing symbiosomes in media reflecting the probable composition of infected cell cytoplasm are required to evaulate PBM permeability under *in planta* conditions. It will also be important to extend our studies to examine the wide range of inorganic compounds which must cross the PBM. Chief among these are iron uptake from the plant cytoplasm and release of haem synthesized in the bacteroids. So far we have barely scratched the surface of the PBM and isolated symbiosomes promise to be a useful tool in the study of nutrient exchange between host and endophyte, for some time to come.

ACKNOWLEDGEMENTS

We thank Dean Price, Susan Young, Susan Walker, Diane Lister and Neda Plovanic for their expert assistance, Keith Herbert for photography, and Joanne Perks for typing the manuscript. Financial assistance was provided by the Australian Research Council.

REFERENCES

(1) Bassarab, S., Mellor, R.B. & Werner, D. (1986) Endocyt. C. Res. **3**, 189-196.
(2) Blumwald, E., Fortin, M.G., Rea, P.A., Verma, D.P.S. & Poole R.J. (1985) Plant Physiol. **78**, 665-672.
(3) Brown, C.M. & Dilworth, M.J. (1975) J. Gen. Microbiol. **86**, 39-48.
(4) Day, D.A., Price, G.D. & Udvardi, M.K. (1989) Aust. J. Plant Physiol. **16**, 69-84.
(5) Dilworth, M.J. & Glenn, A.R. (1984) TIBS **9**, 519-523.
(6) Domigan, N.M., Farnden, K.J.F., Robertson, J.G. & Monk, B.C. (1988) Arch. Biochem. Biophys. **264**, 564-573.
(7) Glenn, A.R. & Dilworth, M.J. (1984) J. Gen. Microbiol. **103**, 1961-1968.
(8) Heldt, H.W. & Flugge, U.I. (1988) in *The Biochemistry of Plants*, ed. D.D. Davies (Academic Press, N.Y.), Vol. 11.
(9) Herrada, G., Puppo, A. & Rigaud, J. (1989) J. Gen Micribiol. **135**, 3165-3171.
(10) Humbeck, C. & Werner, D. (1987) Endocyt. C. Res. **4**, 185-196.
(11) Katinakis, P., Klein-Langhorst, R.M., Louwerse, J., van Kammen, A. & van der Bos, R.C. (1988) Plant Molec. Biol. **11**, 183-190.
(12) Kohl, D.H., Schubert, K.R., Carter, M.B., Hagedorn, C.H. & Shearer, G. (1988) Proc. Natl. Acad. Sci. USA **85**, 2036-2040.
(13) Mellor, R.B. (1989) J. Expt. Bot. **40**, 831-839.
(14) Poole, P.S., Franklin, M., Glenn, A.R. & Dilworth, M.J. (1985) J. Gen Microbiol. **131**, 1441-1448.
(15) Ronson,C.W. (1988) in *Nitrogen Fixation: Hundred Years After*, eds. Bothe, H., de Bruijn, F.J. & Newton, W.E. (Gustav Fischer, Stuttgart), pp. 541-551.
(16) Roth, E., Jeon, K. & Stacey, G. (1988) in *Molecular Genetics of Plant-Microbe Interactions*, eds. Palacios, R. & Verma, D.P.S. (American Phytopath. Soc. Press, St Paul), pp. 220-225.
(17) Streeter, J.G. & Salminen, S.O. (1988) in *Physiological Limitations and the Genetic Improvement of Symbiotic Nitrogen Fixation*, eds. O'Gara, F.O. et al (Kluwer Acad. Pub.), pp. 11-20.
(18) Stumpf, D.K. & Burris, R.H. (1979) Anal. Biochem. **95**, 311-315.
(19) Sze, H. (1985) Ann. Rev. Plant Physiol. **36**, 175-208.
(20) Udvardi, M.K. & Day, D.A. (1989) Plant Physiol. **90**, 982-987.
(21) Udvardi, M.K., Price, G.D., Gresshoff, P.M. & Day, D.A. (1988) FEBS Lett. **231**, 36-40.
(22) Udvardi, M.K., Salom, C.L. & Day, D.A. (1988) Molec. Plant-Microbe Interact. **1**, 250-254.
(23) Udvardi, M.K., Ou Yang, L-J., Young, S. & Day, D.A. (1990) Molec. Plant-Microbe Interact. in press.
(24) Verma, D.P.S., Fortin, M.G., Stanley, J., Mauro, V.P., Purohit, S. & Morrison, N. (1986) Plant Molec. Biol. **7**, 57-61.

Gresshoff, Roth, Stacey & Newton (eds.) Nitrogen Fixation: Achievements and Objectives
Chapman and Hall New York • London 1990

Immunological dissection of the plant-microbe interface in pea nodules

N. J. Brewin, A. L. Rae, S. Perotto, J. P. Knox, K. Roberts, M. F. LeGal, S. S. Sindhu, E. A. Wood and E. L. Kannenberg

John Innes Institute, John Innes Center for Plant Science Research, Colney Land, Norwich NR4 7UH, UK.

Introduction

The development of the nitrogen-fixing legume root nodule involves many forms of cell-to-cell contacts and interactions in the pathway of differentiation. Contacts between one plant cell surface and another lead to the differentiation of a variety of different cell types from the same nodule meristem. Contacts between plant and rhizobial cell surfaces guide rhizobia through the infection thread structures that penetrate plant cells. Finally, contact between the bacterial cell surface and the plant cell plasma membrane lead to endocytosis of bacteria and their development within the cytoplasm into endosymbiotic nitrogen-fixing bacteroids enclosed by a plant-derived peribacteroid membrane.

In the following presentation, we will describe some of the components involved in this series of surface interactions. These have been identified using a range of monoclonal antibodies as specific molecular probes in the study of pea root nodule development (2). In addition to a biochemical characterization of the corresponding antigens, we have also examined the localized expression of these antigens using immunocytochemical techniques (21,22). Because pea nodules have apical meristems, their medium longitudinal axis also represents a time course of differentiation, and hence an immunocytochemical analysis reveals many interesting aspects of nodule development occurring both in space and in time (Fig. 1).

Nodule endodermis and boundary layer

The endodermis is a distinct sheath or cylinder, one cell thick, defining the innermost layer of the nodule cortex and separating it from the inner parenchyma and central infected tissue. Using a monoclonal antibody, we have identified a glycoprotein component of the endodermal cell wall. Before the development of the endodermis, this same epitope is observed to be present in cytoplasmic vesicles in the post-meristematic cells of the nodule apex. The continual self-propagation of the single-celled endodermal sheath poses an interesting model system for the analysis of cell fate determination during plant differentiation. The transfer of this epitope from cytoplasmic vesicles into the wall of a nascent endodermal cell should serve as a useful molecular marker for these developmental phenomena.

Just underneath the endodermis is a layer of cells which Parsons (13) has described as the boundary layer, and between this and the bacteroid-containing cells is a region known as the distribution zone. Together these layers form the uninfected nodule parenchyma.

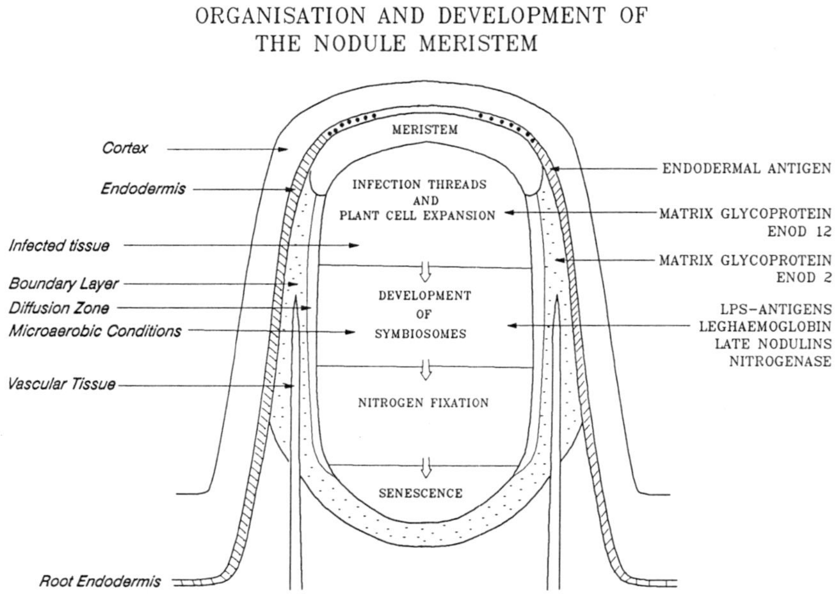

Figure 1. Diagrammatic longitudinal section through a pea nodule. **Left**, organization of cell layers; **center**, tissue invasion by *Rhizobium* and the development of nitrogen fixation, **right**, the development of gene expression identified using antibody or cDNA probes.

In the intercellular spaces of the boundary layer, we have identified a matrix glycoprotein (21) which apparently plugs the network of intercellular air spaces, forcing oxygen to diffuse in aqueous solution through the cells of the boundary layer. As discussed at the previous symposium in Cologne, physiologically induced variations in the quantity, distribution or water-holding capacity of this intercellular matrix glycoprotein could result in rapid changes in the oxygen diffusion barrier of the nodule by changing the total path length for aqueous oxygen diffusion. These ideas are discussed further in the presentation by Minchin *et al.* at this symposium.

Infection threads

Although the small polyhedral cells of the nodule meristem are not themselves invaded by *Rhizobium*, the development of tubular infection threads occurs within 5-10 cell diameters of the proximal face of the meristem. Subsequently, unwalled infection droplets arise, permitting the endocytosis of rhizobia into the plant cytoplasm. In order to study the process of tissue invasion by *Rhizobium* we have used a range of specific antibodies as immunocytochemical probes for immunogold localization studies. MAC 265 reacts with a glycoprotein component of the infection thread matrix (21); MAC 207 recognizes an arabinogalactan protein first identified as an antigen on peribacteroid membranes and subsequently recognized as a component of the plant cell glycocalyx in all flowering plants (7,14,15); JIM5 and JIM7 recognize unesterified and methylesterified pectin respectively (7), and a rabbit antiserum specific for xyloglucan was kindly provided by A. Staehelin (12). A hypothetical model for cell wall growth is presented in Figure 2.

In the uninfected meristematic region, cell walls expanded radially by incorporation of components exported by Golgi vesicles and by biosynthetic activities at the cell surface. Pectin, cellulose and xyloglucan antigens are present in the cell wall while matrix glycoprotein seems to be present as a thin layer between the plasma membrane and the primary wall itself. Underneath the plasma membrane, a microtubular cytoskeleton targets the pectic vesicle into the growing cell walls. Within the plant cell cytoplasm, dual immunogold labelling experiments have revealed that the Golgi apparatus has at least two distinct export pathways, one for pectin-containing vesicles and another for vesicles containing glycocalyx components such as the membrane glycoprotein antigen identified by MAC 207 (Pennell & Roberts in preparation). For reasons discussed below in relation to the formation of infection droplets, it is possible that the Golgi-derived vesicles that contain glycocalyx antigens also harbor the precursors of matrix glycoprotein and perhaps also the xyloglucan precursors. However, neither of these antigens could be convincingly labelled in Golgi vesicles, and dual immunogold labelling experiments have not been undertaken.

The structural components of the infection thread wall appear to be the same as for the meristematic cell wall described above, but these components are organized somewhat differently. The major difference is that deposition of new wall material is localized to a cylindrical invagination which grows apically to produce an inwardly growing cylinder of fairly uniform diameter. Apical cell wall growth is also a characteristic of root hair cells (8), where it is organized and given polarity by a helical array of microtubules (16). It is generally thought that during *Rhizobium*-induced infection of root hair cells the cytoskeletal machinery for apical cell wall growth is subverted, diverted, inverted and converted into an inwardly growing infection thread structure which somewhat resembles an 'inside-out' root hair. Our observations are consistent with this general model. Moreover, we observe that production and secretion of matrix glycoprotein appears to be stimulated in response to *Rhizobium* infection. Most of this matrix material accumulates in the lumen of the infection thread where it is ensheathed by the pectic, xyloglucan and cellulose components of the wall. We estimate that, for peas and other legumes that produce indeterminate nodules, the volume of matrix material in the infection thread lumen is about five-fold greater than the volume of enclosed rhizobia. If most of this matrix glycoprotein is plant-derived, it is possible that growth of the infection thread could be driven as much by secretion of plant matrix glycoprotein at the tip as by growth and division of rhizobia enclosed in the infection thread lumens.

Infection droplets differ from infection threads in that they are not bounded by cell wall cellulose, pectic and xyloglucan components, not it seems are they constrained by the plant cytoskeleton. They grow spherically, forming a shapeless vesicle bounded by plasma membrane bearing the normal glycocalyx components (typified by MAC 207 antigen). Matrix glycoprotein accumulates in the lumen of these infection droplets, together with some disorganized inclusions of wall material. For this reason we hypothesis that matrix glycoproteins and membrane glycocalyx may be packaged and exported in the same class of Golgi vesicle, whereas pectic substances are packaged in a different class of vesicle which can only be targeted to the plasma membrane by an organized system of microtubules. It is clear from these results that a study of tissue

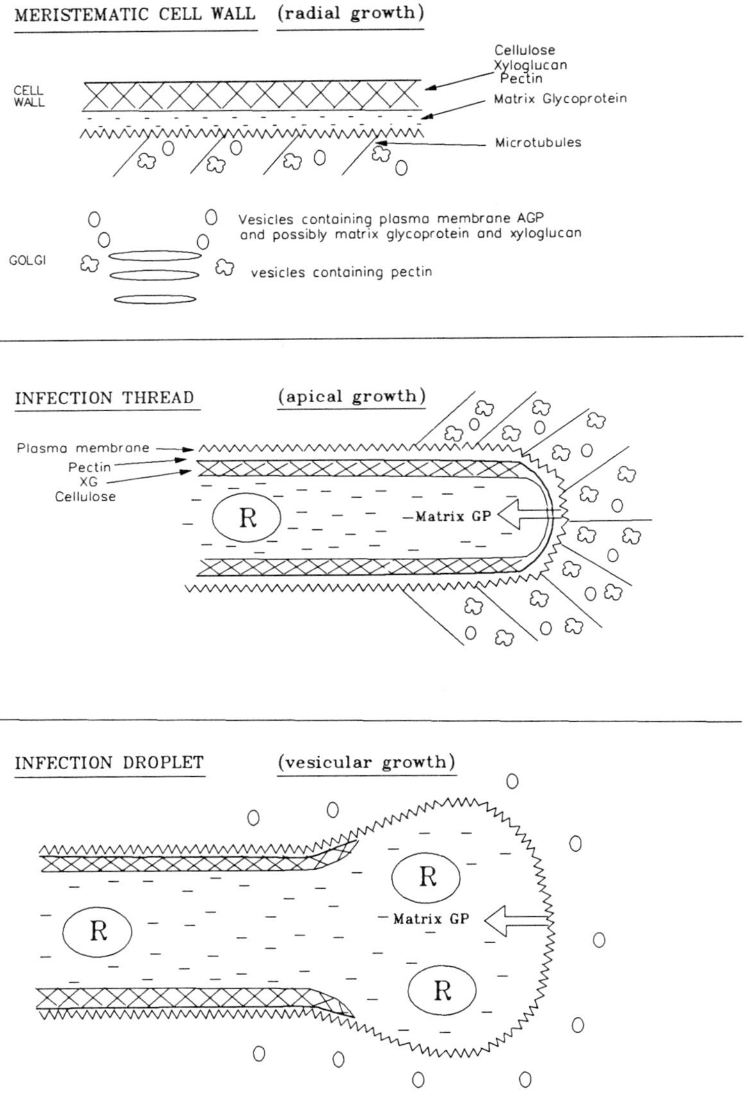

Figure 2. Relationship of matrix glycoprotein to other components involved in cell wall growth. **Top**, meristematic cell; **centre**, infection thread; **bottom**, infection droplet.

invasion by *Rhizobium* will also lead to interesting insights into the normal mechanisms of plant cell wall growth and meristematic activity.

Endocytosis

Endocytosis of bacteria can occur from infection droplets because the bounding plasma membrane is not ensheathed by plant cell wall materials, nor probably underpinned by a cytoskeleton of microtubules. For some time, we have been interested to characterize the nature of molecular interactions occurring at this plant-microbial interface which are probably responsible for uptake of rhizobia into the cytoplasmic space and for division of the peribacteroid membrane in synchrony with the dividing bacteroids (1). These studies have led us to examine the structure and function of *Rhizobium* lipopolysaccharide in nodule development (4,18), and also to characterize the nature of the plant membrane glycocalyx which is associated with the peribacteroid membrane (2,3).

Our molecular studies of *Rhizobium* lipolysaccharide (LPS) have centered round the expression of a particular antigen identified by monoclonal antibody MAC 203 (22). Expression of this antigen within the nodule is apparently induced by low oxygen (or possibly by low pH) growth conditions (5) and may be physiologically adaptive for growth of bacteria under conditions of low pH stress (18). We have recently isolated mutants which lack MAC 203 antigen (24). Many of these mutants are also impaired in the extent of intracellular infection and result in symbiotically defective (Fix^-) nodules. The underlying genetics and biochemistry of *Rhizobium* LPS is currently being investigated in an attempt to relate LPS structure to its function in symbiosis. Currently, there are three general theories concerning the role of LPS. These are: (a) that LPS is needed to stabilize the bacterial outer membrane in response to physiological stresses encountered within the nodule e.g., low oxygen , low pH, salt or osmotic stress (5); (b) that LPS is important in determining overall surface properties such as hydrophobicity or surface change (9); and (c) that particular LPS substituents are involved in specific molecular recognition with components of the plant plasma membrane and peribacteroid membrane (2).

Using monoclonal antibodies, we have identified many carbohydrate antigens associated with plant glycolipid and glycoprotein components of the peribacteroid membrane. These antigens have been characterized biochemically and immunocytologically (Perotto *et al.*, manuscript in preparation). One group of membrane antigens is more abundant on plasma membranes and present in reduced amounts in peribacteroid membranes. another group of glycolipid antigens is particularly associated around mature and senescent bacteroids. Finally one rather exceptional monoclonal antibody identifies a 75kDa protein that is associated both with the peribacteroid and nuclear membranes. Further analysis of these plant membrane components should greatly improve our understanding of the structure and biosynthesis of the peribacteroid membrane and of its functional interactions with the surface of *Rhizobium.*

The peribacteroid space

Plant cell wall components and plant matrix glycoproteins have not been detected in the peribacteroid space which is the lumen of the symbiosome. However, the glycocalyx antigens recognized by antibodies such as MAC 207 are still incorporated into the peribacteroid membrane as it grows in synchrony with the enclosed bacteroids. This observation may imply that, for this particular compartment of the endomembrane system (the symbiosome), delivery of Golgi-derived membrane material carrying glycocalyx antigens is uncoupled from delivery of matrix material in the same vesicles. Alternatively it is possible that matrix glycoprotein is still delivered to the symbiosome but is rapidly degraded.

We have observed that the contents of the peribacteroid space (recovered from isolated intact symbiosomes) are very susceptible to autohydrolysis at low pH. This suggests that acid hydrolases may be present in the peribacteroid space, and we have preliminary evidence that some of the glycoproteins recognized by monoclonal antibody MAC 254 are acid hydrolases. Other characteristics of plant glycoproteins recognized by MAC 254 are described in the poster by VandenBosch *et al.* in this symposium volume.

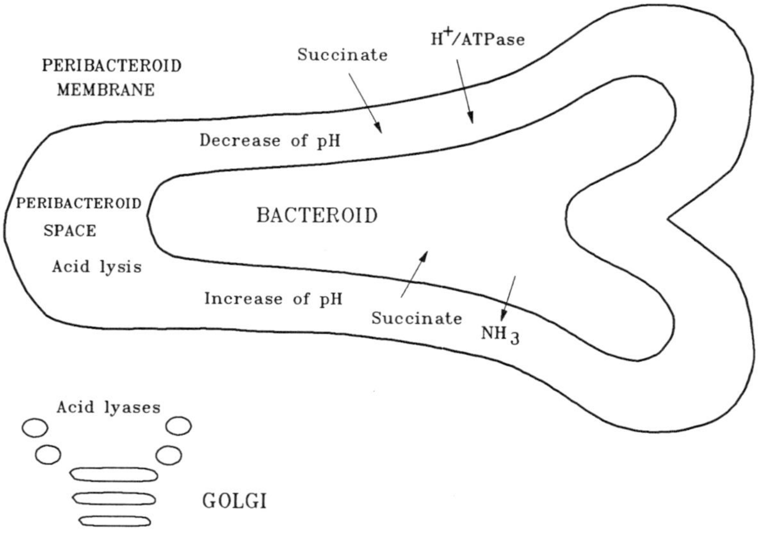

Figure 3. Model for pH balance in symbiosis.

We have recently proposed a model for pH balance in symbiosis (5) which is summarized in Figure 3. The two major transport activities of the plant-derived peribacteroid membrane would progressively acidify the peribacteroid space. Those are dicarboxylic acid transport (19) and a vectorial proton ATPase (20). Moreover, the two major metabolic activities of bacteroids would both tend to increase the pH of the peribacteroid space: these are the uptake and respiration of dicarboxylic acids and the excretion of ammonia as the product of biological nitrogen fixation. Finally, the peribacteroid space contains acid hydrolases which, if they were activated, would degrade any transport proteins present in the peribacteroid space and ultimately would convert the symbiosome into a lytic vesicle (10). It is obvious from the components of this model that the pH of the peribacteroid space is held in dynamic equilibrium. The plant cell tends to acidify this space, while the activity of bacteroids tends to restore pH to neutrality.

The potential activation of acid hydrolases hangs over the whole system like the proverbial sword of Damocles.

This model has a number of interesting consequences. (a) Mutant bacteroids that fail to fix nitrogen would senesce prematurely. (b) When the nitrogenase of bacteroids is used to synthesize ethylene (from acetylene) or hydrogen (for nodules grown under argon) the lack of ammonia production would result in a rapid decline in the pH of the peribacteroid space and hence to an impairment of bacteroid function. This decline in pH could theoretically be counterbalanced if the uptake and respiration of succinate was dramatically increased, for example, if the nodules were transferred to very high oxygen concentrations (9). (c) On the basis of this model, it seems likely that the rate of acidification in very young symbiosomes may be less than in the mature nitrogen fixing state because, when bacteria are first released into symbiosomes, they do not fix nitrogen and their only mechanism to regulate pH is through the uptake and respiration of succinate. Similarly, programmed senescence of bacteroids in nodule development could result from an increased rate of acidification of the peribacteroid space in older symbiosomes beyond the rate that could be counterbalanced by nitrogen-fixing bacteroids. We hope that the predictions of this model will be carefully investigated by nodule physiologists and molecular biologists.

Conclusions

Our description of the *Rhizobium*-legume symbiosis has emphasized mechanisms by which the normal processes of plant growth and development have been redirected to create the symbiotic state. Therefore, we do not confine our attentions to tissue-specific (nodulin) gene functions, nor to flavone-inducible *Rhizobium nod* genes. Early nodulin genes ENOD 2 and ENOD 12 appear to encode cell wall glycoproteins (17,23) localized respectively in the boundary layer and in the sub-apical zone of nodule cell elongation and infection thread development. It is interesting that these two tissues have previously been shown to produce abundant quantities of matrix glycoprotein identified by monoclonal antibody MAC 265 (21). In future, we can hope to see many more examples of complementary interactions between the products of 'nodulin' research and our own immunological analysis of the molecular biology of nodule development.

Acknowledgements

We are grateful for financial support from the UK Agriculture and Food Research Council, the European Commission (SP and MFLeG), the Commonwealth Commission (SSS) and Deutsche Forschungsgemeinschaft (ELK). The figures were drawn by A. L. Davies.

References

1. Bradley, D.J., Butcher, G.W., Galfe, G., Wood, E.A. & Brewin, N.J. (1986) J. Cell Sci. **85**, 47-61.
2. Bradley, D.J., Wood, E.A., Larkin, A.P., Galfe, G., Butcher, G.W. & Brewin, N.J. (1988) Planta **173**, 149-160.

3. Brewin, N.J., Robertson, J.G., Wood, E.A., Wells, B., Larkins, A.P., Galfe, G. & Butcher, G.W. (1985) EMBO J. **4**, 605-611.
4. Brewin, N.J., Wood, E.A., Larkins, A.P., Galfe, G. & Butcher, G.W. (1986) J. Gen. Microbiol. **132**, 1959-1968.
5. Kannenberg, E.L. & Brewin, N.J. (1989) J. Bacteriol. **171**, 4543-4548.
6. Knox, J.P., Linstead, P., Cooper, C., King, J. & Roberts, K. (1990) Planta (in press).
7. Knox, J.P., Day, S. & Roberts, K. (1989) Development **106**, 47-56.
8. Lloyd, C.W., Pearce, K.J., Rawlings, D.J., Ridge, R.W. & Shaw, P.J. (1987) Cell Mot. Cytoskel. **8**, 27-36.
9. de Maagd, R.A., Rao, A.S., Mulders, I.H.M., Goosen-de-Roo, L., van Loosdrecht, M., Wijffelman, C.A. & Lugtemberg, E.J.J. (1989) J. Bacteriol. **171**, 1143-1150.
10. Mellor, R.B. (1989) J. Exp. Bot. **40**, 831-839.
11. Minchin, F.R., Witty, J.F., Sheehy, J.E. & Muller, M. (1983) J. Exp. Bot. **34**, 641-649.
12. Moore, P.J., & Staehelin. L.A. (1988) Planta **174**, 433-445.
13. Parsons, R. (1989) Ph.D. Thesis, Aust. Nat. Univ.
14. Pennell, R.I., Knox, J.P., Scofield, G.N., Selvendran, R.R. & Roberts, K. (1989) J. Cell Biol. **108**, 1967-1977.
15. Pennell, R. I. & Roberts, K. (1990) Nature **344**, 547-549.
16. Ridge, R.W. (1988) Bot. Mag. Tokyo **101**, 427-441.
17. Scheres, B., Van de Wiel, C., Zalenski, A., Horvath, B., Spaink, H. & van Eck, H. (1990) Cell **60**, 281-294.
18. Sindhu, S.S., Brewin, N.J. & Kannenberg, E.L. (1990) J. Bacteriol. **172**, 1804-1813.
19. Udvardi, M.K., Price, D.G., Gresshoff, P.M. & Day, D.A. (1988) FEBS Lett. **231**, 36-40.
20. Udvardi, M.K. & Day, D.A. (1989) Plant Physiol. **90**, 982-987.
21. VandenBosch, K.A., Bradley, D.J., Knox, J.P., Perotto, S., Butcher, G.W. & Brewin, N.J. (1989a) EMBO J. **8**, 335-342.
22. VandenBosch, K.A., Brewin, N.J. & Kannenberg, E.L. (1989b) J. Bacteriol. **171**, 4537-4542.
23. Van de Wiel, C., Scheres, B., Franssen, H., Van Lierop, M.J., Van Lammeren, A., Van Kammen, A. & Bisseling, T. (1990) EMBO J. **9**, 1-7.
24. Wood, E.A., Butcher, G.W., Brewin, N.J. & Kannenberg, E.L. (1989) J. Bacteriol. **171**, 4549-4555.

Gresshoff, Roth, Stacey & Newton (eds.) Nitrogen Fixation: Achievements and Objectives
Chapman and Hall New York • London 1990

Endosymbiosis of *Rhizobium*: Internalization of the "extracellular compartment" and metabolites exchange

Desh Pal S. Verma

Department of Molecular Genetics and Biotechnology Center,
The Ohio State University, Columbus, OH 43210,USA.

Following a successful infection, the *Rhizobium* is released inside the host cell *via* a process resembling endocytosis. A rapid proloiferation of membrane vesicles and their fusion at the tip of the infection thread allows each bacterium to be released inside a membrane-bound vesicle. During the endosymbiotic state bacteria remain enclosed in a membrane envelope (peribacteroid membrane, PBM) which is constituted by the components of the plasma membrane of the host cell (8). The nature of this "extracellular compartment" is, however, not fully understood. It appears to be unique in that it is able to change its properties depending upon the fate of the process ie. symbiotic *vs* pathogenic. Physiological "internalization" of this compartment is a key step in acceptance of the foreign organism inside the host cell. *Rhizobium* enclosed in PBM behaves as an "organelle" having many properties in common with mitochondria. The PBM, while keeping bacteria physiologically "outside" the host cell cytoplasm and avoiding any pathogenic interaction, allows development of a close contact between the plant cell and bacteria for rapid and efficient exchange of metabolites. Recent studies from our laboratory suggest that many specific host gene products (nodulins) are integrated into the PBM.

The PBM nodulins are found to be induced at an early stage of development (6,9). Apparently, transduction of all the signals necessary for development and differentiation of nodules preceed the release of bacteria from the infection thread (10). Using a *Bradyrhizobium japonicum* mutant strain defective in endocytosis, it was observed that the PBM nodulin genes are induced prior to the formation of the PBM compartment (6) and there are apparently differently signals for the induction of different PBM nodulin genes. The mechanism of targeting of nodulins to PBM also differs for each peptide. Nodulin-24 was found to have an amino terminus signal sequence which is cleaved co-translationally (4), while nodulin-26, an integral PBM protein, does not have an amino terminus signal sequence and the signal seems to reside internally towards the carboxy end. The soybean nodulin-26 has homology with an intrinsic protein of the bovine eye lens membrane, MIP-26. The latter is known to form a gap junction type channel. We have determined the topology of nodulin-26 in PBM which suggests that both carboxy and amino ends face the cytoplasmic side. This protein is glycosylated co-translationally as evidenced by Con-A binding. The carboxy end contains several phosphorylation sites which may interact with the acidic lipids present in the PBM.

Nodulin-26 shows sequence similarity with a Pea protein induced as a result of change in turgor pressure (3) and the glycerol facilitator protein (GlpF) of *Escherichia coli* (1). While the subcellular location of turgor regulated pea protein is not known, recently it has been observed that an analog of nodulin-26 may be present in the seed vacuole membrane (M. Chrispeels, personal communication). This suggest that that the peribacteroid compartment may be vacuoler in nature (see ref.5).

In order to decipher the function of nodulin-26 in PBM, we attempted to integrate this protein into *E. coli* membrane and compare the properties of the channel formed by this protein with that formed by the glycerol facilitator of *E. coli* The results suggest that nodulin-26 does not transport glycerol and is unable to complement *glp*F mutation in *E. coli*. However, we discovered that it does form a channel analogous to GlpF that facilitates the translocation of succinate and glycine. A rapid transport of dicarboxylic acids is essential for effective symbiosis. The uptake of these compounds appears to be passive and differs from active transport in bacteroids. A passive transport will allow rapid equilibration of the peribacteroid fluid with respect to the specific metabolites from the host cell cytoplasm which then can be actively takenup by the bacteroids. This also eliminates any need to couple the transport systems across the two different membranes, procaryotic and eucaryotic, which are physically separate from each other. Therefore, recent results obtained on the uptake of dicarboxylic acids using peribacteroid units (7) must be reassessed in light of this data.

The presence of nodulin-26 type channels in the PBM sheds new light on root nodule symbiosis where the "extracellular compartment" enclosing the microsymbiont must be "internalized" or equilibrated with respect to its ionic contents with the host cell cytoplasm and thus bring nitrogen-fixing bacteria closer to the cellular metabolism of the host. Since the peribacteroid space occupies as much or more space then the total cytoplasm in an infected cell such an equilibration becomes even more essential. It is also possible that this channel or other similar channels in the PBM translocate NH_4 from the bacteroids, since an increase in the concentration of this compound in the peribacteroid fluid would directly inhibit nitrogen fixation and NH_4 must be rapidly transported to the host cell, where it is converted to glutamine by the host glutamine synthetase. An ammonia-induced glutamine synthetase is present in the infected cells of soybean root nodules (see Miao et al., this volume). Such channels may also exist in other symbioses between eucaryotic and procaryotic cells for the rapid exchange of metabolites.

We have also observed that the infected cells of the nodules are much lower in oxygen concentration as indicated by the expression of maize alcohol dehydrogenase promoter and GUS reporter gene fusion in a transgenic legume. Moreover, the osmoticum in the infected cells is much higher as indicated by the elevated levels of Δ^1-pyrroline-5-carboxylate reductase gene expression (2) and concomitant proline accumulation which may further reduce the solubility of oxygen. Thus the celluar environment of the nodule cells is very different from that of the normal root cells and specific needs exist for compartmentalizing metabolisms in the infected and uninfected cells to sustain the nodule function (11).

Acknowledgements
This work was supported by research grants from NSF (#DCB 8819399 and DCB 8904101)

References

1. Baker, M.E. and Saier, M.H. (1990) Cell **60**, 185-186.
2. Delauney, A.J. and Verma , D.P.S. (1990) Mol. Gen. Gen., in press.
3. Guerrero, F.D., Jones, J.T. and Mullet, J.E. (1990) Plant Mol. Biol. in press.
4. Katinakis, P. and Verma, D.P.S. (1985) Proc. Natl. Acad. Sci. USA **82**, 4157-4161.
5. Mellor, R.B. and Werner, D. (1987) Symbiosis **3**, 75-100.
6. Morrison, N. and Verma, D.P.S. (1987) Plant Mol. Biol. **7**, 51-61.
7. Udvardi, M.K., Price, G.D., Gresshoff, P.M. and Day, D.A. (1988) FEBS Letters **231**, 36-40.
8. Verma, D.P.S., Kazazian, V., Zogbi, V. and Bal, A.K. (1978) J. Cell Biol. **78**, 919-936.
9. Verma, D.P.S., Fortin, M.G., Stanley, J., Mauro, V., Purohit, S. and Morrison, N. (1986) Plant Mol. Biol. **7**, 51-61.
10. Verma, D.P.S., Delauney, A.J., Guida, M., Hirel, B., Schafer, R. and Koh, S. (1988) In Molecular Genetics of Plant-Microbe Interactions, eds. R. Palacios and D.P.S. Verma (APS Press, St. Paul) p. 315-320.
11. Verma, D.P.S. and Fortin, M.G. (1989) In The Molecular Biology of Nuclear Genes, ed. I.K. Vasil, (Academic Press Inc., New York), pp. 329-353.

Gresshoff, Roth, Stacey & Newton (eds.) Nitrogen Fixation: Achievements and Objectives
Chapman and Hall New York • London 1990

Workshop summary: Compilation of the *nod*, *fix*, and *nif* genes of Rhizobia and information concerning their function

Gary Stacey

Center for Legume Research, Dept. of Microbiology and Graduate Program of Ecology, The University of Tennessee, Knoxville, TN 37996, USA.

As a means of summarizing the Workshop on the Biochemical Function of the *nod, nif,* and *fix* Genes, I have simply listed all of the identified genes below with a few lines describing the information available concerning their function. I apologize to those individuals whose papers I have failed to reference due to a limitation on space.

NODULATION GENES

nodABC -- NodABC are essential for root hair curling and cortical cell division. Nod ABC likely synthesize a phytohormone-like substance (cf., [(1-9]). Such a substance, NodRm-1, produced by *R. meliloti* has recently been identified as a N-acyl-tri-N-acetyl- -1,4-D-glucosamine tetrasaccharide bearing a sulfate group on carbon 6 of the reducing sugar moiety [10].

nodD -- Transcriptional regulator required for the expression of the other *nod* genes (cf., [11-14]).

nodE -- Host range; reported sequence similarity to the *fabB* gene of *E. coli*, condensing enzyme of fatty acid synthase. Membrane protein [15-21]. Mutants lacking NodE are affected in capsular polysaccharide biosynthesis (Dazzo, pers. commun.).

nodF -- Host range; reported sequence similarity to acyl carrier protein [17-22]. NodF carries 4'-phophopantetheine as a prosthetic group. At least two different forms of NodF can be isolated from a NodFE-overproducing strain of *R. leguminosarum* bv. *viciae* (Geiger et al., this volume).

nodG -- Host range; reported sequence similarity to dehydrogenases [17,20,23].

nodH -- Host range; may modify factor produced by NodABC [5,6,17,20]. May function as a sulfate transferase for the synthesis of NodRm-1 (Denarie et al., this volume).

nodI -- Reported sequence similarity to ATP-binding transport proteins [24,25]. *R. leguminosarum* bv. *trifolii* mutants defective in NodI have a decrease in the 3-hydroxybutyrate substitution of the capsular polysaccharide (Philip-Hollingsworth et al., this volume).

nodJ -- Membrane protein; may act in conjunction with *nodI* [25].

nodK -- Reported only in *B.* sp.(*Parasponia*). Function unknown [4].

nodL -- Host range; reported sequence similarity to acetyltransferase [18,26-29].

nodM -- Host range; reported sequence similarity to amido transferase [27,29]. Glucosamine synthase for the synthesis of NodRm-1 (this volume; also A. Kondorosi, pers. commun.).
nodN -- Reported [27]. Involved in biosynthesis of hair deformation factor (Baev, Endre, Kondorosi, pers. commun.).
nodO -- Function unknown; encoded protein is exported. Reported to bind calcium [30,31].
nodP -- Function unknown; gene hybridizes to *E.coli* genomic DNA [32-34].
nodQ -- Host range; reported sequence similarity to GTP-binding proteins, EF-tu and EF-1. May modify factor produced by NodABC [5,32-34]
nodR -- Reported only in *R. leguminosarum* bv. *trifolii*. Host range. Function unknown [29,35].
nodS -- Function unknown. (Gottfert et al., personal communication).
nodT -- Host range. Function unknown [26,36].
nodU -- Function unknown. (Gottfert et al., pers. commun.).
nodV -- Reported only in *B. japonicum*. Important for host range; sequence similarity to membrane sensor family [37].
nodW -- Reported only in *B. japonicum*. Important for host range; sequence similarity to family of transcriptional regulatory proteins [37].
nodX -- reported only in *R. leguminosarum* strains capable of nodulating Afghanistan pea [38].
nodY -- Reported only in *B. japonicum*, but found in all rhizobia by hybridization. Function unknown [39,40].
nodZ -- Host range; required for growth under microaerobic conditions [41] (Schell et al., pers. commun.).
nolA -- Host range. Essential for nodulation of selected soybean cultivars. Reported only from *B. japonicum*. Shows sequence similarity to MerR, a transcriptional regulatory protein [42].
nolD, nolE, nolF, nolG -- Host range in *R. meliloti* (Baev, Banfalvi, Kondorosi, pers. commun.).

<u>FIX</u> GENES

fixABC -- May function in electron transport to nitrogenase [43-45]. *fixC* has been reported to have a signal sequence for membrane insertion [44].
fixGHIS -- The proteins encoded by these genes appear to contain transmembrane sequences. FixI shows sequence similarity with ATP-dependent cation pumps in bacteria. FixG contains cysteine clusters and may be involved in redox reactions. FixGHIS have been proposed to act as a membrane-bound complex coupling a cation pump to redox processes [46].
fixLJ -- Transcriptional regulator of the *nif*A gene [47]. FixL shows sequence homology to a family of proteins which functions as a membrane sensor family. Some members of this family have been shown to act as protein phosphotransferases. FixJ shows sequence similarity to a family of transcriptional regulatory proteins.
fixKN -- Reported in *R. meliloti* [47]. FixK activates transcription

of *fixN* and exerts negative control on *nifA* [48].
fixR -- Function unknown. Reported in *B. japonicum* [49]. Protein sequence has similarity to NAD-dependent dehydrogenases (this volume).
fixX -- Sequence similarity to ferredoxin. Reported in *R. meliloti* [43,44,50] and *R. leguminosarum* [51].

NIF GENES

nifA -- Transcriptional regulator required for the transcription of the *nif* genes and some *fix* genes (reviewed in [52]).
nifB -- FeMoco biosynthesis [53].
nifHDK -- *nifH* encodes Fe protein subunits. *nifDK* encodes MoFe protein subunits (c.f., [54-57]).
nifE -- FeMoco biosynthesis [58,59].
nifN -- FeMoco biosynthesis [60]. *fixF* = *nifN* [61].
nifS -- May be involved in nitrogenase maturation [60].

A description of other genes involved in nodulation or nitrogen fixation can be found in recent reviews (cf. [62-64]).

LITERATURE CITED

1. Rossen, L., Johnston, A.W.B. & Downie, J.A. (1984) Nucl. Acids Res. **12**, 9497-9508.
2. Torok, I., Kondorosi, E., Stepkowski, T., Postfai, J. & Kondorosi, A. (1984) Nucl. Acids Res. **12**, 9509-9524.
3. Egelhoff, T.T. & Long, S.R. (1985) J. Bacteriol. **164**, 591-599.
4. Scott, K.F. (1986) Nucl. Acids Res. **14**, 2905-2910.
5. Faucher, C., Camut, S., Denarie, J. & Truchet, G. (1989) Mol. Plant-Microbe Int. **2**, 291-300.
6. Faucher, C., Maillet, F., Vasse, J., Rosenberg, C., van Brussel, A.A.N., Truchet, G. & Denarie, J. (1988) J. Bacteriol. **170**, 5489-5499.
7. Hollingsworth, R., Squartini, A., Philip-Hollingsworth, S. & Dazzo, F. (1989) in *Signal Molecules in Plants and Plant-Microbe Interactions*, ed. Lugtenberg, B.J.J. (Springer Verlag, Berlin), pp. 387-393.
8. Schmidt, J., Wingender, R., John, M., Wieneke, U. & Schell, J. (1988) Proc. Natl. Acad. Sci. USA **85**, 8578-8582.
9. Banfalvi, Z. & Kondorosi, A. (1989) Plant Mol. Biol. **13**, 1-12.
10. Lerouge, P., Roche, P., Faucher, C., Maillet, F., Truchet, G., Prome, J.C. & Denarie, J. (1990) Nature **344**, 781-784.
11. Mulligan, J.T. & Long, S.R. (1985) Proc. Natl. Acad. Sci. USA **82**, 6609-6613.
12. Gottfert, M., Horvath, B., Kondorosi, E., Putnoky, P., Rodriquez-Quinones, F. & Kondorosi, A. (1986) J. Mol. Biol. **191**, 411-420.
13. Rossen, L., Shearman, C.A., Johnston, A.W.B. & Downie, J.A. (1985) EMBO J. **4**, 3369-3373.
14. Appelbaum, E.R., Thompson, D.V., Idler, K. & Chartrain, N. (1988) J. Bacteriol. **170**, 12-20.
15. Spaink, H.P., Weinman, J., Djordjevic, M.A., Wijffelman, C.A.,

Okker, R.J.H. & Lugtenberg, B.J.J. (1989) EMBO J. **8**, 2811-2818.
16. Bibb, M.J., Biro, S., Motamedi, H., Collins, J.F. & Hutchinson, C.R. (1989) EMBO. J. **8**, 2727-2736.
17. Debelle, F., Rosenberg, C., Vasse, J., Maillet, F., Martinez, E., Denarie, J. & Truchet, G. (1986) J. Bacteriol. **168**, 1075-1086.
18. Djordjevic, M.A., Innes, R.W., Wijffelman, C.A., Schofield, P.R. & Rolfe, B.G. (1986) Plant Mol. Biol. **6**, 389-394.
19. Downie, J.A., Knight, C.D., Johnston, A.W.B. & Rossen, L. (1985) Mol. Gen. Genet. **198**, 591-597.
20. Horvath, B., Kondorosi, E., John, M., Schmidt, J., Torok, I., Gyorgypal, Z., Barabas, I., Wieneke, U., Schell, J. & Kondorosi, A. (1986) Cell **46**, 335-343.
21. Djordjevic, M.A., Schofield, P.R. & Rolfe, B.G. (1985) Mol. Gen. Genet. **200**, 463-471.
22. Shearman, C.A., Rossen, L., Johnston, A.W.B. & Downie, J.A. (1986) EMBO J. **5**, 647-652.
23. Debelle, F. & Sharma, S.B. (1986) Nucl. Acids Res. **14**, 7453-7472.
24. Higgins, C.F., Hiles, I.D., Salmond, G.P.C., Gill, D.R., Downie, J.A., Evans, I.J., Holland, I.B., Gray, L., Buckel, S.D., Bell, A.W. & Hermodson, M.A. (1986) Nature **323**, 448-450.
25. Evans, I.J. & Downie, J.A. (1986) Gene **43**, 95-101.
26. Canter Cremers, H.C.J., Spaink, H.P., Wijfjes, A.H.M., Pees, E., Wijffelman, C.A., Okker, R.J.H. & Lugtenberg, B.J.J. (1989) Plant Mol. Biol. **13**, 163-174.
27. Surin, B.P. & Downie, J.A. (1988) Mol. Microbiol. **2**, 173-183.
28. Downie, J.A. (1989) Mol. Microbiol. **3**, 1649-1651.
29. Weinman, J.J., Djordjevic, M.A., Sargent, C.L., Dazzo, F.B. & Rolfe, B.G. (1988) in *Molecular Genetics of Plant-Microbe Interactions*, eds. Palacios, R. & Verma, D.P.S. (American Phytopathological Society Press, St. Paul, MN), pp. 33-34.
30. Economou, A., Hamilton, W.D.O., Johnston, A.W.B. & Downie, J.A. (1990) EMBO J. **9**, 349-354.
31. de Maagd, R.A., Wijfjes, A.H.M., Spaink, H.P., Ruiz-Sainz, J.E., Wijffelman, C.A., Okker, R.J.H. & Lugtenberg, B.J.J. (1989) J. Bacteriol. **171**, 6764-6770.
32. Schwedock, J. & Long, S.R. (1989) Mol. Plant-Microbe Int. **2**, 181-194.
33. Long, S.R., Schwedock, J., Egelhoff, T., Yelton, M., Mulligan, J., Barnett, M., Rushing, B. & Fisher, R. (1989) in *Signal Molecules in Plants and Plant-Microbe Interactions*, ed. Lugtenberg, B.J.J. (Springer-Verlag, Berlin), pp. 145-151.
34. Cervantes, E., Sharma, S.B., Maillet, F., Vasse, J., Truchet, G. & Rosenberg, C. (1989) Mol. Microbiol. **3**, 745-755.
35. Rolfe, B.G., Sargent, C.L., Weinman, J.J., Djordjevic, M.A., McIver, J., Redmond, J.W., Batley, M., Yuan, D.C. & Sutherland, M.W. (1989) in *Signal Molecules in Plants and Plant-Microbe Interactions*, ed. Lugtenberg, B.J.J. (Springer-Verlag, Berlin), pp. 303-310.
36. Surin, B.P., Watson, J.M., Hamilton, W.D.O., Economou, A. & Downie, J.A. (1990) Mol. Microbiol. **4**, 245-252.
37. Gottfert, M., Grob, P. & Hennecke, H. (1990) Proc. Natl. Acad.

Sci. USA **87**, 2680-2684.
38. Davis, E.O., Evans, I.J. & Johnston, A.W.B. (1988) Mol. Gen. Genet. **212**, 531.
39. Nieuwkoop, A.J., Banfalvi, Z., Deshmane, N., Gerhold, D., Schell, M.G., Sirotkin, K.M. & Stacey, G. (1987) J. Bacteriol. **169**, 2631-2638.
40. Banfalvi, Z., Nieuwkoop, A.J., Schell, M.G., Besl, L. & Stacey, G. (1988) Mol. Gen. Genet. **214**, 420-424.
41. Stacey, G., Schell, M.G. & Deshmane, N. (1989) in *Signal Molecules in Plants and Plant-Microbe Interactions*, ed. Lugtenberg, B.J.J. (Springer-Verlag, Berlin), pp. 394-399.
42. Sadowsky, M.J., Cregan, P.B., Gottfert, M., Sharma, A., Gerhold, D., Rodriguez-Quinones, F., Keyser, H.H., Hennecke, H. & Stacey, G. (1990) Proc. Natl. Acad. Sci. USA, submitted,
43. Gubler, M., Zurcher, T. & Hennecke, H. (1989) Mol. Microbiol. **3**, 141-148.
44. Earl, C.D., Ronson, C.W. & Ausubel, F.M. (1987) J. Bacteriol. **169**, 1127-1136.
45. Gubler, M. & Hennecke, H. (1986) FEBS Lett. **200**, 186-192.
46. Kahn, D., David, M., Domergue, O., Daveran, M.-L., Ghal, J., Hirsch, P.R. & Batut, J. (1989) J. Bacteriol. **171**, 929-939.
47. David, M., Daveran, M.L., Batut, J., Dedieu, A., Domergue, O., Ghai, J., Hertig, C., Boistard, P. & Kahn, D. (1988) Cell **541**, 671.
48. Batut, J., Daveran-Mingot, M.L., David, M., Jacobs, J., Garnerone, A.M. & Kahn, D. (1989) EMBO J. **8**, 1279-1286.
49. Thony, B., Fischer, H.-M., Anthamatten, D., Bruderer, T. & Hennecke, H. (1987) Nucl. Acids Res. **15**, 8479-8499.
50. Dusha, I., Kovalenko, S., Banfalvi, Z. & Kondorosi, A. (1987) J. Bacteriol. **169**, 1403-1409.
51. Gronger, P., Manian, S.S., Reilander, H., O'Connell, M., Priefer, U.B. & Puhler, A. (1987) Nucl. Acids Res. **15**, 31-49.
52. Gussin, G.N., Ronson, C.W. & Ausubel, F.M. (1986) Ann. Rev. Genet. **20**, 567-591.
53. Buikema, W.J., Klingensmith, J.A., Gibbons, S.L. & Ausubel, F.M. (1987) J. Bacteriol. **169**, 1120.
54. Corbin, D., Barran, L. & Ditta, G. (1983) Proc. Natl. Acad. Sci. USA **80**, 3005.
55. Torok, I. & Kondorosi, A. (1981) Nucl. Acids Res. **9**, 5711.
56. Krol, A.J.M., Hontelez, J.G.J., Roozendaal, B. & Van Kammen, A. (1982) Nucl. Acids Res. **10**, 4147.
57. Hennecke, H. (1981) Nature **291**, 354.
58. Puhler, A., Aguilar, O.M., Kapp, D., Reilander, H. & Weber, G. (1985) in *Natural Products Chemistry, 1984*, eds. Zalewski, R.I. & Skolik, J.J. (Elsevier, Amsterdam), pp. 667.
59. Norel, F., Desnoues, N. & Elmerich, C. (1985) Mol. Gen. Genet. **199**, 352.
60. Ebeling, S., Hahn, M., Fischer, H.M. & Hennecke, H. (1987) Mol. Gen. Genet. **207**, 503.
61. Aguilar, O.M., Reilander, H., Arnold, W. & Puhler, A. (1987) J. Bacteriol. **169**, 5393-5400.

62. Martinez, E., Romero, D. & Palacios, R. (1990) Crit. Rev. Plant Sci. **9**, 59-93.
63. Long, S.R. (1989) Ann. Rev. Genet. **23**, 483-506.
64. Appelbaum, E. (1989) in *Molecular Biology of Nitrogen Fixation*, ed. Gresshoff, P.M. (CCRC, Boca Raton, FL), pp. 131-158..

Gresshoff, Roth, Stacey & Newton (eds.) Nitrogen Fixation: Achievements and Objectives
Chapman and Hall New York • London 1990

Inhibition of the 1,3-ß-glucan synthase from *Glycine max* and *Pisum sativum* by exopolysaccharides of *Bradyrhizobium japonicum* and *Rhizobium leguminosarum*

Bettina Ahlborn and Dietrich Werner

FB Biologie/Botanik, Philipps-Universität Marburg, Karl-v.-Frisch-Straße, D-3550 Marburg FRG.

Penetration of the leguminous root hair cell wall by rhizobia -a process involving alteration and degradation of plant cell wall polysaccharides (1)- is expected to be met by plant defense reactions. One of the characteristic responses to pathogenic attacks, the deposition of callose (a 1,3-ß-D-glucan), is missing at the infection-site of compatible *Rhizobium*-legume interactions (1). We were interested in analyzing whether these bacteria have any effect on the callose synthezising enzyme, the 1,3-ß-glucan synthase, localized on the plasma membrane. Since rhizobial exopolysaccharides (EPS) have been asigned an important role in protecting the bacteria against plant defense responses (2), the effect of EPS on the glucan synthase was assayed using two model systems, *Glycine max/B. japonicum* and *Pisum sativum/R. leguminosarum*

A standard in-vitro assay at Ca^{2+}-saturating conditions was employed, based on the incorporation of UDP-[^{14}C]-glucose into an insoluble ß-1,3-glucan (3).
A marked inhibition of the enzyme activity occurred by addition of rhizobial EPS. EPS of *B. japonicum* exhibited the strongest inhibition potency, leaving a rest activity of 35-45% of the control whereas EPS of *R. leguminosarum* reduced the enzyme activity to a level of 60-70% . Since different EPS-batches revealed different inhibitory rates, we assumed that the portion of the effective structure varied, probably due to small differences in growth conditions, culture age of the bacteria and purification procedure of the EPS. Cross experiments of the two model systems demonstrated that the observed inhibition is not host-specific.
The ß-glucan synthase is Ca^{2+}-regulated (3,4). However, a 20-fold increase in Ca^{2+}-concentration did not significantly alter EPS-effectivity, suggesting that the inhibition cannot be due to Ca^{2+}-complexation by acidic functions of rhizobial EPS.

Two polymers of non-symbiotic organisms, xanthan and laminarin, exhibited no significant effect on either of the 1,3-ß-glucan synthases tested, implying that certain structural requirements typical for rhizobia are necessary for a successful inhibition of the activated GS II-system.
In short, our data confirm the widespread assumption that rhizobial surface polysaccharides, or probably oligosaccharides, play an important role in suppressing plant defense reactions, thereby facilitating the infection of the host-legume.

(1) Callaham, D.A. and Torrey, J.G. (1981) Can. J. Bot. **99**:1647-1664
(2) Glazebrook, W. and Walker, G.C. (1989) Cell **56**:661-672
(3) Fink, J., Jeblick, W., Blaschek, W. & Kauss, H. (1987) Planta **172**:101-105
(4) Girard, V. and Maclachlan, G. (1987) Plant Physiol. **85**:131-136

Gresshoff, Roth, Stacey & Newton (eds.) Nitrogen Fixation: Achievements and Objectives
Chapman and Hall New York • London 1990

Toxicity of *Leucaena leucocephala* seed diffusate to *Rhizobium*

M. Alameda and E.C. Schröder

BNF Laboratory, Dept. of Agronomy and Soils, University of Puerto Rico, Mayagüez, PR 00709-5000, USA

Several authors have reported the inhibitory effects of seeds towards Rhizobium growth (1,3,5) and we observed poor nodulation of L. leucocephala growing in test tubes. The objective of this study was to determine if leucaena seeds produce compounds toxic to rhizobia.
Leucaena K8 and strains TA1 and TAL 1145 of Rhizobium were used. Autoclaved seeds received several treatments and were placed in YEM plates where bacteria inhibition zones were measured.
Both Rhizobium strains were inhibited by antibacterial substance(s) produced by leucaena seeds. Scarification facilitated the diffusion of the compound, since larger inhibition zones were observed around scarified seeds. Placing the seeds 24 hrs. before the bacteria and reduced agar concentration produced larger inhibition zones. Separation of seed coats and cotyledons indicated that inhibitory substances were located in the endosperm, while in clovers the antibacterial substance is associated with the seed coat. Filter paper discs impregnated with the sterile aqueous seed extract also showed inhibition zones. This shows that the toxic factor can be separated from the seed and its relative concentration measured by the degree of inhibition.
The above results indicate that leucaena seed produces a toxic factor. Further research is needed to identify the substance. Previous studies have stated that the flavonol myricetin (2) and tannins (4) show antibiotic activity. It might be possible to select resistant Rhizobium mutants or non-toxic genotypes. Seed washing and addition of absorbants might promote better nodulation.

REFERENCES

1. Bowen, G.D. (1961) The toxicity of legume seed diffusates towards rhizobia and other bacteria. Plant and Soil 15: 155-165.
2. Fottrell, P.F., O'Connor, S. and Masterson, C.L. (1964) Identification of the flavonol myricetin in legume seeds and its toxicity to nodule bacteria. Irish J. Agric. Res. 3: 246-249.
3. Masterson, C.L. (1962) Inhibition of Rhizobium species by legume seeds. Irish J. Agric. Res. 1: 343-344.
4. Masterson, C.L. (1965) Studies on the toxicity of legume seeds towards Rhizobium. Ann. Inst. Pasteur, Paris 109: 216-217.
5. Thompson, J.A. (1960) Inhibition of nodule bacteria by an antibiotic from legume seed coats. Nature 187: 169.

Gresshoff, Roth, Stacey & Newton (eds.) Nitrogen Fixation: Achievements and Objectives
Chapman and Hall New York • London 1990

Characteristics of exopolysaccharide-deficient mutants of *Rhizobium* spp. with altered nodulation competitiveness

Richard S. Araujo and Jo Handelsman

Dept. Plant Pathology and Center for the Study of Nitrogen Fixation, University of Wisconsin, Madison, WI 53706, USA.

The major barrier to the use of *Rhizobium* inoculants in commercial bean (*Phaseolus vulgaris* L.) production is the poor nodulation competitiveness of the high-nitrogen-fixing inoculant strains. Previous work (1-3) has implicated the cell surface in nodulation competitiveness of other rhizobia, but little is known about the determinants of competitiveness among bean rhizobia. We are currently using a genetic approach to study the relationship between the ability to produce expolysaccharides (EPS) in culture and the nodulation competitiveness of bean rhizobia.

Rhizobium spp. strains KIM5s, CIAT899 and CE3 were mutagenized with transposon Tn5 and the mutants were screened for non-mucoid colony morphology on high carbon, solid medium. All EPS-deficient mutants induced normal nodules on beans. The nodulation competitiveness of each of 14 EPS-deficient mutants of CE3, 3 of CIAT899, and 6 of KIM5s was determined in experiments in which bean seeds were inoculated with each mutant paired with its parent. Nodulation competitiveness was assessed by determining the proportional representation of each strain in the nodules relative to its proportional representation in the inoculum (4). EPS-deficient mutants of all three strains were consistently less competitive than their parents. One EPS-deficient mutant of strain CE3 was previously isolated and complemented by Diebold & Noel (5) to restore EPS. The mutant was less competitive than the parent whereas the EPS-restored mutant was as competitive as CE3. These results suggest that mutations that affect EPS production may directly or indirectly affect nodulation competitiveness in rhizobia that nodulate beans. A further genetic analysis will enable us to determine the genetic and physiological relationship between EPS production and nodulation competitiveness.

1. Ames, P. & Bergman, K. (1981) J. Bacteriol. **148**, 728-729.
2. Handelsman, J. et al (1984) J. Bacteriol. **157**, 703-707.
3. Mellor, H.Y. et al (1987) Arch.Microbiol. **148**, 34-39.
4. Beattie, G.A. & Handelsman, J. (1989) J. Microbiol Methods **9**, 29-33.
5. Diebold, R. & Noel, K.D. (1989) J. Bacteriol. **171**, 4821-4830.

Gresshoff, Roth, Stacey & Newton (eds.) Nitrogen Fixation: Achievements and Objectives
Chapman and Hall New York • London 1990

Nodulation and N_2-fixation of *M. sativa* under osmotic (PEG) and salt stress conditions

G. Banet, H. Badani, S. Weininger and Y. Kapulnik.

Legume Inoculation Dept. ARO, Volcani Center, Bet-Dagan, Israel.

The symbiosis between legume plants and Rhizobium bacteria is greatly affected by saline (NaCl) conditions under which fewer nodules are formed with no major changes in their specific activity. Increasing levels of salinity impose high toxic ions and low water potential ("osmotic stress") environment. The objective of this study was to distinguish between the osmotic and toxic effects which are normally induced by NaCl application and to relate nodule initiation and function to each of these conditions.
The effect of NaCl, KCl, K_2SO_4, Na_2SO_4 and PEG (polyethyleneglycol) was studied under controlled conditions in modified Leonard jars containing dune sand. N free nutrient solution was adjusted to the same osmolality (mOsm) to obtain equal osmotic pressure of all salts. Nodule activity was determined by the ARA method and ion concentrations were ascertained from wet digestion by standard methods.
All salts tested were found to significantly reduce the number of nodules per plant in a range of 42 and 75%, nodules dry weight, 40 to 67% and plant dry weight, 26 to 41% as compared with untreated controls. No significant reduction in ARA was found in any of the treatments investigated. Although, nodule number was not affected by any of the PEG levels tested (up to 195 mOsm), nodule dry weight was reduced by 40% as compared with untreated controls. It was concluded that nodule number was affected by the toxic and not the osmotic stress induced; nodule activity (ARA) was not affected by both: toxic and osmotic . Based on internal ion concentration, the relative toxic effect of the individual ion was rated as: $Na^+ > Cl^- > K^+ > SO_4^{--}$ for plant yield, nodule number and weight; nodule dry weight was more affected by the osmotic stress than the toxic stress induced by salinity.

Gresshoff, Roth, Stacey & Newton (eds.) Nitrogen Fixation: Achievements and Objectives
Chapman and Hall New York • London 1990

Specific forms of exopolysaccharides from *Rhizobium meliloti* restore symbiotic effectiveness to *R. meliloti* exo-mutants

L. Battisti, C.C. Lee and J.L. Leigh

Dept. of Microbiology SC-42, Univ. of Washinton, Seattle, WA
USA 98195

Rhizobium meliloti strain SU47 produces both high molecular weight (HMW) and low molecular weight (LMW) forms of an acidic exopolysaccharide, succinoglycan (EPSA), composed of repeating octomeric subunits of beta-linked glucose, beta-linked galactose, acetyl, pyruvyl and succinyl substituents (7:1:1:1:1). The LMW form may be further separated by anion exchange chromotography into several distinct subfractions differing in subunit multiplicities (e.g. monomer, trimer, tetramer) and charge. Genetic studies have shown that EPSA is required for nodule invasion; Tn5 mutants of SU47 which fail to produce any form of EPSA (Exo$^-$) exhibit a delay in root hair curling and induce empty non-nitrogen fixing (Fix$^-$) nodules on alfalfa roots. In order to show conclusively that the acidic EPS of R. meliloti has a biologically active role in nodule invasion, a plant bioassay was developed to determine if various forms of EPSA could correct the nodule invasion defect of Exo$^-$ mutants. In addition to EPSA, heterologous EPSs and a cryptic acidic EPS of R. meliloti, EPSB, were also tested in the bioassay. Results from these experiments support the following conclusions: (1) unfractionated LMW EPSA partially restored nodule invasion to various Exo- mutants; heat and protease treatment had no effect on this activity. HMW EPSA, variously charged monomeric subfractions of LMW EPSA as well as depyruvylated LMW EPSA (unfractionated) were not able to complement exoA mutants suggesting that a specific form of LMW EPSA is required for biological activity (2) EPSs from R. trifolii, R. leguminosarum, A. tumefaciens and R. strain NGR234 failed to restore nodule invasion providing further evidence there is structural specificity associated with biological activity (3) EPSB, also partially restored nodule invasion in the bioassay, confirming that the two EPSs are functionally interchangeable (4) preliminary results using QAE derived subfractions appear to indicate that either the larger multimeric forms (trimer, tetramer) alone or all forms of LMW EPSA (a synergistic effect) are required for biological activity.

Gresshoff, Roth, Stacey & Newton (eds.) Nitrogen Fixation: Achievements and Objectives
Chapman and Hall New York • London 1990

Cloning of DNA sequences conferring enhanced competitiveness in *Rhizobium leguminosarum* bv. *phaseoli*

Gwyn A. Beattie and Jo Handelsman

Dept. of Plant Pathology, University of Wisconsin, Madison, WI 53706, USA.

Most investigations into the molecular basis of nodulation competitiveness test for a role of a particular bacterial trait in superior competitiveness using mutant analysis. These studies have shown that motility, cell surface polysaccharides, bacteriocin production, and the rate of nodule induction affect the competitiveness of strains. Although this approach can elucidate factors that are involved in competitiveness, it is limited to the predictive abilities of the researchers. In this study, we directly isolated cloned genes involved in enhanced nodulation competitiveness in *R. leguminosarum* bv. *phaseoli* (Rlp) through selection by the host plant to enrich for highly competitive transconjugants.

Previously, we identified a pair of Rlp strains, KIM5s and CE3, whose relative competitiveness is similar in various environments. KIM5s is consistently 10 to 15 times more competitive than CE3, which makes these strains ideal for a genetic study of competitiveness. To identify genes that contribute to the successful competitiveness of KIM5s, libraries of the KIM5s genome were transferred into CE3 and the transconjugants were screened for enhanced competitiveness.

In order to use the plant as the selective agent for transconjugants with enhanced competitiveness, the CE3 transconjugants were cycled twice through plant nodulation by applying transconjugants in groups of 48 to bean seeds and using the bacteria in the resulting nodules as the inoculum for a second set of seeds. The competitiveness of randomly selected nodule isolates from these plants was evaluated by competing each isolate against KIM5s.

Of 4,000 independent transconjugants applied to plants, at least ten nodule isolates from the final enrichment appear more competitive than the CE3 parent. The cosmids in two, designated pOWR1 and pOWR2, contain inserts that are approximately 20.5 and 19.8 kb, respectively. The cosmid pOWR1 may also confer the ability to produce a dark pigment on *Escherichia coli*. If the pigment is melanin, pOWR1 may contain an insert from the pSym of KIM5s since melanin production is encoded on the pSym of Rlp. These cosmids should provide the means for understanding the genetic and physiological basis of nodulation competitiveness.

Gresshoff, Roth, Stacey & Newton (eds.) Nitrogen Fixation: Achievements and Objectives
Chapman and Hall New York • London 1990

Competition defective mutants of *Bradyrhizobium japonicum*

Arvind A. Bhagwat and Donald L. Keister.

Nitrogen Fixation and Soybean Laboratory, USDA-ARS, BARC-W, Beltsville, MD 20705 USA.

The superior ability of indigenous rhizobial strains to compete for nodulation with inoculated strains is one of the important agronomic problems that must be solved before new strains, enhanced for nitrogen fixation, can be utilized effectively. Competitiveness presently is identified as a strain characteristic and is only vaguely associated with serology and geographic origin of the strain.

Considerable genetic and physiological evidence suggests that rhizobial extracellular polysaccharides (EPS) play an important role in legume root nodule initiation and development. We postulated that EPS may also have a role in intra-strain competition. Tn5 mutants defective in EPS synthesis (Exo$^-$) were isolated and were screened for defects in competitive ability. We adapted and developed a competition assay (McLoughlin et al., J. Bacteriol. 169, 410-413, 1987) in which the test strain is challenged with a nodulating but Nif$^-$ strain of B. japonicum (MC 617). The assay was designed so that the ratio was 10:1 (test strain: MC 617). Four Tn5 mutants of USDA 110 were identified, which were defective in competition (Com$^-$). Each mutant had the Tn5 insertion at a unique site. When inoculated singly, the mutants formed effective nodules, indistinguishable from the wild-type. The DNA segment carrying the com gene for one of the mutations was isolated, cloned and mapped. A clone (pAAB8.0) carrying 9.6 kb wild-type DNA and its subclone (pAAB8.1), carrying a 3.6 kb EcoR1 fragment, both complimented the competition defect of the mutant. Fragment saturation mutagenesis of pAAB8.0 was performed to delimit the gene(s) responsible for the Com$^-$ phenotype. At least two com genes and two fix genes were found on pAAB8.1, and an additional com locus was found on the subclone, pAAB8.2.

Based on available restriction maps in the literature, none of the Tn5 mutations on pAAB8.0 are on the large nod gene-containing region of USDA 110 and thus represent new symbiotic loci.

Two of the Com$^-$ mutants were Exo$^+$ and represent an additional new class of mutants, whose only phenotype recognized at this point, is a defect in competitive ability.

Gresshoff, Roth, Stacey & Newton (eds.) Nitrogen Fixation: Achievements and Objectives
Chapman and Hall New York • London 1990

Genetic complementation of *Rhizobium leguminosarum* bv. *phaseoli* Tn5 mutants defective in nodulation and nodulation suppression

Maria Luz F. Caces, Francoise M. Robert, Dulal Borthakur, and B. Ben Bohlool

University of Hawaii, Honolulu, HI 96822, USA.

A plant signal induced by infecting rhizobia is postulated to regulate nodulation and inhibit subsequent nodulation. Evidence for this is seen in a split-root system where a suppression of nodulation occurs on a half root when the other half has been previously nodulated (1,2,3,4). We investigated the genetics of suppression in *Phaseolus vulgaris* (common bean) using *Rhizobium leguminosarum* bv. *phaseoli* mutants. A highly competitive and effective strain, TAL182, was mutagenized with Tn*5* and screened for symbiotic defects. Two classes of nodulation mutants and four types of auxotrophic mutants (Ade^-, Leu^-, Phe^-, Ura^-) that are blocked at different stages of symbiotic development were identified. When the mutants were used as primary inoculants in split-root experiments, they failed to induce nodulation suppression. Complementation of the nodulation defective mutants with a cosmid library of *Rhizobium* sp. (*Leucaena leucocephala*) TAL1145 restored varying symbiotic capabilities to the mutants. Class I mutants regained effective nodulation and suppression while Class II mutants regained ineffective nodulation but not suppression. Restriction analysis of the overlapping clones complementing the Class I and Class II mutants showed an 8.3 kb region and a 12 kb region in common, respectively.

REFERENCES

1. Caetano-Anolles, G. & Bauer, W.D. (1988) Planta **174**, 385-395.
2. Djordjevic, S.P., Ridge, R.W., Chen, H., Redmond, J.W., Batley, M. & Rolfe, B.G. (1988) J. Bacteriol. **170**, 1848-1857.
3. Kosslak, R.M. & Bohlool, B.B. (1984) Plant Physiol. **75**, 125-130.
4. Sargent, L., Huang, S.Z., Rolfe, B.G., & Djordjevic, M.A. (1987) App. Environ. Microbiol. **53**, 1611-1619.

Gresshoff, Roth, Stacey & Newton (eds.) Nitrogen Fixation: Achievements and Objectives
Chapman and Hall New York • London 1990

Isolation and analysis of the lipopolysaccharides from a symbiotic mutant of *Bradyrhizobium japonicum*

R.W. Carlson, B. Lakshmi, U.R. Bhat and G. Stacey.

Complex Carbohydrate Research Center, University of Georgia, Athens, GA 30602 and Dept. of Microbiology, University of Tennessee, Knoxville, TN 37996 USA.

The lipopolysaccharide (LPS) from a symbiotic mutant of *B. japonicum* 61A101C, J314, was partially characterized by PAGE and glycosyl composition analysis. The mutant forms empty nodules that have central, rather than the normal peripheral vascular tissue. The LPSs are extracted into the phenol layer during phenol/water extraction and further purified by treatment with proteinase K, RNase, DNase and gel-filtration chromatography. PAGE was performed in buffer containing deoxycholate (DOC) and glycosyl components were determined by preparing alditol acetates followed by GC/MS analysis. PAGE analysis shows that the mutants lack the higher molecular weight form of LPS, LPS I. The parent LPS is composed of 2,3-di-O-methylfucose, fucose, rhamnose, 4-O-methylmannose, mannose, galactose, glucose, quinovosamine, fucosamine and 2,3-diamino-2,3-dideoxyglucose. The major sugars in the mutant are 4-O-methylmannose, mannose and glucose in a 1/3/1 ratio. Major sugars in the parent LPS, e.g. 2,3-di-O-methylfucose, fucose, and quinovosamine, are missing in the mutant LPSs. Fatty acids identified in both parent and mutant LPSs are: C12:OH, C12, C14:OH, C14, C16, C18:1, C18, C20:1 and C24. The long-chain fatty acid, C28:27-OH is also detected.

These results indicate that the core region of this LPS consists of 4-O-methylmannose, mannose and glucose. Similar results are reported for *B. japonicum* 61A123 and USDA110 LPSs (1,2) indicating that *B. japonicum* LPSs from divergent serogroups may have the same core structures.

The EPSs from both parent and mutant are the 4-O-methylglucuronsyl/rhamnose containing EPSs and have identical glycosyl linkages and proton NMR spectra. The mutant produces approximately three times the EPS that is produced by the parent. Thus this mutant has a phenotype reminicent of the *R. meliloti* EPS$^-$ strains but is defective in LPS, not EPS.

Supported in part by NIH grant RO1 GM39583-02 and by the DOE/USDA/NSF Plant Science Centers program (project funded by DOE grant DE-FG09-87-ER13810).

1. Puvanesarajah, V., Schell, F.M., Gerhold, D. & Stacey, G. (1987) *J. Bacteriol.* **169**:137-141.

2. Carrion, M., Bhat, U.R., Reuhs, B. & Carlson, R.W. (1990) *J. Bacteriol.* **172**:1725-1731.

Gresshoff, Roth, Stacey & Newton (eds.) Nitrogen Fixation: Achievements and Objectives
Chapman and Hall New York • London 1990

Mutations that alter O-antigen structure of *R. trifolii* ANU843

K. Chapman[*+], S. Pankratz[+], S. Philip-Hollingsworth[+], F. Dazzo[+], and S. Wright[@]

+ Michigan State University, East Lansing, MI 48824
@ USDA-ARS Beckley, WV 25801

We are studying the role of lipopolysaccharide in the establishment of a nitrogen fixing rhizobium-legume symbiosis using mutants with altered LPS. Monoclonal antibodies were raised against R. trifolii ANU843 cells grown in the presence of the flavone apigenin. These antibodies were used to screen kan^r Tn5 mutagenized cells. Mutants lacking one or more cell surface epitopes were found at a frequency of 1 - 2%. These mutants were grouped into classes according to their ability to bind various monoclonal antibodies.
Rough mutants are $nod^+ fix^-$ on all clover varieties tested. Examination of clover root nodules formed by rough mutants show that some bacteria are released from the infection threads but that they usually fail to undergo normal bacteroid development (bad^-).
Several hypotheses could account for the necessity of the O-antigen: 1) it may act as a signal to the plant, 2) it may coat underlying epitopes that would irritate the plant, 3) complete LPS may act as a carrier of a signal molecule, 4) rough mutants may have a pleiotropic loss of outer membrane components that are required for bacteroid development, or 5) involved in a signalling process. Our evidence supports the idea that the O-antigen is involved in a signalling process.
Some observations support the idea that rough mutants are not intrinsically incapable of bacteroid development: 1) rough mutants can assume a bacteroid shape when grown in the presence of succinate, 2) rough bacteroids are rare but they do occur, 3) some rare individual Red Clover plants may tolerate the rough mutants.
One mutant has a reduced ratio of complete vs. rough LPS, but shows no change in nodulation kinetics or nitrogen fixation.
Some mutations eliminate some but not all O-antigen epitopes. These mutants produce chemically distinct LPSs but show no change in nodulation kinetics or nitrogen fixation on Dutch White clover.

Gresshoff, Roth, Stacey & Newton (eds.) Nitrogen Fixation: Achievements and Objectives
Chapman and Hall New York • London 1990

Effect of various N sources on isoflavonoid concentration in wild-type and nodulation-mutant soybean roots

Myeong-Je Cho and James E. Harper

Dept. of Agronomy and USDA/Agricultural Research Service, 1102 S. Goodwin, Univ. of Illinois, Urbana, IL 61801, USA.

The isoflavones, daidzein and genistein, have been isolated and identified as the major inducers of *nod* genes of *Bradyrhizobium japonicum* (3). In addition, coumestrol and daidzein have also been shown to promote the growth of *B. japonicum* (1). It is well known that soil N (primarily NO_3^-) inhibits nodulation and N_2 fixation (2), and Vigue *et al.* (4) showed that NO_3^- is more inhibitory to soybean nodulation than is urea. This study evaluated the effect of various N sources on isoflavonoid concentration in soybean root extracts from selected hypernodulating mutants (NOD1-3, NOD2-4, and NOD3-7) and a nonnodulating mutant (NN5), compared to the Williams wild type.
Soybean seeds were germinated in sterilized sand trays in a growth chamber and 6-d-old seedlings were inoculated with *B. japonicum* (USDA 110) and transplanted to 18-L trays containing a modified Hoagland nutrient solution. Applied N sources were 0, 1.5, and 5.0 mM $NaNO_3$, 2.5 mM urea, and 2.5 mM $(NH_4)_2SO_4$. At 12 d after inoculation and N treatment, 5 g of fresh roots were harvested and prepared for HPLC analysis.
HPLC analyses of soybean root extracts showed that all three hypernodulating mutants and NN5 had markedly higher isoflavonoid concentrations than did the Williams control when grown at O mM N level. N application decreased the concentration of all three isoflavonoid compounds in all soybean lines. Application of urea and $(NH_4)_2SO_4$ resulted in similar decreases in isoflavonoid concentration; resulting in 56 to 86% daidzein, 25 to 67% genistein, and 42 to 90% coumestrol, relative to respective controls without N. Application of NO_3^- was more inhibitory to isoflavonoid concentrations; resulting in 37 to 55% daidzein, 12 to 32% genistein, and 10 to 24% coumestrol, relative to respective controls without N. The degree of inhibition by NO_3^- was concentration dependent. Although these results are consistent with a conclusion that differential NO_3^- inhibition of nodulation may be partially due to changes in isoflavonoid levels, the similar response noted for the NN5 line brings this conclusion into question. On the other hand, nodulation control in the NN5 mutant may be due to factors totally unrelated to isoflavonoids, leaving open the possibility that isoflavonoids play a role in NO_3^- inhibition of nodulation of lines genetically competent to nodulate.

1. d'Arcy Lameta, A. (1987) Plant Soil **101**, 267-272.
2. Harper, J.E. (1987) in *Soybeans: Improvement, Production, and Uses*, eds. Wilcox, J.R. (Amer. Soc. Agron., Madison, WI), pp. 497-533.
3. Kosslak, R.M., Bookland, R., Barkei, J., Paaren, H.E. & Appelbaum, E.R. (1987) Proc. Natl. Acad. Sci. USA **84**, 7428-7432.
4. Vigue, J.T., Harper, J.E., Hageman, R.H. & Peters, D.B. (1977) Crop Sci. **17**,169-172.

Gresshoff, Roth, Stacey & Newton (eds.) Nitrogen Fixation: Achievements and Objectives
Chapman and Hall New York • London 1990

NodF protein carries a 4'-phosphopantetheine

Otto Geiger, Herman P. Spaink, and Eugene P. Kennedy

Department of Biological Chemistry and Molecular Pharmacology, Harvard Medical School, Boston, MA 02115, USA.

In addition to the constitutive acyl carrier protein (ACP), which is required for synthesis of essential cell lipids, recently a flavonoid-inducible ACP-like gene product with a predicted molecular weight of 9946, NodF, has been recognized in Rhizobium. NodF together with NodE participate in generating a host-specific response to the plant host (Spaink et al. (1989). EMBO J. 8, 2811-2818).

As a part of an on-going study of the diverse function of bacterial acyl carrier proteins, the constitutive ACP of Rhizobium meliloti has been isolated and characterized in this laboratory (Platt et al., J. Bacteriol., submitted). We have now carried out a similar study with the NodF ACP of Rhizobium leguminosarum. The nodFE region of R. leguminosarum has been cloned into a multicopy plasmid. A flavonoid-inducible band is effectively labeled with radioactive β-alanine. The flavonoid-induced radioactivity was purified and moves as a single protein band corresponding to a molecular weight of 5 kD on an SDS polyacrylamide gel electrophoresis. Separation of the purified radioactivity in a native polyacrylamide gel electrophoresis displayed two protein bands which migrated much slower than even the constitutive ACP from Escherichia coli. The N-terminal sequences of both bands are as expected for NodF. Pulse-chase experiments with labeled β-alanine indicate that the slower moving band is the precursor of the faster moving one. Treatment of NodF with mild alkali released labeled 4'-phosphopantetheine by β-elimination.

We also have over-expressed a NodF fusion protein under the control of the lac promoter in E. coli. Again NodF is effectively labeled with β-alanine indicating that ACP synthase from E. coli recognizes apo-NodF as a substrate. E. coli ACP phosphodiesterase, however, an enzyme known to hydrolyze 4'-phosphopantetheine from the constitutive ACP in E. coli, does not release any phosphopantetheine from NodF.

(Supported by grants GM19822 and GM22057).

Gresshoff, Roth, Stacey & Newton (eds.) Nitrogen Fixation: Achievements and Objectives
Chapman and Hall New York • London 1990

Antigenic changes in lipopolysaccharide (LPS I) and an outer membrane protein of *Rhizobium leguminosarum* bv. *viciae* strain 248 during the differentiation of bacteria into bacteroids in nodules of *Vicia sativa* ssp. *nigra*

Leentje Goosen-de Roo, Ruud A. de Maagd and Ben J.J. Lugtenberg.

Department of Plant Molecular Biology, Leiden University, Nonnensteeg 3, 2311 VJ Leiden, The Netherlands

Immunochemistry showed that cell surface antigens of R. leguminosarum bv. viciae strain 248 decreased in quantity during differentiation from the free-living state to the mature bacteroid state in nodules (1). Immunoelectronmicroscopy was applied to follow the fate of two of these antigens in detail.

Monoclonal antibodies, MAb 16 which recognizes LPS I (presumably O-antigen containing) and MAb 38 which recognizes group III outer membrane protein antigens (1), were used. For detection of LPS I antigen, 3 or 4-week old nodules of Vicia sativa ssp. nigra were fixed in 1% glutaraldehyde-2% paraformaldehyde-0.1 M sodium cacodylate, pH 7.2, post-fixed in 1% osmiumtetroxide, dehydrated, infiltrated with LR White acrylic resin at room temperature, and polymerized at 50°C for 16 h. For detection of MAb 38 antigen, nodules were fixed without post-fixation at 4°C, dehydrated, infiltrated at -20°C in LR White and polymerized with UV light. Immunogold labeling of the sections was quantified from electron micrographs. Antigen density was estimated as the number of gold particles per um^2 of Rhizobium section surface.

The electron micrographs showed the LPS I antigen only on the outer membrane (OM) of bacteria in infection threads and in infection droplets, i.e., sites of endocytosis where the cell wall of the thread is absent. The MAb 38 antigen was found on the OM of Rhizobium in all differentiation stages. However, antigen density estimates show that MAb 38 antigen density is high on bacteria, less on young bacteroids and low on mature bacteroids. The LPS I antigen decreased abruptly during endocytosis. The MAb 38 antigen diminished gradually after endocytosis during bacteroid differentiation. It is postulated that synthesis of both antigens may stop during endocytosis, upon reception of a signal, presumably in the infection droplet. Subsequently, the LPS I antigen is rapidly and actively modified or degraded, and the MAb 38 antigen remains at the OM and is diluted due to growth of the bacteroids.

1. De Maagd, R.A., De Rijk, R., Mulder, I.H.M. & Lugtenberg, B.J.J. (1989) J. Bacteriol. **171**, 1136-1142.

Gresshoff, Roth, Stacey & Newton (eds.) Nitrogen Fixation: Achievements and Objectives
Chapman and Hall New York • London 1990

Release and modification of flavonoid nodulation factors by alfalfa seeds

U.A. Hartwig[1], C.M. Joseph, and D.A. Phillips

Dept. Agronomy & Range Science, Univ. California, Davis, CA 95616, USA. [1]Present address: Inst. of Plant Sciences, Federal Institute of Technology, Zurich, Switzerland.

Five rhizobial *nod* gene inducers are released by alfalfa (*Medicago sativa* L.) seeds (1, 2) and roots (3). Roots concurrently synthesize and release *nod*-gene inducers (4), but seeds probably release stored *nod*-gene inducers. This conclusion is suggested by the fact that flavonoids from isolated seed coats and intact seeds show the same HPLC patterns and that most flavonoids are washed from seeds within a few hours. Luteolin-7-O-glucoside from seeds can be hydrolyzed to luteolin by β-glucosidase activity in sterile seed exudate and by *R. meliloti* 1021. Luteolin, luteolin-7-O-glucoside, and quercetin-3-O-galactoside from alfalfa seeds, greatly stimulate growth rate of *R. meliloti* on a defined minimal medium (e.g. doubling time decreases from 70 to 15 h). The process occurs on both mannitol and glucose, it does not require *nodD* genes, and it is associated primarily with the 5,7-dihydroxyl substitution pattern present in seed, but not root, flavonoids. These facts suggest that flavonoids released by seeds and roots are modified in the soil to influence *nod* gene induction (1, 2, 3), as well as rhizobial chemotaxis (5) and growth. Interactions between seed and root flavonoids at the top of the primary root may create a region of intense *nod* gene induction (6, 7).

1. Peters, N.K., Frost, J.W. & Long, S.R. (1986) Science **233**, 977-980.
2. Hartwig, U.A., Maxwell, C.A., Joseph, C.M. & Phillips, D.A. (1990) Plant Physiol. **92**, 116-122.
3. Maxwell, C.A., Hartwig, U.A., Joseph, C.M. & Phillips, D.A. (1989) Plant Physiol. **91**, 842-847.
4. Maxwell, C.A. & Phillips, D.A. (1990) Plant Physiol. In press.
5. Caetano-Anolles, G., Crist-Estes, D.K. & Bauer, W.D. (1988) J. Bacteriol. **170**, 3164-3169.
6. Hartwig, U.A., Maxwell, C.A., Joseph, C.M. & Phillips, D.A. (1989) Plant Physiol. **91**, 1138-1142.
7. Hartwig, U.A., Maxwell, C.A., Joseph, C.M. & Phillips, D.A. (1990) J. Bacteriol. **172**, In press.

Gresshoff, Roth, Stacey & Newton (eds.) Nitrogen Fixation: Achievements and Objectives
Chapman and Hall New York • London 1990

Biological characterization of seed and root exudates from nodulation mutant lines of chickpea (*Cicer arietinum*)

J. E. Heard and T. M. Davis

Dept. of Plant Biology, University of New Hampshire, Durham, NH 03824, USA.

Tsr and β-galactosidase bioassays were used to compare rhizobial nod gene induction activities of plant exudates from wild type and seven non-nodulating or ineffectively nodulating mutant chickpea lines.

A β-galactosidase assay (1,2) was used to measure nodABC-lacZ induction on pRmM57 in a heterologous symbiont, Rhizobium meliloti strain 1021 (Sharon Long, Standford Univ.), in response to chickpea seed exudate. No nod-lacZ constructs are available in chickpea Rhizobium. Exudate prepared by soaking seeds overnight in 10X volume distilled water induced nodC-lacZ expression at approximately 20% of the level induced by similarly prepared alfalfa seed exudate. No significant variation was detected between chickpea lines in their ability to induce nodC expression.

Support for this finding was provided in a bioassay developed to monitor nod gene activity through a "thick short root" (Tsr) phenotypic response. The Tsr phenotype has been characterized as a nodABCD-dependent host legume response in Vicia sativa (3). Three-day-old chickpea seedlings with 2-3 cm long primary roots showed root thickening but little to no root elongation after four days incubation in N-free mineral solution containing filtrate from a culture of R. ciceri strain CC1192 grown in presence of wild type or mutant chickpea root exudates (prepared by soaking sterile seedlings in mineral solution for 30 hrs). Control seedlings incubated with either filter sterilized R. ciceri culture or uninoculated culture media resulted in mean primary root elongation of 6.6 cm and 8.0 cm, respectively. Seed exudates of mutant and wild type chickpeas did not differ in their ability to cause the Tsr phenotype by inducing presumed nod gene expression in their microsymbiont, R. ciceri.

1) Miller, J.H. (1972) Experiments in Molecular Biology (Cold Spring Harbor Laboratory, Cold Spring Harbor, NY).
2) Mulligan, J.T. & Long, S.R. (1983) Proc. Natl. Acad. Sci. USA 82, 6609-6613.
3) Zaat, S.A.J., Van Brussel, A.A.N., Tak, T., Lugtenberg, B.J.J., Kijne, J.W. (1989) Planta 177, 141-150.

Gresshoff, Roth, Stacey & Newton (eds.) Nitrogen Fixation: Achievements and Objectives
Chapman and Hall New York • London 1990

Had and Noi signals from *Rhizobium trifolii*

Rawle I. Hollingsworth[1], Andrea Squartini[2], Saleela Philip-Hollingsworth[2], and Frank B. Dazzo[2]

[1]Departments of Biochemistry and Chemistry, and [2]Department of Microbiology, Michigan State University, East Lansing, MI 48824 USA.

Introduction -- We report two excreted metabolites (BF-3 and BF-4) of *R. trifolii* ANU843 which elicit root hair deformation (Had) and initiate nodule-like primordia (Noi) in white clover roots.

Methods -- Had (1) and Noi (2) are scored microscopically in an axenic seedling bioassay. Enhancement of infection is scored in inoculated roots preincubated with metabolites. Sterile -N Fahraeus medium is evaluated as a control in all bioassays.

Results -- BF-3 and BF-4 were purified from the culture supernatant by HPLC, TLC, reverse phase and gel filtration chromatography. Flavone *nod* inducers significantly increased BF-3 & BF-4 accumulation. BF-3 elicits an exaggerated Had response on white clover root hairs (begins 9 h, near maximum at 21 h) and induces foci of cortical cell divisions which develop into nodule-like primordia in axenic seedling roots. Laser scanning confocal microscopy indicates that very young cortical meristems consist of small, dividing cells which form a multilobed mound, suggesting that several (rather than one) cortical cells are triggered to divide in response to *Rhizobium* Noi signals during nodule initiation. 2-D PAGE indicates that at least 12 proteins change significantly in root meristems treated with BF-3: 6 appear and 6 others disappear. BF-4 is a weak Had and Noi signal on white clover. BF-4 synergistically enhances Noi signalling by BF-3. Also, BF-3 enhances root hair infection and nodulation of white clover by wt *R. trifolii* ANU843 and, most importantly, enables a *nodA*::*Tn5* mutant derivative (*R.t.* ANU252 from B. Rolfe, ANU) to produce hyaline "bright spots" on deformed root hairs as an early step of infection thread formation. Neither BF-3 nor BF-4 (alone or together) elicited Had or Noi responses on Empire Birdsfoot Trefoil or Vernal Alfalfa under the same experimental conditions.

1. Hollingsworth, R., Squartini, A., Philip-Hollingsworth, S. & Dazzo, F. (1989) in *Signal Molecules in Plants and Plant-Microbe Interactions*, ed. Lugtenberg, B. (Springer-Verlag, Berlin), pp. 387-393.
2. Truchet, G., Camu, S., deBilly, F., Odorico, R., and Vasse, J. (1989) Protoplasma 149, 82-88.

Gresshoff, Roth, Stacey & Newton (eds.) Nitrogen Fixation: Achievements and Objectives
Chapman and Hall New York • London 1990

The effect of a seed color mutation on *nod*-gene-inducing flavonoids and nodulation in *Phaseolus vulgaris* L.

M. Hungria, C.M. Joseph and D.A. Phillips

Department of Agronomy & Range Science, University of California, Davis, CA 95616 USA.

Common bean (*Phaseolus vulgaris* L.) line PI165426 has black seed coats rich in anthocyanin, an end product of flavonoid metabolism. Recent tests for resistance to fungal pathogens showed that an isogenic white-seeded mutant derived from PI165426 was more strongly affected by pathogens (1), and it was suggested that differences in flavonoid metabolism between the two lines altered plant-microbe interactions. To test whether differences in flavonoids between these bean lines affect interactions with *Rhizobium leguminosarum bv. phaseoli* CIAT 899, we examined their patterns of root nodulation and exudation of *nod*-gene-inducing flavonoids. Seedlings from the two lines had equal shoot and root dry weight and numbers of primary and lateral roots. There were, however, nearly twice as many root nodules on the black-seeded line as on the white-seeded mutant 14 days after planting. Rinsing black seeds for 24 h or removing their seed coats decreased the number of primary root nodules to that of white-seeded plants. Tests with *R.l.* bv. *phaseoli* containing either a *nodA-lacZ* fusion (supplied by Drs. B. Lugtenberg and R. Okker) or a *nodC-lacZ* fusion (from Drs. F. Sanchez and A. Penas) showed that total *nod*-gene-inducing activities of seed and root exudates from the black-seeded line were as much as 10-fold higher than corresponding exudates from white-seeded mutants. HPLC separations of seed exudate flavonoids show that many different *nod* gene inducers are exuded from both plant lines and suggest that two active compounds from the black seeds are anthocyanins with UV/visible absorbance maxima of 282 & 531 and 276 & 526 nm.

1. Dickson, M.H. & Petzoldt, R. (1988) J. Amer. Soc. Hort. Sci. 113, 111-114.

Gresshoff, Roth, Stacey & Newton (eds.) Nitrogen Fixation: Achievements and Objectives
Chapman and Hall New York • London 1990

The role of plasmids in competition between strains of *Rhizobium leguminosarum*

Michael F. Hynes

Agriculture Canada, Lethbridge, AB, Canada T1J 4B1.

The effect of various plasmids in Rhizobium leguminosarum strain VF39 on the strain's ability to compete with other R. leguminosarum strains was examined using derivatives of VF39 cured of each of its six plasmids (Hynes and McGregor, 1990). Preliminary studies showed that the two smallest plasmids, which code for melanin production and for the production of a restriction enzyme, do not affect competition significantly. The four largest plasmids all had a significant effect on competition. Since three of these, pRleVF39c, pRleVF39d, and pRleVF39e, carry genes necessary for symbiosis (Hynes and McGregor, 1990), it was not surprising that they affect competitiveness. Plasmid pRleVF39f, a 600 kb plasmid which is not necessary for symbiosis, was required for competitive nodulation by VF39 on both peas and lentils. This was true both under sterile conditions and in soil. The loss of competitiveness due to the absence of pRleVF39f may be due to nutritional requirements. Strains cured of pRleVF39f grow more slowly than the wild-type (generation time 1.6 times as long), have a leaky auxotrophic phenotype, and are unable to grow on nitrate as a sole nitrogen source. Nutrient depletion studies using plant exudates as a medium for growth showed that strains cured of pRleVF39f were unable to utilise plant exudates to the same extent as the wild-type.

Hynes M.F. & McGregor N.F. (1990) Molec. Microbiol. **4**, 567-574.

Gresshoff, Roth, Stacey & Newton (eds.) Nitrogen Fixation: Achievements and Objectives
Chapman and Hall New York • London 1990

The effect of alfalfa *nod*-gene-inducing flavonoids on *nodABC* transcription in *Rhizobium meliloti* strains containing different *nodD* genes

C.M. Joseph, U.A. Hartwig, C.A. Maxwell and D.A. Phillips

Dept. Agronomy & Range Science, University of California, Davis, CA 95616 USA.

Rhizobial *nodABC* transcription requires a plant inducer and active *nodD* gene product. Three *nodD* genes, *D1*, *D2* and *D3*, with differing specificities for inducing signals from plants are present in *R. meliloti*, and aqueous seed rinses of alfalfa (*Medicago sativa* L.) contain factors that activate each of their gene products (1). Whether compounds activate various *nodD* gene products can be assessed by measuring *nodABC-lacZ* transcription in *R. meliloti* containing extra copies of different *nodD* genes. All *nod* gene inducers purified from alfalfa have been identified by their capacity to activate NodD1 (2,4,5). One of those molecules, 4,4'-dihydroxy-2'-methoxychalcone, which is exuded from roots but not seeds, also activates NodD2 (3). Mixtures of the seed flavonoid luteolin and the root chalcone give a synergistic promotion of *nodABC* transcription in *R. meliloti* with extra *nodD1* copies, but luteolin inhibits the chalcone inducing activity with extra *nodD2* copies. The latter fact suggests that luteolin and the chalcone bind competitively to NodD2. We find one NodD2 activator in alfalfa seed exudate which does not activate NodD1. It does not bind strongly to C_{18} resin in water, and its UV/visible spectrum in methanol shows 204 and 265 nm maxima with a shoulder at 272 nm. All data indicate this NodD2 activator is much more water soluble than any *nod*-gene-inducing flavonoid identified thus far from alfalfa seed or root exudates.

1. Gyorgypal, Z., Iyar, N. & Kondorosi, A. (1988) Mol. Gen. Genet. **212**, 85-92.
2. Hartwig, U.A., Maxwell, C.A., Joseph, C.M. & Phillips, D.A. (1990) Plant Physiol. **92**, 116-122.
3. Hartwig, U.A., Maxwell, C.A., Joseph, C.M. & Phillips, D.A. (1990) J. Bacteriol. **172**, 2769-2773.
4. Maxwell, C.A., Hartwig, U.A., Joseph, C.M. & Phillips, D.A. (1989) Plant Physiol. **91**, 842-847.
5. Peters, N.K., Frost, J.W. & Long, S.R. (1986) Science **233**, 977-980.

Gresshoff, Roth, Stacey & Newton (eds.) Nitrogen Fixation: Achievements and Objectives
Chapman and Hall New York • London 1990

Chemotaxis and *nod* gene activity of *Bradyrhizobium japonicum* in response to hydroxycinnamic acids and isoflavonoids

Rüdiger Kape and Dietrich Werner

Fachbereich Biologie, Philipps-Universität Marburg, Karl-von-Frisch-Str., D-3550 Marburg, FRG.

The importance of rhizobial chemotaxis for the *Rhizobium*-legume symbiosis has been a point of interest for quite a while. Recent work focused in particular on the chemotactic effect of *nod* gene inducing isoflavonoids. For fast growing rhizobia,a more or less close correlation between *nod* gene inducing and chemotactic effects of flavonoids was found (1, 2, 3). The purpose of this work was to investigate whether such a correlation also exists for *B. japonicum*. Besides *nod* gene inducing isoflavonoids, also hydroxycinnamic acids were analysed.
The chemotactic activity of *B. japonicum* 110*spc*4 (wild type) was investigated according to Palleroni (5). The *nod* gene induction in *B. japonicum* 110*spc*4 harboring plasmid pRJ458 (*nod YABC-lacZ*) was determined according to Miller (4).
None of the tested isoflavonoids (genistein, daidzein, and coumestrol) attracted *B. japonicum*. All tested hydroxycinnamic acids (cinnamic acid, coumaric acid, caffeic acid, ferulic acid, and sinapic acid) were strong chemoattractants. Their derivatives (coniferyl alcohol and chlorogenic acid) were chemotactically inactive.
Among the hydroxycinnamic acids/derivatives,weak *nod* gene inducers were identified,in particular,ferulic acid, coniferyl alcohol, and chlorogenic acid.
No correlation between *nod* gene inducing and chemotactic effects similiar to the one described for fast growing rhizobia was found in *B. japonicum*. Chemotaxis towards hydroxycinnamic acids is probably not correlated with their weak *nod* gene inducing effect but may be a result of the importance of phenolic compounds as nutrients for *B. japonicum*.

(1) Aguilar, J. M. M., Ashby A. M., Richards A. J. M. , Loake G. J., Watson M. D. & Shaw H.(1988) J. Gen. Microbiol. **134**, 2741-2746.
(2) Armitage J. P., Gallagher A., & Johnston A. W. B. (1988) Mol. Microbiol. **2**, 743-748.
(3) Caetano-Anollés G., Crist-Estes D. K., & Bauer W. D. (1988) J. Bacteriol. **170**: 3164-3169.
(4) Miller J. H. (1972) *Experiments in Molecular Genetics* (Cold Spring Harbor Laboratory, Cold Spring Harbor, NY).
(5) Palleroni N. J. (1976) Appl. Environ. Microbiol. **32**: 729 - 730.

Gresshoff, Roth, Stacey & Newton (eds.) Nitrogen Fixation: Achievements and Objectives
Chapman and Hall New York • London 1990

Rhizobium meliloti lipopolysaccharide (LPS) saccharidic moiety involved in early symbiotic recognition by alfalfa

A. Lagares and G. Favelukes.

Química Biológica I, Facultad de Ciencias Exactas, Universidad Nacional de La Plata, 1900-La Plata, Argentina

Crude preparations of LPS from *R. meliloti* (*Rme*), but not from heterologous rhizobia, have been previously shown to inhibit specific adsorption of *Rme* to alfalfa roots (1). Chromatography of crude LPS from *Rme* L5-30 (late-exponential phase) on Sepharose CL-4B/EDTA (10 mM)-TEA, pH 7.0, has now separated three components of which only the mid-one eluting as a partially included band (Kd = 0,47), was actively inhibitory in the *Rme* specific root adsorption assay, and had properties expected of *Rme* LPS proper: a high ketodeoxyoctonate:hexose content ratio, 0.37 (w/w), and marked hydrophobicity as detected with the optical probe merocyanin 540 (suggesting the presence of a lipid A moiety). After mild acid hydrolysis (1% acetic acid or 0.1M sulfuric acid, 100 °C, 30 min) of crude LPS from *Rme* L5-30 or 1021 which eliminated the lipid A moiety, the inhibitory activity was preserved and even increased in the remaining soluble product(s), suggesting an active role for the saccharidic region of the original LPS.

Double-diffusion and ELISA assays have revealed strong and extensive cross-immunoreactivity among LPS from different *Rme* strains (L5-30, 1021, 1220, DMG177, AP8, and AK631), but excluded LPS from heterologous rhizobia (*R. trifolii* TA1 and 0403, *R phaseoli* CIAT 144 and CO5, *B. japonicum* USDA 110). When LPS of *Rme* 1021 was selectively removed from a crude preparation by using affinity chromatography (Detoxi-Gel), the remainder had lost the reactivity with the homologous antiserum. These results are independent evidence of the ability of LPS as a common marker of the *Rme* cross-inoculation group.

Our studies support the hypothesis that *Rme* LPS saccharidic chains carry determinants for early symbiotic recognition by alfalfa, which are involved in the process of specific root adsorption during preinfection.
--Supported by CONICET (Argentina) and by SAREC (Sweden).
(1) Lagares,A. & Favelukes,G.(1988) in *Nitrogen Fixation:Hundred Years After*, eds. Bothe,H., de Bruijn,F.J. &Newton,W.E. (Gustav Fischer, Stuttgart & New York), p. 477.

Gresshoff, Roth, Stacey & Newton (eds.) Nitrogen Fixation: Achievements and Objectives
Chapman and Hall New York • London 1990

The AICAR requirement for infection of legumes by *Rhizobium*

Jeffrey D. Newman, Bruce W. Shultz, K. Dale Noel

Marquette University, Milwaukee, WI 53233 USA

Introduction - Purine auxotrophs of Rhizobium leguminosarum bv. phaseoli CFN42 elicit pseudonodules devoid of bacteria and scattered throughout the bean root system. It has previously been reported that supplementation of the root medium with the riboside of a purine precursor, 5-aminoimidazole-4-carboxamide riboside (AICAR), but not inosine, allows infection and enhances nodule development. The resulting nodules are clustered near the top of the root and are the same size as those elicited by the wild type (1). We report here the concentration and temporal dependence of this effect, the ability of inosine to promote infection, and some properties of a new class of purine auxotrophs.

Methods - AICAR or inosine supplements were added to both the plant growth medium and the watering solutions at concentrations ranging from 1 μM - 5 mM. The temporal requirement for AICAR was determined by delaying AICAR addition to mutant inoculated plants and analyzing nodule distribution.

Results and Discussion - Contrary to previous findings (1), inosine has been found to rescue infection, although at a concentration 10-fold higher than the effective concentration of AICAR (0.1 mM). AICAR and inosine served equally well as purine sources ex planta. The higher requirement for inosine is consistent with the hypothesis that AICAR directly promotes infection, whereas inosine acts indirectly by leading to AICAR production via the histidine pathway.

When AICAR addition was delayed for various times following inoculation with the mutant, infection occured in areas of the root developing at the time of AICAR addition, rather than areas developing at the time of inoculation. Nodule distribution was altered after a delay of only 1 day. Delayed AICAR addition to the mutant affected nodule distribution similarly to delayed inoculation with the wild type. These results suggest that AICAR is required within 24 hours of nodule initiation and that to be successful, infection must occur very soon after nodule initiation.

A class of mutants defective in de novo purine biosynthesis both before and after AICAR production has been isolated. Despite their inabilty to use AICAR as a purine source, AICAR is able to promote infection in these mutants. This supports our hypothesis that AICAR plays some role in infection thread development, other than as an intermediate in purine biosynthesis.

References

1. Noel,KD.,Diebold,JR.,Cava,JR.,Brink,BA.(1988)Arch.Microbiol.149:499

Gresshoff, Roth, Stacey & Newton (eds.) Nitrogen Fixation: Achievements and Objectives
Chapman and Hall New York • London 1990

Exopolysaccharide-deficient mutants of *Rhizobium* sp. CIAT899 induce chlorosis in common bean (*Phaseolus vulgaris* L.)

Kevin P. O'Connell, Ricardo S. Araujo, and Jo Handelsman

Dept. of Plant Pathology, University of Wisconsin-Madison, WI 53706, USA.

Expolysaccharide (EPS) production appears not to be required for infection of common bean by strains of Rhizobium leguminosarum bv. phaseoli, since EPS-deficient mutants nodulate and fix nitrogen normally. However, EPS may play a role in nodulation competitiveness. To test this hypothesis, EPS-deficient mutants of Rhizobium strain CIAT899 were induced with Tn5 and with ethyl methane sulfonate.

The EPS-deficient mutants produced non-mucoid colonies on solid media and released 8.1-41.4% of the amount of EPS produced by CIAT899 in broth culture. All EPS-deficient mutants of CIAT899 tested induced an interveinal chlorosis in the first trifoliolate leaf of bean plants, whereas CIAT899 caused slight chlorosis on a small proportion of inoculated plants. Severely chlorotic leaflets were often elongated and displayed stunting and witch's broom-like symptoms. Chlorosis was not reversible by spotting solutions containing iron or Hoagland's micronutrients on leaflets. Two other bean-nodulating strains, R. l. bv. phaseoli CE3 and KIM5s, did not induce chlorosis, nor did EPS-deficient mutants derived from these two strains. CIAT899 and its EPS-deficient mutants all induced pink, effective nodules, indicating that chlorosis was not due to N-deficiency resulting from an ineffective symbiosis. One EPS-deficient mutant, CT9005, when applied to Leucaena leucocephala, induced foliar chlorosis but still nodulated that host. Cultures of chlorosis-inducing strains lost the ability to induce chlorosis in bean plants upon dilution, while retaining sufficient cells to nodulate abundantly, suggesting that nodulation is not sufficient for chlorosis induction. These data suggest that EPS-deficient mutants of CIAT899, and CIAT899 to a lesser extent, produce a chlorosis-inducing factor in broth culture.

Our results also suggest a relationship between the loss of EPS production and acquisition of the chlorosis-inducing phenotype. LaFavre et al. (Can. J. Microbiol. **34**, 1017-1022) described a similar phenomenon. Capsule-deficient variants of Bradyrhizobium japonicum strain USDA76 induce chlorosis on soybeans, whereas capsule-producing variants do not. The phytotoxin rhizobitoxine is responsible for chlorosis induction by USDA76. It is possible that the EPS-deficient mutants of CIAT899 produce a similar substance that induces chlorosis in beans.

Gresshoff, Roth, Stacey & Newton (eds.) Nitrogen Fixation: Achievements and Objectives
Chapman and Hall New York • London 1990

Isoflavonoid-inducible resistance to the phytoalexin glyceollin in soybean rhizobia

M. Parniske, B. Ahlborn and D. Werner

FB Biologie/Botanik; Philipps-Universität; Karl-von-Frisch-Str.; D-3550 Marburg, FRG

The toxic isoflavanoid glyceollin is produced when soybeans are challenged by environmental stress, such as UV-light, heavy metals or the presence of fungi (1). We have to conclude that soybean rhizobia have to cope with the presence of this compound in the rhizosphere, because, even the compatible symbiotic interaction leads to increased levels of glyceollin accumulation during the first 20h. Furthermore, glyceollin accumulation may take place in nodules of incompatible *Glycine/Bradyrhizobium* combinations (2). We, therefore, studied the consequences of glyceollin accumulation for the microsymbiont. Growth of *B. japonicum* exhibited a concentration-dependent lag period. This lag period was followed by logarithmic growth, indicating that *B. japonicum* became resistant during incubation in glyceollin-containing medium. Glyceollin resistance of *B. japonicum* is not due to degradation. The isoflavonoids genistein and daidzein, previously detected in the rhizosphere of *G. max*,are able to induce glyceollin resistance in *B. japonicum.* The flavone luteolin had no such effect. Glyceollin resistance is also inducible by isoflavonoids in a *common-nod*-deletion mutant, indicating another recognition site for these signal substances apart from the known *nodD* gene products. Genistein-inducible resistance to glyceollin was also detected in *R. fredii*, suggesting that this phenotype is a unique feature of soybean rhizobia. A marked loss in living cell counts was observed after exposure of *B. japonicum* to phytoalexin-producing soybean roots, indicating that phytoalexin concentrations in the rhizosphere may reach relevant levels. Resistant cells were isolated at similar rates from healthy and from phytoalexin-producing roots. Glyceollin resistance is, therefore, a considerable advantage for survival of the microsymbiont in the rhizosphere of its host plant.

1. Ebel & Grisebach (1988) TIBS **13**, 23-27.

2. Parniske et al. (1990) Botanica Acta **103**, 143-148.

Gresshoff, Roth, Stacey & Newton (eds.) Nitrogen Fixation: Achievements and Objectives
Chapman and Hall New York • London 1990

Nodule initiation and function influenced by VA-mycorrhiza infection

N.A. Patterson, S. Weininger, Y. Friedman and Y. Kapulnik.

Legume Inoculation Dept. ARO, Volcani Center, Bet-Dagan, Israel.

There is little information on the effect of mycorrhizal establishment on the control of nodulation in legumes. The purpose of this study was to demonstrate the effect of mycorrhizal colonization on nodule initiation and function in alfalfa and trifolium.

Experiments were conducted under controlled growth conditions in pots containing sandy soil mixed with low levels of superphosphate (15 mg/kg soil). Mycorrhizal infection was delayed by introducing three depths of inocolum application (4, 10 and 16 cm). The extent of mycorrhizal infection was determined colorimetrically by the glucosamine assay (1). Rhizosphere bacteria counts were determined by MPN using plant infection technique and the indirect technique of plating. Nodulation gene induction was assayed as ß-galactosidase activity of nodABC-lacZ fusion on plasmid pRmM57 in R meliloti strain 1021 (1) exposed to alfalfa root flavonoids extracts.

Fewer and bigger nodules were formed on mycorrhizal compared with non-mycorrhizal plant roots. Althouh fewer, these nodules had higher dry weight, greater specific nitrogenase activity (ARA) and were specifically patterned on mycorrhizal root system. Number of nodules confined to the aerial part of mycorrhizal compared with non-mycorrhizal root system did not differ significantly, however, number of nodules declined significantly (15%) on the subterranean portion of the mycorrhizal root system. Bacteria number from rhizosphere of different root portions did not differ significantly (10^5 c.f.u./g soil) after 4 and 8 weeks of plant growth. Studies with nodABC-lacZ fusion mutants (2) indicated that 60% of nod gene induction was reduced by mycorrhizal root extractions as compared with non-mycorrhizal plants. This suggests that the establishment of the VAM fungus might have interferred with the production of nod gene inducing compounds in the root thus controlling nodule inititiation on the lower part of the root system.

REFERENCES

1. Hepper, C.M. (1976). Soil Biol. Biochem **9**, 15-18.
2. Peters, N.K., Frost, J.W., and Long, S.R. (1986). Science 233, 977-980

Gresshoff, Roth, Stacey & Newton (eds.) Nitrogen Fixation: Achievements and Objectives
Chapman and Hall New York • London 1990

pSym *nodI* function downregulates 3-hydroxybutyrate substitution in the acidic heteropolysaccharide of *Rhizobium trifolii* ANU843

Saleela Philip-Hollingsworth, Lucette Cargill, and Frank Dazzo
Department of Microbiology, Michigan State University
East Lansing, MI48824

We used a biochemical and dual genetic approach (gain or loss of function) to evaluate the contribution of certain pSym nod gene products in modifying the acidic capsular heteropolysaccharide (CPS) of Rhizobium trifolii. The CPS of wild-type R. trifolii ANU843 was compared to various recombinant strains and Tn5 insertion mutants altered in certain pSym nod genes. Cells were grown with, and in some cases, without 4'-7 dihydroxyflavone to control the level of pSym nod gene expression in defined BIII medium. Quantitative NMR studies of enzyme-cleaved octasaccharide units of CPS showed that they were substituted with less 3-hydroxybutyrate (3-HB) in the wild-type and recombinant strains when grown to express pSym nodI. In contrast, the CPS oligomers contained more 3-HB substitution when produced by recombinant strains lacking nodI or the nodI::Tn5 mutant derivative of wild type ANU843. All strains produced CPS octasaccharide units with the same glycosyl sequence. The data suggest that the products of nodIJ, and especially nodI, modulate the level of 3-HB substitution in the acidic CPS synthesized by R. trifolii ANU843. We propose two possible mechanisms to explain the data. (I) The R. trifolii nodI gene product may directly control the activity of the enzyme(s) which catalyze the incorporation of 3-HB into the CPS. In relation to this hypothesis, it is important to note that a major difference between the CPS structure of several wild-type R. trifolii and R. leguminosarum strains is the extent to which the galactosyl residues are substituted with 3-HB. (II) The activities of the nodI gene product may indirectly interfere with 3-HB incorporation, perhaps by affecting the cellular pool of precursors in the cytoplasmic membrane environment where the octasaccharide repeat unit is synthesized. The symbiotic phenotypes of the nodI and nodJ mutants are a significantly reduced ability to infect root hairs and form root nodules on subterranean clover and inability to form root nodules on certain varieties of white clover.

Gresshoff, Roth, Stacey & Newton (eds.) Nitrogen Fixation: Achievements and Objectives
Chapman and Hall New York • London 1990

Inoculation of *Vicia sativa* ssp. *nigra* with *Rhizobium leguminosarum* bv. *viciae* increases the number of *nod* gene inducing flavonoids released by the roots of the host plant

Kees Recourt, Anton van Brussel, Jan Kijne, Jan Schripsema[*], and Ben Lugtenberg.

Dept. of Plant Molecular Biology and [*]Dept. of Pharmacognosy, Leiden University, Nonnensteeg 3, 2311 VJ Leiden, The Netherlands.

Uninoculated roots of Vicia sativa ssp. nigra release 2 major flavonoids which activate nod genes of Rhizobium leguminosarum bv. viciae (R.l.viciae) (1). Recently, it has been shown that inoculation with R.l.viciae increases the nod gene inducing activity (Ini) of Vicia root exudate. This increase is biovar-specific, and requires the genes nodLEFDABC (Van Brussel et al, this volume). This study concerns the molecular characterization of Ini. Nod gene inducers were purified from root exudate of 4000 Vicia seedlings, and analyzed by HPLC and NMR.
Inoculation with R.l.viciae results in release of 6 flavanones and 2 chalcones, each activating the nod genes of R.l.viciae. Only 4,4'-dihydroxy-2'-methoxy-chalcone (#6) efficiently induces the nod genes of R.meliloti. The released flavonoids are structurally closely related, and can be linked according to known pathways (see figure).

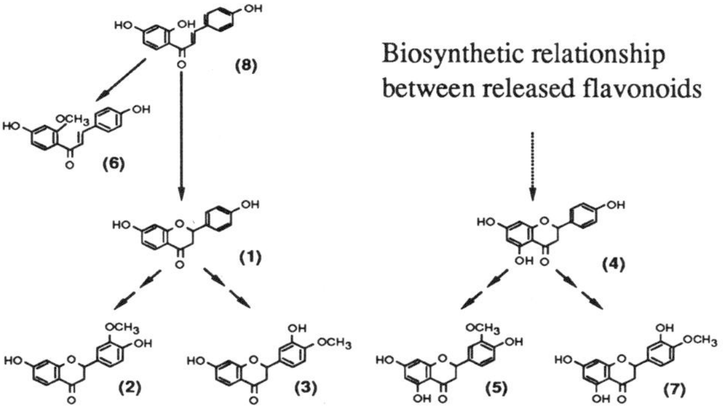

Biosynthetic relationship between released flavonoids

(Non)inoculated roots of Vicia do not contain significant amounts of nod gene inducers. It is therefore likely that inoculation with R.l.viciae enhances the *biosynthesis* of the released flavonoids. This hypothesis is supported by initial studies on Vicia roots, showing that inoculation increases both transcription and activity of key enzymes (PAL,CHS, methyltransferases) from the flavonoid biosynthetic pathway.

1) Zaat,S.A.J., J.Schripsema, C.A.Wijffelman, A.A.N. van Brussel, and B.J.J. Lugtenberg. (1989). Plant Mol. Biol. **13**, 175-188.

Gresshoff, Roth, Stacey & Newton (eds.) Nitrogen Fixation: Achievements and Objectives
Chapman and Hall New York • London 1990

Flavone dependence for *Rhizobium* elicitation of peroxidases from the white clover root surface

J.L. Salzwedel and F.B. Dazzo

Dept. of Microbiology, Michigan State University, East Lansing, MI 48824, USA.

Plant cell walls possess ionically bound peroxidases which are involved with wound or pathogenic resistance responses such as polymerization of suberin and extensin. Systemic increases in peroxidase activity have been observed as an induced resistance response in cucumber (2). It has been suggested that plant defense responses play a role in determining host specificity in the Rhizobium-legume symbiosis (1).

In this work, we used peroxidase activity as a measure of a plant defense response at the first barrier to microbial penetration, the root hair cell wall. We determined whether clover root or root hair peroxidase activity is altered by inoculation with homologous and heterologous rhizobia or culture supernatants.

After clover and pea seedlings were incubated with R. trifolii ANU843 or R. leguminosarum 300, peroxidases were eluted with 1.0 M NaCl and assayed spectrophotmetrically according to standard protocols. For both clover and pea, the heterolgous combination elicited the greater peroxidase specific activity than the homologous combination. The increase in peroxidase activity was also more rapid for the heterologous combinations. This activity was localized at deformed root hair tips. For the homologous combination, the peroxidase activity was concentrated at the point of infection thread initiation.

Only culture supernatant from flavone-grown Rl300 plates also elicits peroxidase activity at clover root surfaces. This excreted eliciting component is heat stable and in an ethanol precipitation, is found primarily in the ethanol fraction.

1. Djordjevic, S.P., Ridge, R.W., Chen, H., Redmond, J.W., Batley, M., and Rolfe, B.G. (1988) J. Bact. 170, 1848-1857.
2. Hammerschmidt, R., Nuckles, E.M., & Kuc, J. (1982) Physiol. Plant Pathol. 20, 73-82.

Gresshoff, Roth, Stacey & Newton (eds.) Nitrogen Fixation: Achievements and Objectives
Chapman and Hall New York • London 1990

Purification of *Bradyrhizobium nod*-gene inducers from peanut

Gerrit Smit[1], Dooley Buckner[1], V. Puvanesarajah[2], Russell W. Carlson[2], and Gary Stacey[1]

Center for Legume Research and Dept. of Microbiology[1], University of Tennessee, Knoxville, TN 37996. Complex Carbohydrate Research Center[2], University of Georgia, 220 Riverbend Road, Athens, GA 30602, USA.

INTRODUCTION

Bradyrhizobium sp. NC92 infects peanut plants in a process leading to nitrogen-fixing nodules on the roots. Although peanut is a major crop, not much is known about the genetics and biochemistry of the nodulation process of these plants. Peanut-*Rhizobium* NC92 harboring a *nodY-lacZ* fusion (Banfalvi et al. 1988) was used to analyze the *nod*-gene inducers from peanut plants. Induction of this strain was compared with *B. japonicum* USDA110 *nod*-gene induction.

MATERIALS AND METHODS

Bradyrhizobium sp. NC92 and *B. japonicum* USDA110, both harboring a *nodY-lacZ* fusion, were used to analyze inducing compounds from peanuts. β-Galactosidase assays were performed as described previously (Banfalvi et al. 1988). Inducers were purified by HPLC on a reverse phase C_{18} column and each fraction was assayed for *nod* gene inducing ability.

RESULTS AND DISCUSSION

A number of flavonoids induced only the *nodY-lacZ* fusion in NC92 but not in USDA110, indicating differences in the NodD-flavonoid specificity within these bradyrhizobia. In both bradyrhizobia, genistein appeared to be the most active inducer of the flavonoids tested. Another difference between these two strains was that although the maximum values of induction were not consistently different between peanut-*Rhizobium* and *B. japonicum* strains, an approximately 50-fold lower concentration of the inducer was required for maximum induction in strain NC92. *Nod*-gene inducing compounds were purified from peanut seeds as assayed by the *nodY-lacZ* fusion in strain NC92. Using HPLC separation, three major and two minor *nod*-gene inducers were purified from the seeds of peanut. Two of the inducers appeared to be predominantly located in the cotyledons, and one was largely located in the seed coat. On the basis of weight of the starting material, it appeared that a relatively high percentage of inducers were located in the seed coat of the peanut seeds. Two of the inducers comigrated with the dyes which give peanut seed its red color. Interestingly, none of the purified peanut inducers showed any inducing activity when tested on the *nodY-LacZ* fusion in *B. japonicum* USDA110. The inducers are currently being chemically characterized.

REFERENCE

Banfalvi, Z., Nieuwkoop, A., Schell, M, Besl, L, and Stacey, G. (1988) Molec. Gen. Genet. **214**: 420-424.

Gresshoff, Roth, Stacey & Newton (eds.) Nitrogen Fixation: Achievements and Objectives
Chapman and Hall New York • London 1990

Bradyrhizobium japonicum nodD1 can be specifically induced by flavonoids which do not induce the *nodYABC* genes

Gerrit Smit[1], V. Puvanesarajah[2], Russell W. Carlson[2], and Gary Stacey[1]

Center for Legume Research and Dept. of Microbiology[1], University of Tennessee, Knoxville, TN 37996. Complex Carbohydrate Research Center[2], University of Georgia, 220 Riverbend Road, Athens, GA 30602, USA.

INTRODUCTION

B. japonicum infects soybean plants in a process leading to nitrogen-fixing nodules on the roots of the host plant. The bacterial genes involved in this process are chromosomally located. In *B. japonicum*, the *nodD1* gene is inducible in contrast to the *nodD1* gene in most other rhizobial species. The main inducers of the *nodYABC* genes were isolated from soybean and identified as genistein and daidzein (Kosslak et al. 1987). These inducers were also found to be *nodD1* inducers (Banfalvi et al. 1988), however, it appeared that there were other compounds present in the seed extract which specifically induced *nodD1*. These specific *nodD1* inducers were purified and characterized.

MATERIALS AND METHODS

B. japonicum USDA135 harboring the *nodD1-lacZ* fusion and the *nodY-lacZ* fusion were used to analyze inducing compounds from *Glycine max* cv. Essex. β-Galactosidase assays were performed as described previously (Banfalvi et al. 1988). Inducers were purified by HPLC on a reverse phase C_{18} column and each fraction was assayed for *nod* gene inducing ability.

RESULTS AND DISCUSSION

Besides genistein and daidzein, which are active inducers of both *nodD1* and *nodYABC* in *B. japonicum*, soybean seed extract was also found to contain a number of compounds which enhance the induction of *nodYABC* genes in the presence of suboptimal genistein concentrations. This synergism was studied in further detail, and it appeared that certain compounds present in seed exudate do not induce *nodYABC*, but do induce *nodD1* as determined by *nodD1*- and *nodY-lacZ* fusions. Therefore, our current hypothesis is that the observed synergism is caused by a specific induction of *nodD1*, leading to a higher amount of nodD protein in the cell. *NodD1*-inducers from soybean SE have been purified by HPLC and characterized chemically. UV and NMR spectroscopy, FAB-MS, acid and enzymatic hydrolysis of inducers revealed that the specific *nodD1* inducers are glycosylated derivatives of daidzein and genistein. Both root exudate and seed exudate appear to contain these compounds.

REFERENCES

Banfalvi, Z., Nieuwkoop, A., Schell, M, Besl, L, and Stacey, G. (1988) Molec. Gen. Genet. **214**: 420-424.
Kosslak, R.M., Bookland, R, Barkei, J, Paaren, H.E., and Applebaum, E.R. (1987) Proc. Natl. Acad. Sci. USA. **84**: 7428-7432.

Gresshoff, Roth, Stacey & Newton (eds.) Nitrogen Fixation: Achievements and Objectives
Chapman and Hall New York • London 1990

Effect of genistein on cytokinin production by *Bradyrhizobium*

B. J. Taller and W. A. Mulroy

Biology Department, Memphis State University
Memphis, Tennessee 38152, USA.

Previously we have shown that Bradyrhizobium japonicum produces three cytokinin-active compounds, tentatively identified as zeatin (Z), ribosylzeatin (ZR) and methylthiozeatin (msZ). Although cytokinin synthesis occurs in defined media, it was of interest to determine whether plant factors affect cytokinin production. When B. japonicum 61A68 was grown in medium with aqueous soybean seed extract (10%), there was a shift in the cytokinin profile. MsZ increased significantly, while Z disappeared. Attempts to reproduce this effect with isoflavones produced equivocal results, presumably because cultures were not consistently induced. Thus, B. japonicum ZB977, containing a nodY:lacZ fusion, was used to monitor both nod gene induction and cytokinin synthesis. ZB977 was grown in RDY medium +/- 2 uM genistein. After 84h, the culture filtrate was applied to an Amberlite XAD-2 column. The XAD eluate was fractionated on Sephadex LH-20 in 35% ethanol, and pooled fractions were tested for cytokinin activity in the tobacco bioassay. Fractions were also examined by cytokinin immunoassay and HPLC. B. japonicum 110 culture medium and RDY medium were similarly analyzed. RDY medium alone contained several cytokinin-active and immunoreactive fractions, which complicated interpretation of some experiments. However, ZB977 did show several cytokinin-active fractions not attributable to RDY. Similar to other strains of bradyrhizobia, both 110 and ZB977 produced ZR and msZ but not isopentenyladenine or its riboside. Genistein qualitatively affected cytokinin synthesis by ZB977. Uninduced cultures contained msZ, identified by cochromatography on LH-20, t-ZR immunoassay and HPLC analysis. Induced cultures contained no msZ, but had instead a later eluting compound which has not been identified. Thus both seed extract and genistein affect cytokinin production, although the change in cytokinin profiles was not the same. This may be due to strain or medium differences. Alternatively, as soybean seed extract contains both genistein and daidzein, daidzein or some combination of inducers may be required to duplicate the effect of soybean extract.

Gresshoff, Roth, Stacey & Newton (eds.) Nitrogen Fixation: Achievements and Objectives
Chapman and Hall New York • London 1990

Lipopolysaccharide epitopes missing from bean bacteroids

Hong Tao and Dale Noel

Dept. of Biology, Marquette University, Milwaukee,WI 53233, USA.

Introduction: Previous work has indicated that Rhizobium leguminosarum must carry abundant lipopolysaccharide (LPS) O-antigen to elicit completely developed and infected nodules. A possible reason for this requirement is that the O-antigen undergoes modifications during legume infection that are required for symbiotic function. Recently, VandenBosch et al. demonstrated that pea bacteroids contained a LPS epitope that is not found in the same strain in the freeliving state. Likewise we found that certain epitopes of the freeliving bacteria are missing from bean bacteroids.

Methods: Four different monoclonal antibodies raised against strain CE3 (Str^r R. leguminosarum CFN42) grown at pH6.7 were obtained from N.J. Brewin. All of them reacted with entire LPS of CE3 grown at pH6.7 by immunoblot. Wild type CE3 bacteroids were harvested 15 days after inoculation. Low pH media was prepared according to Kannenberg et al., 1989. Root exudate induction was performed by the method of Mulligan et al.(1985) using wild type strain CE3 and pSym- strain CE144. SDS extracts pf bacteroids and freeliving under different culture conditions were run on 15% SDS-PAGE followed by immunoblot with four different monoclonal antibodies and stained with peroxidase. Dot immunoassay was done by applying a dot (2μl culture) to the nitrocellulose sheet.

Results: Two epitopes were missing from the LPS of the wild type bacteroids. These epitope changes could be mimicked by culture *ex planta* at pH below 5.0. Mere incubation at this pH in the absence of growth did not result in these changes. Root exudate can also induce changes in LPS but these changes are distinct from those induced by low pH. Unlike the changes observed at low pH, the exudate-induced changes require the presence of Sym plasmid genes.

Discussion: Although low pH causes the same epitopes to be missing in freeliving bacteria compared to bacteroid, SDS-PAGE indicates that the resulting LPS structure under these two conditions are at least somewhat different. And in the presence of pSym genes root exudate can also induce subtle LPS changes. These indicate that bacteroid LPS changes may be due to more than one inducing factor.

References: VandernBosch, D. A., Brewin, N.J., & Kannenberg, E.L. (1989) J. Bacteriol. 171:4537-4542.

Kannenberg, E.L.& Brewin, N.J. (1989) J. Bacteriol. 171:4543-4548.

Mulligan, J.T. & Long, S.R. (1985) Proc. Natl. Acad. Sci. USA.82:6609-6613.

Gresshoff, Roth, Stacey & Newton (eds.) Nitrogen Fixation: Achievements and Objectives
Chapman and Hall New York • London 1990

An extracellular, biovar-specific signal of *Rhizobium leguminosarum* biovar *viciae* induces increased *nod* gene-inducing activity in root exudate of *Vicia sativa* ssp. *nigra*

A.A.N. Van Brussel, K. Recourt, E. Pees, H.P. Spaink, T.Tak, C.A. Wiffelman, J.W. Kijne and B.J.J. Lugtenberg.

Department of Plant Molecular Biology, Leiden University, Nonnensteeg 3, 2311 VJ Leiden, The Netherlands

INTRODUCTION. Root exudates of uninfected V. sativa ssp. nigra (Vsn) plants contain, like those of other leguminous plants, flavonoids which activate the transcription of genes of R. leguminosarum bv. viciae (Rlv) involved in the formation of root nodules (nod genes) (3). In this study, we investigate the influence of the presence of Rhizobium in the rhizosphere of Vsn plants on the level of nod gene-inducing activity in root exudate.

MATERIALS AND METHODS. Bacterial strains, media, plant culture methods and other experimental details have been published elsewhere (1,3).

RESULTS AND DISCUSSION. By using Rhizobium indicator strains, containing a nodD gene and a cloned nodA promotor fused to the lacZ gene of Escherichia coli (3), we found an increased nod gene-inducing activity (Ini) in root exudates of Vsn plants infected with the homologous symbiont Rlv, whereas inoculation with heterologous Rhizobium strains had no effect. The nod genes involved in Ini were the same as those involved in nodulation (2). Thus, Rhizobium mutants exhibiting a delayed and diminished nodulation phenotype also induced delayed and diminished Ini and nod^- mutants also cause an Ini^- phenotype. The nod genes minimally required for Ini were nodLEFDABC(IJ). The nodE gene of Rlv is responsible for the biovar specificity of Ini induction since it could not be replaced for this function by the nodE gene of R. leguminosarum bv. trifolii. Ini was caused by a soluble heat-stable factor, the Ini-factor, which was only produced by Rlv with induced nod genes. The molecular weight, as determined by ultrafiltration, was between 1000 and 10,000 Da.

1. Van Brussel, A.A.N., Recourt, K., Pees, E., Spaink, H.P., Tak, T., Wiffelman, C.A., Kijne, J.W. & Lugtenberg, B.J.J. (1990) J. Bacteriol., submitted.
2. Wiffelman, C.A., Pees, E., Van Brussel, A.A.N., Okker, R.J.H., Lugtenberg, B.J.J. (1985) Arch. Microbiol. **143**, 225-232.
3. Zaat, S.A.J., Wiffelman, C.A., Mulders, I.H.M., Van Brussel, A.A.N. & Lugtenberg, B.J.J. (1988) Plant Physiol **86**, 1298-1303.

Gresshoff, Roth, Stacey & Newton (eds.) Nitrogen Fixation: Achievements and Objectives
Chapman and Hall New York • London 1990

Immunolabeling suggests uptake of a plant glycoprotein by *Rhizobium leguminosarum*

KA VandenBosch*[1], DA Dreyer[1] and NJ Brewin[2]

[1]Dept. of Biology, Texas A & M University, College Station, TX, 77843-3258, USA, and [2]John Innes Institute, Colney Ln, Norwich, NR4 7UH, UK.

In an ongoing characterization of the peribacteroid membrane of nodules of Pisum sativum, we have isolated a monoclonal antibody, MAC 254, that reacted with a distinctive set of peribacteroid proteins. The epitope recognized by MAC 254 was a periodate-sensitive carbohydrate moiety borne on numerous glycoproteins. The antibody recognized several nodule-specific proteins in the molecular weight range 25-35 kDa. The antibody also identified a 50 kDa glycoprotein that occurs in both the peribacteroid fraction and in the bacteroids, but that was absent from cultured rhizobia. Immunocytochemistry demonstrated three prominent locations of the epitope within infected cells: Golgi bodies, the peribacteroid space, and an inclusion between the bacteroids' inner and outer membranes. The latter feature, which developed after the release of rhizobia from infection threads, represents a novel structure that has not been previously described. The observations indicate that the labeled glycoproteins are glycosylated in the plant cell's Golgi bodies, and then transported to the peribacteroid space. The lack of glycoproteins in gram-negative bacteria suggests that the 50 kDa glycoprotein may be of plant origin and that the bacteroids may import this exogenous protein from the peribacteroid space into the bacterial periplasm. Future work will emphasize the isolation and characterization of this protein in order to determine its origin.

Gresshoff, Roth, Stacey & Newton (eds.) Nitrogen Fixation: Achievements and Objectives
Chapman and Hall New York • London 1990

Stimulation of *Rhizobium meliloti* for early adsorption and nodulation of alfalfa roots, by previous symbiont-specific interaction with root exudate protein factor or seed agglutinin

L. G. Wall, M. C. Giménez and G. Favelukes

Química Biológica I, Facultad de Ciencias Exactas, Universidad Nacional de La Plata, 1900-La Plata, Argentina.

A symbiont-specific interaction between *R. meliloti* (*Rme*) L5-30 and an alfalfa root exudate (A-RE) macromolecular factor stimulates rhizobial adsorption to roots (L.G.W. & G.F.,7th Int. Congr.Nit.Fix.,Köln). The stimulatory activity in A-RE involves a protein factor since it is trypsin-sensitive. A short pretreatment of A-RE with just $5x10^5$ Rme L5-30 cells/ml from either a late-exponential or a stationary culture, completely eliminated the stimulatory activity from the filtered A-RE. While this elimination ability of *Rme* (also symbiont-specific) appeared to be constitutive, the slower ensuing stimulation of rhizobial adsorption to roots required late-exponential-phase bacteria, suggesting at least two different, perhaps consecutive, interactive events. A similar preincubation of *Rme* L5-30 with A-RE also stimulated precocious infection and nodulation of alfalfa: while total numbers of nodules per plant were not modified by the pretreatment, the process of infection occurred faster, as shown by the number of nodules above RT mark, by the average position of the uppermost nodule, and by the profile of nodulation along the roots, compared to control rhizobia preincubated either in heat-inactivated A-RE or in mineral medium. Adsorption to roots, and nodulation, were also stimulated by preincubating *Rme* L5-30 with alfalfa seed agglutinin (instead of A-RE) at pH 7.0, using a very low concentration of protein (7 femtog per $5x10^4$ rhizobia). The stimulation of nodulation by a root exudate protein or lectin had been observed already in cowpeas (1) and in soybeans (2). Thus, a root-secreted protein in the rhizosphere (probably the root form of alfalfa agglutinin) and maybe the seed agglutinin proper at an earlier stage in the spermosphere are involved in an early *Rme*-alfalfa recognition event, which stimulates binding to roots and its consequent process of infection and nodulation.

--Supported by CONICET (Argentina) and SAREC (Sweden).

(1) Bhagwat,A.A. & Thomas,J.(1982) Appl.Environ.Microbiol. **43**,800-805; (1983) Arch.Microbiol. **136**,102-105.
(2) Halverson,L.J. & Stacey,G.(1984) Plant Physiol.**74**,84- -89; (1985) ibid. **77**,621-625.

SECTION C:

LIMITING FACTORS IN NITROGEN FIXATION

Invited papers
Poster summaries

Gresshoff, Roth, Stacey & Newton (eds.) Nitrogen Fixation: Achievements and Objectives
Chapman and Hall New York • London 1990

Limiting factors in symbiotic nitrogen fixation: Structure and dynamics of the *Rhizobium* genome

R. Palacios, S. Brom, G. Davila, M. Flores, M.L. Girard and D. Romero

Departamento de Genética Molecular, Centro de Investigación sobre Fijación de Nitrógeno, Apdo. postal 565-A, Cuernavaca, Mor. MEXICO.

In a broad context, the analysis of limiting factors in symbiotic nitrogen fixation should take into account all the elements that participate with either a positive or a negative role in the process and that may limit it at any stage. The survival of strains in the soil, the specific recognition between bacteria and plants, the selection of the most competitive bacteria to establish the interaction, the differentiation of both bacteria and plant cells with its intimate exchange of signals, the supply of energy to nitrogenase from dicarboxilate and aminoacid metabolism and finally the onset of nitrogen fixation resulting from a balanced interaction of regulatory circuits that sense nitrogen and carbon metabolism as well as oxygen concentration, are major events to be consider.

Ultimately, the symbiotic process may be visualized as the eventual interaction between two genomes that are as well suited to sustain the independent growth and survival of the organism. As part of the structure of each of these genomes, elements are present that participate specifically in the interaction process, the symbiotic genome. Most of the research in the field has focussed on these particular DNA sequences and in this regard, our knowledge has enormously grown. We must consider as well that as part of the complete structure these specific elements must follow the general rules of the genome as a whole. Our studies have focussed on some general characteristics of the Rhizobium genome and its influence on the specific elements that participate in the symbiosis.

We have previously reported that the Rhizobium genome contains a large amount of reiterated DNA sequences (1) located in different replicons including the chromosome as well as the plasmids. In Rhizobium leguminosarum biovar phaseoli (R. phaseoli), one such repeated DNA family corresponds to nitrogenase genes and is located in the symbiotic plasmid (4). Two complete nifHDK operons and a third nifH gene are present in different regions of the plasmid. In addition to its specific biological significance, repeated DNA sequences might function as points for recombination generating genomic rearrangements. Direct experiments to detect genomic rearrangements indicated that such events are frequent in R. phaseoli (2). This characteristic of the Rhizobium genome might correlate with the instability of symbiotic properties found in several strains.

Recent efforts from our laboratory have focussed on the structure and dynamics of a particular replicon of the Rhizobium genome, the symbiotic plasmid of R. phaseoli. The physical map of the plasmid has been established (Girard et al, to be published). It contains a large number of reiterated DNA sequences. Some of these elements are reiterated within the plasmid while others are repeated in other replicons. The two nifHDK operons are present as direct repeats separated by 120kb. Several essential nodulation genes were mapped in this region. The third nifH gene is an inverted repeat in relation to the operons and is situated in the region between them.

To study genomic rearrangements in the Rhizobium genome and in particular in the symbiotic plasmid of R. phaseoli a genetic element was constructed (Romero et al, to be published). When introduced into the genome the element allows the positive selection of different types of rearrangements, in particular DNA amplification and deletion. The introduction of this element in the pSym, revealed that the two direct repeats of the nif operons are used as recombination points to generate amplifications or deletions of the 120kb strech of DNA bordered by the repeats (see Romero et al, this volume). The frequency of both types of rearrangements, amplification and deletion, is much higher than the rate of spontaneous point mutations. These rearrangements might be significant for the symbiotic process. In fact, deletions render the strain nod$^-$. Experiments are in progress to determine the relevance of DNA amplification.

Preliminary data suggest that DNA amplification is a frequent phenomenon, not only in the described region, but as well in other regions of the pSym and in other replicons of the genome. We have proposed that Rhizobium strains are collections of similar but not identical organisms (3). Our recent experiments suggest that some of the cells harbor genomes amplified in certain regions. These extra gene copies might provide selective advantages under particular conditions. On the other hand, the knowledge of this characteristic of the genome may be used to select for organisms best fitted for specific functions, one of them might well be the symbiotic process.

Partial financial support for this research was provided by the U.S. Agency for International Development.

References:

1. Flores, M., González, V., Brom, S., Martínez, E., Piñero, D., Romero, D., Dávila, G. and Palacios, R. (1987) J. Bacteriol. **169**, 5782-5788.

2. Flores, M., González, V., Pardo, M.A., Leija, A., Martínez, E., Romero, D., Piñero, D., Dávila, G. and Palacios, R. (1988). J. Bacteriol. **170**, 1191-1196.

3. Martínez, E., Romero, D., and Palacios, R. (1990) Crit. Rev. Plant Sci. **9**, 59-93.

4. Quinto, C., de la Vega, H., Flores., M., Leemans, J., Cevallos, M.A., Pardo, M.A., Azpiroz, R., Girard, M.L., Calva, E., and Palacios, R. (1985) Proc. Natl. Acad. Sci. USA **82**, 1170-1174.

Gresshoff, Roth, Stacey & Newton (eds.) Nitrogen Fixation: Achievements and Objectives
Chapman and Hall New York • London 1990

Oxygen diffusion in the legume root nodule

J.F. Witty and F.R. Minchin

Department of Environmental Biology, AFRC Institite of Grassland and Environmental Research, Welsh Plant Breeding Station, Aberystwyth, Dyfed SY23 3EB, UK.

Introduction

All aerobic nitrogen fixing organisms require O_2 for ATP production but need to protect nitrogenase against inactivation by direct contact with O_2. In legume nodules, O_2 supply to the bacteroid zone is restricted by a variable diffusion barrier and within this zone it is evenly distributed and rapidly consumed by bacteroid respiration.

Evidence for a variable diffusion barrier

Indirect evidence for a variable barrier arose from studies using flow-through gas systems. Exposure of nodulated roots to C_2H_2 causes a reversible decline in production of both C_2H_4 and CO_2 from nodules (23, 21). Also, external O_2 can be increased up to 100% without damage to nitrogenase, despite little-or-no increases in nodule respiration (8, 31, 37, 38). If gaseous diffusion follows Fick's first law, these observations can only be explained by an increase in diffusion resistance (43).

Direct evidence has been provided by advancing an O_2-selective micro-electrode tip into the bacteroid zone until O_2 cannot be detected and then increasing external pO_2 to 40%. A transient increase in internal pO_2 (2 to 3 min) occurs. This can be interpreted as an increase in O_2 flux into the bacteroid zone which is then reduced as the resistance of the barrier increases (44). Fiber optic spectrophotometric measurements show a similar transitory increase in leghemoglobin oxidation state, when external O_2 is increased (14). Also, a 'spike' in $^{14}CO_2$ evolution occurs when soyabean nodules are returned to 21% O_2 following a 60-minute exposure to 3% O_2 (6).

Site of the diffusion barrier

Dark field microscopy (44) and O_2 gradients within nodules (see below) show that the bacteroid zone is permeated with interconnected air spaces (pO_2 ca. 20 nmol; 14), which offer little diffusion resistance. This suggests that the O_2 diffusion barrier must reside in the cortical region outside the bacteroid zone. Light and dark field microscopy (43,44) show an annulus of greatly reduced air spaces in the inner cortex. Micro-electrode studies also show a sharp decrease in pO_2 across the inner cortex and a very low concentration in the bacteroid zone (44).

It might be argued that such profiles do not demonstrate a barrier in the inner cortex but result from high rates of respiration in the infected zone. However, the use of H_2 specific micro-electrodes with Hup^{-ve} symbioses (lupin, pea and soyabean) suggests that H_2 concentrations are reciprocal to those of O_2 (eg, Fig 1). In unstressed symbioses, only a shallow H_2 gradient exists across the infected zone indicating that H_2 can diffuse freely through the interconnected air spaces within this region. A converse

O_2 gradient must also exist to drive the required inward flux of this gas. In the absence of such spaces, the flux needed to support bacteroid respiration at the center of the nodule would require a steep O_2 gradient across the infected zone which could be achieved only with damaging O_2 concentrations at the periphery.

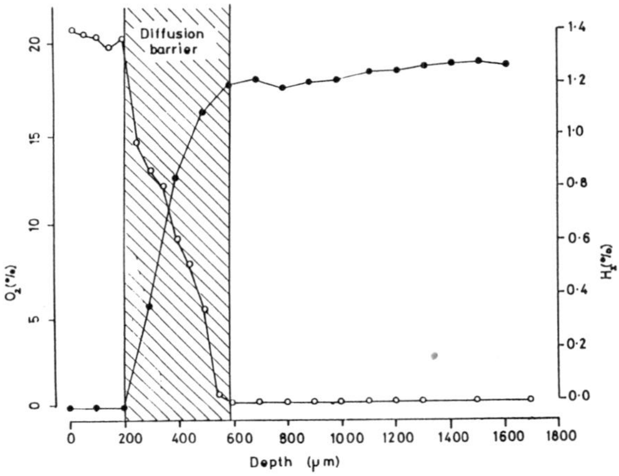

Figure 1. Traces from O_2 and H_2 specific micro-electrodes as they were advanced into a soyabean nodule (Clarke/RCR 3442).

Measurements of diffusion resistance and air spaces

Flux across a barrier is determined by area, resistance and the concentration gradient driving the diffusion process (Ficks' Law). Thus, if these can be determined, resistance can be calculated. This has been applied to intact plants in a flow through system by relating respiratory CO_2 production (which, with an RQ of unity, equals O_2 consumption, 20) to external O_2 over a range of pO_2 (18). The proportion of total root respiration linked to nitrogenase activity (flux across the barrier) can be determined from the regression intercept of CO_2 upon C_2H_4 (41) and internal pO_2 is assumed to be zero.

An alternative approach to estimating resistance is based on measurements of H_2 flux from and concentration within individual nodules. Flux was measured in a flow-through system using a Figaro gas sensor (16), external pH_2 was assumed to be zero and internal concentrations were measured using H_2 specific micro-electrodes.

The micro-electrode technique can also be used to evaluate the volume of air space inside the barrier by exposing nodules to increased pH_2 (ca 1% to avoid H_2 inhibition of nitrogenase) and measuring the rate of decline in internal concentration when H_2 is removed (Fig 2). A logarithmic transformation of 'C' (Fig. 3) produces a straight line (and see 39), the gradient of which is determined by two factors; the diffusion resistance of the nodule and the *amount* of H_2 within the diffusion barrier. Hydrogen is poorly water soluble (0.018 v/v at 20°C) and the amount within the nodule at a given % saturation is highly dependent upon the proportion of air spaces.

Gresshoff, Roth, Stacey & Newton (eds.) Nitrogen Fixation: Achievements and Objectives
Chapman and Hall New York • London 1990

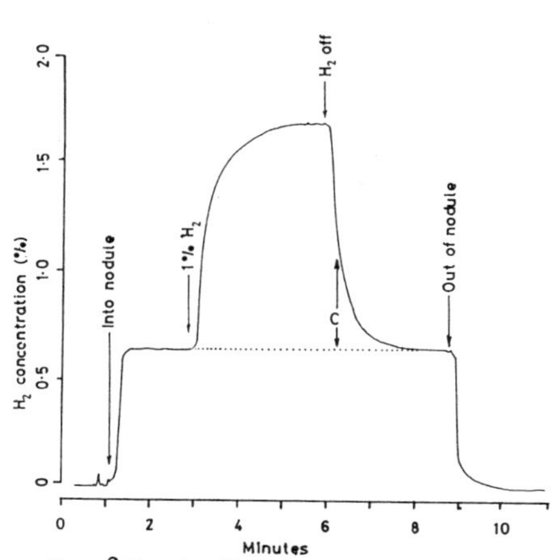

Figure 2 Trace from H_2 specific micro-electrode. The electrode tip was advanced to the centre of the nodule at time 1 min and left in position for 1 min before 1% H_2 was added to the gas stream. Hydrogen was removed from the gas stream after 6 min. The rate of the resulting decline in internal H_2 concentration is determined by nodule size, diffusion resistance and amount of internal air spaces - see text.

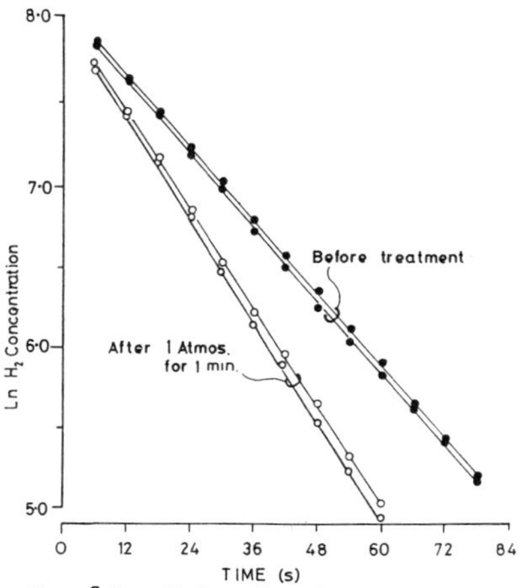

Figure 3 Logarithmic transform of decline data produced as in Fig.2 showing the effect of pressure on detached nodules of soyabean (Clarke/RCR 3442). Paired measurements taken before (•) and after (o) the nodule was exposed to the a pressure of 1 atmosphere for 1 min.

The relationship between concentration and time is determined by the integral of the Ficks' Law equation:

$$\text{Ln } C^t = \text{Ln } C^o - t/RVP \text{...............(1)}$$

where C^t=H_2 concentration above steady state level at time = t (mol m^{-3} see Fig 1)
C^o = H^2 concentration above steady state level at time 0 (mol m^{-3})
V = Nodule volume within diffusion barrier (m^3)
R = Diffusion resistance (sm^{-1})
and P = Proportion of nodule volume equivalent to air space ie P = [A + 0.018 (1-A)] where A = proportion of air space inside the diffusion barrier and 0.018 is the solubility of H_2 in water.

Table 1. Effect of removal of plant top and NO_3^- feeding on H_2 concentration, H_2 flux and diffusion resistance on the Hup^{-ve} soyabean Clarke/RCR 3442 symbioses. (mean of 6 replicates ± S.E.)

	Control		Detopped		NO_3^- (24h)		NO_3^- (48h)	
H_2 within nodule (%)	0.94	±0.047	0.45	±0.022	1.21	±0.077	0.85	±0.070
H_2 flux (mol nod^{-1} min^{-1}x 10^{-12})	11.85	±1.08	1.52	±0.09	7.50	±0.88	3.44	±0.323
Nodule surface area (m^2 x 10^{-6})	33.6	±2.74	34.5	±2.10	35.6	±2.47	35.2	±2.30
Resistance to H_2 (sm^{-1} x 10^6)	1.11	±0.05	4.18	±0.05	2.41	±0.19	3.59	±0.35

Hydrogen concentration within nodules following 24h of NO_3^- were higher than the controls although the H_2 flux was lower (Table 1). This cannot be due to the increased resistance alone. It is possible that the decrease in electron flux through nitrogenase increased the electron allocation to H_2. However, the results in Table 2, and dark field microscopy show a decrease in air pathways traversing the infected zone of nitrate-treated nodules which contribute to the increased H_2 concentration. Micro-electrode measurements in control nodules were characterized by a shallow diffusion gradient through the center, whereas when air spaces are reduced by NO_3^-, this gradient was steepened to give a higher central concentration (Fig. 4).

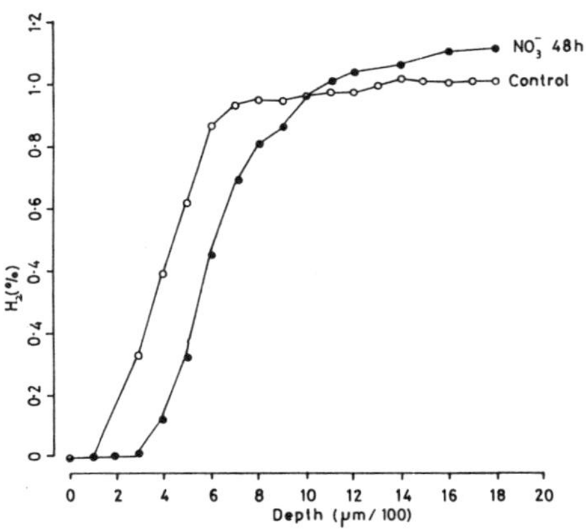

Figure 4. Effect on soyabean nodules (Clarke/RCR 3442) of 48 h exposure to 10 mM NO_3^-. Hydrogen concentration profiles measured using H_2 specific micro-electrodes (o=control, •=+NO_3^-).

The volume of air spaces within nodules of control plants, calculated from the H_2 decay data (Table 2), is similar to that reported in the literature (e.g. 3). However, detopping or NO_3^- caused a marked reduction in air space volume. The relevance of these changes to those of the variable cortical barrier have yet to be determined, but they are not the primary site of rapid changes in resistance such as produced by O_2. This also applies to stress-induced changes in nodule surface features (35, 29).

Structure of the variable diffusion barrier

Progress in understanding the structure of the barrier has been made with both the observation of a "boundary" layer (or layers) of thin-walled cells within the inner cortex which do not appear to be traversed by air spaces (27) and the recognition of glycoprotein occlusions within the intercellular spaces of the inner cortex of pea and soyabean nodules (4, 33). The degree of occlusion can be varied by long-term exposure of nodules to 10, 21 or 40% O_2 (12).

The extent of variability in resistance within the "boundary" layer may be less than that observed for nodules (27 and see above). It is, therefore, tempting to speculate that the main function of this layer is to provide a largely fixed diffusion resistance whilst the capacity for variation resides with the extent of additional air-space occlusion by the glycoprotein. This would be analogous to the theoretical separation of resistance (18) into a minimum component and an adjustable component .

Table 2. Calculation of percentage air spaces within the infected zone of Clarke/RCR 3442 calculated from H_2 relaxation data (mean of 6 replicates ± S.E.).

	Control		Detopped		NO_3^- (24h)		NO_3^- (48h)	
[a]Resistance to H_2 (nodules $sm^{-3} \times 10^9$)	3.3	±0.23	120.7	±10.1	68.5	±5.15	102.0	±9.2
Nodule volume ($m^3 \times 10^9$)	18.31	±1.20	19.03	±1.92	19.90	±2.09	19.60	±0.96
[b]Ln (H_2 decay)	0.034	±0.001	0.016	±0.001	0.027	±0.002	0.025	±0.002
[c]Air space within barrier (%)	3.14	±0.13	0.15	±0.01	0.93	±0.04	0.20	±0.02

[a] Calculated from Table 1 and nodule surface area
[b] Equals 1/RVP - see text
[c] Solving above for P and calculating 'A' (see equations)

Possible mechanisms of operation of the diffusion resistance

Variation in shape or size of cortical cells (27, 43) probably involves changes in turgor pressure. Such changes could occur via active ion movement into or out of cells (26), changes in phloem supply to nodules (35) or pH regulated variation in starch:malate balance (1). All these mechanisms may operate in nodules to produce slow changes in resistance (eg in response to C_2H_2 or detopping) but it may be that only ionic movement could account for very rapid resistance changes in response to increased O_2 (c. 1 min, 42). The involvement of active ion movement across membranes is supported by declines in nodulated root respiration caused by exposure to chloroform vapour, which closely parallel that caused by the acetylene (Figure 5). Anesthetics such as chloroform act by inhibiting ionic transport across membranes (10, 13).

Rapid variation in the degree of air-space occlusion by glycoprotein is unlikely to occur by synthesis or breakdown of this complex compound but could involve water movement into, or out of, the intercellular spaces (11). This could occur by changes in the hydrophillic/hydrophobic nature of the glycoprotein, possibly related to the nature of its ionic binding. X-ray micro-analysis shows that the inner cortical cells of soyabean nodules contain high levels of magnesium and potassium (Thomas and Minchin, unpublished data) and these ions produce opposite shifts in the NMR proton spectrum of a closely related glycoprotein (Gum-arabic, James et al, unpublished data). If water movement into intercellular spaces does occur, it would probably also require changes in turgor pressure of the inner-cortical cells, which may involve one or all of the mechanisms mentioned above.

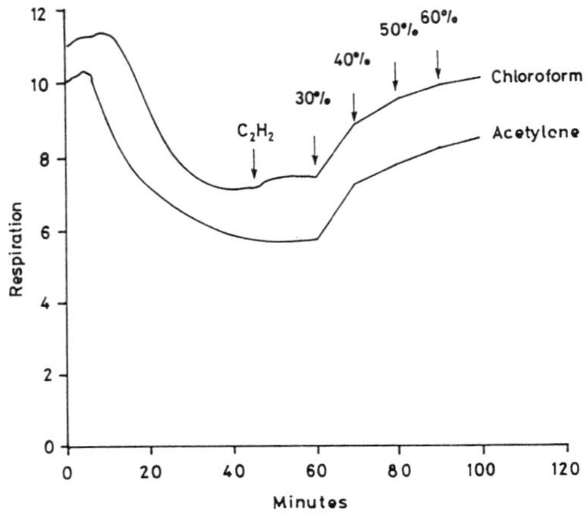

Figure 5 Effect of chloroform vapour or 10% acetylene on nodulated root respiration of intact undisturbed soyabean (Clarke/RCR 3442). An acetylene treatment was superimposed on the chloroform treatment after 40 minutes and in both treatments O_2 was increased in 10% steps after 60 min.

Biochemistry of the diffusion barrier

The dearth of knowledge regarding the biochemistry of the diffusion resistance reflects the novelty of this area of nodule function and emphasises the limited information on the biochemistry of the nodule cortex. It is controversial as to whether the initial steps of sucrose metabolism occur in the cortex (15, 5) but there is evidence that ureide metabolism can occur here (24). These observations support the hypothesis that changes in the import of sucrose or the export of nitrogenous compounds could have direct effects on the cortex, leading to changes in resistance (35). The hypothesis that the acetylene-induced and argon-induced declines reflect the cessation of ammonium production by nitrogenase (23) has also been supported by the production of similar declines following the inhibition of ureide synthesis with allopurinol (2) or glutamine synthetase activity with methionine sulfoximine (26). However, a causal relationship between diffusion resistance and nitrogenase activity has not been found over a range of external NH_3 concentrations (28).

The inner cortex of soyabean is also an area of high starch production (eg, 24) and the balance between starch and malate in these cells could be altered by changes in cell pH, affecting their turgor pressure (1). However, the pH of these cells is likely to be buffered (32) and nodule activity is unaffected by changes in external pH over the range 4.5 to 8.5 (7, Minchin and Fernandez-Pascual, unpublished).

Genetics of the oxygen diffusion barrier

The characteristics of the barrier can be affected by strain of *Rhizobium* (17) and by mutations to the host plant (30). It is possible that the use of the H_2 sensitive micro-electrode (see above) will lead to a rapid screening procedure for identifying further genetically determined variations in resistance characteristics. We are currently developing a micro-electrode system for intact, undisturbed nodules.

The early nodulin transcript ENOD2, which may encode a cell-wall protein related to the diffusion barrier, has been localised to the inner cortex of pea and soybean nodules (34) A significant advancement would accrue if the gene for the intercellular glycoprotein could be found.

Potential applications

It is very difficult to determine the extent of *in situ* O_2 limitation of legume nodules because most techniques used to measure nitrogenase activity invoke changes in the diffusion resistance (19). However, measurements based on either non-saturating levels of C_2H_2 (36) or rapid exposure to Ar/O_2 mixtures (9) suggest that nitrogenase in soybean nodules is operating under O_2 limitation. If this is correct, rates of nitrogen fixation might be increased by "resetting" the diffusion barrier and allowing a greater O_2 flux to the bacteroids. However, such a "resetting" could be detrimental to fixation by field grown crops if increases in resistance are not sufficient to protect nitrogenase during periods of enviromental stress (43). Increases in diffusion resistance are associated with the protection of nitrogenase under a wide range of stresses (22) and stress tolerance or susceptibility may be related to the speed and extent of resistance responses. Thus, an ability to manipulate the diffusion barrier to provide maximum protection for nitrogenase, and hence maximum tolerance, under a range of conditions represents one promising application of this research.

Conclusions

A plethora of evidence establishes the existence of a variable O_2 diffusion barrier located in the inner cortex of legume nodules. The challenge now is to confirm the structural components of this barrier, improve techniques for measuring gaseous diffusion in nodules, advance our knowledge of the underlying physiological and biochemical processes governing operation of the barrier, identify the relevant genes in both partners, and explore the potential applications of this control mechanism.

References

1. Appels, M.A. (1989) *The symbiosis between Rhizobium leguminosarum and Pisum sativum: regulation of the nitrogenase activity* (Ph.D Thesis, University of Wageningen)
2. Atkins, C.A., Sanford, P.J., Storer, P.J. & Pate, J.S. (1988) Plant Physiol. **88**, 1229-1234.
3. Bergersen, F.J. (1982) *Root Nodules of Legumes: Structure and Function* (John Wiley and Sons Ltd, Chichester)
4. Brewin, N.J., Wood, E.A., Bradley, D.J., Harding, S.C., Sindhu, S.S., Perotto, S., Kannenberg, E.L. & VandenBosch, K.A. (1988) in *Nitrogen Fixation: Hundred Years After*, eds. Bothe, H., de Bruijn, F.J. & Newton, W.E. (Gustav Fischer, Stuttgart), pp. 517-521.
5. Copeland, L., Vella, J. & Hong, Z. (1989) Phytochemistry **28**, 57-61.
6. Gordon, A.J., Ryle, G.J.A., Mitchell, D.F. & Powell, C.E. (1985) J. Exp. Bot. **36**, 756-769.
7. Heckman, M-O., Drevon, J-J., Saglio, P. & Salsac, L. (1989) Plant Physiol. **90**, 224-229.
8. Hunt, S., King, B.J., Canvin, D.T. & Layzell, D.B. (1987) Plant Physiol. **84**, 164-172.
9. Hunt, S., King, B.J. & Layzell, D.B. (1989) Plant Physiol. **91**, 315-321.
10. Jackson, P.C. & St.John, J.B. (1984) Planta **162**, 415-421.
11. James, E.K. (1990) *Oxygen Diffusion Barriers in the Root Nodules of Legumes*, (Ph.D Thesis, University of Dundee)

12. James, E.K., Sprent, J.I., Minchin, F.R. & Brewin, N.J. (1990) Plant Cell Environ. (In Press)
13. Kasamo, K. (1988) Plant Cell Physiol. **29**, 215-221.
14. King, B.J., Hunt, S., Weagle, G.E., Walsh, K.B., Potter, R.H., Canvin, D.T. & Layzell, D.B. (1988) Plant Physiol. **87,** 296-299.
15. Kouchi, H., Fukai, K., Katagiri, H., Minamisawa, K. & Tajima, S (1988) Plant Physiol. **73**, 327-334.
16. Layzell, D.B., Weagle, G.E. & Canvin, D.T. (1984) Plant Physiol. **75**, 582-585.
17. Minchin, F.R. & Schuller, K.A. (1988) in *Nitrogen Fixation: Hundred Years After*, eds. Bothe, H., de Bruijn, F.J. & Newton, W.E. (Gustav Fischer, Stuttgart), pp. 529.
18. Minchin, F.R., Sheehy, J.E. & Witty, J.F. (1985) Ann. Bot. **55**, 53-60.
19. Minchin, F.R., Sheehy, J.E. & Witty, J.F. (1985) in *Nitrogen Fixation Research Progress*, eds. Evans, H.J., Bottomley, P.J. & Newton, W.E. (Martinus Nijhoff, Dordrecht), pp. 285-291.
20. Minchin, F.R. & Witty, J.F. (1990) *J. Exp. Bot.* (In Press).
21. Minchin, F.R. Witty, J.F. & Sheehy, J.E. (1983) in *Temperate Legumes: Physiology, Genetics and Nodulation*, eds. Jones, D.G. & Davies, D.R. (Pitman, London), pp. 201-217.
22. Minchin, F.R., Witty, J.F. & Skøt, L. (1988) in *Physiological Limitations and the Genetic Improvement of Symbiotic Nitrogen Fixation* eds. O'Gara, F., Manian, S. & Drevon, J.J. (Kluwer Academic, Dordrecht), pp. 77-85.
23. Minchin, F.R., Witty, J.F., Sheehy, J.E. & Muller, M. (1983) J. Exp. Bot. **34**, 641-649.
24. Newcomb, E.H, Kaneko, Y. & VandenBosch, K.A. (1989) Protoplasma **150**, 150-159.
25. Pankhurst, C.E. & Sprent, J.I. (1975) Protoplasma **85**, 85-98.
26. Parsons, R. (1989) *Physiological Regulation of Nitrogen Fixation in Soybean Root Nodules* (Ph.D Thesis, University of Canberra).
27. Parsons, R. & Day, D.A. (1990) Plant Cell Environ. (In Press).
28. Purcell, L.C. & Sinclair, T.R. (1990) Plant Physiol. **92**, 268-272.
29. Ralston, E.J. & Imsande, J. (1982) J. Exp. Bot. **33**, 208-214.
30. Schuller, K., Minchin, F.R. & Gresshoff, P.M. (1988) J. Exp. Bot. **39**, 865-877.
31. Sheehy, J.E., Minchin, F.R. & Witty, J.F. (1983) Ann. Bot. **52**, 565-571.
32. Smith, F.A. & Raven, J.A. (1979) Ann. Rev. Plant Physiol. **30**, 289-311.
33. VandenBosch, K.A., Bradley, D.J., Knox, J.P., Perotto, S., Butcher, G.W. & Brewin, N.J. (1989) EMBO *J.* **8**, 335-342.
34. van de Weil, C., Scheres, B., Franssen, H., van Lierop, M.J., van Lammeren, A., van Kammen, A. & Bisseling, T. (1990) EMBO *J.* **9**, 1-7.
35. Walsh, K., Canny, M. & Layzell, D.B. (1989) Plant Cell Environ. **12**, 713-723.
36. Weisz, P.R., Denison, R.F. & Sinclair, T.R. (1985) Plant Physiol. **78**, 525-530.
37. Weisz, P.R. & Sinclair, T.R. (1987) Plant Physiol. **84**, 900-905.
38. Weisz, P.R. & Sinclair, T.R. (1987) Plant Physiol. **84**, 906-910.
39. Weisz, P.R. & Sinclair, T.R. (1988) Plant Soil **105**, 69-78.
40. Weisz, P.R. & Sinclair, T.R. (1989) in *Applications of Continuous and Steady-State Methods to Root Biology*, eds. Torrey, J.G. & Winship, L.J. (Kluwer Academic, Dordrecht), pp. 29-53.
41. Witty, J.F., Minchin, F.R. & Sheehy, J.E. (1983) J. Exp. Bot. **34**, 951-963.
42. Witty, J., Minchin, F. Sheehy, J. & Minguez, M.I. (1984) *Ann. Bot.* **53**, 13-20.
43. Witty, J.F., Minchin, F.R., Skøt, L. & Sheehy, J.E. (1986) Oxford Surveys Plant Molecular Cell Biol. **3**, 275-314.
44. Witty, J.F., Skøt, L. & Revsbech, N.P. (1987) J. Exp. Bot **38**, 1129-1140.

Gresshoff, Roth, Stacey & Newton (eds.) Nitrogen Fixation: Achievements and Objectives
Chapman and Hall New York • London 1990

A genetic approach to analyze the critical role of oxygen in bacteroid metabolism

H. Hennecke, M. Bott, T. Ramseier, L. Thöny-Meyer, H.-M. Fischer, D. Anthamatten, I. Kullik and B. Thöny

Mikrobiologisches Institut, Eidgenössische Technische Hochschule, ETH-Zentrum, Schmelzbergstrasse 7, CH-8092 Zürich, Switzerland.

Introduction

There are two essential processes in endosymbiotic bacteroids for which the ambient oxygen concentration is critical: respiration and N_2 fixation. The apparently paradoxical requirements of respiration for oxygen supply and of nitrogenase for oxygen exclusion are met (i) by the induction of a respiratory chain with a high-affinity terminal oxidase that functions at nanomolar free oxygen, and (ii) by a sophisticated redox control system that allows nitrogenase gene expression only at these low O_2 tensions. Our laboratory is engaged in studies aimed at elucidating the molecular genetic basis of both processes (i) and (ii) in the soybean root nodule symbiont, *Bradyrhizobium japonicum*. This article summarizes the progress made recently.

Analysis of genes encoding components of the respiratory chains

Background
Earlier studies by Appleby and coworkers have shown that aerobically grown *B.japonicum* cells contain cytochrome *o*- and *aa*$_3$-type terminal oxidases (for reviews see Refs. 1,12). These oxidases disappear in bacteroids or in cells grown anaerobically with nitrate as terminal electron acceptor, whereas new cytochromes are induced in these conditions. The data suggest, therefore, that the *B.japonicum* respiratory chain may consist of several branches, two functioning in aerobiosis, and at least one working in bacteroids and generating energy even at the low free oxygen concentration prevailing in the infected nodule cells. Little genetic evidence for these inferences has been available until recently.

Genes for the cytochrome bc$_1$ *complex, a novel membrane-bound cytochrome* c, *and cytochrome* aa$_3$
The first rhizobial cytochrome (cyt) genes that were cloned and sequenced are the *B.japonicum fbcF* and *fbcH* genes which are transcribed as an operon, *fbcFH* (18). The *fbcF* gene codes for the Rieske iron-sulfur protein; the *fbcH* gene is unique in that it codes for two proteins: cyt *b* and cyt c_1 (see below). These three electron transfer proteins are the constituent components of the membrane-bound bc_1 complex present in many aerobic bacteria in which they transfer electrons from the ubiquinol pool to a (usually) soluble cyt *c*. From there the electrons are passed on to cyt aa_3 and then to O_2. Mutants defective in *fbcF* and *fbcH* are Fix$^-$ and do not proliferate in symbiosis after release from the infection thread (18); yet, the mutants are able to grow aerobically. An analysis of the membrane frac-

tion revealed that *fbcF* and *H* mutants not only lacked cyt c_1 but also another membrane-bound cyt *c* with an M_r = 20'000. Apparently, its assembly into the membrane is dependent on the presence of the bc_1 complex. It may thus be analogous to a 22K cyt *c* in *Paracoccus denitrificans*, in which it is part of the bc_1-aa_3 supercomplex (4) and transfers electrons from bc_1 to aa_3. The missing 20K cyt *c* in *B.japonicum fbcFH* mutants also explains why these strains are negative in the simple oxidase assay (NADI test or TMPD test). We made use of this trait and isolated and analyzed additional oxidase-negative Tn*5* mutants among which we found one that lacked specifically the 20K membrane-bound cyt *c*. The affected gene was called *cycM*. Another mutant was defective in cytochrome aa_3. The corresponding wild-type gene was cloned and sequenced and shown to be the structural gene (*coxA*) for subunit I of the aa_3-type terminal oxidase. Like the *fbcFH* mutants, the *cycM* and *coxA* mutants were able to grow aerobically, but unlike the *fbcFH* mutants, they were Fix^+ in symbiosis (see Table 1).

Table 1. Genes for respiration in *Bradyrhizobium japonicum*, and phenotypes of mutations in them

Cloned gene	Gene product: apoprotein	Phenotypes of Tn*5* insertion mutations		
		Oxidase test	Aerobic growth	Symbiotic N_2 fixation
fbcF	Rieske FeS protein	–	+	–
fbcH	cyt *b* cyt c_1	–	+	–
cycM	cyt *c*	–	+	+
coxA	cyt aa_3, CoI	–	+	+

These results allow the following important conclusions: (i) Aerobic *B.japonicum* cells contain the respiratory chain 2[H]→Q→FeS protein/cyt bc_1 → cyt *c* (20K) → cyt aa_3 → O_2. (ii) When genes for individual components of this chain are mutated, the mutants can grow aerobically because they possess a second respiratory chain probably using the cyt *o* complex as the terminal oxidase. Like in *Escherichia coli* (9), for example, cyt *o* is likely to function as a quinol oxidase. (iii) The fact that *fbcFH* mutants are Fix^- whereas *cycM* and *coxA* mutants are Fix^+, shows that the bc_1 complex respresents a branchpoint in the electron transport chain. The branch via cyt *c* (20K) to aa_3 is not important for symbiosis. Hence, we propose that the bc_1 complex can also deliver electrons to another transport chain which operates in, and is essential for, a functional, N_2 fixing symbiosis. It is quite possible that this is the branch that leads to the aforementioned high-affinity terminal oxidase. This oxidase and other components of this branch remain to be characterized.

Biogenesis of cytochromes b *and* c_1
The unique gene-protein relationship of the single *fbcH* gene, which codes for two apocytochromes, has prompted us to unravel the biogenesis of the proteins. While the 5' half of the *fbcH* gene encodes the apocytochrome *b* and the 3' half the apocytochrome c_1, we observed that these two segments are connected by an extra piece of DNA encoding a perfectly conserved bacterial signal peptide whose function is to direct protein translocation (in this case cytochrome c_1) across the cytoplasmic membrane. This suggested that the regular cellular signal peptidase could cleave the FbcH precursor protein after the precursor-internal signal sequence. To prove this, we altered the amino acids at the putative signal peptidase cleavage site by site-directed mutagenesis. In fact, it was then possible to detect a large FbcH precursor protein in the membrane, whereas in the wild type the two mature apoproteins *b* and c_1 appeared.

Discovery of genes for the biogenesis of all c*-type cytochromes*
In the course of studying mutants defective in respiration we have described previously the *B.japonicum* mutant 2960 which, apart from being Fix$^-$, also had an oxidase-negative phenotype (13). We observed that all *c*-type cytochromes (both the soluble and the membrane-bound) were missing in this mutant. A wild-type DNA region covering the site of the Tn*5* insertion in mutant 2960 was cloned and sequenced and shown to contain at least two open reading frames coding for putative, membrane-associated proteins. One of the two predicted proteins shared significant sequence similarity with ATP binding proteins belonging to the periplasmic binding protein-dependent transport systems in bacteria (10). While most of these systems are involved in transporting small molecules across the cytoplasmic membrane, some others function in the secretion of specific proteins. We would like to speculate, therefore, that the newly detected ORFs are involved in the translocation of soluble *c*-type cytochromes into the periplasm and of membrane-bound *c*-type cytochromes onto the periplasmic side of the cytoplasmic membrane. Their role in other processes, however, such as the transport of heme or iron, cannot be ruled out at present.

In conclusion, our investigations into the genetics of *B.japonicum* respiration have not only led to a better understanding of the composition and function of the branched respiratory chain but have also resulted in the discovery of genes for cytochrome formation which may be common to all aerobic bacteria.

Regulation of symbiotic nitrogen fixation genes by oxygen

Background
In the rhizobia, oxygen appears to be the only environmental signal that governs the expression of *nif* and *fix* genes, i.e. these genes are expressed only at low partial pressures of O_2 whereas they are not expressed in aerobiosis. At least two levels of control have been identified: one occurs at the level of the expression of the *nifA* gene, which encodes the positive regulator for all other *nif* and *fix* genes, and the other one occurs at the level of the function of the NifA protein. The ways these controls are achieved, however, differ considerably in detail from one rhizobial species to the other (3,5,14,17,19). Figure 1 shows the current model of *nif* and *fix* gene regulation in *B.japonicum*. Details of this scheme will be discussed in the following paragraphs.

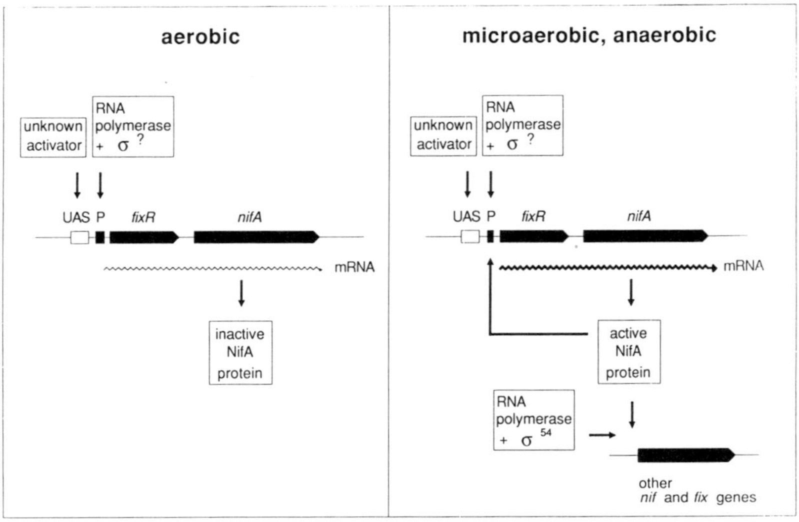

Fig. 1. Current model of *nif* and *fix* gene regulation in *B.japonicum*. The scheme is a modified and updated version of that presented in Ref. (17). For details, see text. Abbreviations: P, promoter of the *fixRnifA* operon; UAS, upstream activator sequence of the *fixR* promoter.

The oxygen- or redox-responsive NifA protein, and a possible function for FixR. In vivo, both in the homologous *B.japonicum* as well as in the heterologous *E.coli* backgrounds, the activity of the *B.japonicum* NifA protein has been shown to be sensitive to high oxygen concentrations (6). By contrast, the *Klebsiella pneumoniae* NifA protein *per se* was not inactivated by oxygen. This functional difference should be reflected in structural differences between the *B.japonicum* and *K.pneumoniae* NifA proteins. In fact, the *B.japonicum* NifA protein has 4 essential cysteine residues in a domain with the following primary structure: cys-X_{11}-cys-X_{19}-cys-X_4-cys (X = any other amino acid) (7,8). Three of these cysteines are missing in the *K.pneumoniae* NifA protein. We have proposed previously that the four-cysteine-motif could function as a metal-binding site, because the *in vivo* activity of *B.japonicum* NifA protein, but not that of the *K.pneumoniae* NifA protein, was inhibited by chelating agents (7). It is thus possible that the *B.japonicum* NifA protein is a metalloregulatory protein, in which the bound metal (perhaps Fe^{2+}) must be in the reduced form in order for NifA to be active. Higher oxygen concentrations or high (positive) redox potentials might lead to oxidation of the bound metal and, hence, to inactivation of NifA. To prove (or disprove) this hypothesis it will be necessary to purify the *B.japonicum* NifA protein and demonstrate its metal dependence in a *nif* transcription assay *in vitro*. Experiments towards this goal are underway (H.M. Fischer, in collaboration with S. Kustu, UC Berkeley).

```
EcGutD                                      MNQVAVVIGGGQTLGAFLCHGLAAEG
BjFixR    MGLDLPNDNLIRGPLPEAHLDRLVDAVNARVDRGEPKVMLLTGASRGIGHATAKLFSEAG
RmNodG                                MFELTGRKALVTGASGAIGGAIARVLHAQG

EcGutD    YRVAVVDIQSDKAANVAQEINAEYGESMAYGFGADATSEQSCLALSRGVDEIFGRVDLLV
BjFixR    WRIISCARQPFDGERCPWEAG...NDDH.FQVDLGDHRMLPRAITEVKKRLAGAPLHALV
RmNodG    AIVGLHGTQIEKLETLATELG...DRVKLFPANLANRDEVKALGQRAEADLEG..VDILV

EcGutD    YSAGIAKAAFISDFQLGDFDRS......LQVNLVGYFLCAREFSRLMIRDGIQGRIIQIN
BjFixR    NNAGVSPKTPTGDRMTSLTTSTDTWMRVFHLNLVAPILLAQGLFDE.LRAA.SGSIVNVT
RmNodG    NNAGITKDGLFLHMADPDWDIV......LEVNLTAMFRLTREITQQMIRRR.NGRIINVT

EcGutD    SKSGKVGSK.HNSGYSAAKFGGVGLTQSLALDLAEYGITVHSLMLGNLLKSPMFQSLLPQ
BjFixR    SIAGSRVHPFAGSAYATSKAALASLTRELAHDYAPHGIRVNAIAPGEI............
RmNodG    SVAGAIGNP.GQTNYCASKAGMIGFSKSLAQEIATRNITVNCVAPGFI............

EcGutD    YATKLGIKPDQVEQYYIDKVPLKRGCDYQDVLNMLLFYASPKASYCTGQSINVTGGQVMF
BjFixR    ...RTDMLSPDAEARVVASIPLRRVGTPDEVAKVIFFLCSDAASYVTGAEVPINGGQHL
RmNodG    ESAMTDKLNHKQKEKIMVAIPIHRMGTGTEVASAVAYLASDHAAYVTGQTIHVNGGMAMI
```

Fig. 2. Amino acid sequence alignments of the *B.japonicum* (Bj) FixR protein to the *R.meliloti* (Rm) NodG protein and the *E.coli* (Ec) glucitol-6-phosphate dehydrogenase (GutD) (2, and further references therein). Identical amino acids and the amino acids of the leucine/isoleucine/valine group are connected by vertical lines. Amino acids that could form a domain for the binding of NAD (15) are circled.

Another, still unsolved problem is whether or not the oxygen- or redox state-dependent inactivation of the *B.japonicum* NifA protein is reversible. In a heterologous *E.coli* system we could show that this inhibition was irreversible, i.e. after shift of an air-exposed culture back to anaerobiosis, *de novo* protein synthesis was necessary to form active NifA (11). Similar experiments in the homologous *B.japonicum* background have not been performed. It cannot be ruled out, therefore, that *B.japonicum* (but not *E.coli*) has a protein that catalyzes the conversion of the oxidized, inactive NifA protein back to the active form under reducing conditions. If such an enzyme exists, the *B.japonicum* FixR protein could be a candidate. This protein is encoded by the *fixR* gene that is located uptream of, and is cotranscribed with, the *nifA* gene (see Fig. 1). Although FixR is not essential for symbiotic N_2 fixation (16), it does appear to play a role in a hitherto unidentified redox process. This assumption is deduced from the observation that the FixR protein has a significant sequence similarity to many NAD-dependent oxidoreductases. As an example, the homology to the *E.coli* glucitol-6-phosphate dehydrogenase is shown in Fig. 2. Interestingly, the *Rhizobium meliloti* NodG protein which has been shown previously to be homologous to ribitol dehydrogenase also belongs to that protein class (Fig. 2). Amino acids known to form a typical pyridine nucleotide-binding site (15) are found to be conserved and at the appropriate positions in all three proteins shown in Fig. 2.

Regulation of nifA *gene expression*
As mentioned, the *B.japonicum nifA* gene is part of the *fixRnifA* operon and thus dependent on transcription from the *fixR* promoter (16). This operon is expressed under aerobic conditions. Under microaerobic conditions or when the cells are grown anaerobically with nitrate, the expression of the *fixRnifA* operon is increased by factor of 5. The increase is *nifA* dependent (17). Thus, at low partial pressures of O_2 NifA appears to activate, directly or indirectly, its own expression (Fig. 1).

The aerobic expression from the *fixR* promoter requires the presence of an upstream activator sequence located around position -66 from the transcription start site (17). This suggests that under aerobic conditions the *fixRnifA* operon is controlled positively. The gene for the postulated aerobic activator has yet to be found. Pursuant to our previously published work (17), several open questions remained of which two have been addressed recently: (i) Is the *fixR* promoter dependent on σ^{54}? We proposed previously that this was likely to be the case because the DNA sequence between -10 and -30 from the start of transcription carried a typical -24/-12 consensus promoter. However, we also noted that mutations in the -24 region did not alter the expression from the *fixR* promoter, a finding that could not readily be explained if this was a σ^{54}-dependent promoter (16). (ii) Is the aerobic expression of the *fixRnifA* operon regulated by *fixLJ*? This question was raised because in *R.meliloti nifA* gene expression was found to be controlled by the FixJ protein (5,19). A conceptual difference of the *B.japonicum* system, however, is that the *fixRnifA* operon is expressed aerobically whereas the FixJ-dependent *nifA* gene expression in *R.meliloti* occurs only at low O_2 concentrations; under aerobiosis the FixL protein modulates FixJ so that it cannot activate *nifA* gene expression. On the basis of this difference one would not expect the *B.japonicum fixRnifA* operon to be regulated by FixJ. The next two paragraph provide the answers to questions (i) and (ii).

Aerobic expression from the fixR *promoter does not appear to depend on* σ^{54}. *B.japonicum* was found to possess two homologs of the σ^{54} gene, *rpoN1* and *rpoN2* (see the article by Fischer *et al.* in this book). Both copies had to be mutated to create a Fix^- phenotype. Consistent with the Fix^- phenotype was the finding that *nifH-lacZ* and *nifD-lacZ* fusions were not expressed in the *rpoN1/rpoN2* double mutant background. Surprisingly, however, a *fixR-lacZ* fusion was expressed aerobically in this strain, similarly as in the wild type. This shows that the *fixR* promoter does not require these two *rpoN* gene products (Fig. 1) even though it has a well conserved -24/-12 consensus sequence at the proper distance from the transcription start site. The theoretical, though very remote, possibility still exists that *B.japonicum* possesses a third σ^{54} that is specific for the *fixR* promoter. Whether or not the anaerobic NifA-dependent autoactivation of the *fixRnifA* operon requires σ^{54} is currently under investigation.

The fixLJ *genes are not involved in the regulation of the* fixRnifA *operon.* *B.japonicum* carries *fixL*- and *fixJ*-like genes (see the article by Anthamatten and Hennecke in this book). In *fixL* and *fixJ* mutant backgrounds, *fixR-lacZ* and *nifA-lacZ* fusions were found to be expressed aerobically at wild-type levels. Thus, as expected, the FixJ protein is not the aerobic activator of the *fixR* promoter. We also could show that the expression of the *fixLJ* genes, both under aerobic and anaerobic conditions was not dependent on *nifA*. This excludes the possibility that the NifA-dependent autoregulation of the *fixRnifA* operon works indirectly via FixJ.

Although FixJ does not activate *nifA* gene expression in *B.japonicum*, as opposed to *R.meliloti*, it is nevertheless important for symbiosis. *B.japonicum fixL* and *fixJ* mutants have a 90% reduced Fix phenotype. Thus, the FixL and FixJ proteins must control other symbiotically essential genes. We are now hunting for these target genes.

Acknowledgements

We are most grateful to Michael E. Baker for pointing out to us the homology of dehydrogenases to the FixR protein. We also thank Michael Göttfert for preparing Fig. 2 and Hélène Paul for typing the manuscript. This work was supported by grants from the Swiss National Foundation for Scientific Research (Berne) and from the Swiss Federal Institute of Technology (Zürich).

References

1. Appleby, C.A. (1984) Ann.Rev.Plant Physiol. **35**, 443-478.
2. Baker, M.E. (1989) Mol.Endocrinol. **3**, 881-884.
3. Batut, J., Daveran-Mingot, M.-L., David, M., Jacobs, J., Garnerone, A.M. & Kahn, D. (1989) EMBO J. **8**, 1279-1286.
4. Berry, E.A. & Trumpower, B.L. (1985) J.Biol.Chem. **260**, 2458-2467.
5. David, M., Daveran, M.-L., Batut, J., Dedieu, A., Dommergue, O., Ghai, J., Hertig, C., Boistard, P. & Kahn, D. (1988) Cell **54**, 671-683.
6. Fischer, H.-M. & Hennecke, H. (1987) Mol.Gen.Genet. **209**, 621-626.
7. Fischer, H.-M., Bruderer, T. & Hennecke, H. (1988) Nucl.Acids Res. **16**, 2207-2224.
8. Fischer, H.-M., Fritsche, S., Herzog, B. & Hennecke, H. (1989) FEBS Lett. **255**, 167-171.
9. Gennis, R.B. (1987) FEMS Microbiol.Rev. **46**, 387-399.
10. Higgins, C.F., Gallagher, M.P., Hyde, S.C., Mimmack, M.L. & Pearce, S.R. (1990) Phil.Trans.Royal Soc.Lond. B **326**, 353-365.
11. Kullik, I., Hennecke, H. & Fischer, H.-M. (1989) Arch.Microbiol. **151**, 191-197.
12. O'Brian, M.R. & Maier, R.J. (1989) Biochim.Biophys.Acta **974**, 229-246.
13. Ramseier, T.M., Kaluza, B., Studer, D., Gloudemans, T., Bisseling, T., Jordan, P.M., Jones, R.M., Zuber, M. & Hennecke, H. (1989) Arch.Microbiol. **151**, 203-212.
14. Ratet, P., Pawlowski, K., Schell, J. & de Bruijn, F.J. (1989) Mol.Microbiol. **3**, 825-838.
15. Rossmann, M.G., Liljas, A., Branden, C.-I. & Banaszak, L.J. (1975) in *The Enzymes* Vol. XI, ed. Boyer, P.D. (Academic Press, New York), pp. 61-102.

16. Thöny, B., Fischer, H.-M., Anthamatten, D., Bruderer, T. & Hennecke, H. (1987) Nucl.Acids Res. **15**, 8479-8499.
17. Thöny, B., Anthamatten, D. & Hennecke, H. (1989) J.Bacteriol. **171**, 4162-4169.
18. Thöny-Meyer, L., Stax, D. & Hennecke, H. (1989) Cell **57**, 683-697.
19. Virts, E.L., Stanfield, S.W., Helinski, D.R. & Ditta, G.S. (1988) Proc.Natl.Acad.Sci. USA **85**, 3062-3065.

Gresshoff, Roth, Stacey & Newton (eds.) Nitrogen Fixation: Achievements and Objectives
Chapman and Hall New York • London 1990

Molecular analysis of terminal oxidases in electron-transport pathways of *Bradyrhizobium japonicum* and *Azotobacter vinelandii*

R. J. Maier, F. Moshiri, R. G. Keefe, and C. Gabel.

Department of Biology, The Johns Hopkins University, Baltimore, MD 21218 (U.S.A.)

INTRODUCTION

Both B. japonicum and A. vinelandii contain branched electron transport pathways, utilizing multiple terminal oxidases. The oxidases display differing affinities for O_2, thus allowing the O_2 concentration to regulate the pathway of electron flow. It is thought that the roles the multiple oxidases play in N_2 fixation may be to permit both coupled ATP-synthesizing, and uncoupled but O_2-scavenging electron transport (respiratory protection, see ref. 20). Indeed, inhibitor studies on both N_2-fixing Azotobacter (9,10,20) and bacteroids of B. japonicum (3-7) suggest the presence of highly-efficient ATP-synthesizing and inefficient O_2-scavenging pathways. The O_2 concentration at the bacterial membrane then is the factor determining the use of the coupled versus uncoupled pathways. It is important to keep in mind that the free O_2 level in root nodules is maintained primarily by leghemoglobins, so that even the highest O_2 levels at the bacteroid surface are much less than fully aerobic. Although inhibitor studies have been useful to demonstrate the multiplicity of O_2 affinity sites and are suggestive of physiological function, data specifically designating roles of each terminal oxidase is lacking. Some of the genes encoding cytochromes involved in O_2 consumption have been cloned (described here); this will allow a molecular analysis of their regulation and help in assigning specific O_2 scavenging or ATP-synthesizing roles to the specific oxidases. Genes encoding components of the B. japonicum cytochrome bc_1 respiratory complex were cloned recently (22) and are also required for B. japonicum bacteroid respiration and symbiotic nitrogen fixation.

TERMINAL OXIDASES OF BRADYRHIZOBIUM JAPONICUM

Free-living *B. japonicum* express cytochromes aa_3 and cytochrome *o* as the major terminal oxidases (2,18), but the cells isolated from root nodules contain a complement of carbon monoxide reactive heme proteins different from those in free-living cells. These include cytochromes P-450, P-420, c-552 and c-554 (1,23), all of which react with carbon monoxide and are thus putative oxidases (see Table 1). It is important to remember that these are putative oxidases and photochemical action spectral data are still needed to confirm or refute O_2-binding capabilities. Efficient respiration by bacteroids occurs at low O_2 tensions and inhibitor studies (5-7) suggest this efficient phase utilizes cytochrome P-450. However, cytochrome P-450 may not be the terminal (O_2-binding) oxidase involved in the coupled electron transport pathway. At high O_2 concentrations (more than 1.0 μm) *B. japonicum* bacteroid respiration is uncoupled from ATP synthesis (6) and does not involve cytochrome P-450 based on inhibitor data (5-7). However, four O_2-affinity states (7) are involved in bacteroid respiration. The highest O_2 affinity state had an apparent K_s value for O_2 of 6 nM, near the estimated O_2 concentration of the root nodule. Other O_2 affinities by bacteroids ranged from 0.015-14 μM. Clearly, bacteroid respiration is a multi-component process reflecting unique and complex relationships to fluctuating O_2 levels.

In addition, an atebrine-sensitive cyanide-reactive flavoprotein with a relatively high redox potential is

Table 1. Putative Terminal Oxidases of *B. japonicum* and *A. vinelandii*

Bacterium	Terminal Oxidase
B. japonicum	
free-living	cytochromes aa_3, cyt *o*
bacteroids	cytochromes c-552, c-554, P-450, flavoprotein cyt *c* oxidase, cyt *o*, cyt aa_3 (some strains)
Azotobacter vinelandii	cytochromes *o* and *d*, possibly a_1

Table 2. Detergent Solubilization of Bacteroid cyt c Oxidase Activity from Membranes

Detergent	Percent solubilized from the membrane	Percent activity remaining in the detergent-treated particulate fraction
CHAPS		
1%	14	84
3%	26	80
Octyl glucoside		
1%	72	17
3%	68	10
Lubrol PX		
1%	30	75
3%	45	70

apparently responsible for a large portion of the cyt c oxidase activity of bacteroid membranes (3,4,15). Comparatively little cytochrome c oxidase activity is associated with the soluble non-particulate bacteroid cell portion. Detergents such as octyl glucoside are effective at removing the cyt c oxidase activity from bacteroid membranes, and any oxidase activity which was not solubilized away from the bacteroid membranes remained with the particulate fraction (see Table 2).

The high turnover rate and the CO-insensitivity of the major bacteroid cytochrome c oxidase activity led Appleby (4) to conclude that it terminated the inefficient branch of electron transport. Evidence that a cyt c oxidase purified from bacteroid membranes is a glycoprotein is presented at this congress (Keefe and Maier, see Abstract). This component can be purified by using a lectin affinity column, and monosaccharide analysis of a sample having cyt c oxidase activity indicated that it contains galactosamine, galactose, glucose, and mannose (Figure 1).

The a-type cytochromes are commonly associated with bacterial cultures incubated in high O_2 environments and are usually not expressed in Rhizobium isolated from nodules or in cultured cells grown in restricted aeration conditions. However, some strains of B. japonicum express cyt aa_3 symbiotically (13,14). Cytochrome aa_3 mutants of B. japonicum derived from a parent strain

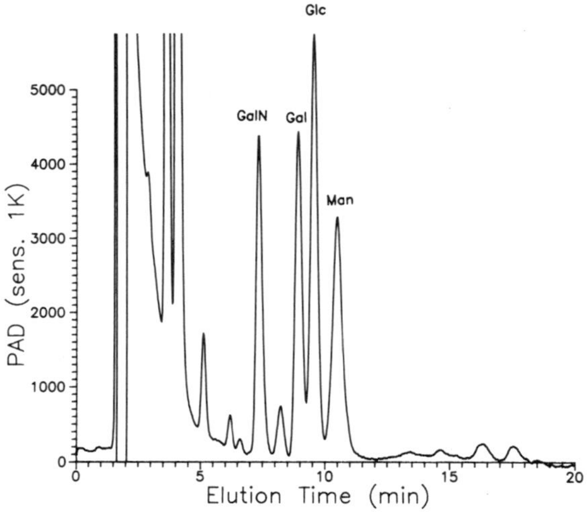

Figure 1. Monosaccharide analysis of pooled fractions from DEAE anion exchange column having cyt *c* oxidase activity

that expresses these cytochromes symbiotically were still symbiotically effective (14). It seems that the low O_2 affinity properties of cyt aa_3 are not useful for O_2 consumption in the low O_2 environment of the root nodule (18,19). However, recently an *R. phaseoli* mutant was described (21) with greater cytochrome aa_3 expression than the parent strain; it showed clearly superior symbiotic properties, suggesting an important function

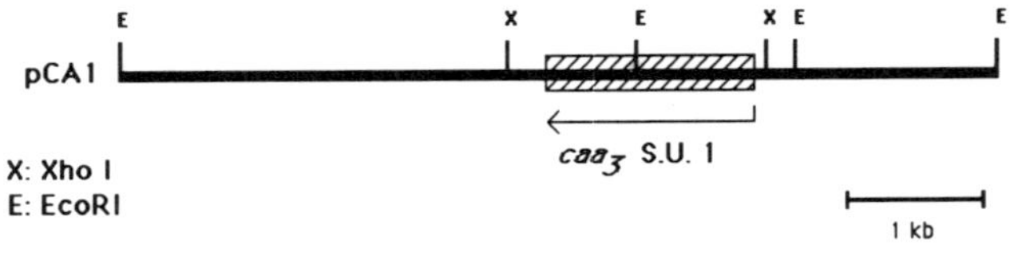

Figure 2. Restriction map of the *B. japonicum* cyt aa_3 gene region showing the cyt aa_3 subunit I gene

for cyt aa_3 in nodules formed by this symbiont. Previously we cloned a B. japonicum cytochrome aa_3 gene; the plasmid pCA1 complemented a cyt aa_3 B. japonicum mutant (strain L0501) both spectrally and functionally (17). Analysis of the nucleotide sequence and the derived amino acid sequences has revealed extensive areas of homology with mitochondrial and procaryotic cyt aa_3 sequences. The sequenced portion is 74% homologous with yeast (S. cerevisiae) COX I or cyt c oxidase subunit 1. A restriction map of the B. japonicum cyt aa_3 gene region with the location of subunit 1 is shown in Figure 2.

B. japonicum cytochrome o seems to be more suited to function under low O_2 tensions than cytochromes aa_3 (16,18). Cytochrome o is expressed symbiotically (16, 18), thus mutants specifically lacking this cytochrome are needed to assess its importance in nitrogen fixation and bacteroid respiration.

TERMINAL OXIDASES OF AZOTOBACTER VINELANDII

The respiratory chain of Azotobacter vinelandii is also branched at the terminal oxidase steps, utilizing cytochromes d and o as terminal oxidases (see Table 1). These oxidases can best be distinguished by their vast differences in cyanide sensitivity, which in turn are a reflection of their differing abilities to bind O_2.

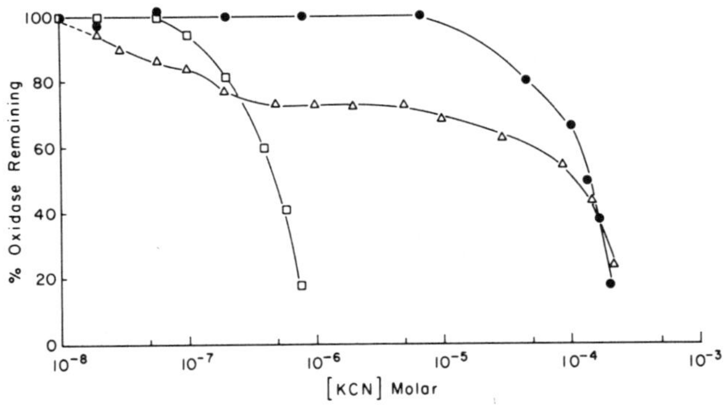

Figure 3. Effect of CN on the oxidation of malate (Δ), TMPD ascorbate (□) and H_2 (●) by A. vinelandii membranes. Reprinted with permission from ref. 24.

Figure 3 shows the effect of CN on the oxidation of malate (uses both cytochromes *d* and *o*), TMPD-ascorbate (uses cytochrome *o*) and H_2 (uses cytochrome *d*)(24,25). Clearly the two oxidases differ in affinity for O_2 and thus play different roles in respiratory metabolism. The substrate specificity of the terminal oxidases in *A. vinelandii* (24,25) suggests that these two oxidases may differ in cellular localization/organization with respect to the dehydrogenases, as well as in affinity for O_2.

Although many studies have suggested that the high respiratory activities of *Azotobacter* during N_2-fixation contribute to protection of nitrogenase from O_2 inactivation (10,20), some workers have observed that irreversible inactivation of nitrogenase by O_2 did not depend on the respiratory activity of the culture (8). Electron flow through the cytochrome *b* → *d* branch is rapid, but is not believed to be coupled to ATP synthesis (9). Therefore, it is postulated that cytochrome *d*

nif H:	TGT-N_{10}-ACA	..........111 bp..........	CTGG-N_8-CTGCA	..........75 bp..........	ATG ->
nif E:	TGT-N_{10}-ACA	..........62 bp..........	CTGG-N_8-TTGCA	..........45 bp..........	ATG ->
nif M:	has no U.A.S.'s		CTGG-N_8-GTGCC	..........53 bp..........	ATG ->
nif F:	TGT-N_{10}-ACA	..........110 bp..........	GTGG-N_8-TTGCT	..........84 bp..........	ATG ->
cyd A:	has no U.A.S.'s		ACGG-N_8-GTGCC	..........84 bp..........	ATG ->

Figure 4. Comparison of upstream sequences of the *A. vinelandii* cytochrome *d* gene with selected *nif* promoters from the major *nif* gene cluster.

may be the component responsible for maintaining a low intracellular O_2 level. Recent analysis of an E. coli cyt d mutant carrying K. pneumoniae nif genes led to the conclusion that one role of the enteric d-type cytochrome is to protect nitrogenase by respiratory protection (11). We have cloned and sequenced the structural gene for cyd of A. vinelandii. Two open reading frames encoding proteins with extensive areas of homology to subunit I and subunit II of the E. coli cyt d complex are present (Moshiri et al., see Abstract). The sequence of the region upstream of the initiation codon for subunit I revealed a ribosome-binding site and a putative promoter region which is similar, but not identical to the canonical nif promoters of A. vinelandii. The nif promoter consensus sequence (12) and the cytochrome d putative promoter sequence are given in Figure 4.

The A. vinelandii cyd can be expressed in E. coli under the control of the lac promoter or as a fusion protein under the control of the E. coli cyd promoter. However the A. vinelandii cyd constructs do not complement an E. coli cyd mutant, thus cytochrome d may play a much different role in Azotobacter than in E. coli.

Acknowledgements

This work was supported by grants from the NIH (GM 40079) , NSF (DCB-8804047), and the USDA (CRCR-1-1406).

References

1. Appleby, C. A. (1969) Biochim. Biophys. Acta 172, 71-87.
2. Appleby, C. A. (1969) Biochim. Biophys. Acta 172, 88-105.
3. Appleby, C. A. (1985) in Nitrogen Fixation and CO_2 Metabolism, eds. Ludden, P. W. and Burris, J. E. (Elsevier, New York), pp. 41-51.
4. Appleby, C. A. (1978) in Functions of Alterntive Terminal Oxidases, eds. Degn, H., Lloyd, D. and Hill, G. (Pergamon Press, Oxford) pp. 11-20.
5. Appleby, C. A., Turner, G., and MacNichol, P. K. (1975) Biochim. Biophys. Acta 387, 461-474.
6. Bergersen, F. J. and Turner, G. L. (1975) J. Gen. Microbiol. 91, 345-354.
7. Bergersen, F. J. and Turner, G. L. (1980) J. Gen. Microbiol. 118, 345-354.
8. Dingler, C. and Oelze, J. (1985) Arch. Microbiol. 141, 80-84.
9. Downs, A. J. and Jones, C. W. (1975) FEBS Lett. 60, 42-46.

10. Haddock, B. A. and Jones, C. W. (1977) Bacteriol. Rev. 4, 47-99.
11. Hill, S., Viollet, S., Smith, A. T., and Anthony, C. (1990) J. Bacteriol. 172, 2071-2078.
12. Jacobson, M. R., Brigle, K. E., Bennet, L. T., Setterquist, R. A., Wilson, M. S., Cash, V. L., Beynon, J., Newton, W. E., and Dean, D. R. (1989) J. Bacteriol. 171, 1017-1027.
13. Keister, D. L., Marsh, S. S., and El Mokadem, M. T. (1983) Plant Physiol. 71, 194-196.
14. Nautiyal, C. S., van Berkum, P., Sadowsky, M. J., and D. L. Keister (1989) Plant Physiol. 90, 553-559.
15. O'Brian, M. R. and Maier, R. J. (1983) J. Bacteriol. 155, 481-487.
16. O'Brian, M. R. and Maier, R. J. (1985) J. Bacteriol. 161, 507-514.
17. O'Brian, M. R. and Maier, R. J. (1987) Proc. Natl.-Acad. Sci. USA 84, 3219-3223.
18. O'Brian, M. R. and Maier, R. J. (1989) Biochim. Biophys. Acta 974, 229-246.
19. O'Brian, M. R., Kirshbom, P. M., and Maier, R. J. (1987) J. Bacteriol. 169, 1089-1094.
20. Postgate, J. R. (1982) The Fundamentals of Nitrogen Fixation (Cambridge Univ. Press, Cambridge).
21. Soberon, M., Williams, H. D., Poole, R. K., and Escamilla, E. (1989) J. Bacteriol. 171, 465-472.
22. Thony-Meyer, L., Stax, D., and Hennecke, H. (1989) Cell 57, 683-697.
23. Tuzimura, K. and Watanabe, I. (1964) Plant Cell Physiol. 5, 157-170.
24. Wong, T.-Y. and Maier, R. J. (1984) J. Bacteriol. 159, 348-352.
25. Wong, T.-Y. and Maier, R. J. (1985 Biochim. Biophys. Acta 807, 320-323.

Gresshoff, Roth, Stacey & Newton (eds.) Nitrogen Fixation: Achievements and Objectives
Chapman and Hall New York • London 1990

Metabolism of C_4-dicarboxylates and amino acids in soybean nodule bacteroids, and their role in energy supply to nitrogenase

Shigeyuki Tajima and Hiroshi Kouchi*

Dept. of Bioresource Science, Kagawa Univ., Miki-cho, Kita-gun, Kagawa, 761-07. *Dept. of Applied Physiology, National Institute of Agrobiological Resources, Tsukuba, Kbaraki, 305 Japan.

INTRODUCTION: Overall carbon metabolism in nodules and bacteroids is already well reviewed (3,9,10), and it is accepted now that bacteroid respiration is mainly dependent on C_4-dicarboxylate supplied from the host cells. The presence of a dicarboxylic acid carrier and absence of a tricarboxylic acid one on the bacteroid membrane suggests a unique nutrient condition surrounding the symbionts.

The bacteroids showed very poor glycolytic activities even though the cultured bacteria actively metabolize disaccharides, suggesting the presence of a regulatory system to suppress glycolytic enzyme expression. Though that phenomenon may be due to the selective permeability of the peribacteroid membrane (3,14), steady-state ^{13}C experiments revealed that sucrose is freely diffusing through the peribacteroid and bacteroid membranes (7). Thus, the molecular mechanisms to suppress sucrose catabolizing enzymes in the symbionts remain to be studied.

Currently, methods to isolate infected cells and uninfected cells from soybean nodules with few cross contaminations have been developed (4,8), and enzyme analyses indicated that those two types of plant cells possess different metabolic features, especially in urate oxidase and anaerobic metabolism enzymes. Furthermore, the uninfected cells contained high levels of glycolytic enzymes, i.e., invertase and sucrose synthase, though the highest specific activities were in the cortex (8,12). These data suggested a symbiotic differentiation of the plant cell and resulting metabolic compartmentation. More effort is necessary to elucidate the role of plant cells surrounding the infected bacteria.

In this chapter, we focus on the metabolic relationships between the amino acid and carbon metabolisms in the bacteroids, and changes in metabolic parameter values during C_4-dicarboxylate oxidation.

RESPIRATORY DICARBOXYLATE OXIDATION IN BACTEROIDS: Organic acid degradation clearly depended on the presence of oxygen in the reaction system, but the maximum oxidation rate was observed when the incubation system contained 100 uM of leghemoglobin (Lb) and was under low O_2 condition. In this condition, succinate stimulated acetylene reduction actively, and the quantity of ethylene formed depended on succinate supplied. Though the stimulation of the respiration and *in vivo* nitrogenase by the addition of Lb to the incubation system was already reported (1) and the metabolic fate of the dicarboxylic acids in the bacteroids has been analyzed by various conditions (9,13), the tracer experiments under low oxygen conditions have rarely been performed (14).

Radiorespirometric experiments using ^{14}C-succinate, -pyruvate and -acetate strongly suggested the rapid oxidation of these substrates through the complete TCA cycle (14). Though the bacteroids were able to release $^{14}CO_2$ from ^{14}C-glyoxylate or -glucose, these activities were low and showed minor effects on the respiration pattern. Effects of malonate (inhibitor of succinate dehydrogenase) and fluoroacetate (inhibitor of aconitase) also supported this conclusion.

In a short-term pulse-chase experiment with [2,3-^{14}C]-succinate, almost all the ^{14}C in the bacteroids was present among succinate, glutamate, polysaccharide and maltose. Spontaneous formation of glutamate from dicarboxylates was also observed when ^{14}C of ^{14}C-malate was incorporated into glutamate (9). Rapid decrease of ^{14}C by forming $^{14}CO_2$ did not occur in glutamate, polysaccharide and maltose, suggesting the accumulation or slow turnover of these compounds.

When the bacteroids were incubated with succinate under 1% O_2, NAD^+/NADH ratios in the bacteroids increased from 1.8 to 3.0. This high value of NAD^+/NADH suggests rapid reoxidation of the NADH formed, resulting in continuous operation of the TCA cycle. The whole soybean nodule also exhibited a very high NAD^+/NADH value of 12. In contrast to the increase of NAD^+/NADH values during the succinate oxidation, $NADP^+$/NADPH ratios in the bacteroids decreased from 0.4 to 0.2. The nodules also showed a low value of 0.28. This difference in the redox state between NAD(H) and NADP(H) systems seems to be a characteristic feature in bacteroid succinate oxidation.

When changes in isolated bacteroid,the adenylate energy charge (AEC = {[ATP]+0.5[ADP]}/{[ATP]+[ADP]+[AMP]} value during the incubation with succinate were monitored to describe the energy state of the ATP-ADP-AMP system, the initial low AEC values (0.3-0.4) increased tentatively to 0.5-0.6. However, the AEC values in the bacteroids responded mainly to oxygen supply (oxygen concentration and presence of Lb) and the presence of substrate in the incubation system produced only minor changes, though the bacteroids showed vivid nitrogenase and CO_2-release activities in response to substrate supply.

Since the low AEC values in the bacteroids might be an artifact resulting from the isolation process, AEC values of the whole nodules were comparable. Table 1 shows that the values were ca. 0.92-0.95, showing that adenine nucleotides in the nodules were fully phosphorylated. To obtain high AEC values, freezing the whole root system bearing nodules by dipping into liquid nitrogen was necessary, after which frozen nodules were powdered and extracted following HPLC assay. When the adenine nucleotides were determined after the nodules were detached from the roots and frozen immediately (in 2-5 sec), the nodules tended to show low AEC values (ca. 0.3-0.4), comparable to the values of isolated bacteroids (0.3-0.4). Nodules detached from young plants (4-week) however showed high AEC values. AEC values of the nodules responded well to oxygen conditions around the root system, suggesting the high AEC values were not artifacts.

Table 1. Adenine nucleotide contents and AEC values in the soybean nodules.

plant age (week)	freezing condition	AMP	ADP	ATP	AEC
		(nmol/g. fresh w.)			
4	detached before freezing	6	19	340	0.94
4	frozen whole roots	0	41	200	0.91
7	detached before freezing	160	56	80	0.37
7	frozen whole roots	10	18	320	0.94
7	frozen whole roots (N_2 gas for 15 min)	170	85	140	0.40
7	frozen whole roots ($5\%O_2/N_2$ for 15 min)	48	70	200	0.76

Since the adenine nucleotides in the bacteroids were supposed to be 30-50% of those of whole nodules, the AEC values in the isolated bacteroids were too low (15). Since AEC values of the isolated bacteroids did not increase more than 0.7 after prolonged incubation with succinate and Lb, more efforts would be essential to isolate the intact bacteroids and elucidate the quantitative relationship between organic acid metabolism and oxidative phosphorylation.

MALIC ENZYMES AS KEY ANAPLEROTIC ENZYMES: The presence of an anaplerotic pathway in the bacteroids is important to supply acetyl-CoA to the TCA cycle. In the soybean nodules high NADP- and NAD-malic enzyme were detected, especially in cortex tissue, but the bacteroids contained much less malic enzyme activities. However, other anaplerotic-pathway enzymes showed negligible activities, suggesting that the malic enzymes are key enzymes for TCA cycle operation in bacteroids (6).

The NADP-malic enzymes in the symbiotic systems were reported in *Bradyrhizobium japonicum*, and some characteristics were observed (2,6). The enzyme was purified to homogeneity from cultures of *Bradyrhizobium japonicum* A1017 by ammonia sulfate precipitation, DEAE-cellulose chromatography, preparative IEF, 2'5'-AdP sepharose 4B and NAD agarose affinity chromatographies. The purified enzyme converted 98% of malate in the reaction system to pyruvate. Gel filtration chromatography and SDS-PAGE of the purified enzyme showed the enzyme to be a dimer of subunit MW 59,000. During the ammonium precipitation, the enzyme was activated by ca. 50%. The activation was not simply due to the effect of ammonium ion, and identification of the activator is under study in our laboratory.

The enzyme required Mn^{2+} for activity. Mg^{2+}, Co^{2+} and Zn^{2+} also stimulated the activity to a less extent.

The optimum pH, the Km values against L-malate and NADP was 7.4, 61 and 4.9 respectively. Since malate concentration in the bacteroids were ca. 30 uM, the high Km value against malate can play a role in TCA cycle operation.

Among the organic acids as metabolic intermediates, OAA, 2-oxoglutarate, succinate, PEP, glyoxylate and malonate inhibited more than 50% of the activity in 1 mM concentration. Though these organic acids, especially malonate, might regulate the enzyme activity, most inhibition may come from high concentrations of ADP and ATP in the bacteroids. Soybean nodules and bacteroids contain ca. 300-400 uM of those nucleotides, and both ADP and ATP inhibited 75% of the enzyme activity in 0.4 mM.

AMINO ACID METABOLISM IN SOYBEAN BACTEROIDS: Currently, glutamate was reported to be an alternative bacteroid respiratory substrate (5,9,11). The presence of an active transport system for the amino acids and natural ^{15}N enrichment in the bacteroids demand more attention to the incorporation of amino acids by the bacteroids and the exudation of NH_3 to plant cytosol (9,16). Though the net and main respiratory substrate for the bacteroids would be C_4-dicarboxylate, the total profile of the relationships between carbon metabolism and amino acid metabolism remains to be elucidated.

When the effect of exogenous glutamate (Glu), aspartate (Asp), glutamine (Gln) and asparagine (Asn) was analyzed, those amino acids significantly stimulated bacteroid nitrogenase activity and CO_2 formation. Ans however has been suggested to be degraded in the medium to Asp by asparaginase released by the bacteroids because much NH_3 and Asp were detected in the medium when Asn was supplied to the bacteroids.

When ^{14}C-(U)-Glu or -Asp were incorporated to the bacteroids, most ^{14}C in both amino acids was in the Glu pool. Only trace amounts of ^{14}C were detected in the Asp pool even when ^{14}C-(U)-Asp was injected into the incubation system. This dynamic degradation of Asp was analyzed using ^{15}N. ^{15}N in the Asp was mainly incorporated into Glu and y-aminobutyrate (GABA), suggesting that the absorbed Asp was immediately deaminated or transaminated with little incorporation into the endogenous Asp pool. Addition of amino-oxyacetate (AOA), a specific inhibitor of transamination, to the incubation system inhibited Glu-dependent stimulation of bacteroid nitrogenase and respiratory activities but not Asp-dependent ones (Table 2). This difference suggests that Asp was mainly deaminated and oxidized through TCA cycle. Glu could be transaminated and utilized as a respiratory substrate as oxoglutarate. Another route by which Glu could be converted to succinate through GABA might at least partly play a role since incorporation of ^{15}N from ^{15}N-Glu to Asp is small. When malate was incorporated into the bacteroids, a large amount of alanine was released into the incubation medium, and when both glutamate and malate were fed to the bacteroids, Asp was released. Such complex relationships between amino acid and organic acid interconversion need more study.

Table 2. ^{15}N atom excess % of amino acids in and NH_3 liberated from bacteroids fed ^{15}N-labelled aspartate or glutamate with or without aminooxyacetate (AOA).

	^{15}N-Aspartate		^{15}N-Glutamate	
	-AOA	+AOA	-AOA	+AOA
Alanine	7.1	3.1	10.1	1.6
Valine	10.4	2.2	10.3	3.3
Isoleucine	12.3	3.9	17.1	4.8
Threonine	4.0	3.9	2.5	0.1
Leucine	6.8	1.7	11.7	4.0
Glycine	1.0	0.5	1.1	0.6
Serine	0.3	0.03	0.7	0.2
γ-aminobutyrate	26.5	4.2	32.8	4.7
Aspartate	5.2	12.5	8.3	3.7
Glutamate	31.8	9.2	35.0	38.7
NH_3 in the medium	30.2	20.8	25.3	6.2

CONCLUDING REMARKS: Overall dicarboxylate and amino acid metabolisms in the bacteroids are proposed in Figure 1. Tracer experiments indicated that soybean nodule bacteroids operate the complete TCA cycle for dicarboxylate oxidation under very low O_2 conditions and that NADP-malic enzyme can be a key enzyme for a continuous supply of acetyl-CoA.

The high NAD^+/NADH and low $NADP^+$/NADPH values in the bacteroids was maintained during the organic acid oxidation, suggesting a characteristic feature of bacteroid respiratory metabolism.

The bacteroids also could utilize Glu and Asp by converting them to organic acids in different pathways, and more effort will be necessary to elucidate the fate of Glu in bacteroids.

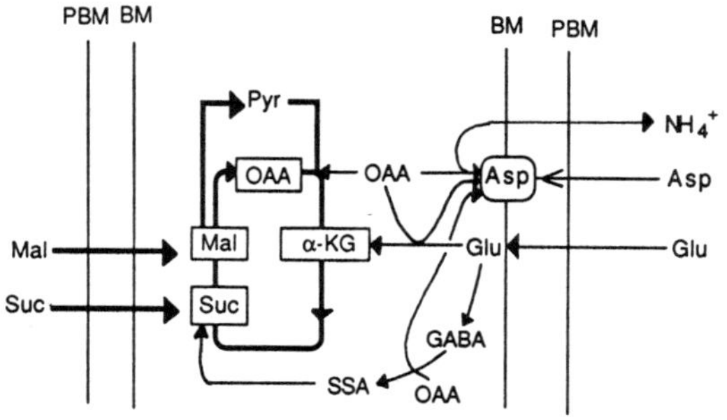

Fig.1 Proposed metabolism of C_4-dicarboxylatesand amino acids in the bacteroids.

REFERENCES

1. Bergersen et al. (1973) Biochim. Biophys. Acta. 292:271-278.
2. Copeland, L. et al. (1989) J. Gen. Microbiol., 135:2005-2011.
3. Emerich, D.W. et al. (1988) In Bothe, H., de Bruijn, F.J. and Newton, W.E. (eds): Nitrogen Fixation: Hundred Years After, 539-546. Gustav Fisher, Stuttgart, New York 1988.
4. Hanks, J.F. et al. (1983) Plant Physiol., 71:869-873.
5. Kahn, M.L. et al. (1985) In Evans, H.J., Bottomley, P.J. and Newton, W.E. (eds): Nitrogen Fixation Research Progress, 193-199. Martinus Nijhoff, Dordrecht, Netherlands.
6. Kimura, I. and Tajima, S. (1989) Soil Sci. Plant Nutr., 35:271-279.
7. Kouchi, H. and Yoneyama, Y. (1984) Annals of Botany, 53, 883-896.
8. Kouchi, H. et al. (1988) Physiol. Plant., 73, 327-334.
9. Salminen, S.O. and Streeter, J.G. (1987) J. Bacteriol., 169, 495-499.
10. Stowers, M.D. (1985) Ann. Rev. Microbiol. 140, 375-379.
11. Streeter, J.G. and Salminen, S.O. (1985) In Evans, H.J., Bottomley, P.J. and Newton, W.E. (eds): Nitrogen Fixation Research Progress, 277-283. Martinus Nijhoff, Dordrecht, Netherlands.
12. Suganuma, N. et al. (1987) Plant Cell Physiol., 28, 113-122.
13. Tajima, S. et al., (1985) Agr. Biol. Chem., 49,3473-3479.
14. Tajima, S. et al., (1990) Agr. Biol. Chem., 54, 891-897.
15. Tajima, S. and Kouzai, K. (1989) Plant Cell Physiol., 30, 589-593.
16. Yoneyama, T et al. (1987) J. Plant Physiol., 127, 251-259.
17. Udvardi, M.K. et al. (1988) FEBS Letters, 231,36-40.

Gresshoff, Roth, Stacey & Newton (eds.) Nitrogen Fixation: Achievements and Objectives
Chapman and Hall New York • London 1990

Measurement of N2 fixation by ^{15}N natural abundance in the management of legume crops: roles and precautions

F.J. Bergersen[1], M.B. Peoples[1], D.F. Herridge[2], and G.L. Turner[1]

[1]CSIRO Division of Plant Industry, GPO Box 1600, Canberra, ACT, 2601, Australia· [2]NSW Agriculture and Fisheries, RMB 944, Tamworth, NSW, 2340, Australia

Introduction: The improved utilization of N_2 fixing components in agricultural production systems is a goal of agricultural research in many countries. Other goals, for example, relate to conservation of biological resources in forests or fragile ecosystems. Evaluation of success in achieving these goals will be aided by more accurate quantitative assessment of the levels of N_2 fixation actually achieved in the field. There are at least 6 reasons to be given in support of this (1). The following relate especially to evaluating N_2 fixation by nodulated legumes. (i) Knowing the magnitude of the input of fixed N [usually as kg N ha^{-1} (growth period)$^{-1}$] is a basic component of evaluating a production system. (ii) Ecological considerations demand understanding of the relative importance of N_2 fixation in the maintenance of N balance.
(iii) Understanding the amount of N_2 fixed by legumes in various cropping sequences allows development of more efficient systems. (iv) When the user is satisfied that nodule development is adequate, measurement of N_2 fixation establishes whether a legume is achieving its potential or establishes success of improvement programs. (v) Measurement allows more reliable assessment of whether N_2 fixation rates are adequate for or a limitation upon crop yield.
(vi) Measurement allows proper assessment of the benefits of fixed N (as distinct from indigenous soil N retained during growth of a leguminous crops) for growth of subsequent crops.

Methods for measuring N_2 fixation have been extensively reviewed and evaluated (e.g. refs. 12, 17). In the present paper we describe the use of the ^{15}N natural abundance method (review ref. 19) in Australia and S.E. Asia, giving examples of results obtained and point to some of the precautions to be observed when using the method.

Methods

Principals: A plant assimilating N simultaneously from air N_2 and soil N, having different isotopic compositions (respectively *c* and *x*) has an isotopic composition of *y*. The proportion (*P*) of N assimilated from air N_2 is given by:

$$P(\%) = \frac{100\,(x-y)}{(x-c)} \qquad \text{- (i)}$$

Plant-available combined N in soil frequently has a slightly higher proportion of ^{15}N than atmospheric N_2, due to isotopic discrimination which occurs during

N-cycle processes (e.g. 15). Such natural differences are small and are conventionally expressed as:

$$\delta^{15}N = \frac{1000\,(R_s - R_{N_2})}{R_{N_2}} \qquad \text{- (ii)}$$

where R_s and R_{N_2} are the isotope ratios of a sample and of atmospheric N_2 (which is very uniform) respectively, measured with a very accurate mass spectrometer. By this definition, $\delta^{15}N$ of atmospheric N_2 is zero but for nodulated plants, entirely dependent upon atmospheric N_2 for growth, $\delta^{15}N$ often is slightly different from zero (e.g. 1, 17) due to small isotopic discriminations during N_2 fixation. Therefore it is advisable to use experimentally-determined values for c in applying expression (i) to the measurement of P using natural ^{15}N abundance (19). Similarly, the term x is usually obtained experimentally from the $\delta^{15}N$ of the total N of a non-N_2-fixing reference plant growing in the same soil, although the $\delta^{15}N$ of mineral N ($NO_3^- + NH_4^+$) extracted from the soil can be used effectively also (e.g. 3). The best correlation between reference plant N and soil nitrogen was obtained with $\delta^{15}N$ of extractable mineral N in the 0-10 cm zone of the soil profile (Bergersen & Turner, unpublished).

Table 1. Precision of field measurement of $\delta^{15}N$.[a]

Sample		n	mean $\delta^{15}N$[b]	s.e.m[c]
Soil				
extracted mineral N (0-10 cm)		21	8.8	0.44
Soybean Shoots				
non-nodulating Clark 63		12	8.0	0.25
uninoculated plants		71	7.4	0.09
maximum nodulation	- pre cropped soil	12	3.0	0.28
	- pre fallowed soil	12	6.3	0.22

[a]Data collected from Trangie, NSW, Australia, March 1986.
[b]105 days after planting. Mean over entire experiment.
[c]Standard error of the mean.

It can be seen from (i) that the statistical significance of estimates of P depends on the variances of $(x\text{-}y)$ and $(x\text{-}c)$ and especially when P is small, it will be an advantage if $(x\text{-}y)$ is as large as possible. Over all, with our collaborators in Australia, Indonesia, Malaysia, Thailand and the Philippines, we have measured values of x in the range 1.0 to 17. Generally, $\delta^{15}N$ of plant-available N has been highest in cropping soils, intermediate in pasture soils and lowest in forest and plantation soils, but in each there was a considerable range.

Requirements: It is desirable that $\delta^{15}N$ of plant-available N at any site is as uniform as possible in space and time. On the areas of many experimental sites the s.e. of mean $\delta^{15}N$ of plant-available-N of most sites was 0.1-0.4 and in no case was it >0.8. The effects of variation are countered by growing reference plants close to, or even paired with the N_2 fixing plants. An example of measurements at one experimental site is given in Table 1. Where $\delta^{15}N$ has been measured in plant-available N at various depths, there is little variation. We found $\delta^{15}N$ of 11.5 ± 1.6 and 9.4 ± 0.7 in extractable mineral N of soil between 0 and 50 cm depth from 2 blocks of a cropping site in NSW, (Turner & Bergersen, unpublished). Although variation in $\delta^{15}N$ of plant-available-N in surface soil has been recorded during growth of a winter cereal (20), generally, changes during measurement periods have been small and gradual (e.g. 3, 14).

Advantages: There are two major advantages of ^{15}N natural abundance methods. As a result of the small variations with time and with depth, the choice of reference plant is much less critical for measurement of N_2 fixation by natural abundance than for ^{15}N-enrichment methods (review 7). Furthermore, the natural abundance method may be applied without disturbance due to additions (however small) of combined N to a cropping system. It may even be used during an experiment not designed specifically for the purpose, by using already existing reference plants such as uninoculated guard rows (5) or weeds (18).

Analytical constraints and precautions

Ten years ago some workers (6) and the senior author of this paper found variable values for $\delta^{15}N$ of soil N and were sceptical about the feasibility of using ^{15}N natural abundance to measure N_2 fixation, as had been suggested (8). In our present view, at least part of the variability in $\delta^{15}N$ measurements which caused these doubts was due to the inadequacy of the analyses. Methods used for ^{15}N-enriched material were applied uncritically to the analysis of ^{15}N at natural abundance. The significance of errors at each step are outlined in Table 2. Here we mention 3 of them in more detail.

Errors due to sampling plant material: The $\delta^{15}N$ of the total N of various plant parts is often different (4). Imperfect grinding and sub-sampling is a major source of error. Coarsely-milled material frequently separates during storage, due to laboratory vibration; great care with mixing and sub-sampling is needed and fine grinding is an advantage.

Loss of N during digestion: Errors arising during Kjeldahl digestion are common. Incomplete digestions may occur if they are not sufficiently prolonged, but more commonly, minute losses of highly-^{15}N-depleted N due to thermal decomposition of $(NH^4{}_4)_2SO_4$, gives rise to elevated and variable ^{15}N values in digests. Control of temperature within the digest is essential. In our experience, 300-310°C is the absolute maximum for digestion of plant material in tubes. This results in digestion times of perhaps 10-15 h.

Loss during drying of distillates: In our laboratory, distillates of Kjeldahl digests are titrated, acidified with one drop of 1% H_2SO_4 and the volume reduced to 5 ml by boiling. If, inadvertently, samples boil dry, it is noted that elevated and variable $\delta^{15}N$ values result. Therefore the practice of evaporation to dryness, advocated by some workers, should be avoided in our opinion. The effect is due to thermal decomposition of ammonium sulphate/borate (even at modest drying temperatures of about 70°C) and consequent loss of minute amounts of highly ^{15}N-depleted nitrogen (Table 3).

Table 2. Sources of variability in measurement of $\delta^{15}N$ due to analytical errors.

Analysis	Status	Cause	Effect on:- N. analysis	δ
Ground[a] Plant sample	imperfect sample	unrepresentative sub-sample: (incomplete, uneven grinding; separation due to vibration)	variable	variable
Digestion	incomplete	dig. too short or presence of recalcitrant N-cpds.	low	variable
	losses	dig. temp too high; insufficient cooling of upper tubes	sl. low or app. normal	high & variable
Distillation[a]	incomplete	insufficient distillate collected	sl. low or app. normal	low & variable
	losses	distillate temp. too high	low	high and variable
Concentration	losses	temp too high drying	sl. low or app. normal	high & variable
Conversion of[a] NH^+_4 to N_2	incomplete	insufficient alk. hypobromite	-	low & variable

[a]At these stages and in field plots, there is always danger from undetected contamination by traces of ^{15}N-enriched materials.

Table 3. Loss of ^{15}N-depleted NH_3 from acidified distillates on drying at 75°C. Concentrated distillates (2.0 ml) in one sidearm of a closed 50 ml Warburg flask, 0.2 ml of 1N H_2SO_4 in the other sidearm: the floor of the flask contained NaOH pellets.

Sample treatment	mg N/sample	$\delta^{15}N$	
		mean	s.e.m.
Not dried	1.843	2.05	0.05
Dried at 75°C	1.824	4.40	0.65
NH_3 lost [a] during drying	0.019	-265[a]	-

[a]NH_3 collected in 0.2 ml of 1N H_2SO_4 in the sidearms of 5 flasks was recovered with distilled H_2O. The diffusates containing 18.5 μg N were pooled for one $\delta^{15}N$ measurement only.

Results: Applications of ^{15}N-natural abundance estimates of N_2 fixation

With our collaborators, we have used the ^{15}N natural abundance method for calculating *P* (expression i), as a means of investigating a wide range of experimental variables on N_2 fixation by nodulated legumes growing in the field in Australia and S.E. Asia. These are summarized in Table 4. In some experiments, detailed validation of the method has been attempted (3, 5) by relating N_2 fixation estimates to nodule development and other measured parameters. Usually, good agreement with the ^{15}N-enrichment method has been achieved (2, 10, 13, 14) and where there have been discrepancies, often due to choice of reference plant, underlying causes have been investigated (13, 14). Recently the natural abundance method has been used to calibrate estimates of N_2 fixation based on the measurement of the relative abundance of ureides in xylem sap of nodulate soybeans grown in the field (Figure 1).

When the ^{15}N natural abundance method has been combined with measurements of nitrogen in crops at several stages of development and in various plant parts and seed, useful information about relationship between N_2 fixation and crop development emerged. For example, for 3 years we have studied a winter cereal, summer soybean rotation under irrigation at Trangie in western, New South Wales, Australia (ref. 3 gives results for year 1). One important finding was that N_2 fixation late in seed development produced high yields of high protein seed, when soybeans were grown with high rates of inoculation in soil depleted in available N by a preceeding winter cereal crop. At 105d these plants, although well-nodulated, had been smaller, and contained less N than plants grown on winter-fallowed (high available N) soil (3). Subsequently, N_2 fixation during seed development was specifically devoted to increasing seed weight and protein content. Table 5 illustrates the distribution of fixed N and total N in the crop at the highest rate of inoculation. The crop grown in the low N soil approached N balance. The high harvest index for N resulted in only a small net loss of N compared with the large negative N-balance from soybeans grown in the winter fallowed soil.

Table 4. Examples of measurements of N_2 fixation made by the $\delta^{15}N$ method in Australia and S.E. Asia.

Legume Country/location (ref.)	Experimental	N_2 fixation variables *P* (%)	Amount (kg/N/ha)	Growth Period days)
Soybean Australia/Leeton, NSW(5)	soil/N inoculation	25-67	79-234	114
Trangie, NSW(3)	" "	9-57	32-149	146
Breeza, NSW(11)	" "	5-70	13-156	131
Thailand/Chiang Mai[a]	variety	78-87	26-57	64-73
Malaysia/Selangor(16)	N-fertilizer	15-68	30-115	90
Ricebean Thailand/Chiang Mai(18)	intercropping	36-86	49-41	90
Green gram Thailand/Chiang Mai[a]	variety	89-90	66-64	57-64
Black gram Thailand/Chiang Mai[a]	variety	95-98	119-140	66
Peanut Australia/Kingaroy, Q[b]	rotation	47-53	85-131	144
	water supply	25-49	42-106	
Lupinus Australia/Wagga, NSW(9,10)	season	49-85	78-225	159-193
	soil N	81-94	85-260	193
Desmodium rensonii Indonesia/Ciawi, Java[c]	cutting height	68-84	n.d.[b]	-
Sub-clover Australia/Canberra(14)	pasture composition	85-86	21-22	25
Lucerne Australia/Canberra(14)	pasture composition	64-81	5-8	25

[a]Sampet and Peoples (unpubl.); [b]Peoples and Bell (unpubl.); [c]Nurhayati and Peoples (unpubl.).

Conclusion

We conclude from our work that the ^{15}N natural abundance method for calculating the proportion of plant-N derived from N_2 fixation, is valid and useful, provided that the precautions noted by (19) and outlined above, are observed.

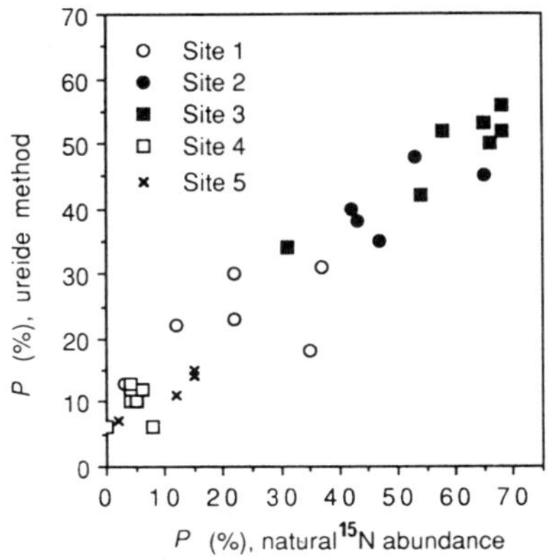

Figure 1. The relationship between P estimated by ^{15}N natural abundance and by analysis of ureides in xylem sap of soybeans grown in the field.

The 5 sites contained various levels of plant-available soil N, site 3 being lowest (11).

Table 5. The distribution of total and fixed nitrogen (kg/ha) in a soybean crop, as affected by soil N status at planting[a].

	Extractable mineral N (mg N/kg soil):			
	--------18.5--------		--------37.6--------	
Crop nitrogen	Total N	from N_2	Total N	from N_2
Total above-ground crop	242	137	316	29
Roots+Nodules	10	4	6	0
Fallen leaves	15	8	26	3
Seeds	206	141	187	26
Crop residue	61	13	151	6
N-Balance[b]	-39.4		-161	

[a]From (ref. 3) and Table 7.
[b]Fixed N in residue and leaf fall and roots and nodules *minus* soil N removed in seeds.

Acknowledgements: Material presented in Tables 3, 4-5 and Fig. 1 arose in part from Project 8800 of the Australian Centre for International Agricultural Research and involved the following collaborators: Drs B. & K. Rerkasem, Cherlampone Sampet and Ampan Bhromsiri, Chiang Mai University, Thailand, Drs Faizah al Wahab, Mohd Noor b. Sudin, Norhayati Moris and Chong Kewi, Rubber Research Institute of Malaysia, Kuala Lumpur and Miss Nurhayati, B.P.T. Ciawi, Java, Indonesia.

References

1. Bergersen, F.J. (1988) in *Biotechnology of Nitrogen Fixation in the Tropics*, eds. Shamsuddin, Z.H., Othman, W.M.W., Marziah, M. & Sundram, J., (UNESCO/Universiti Pertanian Malaysia/Malaysian Soc. Soil Sci), pp. 105-115.
2. Bergersen, F.J. & Turner, G.L. (1983) *Aust. J. Agric. Res.* **34**, 391-401.
3. Bergersen, F.J., Brockwell, J., Gault, R.R., Morthorpe, L., Peoples, M.B. & Turner, G.L. (1989) *Aust. J. Agric. Res.* **40**, 763-780.
4. Bergersen, F.J., Peoples, M.B. & Turner, G.L. (1988) *Aust. J. Plant Physiol.* **15**, 407-420.
5. Bergersen, F.J., Turner, G.L., Gault, R.R., Chase, D.L. & Brockwell, J., (1985) *Aust. J. Agric. Res.* **36**, 411-423.
6. Broadbent, F.E., Rauschkolb, R.S., Lewis, K.A. & Chang, G.Y. (1980) *Soil Sci. Soc. Amer. J.* **44**, 524-527.
7. Chalk, P.M. (1985) *Soil Biol. and Biochem.* **17**, 389-410.
8. Delwiche, C.C. & Steyn, P.L., (1970) *Env. Sci. Tech.* **4**, 929-935.
9. Evans, J., O'Connor, G.E., Turner, G.L. & Bergersen, F.J. (1987a) *Fld. Crops Res.* **17**, 10we9-120.
10. Evans, J., Turner, G.L., O'Connor, G.E. & Bergersen, F.J. (1987b) *Fld. Crops Res.* **16**, 309-322.
11. Herridge, D.F., Bergersen, F.J. & Peoples, M.B. (1990) *Plant Physiol.*, (in press).
12. Ledgard, S.F. & Peoples, M.B. (1988) in *Advances in Nitrogen Cycling in Agricultural Ecosystems*, ed. Wilson, J.R., (CAB International, Wallingford, U.K.), pp. 351-367
13. Ledgard, S.F., Simpson, J.R., Freney, J.R. & Bergersen, F.J. (1985a) *Aust. J. Agric. Res.* **36**, 663-676.
14. Ledgard, S.F., Simpson, J.R., Freney, J.R. & Bergersen, F.J. (1985b) *Aust. J. Agric. Res.* **36**, 247-258.
15. Mariotti, A., Germon, J.C., Hubert, P., Kaiser, P., Letolle, R., Tardieux, A., & Tardieux, P. (1981) *Pl. Soil* **62**, 413-430.
16. Norhayati, M., Noor, S.M., Chong, K., Faizah, A.W., Herridge, D.F., Peoples, M.B., & Bergersen, F.J. (1988) *Pl. Soil* **108**, 143-150.
17. Peoples, M.B., Faizah, A.W., Rerkasem, B., & Herridge, D.F. (eds) (1989) *Methods for Evaluating Nitrogen Fixation by Nodulated Legumes in the Field*, ACIAR Monograph No. 11. (ACIAR, Canberra.), 76pp.
18. Rerkasem, B., Rerkasem, K., Peoples, M.B., Herridge, D. & Bergersen, F.J. (1988) *Pl. Soil* **108**, 125-135.
19. Shearer, G. & Kohl, D.H. (1986) *Aust. J. Plant Physiol.* **13**, 699-756.
20. Turner, G.L., Gault, R.R., Morthorpe, L., Chase, D.L. & Bergersen, F.J. (1987) *Aust. J. Agric. Res.* **38**, 15-25.

Gresshoff, Roth, Stacey & Newton (eds.) Nitrogen Fixation: Achievements and Objectives
Chapman and Hall New York • London 1990

Genetic analysis of *sym* genes and other nodule-related genes in *Pisum sativum*

Norman F. Weeden[1], Barbara E. Kneen[2] and Thomas A. LaRue[2]

[1]Department of Horticultural Sciences, Cornell University, Geneva, NY
[2]Boyce Thompson Institute, Ithaca, NY

INTRODUCTION

The formation of nodules on the roots of legumes is a complex process, involving genes on both the host's and the bacterial genome. We use the garden pea (Pisum sativum L.) as our model system, concentrating on the genes of the host plant that affect nodulation. Two classes of genes are being analyzed: mutants that produce phenotypic changes in nodule number or physiology (referred to as sym mutants) and genes encoding nodulins or other proteins with important roles in nodule biochemistry. Our approach has been to carefully characterize the sym mutants, thereby defining as accurately as possible their role in nodule formation and to map each of the sym and nodulin genes on the pea linkage map. Functionally related genes do not generally form clusters in the genomes of higher eukaryotes, and we expected the genes involved in nodule formation to be widely distributed on the pea linkage map. Correspondance between the map positions of sym and nodulin genes might indicate that the two are genetically related. If the two are coded by the same locus the precise metabolic defect affecting nodule formation in the mutant and, conversely, the mechanism through which a nodulin influences nodulation would be revealed.

Presently 19 sym genes have been defined (6, 14, 15, 20). One of the first non-nodulating mutants to be described in pea was sym-2, a naturally occurring variant that confers strain specificity. Lines homozygous recessive sym-2 will nodulate some Rhizobium leguminosarum strains from the Middle East (e.g. TOM) but not with temperate strains. The locus was mapped to chromosome 1 using both morphological (29) and isozyme (17) markers. Other sym genes have been isolated by treating commercially available lines with various mutagenic agents. We have obtained numerous stable nodulation mutants by exposing the pea cultivar 'Sparkle' to mutagens such as EMS, NMU, gamma and neutron radiation. These induced mutants have been arranged into at least 15 complementation groups based on the phenotypes of the F_1.

A number of nodulins have been identified and characterized in pea and other legume species (1, 10, 12, 23). The best known is leghemoglobin (Lghb), encoded by a small gene family in soybean and common bean (3, 18, 19). Another nodulin is an isozyme or isozymes of glutamine synthetase. Two genes encode nodule-specific glutamine synthetase in garden pea, whereas two other genes code for the cytosolic and plastid-specific forms.

Using morphological and molecular markers, we have mapped many of the sym genes as well as leghemoglobin and glutamine synthetase genes on the pea linkage map. A region has been identified on chromosome 1 where several of these genes cluster. The details of this work are presented here.

MATERIALS AND METHODS

Plant material: Pisum sativum cv 'Sparkle' is a commercial freezer pea. The 'Afghanistan' line homozygous for the sym-2 mutation was obtained from T. A. Lie, Wageningen, the Netherlands. Morphological marker lines were obtained from G.A. Marx, Cornell University, Geneva. Allozyme and RFLP markers lines were identified in material available through the Pisum collection at the Northeast Regional Plant Introduction Station, Geneva, NY.

Generation of mutants: Strains of R. leguminosarum, culture medium and nodulation assay were as described previously (16). Ethylmethane sulfonate (E), nitrosoethylurea (NEU), nitrosomethylurea (NMU), gamma radiation (R) and fast neutrons (N) were used as mutagens. Procedures for mutagenesis and screening the M_2 progeny for nodulation with R. leguminosarum 128C53 have been described (16). Plants with white or very pale pink nodules were assayed for acetylene reduction activity.

Mapping studies: Lines homozygous for the mutant allele were crossed with markers lines yielding F_2 populations segregating for various combinations of 56 marker loci distributed across most of the pea linkage map. Morphological markers were scored according to Blixt (2), allozymes were visualized on horizontal starch gels using methods detailed previously (28), and RFLP analysis was performed essentially as described by Polans et al. (21) except that labeling of probe was accomplished by random priming (9). In nodulation assays parental controls were always planted simultaneously with the progeny. In some cases F_1 seeds were planted with related F_2 progeny to define the heterozygous nodulation phenotype.

RESULTS

The mutants isolated and examined in the study are presented in Table 1. Each sym gene represents a different complementation group, although in several cases the phenotypes are similar. Four genes (sym-2, sym-5, sym-18 and sym-19) are of particular interest with respect to the mapping results and will be described briefly. Sym-2 confers strain specificity, nodules form with TOM but not with 128C53. It was first identified as a natural variation in pea lines from Afghanistan which also contain genes which lower nodule number. We obtained a near-isogenic sym-2 line using 'Sparkle' as the recurrent parent for five cycles and selecting for high nodule number in the F_3 from strain specific F_2 selections. Two mutants, N25 and N27, display similar strain specificity and allelism with the backcross line.

Sym-5 appears to be a mutational 'hot spot' because six mutants were isolated with EMS and one with gamma radiation. There is a constitutive 66 kDA peptide that is altered in four of five mutants tested (7). Furthermore, these mutants provide the first example of reversing a mutated host nodulation phenotype with plant growth regulators. Addition to the rhizosphere of Ag^+ (an inhibitor of ethylene action) or aminoethoxyvinylglycine or Co^{++} (inhibitors of ethylene synthesis) restores nodulation to all sym-5 mutant lines, suggesting that the lines are ethylene sensitivity mutants (8).

The sym-18 gene conditions variable nodule number with different rhizobial strains. Unlike every other mutant examined, the allele is semi-dominant; F_1 progeny have low nodule numbers. The mutant E54 is pleiotropic. It flowers later than 'Sparkle', first producing pods at node 13 or 14 rather than node 10 or 11.

Three mutants have been obtained at sym-19, two with nitroso urea induction of 'Sparkle' and one (K24) by EMS mutagenesis of 'Rondo' (13). The mutants are not nodulated by any Rhizobium strain tested and appear to be resistant to mycorrhizal infection (5).

Table 1. Sym mutants isolated from pea cultivar 'Sparkle'.

Gene	No. mutants isolated	Mutant designations	Phenotype of mutants
sym-2	2	N25, N27	strain specificity
sym-5	7	E2, others	low nodule number
sym-7	2	E69, N12	non-nodulating
sym-8	3	E14, R19, R25	non-nodulating
sym-9	1	R72	non-nodulating
sym-10	1	N15	non-nodulating
sym-11	1	N24	non-nodulating
sym-13	1	E135F	non-fixing
sym-14	1	E135N	non-nodulating
sym-15	1	E151	short lateral roots, few nodules
sym-16	1	R50	short lateral roots, few nodules
sym-17	1	R82	stubby roots and shoots, few nodules
sym-18	1	E54	strain specificity, late flowering
sym-19	2	NEU5, NMU1	non-nodulating
brz	1	E107	iron accumulator, non-nodulating

Of the 15 genes analyzed that altered nodulation phenotype, 14 were located to specific regions of the pea linkage map (Figure 1). Non-random segregation was observed between known marker genes and each of these genes (Table 2). Eleven of the 25 mutants isolated, representing the four sym genes described above, mapped to a small region on chromosome 1 near d and Idh. The precise distance of each of the sym genes from d, the closer of the two markers, could only be approximately determined because the F_2 populations were relatively small and at least in pea the calculated map distance will often vary among crosses using the same two markers. Sym-2, sym-18, and sym-19 appeared to cluster very close together 5 to 10 map units from d. The fourth locus, sym-5, mapped slightly farther from d (Figure 2), but this value may exceed the actual distance. Homozygous sym-5 plants will form a few nodules, and if root temperature is not precisely controlled these same plants will form enough nodules to be indistinguishable from normal plants. Even though we reevaluated sym-5 phenotypes in the F_3 generation some of the homozygous mutants could have been misclassified, thereby increasing the number of recombinants observed between sym-5 and the marker locus.

The remaining sym genes, including the interesting iron-accumulating mutant, brz, were widely distributed with at least one on every chromosome (Figure 1 and Table 2). Sym-7 mapped to chromosome 3 near Lap-1 and sym-8 on chromosome 6 near the morphological marker, Arg. Sym-9 did not display close linkage (<10 map units) with any of the markers tested, although skewing towards the parental genotypes was observed with was on chromosome 4. Sym-10 mapped close to I, the gene responsible for the yellow versus green cotyledon color variation used by Mendel in his original genetic experiments. This locus is on chromosome 1 but assorts independently of Idh and the sym cluster 1.

Two mutants, sym-11 and sym-13, showed linkage to Skdh on chromosome 7. The two mutants appear to be about the same distance from the marker locus, 15% and 21% recombination frequencies being within the range of variation expected between the same two loci in different crosses. However, their distinct phenotypes and inability to complement each other argue against their being mutants of the same locus. Sym-15 is also located on chromosome 7, but assorted independently of the region containing sym-11 and sym-13.

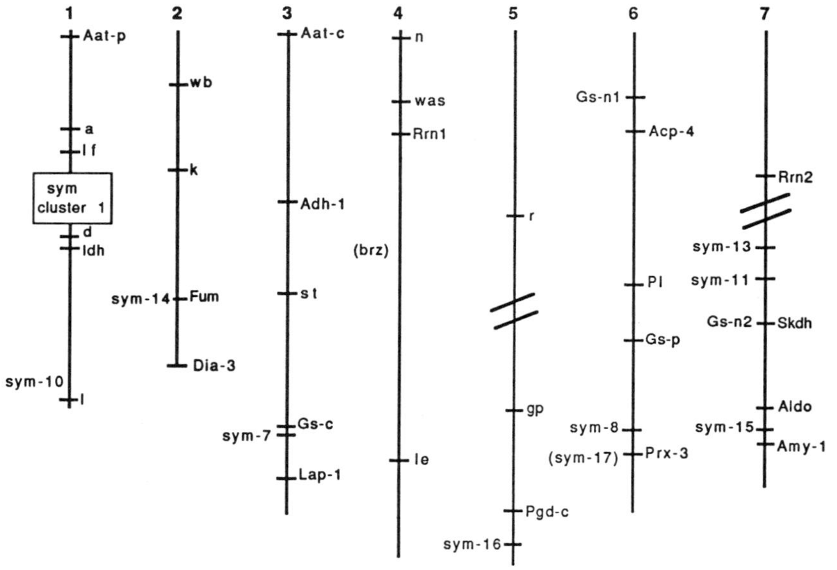

Figure 1. The seven chromosomes of the garden pea genome showing the positions of sym cluster 1, nine additional sym genes, brz, and the four Gs loci. Appropriate marker genes also are located on the map. However, for the purposes of clarity, the positions of most of the markers used in the mapping experiments are not identified. The short, thick diagonal bars on chromosomes 5 and 7 indicate postulated but unconfirmed linkages between sections of the map (see reference 27).

The inheritance of sym-16 correlated closely with markers on chromosome 5, including the gene coh, which also produces stubby roots and short internodes. The change in root morphology produced by coh differs slightly from that produced by sym-16, and coh/coh plants nodulate normally (data not presented). For these reasons we do not believe the two mutants are at the same locus. Sym-17 and brz are only loosely linked to marker loci, and the actual position of each is uncertain. A three-point test was not performed with the former mutant, and it could be on either side of Prx-3. Linkage of brz to a second marker on chromosome 4 is required to verify its weak linkage with was. Preliminary studies with the Rrn1 locus indicate that brz is on this side of was; however relatively few progeny have been examined.

Table 2. Joint segregation data for sym and other nodule-related genes with adjacent marker loci.

Loci	N	No. with designated phenotype[1]						χ^2	Recomb. fraction	Std. error
		+/1	+/h	+/2	-/1	-/h	-/2			
sym-2[2]/Idh	93	27	26	3	2	23	12	23.2**	0.30	0.06
sym-2[3]/Idh	47	10	20	3	0	5	9	16.9**	0.17	0.07
sym-5/Idh	40	11	16	4	1	4	4	5.9*	0.29	0.09
sym-7/Lap-1	74	19	25	4	0	4	22	42.6**	0.10	0.04
sym-8/Arg	65	45	--	3	0	--	17	51.8**	0.05	0.12
Sym-10/I	38	28	--	2	1	--	7	22.8**	0.09	0.16
Sym-11/Skdh	36	6	20	5	0	0	5	15.1**	0.15	0.06
sym-13/Skdh	58	11	20	2	0	12	13	20.3**	0.21	0.07
sym-14/Fum	53	15	27	0	0	0	11	53.0**	<0.02	undef
sym-15/Amy-1	57	12	28	1	1	3	12	34.4**	0.10	0.04
sym-16/Pgd-c	59	11	25	3	1	3	16	31.7**	0.12	0.04
sym-17/Prx-3	50	9	16	9	2	3	11	8.1**	0.29	0.07
brz/was	75	23	--	36	0	--	16	9.0**	0.41	0.09
sym-18/Idh	42	9	19	2	0	8	4	7.7*	0.26	0.08
sym-19/Idh	27	6	7	3	0	5	6	6.6*	0.25	0.09
Lghb/d	35	8	11	2	0	2	12	21.8**	0.10	0.07
Gs-n1/Acp-4	24	1	10	6	4	2	1	7.6*	0.22	0.11
Gs-n2/Skdh	22	4	15	0	0	0	3	19.5**	0.01	0.02

[1]Phenotype designations: + = nodulating, - = mutant, 1 = parent 1, h = heterozygous, 2 = parent 2.

[2]Data from a cross involving the naturally occurring sym-2 in the pea line 'Afghanistan'.

[3]Data from a cross involving the induced sym-2 mutant, N25.

* indicates P<0.05, ** indicates P<0.01.

The segregation of DNA sequences hybridizing to the pALB insert indicated that most of these sequences were located 10-15 map units from d on the side opposite Idh (Table 2). This position is indistinguishable at the current level of resolution from those of sym-2, sym-18 and sym-19. The number of leghemoglobin genes in the cluster has not been determined, but comparison of the band intensity on autoradiographs with those produced by apparently single copy leghemoglobin sequences showing linkage with marker loci on chromosome 7 suggests that at least five genes are present (data not presented).

Four glutamine synthetase genes are known to be present in the pea genome (23). Two of these (Gs-n1 and Gs-n2) encode nodule-specific isozymes. The other two (Gs-c and Gs-p) code for the cytosolic and plastid isozymes, respectively. Each of these genes mapped to a different region of the pea linkage map (Figure 1). Gs-n1 and Gs-p both showed linkage with markers on chromosome 6, but did not show linkage with each other. The other nodule-specific isozyme is coded by a gene mapping to chromosome 7 near Skdh. Two sym genes (sym-11 and sym-13) also appear to map to this region. The relative positions of the three genes could not be determined from the data available. The cytosolic glutamine synthetase is coded by a gene near Lap-1 on chromosome 3. A homologous DNA fragment, designated GSe, was observed on blots of *Eco*RI-cut DNA. RFLP mapping of this sequence indicated it was located on chromosome 1 near d.

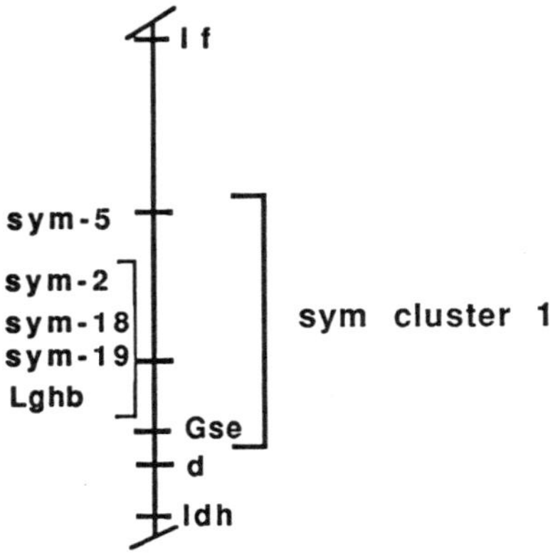

Figure 2. Enlarged view of the sym cluster 1 on chromosome 1. Idh, d, and lf are marker loci flanking the region. Gse is a sequence displaying homology to the gene coding the cytosolic glutamine synthetase. Lghb is the major cluster of leghemoglobin genes. Sym-2, sym-5, sym-18, and sym-19 are genes producing abnormal nodulation phenotypes as described in the text.

DISCUSSION

The most striking result of our mapping study is the clustering of many of the mutants to a restricted region on chromosome 1. Other functionally related genes, such as those involved in anthyocyanin production, carbohydrate metabolism, or chloroplast biochemistry, are dispersed throughout the pea genome (2, 27). Similar, apparently random distribution of such genes has been found in other species (11, 22). The sym mutants appear to follow a distribution pattern more characteristic of seed proteins (4) or other multigene families. Leghemoglobin is coded by such a gene family, and the position of the major Lghb gene cluster near four of the sym genes is intriguing. However, none of these sym mutants show any defect in leghemoglobin synthesis when nodules are formed. Hence, the sym mutants apparently identify other genes involved in nodule formation.

We do not believe the clustering of nodulation mutants is a consequence of our techniques or the use of 'Sparkle' as the parent. Ir. E. Jacobsen (Rijksuniversiteit Gronigen) independently generated nodulation mutants in the cultivar 'Rondo'. One of the nodulation mutants isolated in his program (K24) appears to be an allele of sym-19 (unpublished data). Thus we hypothesize that the Sym-2/Sym-5 region is particularly susceptible to mutation by several types of mutagenic agents. The reason for such susceptibility may be due to the arrangement or number of genes present or to a peculiarity in the chromatin structure in this region.

The size of the region covered by these genes is difficult to determine. Recombination frequencies between stable, easily scored markers in Pisum can fluctuate 2-fold, even in cases where chromosomal rearrangements are not believed to be present. The range of recombination values (10-25 map units) we obtained between Idh and the various mutants could reflect a relatively tight cluster of loci or a segment of chromosome some 15 cM in length.

SUMMARY

Twenty-six Pisum sativum mutants with reduced capabilities to form nodules or fix nitrogen could be partitioned into 15 complementation groups based on F_1 phenotype. Fourteen of the complementation groups were located on the pea linkage map. Nearly half of the mutants (11 out of 26) mapped to a small region on chromosome 1. The other mutants were widely distributed throughout the genome. Genes encoding the nodule specific proteins leghemoglobin and glutamine synthetase also were located on the linkage map, with the major leghemoglobin cluster mapping very near the cluster of sym genes on chromosome 1. The prevalence of mutants on this chromosome suggests that this region may contain a considerable proportion of the genes required for nodulation or that this region is particularly susceptible to mutation.

REFERENCES

1. Bisseling, T., Been, C., Klugkist, J., van Kammen, A. & Nadler, K. (1983) EMBO J. **2**, 961-966.
2. Blixt, S. (1970) in Handbook of Genetics, Vol 2, ed. King, R. C. (Plenum Press, New York), pp. 181-221.
3. Bojsen, K., Abildsten, D., Jensen, E. O., Paludan, K. & Marcker, K. A. (1983) EMBO J. **2**, 1165-1168.
4. Domoney, C., Ellis, T. H. N. & Davies, D. R. (1986) Mol. Gen. Genet. **202**, 280-285.
5. Duc, G., Trouvelot, A. Gianinazzi-Pearson, V. & Gianinazzi, S. (1989) Plant Sci. **60**, 215-222.
6. Engvild, K. J. (1987) Theor. Appl. Genet. **74**, 711-713.
7. Fearn, J. C. & LaRue, T. A. (1990) Plant Mol. Biol. **14**, 207-216.
8. Fearn, J. C. & LaRue, T. A. (1990) in Proc. Penn State Symposium "Polyamines and Ethylene: Biochemistry, Physiology and Interactions (in press).
9. Feinberg, A. P. & Vogelstein, B. (1983) Anal. Biochem. **132**, 6-13.
10. Gloudemans, T., Moerman, M., van Beckum, J., Gundersen, J., van Kammen, A., & Bisseling, T. (1988) in Nitrogen Fixation: Hundred Years After, eds. Bothe, H., de Bruijn, F. J. & Newton, W. E. (Gustav Fischer, Stuttgart) pp. 611-622.

11. Goodman, M. M., & Stuber, C. W. (1983) in Isozymes in Plant Genetics and Breeding, Part B, eds. Tanksley, S. D. & Orton, T. J. (Elsevier, Amsterdam) pp. 1-33.
12. Govers, F., Nap, J.-P., van Kammen, A. & Bisseling, T. (1987) Plant Physiol. Biochem. **25**, 309-322.
13. Jacobsen, E. (1984) Plant Soil **82**, 427-438.
14. Kneen, B. E. & LaRue, T. A. (1984) J. Hered. **75**, 238-240.
15. Kneen, B. E. & LaRue, T. A. (1986) Pisum Newslett. **18**, 33.
16. Kneen, B. E. & LaRue, T. A. (1988) Plant Sci. **58**, 177-182.
17. Kneen, B. E., LaRue, T. A. & Weeden, N. F. (1984) Pisum Newslett. **16**, 31-34.
18. Lee, J. S., Brown, G. G. & Verma,, D. P. S. (1983) Nucleic Acids Res. **11**, 5541-5553.
19. Lee, J. S. & Verma, D. P. S. (1984) EMBO J. **3**, 2745-2752.
20. Lie, T. A. (1984) Plant and Soil **82**, 415-425.
21. Polans, N. O., Weeden, N. F. & Thompson, W. F. (1985) Proc. Natl. Acad. Sci. USA **82**, 5083-5087.
22. Tanksley, S. D. & Rick, C. M. (1980) Theor. Appl. Genet. **57**, 161-170.
23. Tingey, S. V., Walker, E. L. & Coruzzi, G. M. (1987) EMBO J. **6**, 1-9.
24. Weeden, N. F. (1985) in The Pea Crop, eds Hebblethwaite, P. D., Heath, M. C. & Dawkins, T. C. (Butterworths, London) pp. 55-66.
25. Weeden, N. F. & Marx, G. A. (1984) J. Hered. **75**, 365-370.
26. Weeden, N. F. & Marx, G. A. (1987) J. Hered. **78**, 153-159.
27. Weeden, N. F. & Wolko, B. (1990) in Genetic Maps, ed. O'Brien, S. J. (Cold Spring Harbor Press, Cold Spring Harbor) pp. 6.106-6.112.
28. Wendel, J. F. & Weeden, N. F. (1989) in Isozymes in Plant Biology, eds. Soltis, D. E. & Soltis, P. S. (Dioscorides Press, Portland, OR) pp. 5-45.
29. Young, J. P. W. (1985) J. Hered. **76**, 207-208.

This research was supported by a grant from the United States Department of Agriculture (USDA CRGO-86-CRCR-1-2060).

Gresshoff, Roth, Stacey & Newton (eds.) Nitrogen Fixation: Achievements and Objectives
Chapman and Hall New York • London 1990

Host legume control of nodulation by flavonoids

D. A. Phillips, U. A. Hartwig, C. A. Maxwell, C. M. Joseph, J. Wery[1], M. Hungria, and S. M. Tsai

Dept. of Agronomy & Range Science, University of California, Davis, CA 95616
[1]Laboratoire d'Agronomie, INRA-ENSA, Place Viala, Montpellier Cedex, France

Over 4,000 different flavonoids are synthesized by plants (11), but their functions are poorly understood (10). One exception to that generalization is the role that flavonoids from legumes play in the induction of rhizobial *nod* genes (16). Although that phenomenon has been studied in diverse legumes and rhizobia, a wider role for exuded flavonoids in rhizosphere biology has not been established. One factor limiting a broader consideration of how flavonoids affect rhizosphere biology is the lack of detailed data on which compounds are actually released by a particular plant. The primary purpose of this discussion is to show how identifying the chemical structures of flavonoids released naturally from seeds (13) and roots (19) of alfalfa (*Medicago sativa* L.) has increased our understanding of rhizosphere biology. Initial data from an investigation designed to test if conclusions from alfalfa can be generalized to include common bean (*Phaseolus vulgaris* L.) will be mentioned.

Background to Current Work

Lundegardh and Stenlid first reported flavonoids in root exudates in 1944 (17), but actual identification and estimates of exudation rates for specific flavonoids were published only recently (4). During the past few years investigators identified flavonoids released from seeds or roots of many plants, but the number from any particular species is limited: soybean (*Glycine max* (L.) Merr.), two isoflavones (4, 15) and a coumestan (4); lentil (*Lens culinaris* Medic.), three flavones (4); pea (*Pisum sativum* L.), one flavone and a flavanone (7); alfalfa (*Medicago sativa* L.), a flavone (22); white clover (*Trifolium repens* L.), three flavones and an isoflavone (6, 23); *Vicia sativa* L., two flavanones (24). Some, but not all, of those flavonoids function as *nod* gene inducers for homologous rhizobia. Few generalizations can be made from this substantial list of flavonoids because in some cases seed rinses and root exudates were not separated, and usually, no quantitative measures of flavonoid exudation were made. It was against this background of information that we initiated a more comprehensive study of flavonoids released from alfalfa and common bean.

Flavonoids from Alfalfa

Extensive lists of flavonoids identified in organic solvent extracts of alfalfa roots (21) or "Medicago x varia" seeds (8) may not indicate the compounds that are released into an aqueous soil solution. Therefore all studies described here used aqueous rinses of unimbibed seeds or hydroponic solutions collected by bathing 72-h-old roots developing from previously rinsed seeds. The term "flavonoid fraction" refers to compounds that bind to a C_{18} resin in water and can be eluted with methanol. Compounds for which chemical structures are indicated were identified by spectroscopic techniques (UV/visible, NMR, MS) (18) and verified by comparison with authentic standards by NMR and UV/visible spectroscopy and HPLC tests. Initially, the capacity of individual compounds to induce *nod* gene transcription was measured by β-galactosidase induction from a *nodC-lacZ* fusion controlled by *nodD1* in *Rhizobium meliloti* 1021(pRmM57). Tests for activation of the *nodD2* gene product were done in *R. meliloti* strain JM57(pRmM137) and compared with JM57(pRmJ30), which contains extra copies of *nodD1* (14). These methods have now identified 10 flavonoids in seed and root exudates of the alfalfa cultivar 'Moapa 69' (Table 1).

Flavonoids Released from Alfalfa Seeds. The largest peak in an HPLC analysis of the flavonoid fraction from Moapa 69 alfalfa seed rinse after 4 h of imbibition is quercetin-3-*O*-galactoside. The second largest peak also releases quercetin with acid hydrolysis, and together these two quercetin conjugates comprise 46% of the flavonoid fraction. Luteolin-7-*O*-glucoside and the luteolin aglycone represent an additional 22% of that flavonoid fraction. Flavonoid fractions from the 4-h seed rinse contain about 3 nmole per seed of both luteolin-7-*O*-glucoside and 5-methoxyluteolin. A third peak contains an unidentified activator of NodD2, and at least 10 smaller peaks are visible on HPLC chromatograms. Luteolin and the minor component chrysoeriol (3'-methoxyluteolin) are the only compounds in seed exudate that activate *nodD1* gene product in low nM concentrations. Chrysoeriol is three to four times more active than luteolin in NodD1 activation assays. Luteolin-7-*O*-glucoside, 5-methoxyluteolin and 3',5-dimethoxyluteolin show *nod*-gene-inducing activity at higher concentrations, but those results can be attributed to luteolin released from the first compound and colorimetric interference in assays of the latter two molecules. When the appearance of individual flavonoids in the seed rinse is followed over time, most show patterns indicating a rapid release (Table 2). Tests in which seed coats were removed before imbibition indicated that most luteolin, luteolin-7-O-glucoside, and quercetin-3-O-galactoside is released from the seed coat.

Table 1. Traits of flavonoids identified in alfalfa seed and root exudates (13, 14, 19, 20, 22 and unpublished results of U.A. Hartwig). All compounds were tested for *nodC-lacZ* induction with extra copies of *nodD1* and *nodD2* in *R. meliloti* with methods described previously (14).

Flavonoid	Exudate Source	*nod* Gene Product Activated	Growth Effect*	C-5†
Luteolin	Seed	NodD1	yes	yes
Chrysoeriol	Seed	NodD1	nd‡	yes
Luteolin-7-*O*-glucoside	Seed	-	yes	yes
3',5-Dimethoxyluteolin	Seed	-	nd	yes
5-Methoxyluteolin	Seed	-	nd	yes
Quercetin-3-*O*-galactoside	Seed	-	yes	yes
4,4'-Dihydroxy-2'-methoxychalcone	Root	NodD1 NodD2	no	no
4',7-Dihydroxyflavone	Root	NodD1	no	no
4',7-Dihydroxyflavanone	Root	NodD1	no	no
Formononetin	Root (under stress)	-	no	no

*Increases growth rate of *R. meliloti* on minimal medium.
†Substitution on C-5 position (flavone numbering system).
‡nd - not determined

Flavonoids Released from Alfalfa Roots. The flavonoid exuded in largest quantities from sterile 72-h-old alfalfa roots is 4',7-dihydroxyflavone. Smaller amounts of 4',7-dihydroxyflavanone and 4,4'-dihydroxy-2'-methoxychalcone also are released, and together these three flavonoids constitute nearly all of the detectable *nod*-gene-inducing activity (19). The chalcone is unique among alfalfa *nod* gene inducers in that it is a 10-fold stronger activator of NodD1 than luteolin and also activates NodD2 (14). In contrast to the seed flavonoids, all molecules thus far identified in alfalfa root exudate lack C-5 substitutions.

Table 2. Kinetics of release for selected compounds exuded by Moapa 69 alfalfa seeds. Compounds were in a sterile aqueous seed rinse which was changed at the indicated times after imbibition began.

Compound	Period of Release (h)				
	0-1	1-2	2-4	4-8	8-24
	(relative units h^{-1})*				
Luteolin-7-*O*-glucoside	28	130	120	5	0
Quercetin-3-*O*-galactoside	325	800	430	14	0
Luteolin	2	7	10	1	0
NodD2 activator†	110	120	60	11	8

*from HPLC peak area at maximum absorbance (260-400 nm)
†Purity and identity have not been established.

Functions of Flavonoids Released from Alfalfa. Five flavonoids identified thus far in alfalfa root and seed exudates induce *nod* genes (Table 1). When combinations of these molecules are tested for effects on *nodC-lacZ* transcription in *R. meliloti*, several important principles emerge. First, weak inducers like 4',7-dihydroxyflavanone or 4',7-dihydroxyflavone can decrease effects of stronger inducers (12). Previous work showed that a commercially available weak *nod* gene inducer decreased the activity of a strong inducer in clover rhizobia (6), and the data for flavonoids released naturally from alfalfa indicate that such interactions may occur during seedling development. A second principle is that nonsaturating concentrations of luteolin and 4,4'-dihydroxy-2'-methoxychalcone have a synergistic effect on NodD1 activation (12, 14). When extra copies of *nodD2* are present, the synergism is not observed, and luteolin decreases the chalcone induction of *nodC-lacZ*.

In addition to its role as a NodD activator, luteolin released from alfalfa seeds induces positive chemotaxis in *R. meliloti* (2). Whether *R. meliloti* is chemotactic to other *nod* gene inducers has not been reported. The phenomenon may be important, however, because water soluble luteolin-7-*O*-glucoside may diffuse some distance in soil, be hydrolyzed to the less soluble luteolin, and induce rhizobial movement toward the germinating seed.

One striking effect of the major flavonoids released from seeds (luteolin, luteolin-7-*O*-glucoside, quercetin-3-*O*-galactoside) is their capacity to increase growth rate of *R. meliloti* on a minimal medium (Table 1; U.A. Hartwig, unpublished data). The effect saturates at 2 to 5 uM and occurs with two carbon sources: mannitol (0.25-1.0 $g \cdot L^{-1}$) and glucose (1.0 $g \cdot L^{-1}$). Rapid growth of cells on a complex medium is not increased by luteolin, but significant decreases in generation time on a minimal medium can be measured easily (e.g. for luteolin doubling time decreased from 61 to 25 h). All three strains of *R. meliloti* tested gave similar responses, as did a strain mutated at the three known *nodD* loci. Although *Pseudomonas putida* grew faster with luteolin and quercetin, these compounds are not general enhancers of microbial growth because they had no effect on *Bacillus subtilis* or *Agrobacterium tumefaciens* growing on the same medium. A previous report noted that two isoflavonoids released naturally from soybean roots increased growth in one of two soybean rhizobia, but unexplained concentration effects made the results difficult to interpret (5).

Alfalfa flavonoids known to promote rhizobial growth are substituted at the C-5 position (Table 1). Tests indicate that chrysin (5,7-dihydroxyflavone) produced growth promoting effects equal to luteolin, but flavone, 5-hydroxyflavone, and 7-hydroxyflavone did not. That observation is consistent with the fact that 5-deoxyflavonoids released from roots did not promote rhizobial growth under any conditions tested. Thus these results offer the first clear data suggesting that plants create ecochemical zones in the soil by releasing structurally different flavonoids from seeds and roots.

Host Control of Flavonoids Outside the Plant. At least three mechanisms are potentially available for altering the biological effectiveness of flavonoids released from alfalfa plants: synthesis, exudation, and extracellular modification. The rapid release of *nod* gene inducers from seed coats into aqueous solution (Table 2; 13) and the fact that in a related plant the same molecules can be extracted into organic solvents without hydrating the living system (8) suggest that these molecules are synthesized as the seed matures, not as it germinates. In contrast, experiments using ^{14}C-phenylalanine and L-2-aminooxy-3-phenylpropionic acid, an inhibitor of phenylalanine ammonia-lyase (1), give strong evidence that exudation of *nod*-gene-inducing flavonoids from alfalfa seedling roots requires concurrent synthesis (20). The data show that roots can synthesize the relevant flavonoids, but hypocotyls also showed high synthetic activities. Thus synthesis in the hypocotyl followed by translocation to the root cannot be eliminated.

Release of *nod*-gene-inducing flavonoids from alfalfa roots apparently is a selective process. Seedling roots contain significant quantities of formononetin (7-hydroxy-4'-methoxyisoflavone), in addition to *nod*-gene-inducing flavonoids, but only the latter molecules are exuded from sterile, unstressed roots (20). Root cells that exude *nod* gene inducers have not been identified, but the absence of formononetin in normal exudates suggests that a general phenomenon such as the sloughing off of cells during root growth is not responsible for the process.

Plant proteins probably convert inactive *nod*-gene-inducing flavonoids into active forms after they are released from germinating seeds. Recent results show that Moapa 69 alfalfa seeds release a glucosidase that produces luteolin from luteolin-7-*O*-glucoside (U. A. Hartwig, unpublished data). Whether plant enzymes capable of hydrolyzing 5-methoxyluteolin to luteolin are released from seeds is not known, but demethylases from bacteria have been reported (3). Thus the 3 nmoles of luteolin-7-*O*-glucoside and 5-methoxyluteolin released from seeds (13) are potential sources of *nod* gene inducers. The known instability of the 5-methoxyl relative to the 3'-methoxyl group (9) may also help convert 3',5-dimethoxyluteolin to chrysoeriol or 5-methoxyluteolin to luteolin.

In addition to direct effects of synthesis, exudation, and extracellular modification on individual flavonoids, flavonoids may interact among themselves in spatial terms (Figure 1). The region where flavonoids from the seed and root coexist is the most interesting. At that site luteolin from the seed could interact with 4,4'-dihydroxy-2'-methoxychalcone from the root to produce a zone of synergistic *nod* gene induction (12, 14). That region is at the top of the primary root, where most legumes have many nodules. Effects of flavonoid combinations on growth of *R. meliloti* have not been examined. It is reasonable to suggest, however, that because seed flavonoids are released before roots develop, rhizobial populations are stimulated in that region, and the root is exposed to a higher density of those bacteria than would occur if seeds did not release flavonoids. The overall effect may be an increased number of rhizobial cells on the the root after it grows out of the seed zone. Thus flavonoids released from roots and seeds may be important mechanisms for defining ecochemical zones that structure microbial activities in the soil. Whether such interactions occur in soil will require detailed studies of flavonoid movement and metabolism under natural conditions. The presence of flavonoids in soils is suggested by the fact that they can be extracted from samples of commercial vermiculite (S.M. Tsai, unpublished data).

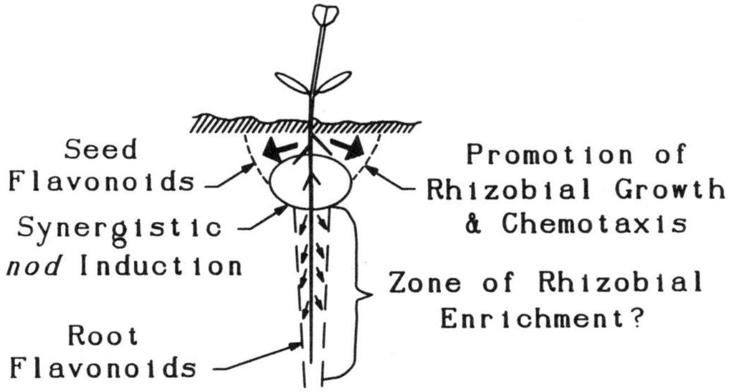

Figure 1. Ecochemical zones of flavonoids released from alfalfa seeds and roots during germination.

Flavonoids from Common Bean

To determine if traits associated with flavonoid synthesis, exudation, and function are unique in alfalfa or occur also in other legumes, we are studying flavonoids released from seeds and roots of common bean. A major role for flavonoids in nodulation of bean is supported by preliminary results which show that a black-seeded cultivar has more primary root nodules and greater *nod*-gene-inducing activity in seed exudate than an isogenic white-seeded mutant (M. Hungria, unpublished data). HPLC data and UV/visible spectral characteristics suggest that there are many more *nod*-gene-inducing flavonoids released from bean than alfalfa seeds and that several inducers exuded from black seeds are anthocyanins, a flavonoid class not previously known to contain *nod* gene inducers. Identifying bean flavonoids and analyzing how they interact with *nodD* gene products in *R. leguminosarum* bv. *phaseoli* will provide important data for comparisons with alfalfa.

Acknowledgments

This work was suported in part by U.S. Department of Agriculture CRGO grants 87-CRCR-1-2552 and 89-37280-4697, as well as grant IS-1348-87 from BARD, the U.S.-Israel Binational Agricultural Research and Development Fund.

References

1. Amrhein, N. & Gödeke, K.-H. (1977) *Plant Sci. Lett.* **8**, 313-317.
2. Caetano-Anolles, G., Crist-Estes, D. K. & Bauer, W. D. (1988) *J. Bacteriol.* **170**, 3164-3169.
3. Cartwright, N. J. & Smith, A. R. W. (1967) *Biochem. J.* **102**, 826-841.
4. D'Arcy-Lameta, A. (1986) *Plant Soil* **92**, 113-123.
5. D'Arcy-Lameta, A. & Jay, M. (1987) *Plant Soil* **101**, 267-272.
6. Djordjevic, M. A., Redmond, J. W., Batley, M. & Rolfe, B. G. (1987) *EMBO J.* **6**, 1173-1179.
7. Firmin, J. L., Wilson, K. E., Rossen, L. & Johnston, A. W. B. (1986) *Nature* **324**, 90-92.
8. Gehring, E. & Geiger, H. (1980) *Zeit. Naturforsch.* **35c**, 380-383.
9. Geiger, H., Casteele, K. V. & Van Sumere, C. F. (1984) *Zeit. Naturforsch.* **39b**, 393-396.
10. Harborne, J. B. (1980) *Encyc. Pl. Physiol.* **8**, 329-402.
11. Harborne, J. B. (1988) *The Flavonoids, Advances in Research Since 1980* (Chapman and Hall, London) p. 621.
12. Hartwig, U. A., Maxwell, C. A., Joseph, C. M. & Phillips, D. A. (1989) *Plant Physiol.* **91**, 1138-1142.
13. Hartwig, U. A., Maxwell, C. A., Joseph, C. M. & Phillips, D. A. (1990) *Plant Physiol.* **92**, 116-122.
14. Hartwig, U. A., Maxwell, C. A., Joseph, C. M. & Phillips, D. A. (1990) *J. Bacteriol.* **172**, 2769-2773.
15. Kosslak, R. M., Bookland, R., Barkei, J., Paaren, H.E. & Appelbaum, E. R. (1987) *Proc. Natl. Acad. Sci. USA* **84**, 7428-7432.
16. Long, S. R. (1989) *Cell* **56**, 203-214.
17. Lundegardh, H. & Stenlid, G. (1944) *Arkiv Bot.* **31A**, 1-27.
18. Mabry, T. J., Markham, K. R. & Thomas, M. B. (1970) *The Systematic Identification of Flavonoids* (Springer-Verlag, New York) p. 354.
19. Maxwell, C. A., Hartwig, U. A., Joseph, C. M. & Phillips, D. A. (1989) *Plant Physiol.* **91**, 842-847.
20. Maxwell, C. A. & Phillips, D. A. (1990) *Plant Physiol.* In press.
21. Olah, A. F. & Sherwood, R. T. (1971) *Phytopathol.* **61**, 65-69.
22. Peters, N. K., Frost, J. W. & Long, S. R. (1986) *Science* **233**, 977-980.
23. Redmond, J. W., Batley, M., Djordjevic, M. A., Innes, R. W., Kuempel, P. L. & Rolfe, B. G. (1986) *Nature* **323**, 632-635.
24. Zaat, S. A. J., Schripsema, J., Wijffelman, C. A., Van Brussel, A. A. N. & Lugtenberg, B. J. J. (1989) *Plant Molec. Biol.* **13**, 175-188.

Gresshoff, Roth, Stacey & Newton (eds.) Nitrogen Fixation: Achievements and Objectives
Chapman and Hall New York • London 1990

Lucerne nitrogen fixation in low rhizosphere oxygen

C. Arrese-Igor[1] and P. M. Aparicio-Tejo[2].

[1]Departamento de Biología Vegetal. Universidad del País Vasco. Apdo. 644. E-48080 Bilbao. Spain. and [2]Departamento de Producción Agraria. Universidad Pública de Navarra. Avenida del Ejército 2. E-31002 Pamplona. Spain.

Introduction

It has been suggested that nitrogen fixation could be limited by oxygen availability, since optimal rates are obtained at supra-ambient O_2 tensions (2). However, in the soil environment of the nodule, O_2 levels greater than 0.21 probably do not occur naturally, while subambient pO_2s exist under various environmental circumstances. Short-term experiments pointed out that soybean nodules were able to adapt rapidly to low oxygen concentrations (1). No explanation concerning the mechanisms of such adaptation was done and these results could reflect an analytical error, since ARA is not a good index for measuring nitrogenase activity in different oxygen environments (3).

Material and methods

Plants of lucerne inoculated with *Rhizobium meliloti* 102-F-51 were maintained to control (0.21 atm) or low oxygen (0.01 atm) partial pressure in the rooting medium. Plants were 12 weeks old and at vegetative stage when used for experiments. ARA was determined by taking samples from the sealed pots after acetylene supply. Respiration was measured as both O_2 uptake using an oxygen-electrode and CO_2 evolution with an IRGA.

Results

Shoot dry weight was unaffected, but root and nodule DW were reduced by 50% in this low oxygen level as related to control plants. FW/DW ratio in roots and nodules was higher at 1% as a result of aerenchyma formation. Specific ARA showed similar values in both sets of plants, although 1% oxygen-grown plants were more sensitive to acetylene supply, showing a dramatic decline in nitrogenase activity after repeated assays. Nodule respiration rates were 5 fold higher in control plants as CO_2 evolution whereas no differences were observed in O_2 uptake between both sets of plants. Transport of nutrients to the shoot was decreased in 1% plants.

Conclusion

It is suggested that nitrogenase activity in low oxygen conditions could be supported by morphological changes, involving aerenchyma formation in the nodule cortex, a decreased nodule gas permeability and by a more efficient recycling of CO_2.

References

(1) Criswell JG et al. (1976) Plant Physiol. **58,** 622-625.
(2) Drevon JJ et al. (1988) Plant Physiol. Biochem. **26,** 73-78.
(3) Witty et al. (1986) *In* Oxford Surveys of Plant Molecular & Cell Biology, vol 3 (Miflin BJ, ed.) pp. 275-314. Oxford Press. Oxford.

Gresshoff, Roth, Stacey & Newton (eds.) Nitrogen Fixation: Achievements and Objectives
Chapman and Hall New York • London 1990

Enzymatic and nonenzymatic mechanisms for ferric leghemoglobin reduction in legume root nodules

Manuel Becana[1,2], Lin Ji[1] and Robert V. Klucas[1]

[1]Dept. Biochemistry, Univ. of Nebraska, Lincoln, NE 68583 USA; [2]E.E. Aula Dei, Apdo. 202, 50080 Zaragoza, Spain.

Evidence is presented for the operation in nodules of three systems for restoring functional ferrous leghemoglobin (Lb^{2+}) from its inactive, ferric form. (i) Reduction of Lb^{3+} by a ferric leghemoglobin reductase. The enzyme is a flavoprotein, has an M_r = 100 kDa and two equally-sized subunits, and exhibits a Km (soybean Lba^{3+} = 9 uM, and K_m(NADH) = 51 uM. NADPH is only 30% (initial velocities) as effective as NADH. Ferric leghemoglobin reductase converts 500 nmol Lb^{3+} to Lb^{2+}.CO (or $Lb^{2+}.O_2$) per min per mg protein, and does not require an exogenous electron carrier. The enzyme shows similar affinity for soybean, bean and cowpea ferrous leghemoglobins, but different V_{max}-values. The reductase is inactive when Lb^{3+} is bound to nicotinate or NO_2^-. (ii) Direct reduction by Lb^{3+} by NAD(P)H, ascorbate and cysteine. Reduction by NAD(P)H is greatly stimulated by trace amounts of metals such as Mn^{2+}. (iii) Reduction of Lb^{3+} by a system NAD(P)H → free flavins → Lb^{3+}. The reaction does not occur via O_2^- or H_2O_2, and thus NAD(P)H-reduced flavins can directly reduce Lb^{3+}. The velocity of the reaction follows the order riboflavin > FMN > FAD. The reaction is also faster and more efficient (less NADH oxidized per Lb^{3+} reduced) in anaerobic conditions, with complete reduction of Lb^{3+} to Lb^{2+}. Therefore, the microaerobic conditions inside the infected region of nodules may greatly facilitate reduction of Lb^{3+} *via* this mechanism.

Gresshoff, Roth, Stacey & Newton (eds.) Nitrogen Fixation: Achievements and Objectives
Chapman and Hall New York • London 1990

Poly-ß-hydroxybutyrate and N_2 fixation of soybean bacteroids

F.J. Bergersen and G.L. Turner

CSIRO Division of Plant Industry, GPO Box 1600, Canberra, ACT, 2601, Australia

Introduction: During supply of malate to bacteroids in flow-chamber experiments at low concentrations (ca 10 nM) of free dissolved O_2, N_2 fixation was partially suppressed but was enhanced when malate was washed out of the chamber by a supply of substrate-free medium (1). Further investigation of this unexpected result is reported.

Methods: Bacteroids retained in a stirred reaction chamber (1), were supplied with a steady flow of medium containing ^{14}C-malate, oxyleghaemoglobin (LbO_2) and dissolved air or the same medium without malate. N_2 fixed to NH_3, $^{14}CO_2$ and residual malate were measured in the effluent solution. Concentration and rate of consumption of O_2 were calculated from spectrophotometry of LbO_2 in the effluent. The location and form of ^{14}C assimilated by the bacteroids was determined.

Results: The principal results are summarized in Table 1. They show that, during supply of ^{14}C-malate, ^{14}C accumulated in the bacteroids and about half was in poly-β-hydroxybutyrate (PHB). At washout of extracellular malate, N_2 fixation increased. After washout, synthesis of ^{14}C-PHB continued. After exhaustion of non-PHB-^{14}C, a new PHB-utilizing steady state was established and N_2 fixation remained high. Low ^{14}C in respired CO_2 reflected the large amount (>65% of bacteroid dry wt.) of PHB already present when the experiment began.

Table 1. Accumulation and utilization of ^{14}C-malate

A. At the start of washout of malate after 106 min of supply. B. At washout of malate, 47 min later. C. At exhaustion of non-PHB-^{14}C, 77 min after washout. D. PHB-utilizing steady state, 153 min after washout. O_2 consumption 4.7 - 5.0 nmol min^{-1}mg^{-1}; O_2 concentration 6 - 13 nM. Basis, bacteroid dry weight.

	^{14}C-malate (mM)	N_2 fixation (nmol NH_3 min^{-1}mg^{-1})	$^{14}CO_2$ (DPM μmole^{-1})	^{14}C-β-hydroxybutyrate (DPM mg^{-1} bacteroids) free monomer	PHB
A.	7.4	0.16	9,400	5,050	4,300
B.	0	0.41	9,100	8,300	7,000
C.	0	1.00	750	<100	11,300
D.	0	0.93	230	280	8,300

Reference: (1) Bergersen, F.J. & Turner, G.L. (1990) *Proc. Roy. Soc. Lond.* **238**, 295-320.

Gresshoff, Roth, Stacey & Newton (eds.) Nitrogen Fixation: Achievements and Objectives
Chapman and Hall New York • London 1990

Generation of reducing power for N_2 fixation by an obligate methanotroph, *Methylosinus trichosporium* OB3b: Properties of formate dehydrogenase

Yung-pin Chen, Michael G. Hardin and Duane C. Yoch

Department of Biological Sciences, University of South Carolina, Columbia, SC, 29208, USA.

The generation of reducing power for driving N_2 fixation in the obligate methane oxidizing bacterium, Methylosinus trichosporium OB3b, involves the oxidation of formate by an NAD-linked formate dehydrogenase (FDH) (ref. 1), and a ferredoxin-NAD reductase (ref. 2). This system, reconstituted from purified components, has a reaction sequence as follows:

FDH → NADH → ferredoxin-NAD reductase → ferredoxin → N_2ase.

FDH was purified and shown to be an Fe/S, FMN-containing enzyme that exists in two molecular weight forms, one approximately 315 Kda and the other 155 Kda. The enzyme has two subunits; the Mr of the α-subunit is 102,600 ± 3,900.and the ß-subunit is 53,800 ± 2,800. The enzyme (Mr = 315,000) is composed of these subunits in an apparent $\alpha_2\beta_2$ arrangement. When the enzyme was isolated aerobically, there was no cofactor requirement for NAD reduction; however, when isolated anaerobically, activity was 80% dependent on the addition of FMN to the reaction mixture. Furthermore, the addition of formate to an active, anoxic solution of formate dehydrogenase rapidly inactivated it in the absence of an electron acceptor. Because this activity could be reconstituted approximately 85% by 50 nM FMN it was concluded that the flavin dissociates from the enzyme under reductive conditions. This dissociation can be reversed and the activity restored by increasing the concentration of FMN in solution. FAD would not replace FMN in reconstituting enzyme activity. The Kms of formate dehydrogenase for formate, NAD and FMN were 146, 200 and 0.02 mM, respectively. 'Pseudomonas oxalaticus' formate dehydrogenase, which has near identical physical characteristics as that of M. trichosporium, was also shown to be inactivted under anoxic conditions by formate and reactivated by FMN. 'P. oxalacticus' is very closely related to Alcaligenes eutrophus (3). The near identical characteristics of their formate dehydrogenases suggest the possibility of lateral gene transfer between these two seemingly unrelated organisms.

1. Chen, Y.P., and Yoch, D.C. (1988) J. Gen. Microbiol. **134**:3123-3128.
2. Chen, Y.P., and Yoch, D.C. (1989). J. Bacteriol. **171**:5012-5016.
3. Jenni, B., Realini, L., Aragno, M., and Tamer, Ü. (1988) System. Appl. Microbiol. **10**:126-133.

Gresshoff, Roth, Stacey & Newton (eds.) Nitrogen Fixation: Achievements and Objectives
Chapman and Hall New York • London 1990

Comparison of O_2 regulation in soybean, pea and lupin nodules

Diaz del Castillo L., S. Hunt, D.B. Layzell.

Department of Biology, Queen's University, Kingston, Ont., Canada.

The O2 concentration in legume nodules is controlled by a barrier to gas diffusion in the nodule cortex. The regulation of this barrier may differ in nodules which have different morphological characteristics. It was the aim of this study to compare diffusion barrier regulation in soybean, lupin and pea symbioses having nodules of different morphology. Nitrogenase activity was measured as H evolution in N2:O2 and in Ar:O2 (total nitrogenase activity, TNA). The potential nitrogenase activity (PNA) was estimated from the maximum activity displayed during a 30 min. ramp from 20-100% O2 in Ar.

It was found that the pea and the lupin nodules were O2 limited (PNA/TNA = 1.3 and 1.4) at ambient O2 concentration, while the soybean nodules were operating close to their PNA under the same conditions (2) (PNA/TNA = 1.1 after a 30 min. ramp from 20-30% O2) (1).

In soybean the step increase in pO2 from 20-30% O2 caused an immediate decline in nitrogenase activity (1). However, in pea and lupin there was a transient increase in nitrogenase activity before the decline occured. This is consistent with a greater O2 limitation in nodules of pea and lupin than in soybean nodules as stated above. In all the three symbioses, the nodules adapted to the increased pO2 and nitrogenase activity returned to initial levels within 30 minutes.

Removal of the shoot from all symbioses caused a reduction in nitrogenase activity to only 20 to 50 % of the initial rate within three hours. In the detopped plants, PNA was 2.3 to 3.1 times TNA and these PNA values were obtained at 60-80% O2 in Ar (3). This indicates that a reduction of phloem sap supply causes nitrogenase inhibition by O2 limitation, presumably through an increase in diffusion barrier resistance.

In the detopped plants the PNA was only 79%, 67% and 52% (for pea, lupin and soybean, respectively) of the PNA in the control plants. This may be attributed to a decrease in the availability of phloem-supplied-carbohydrate for nodule metabolism in these plants.

We can conclude from this study that despite morphological differences in the nodules of all three symbioses, they respond in a similar way to increases in external oxygen concentration and to deprivation of phloem sap supply.

(1) HUNT, S., B. J. KING, D. B. LAYZELL. Plant Physiol. (1989) 91:315-321

(2) KING, B. J. Ph.D. Thesis (1989). Queen's Univ. Kingston, Ont., Canada.

(3) VESSEY, J. K., K. B. WALSH AND D. B. LAYZELL. Physiol. Plant.(1988) 73:113-121

Gresshoff, Roth, Stacey & Newton (eds.) Nitrogen Fixation: Achievements and Objectives
Chapman and Hall New York • London 1990

Measurement of nodule diffusion barrier resistance and respiration rate by spectrophotometry of leghemoglobin

R. Ford Denison[1] and David B. Layzell[2]

[1]USDA/ARS, Beckley, WV 25802, USA
[2]Biology Dept., Queen's Univ., Kingston, ON K7L 3N6, Canada

The responses of legume nodules to such diverse factors as photosynthate supply, soil water status, and soil nitrate all apparently involve changes in nodule gas permeability. A decrease in O_2 permeability results in a reduced O_2 supply for respiration in support of N_2 fixation. Nodule permeability has been estimated from steady-state nodule respiration rate, usually estimated from root+nodule CO_2 production (4), or from unsteady- (5) or steady-state (1) acetylene reduction. Only the latter method has been used in field studies. A new technique, noninvasive fiber-optic spectrophotometry of leghemoglobin (nodule oximetry), measures the O_2 concentration within intact, attached nodules in the laboratory (3) and under field conditions (2). Nodule oximetry can also be used to estimate nodule respiration (O_2 consumption rate inside the cortical diffusion barrier), from the rate of decrease of leghemoglobin (LHG) oxygenation in a nodule surrounded by N_2. The rate of increase in LHG oxygenation after an increase in external O_2 concentration can be used to estimate nodule O_2 permeability, after correcting for O_2 consumption using the respiration data obtained under N_2. The experimental unit for this procedure is a single nodule, which has both advantages and disadvantages. A portable nodule oximeter, to measure nodule respiration and permeability under field conditions, is under development.

1. Denison, R.F., Weisz, P.R. & Sinclair, T.R. (1983) Plant Physiol. 73:648-651.
2. Denison, R.F., Smith, D.L., Legros, T. & Layzell, D.B. (1991) Agron. J. (in review).
3. Layzell, D.B., Hunt, S. & Palmer, G.R. (1990) Plant Physiol. 92:1101-1107.
4. Sheehy, J.E., Minchin, F.R. & Witty, J.F. (1983) Ann. Bot. 52:565-571.
5. Weisz, P.R. & Sinclair, T.R. (1988) Plant Soil 105:69-78.

Gresshoff, Roth, Stacey & Newton (eds.) Nitrogen Fixation: Achievements and Objectives
Chapman and Hall New York • London 1990

Influence of oxygen on the hydrogenase-nitrogenase interaction in soybean

J.J. Drevon and A. Godfroy

Laboratoire de Recherches sur les Symbiotes des Racines, INRA, 2 Place Viala, 34060 Montpellier Cedex, France.

Most legume root-nodules evolve nitrogenase-produced H_2, which is a loss of energy, except when they possess an uptake hydrogenase. This hydrogenase expression limited N_2 fixation in greenhouse and field pluriannual trials that compared the efficiency of isogenic Hup^+ (H_2 uptake positive) and Hup^- *B. japonicum* (Drevon et al., 1987, 1988). To test whether this negative effect of hydrogenase could be due to O_2 limitation, we measured the nitrogenase activity (ARA) of the *B. japonicum* PJ17-1 (Hup^+) - *Glycine max.* cv Kinsoy symbiosis under various external O_2 tensions (pO_2), either on bacteroid suspensions or intact nodulated roots and in the presence or absence of H_2.

1. Influence of hydrogenase on *ex-planta* bacteroid's nitrogenase activity: Hydrogenase activation by an external supply of H_2 modified the *in vitro* ARA response to O_2 of Hup^+ bacteroids in two ways: (1) the optimal pO_2 for ARA was increased from 0.5 kPa without H_2 to about 2 kPa in the presence of H_2; (2) under suboptimal pO_2 (when O_2 was limiting), the ARA in the presence of H_2 was lower than in the absence of H_2. There was no H_2 effect with the isogenic Hup^- bacteroids.

2. Influence of hydrogenase on *in-planta* bacteroid's nitrogenase activity: The response of plants inoculated with Hup^+ rhizobia, as measured with an open-gassing system, was modified in three ways by H_2 addition: (1) the optimal pO_2 for ARA was higher, i.e., 40.3 kPa O_2 in the presence of H_2 as opposed to 35.3 kPa O_2 in the absence of H_2; (2) the ratio of ARA under optimal pO_2 to ARA under 20 kPa O_2 was higher, i.e., 141 in the presence of H_2 as compared to 110 in the absence of H_2; (3) the ARA for suboptimal pO_2 was lower, although this effect of H_2 was less significant than in *ex-planta* experiments. The addition of H_2 had no affect on the ARA response to O_2 with plants inoculated with isogenic Hup^- rhizobia; these gave a similar response to the Hup^+ in the absence of H_2.

The *ex-planta* results show that the hydrogenase reaction limits nitrogenase activity under O_2-limited conditions although it can participate in O_2 protection of nitrogenases. The *in-planta* results suggest that soybean nodule activity is O_2-limited and that in natural habitats where the pO_2 is equal to or less than 21 kPa, the hydrogenase reaction limits nitrogenase activity. However, for an unknown reason, the negative effect of H_2 is slower *in* than *ex-planta*.

Gresshoff, Roth, Stacey & Newton (eds.) Nitrogen Fixation: Achievements and Objectives
Chapman and Hall New York • London 1990

Expression of alfalfa nodule aspartate aminotransferase resulting from divergent selection for nodule enzyme activity

M.W. Farnham*, C.P. Vance, N.R. Degenhart and D.K. Barnes.

Plant Science Research Unit, USDA-Agricultural Research Service and the Department of Agronomy and Plant Genetics, University of Minnesota, St. Paul, MN 55108 USA.

Aspartate aminotransferase (AAT:EC 2.6.1.1) is a key enzyme in the assimilation of fixed nitrogen (N) derived from the legume-Rhizobium symbiosis. Aspartate and asparagine are the major nitrogenous compounds exported from root nodules of alfalfa (Medicago sativa L.). Since most of the N in nodules is incorporated into aspartate and asparagine, AAT must play a primary role in the assimilation of symbiotically fixed N.

Two isozymes of AAT have been identified in alfalfa nodules. AAT-1 is constituitively expressed and AAT-2 is a nodule-enhanced enzyme. Previous research has shown that AAT-1 and AAT-2 are immunologically distinct and the products of two separate genes. Three isoforms of AAT-2 (AAT-2a, 2b, and 2c) occur due to allelic differences at the AAT-2 locus and are evident in all populations examined to date. Conversely, there is no evidence for allelic variation at the AAT-1 locus.

Bidirectional selection for nodule AAT activity in two unrelated alfalfa germplasms (Saranac AR and MNPL10 X MNNC7) resulted in high AAT activity subpopulations with about 20 % more nodule AAT activity on a fresh weight basis than low AAT activity subpopulations. Immunoprecipitation of in vitro nodule AAT activity, quantification of enzyme protein by ELISA, and AAT staining of native isozymes on PAGE gels were used to study the nature of nodule AAT activity in the high and low AAT subpopulations.

High AAT subpopulations from both germplasms exhibited about 5 % more AAT-2 activity as a percent of total activity than did low AAT subpopulations. Additionally, high AAT subpopulations had 18 % more AAT-2 enzyme protein per fresh weight of nodules than low AAT subpopulations, and this accounts for greater total AAT activity and greater AAT-2 activity in nodules of high activity subpopulations. The enhancement of nodule AAT activity by selection resulted due to enhanced expression of AAT-2, confirming the importance of AAT-2 to effective alfalfa root nodule AAT activity. No significant and consistent changes in AAT-2 genotypic makeup of subpopulations appeared to be caused by selection for AAT activity. Since no underlying genotypic changes at the AAT-2 locus were evident, selection must have affected a change(s) at another locus (or loci), other than AAT-1 or AAT-2, which indirectly effected the expression of nodule AAT and more specifically nodule AAT-2.

Gresshoff, Roth, Stacey & Newton (eds.) Nitrogen Fixation: Achievements and Objectives
Chapman and Hall New York • London 1990

O_2 induced changes in metabolite pool sizes in soybean nodules

S.M.Fernando, S.Hunt, R.Smith, D.H.Turpin, C.Atkins [1], D.B.Layzell

Dept. Biology, Queen's Univ., Kingston, Ont, Canada K7L 3N6. [1] Dept. Botany, Univ. Western Australia, Nedlands, W.Australia, 6009.

In recent years, various evidence has accumulated indicating that, in legume nodules, respiration and nitrogenase activity are limited by O_2 at all times (4). However, little is known about the role of O_2 in regulating specific pathways of carbon and nitrogen metabolism in legume nodules. In this present preliminary study, we attempt to identify those biochemical pathways which are sensitive to rapid changes in O_2 by monitering the sizes of metabolite pools following step increase (20% to 30%) or decrease (20% to 10%) in the O_2 concentration surrounding nodulated soybean roots.

The results and conclusions from these measurements are summarized below:

1. In the high O_2 treatment the sugar concentration declined within 2 min. This is consistent with a role for sucrose and other sugars in the osmoregulation of the cortical diffusion barrier(2).
2. The adenylate energy charge (AEC) increased under high O_2. This may be due to a increase in bacterial electron transport chain (ETC) activity at the expense of reductent used by nitrogenase, resulting in nitrogenase inhibition. The maintenance of the malate pool suggests that bacteroids were not carbohydrate-limited under high O_2.
3. The inhibition of nitrogenase activity associated with the low pO_2 treatment was corelated with a reduction in the ATP pool size and AEC and may be attributed to either:
   (a) A decrease in the bacteroid ATP pool required directly by nitrogenase OR
   (b) A decrease in the ATP pool size in the plant fraction required for carbon supply to the bacteroids.

References

1. Day, D.A., Price, G.D. & Udavardi, M.K.(1989) Aust.J.Plant Physiol. **16**, 69-84.
2. Hunt, S., Denison, R.F., King, B.J., Tajima,S., Kouchi, H., Layzell, D.B. (1990) Abstract in this volume.
3. Hunt, S., .King, B.J., Layzell, D.B. (1989) Plant Physiol. **91**, 315-321.
4. Layzell, D.B.,Hunt, S., Palmer, G.R. (1989) Plant Physiol. **91**,1101-1107.

Gresshoff, Roth, Stacey & Newton (eds.) Nitrogen Fixation: Achievements and Objectives
Chapman and Hall New York • London 1990

Characterization of hydrogen-ubiquinone oxidoreductase activity in *Bradyrhizobium japonicum* membranes

D.M. Ferber and R.J. Maier

Department of Biology, The Johns Hopkins University, Baltimore, MD 21218, USA

H_2 oxidation in *Bradyrhizobium japonicum* involves a membrane-bound electron transport chain from H_2 to O_2 that functions in recycling the H_2 evolved by nitrogenase. The proposed sequence involves electron flow from hydrogenase to ubiquinone, then branched electron flow through either cytochromes bc and aa_3 to O_2 or through cytochrome b'(o) to O_2 (1). Since ubiquinone has been proposed as the immediate electron acceptor from hydrogenase we tested for H_2 uptake to ubiquinone with pure hydrogenase and with membranes. Pure hydrogenase, which is active with methylene blue (MB) as a substrate, was not able to catalyze H_2 uptake to ubiquinones Q_0, Q_1, and Q_{10} under a variety of assay conditions. Membrane vesicles were able to catalyze H_2 uptake to these ubiquinones. When exogenous Q_0 was added to H_2-respiring membranes, H_2 uptake increased and O_2 uptake decreased, indicating that Q_0 functions upstream of O_2 as a terminal electron acceptor. This conclusion is supported by the fact H_2-ubiquinone oxidoreductase activity was still observed when O_2 is depleted by respiration and when respiration is inhibited by NaCN. Strong evidence exists that in membranes, ubiquinone is situated upstream of the cytochromes (1). The quinone analog 2-n-heptyl-4-hydroxy quinoline-N-oxide (HQNO), which has been shown to inhibit H_2-O_2 respiration in membranes, was shown here to inhibit H_2-Q_0 oxidoreductase activity but not H_2 uptake to methylene blue (MB), indicating that different sites exist in membranes for ubiquinone reduction and MB reduction. In contrast, NaCN inhibited H_2-O_2 respiration, but did not inhibit H_2-Q oxidoreductase activity, as expected from its known function as a terminal oxidase inhibitor. H_2-ubiquinone oxidoreductase activity was found to be significantly more enriched in membranes than H_2-MB activity, implying that the ubiquinone-reducing component may be physically separable from hydrogenase by cell fractionation. Based on all of these results and other available data, a model was proposed which includes a membrane-bound, non-cytochrome component which functions in electron transport between hydrogenase and ubiquinone in *B. japonicum* membranes.

1. O'Brian, M.R. and R.J. Maier (1988) Adv. Microb. Physiol. **29**, 1-52.

Gresshoff, Roth, Stacey & Newton (eds.) Nitrogen Fixation: Achievements and Objectives
Chapman and Hall New York • London 1990

Hydrogen recycling and nitrogen fixation by *Azorhizobium caulinodans* ORS571: Isolation and characterization of Hup mutants and *hup* genes

U.C.K. Hilgert, H. Stam*, W. de Vries*, M.M.A. Ferdinandy-van Vlerken*, A.H. Stouthamer*, and F.J. deBruijn**

Max-Planck-Institut fur Zuchtungsforschung, 5000 Koln 30, FRG and *Vrije Universiteit Amsterdam, The Netherlands. **Present address: MSU-DOE Plant Res.Lab, East Lansing, MI 48824, USA.

Azorhizobium caulinodans ORS571 which expresses an effective hydrogen uptake (Hup) system in culture and *in planta,* was examined for the energetics of nitrogen fixation and for the benefit of the H_2 recycling capacity. Following random Tn*5* mutagenesis of ORS571, four different Hup mutant strains were isolated (W8, W44, W52, W58; 1). The corresponding transposon insertions were found in four different loci clustered on the ORS571 genome (Fig. 1). The Tn*5* containing regions of three mutants were cloned and homogenotized into the chromosome of the ORS571 wild-type strain. These homogenotes (U8, U44, and U58) were shown to lack hydrogenase activity. Via hybridization, three complementing clones could be isolated from a cosmid gene bank of ORS571 (Fig. 1). These cosmid clones carry a genomic region harboring genes essential for Hup activity in ORS571. In addition, these clones share a high degree of homology with cloned *hup* specific sequences of *R. leguminosarum* (2). Nitrogen fixation by the mutant strains was found to be comparable to that of the wild-type in culture and *in planta,* although the stem nodules they induced turned senescent significantly earlier. Studies on succinate-limited, nitrogen-fixing chemostat cultures of U58 revealed a low H_2/N_2 ratio of 2.4. The energy cost of N_2 fixation under these conditions was $ATP/N_2=51$.

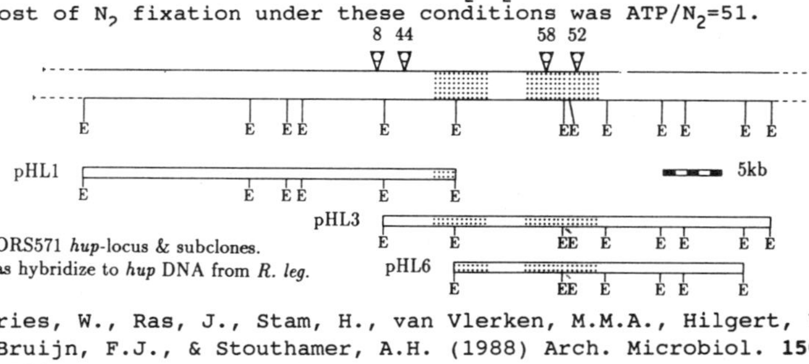

Figure 1. ORS571 *hup*-locus & subclones.
Dotted areas hybridize to *hup* DNA from *R. leg.*

1. DeVries, W., Ras, J., Stam, H., van Vlerken, M.M.A., Hilgert, U., de Bruijn, F.J., & Stouthamer, A.H. (1988) Arch. Microbiol. **150,** 595-599.
2. Tichy, H.V. Scheld, C., Ripke, H.M., Nelson, L.M., Fees, H. & Lotz, W. (1987) in *Molecular genetics of plant microbe interactions,* eds. Verma, D.P.S. and Brusson, N. (Martinus Nijhoff Dordrecht, The Netherlands), pp. 279-281.

Gresshoff, Roth, Stacey & Newton (eds.) Nitrogen Fixation: Achievements and Objectives
Chapman and Hall New York • London 1990

Low [O_2] is the probable cause of trehalose accumulation in legume nodules

I. Hoelzle and J.G. Streeter

Ohio State University/OARDC, Wooster, OH 44691, USA

Trehalose is synthesized by bacteroids and accumulates rapidly at the onset of N_2 fixation in legume nodules (2, 3). We wondered if trehalose accumulation might be related to the microaerophilic conditions found in infected cells and tested this hypothesis by comparing the accumulation of trehalose in cultured rhizobia grown with 1% or 21% O_2 in the gas phase. Liquid cultures of rhizobia were maintained with 1% or 21% O_2, then carbohydrates were extracted with 80% ethanol and analyzed using gas chromatography (4). Rhizobia cultured with reduced O_2 concentration accumulated significantly more trehalose. A similar response was obtained when different C sources were used (glycerol, glucose, succinate or complex media), and with 4 different rhizobia (*Rhizobium leguminosarum* bv. *phaseoli* USDA 2667, *R. l.* bv. *viceae* USDA 2391, *R. fredii* USDA 191, *R. meliloti* USDA 1021). Increases in trehalose concentration (expressed as μg trehalose/mg protein) under low O_2 tension ranged from 8-fold to more than 40-fold, depending on the strain of rhizobia and the culture medium composition. An experiment in which *R. l.* bv. *phaseoli* USDA 2667 cells were grown under 1% O_2 and then switched to 21% O_2 demonstrated the ability of rhizobia to accumulate trehalose under O_2 stress and then rapidly metabolize it when O_2 levels increased. Previous data from legume nodules indicates that bacteroid disaccharidases are not functional (1); the lack of trehalase activity in bacteroids implies that the increased trehalose pool size in free-living rhizobia cultured under 1% O_2 is due to decreased breakdown of the disaccharide. The marked accumulation of trehalose under low [O_2] provides additional evidence that this disaccharide, which is found in all N_2 fixing symbioses, may have some as yet unknown function in the fixation process.

1. Hoelzle, I., and Streeter, J.G. (1990) Can. J. Micro. 36,223-227.
2. Streeter, J.G. (1980) Plant Physiol. 66,471-476.
3. Streeter, J.G. (1985) J. Bacteriol. 164,78-84.
4. Streeter, J.G. (1987) Plant Physiol. 85,768-773.

Gresshoff, Roth, Stacey & Newton (eds.) Nitrogen Fixation: Achievements and Objectives
Chapman and Hall New York • London 1990

Light driven $^{14}CO_2$ fixation, light-decreased O_2 uptake, and acetylene reduction activity by free-living *Rhizobium* strain BTAil

M. Hungria, J.M. Ellis, A.R.J. Eaglesham and R.W.F. Hardy

Boyce Thompson Institute for Plant Research at Cornell University, Ithaca, NY 14853, USA.

BTAil is a strictly aerobic Rhizobium strain (isolated from sand) that nodulates both stems and roots of Aeschynomene, a tropical genus that usually grows in water-logged conditions. Previous experiments have shown that this strain, when grown in aerobic conditions and exposed to light; (1) produces pink colonies, (2) contains bacteriochlorophyll a, (3) has light-increased CO_2 fixation and (4) exhibits light-decreased O_2 uptake, suggesting that it is photosynthetic. $^{14}CO_2$ uptake by BTAil was detected from 1 min of incubation in all growth stages, and the reaction was linear for 60 min. The addition of inhibitors of photosystem II, orthophenantroline, dichlorophenyldimethylurea, and the uncoupler NH_4Cl immediately suppressed both light-driven $^{14}CO_2$ uptake and light-decreased O_2 uptake. Although BTAil showed photosynthetic properties, the strain could not grow in medium with C and it was unable to grow chemoautotrophically in an atmosphere enriched with H_2 and CO_2. In microaerophilic conditions and media deficient in N, free-living BTAil expressed both acetylene-reduction and $^{14}CO_2$-uptake activities. Acetylene-reduction activity was expressed without a lag phase over a two-day period. Higher rates of CO_2-uptake and acetylene-reduction activity occurred at exponential and early stationary growth phase in microaerophilic conditions and low-C medium (1 g l^{-1} of glucose), light increased about two-times over the acetylene-reduction activity rates by BTAil, suggesting that the photosynthesis could provide energy to the N_2-fixation process.

Gresshoff, Roth, Stacey & Newton (eds.) Nitrogen Fixation: Achievements and Objectives
Chapman and Hall New York • London 1990

An osmotic mechanism for diffusion barrier regulation in soybean nodules

S. Hunt, R.F. Denison[1], B.J. King, H. Kouchi[2], S. Tajima[2], D.B. Layzell.
Dept. Biol., Queen's Univ., Kingston, Canada. [1]USDA/ARS, Beckley, WV.
[2] Nat. Inst. Agrobiol. Res., Tsukuba, Japan.

The barrier to O_2 diffusion in soybean nodules is thought to consist of a layer of cells in the inner cortex, the intercellular spaces of which are either open or are plugged with water. The conductance of such a barrier would be determined by the extent of its aqueous phase. The following evidence supports our hypothesis that sucrose acts as an osmoticant to regulate intercellular water content and thereby diffusion barrier conductance: **(a)** When nodules were adapted to high pO_2, their conductance and sucrose concentration declined. The pool sizes of 37 other sugars, amino acids and organic acids did not change significantly. **(b)** After exposure to $Ar:O_2$, nodule conductance declined together with nodule osmolarity and sucrose concentration. **(c)** A reduction in phloem supply to the nodule decreased conductance and reduced sucrose content. **(d)** A mathematical model predicted that the measured changes in sucrose content and nodule osmolarity are sufficient to cause the redistribution of water necessary to produce the observed changes in nodule conductance.
Since most of the nodule's starch is localised in the inner cortex, we propose that the nodule controls diffusion barrier conductance by an osmotic mechanism involving the regulation of starch/sucrose interconversion. The scheme below illustrates our working hypothesis for diffusion barrier regulation.

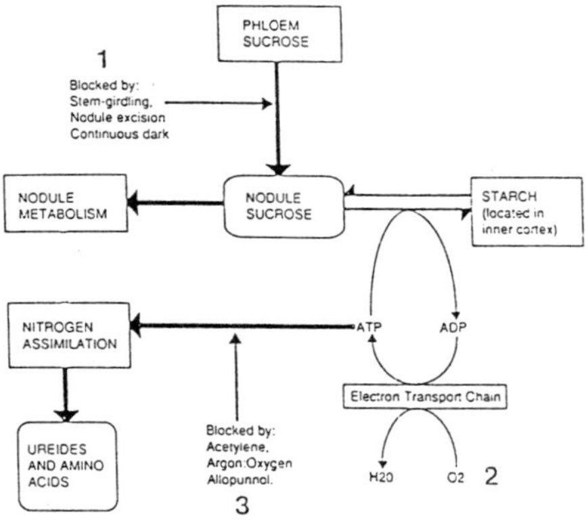

1. Nodule sucrose concentration is lowered by stem girdling, nodule excision and continuous darkness. These treatments also decrease nodule conductance.

2. Low O_2 concentrations inhibit starch formation, increase sucrose concentration and increase nodule conductance.

3. Treatments which block N assimilation increase nodule ATP, favouring starch formation and reducing nodule sucrose concentration. Lower ureide and amino acid concentrations may also contribute to a lower nodule osmolarity and conductance.

Gresshoff, Roth, Stacey & Newton (eds.) Nitrogen Fixation: Achievements and Objectives
Chapman and Hall New York • London 1990

Oxygen requirement for the enzyme-catalyzed NADH oxidation and ferric leghemoglobin reduction

LIN JI[§], MANUEL BECANA[§,¶] AND ROBERT V. KLUCAS[§]

[§]Department of Biochemistry, University of Nebraska Lincoln, NE, 68583-0718,USA; [¶]E.E. Aula Dei, Apdo.202, 50080 Zaragoza, Spain

Ferric leghemoglobin reductase (FLR), an enzyme hypothesized to maintain soybean nodule leghemoglobin in the functional ferrous state, was purified to homogeneity. The pure protein has ferric leghemoglobin reductase and DCPIP reductase activities and utilizes NADH or NADPH (about 30% as effective as NADH) for activity. In the absence of any other exogenous electron carrier, the enzyme could also catalyze NADH oxidation aerobically, functioning as a NADH oxidase with the possible generation of H_2O_2. Enzymic activities for NADH oxidation, Lb^{+3} reduction and DCPIP reduction are completely inhibited by removing O_2 from the reaction system and restored by re-addition of O_2. A study to investigate the affinity of the enzyme for O_2 shows that the enzyme requires at least µM levels of dissolved O_2 for its optimum activities. Peroxidase and catalase at low concentration (about 2 µg/ml) stimulate oxidation of NADH and reduction of Lb^{+3}, suggesting that peroxide intermediates are involved in the reaction. Stoichiometry between the enzyme-catalyzed NADH oxidation and Lb^{+3} reduction is not constant but changes with time as well as with NADH and O_2 concentrations in the reaction solution, indicating that the reactions are not directly coupled and electrons from NADH oxidation are carried by reaction intermediates. The enzyme also shows different kinetics for NADH oxidation and Lb^{+3} reduction, in which the enzyme has a Km of 36.9 µM for NADH and of 9.5 µM for Lb^{+3}, respectively. A sequence of 50 N-terminal amino acids of the purified protein was obtained. Comparisons with known protein sequences showed that the sequence of this enzyme is highly similar to those of the flavin-nucleotide disulfide oxidoreductases, especially dihydrolipoamide dehydrogenase (E_3). The FLR protein sequence also shows remarkable homology to amino acid sequences around the FAD binding domain and disulfide active site of *E. coli*, yeast, pig heart, and human liver E_3s, human glutathione reductase and other flavoproteins.

Gresshoff, Roth, Stacey & Newton (eds.) Nitrogen Fixation: Achievements and Objectives
Chapman and Hall New York • London 1990

Purification of a cytochrome c oxidase from *Bradyrhizobium japonicum* bacteroids by lectin affinity chromatography

Robert G. Keefe and Robert J. Maier

Department of Biology, Johns Hopkins University, Baltimore, MD 21218, USA.

Our previous work directed towards purifying a cytochrome c (cyt c) oxidase from *Bradyrhizobium japonicum* bacteroids showed that this enzyme might be glycosylated. For the most part, purification efforts have been carried out using root nodules from soybeans (*Glycine max* cv. Essex) inoculated with strain LO501, which is derived from USDA DES 122. LO501 is a Tn 5 induced cyt aa_3 structural mutant as a free-living bacterium, but symbiotically, it also has elevated cyt c oxidase activity compared to wildtype. Therefore, it seemed to be an appropriate candidate for use as starting material. Using data from a monosaccharide analysis of a partially purified cyt c oxidase sample, experiments were carried out to see if oxidase activity could bind lectin. Three different lectins, each covalently attached to either sepharose or 4% cross-linked beaded agarose, were tested. Crude protein solubilized from membranes of LO501 bacteroids using octyl glucoside was loaded onto the lectin columns. Of the three, an affinity column made of concanavalin A (con A) covalently attached to sepharose could bind bacteroid cyt c oxidase activity. This activity could be eluted from the con A column with 0.3 M methyl-mannose. When samples from peak activity fractions from a con A column were run on SDS polyacrylamide gels and silver-stained, three protein bands were observed. Their molecular weights were roughly calculated (using SDS PAGE, 15% resolving gel) to be 25.6, 24.4, and 13.5 kdal. The two larger subunits are visible on denatured gels using the periodic acid-Schiff (PAS) stain which is specific for carbohydrate. The 13.5 kdal band did not stain with PAS. Cyt c oxidase activity has been measured in crude membrane protein preparations from free-living *B. japonicum* but that measured in bacteroids is 15-20 times greater per milligram protein. Denatured gels loaded with crude membrane protein from free-living LO501 and LO (wildtype) and stained with PAS show only a doublet (each band) migrating at an approximate molecular weight of 25 kdal. It has not yet been ascertained whether cyt c oxidase activity from free-living cell membranes will bind lectin.

Gresshoff, Roth, Stacey & Newton (eds.) Nitrogen Fixation: Achievements and Objectives
Chapman and Hall New York • London 1990

Nitrogenase function in *Rhizobium meliloti* bacteroids

R.W. Miller

Plant Research Centre, Agriculture Canada, Ottawa, Ontario, Canada K1A OC6

Nitrogenase activity of alfalfa nodules is notably less temperature-dependent than purified nitrogenase proteins (1). Both adenylate energy charge (AEC) and nitrogenase activity of detached nodules are oxygen-limited. Isolated, mature bacteroids display anomalous temperature behavior and oxygen limitation of nitrogenase activity.

Bacteroids were isolated anaerobically and assayed between 5° and 30° for nitrogenase, adenine nucleotides and carbon metabolites after incubation for 15-30 min with varying levels of C_4-dicarboxylic acids, ^{14}C-succinate and oxygen. Bacteroids treated with chlorpromazine were assayed with external reductant and an enzymatic ATP-regenerating system. Nitrogenase was purified from isolated bacteroids to give specific activities for Rm1 and Rm2 of 1250 and 1100 nmoles H_2 $min^{-1}mg^{-1}$.

Only succinate, malate, fumarate or oxaloacetate supported nitrogenase activity of bacteroids. Over 75% of radiolabeled succinate carbon was converted to glutamate, gamma-aminobutyrate and other organic acids during incubation under 4kPa O_2. Oxygen was rate limiting at 20°C and $<$ 10kPa O_2, and AEC declined to suboptimal values below 0.85. AEC values varied inversely with incubation temperature over the range 30°C-5°C, increasing to above 0.9 at the lower temperature.

Chlorpromazine (2-4mM) damaged bacteroid membranes, reduced adenylate pools and uncoupled nitrogenase activity from respiration. Chlorpromazine-treated bacteroids had nitrogenase activity 3-4 times the respiration-supported levels. Hence, only 1/3 of their potential nitrogenase activity is elicited by respiration at 20°C under 4% O_2 and AEC is also suboptimal. Lower temperatures caused AEC levels to increase and a greater fraction of the potential activity to be measured. This temperature compensation mechanism may operate in intact nodules and be responsible for the relative temperature insensitivity of nitrogenase in symbiotic legume-Rhizobium systems.

1. Miller,R.W., Al-Jobore, A. and Berndt, W.B. (1986) Can. J. Biochem. and Cell Biol. **64**, 556-564.

Gresshoff, Roth, Stacey & Newton (eds.) Nitrogen Fixation: Achievements and Objectives
Chapman and Hall New York • London 1990

Effects of acetylene and external oxygen concentration on RQ of legume nodules

F.R. Minchin and J.F. Witty

Department of Environmental Biology, Welsh Plant Breeding Station, Aberystwyth, U.K.

Introduction:

Some methods for determining the 0_2 diffusion resistance of legume nodules involve measurements of respiration as CO_2 output and conversion to 0_2 uptake assuming an RQ of 1. These have been criticised because the use of this RQ value may introduce a substantial error. This poster reports the measurement of RQs of attached nodules of undisturbed plants using an open flow-through gas system.

Results:

Nodulated root respiration was measured as CO_2 production, using an IRGA, and 0_2 uptake, using a Clarke-type electrode. The polarizing unit for this electrode had a reverse current facility which allowed the signal to be backed off to zero. The signal to noise ratio for this system was quite high at 21% 0_2 but decreased progressively up to 60% 0_2. In the absence of acetylene nodulated root respiration showed no response to external 0_2 concentration over the range 10 to 60%. However, the addition of 10% acetylene produced a sharp decrease which was countered by increases in external 0_2. The patterns of respiration determined as 0_2 uptake was essentially similar to that obtained as CO_2 output. These measurements were used to calculate RQ values for both total root respiration (TRR) and nitrogenase-linked respiration (NLR) (based on the percentage of TRR represented by NLR). RQ values for roots not exposed to acetylene were not significantly different over the range 10 to 60% 0_2. Addition of acetylene caused an initial reduction in RQ, as did exposure to 10% 0_2. An essentially similar picture is produced by the RQ values for NLR. The data also provide further evidence that changes in diffusion resistance are largely due to physical, rather than biochemical, processes. Nitrogenase activity was undamaged by increasing external 0_2 from 10 to 60% but 0_2 uptake over this 6-fold range was only increased by 61%. Thus, protection of nitrogenase must be due to 0_2 exclusion rather than consumption.

Conclusion:

The errors involved in calculating nitrogenase-linked 0_2 uptake from CO_2 output, assuming an RQ of 1, are small compared to reported variations in nodule oxygen diffusion resistance. This is the best routine experimental procedure for obtaining comparative estimates of this resistance.

Gresshoff, Roth, Stacey & Newton (eds.) Nitrogen Fixation: Achievements and Objectives
Chapman and Hall New York • London 1990

In vivo regulation of nitrogenase electron allocation

A.H. Moloney, R.D. Guy & D.B. Layzell

Department of Biology, Queen's University, Kingston, Ontario, Canada. K7L 3N6

In legume nodules, the proportion of nitrogenase electron flow used to fix N_2 rather than to produce H_2 (Electron Allocation Coefficient, EAC) varies from 0.75 (the theoretical maximum) to 0.55 or lower. This study used theoretical and empirical approaches to examine the possible role of competitive inhibition by H_2 of N_2 fixation in the regulation of EAC in soybean nodules (Glycine max cv. Harosoy 63 X Bradyrhizobium japonicum str. USDA 16).

Excised soybean nodules were exposed to pH_2 from 0 to 27.5 kPa in a small volume gas-exchange system which could be switched between open and closed modes. Various EAC values were calculated from measurements of H_2 evolution in the presence and absence of N_2, $^{15}N_2$ incorporation and C_2H_2 reduction. As pH_2 increased, EAC declined, but total electron flow through nitrogenase was not affected (Fig 1).

A good fit was obtained between these measured EAC values and those predicted from a mathematical model based upon the measurements of $Km(N_2)$ and $Ki(H_2)$ for nitrogen fixation (1) and the magnitude of a variable barrier to O_2, N_2, and H_2 diffusion (2) (Fig 1). In addition, the model was able to predict the changes which occur in EAC values with varying total nitrogenase activity (3) and following adaptation of legume nodules to various pO_2 and pN_2 (4).

From these results we conclude H_2 inhibition of N_2 fixation is an important determinant of EAC in soybean nodules.

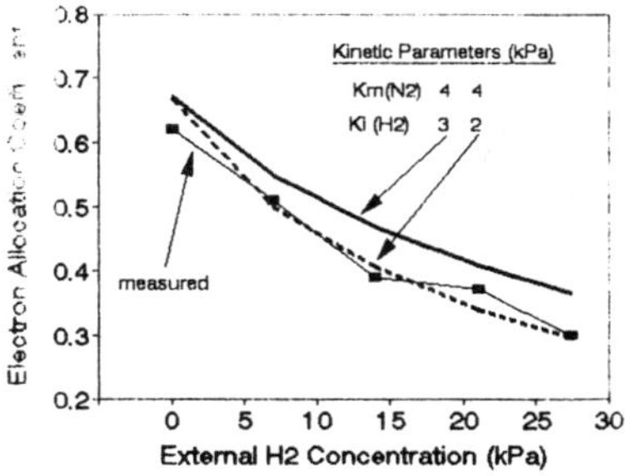

Figure 1: The effect of pH_2 on the measured (■) and predicted EAC of soybean nodules.

REFERENCES

1. Rasche, M.E., & Arp, D.J. (1989) Plant Physiol. **91**, 663-668.
2. Hunt, S., Gaito, S. T. & Layzell, D.B. (1988) Planta **173**, 128-141.
3. Vessey, J.K., Walsh, K.B. & Layzell, D.B. (1988) Physiol. Plant. **74**, 137-146.
4. Vessey, J.K., Walsh, K.B., & Layzell, D.B. (1988) Physiol. Plant. **73**, 113-121.

Gresshoff, Roth, Stacey & Newton (eds.) Nitrogen Fixation: Achievements and Objectives
Chapman and Hall New York • London 1990

Cloning and characterization of the structural gene for cytochrome *d* from *Azotobacter vinelandii*

F. Moshiri*, A. Chawla, and R. J. Maier.

Dept. of Biology, The Johns Hopkins University, Baltimore, MD 21218, USA.

Nitrogen-fixing cells of *Azotobacter vinelandii* have a rich complement of electron transport and respiratory components, giving *Azotobacter* one of the highest respiration rates of all aerobic organisms. This high respiration rate is thought to play a role in maintaining a low intracellular free oxygen concentration, which is an essential condition for the functioning of the nitrogenase components ("respiratory protection"). A *d*-type cytochrome has been proposed to be a major component responsible for the high respiratory rates of *A. vinelandii* under nitrogen-fixing conditions. As a first step towards a more detailed study of the structure and function of the terminal oxidases of *A. vinelandii*, we have cloned the structural gene for cytochrome *d*. The gene was initially cloned based on its homology to the structural gene of cytochrome *d* from *Escherichia coli*. The nucleotide sequence of a 2.9 kb Sac I - Eco RI restriction fragment containing the region homologous to the *E. coli* gene has been determined. Two open reading frames encoding proteins with extensive regions of homology to subunit I and subunit II of the *E. coli* cytochrome *d* complex are present in this region. The sequence of the region upstream of the initiation codon for subunit I reveals a putative promoter region containing an *NtrA* binding site, although further upstream sequences do not contain other *nif*-regulatory sequences (i.e. *NifA binding sites).* Both subunits of the *A. vinelandii* cytochrome *d* complex can be expressed in *E. coli*, either under the control of the *lac* promoter, or as a fusion protein, under the control of the *E. coli cyd* promoter. Despite the high degree of homology between the two oxidases, the *A. vinelandii* enzyme is unable to phenotypically complement *E. coli cyd* or *cyd* + *cyo* mutants. This may be due to the proposed uncoupled nature of electron transport through the cytochrome *d* of *A. vinelandii*.

Gresshoff, Roth, Stacey & Newton (eds.) Nitrogen Fixation: Achievements and Objectives
Chapman and Hall New York • London 1990

Glutamine synthetase may limit nitrogen fixation in *Phaseolus vulgaris*

Raymond S. Pacovsky and Mariangela Hungria

USDA-Western Regional Research Center, Albany, CA 94710, USA
and University of California, Davis, CA 95616 USA.

Nodulated beans (Phaseolus vulgaris L.) typically display a period of N stress between 15 and 20 days after emergence (DAE) when N reserves in the cotyledon have been depleted, but nitrogen fixation and N assimilation have not yet reached significant levels. Although N uptake and acetylene reduction have been measured in nodulated beans, few comparisons between nodule activity (SNA) and glutamine synthetase (GS) or glutamate synthase (GOGAT) acitivites exist. The purpose of the two experiments presented here was to quantify dinitrogen fixation, N assimilation and nodule enzyme activities in a number of Phaseolus cultivar-Rhizobium strain combinations.

In two separate experiments acetylene reduction with hydrogen evolution, dry weight, Kjeldahl N, protein content, nodule GS and GOGAT activities were measured. Specific nodule activity (SNA), based on N-assimilation rates rather than acetylene reduction and hydrogen evolution, were calculated assuming a 14-hour activity period. Cultivars responsive to nodulation assimilated 3 to 4 times more N and had SNA's 2 to 3 times greater than the non-responsive cultivar. Nodule-specific GS activity was significantly elevated in superior strain bean combinations (52% increase) and in cultivars responsive to nodulation (40% increase). Transferase/synthetase ratio was low for GS activity (approximately 36) indicating that synthetase activity was at or near saturation and operating at maximal level since substrate was not limiting. SNA and GS-synthetase activity were significantly correlated ($r = 0.82$, $p < 0.01$) and GS activity was only 10-30% larger than SNA, a situation that would require GS to operate continuously near V_{max} to accommodate the ammonia flux from nitrogen reduction.

Often it has been suggested that the low levels of nitrogen fixation in beans is due to low host photosynthesis or an inefficient bacterial nitrogenase. Our data suggest that the low levels of nodule-specific GS and GOGAT may limit nitrogen fixation in Phaseolus. In some cases, the rhizobial strain found in the nodule affected host-GS activity apparently limited dinitrogen reduction. A search for Phaseolus germplasm containing elevated levels of nodule-specific GS and GOGAT must be initiated in bean-breeding programs that attempt to enhance biological nitrogen fixation.

Gresshoff, Roth, Stacey & Newton (eds.) Nitrogen Fixation: Achievements and Objectives
Chapman and Hall New York • London 1990

Structure of the soybean nodule cortex: In relation to oxygen diffusion and regulation of nodule activity

Richard Parsons[1] and David A. Day[2]

[1]Biological Sciences, University of Dundee, DD1 4HN, UK.
[2]Botany Department, ANU. GPO Box 4 ACT2601, Australia.

Introduction : The maintenance of a microaerobic environment within soybean nodules depends on both the structure of the nodule and rapid oxygen use within the infected zone. The major resistance to oxygen diffusion occurs in the inner cortex, although the question of whether continuous gas connections occur across the cortex, remains.

Results : This drawing shows the cell layers observed in a typical transversesection through a soybean nodule cortex. Discrete layers of cells occur, showing that the nodule structure is comparable in complexity to other plant organs (eg leaves).

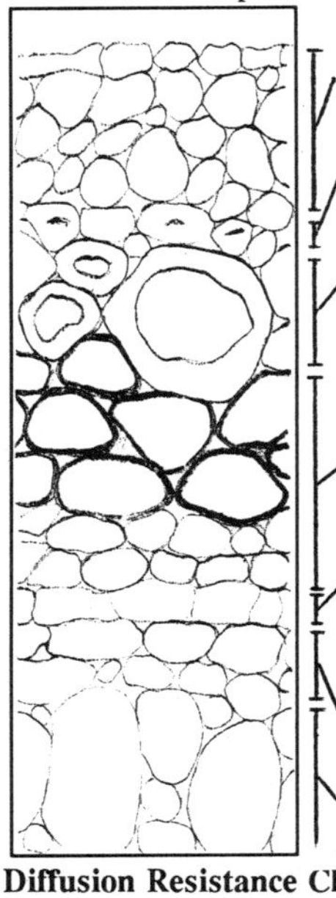

Outer Cortex - In the outer cortex, cells are thin walled and intercellular gas spaces occur at many three cell junctions, providing little resistance to oxygen movement.

Calcium Oxalate Crystals - In a layer of cells immediately external to the scleroid cells, calcium oxalate crystals frequently occur. Gas spaces are present between cells.

Scleroid Cells - These cells form a layer of large thick walled cells with a clumped mosaic pattern encircling the nodule. In tangential sections intercellular gas spaces, although often small, are present between some of the cells in this layer.

Inner Cortex - The first two or three cell layers of the inner cortex are composed of relatively thick walled cells with numerous gas spaces between them. A few of these gas spaces sometimes appear to be occluded with a material which may significantly increase the resistance to gas diffusion.

Boundary Layer - **Gas spaces do not occur** between cells in this boundary layer, which forms a distinct layer which surrounds the infected zone. It is one or two cells thick and made up of thin walled cells without intercellular gas spaces.

Distributing Zone - Internal to the boundary layer is a layer of cells with a large number of intercellular gas spaces.

Infected Zone - Infected and uninfected cells occur in the central infected zone of the nodule. Gas spaces occur at every three cell junction and thus interconnect throughout this region.

Diffusion Resistance Changes : In transverse sections from soybean nodules, the cells forming the boundary layer were frequently observed to be collapsed. These thin walled cells may change in volume and shape, effectively altering the resistance of the cortex to oxygen movement and thus regulating nodule activity. This may occur either by collapse of the cells shortening the distance between gas spaces in the inner cortex and the infected zone, or by the water released on collapse of these cells filling intercellular spaces and increasing the effective water filled space.

Gresshoff, Roth, Stacey & Newton (eds.) Nitrogen Fixation: Achievements and Objectives
Chapman and Hall New York • London 1990

Enhanced nitrogen fixation by genetically improved alfalfa grown under field conditions

D.A. Phillips and L.R. Teuber

Dept. Agronomy & Range Science, University of California, Davis, CA 95616, USA.

M69-33, a population of alfalfa (*Medicago sativa* L.) developed from the agronomic cultivar 'Moapa 69' by three generations of phenotypic recurrent selection for increased forage dry weight and forage N concentration under N_2- and NH_4NO_3-dependent growth conditions, has been tested for several years under field conditions. Previous results showed that M69-33 expressed the desired traits under glasshouse conditions (1, 2). Current tests now permit an initial report on the agronomic performance and the N_2-fixation capacity of M69-33 in the field.

Estimates of N_2 fixation by the ^{15}N-dilution method averaged over six harvests during the first year of tests at Davis show the improved M69-33 derived 7% more forage N from N_2 than the original Moapa 69 plants ($P<0.05$). Although there was no difference in forage yield of the two populations (3,330 kg/ha/harvest), forage N concentration was increased significantly (Moapa 69: 3.24%; M69-33: 3.47%). Those results are supported by 3 years of data from a Central Valley site where forage yields of the two populations did not differ over 22 harvests and M69-33 had 5.9% higher N concentration ($P<0.05$). Forage N concentration of M69-33 also was significantly higher at an Imperial Valley location during 24 harvests taken in 3 years, but total forage yield was less than Moapa 69 at that site. These results indicate that plants selected under controlled conditions do not always show the desired phenotype in field environments, but they suggest improving N_2 fixation can contribute to alfalfa forage quality.

1. Phillips, D.A., Cunningham, S.D., Bedmar, E.J., Sweeney, T.C. & Teuber, L.R. (1985) Crop Sci. **25**, 1011-1015.
2. Teuber, L.R. & Phillips, D.A. (1988) Crop Sci. 28, 599-604.

Gresshoff, Roth, Stacey & Newton (eds.) Nitrogen Fixation: Achievements and Objectives
Chapman and Hall New York • London 1990

Rapid release of $^{14}CO_2$ during incubations of *Bradyrhizobium japonicum* bacteroids with ^{14}C-labeled substrates

S. O. Salminen and J. G. Streeter

Department of Agronomy, Ohio State University/O.A.R.D.C., 1680 Madison Avenue, Wooster, OH 44691, USA.

Previous studies have indicated that significant amounts of $^{14}CO_2$ are released when [^{14}C] substrates are supplied to *Bradyrhizobium japonicum* USDA 110 bacteroids during incubations of 20 to 60 min (1). During shorter incubations (<5 min) generally used for estimating Km and Vmax values, $^{14}CO_2$ evolution has not been considered significant. We tested this assumption in experiments where ^{14}C in bacteroids was estimated along with an independent estimation of $^{14}CO_2$ evolved during incubations of <5 min. We found that release of $^{14}CO_2$ began immediately after mixing bacteroids and ^{14}C substrate. Most of the $^{14}CO_2$ was retained by the reaction mixture and was quantitatively recovered only after acidification of the mixture and shaking for 10 min. As expected, the relative rate of $^{14}CO_2$ release varied among substrates; e.g. 27% of total ^{14}C-glutamate versus 53% of total ^{14}C-malate taken up was recovered as CO_2. Inclusion of $^{14}CO_2$ released in the calculation of kinetic parameters had little influence on Km values but resulted in substantial differences in estimates of Vmax (Table 1). Thus, one would conclude that the Vmax for glutamate is greater than the Vmax for succinate uptake if $^{14}CO_2$ release was not measured, whereas the reverse conclusion would be reached if respired CO_2 were included in the calculations.

Table 1. Km and Vmax values for uptake of [^{14}C] substrates by *B. japonicum* USDA 110 bacteroids in a 5 min incubation.
(Results represent the saturable component of uptake)

	Km (μM)		Vmax[b]	
Substrate	$+CO_2$[a]	$-CO_2$[a]	$+CO_2$[a]	$-CO_2$[a]
Succinate	78	92	9.3	5.0
Glutamate	34	35	8.1	6.4
Glucose	580	635	1.1	0.9
Malate	154	134	27.7	13.1

[a] $+CO_2$ & $-CO_2$ indicate that CO_2 release by bacteroids was or was not considered in the calculation, respectively.
[b]nmole x min^{-1} x mg $protein^{-1}$

1.Salminen,S.O.& Streeter,J.G. (1987) J. Bacteriol.**169:**495

Gresshoff, Roth, Stacey & Newton (eds.) Nitrogen Fixation: Achievements and Objectives
Chapman and Hall New York • London 1990

Release of bacteroid proteins during the preparation of symbiosome membrane and symbiosome space fractions from soybean nodules

J. G. Streeter

Department of Agronomy, Ohio State University/O.A.R.D.C, 1680 Madison Avenue, Wooster, OH 44691, USA.

The symbiosome membrane (SBM, also termed "peribacteroid membrane") and symbiosome space (SBS, also "peribacteroid space"; the space between the SBM and the bacteroids) probably play an important role in transport of metabolites in legume nodules. For studies of SBM and SBS,symbiosomes are first isolated and are then ruptured, generally by osmotic shock, and the SBS, SBM and bacteroid fractions are separated by centrifugation. In this work, symbiosomes were purified on Percoll step gradients (2) and were then divided into equal fractions which were subjected to various rupture techniques. Phosphodiesterase was assayed as a bacteroid periplasmic marker and ß-hydroxybutyrate dehydrogenase was assayed as a bacteroid cytoplasmic marker (3). α-Mannosidase was assayed (4) as a potential marker for the SBS (1).

Treatment of symbiosomes with osmotic shock led to the recovery of as much as 50% of the phosphodiesterase activity and 5% of the ß-hydroxybutyrate dehydrogenase activity in the SBS and SBM fractions. Of the many combinations of treatments tried, resuspension with a Pasteur pipette and rupture of symbiosomes by extrusion through a fine needle is recommended. This approach released a small proportion of the phosphodiesterase and virtually no ß-hydroxybutyrate dehydrogenase activity from bacteroids. Osmotic shock should be avoided because the SBS and SBM fractions will be significantly contaminated with periplasmic and cytoplasmic proteins from bacteroids. α-Mannosidase was found in the SBS fraction only when ß-hydroxybutyrate dehydrogenase was also found; thus, α-mannosidase may not be a useful marker for the SBS. A detailed account of this work will be published (4).

1.Kinnback,A.,Mellor,R.B.& Werner,D.(1987) J.Exp.Bot. **38**, 1373-1377.
2.Price,G.D.,Day,D.A.& Gresshoff,P.M.(1987) J.Plant Physiol. **130**,157-164.
3.Streeter,J.G.(1989) J.Gen.Microbiol. **135**,3477-3484.
4.Streeter,J.G.& Morré,D.J.(1990) Symbiosis (in press).

Gresshoff, Roth, Stacey & Newton (eds.) Nitrogen Fixation: Achievements and Objectives
Chapman and Hall New York • London 1990

C_2H_2 reduction of broadbean nodules and bacteroids under water stress: role of oxygen

Jean-Charles Trinchant, Vincent Guérin and Jean Rigaud

Laboratoire de Biologie Végétale et Microbiologie, URA CNRS 1114, Faculté des Sciences, Nice, France.

Broadbeans (*Ficia faba* L. var minor cv. Soravi) cultivated in a glasshouse on a sandy mixture exhibited decreasing leaf and nodule water potentials throughout the water deprivation period. A linear relation ($r^2 = 0.99$) occurred between these two water potentials which allowed us to estimate the nodule water potential (Ψ_{nod}) from the values determined with leaves, in the range of - 0.5 to - 1.5 MPa. A decrease of 0.2 MPa of Ψ_{nod} induced a 25% inhibition of C_2H_2 reduction activity of attached nodules. This activity continued to depress when water stress increased but the effect was less pronounced since a further decrease of Ψ_{nod} of 0.2 MPA did not cause a 50% decline in activity (1). Increasing O_2 tension up to 60 kPa in the gas phase partially restored (17%) C_2H_2 reduction activity of mildly stressed nodules, suggesting that an O_2 limitation was induced by water stress. In this way, interesting information was given by structural studies of nodule sections where cortex cells appeared dense packed after plant water deprivation. Such alterations, in reducing air spaces, contributed to limit O_2 diffusion to the bacteroids of the central tissue.

Isolation of bacteroids from broadbean nodules led to observe their higher C_2H_2 reduction capacity by comparison to those of bacteroids from *Phaseolus* and *Glycine* (2) or *Sesbania* (3). As in other Legumes, glucose stimulated N_2 fixation of isolated bacteroids but was 2.8 times more efficient than succinate. Moreover, in maintaining a significant level of bacteroid activity, glucose appeared a more appropriate substrate than succinate under water-limited conditions. At the bacteroid level, the lower O_2 requirement for optimal C_2H_2 reduction confirmed the involvement of O_2 in the depressive effect of water stress. In all cases, inhibition of N_2 fixation by water stress was greater when determined with isolated bacteroids than with nodules indicating a protective role played by the bacteroid environment.

(1) Guérin, V., Trinchant, J.C. & Rigaud, J. (1990) Plant Physiol. **92**, 595-601.

(2) Trinchant, J.C., Birot, A.M. & Rigaud, J. (1981) J. Gen. Microbiol. **125**, 159-165.

(3) Trinchant, J.C. & Rigaud, J. (1989) Plant Sci. **59**, 141-149.

Gresshoff, Roth, Stacey & Newton (eds.) Nitrogen Fixation: Achievements and Objectives
Chapman and Hall New York • London 1990

Development of a hydrogen-sensitive micro-electrode for use with legume nodules

J.F. Witty, F.R. Minchin and A. Cresswell

Department of Environmental Biology, Welsh Plant Breeding Station, Aberystwyth, U.K.

Introduction:
There has been controversy regarding the concentration of H_2 within the infected region of legume nodules and whether this leads to partial inhibition of nitrogen reduction by nitrogenase. The obvious way to resolve this question is to directly measure internal concentrations, but techniques for this were lacking. We have now developed a modification of oxygen-sensitive micro-electrodes which allow them to detect hydrogen.

Methodology:
Micro-electrodes were made as previously described (1) but modified by the use of an acidic electrolyte and reversal of the polarity (2). The additional sensitivity of the hydrogen micro-electrode required it to be housed in a perspex "nodule chamber" within a Faraday cage.

Results:
H_2 concentrations within detached nodules of Hup +ve strains are 0 (or below 0.01%). In unstressed nodules of Hup -ve strains they range from 0.4 to 2.2%. Increasing the gas phase 0_2 concentration around pea nodules to 50% can increase internal H_2 to more than 3%, but higher $p0_2$ leads to nitrogenase damage. Similar internal H_2 concentrations can be produced by replacing N_2 in the gas phase with Ar. This approach can also be used as a method of measuring relative efficiency.

Increasing the gas phase pH_2 also produces increases in internal H_2 concentration and at values above 5% the internal concentration shows a "run-away" response. This is probably due to the inhibition of nitrogen reduction by nitrogenase as e^- allocation to H_2 production is increased.

References:
1. Revsbech, N.P. and Ward, D.M. (1983) Applied Environ. Microbiol. 45, 755-759.
2. Wang, R., Healy, F.P. and Myers, J. (1971) Plant Physiol. 48, 108-110.

SECTION D:

NITROGEN FIXATION AND THE ENVIRONMENT

Invited papers
Poster summaries

Gresshoff, Roth, Stacey & Newton (eds.) Nitrogen Fixation: Achievements and Objectives
Chapman and Hall New York • London 1990

Ecological aspects of nitrogen fixation

S.V. Shestakov

The Vavilov Institute of General Genetics, USSR Academy of Sciences, Gubkin St. 3, Moscow 117809, USSR.

Dear colleagues:

Opening this meeting devoted to the ecological aspects of nitrogen fixation, I would like to emphasize the global significance of the problems discussed. The biospheric nitrogen pool is known to be maintained by the activity of nitrogen-fixing bacteria. According to the FAO estimates, 70% of the total planetary nitrogen yield is of biological origin. In recent years, the dynamic equilibrium of nitrogen circulation in nature has been subjected to constantly increasing anthropogenic pressure. This pressure arises from the processes of artificial irrigation, drainage of marshes, extensive application of fertilizers, utilization of nitrogen, humus, peat, coal and eutrophication of the water bodies. Along with this, the release of gases into atmosphere has increased. Of primary concern is nitrous oxide, which is known as one of the main factors enhancing the greenhouse effect and exhausting the atmospheric ozone layer. More intensive use of biological nitrogen fixation through agricultural systems, which are optimized to protect nature, is one of the main ways for preventing disastrous ecological changes resulting from the unbalanced nitrogen budget of the Earth. And release of the gaseous forms of nitrous oxide and ammonia into the atmosphere has the severest ecological consequences.

To cope with these problems, it is important to comprehend the role of various groups of bacteria involved in the different stages of nitrogen circulation. It is important to know the correlations between the processes of nitrogen fixation, autotrophic and heterotrophic nitrification, denitrification, ammonification, and other pathways for nitrogen flow under various ecological conditions. We must also have a clear understanding of the adaptive capacities of natural microbial communities exposed to various stress factors. I believe that many of these problems will be touched upon in the reports of our distinguished speakers.

In my introductory statement, I will mention several principal circumstances. First, I would like to note that the list of microorganisms capable of nitrogen fixation and of developing in various ecological niches has been considerably extended lately. In other words, nitrogen fixation cannot be attributed any longer to a highly-specialized group of bacteria only. From the evolutionary point of view, it is notable that nitrogen fixation is inherent to such ancient groups of archebacteria as obligatory acidophiles and methanogens. Thus, the channels for genetic communication existing in nature must be assumed to be wide enough for the implementation of nitrogen-fixation-gene transfer. Second, apart from the highly-specialized symbiotic systems (including *Rhizobium, Frankia, Cyanobacteria*, etc.), there is quite a number of considerably less specialized associated and free-living communities of nitrogen fixers characterized by enormous ecological versatility. Many diazotrophic microorganisms are capable of denitrification, heterotrophic nitrification, and other processes important for nitrogen circulation. The ecological significant of denitrification is extremely high. This biological method is used to control the nitrogen losses in the form of gaseous nitrous oxide and ammonia. Now, it has become clear that most of the soil bacteria are potential denitrifiers. The fact that one and the same microorganism is capable of implementing two antipodal processes, denitrification and nitrogen fixation, plays an important role in maintaining the nitrogen balance in natural ecosystems. Therefore, of special interest appears to be a study of the metabolic regulation characteristic of mixed populations of various natural ecosystems, as well as that of the population regularities responsible for the high stability and competitiveness of the members of nitrogen-fixation communities.

A large number of up-to-date complimentary analytical methods (e.g., isotopes, inhibitors, chromatography, etc.) make it possible to evaluate quantitatively the impacts of separate microbial groups and communities on the nitrogen balance of soil and aquatic ecosystems. As is known, one of the main factors limiting nitrogen fixation is the energy deficit due to the entry of carbohydrates, i.e., there is a direct correlation between the intensity of photosynthesis and the level of nitrogen fixation. For instance, one of our experiments has shown that a 3-fold increase in the concentration of CO_2 in the greenhouse caused a 5-fold elevation of the bound nitrogen in pea on account of increased nitrogen fixation. Thus, the ecology of nitrogen fixers depends greatly on the degree of their interaction with the plant in its rhizosphere and will respond to all the anthropogenic environmental changes affecting the agricultural and silvicultural ecosystems.

Gresshoff, Roth, Stacey & Newton (eds.) Nitrogen Fixation: Achievements and Objectives
Chapman and Hall New York • London 1990

The ecology of indigenous populations of *Rhizobium leguminosarum* bvs. *trifolii* and *viceae*

P.J. Bottomley, K. Leung, S.R. Strain, K. Yak, N.Dashti and P. Claycomb

Department of Microbiology, and Department of Crop and Soil Science, Oregon State University, Corvallis OR 97331-3804 USA

Introduction. Although competitive nodulation has been studied for many years (1,3,5,7,8), it has been difficult to gain an understanding of what biological and abiological factors control the outcome of this phenomenon in natural soils containing native rhizobial populations. An issue which has received considerable attention is concerned with the debate about whether differences in saprophytic competence exist among members of soil rhizobial populations. If such differences occur, can they affect the outcome of competitive nodulation by influencing the sizes and physiological states of subpopulations? The alternate scenario is that particular rhizobial strains within the soil population are 'preferred' as microsymbionts by the host. As a result, differences in population size and free-living saprophytic attributes are relegated to minor roles in the affairs of competitive nodulation.

In this paper it is my intent to discuss some of our findings on the ecology and nodulating characteristics of soil-borne populations of Rhizobium leguminosarum biovars trifolii and viceae. I hope to convince the reader that our understanding of competitive nodulation and legume-rhizobia interactions in the field will remain incomplete until the composition and ecological behavior of soil rhizobial populations are better understood.

The Composition of Soil Rhizobial Populations. Over a period of fifty years several studies have shown that a tremendous diversity of strains can exist within the same soil rhizobial population (3). Although some methods of strain identification (intrinsic antibiotic resistance and phage typing) have revealed greater diversity within soil populations than other methods, it is quite remarkable that two distinctly different methods (serotyping and multi-locus enzyme electrophoretic typing) have revealed a similar number of subpopulations (6 to 15) within the majority of soil populations examined. Nevertheless, there have been no published reports in which the two methods were compared on the same population. Since each method has its own strengths and weaknesses for ecological work, we compared the two methods on populations of R. leguminosarum bvs. trifolii and viceae recovered from the same soil (Table 1).

Table 1. Electrophoretic types and serotypes within populations of R. leguminosarum bvs. trifolii and vicaea recovered from the same soil.

SEROTYPE	ELECTROPHORETIC TYPE A	B	C
R. leguminosarum bv. viceae	<—	No. of Isolates	—>
AS6	0	1	0
AS6/21	2	13	0
AS21	3	12	7
AR23	15	0	0

SEROTYPE	ELECTROPHORETIC TYPE I	J	K	L	M
R. leguminosarum bv. trifolii	<—	No.	of Isolates	—	—>
AS6	23	0	0	1	0
AS6/21	13	0	0	0	1
AS21	1	20	0	1	0
AG4	0	0	19	0	0

The majority of bv. trifolii isolates from within each of four major nodule-occupying serotypes (AS6, AS6/21, AS21, and AG4) were found to have identical ETs (Table 1). Although serotypes AS6 and AS6/21 are antigenically distinct members of a serogroup, members of both serotypes are represented predominantly by the same ET (I). The latter is different from the predominant ET of either serotype AS21 (J), or serotype AG4 (K). In the case of the bv. viceae population, the three ETs representing the majority of the population are distinctly different from the ETs associated with the same serotypes in the bv. trifolii population. In addition, the association of particular ETs with specific serotypes is not so clear-cut as with bv. trifolii. While all representatives of serotype AR23 were of ET (A), a few isolates of the latter were also found in antigenically dissimilar serotypes. Furthermore, while isolates of ET (C) were restricted to serotype AS21, isolates of ETs A and B were also found within the same serotype. The data emphasize the potential for using multi-locus enzyme electrophoresis to determine the genetic homogeneity within serotypes prior to choosing them for immunofluorescence ecological studies. They also illustrate the difficulties which might be encountered when attempting to compare ecological findings about either different rhizobial types generated with the same method or about the same rhizobial type generated with different methods.

Significance of Diversity within Soil-borne Rhizobial Populations. Although the majority of studies in the literature indicate that soil rhizobial populations are quite diverse, it was reported recently that the diversity within low density populations ($<10^2$/g) of Rhizobium leguminosarum bv. trifolii in acidic soils (≤pH4.5) was less than the diversity found in larger populations ($>10^4$/g) of the same biovar colonizing soils of ≥pH5.7 (9). The complexity of diversity is revealed by our own data where the density and diversity of populations of bvs. trifolii and viceae were compared at two different sites (Table 2). The

Table 2. Distribution of electrophoretic types among the biovars of R. leguminosarum at two sites.

Site	A	B	C	D	E	F	G	H	I	J	K	L	M	N	O	Total
	Electrophoretic types — Number of Isolates															
A⁺.																
bv. viceae	20	26	7	5	1	1	1	1	-	-	-	-	-	-	-	62
bv. trifolii	-	-	-	-	-	-	-	-	42	20	8	3	1	1	-	75
C⁺⁺.																
bv. viceae	6	2	10	1	-	-	1	-	-	-	2	-	-	-	2	24
bv. trifolii	-	1	-	-	-	-	-	-	-	-	-	-	-	2	18	23

⁺Site A is located upon an improved pasture in which subclover (Trifolium subterraneum L.) is the only legume. The population densities of bvs. viceae and trifolii are $<10^2$/g and $>10^5$/g respectively.

⁺⁺Site C is located upon an unimproved oak Savannah hillside in which Vicia spp. contribute >99% of the legume community. The population densities of bvs. viceae and trifolii are $>10^5$ and $<10^2$/g respectively.

relative population densities of R. leguminosarum bvs. trifolii and viceae are mirror images of each other in two similar acidic soils under improved subclover pasture (site A) and unimproved hill land (site C). The larger biovar population at each site corresponds with the legume present in the plant community, i.e. Vicia spp. (site C), and Trifolium spp. (site A). Although the population density of bv. viceae ($\leq 10^2$/g) was several orders of magnitude less than of bv. trifolii (10^5 to 10^6/g) at site A, the extent of diversity revealed by the number of electrophoretic types was similar in both biovars. Although the sharing of similar somatic antigens by the two biovar populations confirms their taxonomic closeness (6,10,13), there were no ETs shared by the two biovar populations at site A. This finding contrasts with those of Young (14) who observed that several ETs were common to populations of both biovars recovered from the same soil. The findings at site C resemble those of Harrison et al. (9) since a low diversity (3 ETs) in the bv. trifolii population coincided with its low population density. A greater diversity (7 ETs) existed within the larger bv. viceae population on the same site. On the issue of ETs common to both biovars, only two of the seven ETs in bv. viceae were shared with bv. trifolii at site C. Both the limited diversity within the bv. trifolii population on site C and the lack of ET overlap between biovars as site A may not hold up after we extend the analysis to populations recovered from more clover species. Nevertheless, they show that limited diversity within one rhizobial type colonizing a soil may not apply to other rhizobia in the same soil, nor to other populations of the same rhizobial type in neighboring soils. Again, the data serve to emphasize the frustration often felt by rhizobial ecologists when attempting to assess the comparative worth of their ecological findings with those generated by other ecologists in different parts of the world.

Significance of the Densities of Subpopulations and their Physiological States. Previously, we established that the population density of serogroup AS6 of R. leguminosarum bv. trifolii was significantly greater than the density of other serotypes of bv. trifolii present in Abiqua soil (4). Subsequently, serogroup AS6 was found to be composed of at least two antigenically distinct serotypes, AS6 and AS6/21. By using adsorbed fluorescent antibodies, we have determined the population densities of each of these two serotypes to be not significantly different from each other in soil sampled from areas in the pasture uncolonized by plants and in nonrhizosphere soil sampled from plant-colonized areas. The soil population of each serotype is significantly greater ($P = 0.05$) by four to seven-fold than the soil populations of two other serotypes AS21 and AP17 (Table 3). In rhizospheres of field-grown plants, however, the situation is somewhat different. With the exception of serotype AG4, the population densities of the four serotypes were 30 to 190-fold greater in subclover rhizospheres than in bare field soil. We have repeatedly observed that the relative subclover rhizosphere responses of AS21 and AS6/21 are greater than of AS6, AP17, and AG4. In contrast, the responses of the serotypes to an orchard grass (Dactylis glomerata) rhizosphere were modest; populations showed a three to six-fold increase in density above bare soil values. These population data confirm our previous findings that differences may exist between serotypes in their saprophytic competence and persistence in bare soil. The novelty of these

Table 3. Rhizosphere effects on population densities of serotypes of R. leguminosarum bv. trifolii.

	Serotypes No. per g of dry soil, 10^5 [b]				
Soil Environment	AS6	AS6/21	AS21	AP17	AG4
Bare soil	2.4	1.4	0.4	0.4	1.2
Nonrhizosphere soil	5.0	2.6	0.5	1.2	2.0
Clover rhizosphere	84 (35)[a]	266 (190)	49 (123)	12.0 (30)	2.6 (2.2)
Grass rhizosphere	12 (5.0)	4 (2.9)	1.8 (4.5)	2.5 (6.3)	3.1 (2.6)

[a]Values in parentheses denote the magnitude of the rhizosphere stimulation relative to bare soil.

[b]The mean population density represents an average of samples collected on two or three occasions during winter and spring of 1990.

findings relates to the different impact of host and nonhost rhizospheres upon the serotype populations. Sufficient selective pressure is created under field conditions to modify the ecological success of serotype populations. Although Bohlool et al. (2) observed that legume rhizospheres stimulated inoculant strains of bv. viceae more than did corn rhizospheres, no strain differences were observed. It is worth remembering that in pasture and legume-intercropping systems, both of which benefit tremendously from N_2 fixation, the plant communities and their associated rhizospheres are mixed. More of rhizobial ecology research effort should be directed at the mixed plant systems.

Substrate Responsiveness of Serotypes. To initiate studies on the physiological state of serotypes within the soil rhizobial populations, we have combined the nalidixic acid-cell elongation assay (direct viable count [DVC] method) with immunofluorescence (4,11). When a rhizobial population is recovered from soil and exposed to yeast extract (200 mg/liter) and nalidixic acid (15 mg/liter), members of serotype AG4 undergo more rapid cell elongation (50% of cells are elongated after 9 h) than those of serotype AP17 (ca. 20% of cells elongated after 20 h). Furthermore, a significantly lower percentage (20 to 35%) of the cells within serotype AP17 was elongated after 24 h than was observed for serotype AG4 (80%). We can conclude that serotype subpopulations are in different states of readiness to respond to substrate appearance. When these findings are considered along with others (4) a tentative relationship is revealed between the fastest responding serotypes (AS6, AG4 and AS21) being the most competitive nodulators of clover species and slow responding serotypes (AP17 and AR23) being poor nodulation competitors.

Competitive Nodulation: Relative Importance of Saprophytic Competence and Host Preference. Despite the presence of at least nine serotypes of bv. trifolii in the Abiqua soil, nodules of field-grown subclover are predominantly occupied by three serotypes AS6, AS6/21, and AS21 (Table 4). Because serotypes AS6

Table 4. Nodule occupancy of field-grown *T. subterraneum* L., cv. 'Nangeela' over three years.

Time of sampling	% nodule occupancy[b]				
	Antigenic types				
	AS6	AS6-21	AS21	AP17	Others[a]
Spring '87	25.7	14.0	44.2	5.2	10.9
Spring '88	23.5	38.6	29.5	ND[c]	8.4
Spring '90	25.0	45.0	10.9	1.7	17.4

[a]Nodules in which the occupants did not react with fluorescent antibodies, AS6, AS21, or AP17.

[b]Nodule occupancy represents the mean percentage values for plants recovered from three separate locations in the field.

and AS6/21 are chromosomally similar, and since their soil populations are significantly greater than of other serotypes and their response to host rhizospheres is vigorous, it is tempting to explain their competitive nodulating success upon their numerical advantage over other serotypes. In contrast, serotype AS21 maintains a soil population which is lower than either AS6 or AS6/21 and which is no greater than the populations of less competitive serotypes (e.g. AP17). A numerical explanation cannot account for its competitive nodulating success. Despite several studies in nonsterile soil showing that numerical dominance is not important to nodulating success by subpopulations (12), I submit that the issue is far from resolved.

HOST PREFERENCE. We have obtained evidence that subclover exerts a strong and unique strain preference for certain members of this soil population. When seven different annual and perennial clover species were planted in replicate samples of Abiqua soil under greenhouse conditions, serotype AS6 was a dominant nodule occupant only of subclover (Table 5). Serotype AG4 was present in substantial numbers of nodules on all clover species (except subclover) regardless of their annual or perennial nature. The remaining indigenous serotypes (including AS6/21 and AS21) did not show evidence of extraordinary competitive behavior on any of the clover species examined. Electrophoretic typing showed that

Table 5. Nodule occupancy by indigenous serotypes on diverse annual and perennial clover species.

Host Species	Serotypes (% nodule occupancy)									
	AR23	AP17	AG4	AS16	AS6	AR21	AS21	AS36	AS27	U+
T. patens Schr.	12	8	24	8	16	8	0	4	4	16
T. parviflorum Erh.	8	0	31	19	6	8	3	11	3	11
T. glomeratum L.	14	0	43	0	14	0	0	0	14	14
T. vesiculosum Savi.	3	0	22	N.D.	3	0	34	6	19	13
T. subterraneum L.	0	14	0	N.D.	46	0	4	0	18	18
T. repens L.	9	1	33	6	4	7	9	0	13	17
T. pratense L.	22	0	34	12	4	4	7	0	5	24

+U designates that the isolates were antigenically unrelated to the serotypes listed.

members of serotype AG4 are chromosomally homogeneous regardless of the clover species from which they were recovered, and their ET is quite distinct from the ETs of serotypes AS6, AS6/21, and AS21. At this time it is unclear how much of the host preference behavior shown by subclover is due to unique features of this particular Trifolium sp. We cannot overlook the fact that this soil has grown subclover for 30 to 40 y and has no prior history of the other clover species used in the study.

Acknowledgments. Financial assistance for these studies was provided by the U.S. Department of Agriculture Competitive Grants Program and the Oregon Agricultural Experiment Station. Carlene Pelroy's excellent word processing skills and patience with the authors are gratefully appreciated. The manuscript represents Technical Paper No. 9260 of the Oregon Agricultural Experiment Station.

References

1. Amarger, N. (1984) in Current Perspectives in Microbial Ecology, eds. Klug, M. J. and Reddy, C. A. (Amer. Soc. Microbiol., Washington, D.C.), pp. 300-305.
2. Bohlool, B. B., Kosslak, R., and Woolfenden, R. (1984) in Advances in Nitrogen Fixation Research, eds. Veeger, C., and Newton, W. E. (Martinus Nijhoff, Dordrecht), pp. 287-293.

3. Bottomley, P. J. (1990) In Press, in The Biology of Nitrogen Fixation, eds. Stacey, G., Burris, R. H., and Evans, H. J. (Chapman and Hall, New York).
4. Bottomley, P. J., and Maggard, S. P. (1990) Appl. Environ. Microbiol. 56, 533-540.
5. Bushby, H. V. A. (1982) in Nitrogen Fixation, Volume 2, Rhizobium, ed. Broughton, W. J. (Clarendon Press, Oxford), pp. 35-75.
6. Crow, V. L., Jarvis, B. D. W., and Greenwood, R. M. (1981) Int. J. System. Bacteriol. 31, 151-172.
7. Date, R. A., and Brockwell, J. (1978) in Plant Relations in Pastures, ed. Wilson, J. R. (CSIRO, East Melbourne), pp. 206-216.
8. Dowling, D. N., and Broughton, W. J. (1986) Ann. Rev. Microbiol. 40, 131-157.
9. Harrison, S. P., Jones, D. G., and Young, J. P. W. (1989) J. Gen. Microbiol. 135, 1061-1069.
10. Kleczkowski, A., and Thornton, H. G. (1944) J. Bacteriol. 48, 661-662.
11. Kogure, K., Simidu, U., and Taga, N. (1979) Can. J. Microbiol. 25, 415-420.
12. Schmidt, E. L., and Robert, F. M. (1985) in Nitrogen Fixation Research Progress, eds. Evans, H. J., Bottomley, P. J., and Newton, W. E. (Martinus Nijhoff, Dordrecht), pp. 379-385.
13. Vincent, J. M., and Humphrey, B. (1970) J. Gen. Microbiol. 63, 533-540.
14. Young, J. P. W. (1985) J. Gen. Microbiol. 131, 2399-2408.

Gresshoff, Roth, Stacey & Newton (eds.) Nitrogen Fixation: Achievements and Objectives
Chapman and Hall New York • London 1990

Molecular analysis of the denitrification system of pseudomonads

Walter G. Zumft

Department of Microbiology, University of Karlsruhe, Kaiserstr. 12, D-7500 Karlsruhe 1, Federal Republic of Germany.

INTRODUCTION

Bacterial denitrification is an alternative form of energy conservation by facultative anaerobic bacteria in the absence of oxygen. Denitrification may be conceived as the assembly of four, not always tightly coupled respiratory processes, which depend as terminal electron acceptors on nitrate, nitrite, nitric oxide (NO) or nitrous oxide (N_2O) (Fig. 1). The process is of importance for ground and waste water management (nitrate removal), agronomy (loss of fixed nitrogen), and atmospheric chemistry (NO and N_2O as reactants in the stratosphere). Denitrification has been described for two genera of the halobacteria and about 50 genera of the eubacteria, totalling nearly 130 denitrifying species (35). The trait occurs in the diazotrophic genera _Aquaspirillum_, _Azospirillum_, _Bradyrhizobium_, _Pseudomonas_, _Rhizobium_, _Rhodobacter_, and _Rhodopseudomonas_, somewhat unexpectedly because antagonistic N-transformations are involved. Denitrification is prevalent among the pseudomonads; they harbor also the dominant denitrifiers worldwide (8). Our objective is the characterization of the process from a single bacterial species. The four enzymes catalyzing the complete denitrification pathway from nitrate to N_2 have been isolated and characterized from _Pseudomonas stutzeri_ strain ZoBell (ATCC 14405), formerly _P. perfectomarina_ (6). The critical threshold levels of O_2, determined for derepression of enzyme synthesis, except NO reductase, have been found to decrease in the order nitrate, N_2O, and nitrite reductase (19).

METHODS

Mutants of _P. stutzeri_ with defects in individual reaction steps of denitrification were isolated from insertional mutagenesis with the transposon Tn_5_ (36, 37). Mutants were selected for lack of gas formation in agar plates (bubble test) in the presence of the respiratory substrates nitrate or nitrite. Recombinant DNA techniques are described in ref. 32, 33, and 40. Nitric oxide reductase was purified to homo-

nar(10)   nir(5)   nor(3)   nos(7)

$$NO_3^- \longrightarrow NO_2^- \longrightarrow NO \longrightarrow N_2O \longrightarrow N_2$$

Fig. 1. Reaction sequence of bacterial denitrification. The genetic notation for each system catalyzing the individual reaction steps is shown above the arrows. The figures give a preliminary account of the numbers of genes involved.

geneity after detergent solubilization (12). Nitrate reductase was purified after heat-dissociation from the membrane by a rapid three-step procedure involving differential centrifugation, DEAE-Sepharose chromatography, and gel filtration (S. Blümle, unpublished). N_2O reductase (4) and nitrite reductase (37) were obtained from the soluble cell fraction.

RESULTS AND DISCUSSION

NITRATE REDUCTASE

Respiratory nitrate reductase, a heterodimer, is characterized best from *Escherichia coli*, a non-denitrifying bacterium. A heme *b*-containing subunit is often observed upon detergent solubilization. Nitrate reductase contains 3Fe and several 4Fe iron-sulfur clusters (18) and bactopterin as cofactor for Mo (1). The enzyme from *P. stutzeri* was resolved in non-denaturing electrophoresis in the presence of 1% Triton X-100 in two catalytically active species (Table 1). The two forms had each a 112-kDa subunit and a 45-kDa subunit, or a mixture of 45 and 60-kDa peptides. The subunit heterogeneity was caused by an endogenous protease which modified the small subunit. A native form of nitrate reductase cannot be isolated from *P. stutzeri* by heat dissociation because of this proteolytic modification. By Western blot analysis it was shown that the enzyme bound to the membrane was composed of only two subunits of 112 and 60 kDa. Random Tn*5* mutagenesis led to 10 classes of mutants with the insertion in distinct *Eco*RI fragments (S. Blümle, unpublished). The nitrate respiratory system of *P. stutzeri* is, therefore, at least as complex as that from *E. coli* for which 15 gene loci have been described so far.

NITRITE REDUCTASE

The tetraheme-dependent cytochrome cd_1 and an alternative copper-containing nitrite reductase are the key enzymes of the denitrification process *sensu stricto*. An immunochemical survey shows the former to be distributed somewhat more widely among denitrifiers than the latter (5). No single bacterium, however, has been found to harbor both enzymes.

Table 1. Enzymes of denitrification from Pseudomonas stutzeri ZoBell

Property	Nitrate reductase[a]	Nitrite reductase[a]	Nitric oxide reductase[b]	Nitrous oxide reductase[c]
EC nr.	1.7.99.4	1.9.3.2	1.7.99._ [d]	1.7.99.6
Reaction	$NO_3^- \longrightarrow NO_2^-$	$NO_2^- \longrightarrow NO$	$NO \longrightarrow N_2O$	$N_2O \longrightarrow N_2$
Location[f]	CM	PPL	CM	PPL
Mol. mass (kDa)				
Gelfiltration	140, 132[g]	120	180[e]	120
Subunits (kDa)	112, 60	61	38, 17	62-74
Quart. structure	aß	a_2	aß	a_2
Metal g-atoms/M_r	~0.8 Mo/172 13 Fe/172	4 Fe/120	~6 Fe/55	~8 Cu/140
S^{2-}	~12/172	none	none	-
Prosthetic groups	Fe_nS_n Bactopterin	heme c heme d_1	heme c heme b	Cu complex
pI	5.6	5.4	-	5.04
Absorbance maxima				
Oxidized form (nm)	280, 290(s) 410, 420(s)	278, 415 465, 535 625	280, 411.5 537	280, 350,480 540, 780
Reduced form (nm)	420	417, 460 522, 554 625	420.5, 522.5 552.5, 560(s)	640
Spec. activity[h]	30	4.2	40[i]	60[j]
pH optimum	7.2	-	4.8	9.8[j]
Electron donors				
Artificial	MV, BV	PMS	PMS	BV, MV, MB
Physiologic	cyt b_{563}?	cyt c_{551}	?	?

[a] Unpublished data; [b] ref. 12; [c] ref. 4 and 25; [d] no valid entry at present; [e] M_r of Triton X-100 micelle; [f] CM, cytoplasmic membrane; PPL, periplasm; [g] by polyacrylamide gel electrophoresis; [h] units in µmol substrate reduced per min per mg protein with artificial electron donors; [i] in the presence of phosphatidylglycerol; [j] base-activated form.

Nitrite reductase is a periplasmic, homodimeric protein of M_r 120 000, with the 60-kDa subunit carrying each a molecule of heme c and heme d_1 (Table 1). The latter is a novel type of 1,3-porphyrinedione (3). Cytochrome cd_1 has been studied extensively as model system for inter- and intramolecular electron transfer (for review see ref. 13). It catalyzes the oxidation of cytochrome c_{551} or azurin, forming NO from nitrite. Cu-containing nitrite reductases with similar properties have been isolated from Alcaligenes xylosoxidans subsp. xylosoxidans (formerly Alcaligenes sp. NCIB 11015) (22) and Pseudomonas aureofaciens (38). The latter is a homodimeric Cu protein with 2 g-atoms of Cu per M_r 80 000. In light of this protein having enzymatic activity it is remarkable that the Cu centers appear to be type I Cu sites. Other Cu-type nitrite reductases have Cu in 2 different environments (7).

NITRIC OXIDE REDUCTASE

Bacteria, unlike eukaryotic cells, metabolize NO as part of the denitrification process. Except of an isolated report concerning Bacillus sp. (24), growth of denitrifying bacteria at the expense of NO has not been achieved. Intact, resting cells, however, can metabolize NO with a high rate to N_2O. Recently, it has been shown in vitro that NO reduction is accompanied by ATP formation (2). An intrinsic, proton-translocating capability of this enzyme has not been documented experimentally, although from studies with intact cells (9, 28) such an activity may be inferred. Nitric oxide-dependent ATP formation attributes an important role to the reductase, rather than only a detoxifying or otherwise auxiliary function.

During NO reduction N-N bonds are being formed; this in essence reverses N_2 fixation. The enzymatic mechanism is still subject to controversy. Free nitroxyl (10) and enzyme-bound oxyhyponitrite (34) were suggested as intermediates. Both hypotheses are based on cytochrome cd_1 as catalyst and preclude the participation of an NO reductase. However, the first representative of an NO reductase purified to homogeneity was now obtained from P. stutzeri (12). The enzyme is a cytochrome bc complex (17- and 38-kDa peptides associated with heme c and heme b, respectively) which is tightly bound to the cytoplasmic membrane (Table 1). This property concurs with its role in energy conservation. The ligand-binding, molecular and spectroscopic characteristics of NO reductase are distinct from that of the cytochrome bc_1 complex or the o-type oxidase. The conclusions from work with isotopes, which seemingly provided strong evidence against a role of NO in denitrification, are no longer upheld (11).

Electronic spectra of the reduced enzyme show maxima at 420.5, 522.5, and 552.5 nm, and a shoulder at 560 nm. Electron paramagnetic resonance spectra at 10 K show a complex set of resonances at $g = 6.38$ of high-spin Fe(III) in rhombic and axial environments, and intense lines at 3.58, 3.02, 2.29 with corresponding high-field lines around 2, 1.58 and 1.44, indicative of low-spin ferric heme b and c. Chemical analysis of NO reductase indicates the presence of nonheme iron; however, no signals for iron-sulfur centers were detected in the reduced or oxidized enzyme.

Solubilization of NO reductase from the membrane by Triton X-100 affects markedly enzymatic activity (12, 29), presumably by displacement of lipids from the complex. The activity of the purified enzyme increases again several-fold in the presence of certain phospholipids or detergents. The following compounds, in increasing order of effectiveness, have proven best for NO reductase: dodecyl-ß-D-maltoside, phosphatidylethanolamine, n-octyl-ß-D-glucopyranoside and its thio-derivative, 1-dodecanoylpropanediol-3-phosphorylcholine ether and phosphatidylglycerol (B. Heiss, unpublished). The maximum increase in NO reductase activity is about four-fold.

NITROUS OXIDE REDUCTASE

N_2O reductase is a multicopper enzyme with approximately 8 g-atoms of Cu per M_r 140 000 (Table 1) (4, 25, 39). The enzyme catalyzes the last step of denitrification and generates dinitrogen from N_2O. It performs, thus, a reaction that is also an activity of dinitrogenase (16). The metal chromophore is a novel type of a mixed valence Cu center (20). The primary structure (33), extended X-ray absorption fine structure and magnetic circular dichroism (27), pulsed EPR (17) and multi-frequency EPR (20) all indicate a Cu_A site similar to that of cytochrome c oxidase. A monomeric enzyme with spectroscopic features similar to the enzyme from P. stutzeri was isolated from Achromobacter cycloclastes (15). From Wolinella succinogenes a Cu and heme-containing enzyme has been described (31).

GENETIC ORGANIZATION

Genes for denitrification are plasmid-coded in Alcaligenes eutrophus (26), which represents until date the only example of this kind. No plasmids were found in P. stutzeri ZoBell using different methods, including those apt to preserve a megaplasmid; attempts of plasmid-curing were also negative. A chromosomal location of genes for denitrification is, therefore, assumed. Physical mapping of transposon Tn5-induced mutations defined a gene cluster of ca. 8 kbp for the respiratory utilization of N_2O by P. stutzeri (32). The underlying mutant phenotypes suggested defects within a regulatory function, the structural gene, and functions for metal-processing. The nos-specific DNA region was cloned by plasmid integration and excision, and the structural gene (nosZ) for N_2O reductase identified by deletion analysis and expression in E. coli (33). Sequence comparisons with nosZ

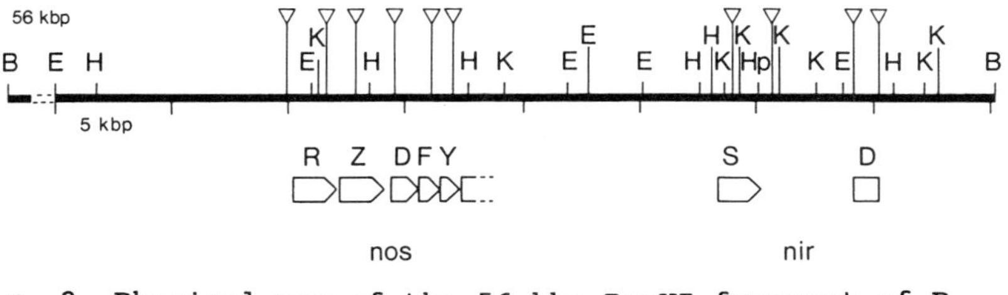

Fig. 2. Physical map of the 56-kbp BamHI fragment of P. stutzeri carrying genes for nitrous oxide (nos) and nitrite (nir) respiration. The flags indicate insertion sites of Tn5 causing Nos^- and Nir^- phenotypes. Arrows indicate transcriptional directions. The size of nirS is given by its gene product; that of nirD by the corresponding Tn5 insertions (A. Jüngst and C. Braun, unpublished).

from A. eutrophus and Pseudomonas aeruginosa revealed extended regions of similarity and a number of conserved histidine residues as potential Cu ligands (S. Löchelt, unpublished).

The structural gene for nitrite reductase (nirS) was identified by mapping the Tn5 insertion site of mutant MK202 which carries its defect within nirS, isolating clones immunoreactive for cytochrome cd_1 from a lambda gt11 expression library, complementation analysis, and N-terminal sequencing of the purified protein and partial DNA-sequencing. A mutation causing overproduction of cytochrome c_{552} (37), mapped within 1,300 base pairs of nirS. Two mutations affecting nirD, which caused the synthesis of an inactive cytochrome cd_1 lacking heme d_1, mapped within a 10.5-kbp EcoRI fragment contiguous to the nirS-carrying fragment. Three other mutants mapped independently on two different EcoRI fragments. Cosmid-mapping and Southern hybridization revealed a close linkage of the genes nosZ, nirS, and nirD. The nirS and nosZ genes are separated by only 16 kbp and transcribed in the same direction. Genes for Cu-processing (vide infra) cover part of the intervening region. Fig. 2 shows the current map of nos and nir genes of P. stutzeri. Downstream of nirS of P. aeruginosa (30) the cytochrome c_{551}-gene was found (23). Whether this is the case in P. stutzeri remains to be seen.

COPPER-PROCESSING

The biosynthesis of metalloproteins and their cofactors involves within the context of the denitrification process metal uptake, the synthesis of organic cofactors, protein insertion into or translocation across the cytoplasmic membrane, and the temporal and spatial control of metal and cofactor incorporation in relation to enzyme export.

Insertional mutagenesis of the DNA region downstream of nosZ generated apoenzyme-producing strains, which argues for the existence of functions for copper acquisition or processing, or both. The relevant region was found to consist of three genes, nosDFY, presumably within a single transcriptional unit (Fig. 2). The expression of nosD and nosF was tested by generating mini-Mulac fusions and found to respond to anaerobiosis and the presence of nitrogenous oxides (H. Cuypers, unpublished). The gene products show structural and organizational similarities to bacterial uptake systems for small molecules. The nosF product comprises a new member of the transport family (14) of ATP/GTP-binding proteins; NosY is an integral membrane protein with 6 predicted transmembrane helices. NosD is suggested from 2-dimensional electrophoretic patterns of the wild type and mutant strain MK402 to be a periplasmic protein. In addition to these gene products an outer membrane Cu-containing protein with channel-forming properties has been identified in P. stutzeri JM300 (21). Mutational loss of this protein led also to apoenzyme-producing strains. Our current data support an assembly of the Cu chromophore within the periplasm upon translocation of the reductase into this cell compartment.

ACKNOWLEDGMENTS

I thank my coworkers S. Blümle, C. Braun, H. Cuypers, B. Heiss, A. Jüngst, S. Löchelt, H. Körner, and A. Viebrock for their many contributions to this work. The support of the Deutsche Forschungsgemeinschaft and the Fonds der Chemischen Industrie is gratefully acknowledged.

REFERENCES

1. Amy, N.K. & Rajagopalan, K.V. (1979) J. Bacteriol. 140, 114-124.
2. Carr, G.J., Page, M.D. & Ferguson, S.J. (1989) Eur. J. Biochem. 179, 683-692.
3. Chang, C.K., Timkovich, R. & Wu, W. (1986) Biochemistry 25, 8447-8453.
4. Coyle, C.L., Zumft, W.G., Kroneck, P.M.H., Körner, H. & Jakob, W. (1985) Eur. J. Biochem. 153, 459-467.
5. Coyne, M.S., Arunakumari, A., Averill, B.A. & Tiedje, J.M. (1989) Appl. Environ. Microbiol. 55, 2924-2931.
6. Döhler, K., Huss, V.A.R. & Zumft, W.G. (1987) Int. J. Syst. Bacteriol. 37, 1-3.
7. Dooley, D.M., Moog, R.S., Liu, M.-Y., Payne, W.J. & LeGall, J. (1988). J. Biol. Chem. 263, 14625-14628.
8. Gamble, T.N., Betlach, M.R. & Tiedje, J.M. (1977) Appl. Environ. Microbiol. 33, 926-939.
9. Garber, E.A.E., Castignetti, D. & Hollocher, T.C. (1982) Biochem. Biophys. Res. Commun. 107, 1504-1507.
10. Garber, E.A.E. & Hollocher, T.C. (1982) J. Biol. Chem. 257, 4705-4708.
11. Goretski, J. & Hollocher, T.C. (1990) J. Biol. Chem. 265, 889-895.
12. Heiss, B., Frunzke, K. & Zumft, W.G. (1989) J. Bacteriol. 171, 3288-3297.
13. Henry, Y. & Bessières, P. (1984) Biochimie 66, 259-289.
14. Higgins, C.F., Gallagher, M.P., Mimmack, M.L. & Pearce, S.R. (1988) BioEssays 8, 111-116.
15. Hulse, C.L. & Averill, B.A. (1990) Biochem. Biophys. Res. Commun. 166, 729-735.
16. Jensen, B.B. & Burris, R.H. (1986) Biochemistry. 25, 1083-1088.
17. Jin, H., Thomann, H., Coyle, C.L. & Zumft, W.G. (1989) J. Am. Chem. Soc. 111, 4262-4269.
18. Johnson, M.K., Bennett, D.E., Morningstar, J.E.,

Adams, M.W.W. & Mortenson, L.E. (1985) J. Biol. Chem. 260, 5456-5463.
19. Körner, H. & Zumft, W.G. (1989) Appl. Environ. Microbiol. 55, 1670-1676.
20. Kroneck, P.M.H., Antholine, W.A., Riester, J. & Zumft, W.G. (1988) FEBS Lett. 242, 70-74.
21. Lee, H.S., Hancock, R.E.W. & Ingraham, J.L. (1989) J. Bacteriol. 171, 2096-2100.
22. Masuko, M., Iwasaki, H., Sakurai, T., Suzuki, S. & Nakahara, A. (1984) J. Biochem. 96, 447-454.
23. Nordling, M., Young, S., Karlsson, G. & Lundberg, L.G. (1990) FEBS Lett. 259, 230-232.
24. Pichinoty, F., Mandel, M., Greenway, B. & Garcia, J.L. (1977) Ann. Microbiol. (Inst. Pasteur) 128A, 75-87.
25. Riester, J., Zumft, W.G. & Kroneck, P.M.H. (1989) Eur. J. Biochem. 178, 751-762.
26. Römermann, D. & Friedrich, B. (1985) J. Bacteriol. 162, 852-854.
27. Scott, R.A., Zumft, W.G., Coyle, C.L. & Dooley, D.M. (1989) Proc. Natl. Acad. Sci. USA 86, 4082-4086.
28. Shapleigh, J.P., Davies, K.J.P. & Payne, W.J. (1987) Biochim. Biophys. Acta 911, 334-340.
29. Shapleigh, J.P. & Payne, W.J. (1985) J. Bacteriol. 163, 837-840.
30. Silvestrini, M.C., Galeotti, C.L., Gervais, M., Schinina, E., Barra, D., Bossa. F. & Brunori, M. (1989) FEBS Lett. 245, 33-38.
31. Teraguchi, S. & Hollocher, T.C. (1989) J. Biol. Chem. 264, 1972-1979.
32. Viebrock, A. & Zumft, W.G. (1987) J. Bacteriol. 169, 4577-4580.
33. Viebrock, A. & Zumft, W.G. (1988) J. Bacteriol. 170, 4658-4668.
34. Weeg-Aerssens, E., Tiedje, J.M. & Averill, B.A. (1988) J. Am. Chem. Soc. 110, 6851-6856.
35. Zumft, W.G. (1991) in The Prokaryotes. A Handbook on the Biology of Bacteria, Ecophysiology, Isolation, Identification, Applications, eds. Balows, A., Trüper, H.G., Dworkin, M., Harder, W. & Schleifer, K.H. (Springer Vlg., New York), 2nd ed., in press.
36. Zumft, W.G., Döhler, K. & Körner, H. (1985) J. Bacteriol. 163, 918-924.
37. Zumft, W.G., Döhler, K., Körner, H., Löchelt, S., Viebrock, A., & Frunzke, K. (1988) Arch. Microbiol. 149, 492-498.
38. Zumft, W.G., Gotzman, D.J. & Kroneck, P.M.H. (1987) Eur. J. Biochem. 168, 301-307.
39. Zumft, W.G. & Matsubara, T. (1982) FEBS Lett. 148, 107-112.
40. Zumft, W.G., Viebrock-Sambale, A. & Braun, C. (1990) Eur. J. Biochem. in press.

Gresshoff, Roth, Stacey & Newton (eds.) Nitrogen Fixation: Achievements and Objectives
Chapman and Hall New York • London 1990

Production of nitrite and N_2O by the ammonia-oxidizing nitrifiers

Alan B. Hooper, David M. Arciero, Alan A. DiSpirito*, James Fuchs, Matthew Johnson, Frank LaQuier, Gary Mundfrom and Hugh McTavish.
Dept. Genetics and Cell Biology, Univ. Minnesota, St. Paul, MN 55108 USA
*Present address Dept. Biology, U. TX, Arlington TX 76019.

Introduction. *Nitrosomonas* is an obligately autotrophic and aerobic bacterium which produces energy for growth from the following reactions: $NH_3 + O_2 + 2e^- \longrightarrow NH_2OH$ (ammonia monoxygenase, "AMO") (8); $NH_2OH + H_2O \longrightarrow 4 e^- + 4H^+ + HNO_2$ (hydroxylamine oxidoreductase, "HAO") (3) and $1/2\ O_2 + 2e^- + 2H^+ \longrightarrow H_2O$ (terminal oxidase).

Origin of N_2O produced from NH_3. At low concentrations of O_2, cells produce significant amounts (5-10%) of NO and N_2O in place of HNO_2 (7). One isolate is reported to produce a small amount (0.05%) of N_2 from ammonia (18). On a per-cell basis, production of N_2O by the nitrifying bacteria can, in fact, exceed production by denitrifying bacteria (2). It is not known with certainty whether this gas is produced by oxidation and/or by reduction of N- compounds. Fig. 1 shows potential means of production of nitrous oxide.

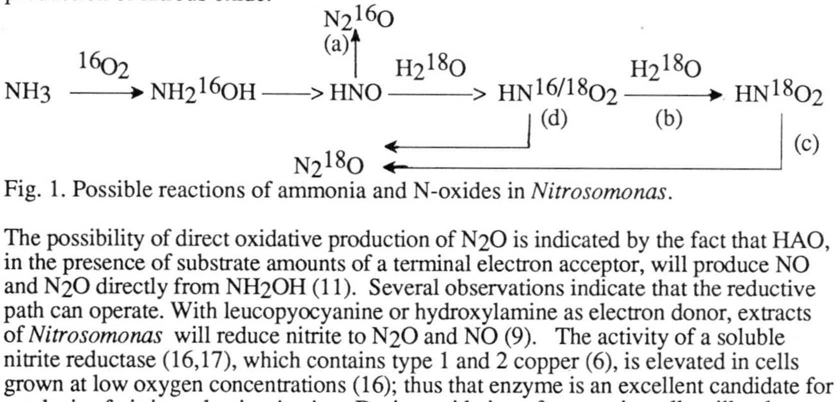

Fig. 1. Possible reactions of ammonia and N-oxides in *Nitrosomonas*.

The possibility of direct oxidative production of N_2O is indicated by the fact that HAO, in the presence of substrate amounts of a terminal electron acceptor, will produce NO and N_2O directly from NH_2OH (11). Several observations indicate that the reductive path can operate. With leucopyocyanine or hydroxylamine as electron donor, extracts of *Nitrosomonas* will reduce nitrite to N_2O and NO (9). The activity of a soluble nitrite reductase (16,17), which contains type 1 and 2 copper (6), is elevated in cells grown at low oxygen concentrations (16); thus that enzyme is an excellent candidate for catalysis of nitrite reduction *in vivo*. During oxidation of ammonia, cells will reduce $^{15}NO_2^-$ to $^{15}N_2O$ (19,20). A kinetic analysis of production of nitrous oxide by cells in the presence of ^{14}N-ammonia and ^{15}N-nitrite has indicated to Poth and Focht (19) that all ammonia-N is oxidized to nitrite prior to formation of N_2O.

We have employed an alternate means of answering this question. During the oxidation of ammonia, cells catalyze the rapid exchange of oxygen between nitrite and water (3,4) (Fig. 1, b). This exchange reaction should replace the nitrite-O which originated from O_2 in the AMO reaction with O from water. N_2O produced by subsequent reduction of nitrite (reaction c) should therefore have the isotopic composition of water. If ammonia is oxidized directly to nitrous oxide (Fig. 1 a), nitrite would not be a precursor of N_2O, so that the O_2-derived O should be retained in the product. The approach has the advantage of direct measurement of N_2O arising from NH_3 rather than from exogenously added NO_2^- (19,20). The conclusion depends on the assumption that nitrite is the only N-oxide which exchanges oxygen with water.

Table 1 shows data resulting from the oxidation of ^{14}N- or ^{15}N-ammonia to nitrous oxide in the presence of $^{16}O_2$ and 55% $H_2{}^{18}O$: 45% $H_2{}^{16}O$. Assuming that, in the direct oxidation of NH_3 to the hypothetical intermediate, HNO, by AMO and HAO, the O came from dioxygen (8), none of the N_2O produced directly from HNO should have contained ^{18}O. If all N_2O had been produced by the oxidation of ammonia-N to nitrite followed by the direct reduction of nitrite and if exchange reactions had not occurred, the intermediate nitrite and thus the product N_2O would have contained 28 % ^{18}O. On the other hand, if all ammonia had been oxidized to NO_2^-, all nitrite had exchanged oxygen with water and nitrite been reduced to nitrous oxide, the latter would have consisted of 55% $N_2{}^{18}O$. In the two experiments shown in Table 1, the product N_2O contained 38 or 43% ^{18}O. Assuming that the water - O exchange occurs only with nitrite, the presence of ^{18}O in N_2O is the first direct evidence that nitrite is an intermediate in the production of nitrous oxide from ammonia. The fact that the observed values of % $N_2{}^{18}O$ are less than 55%, indicates that either (a) exchange was incomplete or (b) some N_2O is produced by direct oxidation of ammonia. The latter would be inconsistent with the conclusions of Poth and Focht (19).

Within the autotrophic nitrifier, *Nitrosomonas*, the nitrapyrin-sensitive oxidation of ammonia is required as both N-source (nitrite) and electron source for the "denitrifying" production of N_2O. In soils, inhibition of AMO of *Nitrosomonas* by nitrapyrin may retard production of N_2O by *Nitrosomonas* and/or production of N_2O by by heterotrophic bacteria because the source of nitrite is diminished.

Together with the proton-utilizing AMO reaction and proton-producing, periplasmic HAO reaction, the nitrite reductase (NiR) activity: $HNO_2 + 2e^- + 2H^+ \rightarrow 0.5\ N_2O + 1.5\ H_2O$, could, in theory, contribute to the production of an energy-transducing proton gradient. If the AMO reaction is on the cytoplasmic surface of the membrane, a net translocation of 4 or 2 protons would occur depending on whether the NiR reaction was cytoplasmic or periplasmic, respectively.

Evolution of N-oxidoreductases. By determining the amino acid sequence of trypsin- and chymotrypsin-generated heme-peptides, we have shown that the 63 kD subunit of HAO contains 7 *c*-hemes (5). One *c*-heme and the chromophore "P-460" appear to be involved in the active site (15). The other hemes are thought to be involved in electron transfer or redox poising of the enzyme. The "cytochrome *c* 552" respiratory nitrite reductases from *E.coli, Wolinella, Desulfovibrio* and *Vibrio* catalyze the reaction; $HNO_2 + 6e^- + 4H^+ \rightarrow NH_3 + 2H_2O$ (14). Interestingly, each subunit (M_r 56,57,63,

Table 1. Fate of $^{18}O_2$ or $H_2{}^{18}O$ During Production of N_2O from Ammonia by *Nitrosomonas europaea*[1]

	Mass and isotopic composition of products				
	44	45	46	47	48
Substrate	$^{14}N_2{}^{16}O$	$^{14/15}N_2{}^{16}O$	$^{14}N_2{}^{18}O$* or $^{15}N_2{}^{16}O$**	$^{15/14}N_2{}^{18}O$	$^{15}N_2{}^{18}O$
	(% of each species observed)				
$^{14}NH_3$ $^{16}O_2$ 55% $H_2{}^{18}O$	62	0	38*	0	0
Expected:					
Oxidation;	100	0	0	0	0
Reduction; no exchange	72	0	28	0	0
complete exchange	45	0	55	0	0
$^{15}NH_3$ $^{16}O_2$ 55% $H_2{}^{18}O$	0	0	57**	3	40

[1]Growth and harvest of cells were as described previously (3). Cells were incubated in 35 ml Hypo Vials (Pierce Chemical co.) sealed with Teflon-silicon discs. The 5 ml volumes of 50 mM phosphate, pH 7.5, solution were in equilibrium with 1% O_2 in the gas phase. Cells (1 mg protein/ml) were incubated in 55% $H_2{}^{18}O$ with either $^{14}NH_3$ or $^{15}NH_3$ (20 mM) as substrate. 10 - 15 mM nitrite were produced in the 1 hr incubation at 28°. N_2O in the head space was analyzed by tandem GC/MS (11).

66 kD, respectively), contains 6 *c* hemes. We note that the cytochrome *c* 552 nitrite reductase is similar in amino acid content to HAO (Table 2). Given that HAO and the hexaheme nitrite reductase catalyze much the same reaction (though in the reverse direction) and have similar numbers of *c*-hemes, subunit size and amino acid composition, we speculate that the two enzymes may have a common evolutionary ancestor. With this in mind, we intend to compare the sequence of HAO of *Nitrosomonas* with the NiR of *E.coli* . The latter has been purified to homogeneity by the published procedures (13). N-terminal sequence analysis of the intact cytochrome failed, presumably because of a blocked n-termini. Therefore a heme-containing peptide was prepared by tryptic digestion, isolated by HPLC, and then sequenced. A probe has been synthesized for each cytochrome and the complementary DNA cloned. The determination of sequence is in progress. We speculate on the function of the putative common ancester. The early biosphere is thought to contain ammonia but not oxygen and thus not nitrite. Therefore nitrite might not have been available to the biosphere until the production of dioxygen by oxygenic photosynthesis and the subsequent evolution of ammonia-oxidizing bacteria. By this scenario, the primitive enzyme might have been involved in oxidation of ammonia or hydroxylamine.

Table 2. Comparison of the Amino Acid Composition of HAO of *Nitrosomonas* (10), and Hexaheme Respiratory NiR of *E.coli, Wolinella, succinogenes, Desulfovibrio desulfuricans* and *Vibrio fisheri* (14).

	HAO	NiR
asp/asn	13.0	10.0
thr	6.6	5.3
ser	4.4	3.6
glu/gln	12.0	13
pro	4.6	4.6
gly	7.2	6.8
ala	8.7	10.0
val	5.7	6.0
met	3.0	3.2
ile	3.0	4.0
leu	6.3	6.7
tyr	3.3	3.5
phe	1.7	3.3
his	2.4	3.5
lys	8.4	9.1
arg	4.6	4.7
cys	3.1	2.1
trp	4.8	1.1

Multiple Copies of Genes for the 63 kD Subunit of HAO in *Nitrosomonas* We have reacted genomic digests of *Nitrosomonas* with two polynucleotide probes for the 63 kD subunit of HAO: a 32-redundant 20-mer and a 64-redundant 29-mer based on the amino acid sequence of one of the heme-containing peptides. DNA of *Nitrosomonas* was digested by the restriction enzymes Bcl, Sal-I and Xho-I and the fragments separated by elecrophoresis on agarose gels . Following electrophoretic transfer to nylon membranes, the restriction fragments were hybridized with the probes. Unexpectedly, using either probe we find that the genome of *Nitrosomonas* appears to contain three copies of genes for the 63 kD protein of HAO. Fig. 1 shows that the 20-mer reacts with three copies of DNA produced by separate digestion with three restriction enzymes. The sequence appears to occur in equal numbers and react with equal stringency. Because the genes are found in equal quantity, it seems unlikely that the culture contains three strains of bacteria, a trivial explanation for the result. Plasmids were not detected in DNA isolated from *Nitrosomonas* and analysed by the procedure of Anderson and McKay (1). HAO is an enzyme central to the energy metabolism of *Nitrosomonas* and is found in high concentration in the cell; it accounts for approximately 5% of the total soluble protein (6). It is possible that cells have three identical copies of the gene to facilitate rapid transcription. However, since *Nitrosomonas* is a rather slow-growing organism and, as an autotroph,must synthesize all components from CO_2 and NH_3, it does not seem likely that having only one copy of the gene would provide a rate-limiting barrier to cell growth. One may speculate that three, slightly different forms of the 63 kD protein form the active multimeric HAO. It is also possible that three, slightly different forms of the enzyme have three different

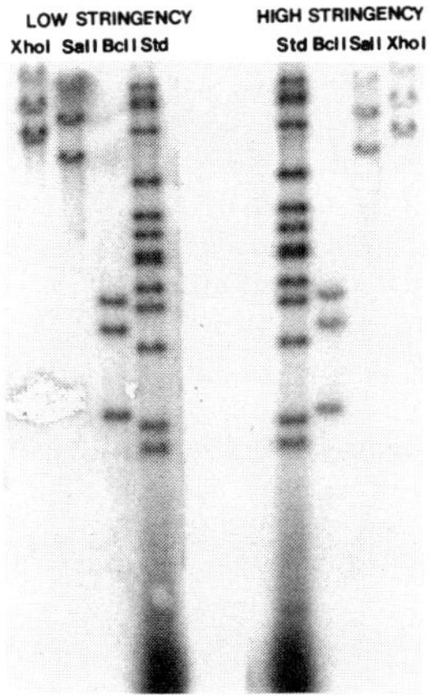

Fig. 3. Reaction of 20-mer Probe of HAO 63 kD Protein with Separate Restriction Enzyme Digests of DNA from *Nitrosomonas*. Hybridization wash was for 30 min in 0.1X standard sodium phosphate EDTA, 0.1 % SDS at 37°. Purification of DNA, restriction enzyme digestion, electrophoresis, transfer and hybridization were carried out by standard methods (21).

biochemical roles. For example a unique form may involved depending on whether the 4 electrons yielded in the oxidation of hydroxylamine terminate in (a) the aa_3 terminal oxidase, (b) AMO or (c) energy-dependent reduction of pyridine nucleotide . Finally, the multiple forms might be suited to particular responses to growth conditions.

Supported by NSF/DMB 87-19338, USDA/CRGO-88-37120-3957. and EPA/CR-816165. We thank Tom Krick for mass spectrometer analysis.

LITERATURE CITED

1. Anderson, D.G. & McKay, L.L. (1983) Appl. Envt. Microbiol. **46**, 549-552.
2. Anderson, I.C. & Levine, J.S. (1986) Appl. Envt. Microbiol. **51**, 938-945.
3. Andersson, K.K. & Hooper, A.B. (1983) FEBS Lett. **164**, 236-240.
4. Andersson, K.K., Philson, S.B., & Hooper, A.B. (1982) Proc. Nat. Acad. Sci. USA **79**, 5871-5875.
5. Arciero, D.M. & Hooper, A.B. (1989) J. Cell Biol. **107**, 620a.
6. DiSpirito, A.A., Taaffe, L.R., Lipscomb, J.D., & Hooper, A.B. (1985) Biochim. Biophys. Acta **827**, 320-326.
7. Goreau, T.J., Kaplan, W.A., Wofsy, F.C., McElroy, M.B., Valois, F.W., & Watson, S.W. (1980) Appl. Env. Microbiol. **40**, 526-532.
8. Hollocher, T.C., Tate, M.E., & Nicholas, D.J.D. (1981)J. Biol. Chem. **256**, 10834-10836.

9. Hooper, A.B.(1968) Biochim. Biophys. Acta **162**, 49-65.
10. Hooper, A.B., Maxwell, P.C., & Terry, K.R. (1978) Biochem. **17**, 2984-2989.
11. Hooper, A.B., & Terry, K.R.(1979) Biochim. Biophys. Acta **571**, 12-20.
12. Hooper, A.B., & Terry, K.R.(1973) J. Bacteriol. **115**, 480-485.
13. Kajie, S.-I. & Anraku, Y. (1986) Eur. J. Biochem. **154**, 457-463.
14. Liu, M-C., Bakel, B.W., Liu, M-Y, & Dao, T.N. (1988) Arch. Biochem. Biophys. **262**, 259-265.
15. Logan, M.S.P. & Hooper, A.B. (1989) J. Inorg. Biochem. **36**, 217.
16. Miller, D.J. & Nicholas, D.J.D.(1985) J. Gen. Microbiol. **131**, 2851-2854.
17. Miller, D.J. & Wood, P.M. (1983) J. Gen. Microbiol. **129**, 1645-1650.
18. Poth, M. (1986) Appl. Envt. Microbiol. **52**, 957-959.
19. Poth, M. & Focht, D.D. (1985) Appl. Envt. Microbiol. **49**, 1134-1141.
20. Ritchie, G.A.F. & Nicholas, D.J.D. (1972) Biochem. J. **126**, 1181-1191.
21. Sambrook, J., Fritsch, E.F., & Maniatis, T. (1989) Molecular Cloning: A Laboratory Manual, Second Edition, (Cold Spring Harbor.Laboratory, Cold Spring Harbor, NY).

Gresshoff, Roth, Stacey & Newton (eds.) Nitrogen Fixation: Achievements and Objectives
Chapman and Hall New York • London 1990

Planned release of genetically modified organisms: Scientific and regulatory issues

Dr. Sivramiah Shantharam

USDA, APHIS, BBEP, BP/6505 Belcrest Road, Hyattsville, MD, USA.

There is a great deal of excitement with the advent of recombinant DNA technology and biotechnology. These disciplines open up a Pandora of possibilities for genetically manipulating plants, animals, and microbes for release into the environment to be used as "work horses." There is little doubt that these genetically engineered organisms will have potential impacts on physical, biological and human components of the environment. The potential effects have to be monitored and assessed in order that careful and prudent releases are conducted. Genetically engineered nitrogen fixing organisms are no exception.

The important scientific concerns are the potential of horizontal transfer of the introduced gene(s) between organisms, potential impacts to animal, plant, and human health. Most of these concerns are dependent upon the survivability (persistence) of the introduced organisms in the environment. Microorganisms, once released, will freely disperse in the environment without any control. It is, therefore, imperative to put in place effective monitoring protocols involving state of the art methodology and allay fears in the public mind. The microorganism in question and the introduced gene(s) all need tagging or labeling so that they can be readily detected and monitored. The gene transfer data sought is more important than ever before. Gene transfer takes place in all ecological niches. It is important to obtain clear cut data regarding the frequency of gene transfer and kinds of gene transfer and the conditions that facilitate gene transfers. Such data, when obtained rigorously, is of critical impportance in developing "prediction" models for the fate of microorganisms. Some of the critical factors that influence gene transfer are the gene(s) itself, especially the trait coded by it, the vector, the hosts mating process, selection pressure, and the environmental factors. Gene transfer is readily demonstrated by the wide dissemination and prevalence of resistance to antibiotics, metals, toxins, and catabolic traits. While the low frequency transfer of genes between microorganisms has contributed to diversity and evolution, the same type of transfer at a high frequency, in localized ecological niches invites potential disaster. There is a significant level of horizontal gene transfer in natural settings as evidenced by a wealth of experimental data. There is a clear cut demonstration of antibiotic resistance gene transfers in enteric bacteria via R-plasmids between patients in hospital settings, multiple drug resistance transfers between animals and humans in the farm environment. R-plasmids have been demonstrated to be transferred in human and animal gut, urinary tracts and bladder, and respiratory tracts of birds, mammals, and humans. Transfer of discrete genetic elements have been shown to occur between free living bacteria in waste treatment plants, soil, and water, on plant surfaces, and between plants and microbes.

Gene transfer via conjugation was the only mode of gene transfer that merited serious study, until now. However, transduction and transformation have been shown to be potent methods of gene transfer in soil and water. The seemingly random processes by which gene transfer takes place raise new questions regarding the origin of genes and their functions in the new host. The classical example is that of transposon Tn5 that codes for kanamycin resistance in *Escherichia coli* and for streptomycin and bleomycin in many gram-negative bacteria. This phenomenon potentially affects ecology and fate of the genetic elements in the environment and as such warrants critical study.

For rhizobiologists, the potential of gene transfer in soil (rhizosphere) and inside the plant (nodules) and on plant surfaces (leaves, stems, and roots) are of critical importance in evaluating the risks of horizontal gene transfer. Most of the factors that influence gene transfer do not act singularly, but in concert. There are also cascade effects as a consequence of changes in one or more factors. The degree of complexity of the influencing factors governing gene transfer in soil is so complex that an accurate prediction of the risk is very difficult. There have been several reports of gene exchanges between rhizobia in the rhizosphere but the data is very meager to form meaningful conclusions. Gene transfers do not occur routinely and successfully in soil. A major vacuum in the knowledge exists in the area of potential effects of recombinant bacteria on structure and function of soils relative to the number of cells and kinds of genes per unit volume of soil per unit time period. Without such a long term study, it will be very difficult to assess the severity of impacts. It is equally important to look out for the pleiotropic effects caused by the introduced genes to account for unpredictable effects before constructing and releasing genetically engineered organisms. This underscores exercising a great deal of caution during initial release experiments on a case-by-case basis.

Genetically engineered strains have found a renewed resurgence for their increased study in microbial ecology of rhizosphere and phyllosphere. Most of the indigenous plasmids found in the plant associated bacteria have no selectable markers but, are known to transfer. This has made the precise knowledge of their transfer properties difficult to determine. Several cryptic plasmids are also known to be conjugal. Transposon tagging has been a useful tool for assessing such plasmid transfers in planta.

There are several gram-negative bacteria that interact with plants to cause pathogenesis, symbiosis, plant growth promotion, and free-living nitrogen fixation. The advances in molecular biology of these bacteria has given very precise insight into the genetics of their systems. Clear cut evidence exists to show the transfer of T-DNA by the Ti-plasmid of Agrobacterium tumefaciens results in crown gall and nodulation, host specificity, and competition genes of Rhizobium species and strains are plasmid borne. In case of Bradyrhizobium japonicum, the symbiotic genes are on chromosonal DNA. As yet, there has been no recorded instance of DNA transfer between Rhizobium and leguminous plants.

Plasmids of Agrobacterium radiobacter coding for agrocins have been shown to be conjugal, symbiotic plasmids of Rhizobium leguminosarum and R. trifolii has been shown to be conjugally transferred. There have been reports of the transfer of megaplasmids between strains within root nodules. All these facts clearly demonstrate the transfer potential of plasmids carrying important traits. Most of these transfers are controlled by stringent transfer mechanisms. If the transfer mechanisms are clearly understood it would be easy to engineer biologically safe recombinant strains for field testing. Although gene transfer from Rhizobium to plants has not been demonstrated, there is a great deal of speculation in light of the fact the presence of vir genes in trans can facilitate transfer and integration of genes other than those found in T-DNA. It is equally important to develop effective sampling methods for accurate monitoring of horizontal, vertical, and aerial dispersal of bacteria.

U.S. Department of AGriculture-Animal and Plant Health Inspection Service (USDA, APHIS) regulates genetically engineered plants and microbes, if they present a plant pest risk under Federal Plant Pest Act (FPA) and Federal Quarantine Act (FQA) and EPA regulates genetically engineered microorganisms under Toxic Substances Control Act (TOSCA) and Federal Insecticide, Fungicide and Rodenticide Act (FIFRA).

At present, USDA, APHIS is basically concerned with the potential of plant pest risk(s) presented by the recombinant strains of Rhizobium and Bradyrhizobium. An opinion letter attesting whether or not the genetically engineered microorganism is a "plant pest" is issued. If it turns out to be a plant pest, then a full blown environmental assessment (EA) as mandated by National Environmental Policy Act (NEPA) will be conducted.

Reference:

Stuart B. Levy., Robert V. Miller (Eds). 1989 Gene Transfer in the Environment. McGraw-Hill Publishing Company.

Gresshoff, Roth, Stacey & Newton (eds.) Nitrogen Fixation: Achievements and Objectives
Chapman and Hall New York • London 1990

Field release of genetically-engineered *Rhizobium meliloti* and *Bradyrhizobium japonicum* strains

C.W. Ronson, *A. Bosworth, M. Genova, S. Gudbrandsen, T. Hankinson, R. Kwiatkowski, H. Ratcliffe, C. Robie, P. Sweeney, W. Szeto, M. Williams and *R. Zablotowicz.

BioTechnica International Inc. 85 Bolton Street, Cambridge, MA 02140; *Plant Science Research, Inc. 6494 Highway N, Sun Prairie, WI 53590 USA.

Introduction

The objectives of BioTechnica International's research and development program in nitrogen fixation are to develop seed-coating products for alfalfa and soybean that improve crop productivity. The seed coatings will contain genetically engineered strains of *Rhizobium meliloti* or *Bradyrhizobium japonicum*, respectively, with improved nitrogen-fixing and competitive abilities. To accomplish these objectives, several technical and regulatory hurdles must be overcome. Strategies to improve nitrogen fixation and competitive ability must be devised and implemented, the resultant strains must be tested under greenhouse and field conditions, and an appropriate delivery system developed.

In this paper, we discuss briefly the genetic strategies adopted to develop improved rhizobial strains, the procedures followed to secure permission to field-test the strains, and the results of ecological studies designed to determine the impact of the genetic engineering on the strains' behavior under field conditions.

Genetic Engineering Strategies

We have followed two strategies to improve the nitrogen-fixing and competitive abilities of rhizobial strains. One strategy involved modifying the expression of the positive regulatory gene, *nifA*, which, if the level of *nifA* product is limiting, will result in increased expression of the *nif* genes and hence an increased level of the enzyme system required for nitrogen fixation. We hypothesized that there would be an optimal level of *nifA* expression. Hence, promoters that gave varying levels of expression were fused to the *nifA* gene. We also hypothesized that as the nitrogen-fixing capacity of the rhizobia was increased, other factors such as energy supply might become limiting. Since C_4–dicarboxylic acids are thought to be the primary energy substrates supplied to nitrogen-fixing bacteroids, we have also investigated the effect of modifying the expression of the C_4–dicarboxylate transport (*dct*) genes on nitrogen fixation.

Earlier studies with *R. meliloti* where the engineered genes were introduced in plasmids showed that many plasmids were unstable and that some plasmids had a

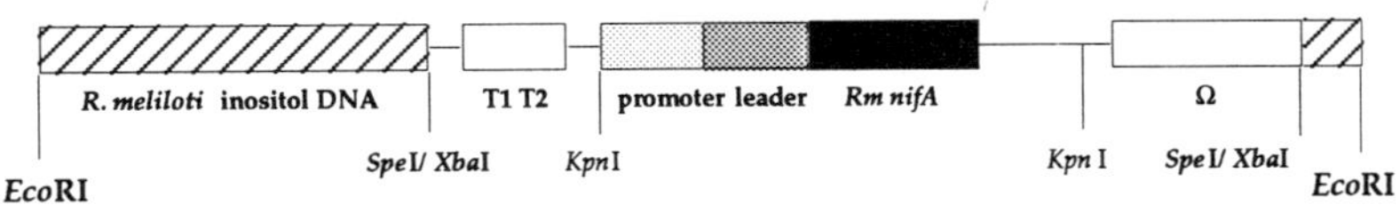

FIGURE 1. Structure of the gene cassette integrated at the ino *site.*

negative effect on the nitrogen-fixing ability of the strain. Furthermore, plasmids are generally unstable in *B. japonicum.* It was therefore decided to integrate the recombinant genes into the genome of the host strains. Considerable efforts were made to select genomic integration sites that, when interrupted by an inserted gene, were not expected to affect the strains' physiological or symbiotic properties. The two sites chosen for *R. meliloti* were in the inositol locus (*2*) and a neutral site in the *nif* region, while a site in repeat sequence RSα9 about 2 kb upstream from the *nifD* gene was chosen in *B. japonicum* (*3*). To facilitate selection of the integration events, an antibiotic-resistance gene, usually the *aadA* gene (conferring streptomycin and spectinomycin resistance) carried on the "Ω" fragment (*1*) was coupled to the gene to be integrated. This marker gene also facilitated environmental monitoring of the strain by adding an assayable resistance that was used in conjunction with intrinsic antibiotic resistances to follow the strains in the field. Furthermore, strains containing only the antibiotic-resistance gene integrated were constructed to provide marker strains for testing whether integration of the marker gene itself affected the strains' ecological or symbiotic properties, and to provide control strains against which to assess the potentially enhanced strains. The structure of the integrated DNA in one potentially enhanced strain is shown in **Figure 1**; note that the Ω fragment and the T_1T_2 fragment provide transcriptional terminators to ensure that transcription does not read into or out of the "enhancement" cassette.

1989 Field Tests

BioTechnica conducted two field tests of recombinant rhizobia in 1989: one of *R. meliloti* at its research station in Sun Prairie, Wisconsin, and one of *B. japonicum* at Baton Rouge, Louisiana in collaboration with Dr. G. Brietenbeck, Department of Agronomy, LSU. The objectives of each test were to *a)* determine whether integration of genes into the selected genomic sites affected the strains' ecological or symbiotic properties; *b)* compare the competitiveness of several host strains; *c)* determine whether alfalfa or soybean yield could be increased through the use of a strain with enhanced nitrogen-fixing ability; and *d)* demonstrate that the population dynamics of the recombinant rhizobia could be monitored under field conditions. Each test consisted of two trials: a large plot "comparison" trial where plots inoculated with a single strain were assessed for yield and were subject to extensive microbiological monitoring; and a small-scale "competition" trial where inocula were mixed pair-wise and only nodule occupancy was monitored to determine the competitive ability of the engineered strains compared to their wild-type parents. The strains that were tested are listed in **Table 1**.

TABLE 1. Strains field-tested in 1989.

Rhizobium meliloti	
RCR2011	Wild-type control
RmSF38	Antibiotic-resistant derivative of RCR2011
UC445	Wild-type control
PC	Wild-type control
RMB7101	Marked, engineered derivative of RCR2011
RMB7103	Marked, engineered, enhanced derivative of RCR2011
RMB7201	Marked, engineered derivative of PC
RMB7401	Marked, engineered derivative of UC445
Bradyrhizobium japonicum	
BJB1000	Wild-type control
BJB2000	Wild-type control
BJB1001	Marked, engineered derivative of BJB1000
BJB1003	Marked, engineered, enhanced derivative of BJB1000
BJB2001	Marked, engineered derivative of BJB2000
BJB2003	Marked, engineered, enhanced derivative of BJB2000

Regulatory Process

Under the coordinated framework, the Environmental Protection Agency (EPA) generally has primary responsibility for reviewing field-test applications of genetically engineered microbial strains. Microbials such as *Rhizobium* species, that are not pesticides, are considered under the Toxic Substances Control Act (TSCA) through pre-manufacture notifications (PMNs). Small-scale field tests of such organisms are covered only if they are "commercial research". The review process includes a mandatory 90–day public comment period. The PMN submission details risks and benefits. It contains a thorough description of the molecular biology of the strains, a discussion of the benefits of the proposed field tests, data comparing the genetically engineered strains to their wild-type parents (*e.g.*, growth rates, intrinsic antibiotic-resistances, effects on target and non-target legumes, persistence in soil and rhizosphere), a discussion of any other possible effects of the genetic engineering (*e.g.*, are any effects such as a disruption of an open reading frame expected from the integration event; are any alterations to host range expected?), and a discussion of any possible environmental effects of the strains. These include potential for gene transfer, exposure to humans and animals, and dissemination from the field site. The PMN submission also contains detailed protocols for the field tests. These include details on the location and design of the field tests, growth of cultures, implementation and supervision of the field tests, containment procedures, and last, but not least, detailed plans for the microbiological monitoring of the strains. Permission to field-test the strains is granted in the form of a Consent Order, a legally binding document that details EPA–approved protocols and reporting responsibilities. Consent Orders for the 1989 field tests were received about 120 days after submitting the PMNs. The PMNs were also reviewed by USDA/APHIS, who determined that the strains were not plant pests, and by state regulatory agencies. The *R. meliloti* field test was implemented on May 24, 1989 and the *B. japonicum* field test on June 16–17, 1989. Information on the progress of the trials, including crop inspection, yield and weather data, and data from the microbiological monitoring program were reported to EPA, with copies to the other agencies, on a quarterly basis.

Microbiological Monitoring

A comprehensive microbiological monitoring program was carried out in association with each field test. Where possible, the behavior of recombinant strains was compared to that of their wild-type parents. In the case of *R. meliloti*, a spontaneous streptomycin-resistant mutant of wild-type strain RCR2011 was also used as a control. Inoculant rhizobia were enumerated both by plating on selective antibiotic-amended media and by Most-Probable-Number (MPN) analyses using laboratory plant nodulation assays. In both cases, strain identity was confirmed using immunofluorescence microscopy. Parameters that were monitored included aerial dispersal on the day of implementation, establishment and persistence of the inoculant in the rhizosphere, vertical migration of the inoculant through the soil profile, lateral migration of the inoculant from the point of inoculation, and nodule occupancy by the inoculant strains. Both sites contained low levels of indigenous rhizobia that provided competitive pressure. Nodule occupancy was sampled from all plots on one or two occasions during the field season, while samples for monitoring inoculant behavior were taken on a monthly basis. These samples were taken in triplicate from four replicate blocks for each strain monitored. This resulted in 108 *R. meliloti* samples and 92 *B. japonicum* samples *per* time point, excluding nodule occupancy studies.

No lateral migration of *R. meliloti* or *B. japonicum* inoculant populations was detected, suggesting that the inoculant remained within six inches of the root zone throughout the field season. A small amount of vertical migration was detected, most likely reflecting the depth of the root system at the time of sampling. Wind-borne dissemination during inoculant application (as a spray in the *R. meliloti* trial and as a seed coating in the *B. japonicum* trial) was negligible in both cases.

Inoculant population profiles in the inner rhizosphere for *R. meliloti* and *B. japonicum* are presented in **Figures 2 and 3**, respectively. The *R. meliloti* populations increased from 100 to 1 x 10^7 cells *per* gram of inner rhizosphere soil by or before the first month after inoculation, shifting to steady state growth with root development throughout the remainder of the season (**Figure 2**). There were no differences in population behavior between strains, suggesting that the genetic engineering did not affect the rhizosphere competence of the strains.

Nodule occupancy data from the *R. meliloti* strain comparison trial showed that both total nodule number and percent nodule occupancy by the inoculant decreased with increasing distance from the crown. Nodule occupancy by the inoculant strain at the crown ranged between 71 and 89%, while at the most distal portion of the root it

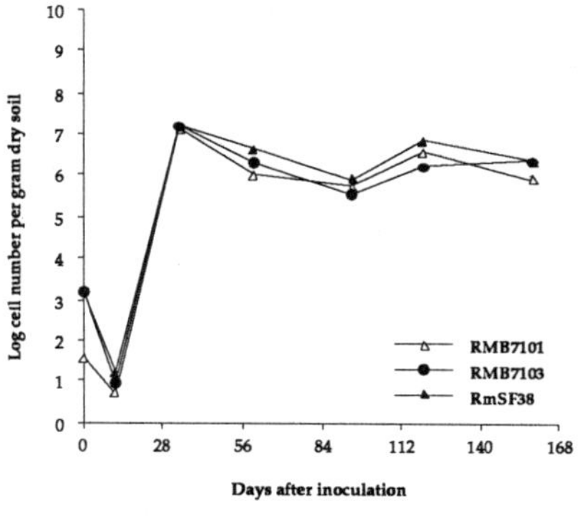

FIGURE 2. Persistence of R. meliloti *inoculant strains in the inner rhizosphere.*

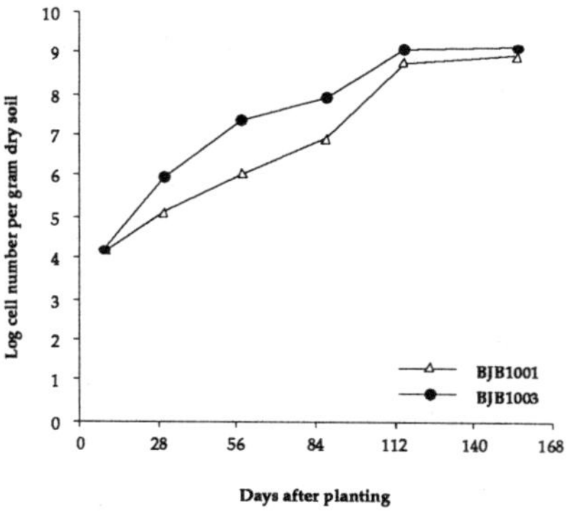

FIGURE 3. Persistence of B. japonicum *inoculant strains in the inner rhizosphere.*

ranged between 2 and 39% (**Table 2**). There were no statistically significant differences in mean nodule occupancy among the strains. These data indicate that the genetic engineering did not affect the ability of the strains to compete for nodule occupancy with the indigenous population. However, data from the strain competition trial, where strains were inoculated in pair-wise combinations, suggested that the genetically engineered strains were slightly less competitive than their wild-type parents (**Table 3**). Nevertheless, recombinant strain RMB7201, derived from host strain PC, was more competitive than the other wild-type strains in the

TABLE 2. Nodule occupancy in the 1989 R. meliloti *strain comparison trial.*

	% Nodule occupancy				
Treatment	Top section	2nd section	3rd section	Bottom section	Average % occupancy
RmSF38	86	65	67	39	70.3
RCR2011	86	60	35	20	63.5
RMB7101	89	60	35	36	61.2
RMB7103	87	50	25	16	55.1
RMB7201	87	48	28	20	54.9
Commercial strains	28	49	63	62	49.6
UC445	71	30	23	13	40.0
RMB7401	79	34	11	2	39.2
Mean	77	49	36	26	54.0

Alfalfa roots were divided lengthwise into four equal sections. All nodules in each section were counted, sterilized and crushed.

TABLE 3. Nodule occupancy in the 1989 R. meliloti *strain competition trial.*

Challenge inoculum	*Column strain % occupancy/Row strain % occupancy*						
	RCR2011	*RmSF38*	*RMB7101*	*RMB7103*	*RMB7201*	*UC445*	*RMB7401*
No challenge	1.78	2.03	0.92	1.27	2.57	1.56	1.63
RCR2011		2.27	0.71	0.22	3.85		0.62
RmSF38	0.44		0.38	0.38			
RMB7101	1.41	2.60				1.25	
RMB7103	4.45	2.63					
RMB7201	0.26					0.23	
UC445			0.80		4.33		0.55
RMB7401	1.62					1.83	

Results for the single-inoculant treatments ("no challenge") are calculated as the percentage of nodules occupied by the inoculum divided by the percentage of nodules occupied by the indigenous strain. Results from co-inoculant treatments are calculated as the percentage of nodules occupied by the "column" strain divided by the percentage of nodules occupied by the "row" strain.

test, RCR2011 and UC445 (**Table 3**). Thus, the degree of impairment to competitive ability caused by integration or expression of the marker gene was minimal.

The *B. japonicum* populations in the soybean rhizosphere increased throughout the course of plant development (**Figure 3**). Populations increased from 1 x 10^4 to 1 x 10^9 cells *per* gram dry inner rhizosphere soil between Days 9 and 115 following planting, slowing considerably after Day 115 as plants senesced during the R7 stage. Similar trends were also observed in the outer rhizosphere, although the growth curves were considerably less linear, and the late-season shift in population growth was not evident (*data not shown*). The mean population density in the inner rhizosphere of BJB1003 on Days 30, 58 and 88 was significantly greater than that of BJB1001 at the 0.05 level of significance. The difference between the BJB1001 and BJB1003 population profiles in the inner rhizosphere may be attributed to more rapid establishment by BJB1003 during the first 58 days of plant development. Following Day 58 the growth rates of the strains were more similar, and the populations densities were similar at the time of crop harvest. Average nodule occupancy by the *B. japonicum* inoculant strains varied between 17% and 59%. In three of four comparisons, the enhanced strain had almost two–fold higher nodule occupancy than the control strain, while in the fourth case, it had equal nodule occupancy (**Table 4**). While these differences were not statistically significant, the results taken in conjunction with those on rhizosphere establishment suggest that the genetically engineered gene integrated in strains BJB1003 and BJB2003 may improve the competitive ability of the strain. Similar results have also been observed in greenhouse studies. Based on these results, BioTechnica is currently conducting another small-scale field test of the recombinant *B. japonicum* strains at Sun Prairie, Wisconsin, to determine whether the gene significantly improves the rhizosphere competence and competitive ability of the host *B. japonicum* strain.

TABLE 4. Percent nodule occupancy in the 1989 B. japonicum *strain comparison trial.*

Strain	*Soybean variety*	*Average % nodule occupancy*
BJB1001	Braxton	39
BJB1001	Centennial	17
BJB1003	Braxton	38
BJB1003	Centennial	28
BJB2001	Braxton	34
BJB2001	Centennial	29
BJB2003	Braxton	56
BJB2003	Centennial	59

Yield Analysis

In both harvests of the *R. meliloti* trial, the enhanced strain RMB7103 out-yielded the other treatments. The differences were not significant since there was considerable variability in the data. Alfalfa stands are generally more uniform in the second year; hence, second-year data are being collected from the trial and may allow more accurate determination of the efficacy of the treatments. Further-improved strains are also being tested in new trials in 1990.

Very high yields (average: 55 bushels *per* acre) were obtained in the *B. japonicum* test, but no effects of inoculation or nitrogen amendment were seen, indicating that nitrogen deficiency was not a significant factor influencing yield. Hence, a conclusive assessment of the effectiveness of the experimental inocula was not possible. The lack of an effect may reflect the excessive rainfall (45 inches above normal) during the season.

Acknowledgements

We gratefully acknowledge Drs. F. Cannon and J. Beynon who established this program at BioTechnica, Drs. D. Glass and E. Owens for their tireless efforts in regulatory affairs, Dr. G. Brietenbeck for carrying out the *B. japonicum* field test and Ms. A. Huot for her extraordinary computer support.

References

1. Prentki, P. and Krisch, H.M. (*1984*) Gene **29**, 303–313.
2. Williams, M.K., Cannon, F.C., McLean, P. and Beynon, J. (*1988*) in *Molecular Genetics of Plant–Microbe Interactions*, eds. R. Palacios and D.P.S. Verma (American Phytopathological Society, St. Paul), pp. 198–199.
3. Yun, A.C., Noti, J.D. and Szalay, A.A. (*1986*) J. Bacteriol. **167**, 784–791.

Gresshoff, Roth, Stacey & Newton (eds.) Nitrogen Fixation: Achievements and Objectives
Chapman and Hall New York • London 1990

Assessment of genetically engineered microorganisms under the Toxic Substances Control Act: Considerations prior to small-scale release

Philip G. Sayre

Office of Toxic Substances, U.S. Environmental Protection Agency, Washington, D.C. 20460 USA

TSCA Authority and Products Reviewed

The "Coordinated Framework for Regulation of Biotechnology" (7) explained the EPA TSCA (Toxic Substances Control Act) authority which allows the review of certain classes of biotechnology products. Under the Coordinated Framework, biotechnology products are regulated in accordance with the use of each product (9). Many agricultural uses of microorganisms, plants and animals are regulated by the Department of Agriculture (USDA); foods, drugs, cosmetics, and biologics are regulated by the Food and Drug Administration (FDA); microbial pesticides are regulated by the EPA's Office of Pesticide Programs (OPP) under the Federal Insecticide, Fungicide, and Rodenticide Act (FIFRA). Uses of microorganisms not covered by other existing authorities are reviewed by the EPA Office of Toxic Substances (OTS) under TSCA.

TSCA thus serves as a "gap filling" environmental law, relative to the authorities of other U.S. federal regulatory bodies (16). TSCA's applicability to reviewing microbial biotechnology products is based on the interpretation that microbes are chemical substances under TSCA. Under this authority, OTS can require information on microbial biotechnology products in order to identify potential hazards and exposures. OTS can also require testing of a microbial biotechnology product that may present an unreasonable risk of injury to human health or the environment, or that is produced in substantial quantities and may result in substantial environmental release or substantial human exposure. Finally, OTS can review the production, processing, distribution, use, and disposal of a microbial biotechnology product if it presents an unreasonable risk of injury to human health or the environment. There is some overlapping of responsibilities with the USDA for some products released in the environment.

According to the 1986 policy statement issued as part of the Coordinated Framework (7), the types of microorganisms reviewed under TSCA are defined by four factors in addition to the type of use: the application environment, the pathogen status of the microbe, the genetic material introduced, and the commercial intent of the product. The microorganisms reviewed under TSCA fall into both of the application environments typically identified with biotechnology products -- closed system and environmental release. Some microorganisms are used in closed systems such as fermentors to produce a product which is generally a protein. Examples of such closed system products uses which would fall under TSCA purview include enzymes for use in detergents and the textile industry and hormones used in cell culture medium. Other microorganisms are produced for intentional release to the environment, and the product often includes the microorganism itself. Examples of product uses for environmental release which would fall under TSCA review include microorganisms used for increased nitrogen fixation in legumes, waste treatment, metal mining, and biomass conversion.

As noted, the type of genetic material contained in the microorganism is a factor which determines whether a microorganism product is subject to TSCA review. Microorganisms which through deliberate human intervention were altered to contain genetic material from dissimilar source organisms are candidates for review. EPA has defined "dissimilar" as organisms from different taxonomic genera; the organisms resulting from such a combination are called "intergeneric" (16). Manufacturers and importers of these intergeneric microorganisms are required to submit a Premanufacture Notice (PMN) if the new microorganism is intended for commercial use. Section 5 PMN submissions currently make up the majority of the genetically engineered microorganisms reviewed under TSCA. Section 5 is the heart of TSCA, and implements the act's goal of identifying potentially hazardous new chemical substances before they enter commerce. This section requires the submission of a PMN to EPA at least 90 days before manufacture or import is to begin. EPA has 90 days to prohibit or regulate the production, processing, distribution or import and disposal of an intergeneric microorganism, otherwise manufacture or import may begin.

Finally, TSCA only applies to products developed for commercial purposes. Although "commercial purposes" is not yet specifically defined by EPA, investigators involved in pure academic research are not subject to TSCA. This statement applies to both contained system and environmental release applications.

Currently, EPA is requesting that industry voluntarily comply with the PMN reporting requirements for any commercial research and development field test release which involves an intergeneric microorganism in a use category subject to TSCA. As a result of this request, PMN submissions such as those for the Monsanto recombinant Pseudomonas and Biotechnica Agriculture, Inc. recombinant Rhizobium and Bradyrhizobium strains have been reviewed prior to small-scale testing. A new Rule is in development which implements the use of TSCA for the regulation of genetically engineered microorganisms. This rule may ease reporting requirements for commercial research and development field tests subject to TSCA.

The OTS Risk Assessment Process

Under Section 5 of TSCA, the premanufacture review program in OTS assesses the potential for risk to health or the environment presented by a new chemical substance before significant human exposure or environmental release occurs (15). This premanufacture review program was initially developed for chemicals, and has been adapted for review of microorganisms. OTS conducts a risk assessment which forms the key element of a PMN review.

A risk assessment is an evaluation which involves an integration of two components: a hazard assessment and an exposure assessment (15). A hazard assessment is an assessment of a substance's potential to cause deleterious effects in humans, other living organisms, or important nonliving environmental components. An exposure assessment is the assessment of a substance's prospects for human exposure or environmental release. The exposure assessment would include an estimation of the partitioning of a substance into various environmental media and the concentrations they could (or do) achieve in these media (13). The exposure assessment provides substance fate information which is specifically intended to aid in predicting the occurrence of the potential effects noted in the hazard assessment.

The "Points to Consider for Environmental Testing of Microorganisms" developed by the NIH Recombinant DNA Advisory Committee (8) provided guidance for the assessment of hazards and exposures which could result from field testing of GEMs. Criteria have also been developed which assist in determining the degree of regulatory oversight by examining characteristics of the engineered trait, the parent organism, and the environment in which the organism is released; these criteria are hazard- and exposure-based (14). The Office of Toxic Substances has guidance documents for both environmental release (18) and closed system (19) TSCA notifications which assist those preparing notifications such as PMN submissions. Both documents specify the following information categories: (1) descriptions of the recipient and donor microorganisms, (2) genetic construction of the PMN microorganism, (3) characteristics of the PMN microorganism, and (4) fermentation process and human exposure information. The environmental release document addresses specific information needed for the OTS review of field test protocols. For closed system PMN applications, information on the environmental behavior of the PMN strains is needed only if significant environmental releases are anticipated.

The OTS risk assessment process for environmental release PMN submissions involves an examination of the five PMN information categories noted above, relative to the five elements which comprise the OTS review (Table). These review elements are: (1) potential environmental hazards, (2) potential human health hazards, (3) exposure to humans and the environment at the fermentation and field test sites, (4) risk assessment, and (5) review of the field test protocol to determine its ability to measure the endpoints of interest to the PMN submitter and OTS. The risk assessment completed for each environmental release PMN submission involves a consideration of the potential hazards in light of the potential for exposure to the PMN microorganism. Additional considerations that form part of a PMN review include an economics analysis, and an assignment of a name to the PMN microorganism and/or product of the recombinant (such as an enzyme) for possible inclusion on the TSCA Inventory.

Table: Listing of specific items covered in a PMN review for a recombinant microorganism intended for environmental release, the information categories relevant to addressing the risk concerns, examples of data submitted for each information category, and computer software used by OTS for risk assessment.

OTS Review Elements	PMN Information Categories	Examples of PMN Information	Software
1. Environmental effects	a. Donor & host characteristics	a. Taxonomy	a. MICRO-IS
2. Human health effects	b. Construct information	b. Probe verification	b. Sequence analysis software
	c. Recombinant characteristics	c. Survival, greenhouse & microcosm tests	
3. Exposure at fermentor site	d. Fermentation process	d. Physical containment	
Exposure at field site			e. MICROBE-SCREEN
4. Risk assessment			
5. Field test protocol review--exposure & effects	e. Field test protocol	e. Occurrence of marker in native microbes	

The OTS Review of Proposed Field Tests with Intergeneric Rhizobium and Bradyrhizobium Strains

Issues that have arisen during the OTS review of recombinant rhizobia will be used to illustrate the review process for environmental release PMN submissions. It is important to note that the small-scale field tests of recombinant Bradyrhizobium sp. and Rhizobium sp. reviewed by OTS were found to pose low risk to human health and the environment. EPA has reviewed such submissions in 1987, 1988, 1989 and 1990. Field data from some of these proposed field tests have also been reviewed. Further, the Biotechnology Science Advisory Committee (BSAC) suggested that it was not necessary for the BSAC to offer advice on future proposed R. meliloti small-scale field tests: in light of both the long history of safe use of, and extensive data on, the species R. meliloti, OTS review of such field tests should suffice (22). The BSAC is a committee which is composed of nine scientific experts (usually from academia) and two representatives of the general public. The committee provides expert scientific advice to the Administrator and the Assistant Administrators of EPA concerning issues related to the risks and other effects of applications of modern biotechnology.

The five listed risk OTS review elements (Table) form the backbone of the OTS review for an environmental release PMN submission. Each consideration will be discussed individually and linked to the relevant PMN information categories. Note that there is a transition from laboratory to microcosm to greenhouse to field testing in order to gather data needed for GEMs intended for environmental release. By contrast, PMN submissions for closed system applications have not contained data derived from greenhouse and field tests.

Environmental Effects Considerations

The first consideration noted in the Table addresses adverse environmental effects. These effects include (1) effects on other microorganisms, (2) effects on target plants and animals, (3) effects on nontarget plants and animals, and (4) effects on non-living resources. The first three information categories listed in the Table are used when addressing these four areas.

In order to identify the PMN microorganism, descriptions of the donor and recipient microorganisms are examined, as well as the techniques used to construct the PMN microorganism. The taxonomic derivation of the recipient microorganism is one key identification element. In the case of the R. meliloti recipient strains reviewed by OTS, most have been derived from strain a well-characterized strain -- SU47 -- isolated from M. sativa in 1939, at Bathurst, Australia (10). The strains of R. meliloti isolated from perennial medics, including alfalfa (Medicago sativa), form a tight taxonomic cluster relative to the genetic diversity of the strains (2, 4). Hence, any one strain, such as SU47, is expected to express nonsymbiotic phenotypic characteristics that are similar to any other strain that fixes nitrogen in alfalfa. In addition, the host range is expected to be restricted to three plant genera: Medicago, Melilotus, and Trigonella. The strains of B. japonicum reviewed by OTS are also well-characterized and, belong to the 110 serotype, and are often derivatives of USDA strain I-110. The phenotypic traits of most GEMs reviewed in OTS are encoded and analyzed using a PC-microcomputer version of the "Micro-IS" software package; this data system was originally developed by NIH (12).

The final step for GEM identification involves verification of the GEM through verification of the introduced DNA of interest and vector DNA remaining. This analysis is based on PMN submission information which usually includes the following: (1) construction of the DNA cassette which codes for traits such as enhanced nitrogen fixation, (2) a complete description of the integration site in the R. meliloti or B. japonicum genome, (3) construction of the vector containing the cassette, (4) introduction of the vector carrying the cassette into the recipient microorganism, and (5) final construct and genetic stability of the PMN microorganism. DNA sequence databases and software may also be used by OTS to provide further construct information.

Effects on other microorganisms are examined, and consideration is given to whether the new microorganism will outcompete indigenous microorganisms which perform functions similar to the GEM in the environment. Increased competitiveness related to a rhizobium or bradyrhizobium strain is not in itself detrimental, unless linked with effects such as decreased target plant seed production or top growth biomass. For example, the introduction of B. japonicum serogroup 123 into agricultural fields has been considered detrimental since its presence makes it more difficult to introduce more effective rhizobia (14). Often, PMN submissions include data from tests on the survival and nodulation ability in field soils of the recombinant rhizobium relative to its parental strain. Such data are indicative of the competitiveness of the rhizobia.

For recombinant rhizobia, the testing sequence that has been followed proceeds from the laboratory to the greenhouse, and finally to the field. For GEMs other than recombinant rhizobia, microcosms could be used to examine the effects of a GEM on community interactions and on ecosystem processes. Microcosm tests may be appropriate if hazards identified in laboratory tests indicate that it may be unsafe to go directly to the field or greenhouse from the laboratory (5).

Determination of the potential adverse effects of a GEM on target plants and animals forms a second component of the environmental assessment. With rhizobia, the considerations center on decreased productivity in alfalfa and soybeans (Glycine max), as seen in greenhouse tests. Top growth, nodule nitrogenase activity, and/or bean yield are some of the endpoints of interest for determination of adverse effects of recombinant rhizobia in agroecosystems. Greenhouse test data for alfalfa are usually part of the PMN submission. These greenhouse tests should be run for a minimum of four to six weeks to determine top growth, and the soil used should be from the field test site. Typical increases in alfalfa yield due to recombinant rhizobia range from 0 to 7.5% in submitted studies, as compared with parental strains. Greenhouse studies with B. japonicum conducted to examine soybean top growth biomass should be run at least until plants are beyond flowering: one of the highest demands for nitrogen occurs during early pod filling.

Rhizobitoxine production is an additional example of a target legume concern. Rhizobitoxine production by some strains of B. japonicum induces chlorosis in soybeans. However, the recipient bradyrhizobia strains reviewed by OTS belong to serotype 110, which is not known to produce the toxin. Also, greenhouse studies involving recombinant bradyrhizobia can be used to screen for chlorosis and stunted growth effects indicative of rhizobitoxine-induced chlorosis.

The inactivation of genes or the alteration of gene products in the recipient rhizobia may also affect the target legume. These rhizobia-associated changes could result from insertion of the genes for enhanced nitrogen fixation. Hence, recipient insertion sites for enhanced nitrogen fixation cassettes are chosen with the goal of altering the phenotype of the rhizobium as little as possible. Disruption of genes at the insertion site should, in particular, not alter the symbiotic properties of the rhizobium-legume interactions. Cassettes for enhanced nitrogen fixation have been inserted into the ino region of the pRmSU47b megaplasmid to produce several recombinant rhizobia reviewed by OTS. In this case, altered host range would not be a concern, since most of the genes responsible for host range are located on a different megaplasmid -- the pRmSU47a megaplasmid.

A third consideration addressed in environmental hazard assessments is the potential of the recombinant microorganism to cause nontarget plant and animal effects. Nontarget organisms are those which are not intended to be affected directly by the GEM. For R. meliloti, identification of the plant host species for the recombinant also identifies the cross-inoculation group for the recipient strain. The construct information is examined to determine whether or not disruption at the insertion site could affect host range genes such as the nod genes, and to determine if the insert genes code for any host range functions.

Increased biomass production in weedy legumes and decreased biomass production in other legumes of agronomic or ecological significance are considered when nontarget legumes within the cross-inoculation group of a given rhizobium are considered. For field tests in Wisconsin, the potential adverse effects of recombinant rhizobia carrying enhanced nitrogen fixation on the nontarget legume Melilotus (sweet clover) is considered since Melilotus is a pest plant species in the Midwest. Since no greenhouse or field tests have been done to determine the effects on sweet clover, any sweet clover within the area likely to be exposed to the recombinant rhizobia should be removed. This prevents the establishment of a plant reservoir for the recombinants and eliminates any potential for the establishment of a fast-growing sweet clover population in the area of the field test. With some rhizobia, there may be uncertainty as to whether all the legumes in a cross-inoculation group have been identified.

Data have also been submitted on nontarget legumes known to be outside the cross-inoculation group of R. meliloti, such as peas. Also, greenhouse data have been submitted on nontarget nonlegume crops grown in rotation with target legumes --data on corn and ryegrass were submitted for possible effects of R. meliloti. It is unlikely that any effects would be noted since such nonlegume roots are incapable of promoting adherence of the rhizobia and triggering nod gene expression.
The potential for adverse effects on animals is determined by an examination of the recipient microorganism and any insert or vector DNA that may code for products such as toxins. For the small-scale field testing of recombinant rhizobia examined thus far, the only possible issue considered was transfer of antibiotic resistances from the recombinants to other microorganisms pathogenic to animals.

The fourth consideration noted in the environmental hazard assessments of biotechnology PMNs is that of effects on non-living resources. There is limited concern for effects such as altered biogeochemical cycling prior to small-scale field tests. If survival and dissemination of the GEM in the small-scale field test are limited, it would then be more appropriate to consider such effects prior to commercial, or large-scale, release. Considerations for large-scale release related to non-living resources include soil nitrogen enrichment resulting from enhanced legume productivity. Elevated soil nitrogen may allow the establishment of new weeds, increase the leaching of nitrates, and increase the flux of nitrogen oxides into the atmosphere (14).

Human Health Considerations

Human health considerations are addressed primarily through the examination of available information on the recipient microorganism, and construct information. Thus far, no major human health issues have arisen as a result of proposed releases of genetically engineered rhizobia. The only human health concern is that of antibiotic resistance gene transfer from the recombinant rhizobia to other microorganisms which may be pathogenic to man. Examples of the antibiotic resistance markers added to recombinant rhizobia reviewed by OTS for small-scale field tests include those for kanamycin and streptomycin resistance.

Antibiotic resistance genes are often introduced into recombinant rhizobia to serve as selectable markers for tracking the fate of the microorganisms during small-scale field testing. A meeting of the BSAC subcommittee on antibiotic resistance genes held on 19 January, 1989 (20) concluded that ideally, GEMs for release should not contain exogenously acquired genes bearing resistance to antimicrobial agents such as the beta-lactam antibiotics (penicillins, cephalosporins). If use of a resistance gene cannot be avoided, it was recommended that the gene be chromosomally located and non-transposable. The resistance should be against a narrow spectrum antimicrobial family with no, or limited, clinical use. The concern was primarily for antibiotic resistance markers used for commercial- scale environmental releases. However, small-scale field tests with clinically useful antibiotic determinants that have high genetic mobility were discouraged. The topic of gene mobility was the subject of a separate BSAC subcommitte meeting discussed under the exposure considerations below.

Exposure to Humans and the Environment at the Fermentation and Field Test Sites

Human and environmental exposures to a GEM that is proposed for use in a small-scale field test generally could occur in three geographical locations: at the site of production of the GEM, on the transportation route from the fermentation facility to the field site, and at the field test site itself. Exposure considerations may be restricted to simple exposure to the microorganism, which is controlled by survival of the microorganism and its ability to disseminate from the field test site. There is an attempt to approximate the survival and spread of the recombinant microorganism after release at the test site. In addition, if there is a high likelihood for genetic transfer of the introduced DNA, human and environmental exposure to the gene products of introduced vector and insert DNA may also be considered.

Human and environmental exposures to the recombinant rhizobia at the fermentation facilities have been considered low. The amount of microorganisms produced for a small-scale field test is low (on the order of 10^{12} cells per strain). Those microorganisms that survive chemical and physical treatments of waste streams from the production facility often have diminished survival in air, water, and/or soil.

In order to determine the fate of the recombinant rhizobia after release at the field test site, information is gathered from the literature on the survival of the recipient rhizobium in soil and water. Laboratory or greenhouse survival tests in field site soil and nodule occupancy tests indicate the survival potential of the recombinant rhizobia during the field test. Most recombinant rhizobia populations have decreased in field site soil samples when tested for approximately one month. Finally, PMN submissions should have data on the efficiency with which the recombinant rhizobium can be separated from the indigenous microflora present in samples taken from soil, water, and air.

While absolute containment of rhizobia released at a test site is not possible, the dispersal of the PMN microorganisms after application is usually considered minimal due to the physical characteristics of the test site, and the field test designs and protocols used. For example, fields with low slope which are isolated from surface waters are usually used; soil erosion control procedures are used; weed control procedures initiated prior to the beginning of the tests eliminate potential nontarget legume hosts for the rhizobia in the test area; buffer zones surround field test plots; access to the test site is limited; and post-harvest treatment of the test site decreases the survival potential of remaining recombinant rhizobia.

If there is a high likelihood for genetic transfer of the introduced DNA, exposure considerations may extend beyond an examination of the fate of the microorganism alone. If genetic transfer is likely, then human and environmental exposures to the introduced vector and insert DNA are also considered. The potential impact of the gene products of the introduced DNA as expressed in recipient microorganisms other than the PMN microorganism would also be a concern (see example of antibiotic resistance determinant transfer above). Insert DNA introduced into the genome of a bacterium on a transposon or plasmid would have a higher likelihood of being transferred to other bacteria in the environment than if the introduced DNA were chromosomally located and lacked closely associated genes that could subsequently mobilize it. In order to better assess potential gene transfer by bacterial plasmids, transposons, and viruses, OTS convened the BSAC subcommittee on Mobile Genetic Elements in July, 1989 (21, 11).

There is an OTS computerized multimedia fate and transport model MICROBE-SCREEN which is designed to assess the potential fate of passively dispersed GEMs released to air, surface water, or soil (17, 1). The model may be used as a first order approximation screening tool for prediction of the fate of GEMs during small-scale field tests. The model couples microbial processes (growth, decay, and survival) to transport components of the chemical fate model TOX-SCREEN (6). For example, the transport of microbes such as rhizobia in soil is modelled using factors which include: microbial growth and death, interaction of the microbe with soil particles as if the bacterium were a water-borne inert particle, surface water runoff estimates, water filtration estimates, and water recharge to the surface. This model will undergo validation by comparing its microbial fate predictions to the fate results obtained in field tests of a recombinant pseudomonad (3) and several recombinant rhizobia.

Risk Assessment

The risk assessment for a proposed small-scale field test involves an integration of the human and environmental hazard considerations with the exposure assessment. The intent of the risk assessment is to determine which adverse effects are most likely to occur, given the exposure pattern predicted for the GEM. The predicted exposure is usually limited to the immediate test site (thus limiting exposure to humans and the environment). Because human health hazards associated with the rhizobia (aside from antibiotic resistance transfer) have not been identified, risk to humans is considered low. Environmental risk of the small-scale field test is considered low because although some environmental hazards exist, the exposure to the GEMs beyond the field test site is significantly diminished. In order to decide whether to allow the field test, the potential risk is then balanced against the potential benefits in a process referred to as risk management. The potential benefits usually involve the gathering of data by a company to determine if the rhizobium will be commercially viable. In addition, the exposure and effects data gathered from the small-scale field tests will be used to assess the impacts of broader scale releases.

Field Test Protocol Review

In addition to the hazard, exposure, and risk assessments, a field test protocol review is conducted. This review involves two components: an assessment of whether monitoring endpoints and effects endpoints are adequate. Monitoring endpoints are designed to measure the survival and dissemination of the PMN strain during the field test. Typical monitoring endpoints include assessment of population survival and dissemination in different environmental media. For rhizobia, samples of the populations in the field site soil, rhizosphere, and air are examined. Effects endpoints are generally designed to indicate adverse effects on target and/or nontarget organisms. In the case of R. meliloti tests, effects on alfalfa top growth are examined, while soybean bean yield may also be determined for B. japonicum tests.

In order to determine if adequate data will be gathered on the monitoring and effects endpoints in the field test protocol, certain general information should be provided in the PMN submission. The nature of the field test site and the field test design should be described: statistics to be employed, diagrams of the plot layout, and controls and treatments to be used are examined. On-site containment practices such as those noted above should be provided, as well as application methods and worker safety procedures. For the purposes of monitoring, the techniques used to isolate the PMN microorganism from test samples, data on the selectivity or specificity of the monitoring technique, descriptions of the sensitivity and reliability of the isolation and identification procedure should be provided. Also, a thorough description of the sampling procedures for effects and monitoring endpoints should include such considerations as how and where the samples were taken and the techniques used for sample analysis. Routine and emergency termination and mitigation procedures should also be provided. No monitoring or effects endpoints have been proposed due to human health concerns associated with small-scale tests of rhizobia.

References

1. Chiu, N., Wickramnayake, G., Evers, D., & Reichenbach, N. (1990) In preparation for Environ. Toxicol. & Chem.

2. Crow, V.L., Jarvis, B.D.W., & Greenwood, R.M. (1981) Int. J. Syst. Bacteriol. **31**, 152-172.

3. Drahos, D. J., G. F. Barry, B. C. Hemming, E. J. Brandt, H. D. Skipper, E. L. Kline, D. A. Kluepfel, T. A. Hughes, & D. T. Gooden. (1988) in The Release of Genetically-Engineered Micro-Organisms, eds. Sussman, M., Collins, C., Skinner, F., & Stewart-Tull, D. (Academic Press, London), pp. 181-191.

4. Eardly, B.D., Materon, L.A., Smith, L. H., Johnson, D.A., Rumbaugh, M.D., & Selander, R.K. (1990) Appl. Environ. Microbiol., **56**, 187-194.

5. Harrass, M. and Sayre, P. (1989) in Aquatic Toxicology and Hazard Assessment: 12th Volume (STP 1027), eds. Cowgill, U.M. & Williams, L.R. (American Society for Testing and Materials, Philadelphia), pp. 204-223.

6. Hetrick, D. M. and McDowell-Boyer, L. M. (1984) User's Manual for TOX-SCREEN: A multimedia screening level program for assessing the potential fate of chemicals released to the environment (Oak Ridge National Laboratory, Oak Ridge, TN).

7. Office of Science and Technology Policy. (1986) Coordinated Framework for Regulation of Biotechnology; Announcement of Policy and Notice for Public Comment. Federal Register **51**, 23302-23393.

8. Milewski, E. (1985) Recombinant DNA Technical Bulletin **8**, 102-108.

9. Milewski, E. (1990) in Biotechnology of Plant - Microbe Interactions, eds. Nakas, J.P., & Hagedorn, C. (McGraw-Hill Publ. Co., NY), pp. 319-340.

10. Robson, A., and Loneragan, J. (1970) Aust. J. Ag. Res., **21**, 427-424.

11. Sayre, P. and Miller, R. (1990) In preparation for Plasmid.

12. Segal, M. (1988) in The Impact of Chemistry on Biotechnology, eds. Shoemaker, S., Middlekauf, R., & Ottenbrite, R. (American Chemical Society, Washington, D.C.), pp. 386-396.

13. Stern, A. (1980) Ecotox. & Environ. Safety, **4**, 404-414.

14. Tiedje, J.M., Colwell, R.K., Grossman, Y.L., Hodson, R.E., Lenski, R.E., Mack, R.N., & Regal, P.J. (1989) Ecology **70**, 298-315.

15. US EPA. (16 March 1979) Toxic Substances Control: Discussion of Premanufacture Testing Policy and Technical Issues; Request for Comment. Federal Register **44**, 16243-44.

16. US EPA. (1987) Report of the Biotechnology Science Advisory Committee of the U.S. Environmental Protection Agency, **1**, 5-6.

17. US EPA. (1987b) User's guide to MICROBE-SCREEN execution in GEMS (Contract No. 68-02-4281; Russell Kinerson, Project Officer), (General Sciences Corporation, Laurel, MD).

18. US EPA. (December, 1989) "Points to Consider in the Preparation and Submission of TSCA Notifications for Microorganisms: Environmental Release" (Draft). US EPA Office of Toxic Substances, Chemical Control Division, Washington, D.C.

19. US EPA. (December, 1989) "Points to Consider in the Preparation and Submission of

TSCA Notifications for Microorganisms: Closed System, Fermentation Use" (Draft). US EPA Office of Toxic Substances, Chemical Control Division, Washington, D.C.

20. US EPA. (December 20, 1989) in Materials distributed to the Biotechnology Science Advisory Committee for the 20 December 1989 meeting, US EPA (Washington, D.C.), Tab 67.

21. US EPA. (December 20, 1989) in Materials distributed to the Biotechnology Science Advisory Committee for the 20 December 1989 meeting, US EPA (Washington, D.C.), Tab 68.

22. US EPA. (December, 1989) in Meeting of the 20 December 1989 Biotechnology Science Advisory Committee, US EPA (Washington, D.C.), Tab 65.

Gresshoff, Roth, Stacey & Newton (eds.) Nitrogen Fixation: Achievements and Objectives
Chapman and Hall New York • London 1990

Characterization of *Rhizobium trifolii* bacteriophage isolated from Iraqi soil

F.Y. Al-Ani, A.K. Al-Sultani and I.A. Al-Khafagi

Department of Biology, Science College, University of Baghdad, Iraq

Bacteriophages of Rhizobia are of particular interest, they are likely to have a significant role in the ecology of their economically important host and they can affect the outcome of competition for nodulation between Rhizobial strains (6,7). These phages were isolated from most Rhizobium species (4,5).

Bacteriophage RT5 was isolated from soil using eighteen strains of Rhizobia as indicators (nine R. leguminosarum, five R. trifolii and four R. meliloti). Seven soil samples were taken from an agricultural area in Baghdad, suspended in yeast-mannitol medium (8) and shaken for 30 min at 30°. After centrifugation, each supernatant was filtered (0.45 µm membrane) and tested for plaque formation on (1). Phage was purified by four single plaque passages.

Morphologically, RT5 seems to be a Bradly type A structure (2,3) with a hexagonal head (86 nm length) and long contractile tail (101 nm length). Its plaques were clear and rounded with diameter ranging from 1-4 mm. RT5 had a latent period of 110 min with a rise time of 90 min and average burst size of 100. The host range of RT5 appeared to be limited since it lysed a single strain of R. meliloti (M13).

Thermostability studies showed that the phage remained stable at 40°C for 1 hr, while 94% of the phage particles were inactivated after heating for 1 hr at 50°C. Moreover, the phage was stable between pH 5-10. RT5 was found to be unaffected by choloroform and diethyl ether. Mitmyocin C experiment showed that RT5 is DNA phage.

1. Adams, M.H. 1959. Bacteriophages. Interscience, New York.
2. Bradley, D.E. 1967. Bacteriol. Rev. 31:230-314.
3. Bradley, D.E. & Rutherford, E.L. 1975. Can. J. Microbiol. 21: 152-163.
4. Delajudie, P. & Bogusz, D. 1984. Can. J. Microbiol. 30:521-525.
5. Dhar, B. & Ramkrishna, K. 1987. Arch. Microbiol. 147:121-125.
6. Lawson, K.A., Barnet, Y.M. & McGilchrist, C.A. 1987. App. Environ. Microbiol. 53:1125-1131.
7. Ugalde, R.A., Handelsman, J. & Brill, W.J. 1986. J. Bacteriol. 166: 148-154.
8. Vincent, J.M. 1970. A manual for practical study of root nodule bacteria. I.B.P. Handbook Nr. 15, Blackwell,Oxford & Edinburgh.

Gresshoff, Roth, Stacey & Newton (eds.) Nitrogen Fixation: Achievements and Objectives
Chapman and Hall New York • London 1990

The effect of root exudates on the growth and activity of nitrogen-fixing bacteria in the rhizosphere of pearl millet

S.L. Albrecht and J.R. Milam

USDA-ARS, Agronomy Department and the Department of Microbiology and Cell Science, University of Florida, Gainesville, FL 32611, USA.

Nitrogen-fixing bacteria in the rhizosphere may make significant contributions to the nitrogen economy of many plants. Although rates are low in comparison to legumes, nitrogen fixation by bacteria on the rhizoplane or in the rhizosphere offers some promise for supplying nitrogen to grass crops. There is the possibility of using plant breeding and genetic techniques to enhance the nitrogen-fixing capacity of these associations. Pearl millet breeding lines able to support a range of nitrogenase activities in seedling agar tubes have been established. In this report an explanation why some breeding lines can continually support elevated nitrogenase activities by their associated bacteria is offered. Pearl millet breeding lines, which support different levels of nitrogenase activity in their rhizosphere, were grown in a greenhouse, labeled with ^{14}C and the amount of the label in root exudates was determined. Rhizosphere nitrogenase activity in seedling agar tubes and the populations of inoculated nitrogen-fixing bacteria were also monitored. The populations of nitrogen-fixing bacteria and associated nitrogenase activity increased during the course of the experiment, however, the specific nitrogenase activity remained relatively constant. Bacterial growth was slightly enhanced, but never inhibited, when root exudates from the differing pearl millet lines were added to cultures of *Azospirillum brasilense* strain JM 125A2 or 13t. This observation was equally valid for growth on solid media or in liquid culture. This indicates that the difference in nitrogenase activity between lines is not due to an inhibitory mechanism or the potential to increase the specific nitrogenase activity of associated bacteria, but the capacity to support more rhizosphere bacteria per plant. The root exudates of pearl millet breeding lines that support a range of nitrogenase activities were evaluated. There were strong positive correlations between the loss of recent photosynthate by the roots and nitrogenase activity and the populations of nitrogen fixing-bacteria in the plant rhizosphere. It is proposed that, under the conditions of this experiment, the mechanism that allows certain pearl millet breeding lines to support greater amounts of rhizosphere nitrogen fixation was the ability to sustain larger populations of nitrogen-fixing bacteria, maintained by increased quantities of photosynthate partitioned to the root and eventually lost to the rhizosphere.

1. Bouton, J.H. & Books, C.O. (1982) Crop Sci. **22**, 680-682.

Gresshoff, Roth, Stacey & Newton (eds.) Nitrogen Fixation: Achievements and Objectives
Chapman and Hall New York • London 1990

Cyanobacterial nitrogen fixation in saline environments

Shree Kumar Apte and Tonina A. Fernandes

Molecular Biology & Agriculture Division, Bhabha Atomic Research Centre, Trombay, Bombay - 400 085 India.

The photoautotrophic nitrogen-fixing cyanobacteria exhibit a considerable tolerance to a variety of environmental stresses. Salt (NaCl) stress, however, inhibits their nitrogen-fixation efficiency and reduces diazotrophic growth. The reasons for the salt sensitivity of nitrogenase activity have been investigated. Exposure to salt inhibited nitrogen fixation and growth in a salt-sensitive strain *Anabaena* L-31, while a salt-tolerant strain *A. torulosa* initially responded with higher rates of fixation and growth. Higher salinities, however, caused inhibition. In contrast, growth in combined nitrogen-supplemented media showed very little sensitivity to salt stress. The effects of ionic and osmotic components of salt stress on nitrogen fixation were experimentally distinguished. Both the synthesis and activity of nitrogenase were completely insensitive to osmotic stress (sucrose), while ionic stress in the form of Na^+ severely inhibited nitrogenase expression in both *Anabaena* strains. Environmental factors, such as alkaline external pH, high external K^+, or exogenous combined nitrogen, effectively curtailed Na^+ influx and alleviated ionic stress, substantially enhancing nitrogen fixation by both *Anabaena* strains irrespective of their inherent salt sensitivities (2). Adaptation to salt/osmotic stress in the two *Anabaena* strains involves synthesis of specific stress-induced proteins which also occurs in the presence of combined nitrogen (1). Some of the stress proteins are induced by combined nitrogen *per se* even in the absence of salt/osmotic stress.

The data suggest that diversion of energy from diazotrophy to osmoregulation and to sodium extrusion in particular is mainly responsible for sensitivity of nitrogen fixation to salinity stress. Factors, which minimize or eliminate the energy requirement of Na+ extrusion, enhance nitrogen fixation under salinity stress. Exogenous combined nitrogen confers maximum osmoprotection on cyanobacteria by eliminating the energy requirement for both osmoregulation and nitrogen fixation (3).

1. Apte, S.K. & Bhagwat, A.A. (1989) J. Bacteriol. **171**, 909-915.
2. Apte, S.K., Reddy, B.R. & Thomas, J. (1987) Appl. Env. Microbiol. **53**, 1934-1939.
3. Reddy, B.R., Apte, S.K. & Thomas, J. (1989) Plant Physiol. **89**, 204-210.

Gresshoff, Roth, Stacey & Newton (eds.) Nitrogen Fixation: Achievements and Objectives
Chapman and Hall New York • London 1990

Screening for bacterocin-producing strains among collections of *Rhizobium meliloti* and *Bradyrhizobium japonicum*

B. T. Briel and E. W. Triplett

Dept. of Agronomy and the Center for the Study of Nitrogen Fixation, University of Wisconsin-Madison, Madison, WI 53706-1597 U. S. A.

Strains of *R. meliloti* and *B. japonicum* in the collections of the USDA and Lipha Chemical Co. were screened for bacteriocin production on minimal media using a plate assay. The assays for the *R. meliloti* and *B. japonicum* strains were conducted at pH 6.8 and 6.0 which are the optimum levels for growth of alfalfa and soybean, respectively.

Of 195 *R. meliloti* strains tested, seven inhibited more than 30% of the *R. meliloti* strains. One strain, 102L1, inhibited 70% of all the strains in the collections and inhibited 81% of those strains isolated from East, Southeast, and Midwest regions of the United States. Strain 102F69 was very inhibitory toward 95% of the strains of *R. leguminosarum* bvs. *phaseoli, viceae*, and *trifolii* in our collection but had no effect on most strains of *R. fredii* and *R. meliloti*. Strain 102F61 inhibited 100% of *R. l.* bv. *viceae*, 63% of *R. l.* bv. *phaseoli*, 39% of *R. meliloti*, 33% of *R. fredii* and none of the *R. l.* bv. *trifolii* strains tested.

The antibiotic-producing strains 102F69, 102F61, 102F84, 104A15, 102N1, and 1047 induced effective nodules on alfalfa roots while 102L1 was ineffective. All of the antibiotic-producing strains of *R. meliloti* were bacteriostatic, not bacteriocidal, toward *R. meliloti* 102F74. The antibiotic produced by 102F69 inhibited *Salmonella* sp., *Staphylococus aureaus*, *Pseudomonas solonacearum,* and *Agrobacterium tumefaciens* but had no effect on *Escherichia coli*, *Klebsiella pneumoniae*, *Erwinia carotovora*, *Streptomyces griseus*, *Corynebacterium* sp., *Micrococcus luteus* and *Bacillus subtilis*. The 102F69 antibiotic was heat stable and greater than 8000 in molecular weight.

Of 366 strains of *B. japonicum*, 51 (14%) were found to be antagonistic toward USDA110 and USDA123. Of the antagonistic strains tested, none inhibited USDA strains 46 and 83, most inhibited USDA strains 127, 129, and 138, while USDA98 showed variable resistance.

Gresshoff, Roth, Stacey & Newton (eds.) Nitrogen Fixation: Achievements and Objectives
Chapman and Hall New York • London 1990

Construction of acid tolerant, nitrogen-fixing strains of *Rhizobium leguminosarum* bv. *trifolii*

Hancai Chen, Elena Gärtner and Barry G. Rolfe

Plant Microbe Interactions Group, Research School of Biological Sciences, Australian National University, P.O. Box 475, Canberra, A.C.T. 2601, Australia.

INTRODUCTION AND RESULTS

Acidification of soils beneath clover-based pastures is a serious problem in southern Australia. Although clover plants themselves are moderately tolerant to acid conditions, *R.l.* bv. *trifolii* and clover nodulation are sensitive to low pH and associated acid soil factors (1). Most commercial inoculant strains of *R.l.* bv. *trifolii* are acid-sensitive strains. Acid-tolerant strains of *R.l.* bv. *trifolii* nodulate clover plants at low pH, but most of them are poor nitrogen-fixers (1). Therefore, we are seeking to construct acid tolerant *R.l.* bv. *trifolii* strains with enhanced nitrogen-fixing capacity.

Strain ANU1173 is an acid-tolerant *R.l.* bv. *trifolii* strain able to nodulate clover plants at pH 4.4. At pH 6.5, its symbiotic effectiveness was about 80% of that of commercial inoculant strains of *R.l.* bv *trifolii*. Strain ANU1173 contains four plasmids, the smallest plasmid being the Sym plasmid. When the Sym plasmid of strain ANU1173 was mobilized into a Nod$^-$ strain of *R.l.* bv. *viciae*, it continued to confer a poor nitrogen-fixing phenotype on clovers. These results suggested that the low symbiotic effectiveness of strain ANU1173 was due to the poor efficiency of genes on the Sym plasmid. To improve the symbiotic effectiveness of strain ANU1173, it was cured of its Sym-plasmid (strain ANU1184) and then replaced with another *R.l.* bv *trifolii* Sym plasmid, pBRIAN which confers more efficient nitrogen fixation (2). Subterranean clover plants inoculated with the hybrid strain ANU1184(pBRIAN) had a dramatic increase in both acetylene reduction activity and dry weight when grown at pH6.5 or pH4.4 compared with plants inoculated with strain ANU1173 and grown under the same pH conditions. When strain ANU1173 was cured of its Sym plasmid and the second largest plasmid (p1173-2) it became acid-sensitive (strain P22) and had a rough colony phenotype on TY agar medium. Mobilization of the plasmid p1173-2 back to strain P22 restored its acid-tolerant and mucoid phenotypes. A series of Tn5-induced acid sensitive mutants of strain ANU1173 have been isolated and the genes involved in the acid-tolerant phenotype have been cloned. Some acid-tolerant loci were non-Sym-plasmid located and the others were located on the chromosome.

1. Richardson, A.E. et al (1989) Aust. J. Physiol. **16**, 117-29.
2. Djordjevic, M.A. et al (1983) J. Bacteriol. **156**,1035-45.

Gresshoff, Roth, Stacey & Newton (eds.) Nitrogen Fixation: Achievements and Objectives
Chapman and Hall New York • London 1990

Symbiotic efficiency of two *Rhizobium meliloti* isolates of the same phage type with different denitrification activities

Yiu-Kwok CHAN

Plant Research Centre, Agriculture Canada, Ottawa, Ontario,
Canada K1A 0C6

INTRODUCTION R. meliloti strains vary considerably in their denitrification activities (3). Such variations exist within phage types and amongst isolates which are of the same phage type, with the same plasmid profile (2) and originate from the same cultivaiton site (1). Both denitrification and nitrogen fixation can occur inside the microaerobic nodule depending on the nitrate concentration. To test for nitrate tolerance of the alfalfa symbiotic system, the efficiency of two isolates (123 and 250) of R. meliloti belonging to phage type 33 were compared in relation to their denitrification activities.

METHODS Whole-cell denitrification was assayed by the acetylene inhibition method (5). Duplicate tube-cultures were prepared in Maniatis M9 medium under 5% acetylene in argon with mannitol as carbon source and either nitrate or nitrite as terminal electron acceptor. Seeds of Medicago sativa cv. Saranac were prepared, germinated and planted into pouches and inoculated according to Sirois & Peterson (4). Five replicate pouches, each containing 5 plants, were used for each inoculant/treatment. Nitrate was added and maintained in the plant nutrient solution for a week after nodule formation. Whole-plant acetylene reduction was assayed by FID-GC after 1 h incubation and nitrous oxide accumulation by ECD-GC after 24 h.

RESULTS & DISCUSSION Denitrifying cultures of isolate 123 reduced nitrate twice as fast as isolate 250 while the latter reduced nitrite an order of magnitude faster. Treatment of the nodulated plants with nitrate up to 10 mM for a week had no effect on plant dry weight or nodule growth in all cases. Acetylene reduction by plants nodulated by isolate 250 was inhibited 50 and 75 % by nitrate at 5 and 10 mM respectively, but the activity of those nodulated by isolate 123 was not affected. At 10 mM nitrate, the simultaneous nitrous oxide accumulation by plants nodulated by isolate 123 was low but significantly higher than by those nodulated by isolate 250. Hence, whole-plant denitrification was probably regulated by nitrate availability to the bacteroids and related to the denitrification activity of the microsymbiont. The role of plant cultivar in forming a nitrate-tolerant symbiotic system needs to be investigated.

REFERENCES

1. Bromfield, E.S.P., Sinha, I.B. & Wolynetz, M.S. (1986) Appl. Environ. Microbiol. **51**, 1077-1084.
2. Bromfield, E.S.P., Thurman, N.P., Whitwill, S.T. & Barran, L.R. (1987) J. Gen. Microbiol. **133**, 3457-3466.
3. Chan, Y-K., Barran, L.R. & Bromfield, E.S.P. (1989) Can. J. Microbiol. **35**, 737-740.
4. Sirois, J.C. & Peterson, E.A. (1982) Can. J. Microbiol. **28**, 265-268.
5. Yoshinari, T. & Knowles, R. (1976) Biochem. Biophys. Res. Commun. **69**, 705-710.

Gresshoff, Roth, Stacey & Newton (eds.) Nitrogen Fixation: Achievements and Objectives
Chapman and Hall New York • London 1990

Heat and cold shock in Proteins arctic and temperate strains of *rhizobia*

J. CLOUTIER[1], D. PRÉVOST[2], P. NADEAU[2], and H. ANTOUN[1]

[1]Dép. des sols, Université Laval, Québec, Canada G1K 7P4
[2]Agriculture Canada, Sainte-Foy, Québec, Canada G1V 2J3

The exposure of bacteria above normal growth temperatures induced heat shock proteins (hsps). Only few studies have reported cold shock proteins (csps) synthesis in bacterial cells exposed to low temperature. Arctic rhizobia, isolated from Astragalus and Oxytropis spp., have shown cold adaptation for growth and nitrogenase activity. The purpose of the present study was to compare the hsps and csps induced in three arctic (psychrotrophic) and three temperate (mesophilic) strains of rhizobia (but only one of each are presented here) in order to identify some of the mechanisms involved in thermo-adaptation. Bacterial cells were grown at their optimal temperatures (arctic 25°C, temperate 30°C). Log phase cells were pulse labelled with ^{35}S-methionine and exposed to different selected shock temperatures allowing comparison of protein synthesis of all strains for a given temperature and a given ΔT (T° shock - T° optimal). Analysis by SDS-gel electrophoresis and fluorography revealed that the arctic strain N-44 synthesized seven hsps at all shock temperatures (36.4, 41.4 and 46.4°C) except at 29°C (2°C over maximal growth temperature) with only three hsps. When the temperate strain SM-2 was subjected to a shift-up of 2°C over its maximal growth temperature (38.4°C) and to a shock at 41.4°C, six hsps were induced. Only one protein band was different in M.W. at 38.4°C (32.3 Kd) and at 41.4°C (30.0 Kd). There was no synthesis of hsps in the temperate strain at 46.4°C. Three hsps of high M.W. (106.9, 83.1, 59.5 Kd) were present in both strains at all heat shock treatments when induced. Cold shock results showed that the arctic and temperate strains of rhizobia synthesized eight and twelve csps respectively when shifted-down to 2°C below their minimal temperature (arctic 0°C, temperate 5.5°C). When the temperate strain SM-2 was shifted-down to 0°C, the same protein pattern [with two additional protein bands (54.1 and 32.8 Kd)] of the previous treatment was obtained concomitant with its maximal induction. The number of csps were considerably decreased at -5°C for both strains of rhizobia. There were no csps induced at -10°C for the temperate strain whereas the arctic strain synthesized four csps. The temperate strain SM-2 responded more vigorously to cold shock treatments than the arctic strain by synthesis of a higher number of csps. This first report on hsps and csps in rhizobia shows that the cold adapted arctic strain (N-44) is able to synthesize heat and cold shock proteins at extreme limits of temperatures (46.4 and -10°C) whereas the temperate strain SM-2 is unable to do so.

Gresshoff, Roth, Stacey & Newton (eds.) Nitrogen Fixation: Achievements and Objectives
Chapman and Hall New York • London 1990

Symbiotic performance and competition ability of some spontaneous mutant of *Rhizobium meliloti*

R. Di Bonito, L. Gatta, and M.L. Belli

Dipartimento Agrobiotechnolgie, E.N.E.A. Casaccia, via Anguillarese 301, 00100 Roma, Italy.

To select strains with good competitiveness, 36 spontaneous mutants of Rhizobium meliloti strains, characterized by antibiotic resistance, were evaluated both for their ability to nodulate in agar tubes and fix nitrogen on six-week-old plants. Some strains, selected for their early nodulation and the high number of nodules formed, were analyzed for competitiveness in soil with indigenous populations. Symbiotic performance of the strains under aseptic conditions was correlated with competitiveness in the soil.

The strains used were 1107 and 1025 furnished by Dr. Griffin, USDA, Beltsville and 102F97 and 102F51a from Dr. Smith, Nitragin Company. The strains 1025-3P, 102F51a-3P, 1025-3T and 102F51a-3T were spontaneous mutants of 1025 and 102F51a resistant respectively to spectinomycin and streptomycin.

We have conducted a pot trial in a growth room on a neutral soil with an indigenous population of R. meliloti of 0.7×10^3 cells/g. Either a single-strain treatment (with 1025-3P, 1025-3T, 102F51a-3P or 102F51a-3T) or a double treatment (1025-3P/1025-3T or 1025-3P/102F51a-3T) of liquid inoculum was added to the soil before sowing to a final concentration of either 2×10^4 or 2×10^6 cells/g of soil. Seeds of cultivar, Romagnola, were surface sterilized, sowed and incubated as previously described. Three replicates for each treatment were performed including a not-inoculated control. After three and four weeks, ten plants were harvested and the number of nodules counted. The nodule occupancy was evaluated by plating on TY agar supplemented with the appropriate antibiotic.

Although some strains exhibited significant precociousness of nodulation or increased nodule number compared to wild type, their acetylene reduction activity was unchanged. After 3 weeks in the single strain inoculations, all strains increased the number of nodules compared with the control, but only 1025-3P did so at both inoculum rates. Nodule occupancy increased significantly for 1025-3P and 102F51a-3P at 2×10^6 inoculum rate. After 4 weeks, only 1025-3P exhibited a significant increase for number of nodules and nodules occupancy against the indigenous populations. In the combined inoculations, strain 1025-3P showed a significant competition ability against 1025-3T and 102F51a-3T at the higher rate after 4 weeks.

Gresshoff, Roth, Stacey & Newton (eds.) Nitrogen Fixation: Achievements and Objectives
Chapman and Hall New York • London 1990

Effect of biological nitrogen fixation on the elemental concentration of *Medicago sativa* grown under salinity levels

Catherine E. Douka, Alex. C. Xenoulis and Th. Paradellis

NRCPS "Demokritos" Agia paraskevi 15310 Athens Greece

Effect of inoculation on the elemental composition of plants has been previously identified[1]. In the present study the growth of Medicago sativa on Nitrogen-free agar medium containing 0.0, 0.2, 0.4, 0.8 and 1.2% NaCl inoculated with a salinity tolerant Rhizobium meliloti strain permitted the identification of a linear correlation between elemental concentration and parameters such as plant yield or nodulation which monitor the induced B.N.F. The comparison between the composition of inoculated and uninoculated biomass, which was determined using the X-ray fluorescence technique, demonstrates that in the inoculated plants the concentration of Mo, Mn, Sr, Cu, Zn decreases with increasing nodulation and plant yield and is significantly lower than that in the uninoculated. The concentration of other elements increases (Rb, K) or is not affected (Ca, Ni) by increasing nodulation and plant yield and is the same (Ca, Ni, K) or higher (Rb) than in the uninoculated. Clearly, the concentration reductions of Mo, Mn, Sr, Cu, Zn are directly correlated with the degree of B.N.F. which has occurred. The relevant equations for Mo are:

(Mo) = 45.1 - 1.9 (plant yield) (r^2 = 0.99)

(Mo) = 45.8 - 3.0 (nodulation) (r^2 = 0.81)

1. Douka, C.E. et al. (1986) Biol. Fert. Soils 2, 113-118; (1986) Folia Microb. 31, 207-214; (1987) Biol. Agr. Hort. 5, 25-39.

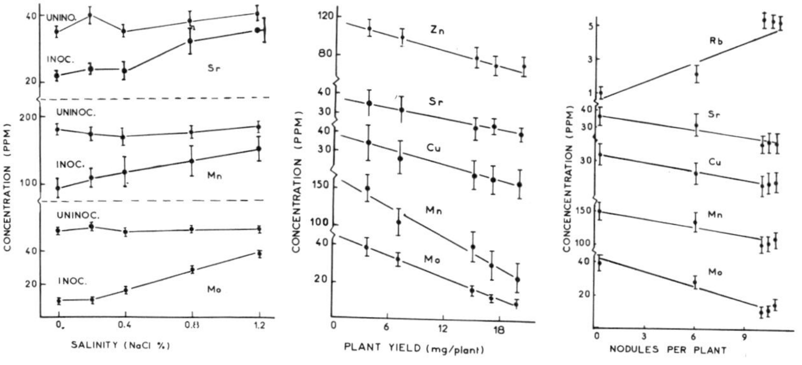

Gresshoff, Roth, Stacey & Newton (eds.) Nitrogen Fixation: Achievements and Objectives
Chapman and Hall New York • London 1990

Inoculation study of soybean under Iraqi conditions

Talib A. Essa, B. Al-Ithawi and H. Al-Dulami

University of Baghdad, Iraq.

Two experiments were conducted to determine soybean response to inoculation and nitrogen fertilizer. Five nitrogen levels (0,40,80, 120, and 160 kg/ha) were used with or without inoculum.

Inoculation increased ($P \leq 0.05$) seed weight, number of pods/plant, protein content, root dry weight and plant dry weight, while the percentage of shriveled seeds and oil content decreased in both years. Increasing N level from zero to 160 kg/ha had increased seed yield, number of pods/plant and seed weight but decreased the percentage of shriveled seeds in the first year, but had no effect on these characteristics in the second year.

Previous inoculation increased seed yield, pod number, seed weight and plant height, and reduced the percentage of shriveled seeds. Nodules were formed only on inoculated plants or in previously inoculated areas. Increasing N levels from zero to 160 kg/ha in the study of previous inoculation also had a great influence on seed yield and other traits.

It was concluded that *Rhizobium* is capable of surviving under environmental Iraqi conditions, and could bring about effective nodulation and normal soybean growth of good productivity. It was recommended that 160 ka/ha of nitrogen should be applied when planting soybean in an area without inoculum, and 40 kg/ha when using inoculum.

Gresshoff, Roth, Stacey & Newton (eds.) Nitrogen Fixation: Achievements and Objectives
Chapman and Hall New York • London 1990

Heavy metal toxicity and tolerance in *Rhizobium leguminosarum* bv. *trifolii*

K.E. Giller, A. Chaudri and S.P. McGrath[2],

Wye College, University of London, Wye, Ashford, Kent TN25 5AH. UK. Institute of Arable Crops Research,[2] Rothamsted Experimental Station, Harpenden, Herts. AL5 2JQ. UK.

Introduction: N_2-fixation and yield of clover (*Trifolium repens* L.) were reduced drastically in field plots where the soil is contaminated with heavy metals at concentrations close to those permitted under current guidelines for environmental protection. (McGrath *et al.*, 1988). The failure of N_2-fixation in metal-contaminated soil was due to the survival of just a single, ineffective *Rhizobium* strain (Giller *et al.*, 1989). If effective strains of *Rhizobium* were inoculated into metal-contaminated soil (10^5 cells/g soil) none survived after two months of incubation. In this work we examined tolerance of our ineffective and effective *Rhizobium* isolates to heavy metals in solution culture.

Materials and Methods: Isolates of clover *Rhizobium* were grown in yeast mannitol glutamate medium (Johnson & Wood, 1987) for 3 days. The cells were then spun down, washed in distilled water and resuspended in dilute solutions of 0, 0.2, 0.4, 0.8 and 1.0 mgl^{-1} $ZnSO_4$ or $CdSO_4$. Numbers of viable cells surviving were estimated after 0, 5, 10, 27 and 72 hours by plate counts.

Results and Discussion: Survival tests of these strains in dilute solutions of Zn and Cd have shown that the ineffective strain isolated from metal-contaminated soil survives ($>10^2$ cells/ml) in concentrations of Zn and Cd up to 1.0 mgl^{-1} for at least 72 hours. Isolates from uncontaminated soil were all killed within 27 hours in 0.8 mgl^{-1} Zn and 0.4 mgl^{-1} Cd. This probably explains the ability of the ineffective strain to survive in soil contaminated with Zn and Cd as these concentrations of metals are similar to those in the soil solution. We have now studied survival of *Rhizobium leguminosarum* bv. *trifolii* in many UK soils and find that there are generally no strains which can survive in soils contaminated with heavy metals when clover is absent.

References:

Giller,K.E.,McGrath,S.P. & Hirsch,P.(1989) Soil Biol.Biochem. **21**,841-848.
McGrath,S.P.,Brookes,P.C. & Giller,K.E.(1988) Soil Biol.Biochem. **20**,415-424.
Johnson,A.C. & Wood,M. (1987) Lett. App. Microb. **4**, 137-139.

Gresshoff, Roth, Stacey & Newton (eds.) Nitrogen Fixation: Achievements and Objectives
Chapman and Hall New York • London 1990

Competition between two effective strains of *Bradyrhizobium sp* (cowpea) in cowpea (*Vigna unguiculata* (L.) Walp var. Tuy) plants inoculated at two different root levels: nodule ureide content, nodule mass and plant dry matter

María Luisa Izaguirre-Mayoral and Margarita Sicardi de Mallorca

Instituto Venezolano de Investigaciones Científicas (IVIC), Laboratorio de Biotecnología y Virología Vegetal, Centro de Microbiología y Biología Celular, Apartado 21827, Caracas 1020 A, Venezuela.

The effective strains of Bradyrhizobium sp (cowpea) I-125 and I-700 were examined for nodulation and ureide production in single, dual or mixed inoculation treatments. Seeds of cowpea (Vigna unguiculata (L.) Walp var. Tuy) were planted in Leonard jars containing a 1:1 mixture of sand:vermiculite and 0.8 L of Fahraeus solution devoid of N. One gram of gamma irradiated peat (1.8 x 10^9 viable B. cowpea cells), was allocated at 5 cm (up) and/or 15 cm (down) from the upper level of the vermiculite according to the following treatments: T1) I-125 up; T2) I-125 down; T3) I-700 up; T4) I-700 down; T5) I-125 up and down; T6) I-700 up and down; T7) I-125 up and I-700 down; T8) I-700 up and I-125 down; T9) I-125 + I-700 up; T10) I-125 + I-700 down. Plants were sampled 1 month after germination and nodules from each strain recognized by their external color. Results of T1, T2, T3 and T4 indicate that the strain I-125 produced 2-3 times more nodule mass than I-700. However, when the strain I-125 was as in T5 the nodule mass in each root area was reduced by 50%. The strain I-700 when single inoculated produced equal nodule mass in T3 and T4, but in dual inoculation (T6) nodulation at 15 cm was 75% lower than in the upper root region. Nevertheless, the total nodule mass per plant in T5 and T6 was similar to that of T1, T2 and T3, T4, respectively. Reduced nodulation of I-700 at 15 cm in the root system was also observed in T7. The presence of I-700 in T8 also decreased the nodule mass of I-125 at 15 cm. In mixed inoculums the inherent competitive ability of I-700 over I-125 was observed at 5 cm (T9) but not at 15 cm in the roots (T10). For both strain a negative relationship was detected between nodule mass and nodule ureide content. High ureide content x total nodule mass^{-1} x plant^{-1} was measured in those plants harbouring the I-700 strain. On the other hand, mixed inoculation caused a reduction in the ureide content in I-700 but not in I-125 nodules (T9, T10). In each individual treatment the ureide content in the petiole of the 2nd trifoliolate leaves resemble the nodule ureide concentration. However, contrasting position of nodules, nodule mass and nodule ureide content were not translated into different plant dry matter. These data suggest that nodules formed by each strain at different inoculum position are able to compensate for ureide production in spite of the differences observed in nodule mass. Under growth controlled conditions movement of the Bradyrhizobium cells was very limited.

Gresshoff, Roth, Stacey & Newton (eds.) Nitrogen Fixation: Achievements and Objectives
Chapman and Hall New York • London 1990

Response of four alfalfa cultivars to salinity under N_2-dependent conditions

Y. Kapulnik[1] and Bruria Heuer[2]

[1]Legume Inoculation Dept. and [2]Soils and Water Dept, ARO, Volcani Center, Bet Dagan, Israel.

Alfalfa (Medicago sativa L.) is an important forage legume crop, moderately sensitive to salinity. Due to the restricted information available on its response to salinity and the suitability of alfalfa to grow in saline habitats, a study was initiated to determine growth and nitrogen accumulation in four alfalfa cultivars exposed to 170 mM NaCl in their rootzone as well as the recovery response upon removal of the stress conditions.
Four alfalfa cultivars, Ghilboa, Mohapa 69 (M69), Hairy Peruvian (HP) and Alioc (TUR) were grown under N_2-dependent conditions in styrofoam containers filled with a vermiculite- sand mixture (1:1). Salt treatments were initiated in half of the containers (while the others served as controls) following the first regrowth, by gradually adding NaCl up to a final concentration of 170 mM. The recovery treatment of the third regrowth was obtained by extensive washing of the rootzone with tap water.
Salinity significantly reduced the total N content in the shoots of all cultivars but cv. TUR. This reduction ranged between 40 to 58%. At the same time, N concentration was not significantly affected by salinity, and never dropped below 3.2%. Forage yield was similarly affected by salinity. cv. Ghilboa produced twice as much forage as all other cultivars in non-saline conditions, and although it was drastically inhibited by salinity, its forage weight was higher than that of TUR by 20%, the salt-resistant cultivar. Although biomass production is entirely derived from photosynthates, no direct corelation was found between growth and CO_2 fixation. Photosynthesis was reduced between 48 to 67% in all cultivars, suggesting a direct inhibition of the photosynthetic mechanism by salt stress. An elevated level of proline was detected in all cultivars with a significant increase only in cv. TUR (by 34%). It is possible that proline partly contributed to the osmotic adjustment observed besides ion and sugar accumulation.
Reduction of salinity level in the root zone resulted in almost full recovery of forage production and nitrogen content and in cultivar TUR it exceeded the controls. Nevertheless, CO_2 fixation in cv. Ghilboa and cv. Mohapa 69 was still inhibited by 28%. Proline levels were more or less the same in all treatments and cultivars.

Gresshoff, Roth, Stacey & Newton (eds.) Nitrogen Fixation: Achievements and Objectives
Chapman and Hall New York • London 1990

Autecological characterization and competitive nodulating behavior of indigenous *Rhizobium leguminosarum* bv. *trifolii*

K. Leung[1] and P. J. Bottomley[1,2]

Departments of Microbiology[1], and Crop and Soil Science[2],
Oregon State University, Corvallis, OR 97331-3804

Introduction. In order to enhance our understanding of the relationship between the competitive nodulating ability of field rhizobia and their rhizosphere population densities, experiments were designed to determine: (1) the field nodule occupancy of subclover; (2) the serological and multi-locus enzyme electrophoretic characteristics of nodule isolates; and (3) the sub-population densities of rhizobia in bulk soil, subclover and grass rhizospheres under field conditions.

Results and discussion. Although eight serotypes had been identified within the soil population of R. leguminosarum bv. trifolii, eighty-three to 92% of the nodules were occupied by serotypes AS6, AS21 and AS6-21 in three separate years. Representatives of our subclover nodule isolates were also analysed by multi-locus enzyme electrophoresis. Only two major ETs represented the isolates from subclover nodules. One ET represented both competitive serotypes AS6, AS6-21 and the noncompetitive serotype AP17. The other ET was restricted to serotype AS21.

The population densities of AS6 and AS6-21 (1.4×10^5 to 5.0×10^5 cells/gm dry soil) were greater than the populations of AS21 and AP17 (4.0×10^4 to 5.0×10^4 cells/gm dry soil) in bulk soil and unplanted soil. Although both subclover and grass rhizospheres could stimulate the growth of rhizobia, our data showed that a subclover rhizosphere maintained a greater population than did a grass rhizosphere. Differential stimulation of different serotypes of R. leguminosarum bv. trifolii was observed in the subclover rhizosphere. Although the population density of AS21 was approximately 10-fold lower than AS6 in the bulk soil or unplanted soil, the population of AS21 was only 50% lower than AS6 in the host rhizosphere. In contrast, the differences between the populations of AS6 and AP17 were similar in both the unplanted soil and the host rhizosphere. AS21 was the most stimulated serotype in the host rhizosphere, and possessed the greatest population density (2.7×10^7 cells/gm dry soil)

Gresshoff, Roth, Stacey & Newton (eds.) Nitrogen Fixation: Achievements and Objectives
Chapman and Hall New York • London 1990

Detecting *Frankia* in soil using PCR

D. D. Myrold, A. B. Hilger, and S. H. Strauss

Oregon State University, Corvallis, OR 97331, U.S.A.

Frankia are present in low numbers in most soils, at least as determined using a plant bioassay method. Consequently, to detect *Frankia* in soil and follow its population dynamics, it was necessary to develop a technique that was capable of detecting only a few *Frankia* per gram of soil. Our objectives were to adapt and optimize the PCR (polymerase chain reaction) to selectively amplify *Frankia nifH* DNA directly extracted from soil.

Frankia ArI6 was added to a sterile sandy loam soil in ten-fold increments from 12 to 121,000 CFU (colony forming units) g^{-1} soil. DNA was extracted from soil using a direct lysis procedure (1), amplified using the PCR, and probed using a ^{32}P-labelled *nifHDK* probe. *Frankia* ArI6 was detected down to 1210 CFU g^{-1} soil, which corresponds to 209 IU (infective units) as assayed by an *Alnus rubra* MPN bioassay. This population is similar to the low natural populations often reported in the literature. The PCR amplification procedure used in this experiment produced considerable amounts of non-target DNA products, which led to several modifications of the procedure.

High background DNA levels interfered with the amplification of low levels of *Frankia nifH*. These non-specific PCR reactions were minimized by using booster PCR (2), which allowed us to measure 90 copies (1 pg) of *nifH* against a background of 0.4 μg DNA compared with 900 copies when using the standard PCR procedure. Sensitivity and specificity were improved by an additional factor of ten by optimizing the Mg^{2+} concentration of the PCR reaction mixture.

These results suggest that it should be possible to dectect *Frankia* in soil at population levels below two IU g^{-1}, which will enable the study of important ecological questions regarding the growth and survival of *Frankia* in soil.

References

1. Hilger, A.B. & Myrold, D.D. (1990) Agric. Ecosys. Environ. (in press)
2. Ruano, G. & Kidd, K.K. (1989) Amplifications 4:12-13.

Gresshoff, Roth, Stacey & Newton (eds.) Nitrogen Fixation: Achievements and Objectives
Chapman and Hall New York • London 1990

Risk evaluation for genetically modified microbial inoculants

F. O'Gara[1], B. Boesten[1], K. Birkenhead[1], P. Stephens[1], M. Morris[1], P. Ross[1], S. Condon[1], M.P. Nuti[2] and A. Giacomini[2].
[1]Microbiology Department and National Food Biotechnology Centre, University College, Cork (Ireland).
[2]Dipartimento di Biotecnologie Agrarie, Universita di Padova, via Gradenigo 6, 35131 Padova (Italy).

This work is aimed at developing appropriate biological material and tools to monitor survival, persistence and competition of Rhizobium, Bradyrhizobium and fluorescent Pseudomonas strains in the soil and rhizosphere and to evaluate potential gene transfer unambiguously and on a routine basis. This involves the introduction of the lacZ gene as a marker system into these microbial strains.

In order to obtain clearly detectable ß-galactosidase activity in R. leguminosarum, a strong synthetic promoter was used. Although this 59mer was constructed as an E. coli "consensus" sequence, when cloned in front of the promoterless lacZ gene of the IncP vector pMP220, this stable construct gives rise to about 12,000 ß-gal Miller units in R. leguminosarum and produces dark blue colonies on X-gal containing plates.

As an alternative to the traditionally used antibiotic resistance genes, we employed a thymidylate synthase (TS, E.C.2.1.1.45) derived from Lactococcus lactis subsp. lactis (1), to provide a positively selectable marker in Rhizobium meliloti. This also required an R. meliloti strain which is deficient in TS activity. Such a strain was obtained by selecting colonies that were resistant to the antifolate drug aminopterin in the presence of thymidine. One TS deficient strain (CM21) was used for further evaluation.

CM21 was found to be severely impaired in its ability to persist in the rhizosphere environment and to associate with the Alfalfa host plant. These defects were overcome by complementation with the cloned L. lactis TS gene. This feature was exploited to ensure stable maintenance of IncP plasmids in R. meliloti in the absence of antibiotics.

(1) Ross, P. (1989). PhD Thesis (National University of Ireland).

Gresshoff, Roth, Stacey & Newton (eds.) Nitrogen Fixation: Achievements and Objectives
Chapman and Hall New York • London 1990

Xanthobacter as a rhizosphere organism of rice

H.Keith Reding[1], Peter G. Hartel[2] and Juergen Wiegel[1]

Departments of Microbiology[1] and Agronomy[2], The University of Georgia, Athens, Georgia, USA.

The genus Xanthobacter is composed of gram-type negative, pleomorphic bacteria capable of fixing dinitrogen while growing chemolithotrophically. Isolates of Xanthobacter have been obtained from marine and freshwater sediments, sewage samples, street ditches and soil of meadows. Recently, several strains were isolated from the rhizosphere of rice in Japan; however, Xanthobacter's role as a rhizosphere organism was not investigated. We were able to isolate Xanthobacter from the soil and rhizosphere of rice from a field in southeastern Arkansas, U.S.A. by plating samples on a mineral salts medium with a headspace of 10% H_2, 10% CO_2, 15% air, and 65% N_2.

Selective enrichment for Xanthobacter yielded 1.38 x 10^4 CFU/ g dry soil and 7.8 x 10^5 CFU/ g dry root (rhizosphere:soil = 57). After successive washings of the roots for 1 min, 5 min, and 10 min using a stomacher blender, 2.2 x 10^3 CFU/ g root remained attached to the root indicating that Xanthobacter is closely associated. In studies using gnotobiotic assemblies, Xanthobacter isolate KR-JW1 was able to bind to the root hairs and root surface via structures which are probably bacterial exopolysaccharides. Both rod-shaped and coccal Xanthobacter were bound to the root surface. Similar morphologies can be produced in vitro by growing the bacteria on specific carbon sources e.g. propanol produces cocci and H_2/CO_2 produces rod. Xanthobacter was also able to multiply when cultured gnotobiotically with rice. Bacterial numbers increased from 10^5 CFU/mg root to $10^{7.5}$ CFU/mg root in one day and remained at that concentration throughout the 7-d study. The Xanthobacter rice isolates were motile when cultured on alcohols, especially propanol, or when cultured with rice. No motile strains of Xanthobacter autotrophicus nor Xanthobacter flavus have been reported; and although all strains of Xanthobacter agalis are motile, it is likely that the rice isolates are a new species of Xanthobacter.

Gresshoff, Roth, Stacey & Newton (eds.) Nitrogen Fixation: Achievements and Objectives
Chapman and Hall New York • London 1990

Diazotrophs repeatedly isolated from roots of kallar grass form a new genus, *Azoarcus*

B. Reinhold-Hurek, T. Hurek, M. Gillis, B. Hoste, K. Kersters and J. De Ley.
Laboratorium voor Microbiologie en Microbiële Genetika, Rijksuniversiteit Gent, Ledeganckstraat 35,
B-9000 Gent, Belgium.

Kallar grass (*Leptochloa fusca* (L.) Kunth) is grown as a pioneer plant on salt-affected, low-fertility soils in the Punjab region of Pakistan without addition of any nitrogenous fertilizer. We were interested in possible bacterial partners of a nitrogen-fixing association. Previously, *Azospirillum lipoferum* and *A. halopraeferens* were found predominating on the root surface, and unidentified rods dominating in the endorhizosphere (Reinhold et al., 1986). During a second survey of the most abundant diazotrophs in roots of Kallar grass carried out on the same field in 1988, we obtained similar diazothrophic rods. Here we propose that this group belongs in a new genus, *Azoarcus*.

Enrichment and isolation were carried out at 30°C on malate media, either salt-affected (SSM) or not (SMV) (Reinhold et al., 1986). For references of methods for identification see Willems et al., 1989.

Eleven diazotrophic isolates, including strain BH72 isolated in 1984, formed five subgroups of 1-3 members. Grouping was consistent according to electrophoretograms(SDS-PAGE) of cellular proteins, gaschromatographic analysis of cellular fatty acids, and, less discriminative, according to nutritional and enzymatic features tested in API galleries. DNA-DNA hybridizations indicated that subgroups differed at the species level, but we think they have too few members and too weak discriminative phenotypic features to propose them as different species. DNA-rRNA hybridizations showed no close relationship to any other known genus, the isolates constituting a separate rRNA branch in rRNA superfamily III at the branching level of 68°C $T_{m(e)}$; on this branch we found two clusters with $T_{m(e)}$ values around 80°C and 72°C, respectively. We propose a new genus, *Azoarcus* ("nitrogen arch") for this group. Cells are gram-negative, slightly curved, yellowish rods fixing N_2 microaerobically, growing well on organic acids but not on carbohydrates, having a strictly respiratory type of metabolism.

Reinhold,B., Hurek,T., Niemann,E.-G., & Fendrik,I. (1986) Appl. Environ. Microbiol. **52**, 520-526.
Willems,A. et al. (1989) Int. J. Syst. Bacteriol. **39**, 319-333.

Gresshoff, Roth, Stacey & Newton (eds.) Nitrogen Fixation: Achievements and Objectives
Chapman and Hall New York • London 1990

Ecotoxicological assessment of heavy metals with melanin producing (Mel+Fix+) *Rhizobium* strains

Minocher Reporter#, Wanda Baune#, & Lawrence Kapustka*

#NSI-EL; *U.S. EPA, Environmental Research Laboratory, Corvallis, OR 97333, USA.

Melanin production in three strains of Rhizobium, R. meliloti, R. fredii, and R. leguminosarum bv. phaseoli obtained from J. Beringer (1). Melanin was released from these strains within three hours by SDS in the presence of tyrosine and copper. All strains produced melanin when copper was substituted by cadmium, lead or zinc; controls did not produce melanin. R. meliloti strain GRO15 gave the most consistent melanin response to metal exposure. Melanin production and growth in culture were evaluated by developing methods to assess biologically available heavy metals. In developing the overnight assays, mid-log phase Mel^+ cultures were used. Companion tests, using acetylene reduction and melanin production by nodulated plants, were also studied since one mel gene of each Rhizobium strain is located on the respective symbiotic plasmid (1) and is a regulatory gene which may be linked to the regulatory nifA gene in certain symbiotic bacteria (2). Melanin is, however, not required for nitrogen fixation (2). Melanin production has not been detected in Bradyrhizobium (1). Nodules of Mel^+ bacteria were incubated with heavy metals overnight and examined after ARA indicated melanin in some cortical cells near vascular bundles.

The figure shows an adaptation of the assay for a microplate reader. Site samples for biologically available heavy metals can be started in the fields and read later in the laboratory.

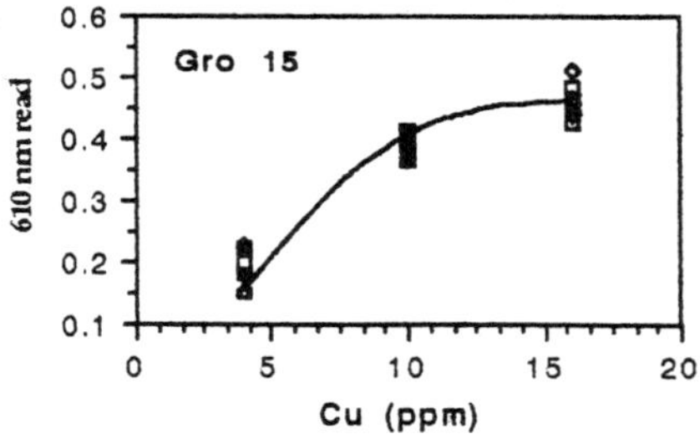

1. Cubo, T., Buendia-Claveria, A. Beringer, J. & Ruiz-Sainz, J. (1988) Appl. & Envir. Micro **54**, 1812-1817.
2. Borthakur, D., Lamb, J. & Johnston, A. (1987) **207**, 155-160.

Gresshoff, Roth, Stacey & Newton (eds.) Nitrogen Fixation: Achievements and Objectives
Chapman and Hall New York • London 1990

Geographic distribution and environmental interactions that affect the abundance of *Rhizobium leguminosarum* biovar *phaseoli* in Costa Rica

Gabriela Soto and Linda K. Blum

Department of Environmental Sciences, University of Virginia, Charlottesville, Va 22903 USA.

The objective of our research was to determine the environmental interactions that influence the abundance of native populations of *Rhizobium leguminosarum* biovar *phaseoli* (Rlp) in Costa Rica. Soil samples were collected from 47 sites representing a wide range of climatic conditions and soil types, and the abundance of Rlp at each site was determined. At each sampling location, two soil samples were collected, one from a field planted to bean, and the second from a field where no legumes had been planted for at least three years. For each field, the agricultural history was obtained from the farmer. Climatic data were obtained from the Inst. Metereolog. Nac. and the Inst. Costarric. the Electric. A complete soil analysis, including texture, organic matter content, pH, P, Ca, Mg, K, Mn, Fe, Cu, Zn, and Al was done for each sample. The numbers of Rlp were determined by a most probable number technique using the bean variety Negro huasteco. Separate principal components analyses were done to relate the abundance of Rlp to soil characteristics and to climatic characteristics.

Low numbers of Rlp (below the mean: 2.46 log cell/g dry soil) were correlated with high concentrations of K (mean = 0.51 meq/100 g). Low numbers of Rlp were also observed in soils with a high percentage of organic matter (mean = 9.6%), high concentrations of Fe (mean = 143 ppm) and Mn (mean = 146.5 ppm). The variation in the numbers of Rlp was not strongly correlated with any of the climatic variables; however, approximately 30% of the variation in Rlp abundance was correlated with temperature-related factors. The effect of temperature differed between samples collected in legume and non-legume fields. Areas with a high temperature (average, maximum, and minimum, with respective averages: 24.6, 29.0 and 18.8 C) presented low numbers of Rlp in the non-legume fields (average 2.2 log cells/g dry soil), and high numbers of *Rhizobium* in the fields where beans were growing (average 2.7 log cells/g dry soil).

A striking geographical distribution of Rlp in Costa Rica was demonstrated, and that distibution was strongly related to the distribution of soil and climate factors in that country.

Gresshoff, Roth, Stacey & Newton (eds.) Nitrogen Fixation: Achievements and Objectives
Chapman and Hall New York • London 1990

Relationships between *Rhizobium leguminosarum* bvs. *trifolii* and *viceae* in soil

S. R. Strain and P. J. Bottomley

Department of Microbiology, Oregon State University
Corvallis, OR, 97331-3804 USA

Multi-locus enzyme electrophoresis has gained recent popularity with some researchers as a method for determining the structure of rhizobial soil populations. Immunofluorescence, however, remains the only practical method for studying the autecology of indigenous soil rhizobia (Bohlool and Schmidt, 1980). The relationship between the delineations made by the two methods remains largely unexplored. There is no convincing evidence to suggest that serotypes within a soil population of *Rhizobium* represent distinct ecotypes.

Isolates of bvs. *viceae* and *trifolii* were recovered from two field sites where population sizes of the biovars differed. Population sizes of bvs. *viceae* and *trifolii* on Site A are $<10^2$ and $>10^5$, respectively. Reciprocal densities are found on Site C. Standard methods (Vincent, 1970) were used to characterize isolates serologically with nine antisera. Diversity was assessed electrophoretically at four loci by the methods of Selander et al. (1986).

Isolates of bv. *viceae* were recovered from four different hosts to trap as diverse a population as possible. Even so, approximately 70-100% of the isolates could be identified serologically by four antisera raised to bv. *trifolii* isolates suggesting the two biovars are related. However, the two biovars shared no common ETs suggesting a lack of relatedness. Future analysis of isolates recovered from other *Trifolium* spp. will help to resolve this contradiction. Evidence was seen for selection by some hosts of certain ETs and serotypes over others. *Pisum sativum* showed strong preference for two of the four serotypes and for one of three ETs. *Vicia sativa* demonstrated a preference for one ET but showed virtually no selection with respect to serotype.

Bohlool, B.B. & Schmidt, E.L. In: Advances in Microbial Ecology, Alexander, M., ed. (Plenum Press, New York). Vol. 4, pp. 203-241.
Selander et al. 1986. Appl. Environ. Microbiol. 51:873-884.
Vincent, J.M. 1970. IBP Handbook No. 15. (Blackwell, Oxford).

This material is based upon work supported under a National Science Foundation Graduate Fellowship awarded to S. R. Strain.

Gresshoff, Roth, Stacey & Newton (eds.) Nitrogen Fixation: Achievements and Objectives
Chapman and Hall New York • London 1990

Survival and nodulation by inoculant strains of *Rhizobium leguminosarum* bv. *phaseoli*

W. Streit, A.B. Wolff, J.A. Kipe-Nolt, H. Vargas and D. Werner

International Center for Tropical Agriculture (CIAT), A.A. 6713, Cali Columbia & Univ. of Marburg, Dept. of Biology, D-3550 Marburg, FRG.

The inability of inoculant Rhizobium bv. phaseoli strains to form a significantly large proportion of the nodules is a frequently cited explanation for the lack of plant responses to inoculation. Three field trials were conducted under diverse climatic and soil conditions to determine actual levels of nodule occupancy, compare different strains and examine the effect of different bean varieties. Survival of two of the strains was studied in two of the field sites to evaluate the importance of this character in determining nodule occupancy over time.

When plants were harvested at the 1st trifoliate leaf stage, % nodule occupancy ranged from highs of 24%-51% in the fertile temperate soil (alfisol) to lows of 7%-26% in the acid tropical soil (orthoxic palehumult). Interestingly, the lowest nodule occupancy was observed in the soil with the lowest native population counts. Significant differences between strains were observed and the strain by location interaction was also significant. However, strain KIM5 showed high nodule occupancy in all three sites and it was only in the acid (Santander) soil that it was outcompeted by CIAT 151. Thus it is possible to identify broadly adapted, competitive strains. There was a decline in % nodule occupancy over time for most of the strains, although a few strains maintained a constant low level. Although 3 contrasting plant genotypes were used, there was no strain by genotype interaction for % nodule occupancy. There were no significant strain effects on nodule number, acetylene reduction activity, plant dry weight or total plant N. Red nodule number was correlated with acetylene reduction activity, and the genotype and genotype by location interactions were significant for these traits.

The immunofluorescence studies used to follow survival of KIM5 and CE3 in the field showed no significant differences between the two strains, or between the two soils. After an initial decline from around 8.0 log to 7.0 log cells per g soil in the first few days, the counts stabilized.

Gresshoff, Roth, Stacey & Newton (eds.) Nitrogen Fixation: Achievements and Objectives
Chapman and Hall New York • London 1990

Soil acidity-factor effects on *Bradyrhizobium japonicum* and soybean symbiosis

A. A. T. Vargas[1], N. D. Denardin[1], L. Evans Roth[2], and G. L. Sexton[2]

[1]Centro de Energia Nuclear na Agricultura, CENA-USP, CP 96, Piracicaba 13400 SP, Brazil and [2]Center for Legume Research, University of Tennessee, M313 Walters Building, Knoxville, TN 37996-0810, USA.

Soil acidity-factors such as acid pH, aluminum toxicity and low phosphorus concentration affect nitrogen fixation by limiting growth of legumes and the establishment of symbiosis with rhizobia. Because of reaction between P and Al, a modified Leonard Jar system was used in which the only function of the lower container was to retain nutrient solution voided from the upper chamber. Jars containing the soybean cultivars IAC-9 and IAC-13 were arranged in the greenhouse and inoculated with 10^9 cells/jar of CM-1529, an acid-tolerant strain of *Bradyrhizobium japonicum*. Eight plant nutrient solutions were used, and the levels of acidity were pH 4.5 and 6.5, aluminum at 0, 50 and 100 μM, and phosphorus at 50 and 500 μM. Sterile nutrient solutions were added to the upper part of the jar each morning. Four treatments containing aluminum were watered with 50 ml solution of 50/100 μM Al at pH 4.5 in the evening to minimize P x Al reaction. Solution retained in the lower container was discarded every two days.

IAC-9 and IAC-13 contained up to 10^9 and 10^{10} bacteroids/g nodule 26 DAP based on direct and pour-plate counts, respectively. The lowest value for bacteroid viability was from nodules of cultivar IAC-13 at pH 4.5 and 100 μM Al. Acid pH associated with aluminum toxicity and low phosphorus concentration reduced nodulation and nitrogen fixation of both cultivars but did not affect top-plant dry weight of IAC-9. Concentrations of N, P and Al in shoots and P and Al in roots were also affected by acid treatments, with different responses for both cultivars. The jar system used was suitable to detect differences between treatments for most of the parameters evaluated.

Electron microscopy indicates that the variables to be examined are the extent of digestive vacuolation, number of bacteroids/symbiosome, disruption of cytoplasmic organization, polyphosphate bodies in bacteroids, and the occurrence of starch in the plastids of infected cells.

Gresshoff, Roth, Stacey & Newton (eds.) Nitrogen Fixation: Achievements and Objectives
Chapman and Hall New York • London 1990

Metabolic and structural condition of *Frankia* in root nodules of *Alnus incana* exposed to prolonged darkness

Per-Åke Vikman, Per-Olof Lundquist and Kerstin Huss-Danell

Department of Plant Physiology, University of Umeå, S-901 87 Umeå, Sweden.

Symbiotic *Frankia* is dependent on photosynthetic assimilates from the host plant for the supply of energy. Nitrogenase activity was decreased when the flow of assimilates was disturbed, and when investigated, the amount of active nitrogenase was also reduced[1]. Does prolonged darkness affect nitrogenase specifically, or *Frankia* metabolism in general?

Cuttings of *Alnus incana* (L.) Moench in symbiosis with a local source of *Frankia* were grown under controlled climate conditions, exposed to darkening for 2 days and then transferred back to ordinary day/night conditions. At various times *Frankia* vesicle clusters were prepared from the root nodules. The respiratory capacity with four different substrates were used to measure the condition of *Frankia*. Nitrogenase activity in intact plants and in the vesicle cluster preparations were also followed.

After 2 days in darkness the respiratory capacity of *Frankia* had decreased to ca. 40% compared to the control in ordinary day/night conditions, and 2-3 days after the dark treatment the respiratory capacity continued to decrease to ca. 25%. This indicates that the general condition of *Frankia* declined by the dark treatment. The following recovery was slow and not completed within the 8 days studied. TEM studies of ultrastructure in the root nodules revealed some disturbances after 2 days in darkness, mainly in young *Frankia* vesicles. Two days after ordinary day/night conditions were restored, structural degradation was pronounced in all forms of *Frankia* and in the host cytoplasm, while 2 days later young *Frankia* vesicles were dominating the infected zone of the root nodules. The amount of active nitrogenase decreased rapidly to ca. 10% of the control after 2 days in darkness. In contrast to respiratory capacity, recovery of nitrogenase in the vesicle cluster preparations started the first day after the dark treatment and was completed after 4 days. Still, whole plant expression of nitrogenase activity did not recover compared to control during the 8 days studied.

We conclude that Frankia is degraded in response to prolonged darkness, but some cells remain intact. Nitrogenase is specifically inhibited but recovers in structurally intact cells. The main recovery is due to synthesis of new *Frankia* cells, probably associated with nodule growth.

1) Huss-Danell, K., Vikman P-Å., Lundquist P-O. (1990) Physiology of symbiotic *Frankia* with emphasis on *Alnus* nodules. In: Endocytobiology IV. 4th Int. Coll. on Endocytobiology and Symbiosis, pp. 39-44, Nardon et al. eds. INRA, Paris,

Gresshoff, Roth, Stacey & Newton (eds.) Nitrogen Fixation: Achievements and Objectives
Chapman and Hall New York • London 1990

Nitrogenase, nitrate reductase and photosynthetic activities of black alder exposed to cold temperatures

Christoph S. Vogel* and Jeffrey O. Dawson

Department of Forestry, University of Illinois, 1301 W Gregory Dr., Urbana, IL 61801, USA.

Previous studies have shown that nodulated black alders (*Alnus glutinosa* [L.] Gaertn.) receiving NH_4NO_3 fertilization exhibited net decreases of up to 38% in foliar N concentration during autumnal senescence while non-nodulated alders receiving the same N fertilization showed no decrease in foliar N concentration (Côté *et al*. 1989). This range of values for alder is below the range for most woody temperate deciduous angiosperms. In the present study, exposure of NO_3^- fertilized black alder seedlings to 2.5 h of chilling temperatures of approximately 0° C decreased nodular acetylene reduction rates by an average of 46%. Partial to full recovery of acetylene reduction activity occurred within 7 days after chilling treatment. Nitrate reductase activities of leaves and fine roots, however, were not significantly reduced by chilling treatments, but instead tended to increase immediately after chilling followed by decreasing nitrate reductase activities as acetylene reduction activities recovered. Photosynthetic rates declined by 10-25% after cold treatment and did not recover within 7 days after treatment. These results suggest that as the energy demanding process of nitrogen fixation is inhibited by cold temperatures, photosynthate is available for other enzymatic processes not affected by cold temperatures, such as nitrate reduction. These results are also consistent with the idea that alder net foliar N resorption is greater in alders that derive the greatest proportion of their nitrogen from chilling sensitive symbiotic nitrogen fixation. It may be possible to approximate the proportion of nitrogen derived from the atmosphere (NDFA) by alder trees using the percentage of net autumnal N resorption and %NFDA determined via ^{15}N enrichment.

Côté, B., C.S. Vogel and J.O. Dawson. (1989) Autumnal changes in tissue nitrogen of autumn olive, black alder and eastern cottonwood. Plant Soil 118:23-32.

Gresshoff, Roth, Stacey & Newton (eds.) Nitrogen Fixation: Achievements and Objectives
Chapman and Hall New York • London 1990

Rhizobium population genetics: Diversity of plasmid profiles in *Sinorhizobium fredii* strains isolated from soybean cultivar Aijiaozao

Fusheng Wang, Weilin Li, Wenli Chen and Fudi Li

Dept. of Soils and Agrochemistry, Huazhong (Central China) Agricultural University, Wuhan, Hubei 430070, P. R. China.

Since the first report in 1982 that fast-growing soybean rhizobia were isolated from nodule samples collected in China(1), this type of rhizobia has attracted great interests of research (2, 3). Extensive investigations have been conducted by a number of laboratories in the world. However, there is no report available on its genetic diversity so far. This summary presents data obtained from part of our work on the population genetics. Sinorhizobium fredii strains were isolated from soybean (Glycine max L. Merr.) cultivar Aijiaozao grown in a soil planted with soybean for at least two decades without inoculation in central China. In order to avoid missing predominant strains and obtain much more information, we collected all nodules from each sampled plant instead of at random as people commonly did and took one rhizobial isolate from each nodule. Ninety-three fast-growing soybean rhizobium isolates were examined for plasmid content by a modified Eckhardt method of horizontal gel electrophoresis (4). One to five plasmids were reproducibly detected in the isolates. The sizes of plasmids ranged from approximately 30MD to 500MD. According to the size and number of plasmids, it was possible to classify the 93 isolates into 22 different groups. The dominant group contained 16% of the isolates, which had only one plasmid. This diversity of plasmid profiles demonstrates the existence of a great heterogeneity of plasmid pools within newly isolated fast-growing soybean rhizobia.

This work was supported in part by a grant from the International Foundation for Science and by a grant from the National Natural Science Foundation of China.

(1) Keyser,HH.et al.(1982) Science 215:1631-1632.
(2) Dowdle,SF & Bohlool,BB (1985) Appl.Environ. Microbiol. 44:960-965.
(3) McLoughlin,TJ.et al.(1986) Can.J.Microbiol.32:183-86.
(4) Harrison,SP.et al.(1988) J.Gen.Microbiol.134:2721-30.

Gresshoff, Roth, Stacey & Newton (eds.) Nitrogen Fixation: Achievements and Objectives
Chapman and Hall New York • London 1990

Growth at low carbohydrate concentrations, survival in soil and competition of *Rhizobium leguminosarum* bv. *phaseoli* strains

A.B. Wolff, W. Streit, J.A. Kipe-Nolt and D. Werner

International Center for Tropical Agriculture (CIAT) A.A. 6713, Cali Colombia. & Univ. of Marburg, Dept. of Biology, D-3550 Marburg, FRG.

To form nodules, inoculant strains of Rhizobium have to pass two steps of competition. First, survival in soil exposed to environmental factors with competition for nutrients. Second, competition for root hair penetration, nodule formation etc. During the first stage, the strains are exposed to many well described stress factors like high temperature, drought, extremes of pH, Al and Mn toxicity. Until seed germination and root rhizosphere development, carbohydrates are a limiting factor and concentrations are in the uM range.

To compare growth rates of bacteria at carbohydrate levels of 1 uM, batch cultures and fermentors are not very useful. Therefore, a dialysis culturing system was developed which allows the comparison of strains at very low concentrations of carbohydrate (Humbeck et al., Arch. Microbiol., 142,223-228 (1985). Growth rates obtained in this system were compared to other important stress factors, to the survival of strains in soil and to nodule occupancy under field conditions.

Differences between strains of R. phaseoli in growth on a range of sugars and amino acids, at low concentrations, were observed. In general, strains with low generation times under carbohydrate limitation showed high nodule occupancy in the two fertile (one temperate, one tropical) soils evaluated, but this relationship was apparently overridden by other stress factors in the acid infertile tropical soil.

No significant differences were found in the survival of five strains when incubated in the greenhouse in soils from the different sites. Survival of the strains was not related to either nodule occupancy or growth rates under carbohydrate limitation.

These results suggest that competitiveness of inoculant strains is more related to the interaction of plant and microsymbiont than to survival in soil during the first weeks after inoculation.

Gresshoff, Roth, Stacey & Newton (eds.) Nitrogen Fixation: Achievements and Objectives
Chapman and Hall New York • London 1990

Characteristics of plasmids in rhizobia isolated from *Leucaena leucocephala*

Mo Xi-mu, Huang Zhi-jian, Cheng Shuang-qi

Department of Biology, South China Normal University, Guangzhou 510631, People's Republic of China.

Rhizobium symbiotic with Leucaena leucocephala have been considered as fast growers, and the occurrence of plasmids has not yet been reported. We collected 7 strains of Rhizobium from nodules of L. leucocephala cultivated in different localities.

Using modified Hirsch's method to isolate large cccDNA, plasmids were detected by 7% gel electrophoresis, and the molecular weight was estimated in comparison to Rhizobium fredii strain USDA 191.

The presence of plasmids was demonstrated in the following rhizobial cell types. Strain TAL 1145 contains 3 plasmids; strains MXLL6 and CB81 contain only one plasmid. The molecular weight of these plasmids ranges from 95Md-195Md. In our experiment, Inc P1 group plasmid RP1::Tn501 can be transferred into strain TAL 1145 and from TAL1145 into Escherichia coli strain HB 101 at frequencies of 10^{-5} recipient cell. The drug resistance patterns of RP::Tn501, namely $Hh^RTc^RKm^RCb^R$, were fully expressed in TAL1145 and HB101. In order to address the question of whether these strains of Rhizobium bore the nod or nif genes, radioactively labelled pIJ1216 and pSA30 DNA were hybridized to plasmid DNA prepared from strains MXLL2 and MXLL7. The homologous sequences of nod and nif genes were found in the plasmids isolated from these two strains.

In our experiment, strains MXLL2, MXLL7 and TAL1145 were tested for plasmid curing by elevating culture temperature 37°C for 7 days. The detection of colonies and nod growth showed neither nod mutation or plasmid deletion had occurred. It is probable that these strains are high-temperature tolerant ones, so it is difficult to obtain HC mutants.

Gresshoff, Roth, Stacey & Newton (eds.) Nitrogen Fixation: Achievements and Objectives
Chapman and Hall New York • London 1990

Ontogenetic changes in natural ^{15}N abundance in the tissues of stem-nodulating legumes and estimates of N_2 fixation by these plants.

T. Yoneyama

National Agriculture Research Center, Tsukuba, Ibaraki, 305 Japan.

Two stem-nodulation (SN) legumes (*Sesbania rostrata* and *Aeschynomene indica*) were grown during summer seasons under upland and paddy field conditions without fertilizer nitrogen. The accumulation of N and variations of natural ^{15}N abundances ($\delta^{15}N$) in plants parts (1) were investigated. The non-nodulating soybean in the upland field and weeds in the paddy field were used as non-fixing reference plants.

Under both soil conditions, the two species nodulated on stem and/or roots. The $\delta^{15}N$ values of the leaves plus stems and fruits of non-nodulating plants were elevated, although the values in the paddy plants were much higher than upland plants. The $\delta^{15}N$ values of the SN plants were slightly negative (form -0.7 to-0.1%) under both soil conditions; $\delta^{15}N$ values of stem and root nodules were high (from +6 to +11%). ^{15}N was enriched in stem nodules in hydroponically grown *S. rostrata*, and in *A. indica*, *A. afraspera*, *A. schimpeni*, and *A. sensitiva* from Thailand, compared to leaves. The $\delta^{15}N$ values of the bacteroid fractions in all nodules from *S. rostrata* and *A. indica* were higher (+8 to +15%) than those of cytosol and cortical fractions. In *S. rostrata* and *A. indica*, amides are the major form of N transport, whereas nodulating soybeans were ureide-exporters. The site of ^{15}N enrichment in pea and broad bean, amide-exporters, was soluble N in the cytosol (2).

The percent N from N_2 fixation (%Ndfa) equals $(\delta^{15}Nn - \delta^{15}Nf)/(\delta^{15}Nn - \delta^{15}Na) \times 100$, where $\delta^{15}Nn$, $\delta^{15}Nf$, and $\delta^{15}Na$ are $\delta^{15}N$ values of non-fixing reference plants, fixing plants, and fixed N, respectively. The $\delta^{15}Na$ values for *S. rostrata* and *A. indica* were -1.0 and -0.7%, respectively. The %Ndfa in tops of *S. rostrata* and *A. indica* were 65; almost 100% in upland fields, and 78 and 93% in paddy fields, respectively. Fixed N in *S. rostrata* and *A. indica* at the final harvest was 15-28 and 8-13 g/m^2 respectively.

1. Yoneyama, T. (1987) Bull. Natl. Inst. Agrobiol. Resour. **3**, 59-87.
2. Yoneyama, T. (1988) J. Plant Physiol. **132**, 59-62.

Gresshoff, Roth, Stacey & Newton (eds.) Nitrogen Fixation: Achievements and Objectives
Chapman and Hall New York • London 1990

Lipopolysaccharide and protein patterns of rhizobia isolated from arid and salt-affected lands of Egypt and the Sudan

HAMDI H. ZAHRAN AND KRISTINA LINDSTRÖM

DEPARTMENT OF MICROBIOLOGY, UNIVERSITY OF HELSINKI, SF-00710 HELSINKI, FINLAND

PURPOSE

The effects of salt, osmotic and heat stress on the synthesis patterns of lipopolysaccharide (LPS) and protein of rhizobia were studied because it has been suggested that a complete LPS structure is essential to establish an effective rhizobium-legume symbiosis. The rhizobia studied here are ecologically important for nodulation of legumes grown in arid lands of Egypt and the Sudan.

METHODS

60 strains isolated from arid and salt-affected lands were screened. Rhizobia were grown in YEM supplemented with NaCl or sucrose at 28 °C or in YEM at 35 °C and 40 °C. Electrophoretic patterns of LPS and protein were analyzed using SDS-PAGE. LPS was detected in proteinase K-treated cell lysate and silver stained. Proteins from whole cell lysates were stained with coomassie brilliant blue.

RESULTS

Rhizobia screened in this study can be divided into three groups according to their salt tolerance: **1**- Sensitive strains which could not maintain complete LPS and protein structure at 1 % NaCl. **2**- Tolerant strains which grew at 3 % NaCl and clearly modified their LPS and protein patterns. **3**- Super tolerant strains which did not alter their LPS and protein patterns in a range of NaCl of 3-10 %. Generally, salt and heat stress altered LPS and protein patterns more than sucrose.

CONCLUSIONS

1- Salt tolerance of rhizobia was not related to any specific LPS type. **2**- The effect on proteins was more specific, since one 65, 000 da protein band was altered by different types of stresses. **3**- Alterations of synthesis patterns of LPS and protein might be a mechanism for adaptation to salt stress. **4**- Selection of effective, salt-tolerant strains is suggested to improve performance of Rhizobium-legume symbiosis under saline conditions.

Gresshoff, Roth, Stacey & Newton (eds.) Nitrogen Fixation: Achievements and Objectives
Chapman and Hall New York • London 1990

A study on the rhizobiophages of soybean

Wang Zifang, Xu Heng and Cen Yinghua

Wuhan Institute of Virology, Chinese Academy of Sciences, Wuhan 430071, P.R. China.

Rhizobiophages occur commonly in the rhizosphere of legumes, on which a volume of literature has been published. However, there are few studies on the rhizobiophage of fast-growing soybean rhizobia. This work explored the feasibility of using rhizobiophage as a vector in genetic studies of soybean rhizobia. We isolated a number of soybean rhizobiophages of both fast- and slow-growing hosts from soybean-grown soils in China and conducted comparative studies of seven phages, including morphology, host range, G+C mol%, patterns of both protein and RE (restriction endonuclease) on electrophoresis, and DNA homology, as well a generalized transduction of phages between different rhizobial strains. It was shown by electron microscopy that all phages isolated could be classified into four morphotypes different in REP. The molecular weight of the rhizobiophages ranged from 40.0 to 70.7 Kb. The G+C content of the DNAs varied between 41.5 and 72.1 mo 1% (Tm). All seven strains had narrow host ranges within the 14 rhizobial strains tested, but DNA-DNA hybridization between them showed that the seven phage genomes had some homology. The seven phages were divided into three groups by SDS-PAGE. Five out of the seven phage strains, i.e., psx, psc, pfm, pfc, and pfx, had the capacity of generalized transduction. Rhizobiophages psc and psx could transduce Tn_5_-induced auxotrophy and kanamycin resistance was carried on Tn_5_ in strains of _Bradyrhizobium japonicum_ USDA110. Transduction of Tn_5_-induced auxotrophy and kanamycin resistance in strains of _R. fredii_ MD_3 by bacteriophage pfm was achieved. We would transfer an antibiotic-resistance gene and _nod_ gene located on a plasmid from ANU_{240} to ANU_{265} by pfc and pfx. The frequency of transduction was 1×10^{-6} to 10^{-7} per plaque-forming unit.

SECTION E:

BACTERIAL SYSTEMS

Invited papers
Poster summaries

Gresshoff, Roth, Stacey & Newton (eds.) Nitrogen Fixation: Achievements and Objectives
Chapman and Hall New York • London 1990

Regulation of the C_4 -dicarboxylate transport in free-living and symbiotic *Rhizobium meliloti*

Alfred Pühler, Thomas Engelke, Doris Jording and Parveen K. Sharma.

Lehrstuhl für Genetik, Fakultät für Biologie, Universität Bielefeld, Postfach 8640, D-4800 Bielefeld 1, Federal Republic of Germany.

Nitrogen fixation in the Rhizobium-legume symbiosis is an energy demanding process. The plant supplies energy to the bacteroid and receives in return, fixed nitrogen. C_4-dicarboxylates are considered to be the compounds which are provided by the plant. These C_4-dicarboxylates have to pass two barriers, the peribacteroid and the bacteroid membrane. In recent years, the investigations were focussed on the transport process through the bacteroid membrane. This transport process is particularly well investigated in Rhizobium leguminosarum (4,5,6) and Rhizobium meliloti (1,7,8,9).

So far, three dct genes organized in one cluster, were analyzed in R.leguminosarum and R.meliloti. The entire region was sequenced in both organisms (2,3,6). It contains the structural gene dctA. This gene is believed to encode the C_4-dicarboxylate carrier, since the derived amino acid sequence shows several hydrophobic regions which are proposed to represent membrane passages (2). The cluster also contains the regulatory genes dctB and dctD which form a two-component regulatory system. It is thought that in the free-living state an extracellular signal (i.e. succinate) activates the sensor in the membrane (DctB). Subsequently, the intracellular mediator DctD is activated. The activated DctD protein together with the sigma factor NtrA may cause the transcription of the dctA promoter (6). A consensus sequence for NtrA protein interaction was found upstream of the dctA coding region (2,3,4). In addition, a consensus sequence, thought to interact with the NifA protein, was also identified in this region. It therefore, was suggested that the dctA promoter, which is activated in the free-living state by the two-component regulatory system DctB and DctD, may be activated in the symbiotic state by the regulatory NifA protein (2,4). We now report the analysis of the R.meliloti dct cluster by transposon mutagenesis and the investigation of the role of the nifA gene in the symbiotic expression of the C_4-dicarboxylate carrier.

After Tn5 mutagenesis, we isolated R.meliloti mutants carrying Tn5 insertions in the genes dctA, dctB, dctD and ntrA. The dctA and the ntrA mutants are unable to take up C_4-dicarboxylates in the free-living as well as in the symbiotic state. Consequently, they induce ineffective nodules. In contrast, dctB and dctD mutants, which also are transport-deficient in the free-living state, are able to transport C_4-dicarboxylates at a reduced rate in the symbiotic state. Nodules induced by dctB and dctD mutants show a reduced fixation capacity. These results can be interpreted as follows: The sigma factor NtrA is essential for dctA activation in both free-living and symbiotic states. The DctA protein is the carrier for C_4-dicarboxylates in both states. The two-component regulatory system consisting of DctB and DctD, is essential for C_4-dicarboxylate transport activity in the free-living state and is also necessary for optimal transport and nitrogen fixation rates in the symbiotic state.

In order to analyze the role of the regulatory nifA gene in the symbiotic expression of the C_4-dicarboxylate carrier, we carried out uptake experiments with bacteroids isolated from alfalfa nodules induced by nifA, by dctB and by nifA dctB R.meliloti mutants. The results showed that the transport rate was clearly reduced in nifA bacteroids, indicating that nifA plays a role in the symbiotic transport of C_4-dicarboxylates. In addition, we found that nifA dctB bacteroids transported C_4-dicarboxylates at approximately the same rate as bacteroids mutated only in dctB. From these results it can be concluded that the regulatory gene nifA is not responsible for the remaining C_4-dicarboxylate transport activity in dctB bacteroids. This also means that the NifA protein does not directly activate the dctA promoter.

1) Engelke, T., Jagadish, M.N. & Pühler, A. (1987) J. Gen. Microbiol. **133**, 3019-3029
2) Engelke, T., Jording, D., Kapp, D. & Pühler, A. (1989) J. Bacteriol. **171**, 5551-5560
3) Jiang, J., Gu, B., Albright L.M. & Nixon, B.T. (1989) J. Bacteriol. **171**, 5244-5253
4) Ronson, C.W. (1988) Nitrogen Fixation: Hundred Years After, eds. Bothe, de Bruijn & Newton (Gustav Fischer Verlag, Stuttgart New York), pp. 547-551
5) Ronson, C.W., Astwood, P.M. & Downie, J.A. (1984) J. Bacteriol. **160**, 903-909
6) Ronson, C.W., Astwood, P.M., Nixon, B.T. & Ausubel, F.M. (1987) Nucl. Acids Res. **15**, 7921-7938
7) Wang, Y.-P., Birkenhead, K., Boesten, B., Manian, S. & O'Gara, F (1988) Gene **85**, 135-144
8) Watson, R.J., Chan, Y.-C., Wheatcroft, R., Yang A.-F. & Han, S. (1988) J. Bacteriol. **170**, 927-934
9) Yarosh, O.K., Charles, T.C. & Finan, T.M. (1989) Mol. Microbiol. **3**, 813-823

Gresshoff, Roth, Stacey & Newton (eds.) Nitrogen Fixation: Achievements and Objectives
Chapman and Hall New York • London 1990

Transcriptional activation of nitrogen fixation genes in *Klebsiella pneumoniae*

Martin Buck

AFRC-IPSR Nitrogen Fixation Laboratory, The University of Sussex, Falmer, Brighton, Sussex, BN1 9RQ, U.K.

INTRODUCTION

The regulation of nif gene expression has been extensively studied in Klebsiella pneumoniae. It remains the model organism for studying many of the molecular mechanisms which underly the transcriptional control of nif in diazotrophs. The basic model for K. pneumoniae nif control is depicted in Figure 1. The rpoN product is a sigma factor (σ^{54}) which complexes with core RNA polymerase (E) to form an RNA polymerase holoenzyme ($E\sigma^{54}$) which recognises the -12, -24 downstream promoter element (DPE) of nif promoters. The interaction of $E\sigma^{54}$ with the DPE is itself non-productive with respect to the initiation of transcription, and activation of nif transcription requires the positive control proteins NtrC (which activates nifLA transcription) and NifA (which activates the other nif operons). The activity of NifA is controlled by NifL in response to nitrogen and oxygen status, that of NtrC by phosphorylation by NtrB in response to nitrogen status, ensuring expression of nif occurs only under nitrogen- limiting anaerobic or microaerobic conditions.

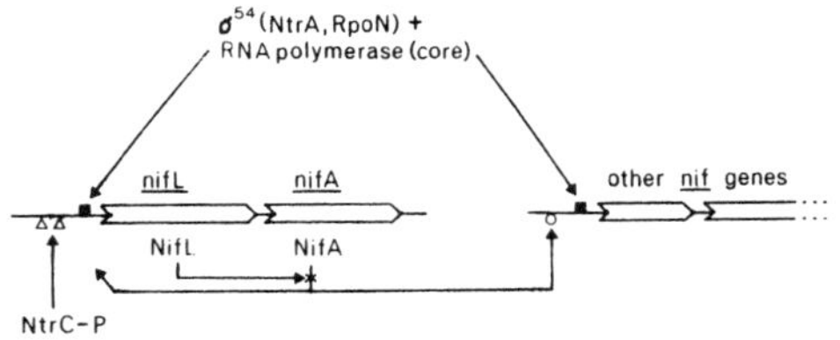

Figure 1. NtrC and NifA bind to sites upstream of the DPE in nif promoters. The phosphorylated form of NtrC activates transcription. NifL protein antagonises NifA- mediated transcription in the presence of fixed N or O_2.

Recent studies addressing the role of NtrC in transcriptional activation in *Escherichia coli* (1,2) and *Salmonella typhimurium* (3,4) are generally applicable to *K. pneumoniae*, but I will focus on recent findings pertinent to the mechanisms controlling *nif* transcription in *K. pneumoniae*.

The $E\sigma^{54}$ recognition sequence

The *nif* promoters of *K. pneumoniae* are amongst a class of promoters recognised by $E\sigma^{54}$ (reviewed in 5). The consensual recognition sequence is shown in Figure 2. The binding of $E\sigma^{54}$ to the DPE of the *glnA*p2 promoter has been demonstrated both *in vivo* and *in vitro* and is readily detected (5,6). In contrast, in the absence of activator the complex between $E\sigma^{54}$ and the DPE of the *nifL*(6), *nifH*, (7,8) and *nifU* (Cannon and Buck, unpublished) promoters is less readily detected, indicating that a relatively weaker closed promoter complex forms. Mutational analysis of the *nifH* promoter has shown that the sequence from -17 to -15 can be altered from CCC to TTT to create a tighter binding site for $E\sigma^{54}$ (8,9). The sequences from -17 to -15 may help to configure the -12, -24 sequences so as to allow $E\sigma^{54}$ binding to the DPE, and/or establish direct contacts with $E\sigma^{54}$ which is thought to occupy two major grooves that span the -12, -24 sequences (10). The products of two open reading frames linked to *rpoN* influence the activity of several σ^{54}-dependent promoters in accord with increasing the occupancy of the DPE by $E\sigma^{54}$ or increasing the activity of $E\sigma^{54}$ (11).

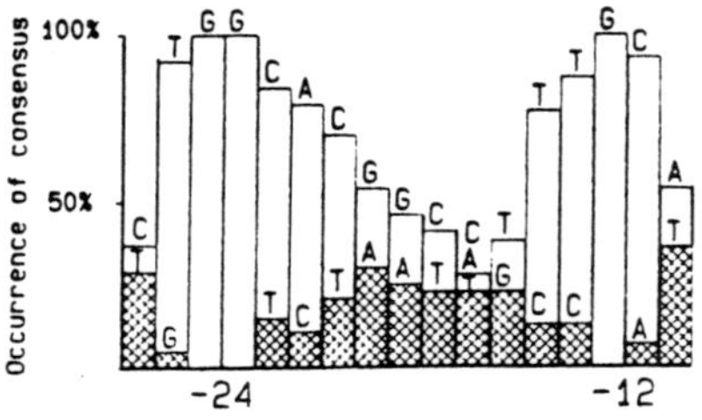

Figure 2. Consensus DPE sequence from sequences cited in (5).

Activator binding to nif promoters

Both NtrC and NifA are DNA binding proteins and a H-T-H motif necessary for DNA recognition lies in the C-terminus of each protein (12,13). In the *nifLA* promoter two binding sites for NtrC are present which are centred at -142 and -162 and to which NtrC binds cooperatively. The active, phosphorylated form of

NtrC binds to these sites more readily than the unphosphorylated form (14); at high concentrations it also induces hypermethylation of the guanine at position -23 (6). Thus some weak binding of NtrC-P in the region of the nifL DPE may occur.

Binding of NifA to the Upstream Activator Sequences (UAS's) of the nifH and nifJ promoters has been demonstrated *in vivo* (7,13), confirming that the TGT-N_{10}-ACA sequence is recognised by NifA, and that NifA (like NtrC) binds to sites upstream of the $E\sigma^{54}$ recognition sequence. DNase 1 protection studies have also demonstrated that the DNA-binding domain of NifA recognises the UAS of the nifH promoter *in vitro* (Buck and Cannon, unpublished). Results obtained with the nifU promoter indicate multiple NifA binding sites upstream of the DPE (see Figure 3). The most upstream binding site comprises two overlapping TGT-N_{10}-ACA sequences similar in arrangement to those present in the *Azospirillum* nifH promoter, the other appears to be UAS half-site at -90. The DPE-proximal UAS at -72 in the nifU promoter is not usually highly occupied by NifA, apparently because it overlaps a binding site for integration host factor (IHF) protein (15). In contrast the UAS at -73 in the nifE promoter is bound by NifA, indicating IHF and NifA do not compete for binding at this promoter (Charlton and Buck, unpublished).

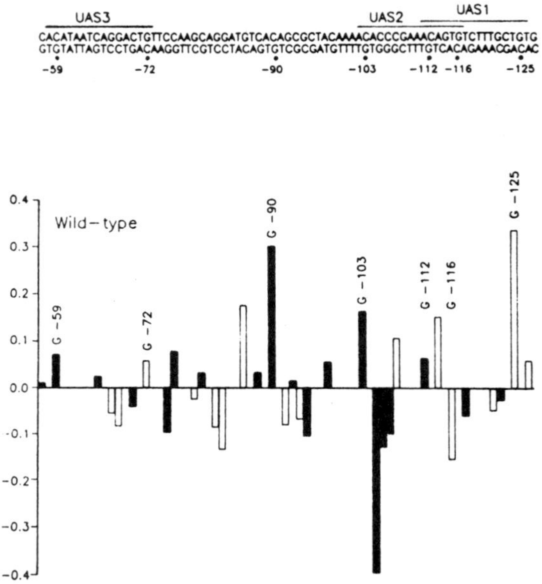

Figure 3. Changes in G reactivity to dimethysulphate in the presence of NifA show the UAS1,2 region and -90 region of the nifU promoter bind NifA but that UAS3 does not normally (15). Positive values- protection, negative values- hypermethylation.

The significance of multiple NifA binding sites in the nifU promoter is not clear, but could indicate that cooperative interactions between NifA dimers or higher order oligomers occur. The fractional occupancy of each site at a given NifA concentration may not be equivalent, and each site may be located relative to the DPE so that different transcriptional activation efficiencies at the nifU promoter result depending upon which sites are bound by NifA.

The interaction of NifA with the nifH promoter UAS was examined *in vivo* in the presence and absence of NifL (16). Under derepressing growth conditions the interaction of NifA with the UAS was found to be unchanged by NifL. However, under repressing (+N) growth conditions the interaction of NifA with the UAS was diminished in the presence of NifL. Thus NifL-mediated repression of nif transcription appears to result from reduced occupancy of nif promoters by NifA. The repressive effect of NifL was also detected in an *in vitro* assay (17). Partially purified preparations of NifA were competent in transcriptional activation assays of the nifH promoter *in vitro*, but not if NifA was isolated from cells which had also synthesised NifL.

Immunochemical analysis indicates NifL and NifA can form a complex, consistent with their stoichiometric production (18). Thus it is possible that NifL acts upon NifA directly rather than modifying it, for example, by phosphorylation. Analysis of isolated domains of NifA and NtrC has shown that these proteins clearly differ at the mechanistic level in their response to their partner proteins NtrB and NifL (19). The amino terminal domain of NifA is dispensable for activity, whereas this domain of NtrC must be phosphorylated by NtrB for NtrC activity. Furthermore, NifA lacking its amino terminal domain remains sensitive to the repressive effect of NifL (19). The PII protein appears not to be involved in nif-specific responses to changes in nitrogen status mediated by NifL (20).

Action at a distance and transcriptional activation

The binding sites for NtrC in the nifLA promoter and for NifA in the nifH and other nif promoters are usually located >100bp upstream of the transcription start site. In the case of the nifL and nifH promoters placing the activator binding site(s) out of phase with the DPE by introducing half-helical DNA turns (+5, +15 bp) hinders activation, consistent with a DNA looping requirement to bring the activator and $E\sigma^{54}$ within close proximity of each other (6,21). At certain nif promoters (nifH, B, U and E), binding of a factor, recently identified as IHF, adjacent to the DPE occurs (22,23,15) which may help to stabilise loop formation via the DNA bending activity of IHF (22,24). At the nifH promoter, mutations which allow a closed $E\sigma^{54}$ promoter complex to form in the absence of activator relax the requirement for upstream bound NifA and therefore DNA loop formation and IHF (Figure 4, ref. 8,9, and Buck and Cannon, unpublished data). Thus one role of NifA (and NtrC) might be to stabilise a normally weak DPE-$E\sigma^{54}$ closed promoter complex by virtue of a high local NifA (or NtrC) concentration afforded by DNA loop formation (6,7). Alternatively the high concentration of topologically constrained UAS-bound NifA might ensure that an increased fraction of $E\sigma^{54}$-DPE interactions were productive, thus leading to transcription. Changes in the reactivity of nifH promoter DNA to $KMnO_4$ and dimethylsulphate have shown that NifA clearly catalyses the isomerisation of a closed $E\sigma^{54}$ promoter complex to that of the open, transcriptionally active promoter complex (8). These open complexes appear as stably bound $E\sigma^{54}$ molecules at the DPE. They appear to contact the guanine residues at -24 and -12 which are important for closed promoter complex formation but at the same time contain denatured DNA sequences downstream of the DPE. The presence of these locally denatured or "melted" regions is absolutely dependent upon activator, and the formation of open promoter complexes is the catalytic function in transcriptional activation fulfilled by NifA and NtrC. Their formation is likely to be ATP-dependent given the presence of a putative ATP binding motif in the central or catalytic domains of activators functioning with $E\sigma^{54}$ and the demonstrated ATP requirement for NtrC-dependent open complex formation at the S. typhimurium glnAp2 promoter (3).

Activation of transcription of the nifL promoter is sensitive to the DNA topology of the promoter and is highly dependent upon negative supercoiling of the DNA (6,25). Since the interaction of NtrC-P with the upstream binding sites of the promoter is not dependent upon negative DNA supercoiling, the influence of DNA topology upon the function of the nifL promoter is likely to involve some aspect of the $E\sigma^{54}$ interaction at the DPE, whether it be the initial recognition or the formation of the open complex.

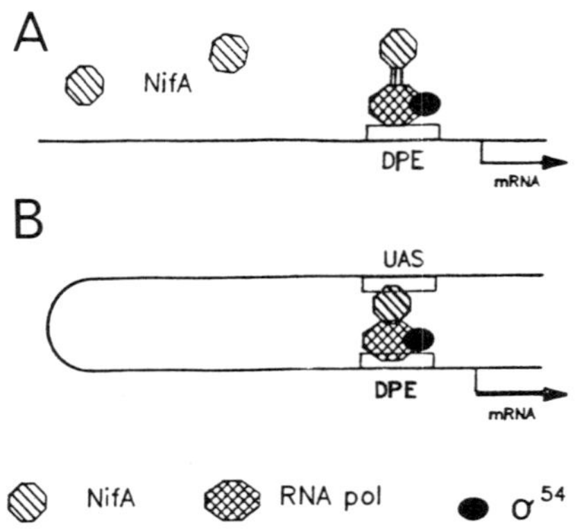

Figure 4. Model for activation of the nifH promoter by NifA. (A) Formation of a stable closed DPE-Eσ^{54} complex at mutant nifH promoter allows transcriptional activation in the absence of a NifA-UAS interaction (8,9). (B) DNA loop formation allows UAS-bound NifA to efficiently activate transcription by increasing its concentration and constraining it within the region of Eσ^{54} which in the wild-type nifH promoter forms a weak closed complex (8). UAS-bound NifA may therefore stabilise closed complex formation prior to open complex formation.

Acknowledgements
I wish to thank Ray Dixon and Mike Merrick for their valuable comments on the manuscript, Sydney Kustu for communicating results of the IHF binding prior to publication and the OECD for financial support to attend the Congress.

References

1. Ninfa, A.J., Brodsky, E and Magasanik, B. (1989) in DNA-Protein Interactions in Transcription. J. Gralla (ed.) UCLA-Symposium (Alan, R. Liss, New York) pp.43-52.
2. Weiss, V. and Magasanik, B. (1988) Proc. Natl. Acad. Sci. USA, 85, 8919-8923.
3. Popham, D.L., Szeto, D., Keener, J. and Kustu, S. (1989) Science 243, 629-635.

4. Keener, J. and Kustu, S. (1988) Proc. Natl. Acad. Sci. USA, 85, 4976-4980.
5. Kustu, S., Santero, E., Keener, J., Popham, D. and Weiss, D. (1989) Microbiol. Rev. 53, 367-376.
6. Minchin, S.D., Austin, S. and Dixon, R.A. (1989) EMBO J., 8, 3491-3499.
7. Morett, E. and Buck, M. (1988) Proc. Natl. Acad. Sci. USA, 85, 9401-9405.
8. Morett, E. and Buck, M. (1989) J. Mol. Biol., 210, 65-77.
9. Buck, M. and Cannon, W. (1989) Nucl. Acids. Res., 17, 2597-2612.
10. Sasse-Dwight, S. and Gralla, J.D. (1988) Proc. Natl. Acad. Sci. USA, 85, 8934-8938.
11. Merrick, M.J. and Coppard, J. (1989) Mol. Microbiol., 3, 1765-1775.
12. Contreras, A. and Drummond, M. (1988) Nucl. Acids. Res., 16, 4024-4039.
13. Morett, E., Cannon, W. and Buck, M. (1988) Nucl. Acids. Res., 16, 11469-11488.
14. Minchin, S.D., Austin, S. and Dixon, R.A. (1988) Mol. Microbiol., 2, 433-442.
15. Cannon, W., Kreutzer, R., Kent, H., Morett, E. and Buck, M. (1990) Nucl. Acids. Res., in press.
16. Morett, E., Kreutzer, R., Cannon, W. and Buck, M. (1990) Mol. Microbiol. in press.
17. Austin, S., Henderson, N. and Dixon, R. (1990) Eur. J. Biochem., 187, 353-360.
18. Henderson, N., Austin, S. and Dixon, R. (1989) Mol. Gen. Genet., 216, 484-491.
19. Drummond, M., Contreras, A. and Mitchenall, L.A. (1990) Mol. Microbiol., 4, 29-37.
20. Holtel, A. and Merrick, M.J. (1989) Mol. Gen. Genet., 217, 474-480.
21. Buck, M., Cannon, W. and Woodcock, J. (1987) Mol. Microbiol., 1, 243-249.
22. Santero, E., Hoover, T. and Kustu, S. (1989) Proc. Natl. Acad. Sci. USA, 86, 7346-7350.
23. Beynon, J., Cannon, M., Buchanan-Wollaston, V. and Cannon, F. (1983) Cell, 34, 665-671.
24. Robertson, C.A. and Nash, H.A. (1988) J. Biol. Chem., 263, 3554-3557.
25. Dixon, R.A., Henderson, N.C. and Austin, S. (1988) Nucl. Acids. Res., 21, 9933-9946.

Gresshoff, Roth, Stacey & Newton (eds.) Nitrogen Fixation: Achievements and Objectives
Chapman and Hall New York • London 1990

Mechanism of transcription from *nif* promoters: involvement of IHF

E. Santero, T. Hoover & S. Kustu

Departments of Plant Pathology and Molecular and Cellular Biology.
University of California, Berkeley, CA 94720, USA.

In a wide variety of gram-negative bacteria belonging to the phylum "Purple bacteria", transcription of nitrogen fixation (*nif*) genes depends on an alternative holoenzyme form of RNA polymerase carrying σ54 as the sigma factor. Sigma 54 confers on core RNA polymerase a unique promoter recognition specificity that consists of an essential GG-N10-GC minimal consensus sequence located 11 to 14 base pairs upstream of transcriptional startsites. This consensus sequence constitutes a binding site for σ54-holoenzyme. In addition of *nif* genes and genes whose products function in nitrogen metabolism, σ54-holoenzyme participates in transcription of genes whose products have diverse physiological roles (reviewed in 5). In all cases transcription appears to occur by a common mechanism: a specific activator protein catalyzes isomerization of closed recognition complexes between σ54-holoenzyme and a promoter to transcriptionally productive open complexes in which the DNA strands are locally denatured around the transcription startsite and the correct strand

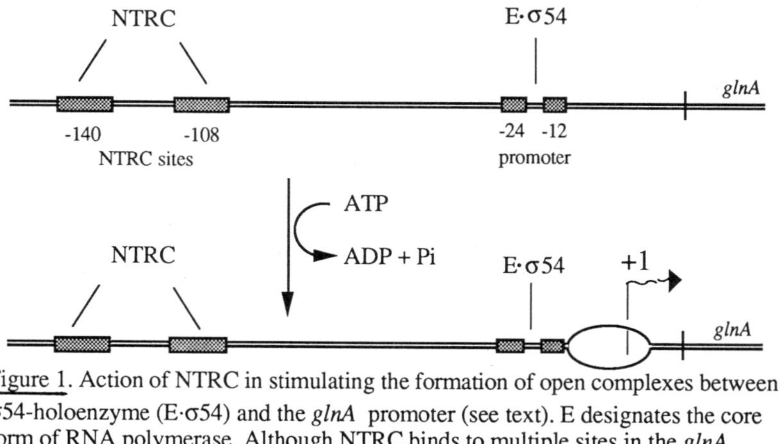

Figure 1. Action of NTRC in stimulating the formation of open complexes between σ54-holoenzyme (E·σ54) and the *glnA* promoter (see text). E designates the core form of RNA polymerase. Although NTRC binds to multiple sites in the *glnA* promoter-regulatory region, the important ones are the enhancer-like sites that are centered at -140 and -108 with respect to the transcriptional start site (10). NTRC must be phosphorylated to stimulate open complex formation (not shown).

can serve as a template for messenger RNA synthesis. The isomerization reaction appears to depend upon hydrolysis of ATP (5, 9)(Fig. 1). NIFA, the specific activator for *nif* genes, and NTRC, the activator for genes whose products participate in general nitrogen metabolism, are examples of activators that catalyze the isomerization reaction.

The best studied activator for σ54-holoenzyme is NTRC. Its function has been studied in detail at the major promoter for the *glnA* gene of enteric bacteria, which encodes glutamine synthetase. NTRC binds to DNA sites in the *glnA* promoter-regulatory region that are located more than 100bp upstream of the transcriptional startsite. These NTRC binding sites have the properties of eukaryotic transcriptional enhancers -- they work efficiently at large distances from the promoter, both downstream and upstream of it, and in either orientation . The binding sites function to tether NTRC to the *glnA* promoter, thereby increasing the local concentration of the activator in the region where it works. The available evidence indicates that NTRC contacts closed complexes between σ54-holoenzyme and the *glnA* promoter via a DNA loop (11,13). Loops formed by contact between these two proteins have been visualized in the electron microscope (14). To activate transcription, but not to bind to DNA, NTRC must be phosphorylated. Phosphorylation is catalyzed by NTRB and is increased under nitrogen-limiting conditions in vivo.

NIFA, the activator for *nif* operons, is structurally and functionally related to NTRC but is not as well characterized biochemically since it has not been purified in active state. It has been demonstrated in vivo that the TGT-N10-ACA motif found upstream of *nif* promoters constitutes a binding site for NIFA (6) and that NIFA is required for the isomerization of closed complexes between σ54-holoenzyme and *nif* promoters to open complexes (7). Thus, the mechanism of initiation of transcription at *nif* promoters appears to be similar to that at *glnA*. However, expression from *nif* promoters involves an additional factor. Beynon et al (1) reported that a protein factor in crude cell extracts of *K. pneumoniae* bound to several *nif* promoter-regulatory regions. This factor binds between the upstream binding site(s) for NIFA and the promoter and we have identified it as the integration host factor "IHF" (12). IHF from *E. coli* is a small heterodimeric protein that binds to DNA in a sequence-specific manner and induces a sharp bend at its binding sites. IHF is involved in several cellular processes that include integration of phage lambda into the bacterial chromosome, plasmid replication, transposition, and regulation of gene expression (reviewed in 3).

We report here that we have obtained NIFA-mediated activation of transcription from the *nifH* promoter of *K. pneumoniae* in vitro and that IHF dramatically stimulates this activation. We propose a model for initiation of transcription at the *nifH* promoter in which an IHF-induced bend brings NIFA and σ54-holoenzyme close together, thereby facilitating productive interactions between them. We also show that IHF binds to *nif* promoter- regulatory regions from a variety of free-living and symbiotic diazotrophs, from which we infer that the involvement of IHF in transcription of *nif* genes is general.

In vitro activity of NIFA present in crude cell extracts: IHF effect on transcription from the *nifH* promoter. When NIFA is overproduced, most of it aggregates and is inactive, as has been reported by others. Upon very high levels of overproduction of NIFA (20% of total cell protein), a small fraction remains soluble after high speed centrifugation of cell extracts and is active in a coupled transcription-translation system (Table 1). NIFA-mediated activation of expression from the *K. pneumoniae nifH* promoter depends on the alternative sigma factor σ54 and is very heat labile (Table 1), two characteristics of NIFA-mediated activation in vivo. Although NTRC has no specific binding sites in the *nifH* promoter-regulatory region, phosphorylated NTRC (NTRC-P) activates transcription from the *nifH* promoter weakly. This activation is also dependent on σ54 but is not heat-labile. Activation of transcription from the *glnA* promoter by NTRC-P is similarly σ54-dependent and is not sensitive to temperature.

Activated gene fusion	Activator	ß-Galactosidase activity nmol. ONP/min. 50 µl reaction 30°C -σ54	30°C +σ54	37°C +σ54
nifH-lacZ	None	1.4	.8	1.1
nifH-lacZ	NIFA	61.0*	605.0	2.6
nifH-lacZ	NTRC-P	1.8	15.5	21.3
glnA-lacZ	NTRC-P	24.3**	659.0	791.0

Table 1. Effects of σ54 and temperature on activation of transcription from the *nifH* promoter of *K. pneumoniae*. Transcription was assessed by measuring the amount of β-galactosidase produced from a *nifH-lacZ* fusion (or in control experiments a *glnA-lacZ* fusion) in a coupled transcription-translation system prepared from an *E. coli* strain that lacks σ54. When present, σ54 was added to 110nM. NIFA (2µl) was provided from crude cell extracts of a strain that overproduced it (4). NTRB (47nM) was provided together with NTRC (154 nM) to phosphorylate it. Plasmid templates were at 15nM. Coupled reactions were run at 30°C or 37°C as indicated.

* The low activity is due to contaminating σ54 in the NIFA crude extract.

** The low activity is due to expression from a weak σ70 dependent promoter.

To allow investigation of the role of IHF in transcription from the *nifH* promoter, both the coupled transcription-translation system and crude cell extracts containing NIFA were prepared from *E. coli* strains that lacked IHF. As shown in Fig. 2A, NIFA had some ability to activate expression from the *nifH* promoter even in the absence of IHF (Y intercept at zero IHF concentration). However, purified IHF stimulated NIFA-mediated activation by up to 20-fold. By contrast, IHF slightly inhibited, weak NTRC-mediated activation of transcription from the *nifH* promoter (Fig. 2B; note change in scale). NTRC-mediated activation may depend upon binding of NTRC to secondary or non-specific sites in the *nifH* promoter-regulatory region, a view we favour (9), or may occur directly from solution. The specificity of the IHF-mediated activation could be accounted for if IHF induced a bend that brought the NIFA binding site close to the promoter.

IHF-induced bending of the *nifH* promoter region. The large mobility shift caused by binding of IHF to a *nifH* promoter fragment from *K. pneumoniae* and the large regions protected by IHF from DNase I (12) or hydroxyl radical cleavage (4) corresponded to those at other IHF-binding sites and were diagnostic of an IHF-induced bend (16). To obtain direct evidence for such bends we visualized them in the electron microscope. Measurements from the apex of the bend to the ends of two different *nifH* promoter fragments indicated that the location of the IHF-induced bend was coincident with the position of the binding site for IHF as assessed in protection assays (4).

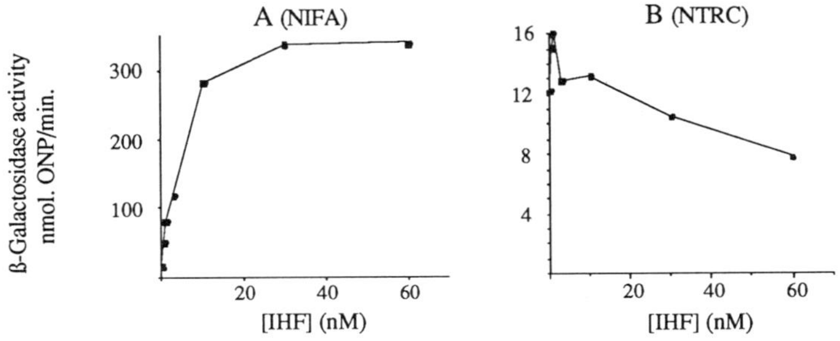

Figure 2. Effect of different concentrations of IHF on NIFA- or NTRC-mediated activation of transcription of a *nifH-lacZ* fusion. NIFA (2μl crude cell extract) was added to the coupled transcription-translation system for panel A and NTRC (154 nM) plus NTRB (47 nM) was added for panel B. Sigma 54 and plasmid template (pVSA2) were present for both panels at 110 nM and 15 nM, respectively.

Effect of promoter strength on IHF stimulation. We compared the effect of IHF on NIFA-mediated activation of transcription from the wild type *nifH* promoter to that from two stronger mutant promoters, which were present on plasmids p318 and p319 (8)(Fig. 3). The mutant promoters carry one or two mutations, respectively, that improve binding of σ54-holoenzyme in closed complexes both in vivo (2,7) and in vitro (unpublished). In the absence of IHF, levels of NIFA-mediated activation increased with promoter strength [Fig. 3A, B and C (circles)]. Whereas IHF markedly stimulated expression from the wild-type promoter (Fig. 2A and Fig. 3A), it had succesively less stimulatory effect on expression from the stronger mutant promoters (Figs. 3B and C). The maximal level of expression from each promoter in the presence of IHF was approximately the same (Figs. 3A, B and C (squares)].

We interpret decreased stimulation by IHF at mutant promoters as follows. When σ54-holoenzyme shows increased occupancy of a promoter in closed complexes, as it does at stronger mutant promoters, NIFA has a greater probability of contacting it by means of transient, and presumably random, conformational changes in the DNA. Consequently, the IHF-induced bend, which stabilizes a productive DNA conformation, has less of a stimulatory effect at stronger mutant promoters. An alternative explanation for the results is that IHF (or the IHF-induced bend) directly stimulates binding of σ54-holoenzyme in closed complexes and is less stimulatory at stronger mutant promoters because polymerase binding has been increased by changes in promoter sequence. There are several lines of evidence against the latter explanation.

If IHF stimulated binding of polymerase in closed complexes, it should stimulate NTRC-mediated activation from the *nifH* promoter as well as NIFA-mediated activation. However this was not the case (Fig. 2B and Figs. 3A, B, and C). Although activation of transcription from the *nifH* promoter by NTRC alone increased with promoter strength as expected [Figs. 3A, B and C (circles)], IHF did not stimulate activation at any promoter but rather was slightly inhibitory [Fig. 2B and Figs. 3A, B and C (squares)]. Moreover, we could not detect IHF-mediated stimulation of the binding of σ54-holoenzyme at either the wild-type *nifH* promoter or the strongest mutant promoter (p319) in direct DNase I protection assays (unpublished).

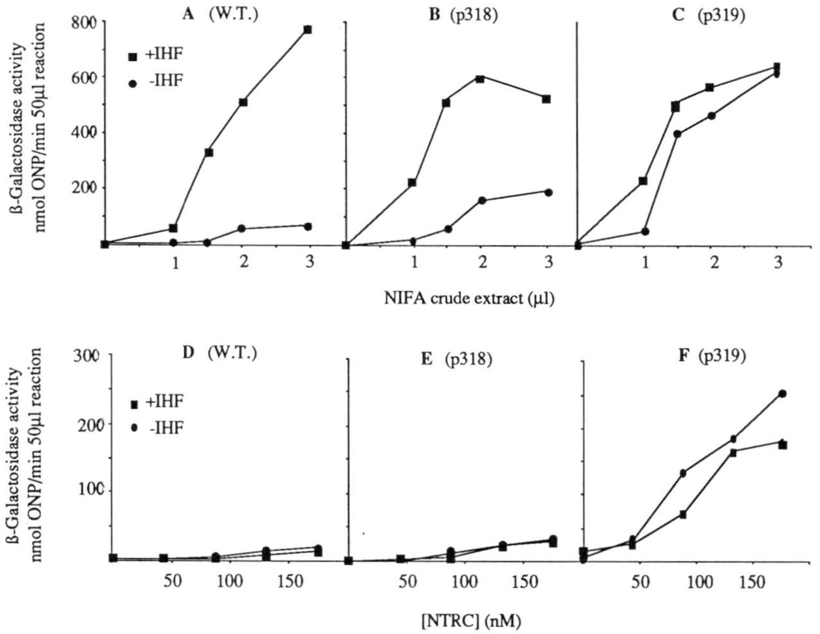

Figure 3. Effect of IHF on activation of transcription from wild-type and mutant *nifH* promoters by NIFA (panels A-C) or NTRC (panels D-F). DNA templates (15 nM) were pVSA2 for panels A and D, p318 (8) for panels B and E and p319 (8) for panels C and F. When present, IHF was at 60nM. Other conditions were as in the legend to Fig. 2.

Binding of IHF to many *nif* promoter regions. We have found that IHF binds to a large number of *nif* promoter-regulatory regions in addition to the *nifH* region of *K. pneumoniae*. It binds to the *K. pneumoniae nifJ*, *nifE*, *nifU*, and *nifB* promoter regions (but not to the *nifF*-*nifL* intergenic region) and to the *nifH* promoter regions from *A. vinelandii*, *R. meliloti*, *B. japonicum*, *R. capsulatus*, *R. rubrum*, and *T. ferrooxidans* (Fig. 4). (For organisms other than *K. pneumoniae* the *nifH* promoter region was the only one we checked). [IHF failed to bind to the *glnA* promoter region from *S. typhimurium* (Fig. 4).] In all cases the binding site for IHF was located between an upstream binding site(s) for NIFA and the promoter, as it was in the *K. pneumoniae nifH* promoter-regulatory region (4). Patterns of protection from DNAse I and hydroxyl radical cleavage were similar to those at other sites and thus indicated that IHF bent the *nif* DNA's. Inspection of sequences in the promoter regions for a large number of additional *nif* and *fix* genes transcribed by σ54-holoenzyme revealed potential IHF-binding sites between upstream sites for NIFA and the promoter in many of them (4).

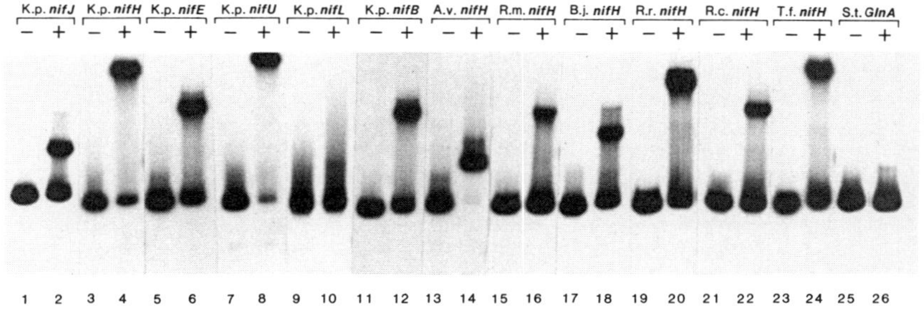

Figure 4. Mobility shifts produced by binding of IHF to DNA fragments carrying different *nif* promoter-regulatory regions. DNA fragments in the absence of IHF (free DNA) have been aligned. Kp: *Klebsiella pneumoniae*. Av: *Azotobacter vinelandii*. Rc: *Rhodobacter capsulatus*. Rr: *Rhodospirillum rubrum*. Rm: *Rhizobium meliloti*. Bj: *Bradyrhizobium japonicum*. Tf: *Thiobacillus ferrooxidans*. St: *Salmonella typhimurium*.

We infer that the stimulatory effect of IHF on expression from the *nifH* promoter of *K. pneumoniae* is not an isolated case. Rather, it appears that the stimulatory effect of IHF on NIFA-mediated activation of transcription is likely to be general for *nif* promoters. We also infer that IHF is not restricted to enteric bacteria but is probably widely spread among gram-negative bacteria in the large phylum "Purple bacteria and their relatives" (15).

A model for initiation of transcription at *nif* promoters: rationalization of the participation of IHF. We propose a model for initiation of transcription at *nif* promoters in which IHF stabilizes a DNA conformation that allows frequent contact between NIFA and σ54-holoenzyme bound to their respective sites (Fig. 5). The IHF-induced bend would thereby stimulate the formation of open complexes at the *nifH* promoter. At the *glnA* promoter (Fig. 5) NTRC bound to its upstream sites contacts polymerase at the promoter by means of random and transient looping of the intervening DNA (11, 13).

Why might the bacterial cell employ IHF at *nif* promoters and not at the *glnA* promoter? We have evidence for one possible reason and have considered several others that we have not yet tested experimentally. If the *nifH* promoter were "strong" like the *glnA* promoter, NIFA-mediated activation would not depend upon IHF (Fig. 3C). However, the cell employs a "weak" *nifH* promoter, for which high levels of NIFA-mediated activation do depend on IHF (Fig. 3A and Fig. 2A). A disadvantage to use of a strong *nifH* promoter is that it decreases the fidelity of activation by NIFA, the physiological activator. NTRC, a non-physiological activator, can activate transcription from a strong *nifH* promoter even though it lacks specific binding sites in the *nifH* promoter-regulatory region (Fig. 3F). This activation occurs in vivo as well (8). It would presumably be detrimental because NTRC would activate under nitrogen-limiting conditions even when molecular oxygen was present, conditions under which products of many *nif* genes would be rapidly inactivated because they are extremely oxygen labile. At the "weak" wild-type *nifH* promoter, activation by NTRC is very poor (Fig. 3D and Fig. 2B). Activation by NIFA, the physiological activator whose function is controlled by molecular oxygen, is also relatively poor at the weak wild-type promoter. However, IHF selectively stimulates NIFA-mediated activation due to the location of the upstream binding site for NIFA with respect to the IHF-induced bend. Use of the weak promoter in conjunction with IHF thereby allows high fidelity of activation by NIFA without a sacrifice in efficiency.

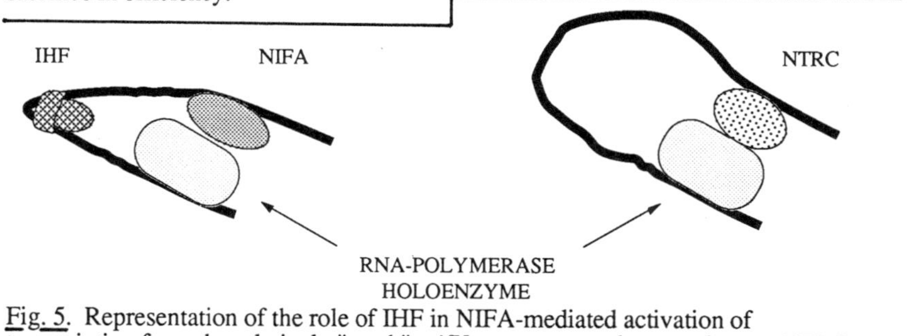

Fig. 5. Representation of the role of IHF in NIFA-mediated activation of transcription from the relatively "weak" *nifH* promoter and comparison to NTRC-mediated activation from the "strong" *glnA* promoter.

We have considered several other possibilities for the role of IHF at *nif* promoters 1) If the amounts of IHF were controlled (e.g. increased in nodules or in free-living diazotrophos under nitrogen-fixing conditions) this would provide yet another means of regulating *nif* expression. 2) In the model of Fig. 5 the position of the IHF-induced bend relative to the binding sites for NIFA and σ54-holoenzyme is important for maximal efficiency of open complex formation. In vivo, levels of expression from different *nif* promoters are known to be very different. Changes in placement of the IHF-induced bend in different *nif* promoter-regulatory regions might account for some of this difference in expression. 3) The seven or eight *nif* operons of *K. pneumoniae* are clustered in a region of ~ 20kilobases. The *nifH* and *nifJ* promoter-regulatory regions are adjacent and these operons are transcribed divergently . Each of the *nif* operons has its own σ54-dependent promoter and, with the exception of the *nifLA* operon, its own binding site(s) for NIFA. Another possible role for IHF would be to direct activation of a particular *nif* promoter by NIFA bound to a particular site or, conversely, to prevent efficient activation of this promoter by NIFA bound to other nearby sites. Such a role might be important in regions such as the *nifH-nifJ* intergenic region where NIFA binding sites are located close together and levels of expression of the two operons are very different.

Acknowledgements. We thank Dennis Dean, Jeffrey Gardner, Arthur Landy, Howard Nash, David Ow and Gary Roberts for providing materials and thoughtful criticisms. This work was supported by Life Sciences Research Foundation fellowship from the Burroughs Wellcome fund to TH, EMBO fellowship ALTF266-1987 to ES, and National Science Foundation Grant 8714761 to SK.

REFERENCES

1) Beynon, J., Cannon, M., Buchanan-Wollaston, V. and Cannon, F. (1983). Cell **34**, 665-671.
2) Buck, M. and Cannon, W. (1989). Nucl. Acids Res. **17**, 2597-2612.
3) Friedman, D. I. (1988). Cell **55**, 545-554.
4) Hoover, T. Santero, E., Porter, S. and Kustu, S. (1990). Cell (in press)
5) Kustu, S., Santero, E., Keener, J., Popham, D., and Weiss, D. (1989). Microbiological Reviews **53**, 367-376.
6) Morett, E. and Buck, M. (1988). Proc. Natl. Acad. Sci. USA **85**, 9401-9405.
7) Morett, E. and Buck, M. (1989). J. Mol. Biol. **210**, 65-77.
8) Ow, D., Xiong, Y., Gu, Q. and Shen, S. (1985). J. Bacteriol. **161**, 868-874.
9) Popham, D., Szeto, D., Keener, J. and Kustu, S. (1989). Science **243**, 629-635.
10) Reitzer, L.J. and Magasanik, B. (1986). Cell **45**, 785-792.
11) Reitzer, L.J., Movsas, B. and Magasanik, B. (1989). J. Bacteriol. **171**, 5512-5522.
12) Santero, E., Hoover, T., Keener, J. and Kustu, S. (1989). Proc. Natl. Acad. Sci. USA **86**, 7346-7350.
13) Wedel, A., Weiss, D., Popham, D., Droge, P. and Kustu, S. (1990). Science **248,** 486-490
14) Wen Su, Porter, S., Kustu, S. & Echols H. (1990). Proc. Natl. Acad. Sci. USA (submitted)
15) Woese, C.R. (1987). Microbiological Reviews **51**, 221-227.
16) Yang, C.C. and Nash, H.A. (1989). Cell **57**, 869-880.

Gresshoff, Roth, Stacey & Newton (eds.) Nitrogen Fixation: Achievements and Objectives
Chapman and Hall New York • London 1990

Organization and regulation of nitrogen fixation genes in *Rhodobacter capsulatus*

W. Klipp

Lehrstuhl für Genetik, Universität Bielefeld,
Postfach 8640, D-4800 Bielefeld, FRG.

The analysis of nitrogen fixation genes (nif) was first carried out in the facultative anaerobe Klebsiella pneumoniae and this organism has become a paradigm. Within the last few years the application of DNA sequencing and molecular genetics in another free-living nitrogen fixing organism, namely the aerobic soil bacterium Azotobacter vinelandii, has allowed comparative studies on nif genes. With the exception of nifJ and nifL, homologues to all K.pneumoniae nif genes are present in A.vinelandii. Although there are spatial differences, the A.vinelandii nif genes have the same sequential arrangement as the corresponding genes from K.pneumoniae (Fig.1). The open reading frames, which are interspersed among the "common" nif genes, have no counterpart in the K.pneumoniae nif gene cluster (13,15). In addition to the conventional molybdenum nitrogenase, A.vinelandii harbors two further genetically distinct nitrogenase systems (for review see ref.14,27): nitrogenase-2, a vanadium-containing enzyme, and nitrogenase-3, an enzyme that does not apppear to contain either Mo or V. The genetics of nitrogen fixation was also analyzed in great detail in a third diazotroph, the phototrophic purplebacterium Rhodobacter capsulatus. In this article I will summarize data on the organization of nif genes in R.capsulatus and will present evidence, that this organism, like Azotobacter, contains at least one alternative nitrogenase system.

1. R.capsulatus genes homologous to nif genes from K.pneumoniae

As summarized in Fig.1 and Tab.1 the current nif map of R.capsulatus includes 36 coding regions most of which have been sequenced. Out of these, 14 have their counterparts in K.pneumoniae. The molecular weights of the R.capsulatus nif gene products homologous to nif proteins from K.pneumoniae are comparable and the homology on the amino acid sequence level is distributed over the entire lengths of the polypeptides. An interesting exception to this is the R.capsulatus nifU gene product (Klipp et al., this volume). Only the C-terminal part of R.capsulatus NifU shows homology to the corresponding gene products of other organisms. The large N-terminal domain, which is highly conserved between NifU from K.pneumoniae (3,5), A.vinelandii (5,13) and Anabaena (25), is not present in the nifU gene product of R.capsulatus.

Homology search in a data bank revealed a second putative nifU gene located between R.capsulatus nifK and nifR4. Since insertion mutants in both copies of nifU and also the corresponding double mutant exhibit a Nif^{+} phenotype, it could be speculated, that these genes are only evolutionary remnants. However, the high degree of homology in the C-terminal part of both NifU copies contradicts this hypothesis.

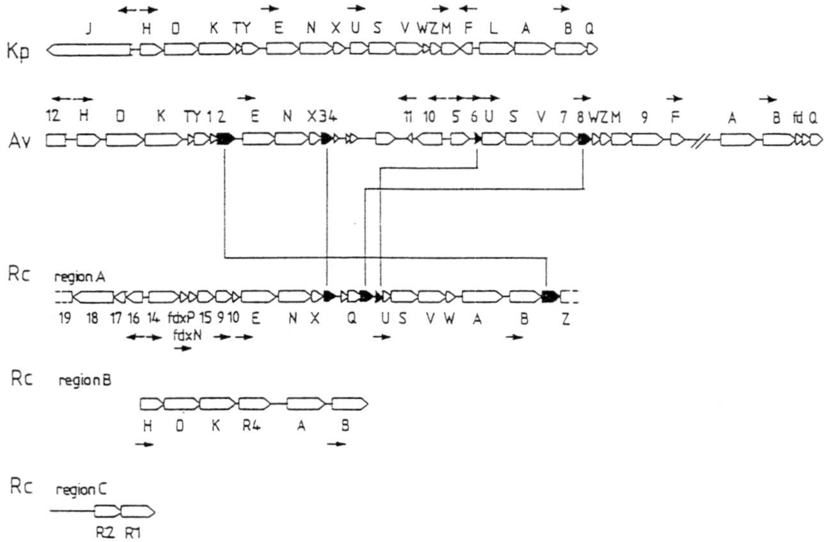

Fig. 1 Organization of nif genes in K.pneumoniae (Kp), A.vinelandii (Av) and R.capsulatus (Rc)

Open reading frames which are conserved between R.capsulatus and A.vinelandii are symbolized by filled arrows. Arrows above the maps of K.pneumoniae and A.vinelandii and below the maps of R.capsulatus mark the position and direction of nif specific promoters.

gene	MW(a)	Nif(b)	homologues in Kp	Av(d)	characteristics. function(c)	reference
ORF19	n.c.	-	-	-	?	(e)
ORF18	55.588	-	-	-	Fe-S-protein ?	(e)
ORF17	19.090	-	-	-	Fe-S-protein ?	(e)
ORF16	20.424	-	-	-	membrane protein ?	(e)
ORF14	41.247	+	-	-	?	(e)
fdxP	10.025	nt	-	-	ferredoxin. plant type ?	(30).(e)
fdxN	6.728	(+)	-	+	ferredoxin. bacterial type ?	(30).(31).(e)
ORF15	27.633	-	-	-	?	(e)
ORF9	21.787	-	-	-	?	(e)
ORF10	19.988	+	-	-	?	(e)
nifE	51.188	-	+	+	FeMoco synthesis (28)	(23)
nifN	49.459	-	+	+	FeMoco synthesis (28)	(23)
nifX	17.459	+	+	+	regulation (8)	(23)
ORF4	17.472	+	-	+	?	(23)
ORF5	10.821	+	-	-	ferredoxin. bacterial type ?	(24)
nifQ	19.524	Q	+	+	processing of Mo (12)	(24)
ORF8	19.891	+	-	+	?	(f)
ORF6	11.019	+	-	+	?	(f)
nifUI	33.274	2x+	C-ter	C-ter	?	(f)
nifS	40.289	-	+	+	?	(f)
nifV	40.624	(+)	+	+	synthesis of homocitrate (11)	(f)
nifW	12.550	+	+	+	?	(f)
nifAI	59.653	2x-	+	+	regulation. oxygen-sensitive	(22).(f)
nifBI	49.453	2x-	+	+	FeMoco synthesis (10).(28)	(22)
ORF1	27.721	+	-	+	periodic protein ?	(22)
nifZ	n.c.	+	+	+	?	(22)
nifH	32.412	-	+	+	nitrogenase reductase	(16).(32)
nifD	56.113	-	+	+	nitrogenase α-subunit	(32)
nifK	n.s.	-	+	+	nitrogenase β-subunit	(4)
nifUII	n.c.	2x+	C-ter	C-ter	?	(17).(f)
nifR4	39.900	-	(+)	(+)	regulation. NtrA-like	(2).(17)
nifAII	n.c.	2x-	+	+	regulation	(22)
nifBII	n.c.	2x-	+	+	FeMoco synthesis (10).(28)	(22)
nifR1	49.400	-	(+)	(+)	regulation. NtrC-like	(17)
nifR2	38.200	-	(+)	(+)	regulation. NtrB-like	(17)
nifR5	12.300	c	+	?	regulation. GlnB	(21)

Tab. 1 The nif genes of R.capsulatus

a) n.c. : not completely sequenced: n.s. : not sequenced
b) (+) : slow diazotrophic growth: Q : diazotrophic growth only with high Mo concentrations: 2x+ : single and double mutants Nif$^+$· 2x- : only double mutants Nif$^-$: c : constitutive expression of nif genes: n.t. : not tested
c) the same function as for the corresponding K.pneumoniae gene products is assumed: ? : exact function not known
d) C-ter : only the C-terminal part of R.capsulatus NifU is homologous: (+) : sequence homology but different regulatory function
e) Schmehl et al., this volume f) Klipp et al., this volume

The R.capsulatus nifA and nifB genes are also present in duplicate (18,22). In this case however, double mutants either in nifA or nifB resulted in a Nif$^-$ phenotype. The nifA/nifB duplication seems to be characteristic for R.capsulatus, since 12 independently isolated wildtype strains contain this reiteration.

2. R.capsulatus genes homologous to nif-associated open reading frames from A.vinelandii

As shown in Fig.1, four R.capsulatus coding regions are homologous to nif-associated open reading frames from A.vinelandii, which are not present in the nif gene cluster from K.pneumoniae. Homologues to A.vinelandii ORF3 (23) and ORF6 are located at identical positions in R.capsulatus, whereas homologues to A.vinelandii ORF2 (22) and ORF8 belong to different transcriptional units. The analysis of the corresponding R.capsulatus mutants revealed no obvious phenotype for these four coding regions. In addition, no transcription of ORF8 could be observed under nitrogenase derepressing conditions. This might be an indication that ORF8 is only an evolutionary relict. Another interpretation could be, that the gene products of ORF8 and the other nif-associated open reading frames are only necessary under not yet identified physiological conditions or are responsible only for a minor increase in nitrogenase activity/stability, which could not be detected under experimental conditions but is advantegous in nature.

3. R.capsulatus genes encoding ferredoxin-like proteins

Three open reading frames of R.capsulatus nif region A encode low molecular weight proteins containing cysteine motifs typical for ferredoxins:

fdxP codes for a protein, which resembles plant type [2Fe-2S] ferredoxins (30).

fdxN (30,31) shows a high degree of homology to nifB-associated genes of A.vinelandii (15), R.meliloti (19,26) and B.japonicum (7) and to a gene located downstream of nifH*, which encodes the vanadium nitrogenase reductase of A.chroococcum (29). The ferredoxin-like proteins of all fdxN genes are characterized by two different cysteine motifs. The first motif C-X_2-C-X_2-C-X_3-C is identical to the arrangement of cysteines found in typical bacterial [4Fe-4S] ferredoxins, e.g. a ferredoxin from Peptococcus aerogenes (1), whereas the arrangement of cysteine residues in the second cluster $C-X_2-C-X_8-C-X_3-C$ is different.

ORF5 (24) which is located immediately upstream of nifQ, contains two cysteine motifs, which match perfectly the consensus sequence of typical [4Fe-4S] ferredoxins. The large spacing between these two cysteine clusters of R.capsulatus ORF5 is also found in a ferredoxin from the archaebacterium Thermoplasma acidophilum (9).

The role of these ferredoxin-like proteins in the nitrogen fixation process is still unclear. R.capsulatus fdxN mutants and fdxN/ORF5 double mutants show only slow diazotrophic growth (Schmehl et al., this volume), whereas ORF5 mutants exhibit no obvious phenotype (24). A clear Nif^- phenotype which allows a detailed biochemical analysis of the possible function of nif-associated ferredoxin-like proteins, was only observed for R.meliloti fdxN mutants (19) and the analysis of this mutant is presented by Masepohl et al. (this volume).

4. R.capsulatus nif genes exhibiting no homology to known genes

Eight coding regions were identified in R.capsulatus nif region A, which exhibit no homology to known genes (Schmehl et al., this volume). Six of these coding regions were shown to be essential for nitrogen fixation, whereas mutants in ORF10 and ORF14 remained Nif^+. The deduced gene products of ORF17 and ORF18 (19kDal and 56kDal, respectively) contain cysteine motifs normally found in low molecular weight ferredoxins. Therefore, it could be speculated, that these putative iron sulfur proteins might play a role in electron transport to nitrogenase in R.capsulatus. In particular, ORF17 and ORF18 might encode oxidoreductases similar to K.pneumoniae NifJ, a pyruvate-flavodoxin-oxidoreductase, which also contains two cysteine motifs of this form (3,6). However, the substrate used as an electron donor in R.capsulatus is yet unknown. In addition, it is still an open question if there might be a correlation between the photosynthetic electron flux and the electron transport to nitrogenase. An indication for this hypothesis is the structure of the ORF16 gene product, which contains six potential integral membrane helices, suggesting that some reactions of the nitrogen fixation process in R.capulatus are membrane-associated.

5. Regulation of nif genes in R.capsulatus

As in K.pneumoniae, the R.capsulatus nitrogen fixation regulatory circuit involves two levels of control. The first level sensing the nitrogen status of the cell contains the genes nifR1 (ntrC), nifR2 (ntrB), nifR4 (ntrA) and nifR5 (glnB). This system is similar to the ntr-system of E.coli and K.pneumoniae. However, significant differences to the ntr-system of enteric bacteria exist. The R.capsulatus ntr-sytem seems to be "nif-specific", since the corresponding mutants show no typical Ntr phenotype, only the expression of nif genes and methylammonium transport is impaired. In addition, the nifR1 (ntrC) gene product seems to be dependent on an alternative RNA polymerase sigma factor different from nifR4 (for a recent review see ref.21).

The second level of nif gene regulation in R.capsulatus is similar to that found in most nitrogen fixing organisms. The gene products of nifA and nifR4 (ntrA) are necessary for the activation of ntrA-dependent promoters. Nine -12/-24 promoters preceding R.capsulatus nif genes were identified (Fig.1). In contrast to these promoters, which are highly homologous to the consensus sequence, putative NifA-binding sites, which match perfectly the TGT-N_{10}-ACA motif, were only found in front of nifH and both copies of nifB. Between the divergently transcribed promoters of ORF14 and ORF16 only one motif could be identified and all other R.capsulatus nif promoters are preceded only by imperfect putative NifA-binding sites. However, the analysis of lacZ fusions demonstated, that these R.capsulatus nif promoters are dependent on nifA and nifR4 (ntrA).

The domain structure of the R.capsulatus nifA gene product resembles more the nifA proteins from organisms that fix nitrogen in symbiosis with legumes than NifA from the free-living nitrogen fixing species K.pneumoniae and A.vinelandii. R.capsulatus NifA also contains the interdomain linker region which might be responsible for the oxygen-sensitivity of this type of NifA. To test if R.capsulatus NifA is also oxygen-sensitive, plasmids were constructed that allowed the constitutive expression of R.capsulatus nifA in E.coli. The analysis of a transcriptional fusion between the R.capsulatus nifB promoter and lacZ demonstrated that R.capsulatus NifA is able to activate nif promoters only under anaerobic conditions. In addition to the requirement of DNA supercoiling (20) the oxygen-sensitivity of NifA might be a second point of oxygen control. This could explain why no nifL analogous gene could be identified in R.capsulatus.

6. Identification of an alternative nitrogenase system in R.capsulatus

To analyze if R.capsulatus contains alternative nitrogenase systems, a nifHDK deletion mutant was constructed. A 4kb BglII-HindIII fragment containing nifK, nifD and most of the nifH coding region was substituted by an interposon coding for kanamycin resistance. The corresponding R.capsulatus mutant is unable to fix nitrogen in the presence of molybdenum. Slow diazotrophic growth, which could not be stimulated by vanadium, was observed in molybdenum-free media. The alternative nitrogenase is able to reduce acetylene to both ethylene and ethane. The partially purified component I, which appeared to be ESR-silent, contains 20 Fe atoms and negligible amounts of Mo, V, W and Re (Schneider and Müller, this volume). These results indicate, that the alternative nitrogenase from R.capsulatus might correspond to nitrogenase-3 from A.vinelandii.

ACKNOWLEDGEMENTS

The author wishes to thank all his colleagues from the University of Bielefeld, W.Paul and M.Merrick (University of Sussex) and M.Jacobson and D.Dean (University of Blacksburg) for their help. This work was supported by grants from the Deutsche Forschungsgemeinschaft (Pu28/14-2) and Bundesministerium für Forschung und Technologie (0319342A).

REFERENCES

1. Adman, E.T., Sieker, L.C. & Jensen, L.H. (1973) J. Biol. Chem. **248**, 3987-3996.
2. Alias, A., Cejudo, F.J., Chabert, J., Willison, J.C. & Vignais, P.M. (1989) Nucl. Acids Res. **17**, 5377.
3. Arnold, W., Rump, A., Klipp, W., Priefer, U.B. & Pühler, A. (1988) J. Mol. Biol. **203**, 715-738.
4. Avtges, P.P., Scolnik, P.A. & Haselkorn, R. (1983) J. Bacteriol. **156**, 251-256.
5. Beynon, J., Ally, A., Cannon, M., Cannon, F., Jacobson, M., Cash, V. & Dean, D. (1987) J. Bacteriol. **169**, 4024-4029.
6. Cannon, M., Cannon, F., Buchanan-Wollaston, V., Ally, D., Ally, A. & Beynon, J. (1988) Nucl. Acids Res. **16**, 11379.
7. Ebeling, S., Noti, J.D. & Hennecke, H. (1988) J. Bacteriol. **170**, 1999-2001.

8. Gosnik, M.M., Franklin, N.M. & Roberts, G.P. (1990) J. Bacteriol. **172**, 1441-1447.
9. Guddat, L.W., Cowan, S.W., Machin, K.J., Isaacs, N.W. & Cotton, R.G.H. (1985) J. Mol. Biol. **186**, 481-482.
10. Hawkes, T.R. & Smith, B.E. (1983) Biochem. J. **209**, 43-50.
11. Hoover, T.R., Imperial, J., Ludden, P.W. & Shah, V.K. (1989) Biochemistry **28**, 2768-2771.
12. Imperial, J., Ugalde, R.A., Shah, V.K. & Brill, W.J. (1984) J. Bacteriol. **158**, 187-194
13. Jacobson, M.R., Brigle, K.E., Bennett, L.T., Setterquist, R.A., Wilson, M.S., Cash, V.L., Beynon, J., Newton, W.E. & Dean, D.R. (1989) J. Bacteriol. **171**, 1017-1027.
14. Joerger, R.D. & Bishop, P.E. (1988) Crit. Rev. Microbiol. **16**, 1-14.
15. Joerger, R.D. & Bishop, P.E. (1988) J. Bacteriol. **170**, 1475-1487.
16. Jones, R. & Haselkorn, R. (1988) Nucl. Acids Res. **16**, 8735.
17. Jones, R. & Haselkorn, R. (1989) Mol. Gen. Genet. **215**, 507-516.
18. Klipp, W., Masepohl, B. & Pühler, A. (1988) J. Bacteriol. **170**, 693-699.
19. Klipp, W., Reiländer, H., Schlüter, A., Krey, R. & Pühler, A. (1989) Mol. Gen. Genet. **216**, 293-302.
20. Kranz, R.G. & Haselkorn, R. (1986) Proc. Natl. Acad. Sci. USA **83**, 6805-6809.
21. Kranz, R.G., Pace, V.M. & Caldicott, I.M. (1990) J. Bacteriol. **172**, 53-62.
22. Masepohl, B., Klipp, W. & Pühler, A. (1988) Mol. Gen. Genet. **212**, 27-37.
23. Moreno-Vivian, C., Hennecke, S., Pühler, A. & Klipp, W. (1989) J. Bacteriol. **171**, 2591-2598.
24. Moreno-Vivian, C., Schmehl, M., Masepohl, B., Arnold, W. & Klipp, W. (1989) Mol. Gen. Genet. **216**, 353-363.
25. Mulligan, M.E. & Haselkorn, R. (1989) J. Biol. Chem. **264**, 19200-19207.
26. Mulligan, M.E., Buikema, W.J. & Haselkorn, R. (1988) J. Bacteriol. **170**, 4406-4410.
27. Pau, R.N. (1989) TIBS, 183-186.
28. Roberts, G.P., MacNeil, T., MacNeil, D. & Brill, W.J. (1978) J. Bacteriol. **136**, 267-279.
29. Robson, R., Woodley, P. & Jones, R. (1986) EMBO J. **5**, 1159-1163.
30. Saeki, K., Miyatake, Y., Young, D.A., Marrs, B.L. & Matsubara, H. (1990) Nucl. Acids Res. **18**, 1060.
31. Schatt, E., Jouanneau, Y. & Vignais, P.M. (1989) J. Bacteriol. **171**, 6218-6226.
32. Schumann, J.P., Waitches, G.M. & Scolnik, P.A. (1986) Gene **48**, 81-92.

Gresshoff, Roth, Stacey & Newton (eds.) Nitrogen Fixation: Achievements and Objectives
Chapman and Hall New York • London 1990

The genetics and biochemistry of the reversible ADP-ribosylation systems of *Rhodospirillum rubrum* and *Azospirillum lipoferum*

G.P. Roberts*, P.W. Ludden, R.H. Burris, W.P. Fitzmaurice, H.-A. Fu, G. Nielsen, J.-H. Liang, L. Lehman, D. Woehle, D. Lies, H. Wirt, S. Montgomery, R. Davis, and Y. Bao

Departments of Bacteriology and Biochemistry, and the Center for the Study of Nitrogen Fixation, University of Wisconsin, Madison, WI 53706 USA

In 1949, Gest and Kamen observed nitrogen fixation by *Rhodospirillum rubrum*, as well as the inhibition of that activity by ammonium (5). While the specific mechanism of that inhibition, the ADP-ribosylation of dinitrogenase reductase, has been known for several years, it is only recently that we have gained some insight into the regulation of that process. The presence of such a reversible ADP-ribosylation system has also been established in *Rhodobacter capsulatus* (7) and *Azospirillum spp.* (6), while suggestive evidence exists for other photosynthetic diazotrophs as well as *Azorhizobium caulinodans* (see ref. 14 for a more extensive review).

Dinitrogenase reductase ADP-ribosyl transferase (DRAT). DRAT is the enzyme responsible for the covalent modification of dinitrogenase reductase. It is an O_2-stable soluble, monomer of 30,000 kd of exceedingly low abundance, requiring a 20,000-fold purification to homogeneity (10). DRAT is extremely specific for native dinitrogenase reductases, modifying them with an ADP-ribose at the highly conserved arginine residue (arg^{101} of *R. rubrum*) adjacent to the cysteine ligand of the iron-sulphur center (cys^{98}). The region around this arginine is also highly conserved in all sequenced dinitrogenase reductases (17) and seems to be involved in their interaction with dinitrogenase, since a mutant affecting the arginine residue is unable to bind with or transfer electrons to dinitrogenase (9). The modification reaction is NAD-dependent and requires MgADP. This latter molecule probably binds to dinitrogenase reductase and affects its conformation in a way conducive to modification.

Dinitrogenase reductase activating glycohydrolase (DRAG). This enzyme was originally termed "activating factor" because of its ability to stimulate nitrogenase activity in crude extracts of *R. rubrum* (12,15). This activation has been shown to reverse the ADP-ribosylation of dinitrogenase reductase by DRAT which occurs during cell harvesting. DRAG is found associated with chromatophores in *R. rubrum* and is removed from them with a 0.5 M NaCl wash. As with DRAT, it is an extremely non-abundant enzyme, requiring 12,000-fold purification and has a similar monomer size of 32,000 kd (19). In contrast with DRAT it is extremely O_2-labile with a half-life in air of 90 sec, and it is likely that this lability reflects a role of redox in the regulation of the enzyme's activity. DRAG is rather less specific than DRAT, in that it will hydrolyse a range of α-N-glycosidic bonds between ADP-ribose and arginine-containing groups (16). The reaction also has a requirement for MgATP and free divalent metal ions, and the role of the former appears to be in inducing a conformational change in dinitrogenase reductase, allowing its recognition by DRAG (18).

Dinitrogenase reductase as substrate. As noted above, the region around the modified arginine is critical for ADP-ribosylation. Only one of the two arginyl residues of the dinitrogenase reductase dimer is modified to effect inactivation and it appears that the dimer cannot be doubly ADP-ribosylated. The modified enzyme is unable to bind to or reduce dinitrogenase *in vitro*. *In vivo* acetylene reduction activity is reduced 90% in response to darkness, but this probably reflects incomplete modification of the pool of dinitrogenase reductase rather than residual activity of the modified proteins. The reason for this incomplete modification is unclear, but may reflect sequestering of dinitrogenase reductase from the DRAG/T system. Through the use of mutant strains, it has been shown that both oxidized and reduced dinitrogenase reductase are substrates for DRAT *in vivo* (13, H. Wirt, unpublished results). The expression of DRAG/T in *Klebsiella pneumoniae* has allowed the analysis of the effects of mutations in other *nif* genes on the suitability of dinitrogenase reductase as a DRAT substrate. The result indicates that all mutants that accumulate dinitrogenase reductase protein, including $NifM^-$ strains, allow the modification of dinitrogenase reductase (H. Wirt, unpublished results).

Genetics of the DRAG/T system. The genes encoding DRAG (*draG)* and DRAT (*draT*) have been cloned from both *R. rubrum* and *A. lipoferum* (3,4). In both cases, *draT* and *draG* are immediately adjacent to each other and rather close to the *nifHDK* operon. The region has been sequenced from *R. rubrum* and the divergently transcribed *nifH* and *draT* genes are separated by only a 400 bp, which presumably contains the regulatory regions for both operons. While this organization certainly opens the possibility of some coordinate transcriptional regulation of the two regulons, this has not been experimentally established. Preliminary results have suggested that the transcription of the *draTG* region is complex: there appears to be a small transcript containing only *draT*, while a much larger transcript contains *draT, G*, and the region on the *nif*-distal side of *dra*. These two transcripts also seem to be affected differentially by environmental conditions (S. Montgomery, unpublished results). This apparent transcriptional regulation of *draTG* is a bit surprising since it has been known for some time that the activities are detectable in cells grown under a wide range of conditions. Quite possibly the activity assays were not sensitive enough to detect modest changes in protein levels.

Insertional mutations have been generated in cloned versions of the *dra* region, but initial attempts to introduce these into the *R. rubrum* chromosome using incompatible RK2 plasmids were unsuccessful (2). This failure has been traced to the combination of reduced plasmid incompatibility and relatively poor recombination frequencies in *R. rubrum* (D. Lies, unpublished results). A different introduction system, based on the pSUP vectors, has been more successful, allowing the construction of both polar and non-polar insertions in *draT, G,* and in the region to the 3' side of *draG* (J. Liang, unpublished results). The characterization of these mutants is described below.

Functionality of DRAG/T in *K. pneumoniae*. The *draTG* genes from *A. lipoferum* and *R. rubrum* have been cloned into pKK223-3, an expression vector carrying the *tac* promoter (3,4). When these vectors are introduced into *K. pneumoniae*, appropriate enzyme activities are detectable in crude extracts of vector-containing cells. More surprisingly, *K. pneumoniae* cells with both *draT* and *draG* show reversible inhibition of nitrogenase activity *in vivo* in response to added fixed nitrogen. This inhibition has been correlated with the ADP-ribosylation of dinitrogenase reductase in extracts of these cells. The physiological implications of these results are discussed below, but at present the nature of the signal transduction pathway involved in this reversible regulation is unknown.

Physiology of the reversible regulatory system. The DRAG/T system inhibits nitrogenase activity of N_2- or glutamate-grown *R. rubrum* in response to a variety of stimuli, including fixed nitrogen, darkness and the oxidizing dye, phenazine methosulfate. In all of these cases the system acts in an "all or none" fashion, with dinitrogenase reductase either being fully active or 90% inactive. Cells that are starved for nitrogen fail to respond to the stimuli (1), even though both DRAT and DRAG are present (11). A critical question concerns the mechanism by which the DRAG/T activities are modulated to achieve regulation: are the activities of each enzyme inversely regulated in response to these conditions, or is the activity of one held constant while the other is regulated? The latter case would imply a futile cycle of NAD hydrolysis under at least some conditions. As the experiments described below demonstrate, both activities appear to be regulated in *R. rubrum* and at least the DRAT activity is regulated in the *K. pneumoniae* containing appropriate expression vectors.

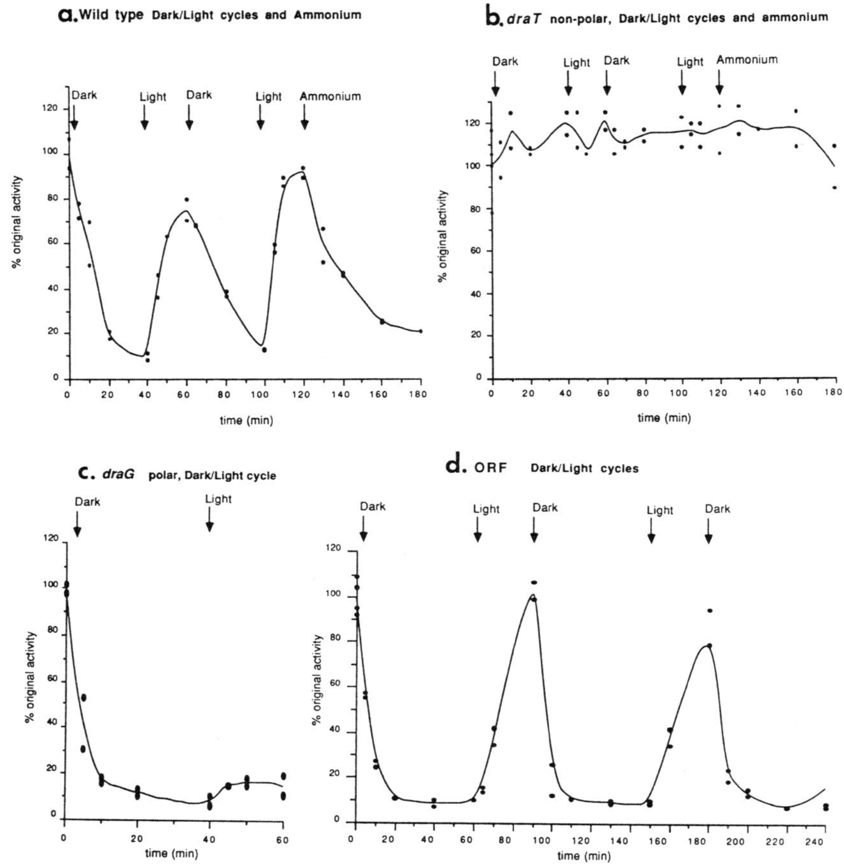

Figure 1. Physiological effects of dark/light cycles and ammonium on nitrogenase activity in wild type *R. rubrum* and selected mutants. Cells were grown to approximately 1 O.D. in glutamate/malate medium anaerobically in the light. Conditions were altered at indicated times and two ml samples were assayed for acetylene reduction activity. a. Strain UR2 (*dra*$^+$). b. Strain UR213 (*draT3*). c. Strain UR214 (*draG4*). d. Strain UR216 (*dod-1*).

Regulation of DRAG. The regulation of DRAG activity *in vivo* was demonstrated by the determination of the half-life of the AD^{32}P-ribose moiety on dinitrogenase reductase during a cold chase in the presence of fixed nitrogen (8). The result clearly showed that this modification is long-lived under inhibition conditions, indicating that there was negligible DRAG activity under these conditions. Since DRAG activity was detectable in crude extracts of these cells, its activity *in vivo* must have been blocked in some fashion. It is unclear if this regulation of DRAG activity is due to a direct effect on DRAG itself or indirectly through a sequestering of dinitrogenase reductase.

Regulation of DRAT. The regulation of DRAT activity *in vivo* has been demonstrated in both *R. rubrum* and *K. pneumoniae*. In the former, a polar cassette insertion was introduced into the *draG* gene, leaving only the *draT* gene intact (G. Nielsen and J. Liang, unpublished results). Under derepressing conditions, this strain displayed normal nitrogenase activity, but upon shifting the culture to dark or ammonium, the activity rapidly disappeared (see Fig. 1c). This result indicates that the DRAT activity was present, but inactive, prior to the shift. A similar result was also obtained when cells were exposed to a protein synthesis inhibitor 30 minutes prior to the shift, indicating that *de novo* DRAT synthesis was not required. In a *K. pneumoniae* strain containing an expression vector with only *draT*, derepression was normal, but dinitrogenase reductase was ADP-ribosylated in response to added ammonium (4). These two results clearly demonstrate that DRAT activity is regulated *in vivo*, but as in the case of DRAG, the biochemical nature of this regulation is unclear.

Other relevant genes in the *dra* region. When the *draTG* region of *R. rubrum* was sequenced (2), an open reading frame with reasonable codon usage was noticed immediately downstream of *draG*. A polar cassette insertion has been introduced into this region and the resulting strain analyzed for its response to darkness and ammonium with the result shown in Fig. 1d. The result appears superficially rather similar to that obtained in the case of wild type, but there appears to be a reproducible difference, namely the response to inhibiting conditions is much faster than is the response of wild type. A similar effect is seen in the case of the polar *draG* mutant (Fig. 1c), raising the possibility that both of these mutations eliminate a gene product that in some way modulates the cell's response to these stimuli. These results may begin to address the central question in this regulatory system, namely what are that signal transduction pathways for these various stimuli. The *K. pneumoniae* results should also provide insight into this question since there is apparently a factor or set of factors in that organism capable of communicating with DRAG/T.

Conclusions. The results presented here support the following conclusions (i) The DRAG/DRAT proteins are necessary for the posttranslational regulation of the activity of the nitrogenase system through the reversible ADP-ribosylation of dinitrogenase reductase. (ii) The activities of DRAG/DRAT proteins are themselves both targets of posttranslational regulation. (iii) The region of DNA on the *nif*-distal side of *draTG* may encode products involved in the regulatory system. (iv) The *draTG* genes in *R. rubrum* are transcriptionally regulated. (v) When expressed in the heterologous background of *K. pneumoniae*, the DRAG/DRAT proteins allow a regulatory response to fixed nitrogen that is remarkably similar to that seen in *R. rubrum*.

ACKNOWLEDGEMENTS

This work was supported by the University of Wisconsin college of Agricultural and Life Sciences, by U.S. Department of Agriculture Competitive Grants Program Award 89-37262-4740 to GPR and PWL, by National Science Foundation grant DCB 8821820 to PWL, and by U.S. Department of Energy grant DE-FG02-87ER13707 to RHB.

1. Carithers, R.P., Yoch, D.C., & Arnon, D.I. (1979) *J. Bacteriol.* **137**, 779-789.
2. Fitzmaurice, W.P., Saari, L.L., Lowery, R.G., Ludden, P.W., & Roberts, G.P. (1989) *Molec. Gen. Genet.* **218**, 340-347.
3. Fu, H.-A., Fitzmaurice, W.P., Roberts, G.P., & Burris, R.H. (1990) *Gene* **86**, 95-98.
4. Fu, H,-A., Burris, R.H., & Roberts, G.P. (1990) *Proc. Natl. Acad. Sci. USA* **87**, 1720-1724.
5. Gest, H. & Kamen, M.D. (1949) *Science* **109**, 558-559.
6. Hartmann, A., Fu, H.-A., & R.H. Burris (1986) *J. Bacteriol.* **165**, 864-870.
7. Jouanneau, Y, Roby, C., Meyer, C.M., & Vignais, P.M. (1985) *Biochemistry* **28**, 6524-6530.
8. Kanemoto, R.H. & Ludden, P.W. (1987) *J. Bacteriol.* **169**, 3035-3043.
9. Lowery, R.G., Chang, C.L., Davis, L.C., McKenna, M.C., Stephens, P.J., & Ludden, P.W. (1988) *Biochemistry* **28**, 1206-1212.
10. Lowery, R.G. & Ludden, P.W. (1988) *J. Biol. Chem.* **263**, 16714-16719.
11. Lowery, R.G., Saari, L.L., & Ludden, P.W. (1986) *J. Bacteriol.* **166**, 513-518.
12. Ludden, P.W. & Burris, R.H. (1976) *Science* **194**, 424-426.
13. Ludden, P.W., Lehman, L., & Roberts, G.P. (1989) *J. Bacteriol.* **171**, 5210-5211.
14. Ludden, P.W. & Roberts, G.P. (1989) *Curr. Top. Cell. Regul.* **30**, 23-56.

15. Nordlund, S., Eriksson, U., & Baltscheffsky, H. (1977) *Biochim. Biophys. Acta* **462**, 187-195.
16. Pope, M.R., Saari, L.L., & Ludden, P.W. (1986) *J. Biol. Chem.* **261**, 10104-10111.
17. Pretorius, I-M., Rawlings, D.E., O'Neill, E.G., Jones, W.A., Kirby, R., & Woods, D.R. (1987) *J. Bacteriol.* **169**, 367-370.
18. Saari, L.L., Pope, M.R., Murrell, S.A., & Ludden, P.W. (1986) *J. Biol. Chem.* **261**, 4973-4977.
19. Saari, L.L., Triplett, E.W., & Ludden, P.W. (1984) *J. Biol. Chem.* **259**, 15502-15508.

Gresshoff, Roth, Stacey & Newton (eds.) Nitrogen Fixation: Achievements and Objectives
Chapman and Hall New York • London 1990

Nitrogen fixation genes of *Clostridium pasteurianum*

Jiann-Shin Chen, Shu-Zhen Wang, and John L. Johnson

Department of Anaerobic Microbiology, Virginia Polytechnic Institute and State University, Blacksburg, VA 24061 USA

Clostridium pasteurianum W5 is an obligately anaerobic, free-living nitrogen fixer. It was one of the first nitrogen-fixing organisms isolated (16, 26), and it has been studied in the laboratory for almost 100 years now. Consistent nitrogen fixation by cell-free preparations was first obtained with this anaerobe (4), which marked a turning point in studies of the biochemistry of nitrogen fixation (2). Although nitrogenase from C. pasteurianum has been very extensively studied and it possesses a number of unique properties (5), the nitrogen-fixation (nif) genes of C. pasteurianum have been studied only in recent years.

Genetic studies with C. pasteurianum have been hampered by a lack of useable genetic markers and of procedures for genetic manipulations. Since nitrogenase structural genes are conserved among nitrogen-fixing organisms (15, 20), the cloned nifHDK genes of Klebsiella pneumoniae became useful probes for the detection and isolation of nif genes from other organisms. Our initial cloning of nitrogenase structural genes from C. pasteurianum utilized K. pneumoniae nifHD genes as a probe. However, because of a significant difference in codon usage between C. pasteurianum and K. pneumoniae, an exact match over 10 nucleotides in length is rare between nitrogenase genes of these two organisms, which made the initial detection of nif genes in C. pasteurianum difficult (5).

Following the initial cloning of nifH1 and part of nifD from C. pasteurianum, we have expanded from this point to clone, sequence, and identify genes corresponding to nifK, nifE, nifN, nifB, and nifV. The organization of these genes differs significantly from those found in other organisms. Furthermore, a gene possibly involved in molybdenum transport, which is not present in the nif cluster of K. pneumoniae, is present in the nif cluster of C. pasteurianum. In addition to the better characterized nif genes, there are also five nifH-like genes (nifH2 through nifH6), whose functions and genomic locations (nifH3 through nifH6) have not yet been determined (23). The salient features of these nif genes are described below.

METHODS

Clostridium pasteurianum strain W5 (ATCC 6013) was used in this study. Isolation of high molecular weight DNA from C. pasteurianum, cloning and sequencing of DNA, and analysis of nucleotide and amino acid sequences were as described (23, 24). pBR325, pUC18, and pUC19 were used in the construction of pCP114, pCP124, pCP134, pCP144, and pCP154, which have inserts covering the contiguous nif genes and ORFs that constitute three possible nif operons (Fig. 1). Additional nifH-like sequences (nifH3 through nifH6) were cloned into pBR322 or pUC plasmids giving pCP3, pCP300, pCP400, pCP500, and pCP600 (5, 23).

A 1.17-kb fragment covering the C-terminal 298 amino acid residues of the deduced nifN-B product was subcloned and expressed under the control of the tac promoter in E. coli strain JM101. The expressed protein was located on polyacrylamide gel after electrophoresis in the presence of SDS, and gel slices containing the protein band were homogenized, emulsified with Freund's complete or incomplete adjuvant, and used in the immunization of rabbits for antibody production. The crude antiserum was preabsorbed with proteins in lysates of E. coli JM109 (harboring the vector plasmid). The treated antiserum was used in Western blot analysis of SDS-PAGE separated proteins from ammonia grown, nitrogen-fixing, and transition phase C. pasteurianum cells.

The function of nifVα and nifVω was established in a nifV deleted strain (DJ388) of A. vinelandii (Wang, Dean, Chen and Johnson, this Proceedings). Plasmids containing nifVα or nifVω or both, with the insert under the control of an A. vinelandii nif promoter, were introduced into the nifV deleted strain. The transformed cells were tested for ability to grow under nitrogen-fixing conditions.

RESULTS AND DISCUSSION

Organization of nif genes in C. pasteurianum.

Characterized nif genes and flanking open reading frames (ORFs) that occur in a cluster in C. pasteurianum are shown in Fig. 1. In a region spanning 11.6 kb of DNA, we have identified genes that correspond to nifH, nifD, nifK, nifE, nifN, nifB, and nifV of K. pneumoniae and other organisms. The nitrogenase structural genes (nifH1, nifD, and nifK) of C. pasteurianum were positively identified from a comparison of the deduced amino acid sequences with protein sequences (5, 21, 24). Additional nif genes (nifE, nifN-B, nifVω, and nifVα) of C. pasteurianum were identified through comparison of the deduced amino acid sequences with those of Azotobacter vinelandii, K. pneumoniae, and other organisms. More recently, an ORF preceding nifVω of C. pasteurianum was found to have sequence similarity to that of chlJ of Escherichia coli, and we designated this ORF nifC.

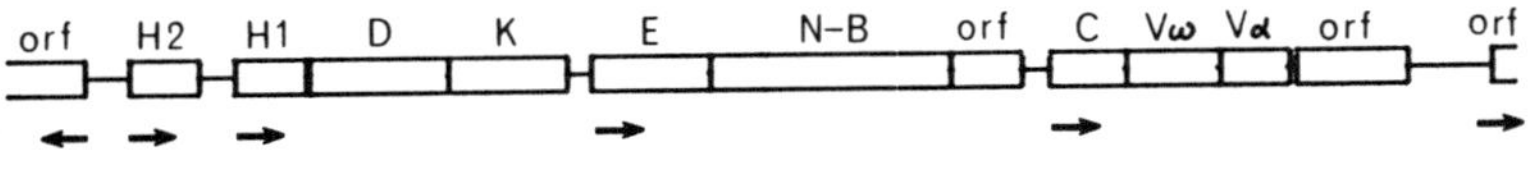

Figure 1. Physical map of the nif cluster of C. pasteurianum.

Despite a conservation in the amino acid sequences encoded by the corresponding genes, the organization of nif genes in C. pasteurianum differs significantly from those found in several Gram-negative organisms (e.g., 1, 8, 9). nif genes of C. pasteurianum seem to be organized according to their functions. Clustering of genes involved in the synthesis of FeMoco is apparent. In C. pasteurianum, three consecutive groups (operons) of nif genes show the following arrangements: (a) the first group consists of structural genes (nifH1DK) for nitrogenase, (b) the second group contains genes (nifEN-B) required for FeMoco synthesis, where nifN and nifB are fused into one gene nifN-B, and (c) the third group contains a gene (nifC) possibly involved in molybdenum transport and additional genes (nifVωVα) for FeMoco synthesis, where the conserved amino acid sequence of nifV is encoded by two separate genes. ORFs with unknown functions are also present in groups 2 and 3. Properties of nif genes and ORFs in a contiguous region in C. pasteurianum (Fig. 1) are listed in Table 1.

Table 1. Properties of nif genes and ORFs in a cluster in Clostridium pasteurianum.

Gene	Coding region	Start codon	Stop codon	Encoded polypeptide: No. of amino acid residues	Encoded polypeptide: M_r
ORF	887-1	AUG	-	(Partial)	-
H2	1395-2213	AUG	UAA	272	29,580
H1	2623-3444	AUG	UAA	273	29,666
D	3486-5087	GUG	UAA	533	58,990
K	5086-6462	AUG	UAA	458	50,115
E	6747-8117	AUG	UAG	456	50,397
N-B	8140-10929	UUG	UAA	929	103,088
ORF	10892-11734	AUG	UAG	280	30,516
C	12084-12944	AUG	UAG	286	31,681
Vω	12962-14020	AUG	UGA	352	41,574
Vα	14030-14839	AUG	UAA	269	29,862
ORF	14953-16224	AUG	UAA	423	47,010
ORF	17207-17471	AUG	-	(Partial)	-

Structural genes for nitrogenase.

The three structural genes, nifH1, nifD, and nifK, constitute a transcription unit. The amino acid sequence deduced from nifH1 matches perfectly the protein sequence (5, 21), whereas amino acid sequences deduced from nifD and nifK have several differences from the protein sequences (24). C. pasteurianum also has five nifH-like genes (nifH2 through nifH6) that have been cloned and sequenced (see below). nifH2 and nifH1 occur in tandem, but nifH2 is not part of the nifH1DK operon (Fig. 1; 23).

The Fe protein of C. pasteurianum has 273 amino acid residues, which is about 20 residues shorter than most nifH-encoded sequences studied to date. The nifH-encoded sequence of Desulfovibrio gigas (Dg H; 14) is only one residue longer than that of C. pasteurianum. These two polypeptides have a similarity coefficient (S_{AB}) of 0.71, which is only slightly higher than the next highest value (0.69) found between Cp and Av nifH1 genes (with the extended C-terminal region of Av NifH excluded from comparison). However, Cp nifH1 and Dg nifH encode long segments (up to 28 residues) of identical amino acid sequences, which include a region (residues 56 to 70) that contain a shared gap of two positions (between residues 64 and 65 when aligned with a nifH reference sequence) and is characteristic of these two sequences (Fig. 2). Both Cp and Dg are obligately anaerobic soil organisms. A study of relatedness concerning the nif and 16s rRNA sequences of these two organisms may shed new lights on the distribution and evolution of nif genes among diverse phylogenetic groups. It should also be informative to compare properties of nitrogenases purified from the two organisms, as the Cp nitrogenase has a number of unique properties (5, 24).

```
              40         50         60          70
Cp H1:   MVVGCDPK ADSTRLLLGG LAQKSVLDTL REEG##EDVELD SI
Dg H:    MVVGCDPK ADSTRLLLGG LSQRTVLDTL REEG##EDVDLD DI
Ref:     -IVGCDPK ADSTRLiLh- KaQ-tvl-lA Ae-GsVEDLEle dV
```

Figure 2. Comparison of a region of conserved amino acid sequences encoded by C. pasteurianum nifH1 (Cp H1; 5) and D. gigas nifH (Dg H; 14). The numbering refers to the C. pasteurianum sequence. Identical residues are underlined. Gaps (#) between residues 64 and 65 are introduced to align the sequences with a nifH reference sequence derived from the following eight organisms: Anabaena 7120 (see 5 for references), A. vinelandii (5), Bradyrhizobium japonicum (5), Frankia strain Ar13 (17), K. pneumoniae (5), Rhizobium maliloti (5), Rhodobacter capsulatus (12), and Thiobacillus ferrooxidans (19). For A. vinelandii, the nifH for the Mo-nitrogenase is used. For the reference sequence, upper case letters denote residues found in all eight organisms, whereas lower case letters denote those found in at least five organisms.

The nifD gene is located 41 base pairs (bp) downstream from nifH1. There are several remarkable features about the nifD gene and its product: (a) it uses the less common start codon GUG, (b) its stop codon (UAA) overlaps with the start codon (AUG) for nifK by one base, and (c) there is an extra stretch (in comparison to the other nifD-encoded sequences) of about 50 amino acid residues in the 380 to 430 region and an apparent periodicity in amino acid sequence and hydrophobicity is observed in this stretch of residues (24).

The nifK gene product of C. pasteurianum is about 60 amino acid residues shorter than the protein deduced from nifK of other organisms, and the shortened region is mostly at the N-terminus (24). The shortness of Cp nifK may be related to the proximity between nifD and nifK genes, which overlap by one base.

nifK is separated by 279 bp from the next ORF (nifE). There are thus no genes equivalent to nifTY next to nifK in C. pasteurianum. Between nifK and nifE, there are two regions of inverted repeats, involving 16 and 10 bp each, that may form stem-and-loop structures.

Genes involved in FeMoco synthesis or Mo transport.

1. The group of nifE-nifN-B-(ORF). The Cp nifE-encoded polypeptide has 456 amino acid residues (25). It shows an S_{AB} of 0.33 with that of A. vinelandii and an S_{AB} of 0.3 with the Cp nifD product. Conserved residues are found over the entire length of the polypeptides. Cys-79, Cys-140, and Cys-255 of Cp NifE correspond to Cys-79, Cys-144, and Cys-271 of Cp NifD.

The Cp nifN-B gene occurs 22 nucleotides downstream of nifE, with the sequence

```
(---- GTT TAG) AATATAAGTGGAGGTGAATATA (TTG AAT AGT -----)
 nifE-Val ***                          Met Asn Ser-nifN-B
```

occurring between the two genes. A potential ribosome-binding site (GGAG) occurs at -13 to -10 nucleotides from the predicted start codon UUG. nifN-B has a coding capacity of 929 amino acid residues (M_r = 103,088); it is so named because the first 480 residues of the predicted polypeptide has sequence similarity to the predicted nifN product of other organisms, whereas the remaining 449 residues has sequence similarity to the predicted nifB product of other organisms (data not shown). The NifN domain has an S_{AB} of 0.29 with either Av NifN or Cp NifK. Cys-17, Cys-42, and Cys-100 of the NifN domain of NifN-B correspond to Cys-23, Cys-48, and Cys-106 of Cp NifK. The NifB domain of Cp NifN-B has an S_{AB} of 0.385 with the predicted Av NifB. Twelve Cys residues are conserved between the Cp NifB domain and Av NifB.

Using antiserum raised against a 33,000-dalton polypeptide (corresponding to the C-terminal 298 residues of Cp NifN-B), Western blot analysis detected a band with a M_r =102,000 in nitrogen-fixing but not in ammonia-grown C. pasteurianum cells, indicating the presence of an intact NifN-B product. The fusion of nifN and nifB genes into nifN-B and the presence of the NifN-B product suggest that the nifB product interact directly with the NifEN complex in other organisms.

The C-terminal 143 residues of the predicted NifN-B also has sequence similarity to the predicted nifX products of other organisms (not shown). The function of this nifX-like domain and the ORF downstream of nifN-B is not known.

2. The group of nifC-nifVω-nifVα-(ORF). The first ORF (nifC) of this cluster (Fig. 1) occurs 330 nucleotides downstream of the ORF of the nifE-nifN-B-(ORF) cluster. Inverted repeats, which may form a stem-and-loop structure with a stem of 23 bp, are present 22 to 77 nucleotides downstream of nifEN-B(ORF).

The predicted polypeptide (286 residues) from nifC has sequence similarity to that of chlJ of Escherichia coli (data not shown). Because the region of chlJ cloned and sequenced covers only the C-terminal 200 residues (11), the comparison is limited to this region, and it has an S_{AB} of 0.325. chlJ of E. coli is part of the chlD locus, and the chlD locus is involved in molybdenum transport. Mo uptake by C. pasteurianum and by K. pneumoniae is coregulated with nitrogen fixation (7, 18). Sequences similar to the proposed consensus nif promoter and upstream sequences of C. pasteurianum precede the nifC gene. We thus propose that nifC is involved in Mo uptake and is coregulated with nitrogen fixation.

Although the nif cluster of K. pneumoniae does not contain a gene equivalent to nifC, nitrogen fixation in K. pneumoniae requires the function of a mol locus, which is equivalent to the chlD locus of E. coli (22). Furthermore, Kennedy and Postgate (13) showed that for the nif genes of K. pneumoniae to function in E. coli grown at low Mo concentrations, a functional chlD locus is required in E. coli. Thus, a nifC- or chlJ-like gene may be expected to exist in K. pneumoniae, although it is not present in the nif cluster.

The two genes nifVω and nifVα are separated by 9 nucleotides, with nifVω occurring 17 nucleotides downstream of nifC. nifVω has a coding capacity of 352 amino acid residues, whereas nifVα has a coding capacity of 269 amino acid residues. These two genes are so named because (a) the C-terminal 195 residues of nifVω are similar to the C-terminal region of the predicted NifV sequence of A. vinelandii and K. pneumoniae, and (b) the predicted sequence of NifVα is similar to the N-terminal region of the predicted NifV of Av and Kp. In addition, a plasmid containing nifVω and nifVα , but not either alone, rescued a nifV-deleted strain of A. vinelandii (see Wang et al. in this Proceedings). Studies of the enzymic properties of the products of nifVα and nifVω should further our understanding of the role of the nifV product in FeMoco synthesis. The function of the ORF following nifVα is not known.

Multiple nifH-like genes.

In C. pasteurianum, there are five nifH-like genes (nifH2 to nifH6) besides nifH1, which encodes the Fe protein (23). nifH2 is upstream of nifH1 (Fig. 1), but they belong to different transcriptional units. The genomic locations of nifH3 through nifH6 have not been determined. Among the nifH-like genes, the predicted polypeptides differ from each other to different degrees: the differences range from one to 23 residues (S_{AB} values range from 0.99 to 0.92), except for nifH3 which differs from nifH1 by 95 residues (S_{AB} = 0.65). mRNA for these nifH-like genes, with the exception of nifH3, has been detected in nitrogen-fixing C. pasteurianum cells grown with normal amounts of Mo (23), but the function of these nifH-like genes remains to be determined.

A striking similarity has been observed between the predicted amino acid sequences from nifH3 of C. pasteurianum and nifH3 (anfH) of A. vinelandii, with an S_{AB} of 0.82 (10). Because C. pasteurianum can grow under nitrogen-fixing conditions with no or low Mo or with the supplement of V (3, 6; our unpublished results), it is likely that the Cp nifH3 product is involved in an alternative nitrogenase system.

nif promoter and upstream sequences.

Following the determination of the transcription start site for nifH1, nifE, and nifH-like genes (excluding nifH3), apparently similar nucleotide sequences were observed in the -100 region relative to the transcription start (23). The proposed consensus sequence is ATCAatat-N_{6-10}-ATGGattc, which appears to be orientation-independent. The -35 and -10 regions are more heterogeneous, with sequences similar to TTG and TATAAT preceding some nif operons (23). The regulatory sequences for nif operons in C. pasteurianum will require further studies.

Concluding remarks.

The organization of nif genes in C. pasteurianum has a number of distinct features. They include (a) an overlap between nifD and nifK, (b) the fusion of nifN and nifB into nifN-B, (c) the division of nifV into nifVα and nifVω, and (d) the presence of nifC upstream of nifVω. The fused and divided nif genes offer a unique opportunity for studies of the structure-function relationship in these gene products. Also, there appears to be a clustering of genes involved in FeMoco synthesis and possibly Mo uptake in C. pasteurianum, which, in conjunction with a proximity of these genes to nifH1DK, may be an important aspect for studies concerned with the evolution and distribution of nif genes and with the maximization of the efficiency of nitrogen fixation.

ACKNOWLEDGMENTS

This work was supported by USDA competitive grants and by projects from the Commonwealth of Virginia.

REFERENCES

1. Arnold, W., Rump, A., Klipp, W., Priefer, U. B. & Puhler, A. (1988) J. Mol. Biol. **203**, 715-738.
2. Burris, R. H. (1988) in Nitrogen Fixation: Hundred Years After, eds. Bothe, H., de Bruijn, F. J. & Newton, W. E. (Fisher, New York), pp. 21-30.
3. Cardenas, J. & Mortenson, L. E. (1975) J. Bacteriol. **123**, 978-984.
4. Carnahan, J. E., Mortenson, L. E., Mower, H. F. & Castle, J. E. (1960) Biochim. Biophys. Acta **44**, 520-535.
5. Chen, K. C.-K., Chen, J.-S. & Johnson, J. L. (1986) J. Bacteriol. **166**, 162-167.
6. Dilworth, M. J., Eady, R. R., Robson, R. L. & Miller, R. W. (1987) Nature (London) **327**, 167-168.
7. Elliott, B. B. & Mortenson, L. E. (1976) J. Bacteriol. **127**, 770-779.
8. Jacobson, M. R., Brigle, K. E., Bennett, L. T., Setterquist, R. A., Wilson, M. S., Cash, V. L., Beynon, J., Newton, W. E. & Dean, D. R. (1989) J. Bacteriol. **171**, 1017-1027.
9. Joerger, R. D. & Bishop, P. E. (1988) J. Bacteriol. **170**, 1475-1487.
10. Joerger, R. D., Jacobson, M. R., Premakumar, R., Wolfinger, E. D. & Bishop, P. E. (1989) J. Bacteriol. **171**, 1075-1086.
11. Johann, S. & Hinton, S. (1987) J. Bacteriol. **169**, 1911-1916.
12. Jones, R. & Haselkorn, R. (1988) Nucleic Acids Res. **16**, 8735.
13. Kennedy, C. & Postgate, J. R. (1977) J. Gen. Microbiol. **98**, 551-557.
14. Kent, H. M., Buck, M. & Evans, D. J. (1989) FEMS Microbiol. Lett. **61**, 73-78.
15. Mazur, B. J., Rice, D & Haselkorn, R. (1980) Proc. Natl. Acad. Sci. **77**, 186-190.
16. McCoy, E., Fred, E. B., Peterson, W. H. & Hastings, E. G. (1930) J. Inf. Diseases **46**, 118-137.
17. Normand, P., Simonet, P. & Bardin, R. (1988) Mol. Gen. Genet. **213**, 238-246.
18. Pienkos, P. T. & Brill, W. J. (1981) J. Bacteriol. **145**, 743-751.
19. Pretorius, I.-M., Rawlings, D. E., O'Neill, E. G., Jones, W. A., Kirby, R. & Woods, D. R. (1987) J. Bacteriol. **169**, 367-370.
20. Ruvkun, G. B. & Ausubel, F. M. (1980) Proc. Natl. Acad. Sci. USA **77**, 191-195.
21. Tanaka, M., Haniu, M., Yasunobu, K. T. & Mortenson, L. E. (1977) J. Biol. Chem. **252**, 7093-7100.
22. Ugalde, R. A., Imperial, J., Shah, V. K. & Brill, W. J. (1985) J. Bacteriol. **164**, 1081-1087.
23. Wang, S.-Z., Chen, J.-S. & Johnson, J. L. (1988) Nucleic Acids Res. **16**, 439-454.
24. Wang, S.-Z., Chen, J.-S. & Johnson, J. L. (1988) Biochemistry **27**, 2800-2810.
25. Wang, S.-Z., Chen, J.-S. & Johnson, J. L. (1989) Nucleic Acids Res. **17**, 3299.
26. Winogradsky, M. S. (1895) Arch. Sci. Biol. (St. Petersburg) **3**, 297-352.

Gresshoff, Roth, Stacey & Newton (eds.) Nitrogen Fixation: Achievements and Objectives
Chapman and Hall New York • London 1990

Azospirillum associations

M. Eyers, K. Michiels, E. Van Bastelaere, C. Croes, P. van Rhijn, A. Vande Broek, A. Milcamps, R. De Mot and J. Vanderleyden
F.A. Janssens Memorial Laboratory of Genetics, University of Leuven, W. de Croylaan 42, B-3030 Heverlee, Belgium

INTRODUCTION

Since its rediscovery in 1975 by Döbereiner and collaborators (18), *Azospirillum* has been the subject of much controversy in the field of biological nitrogen fixation in relation to plant nutrition. This controversy stems from non-reproducible results of *Azospirillum* inoculations in pot and field trials. Based on the natural abundance of diazotrophic Azospirilla in the rhizosphere of tropical vegetations on low N-content soils, it was postulated that these bacteria actively contribute to the nitrogen demand of plants. Moreover since these tropical vegetations are often composed of non-leguminous plants, a new niche for research on biological nitrogen fixation was created, the so called associative nitrogen fixation. With the introduction of N balance and ^{15}N techniques, the contribution of biological nitrogen fixation in the plant growth promotion effect of *Azospirillum* inoculations had to be reconsidered. The uptake of biologically fixed N_2 by the plant turned out to be less than expected, and insufficient to account for the increased N content of inoculated plants versus non-inoculated plants (2,7). However, it is important to stress that an associative interaction versus a symbiotic interaction is more difficult to reconstruct. There are no such things like time of emergence, numbers and size of nodules that can be used to monitor a successful interaction. Therefore it is crucial that for *Azospirillum* the molecular genetics of both nitrogen fixation and plant interaction are developed.

ORGANIZATION OF *A. BRASILENSE NIF* GENES

Genes involved in nitrogen fixation in *A. brasilense* ATCC29145 have been identified either by DNA hybridizations with DNA from *Klebsiella pneumoniae, Azorhizobium caulinodans*, *Bradyrhizobium japonicum*, or by Tn*5* mutagenesis. The organization of five regions, *nifHDK*, *nifE*, *nifUS*,*fixABC*, and an unidentified open reading frame between *nifE* and *nifUS* has been presented by Elmerich and coworkers (5). Recently, transcription of the *nifHDK* operon of *A. brasilense* has been analysed (3) and the nucleotide sequence of *nifH* has been determined (3,4). The 5' region contains an NtrA-dependent promotor and upstream activator sequences that show striking homology with UAS for NifA-mediated activation in other diazotrophs (3). The occurrence of a *nifA* type gene in *A. brasilense* was already suggested by Pedrosa and Yates (12) and Singh *et al.* (14). All the *nif* and *fix* genes identified so far in *Azospirillum* are located on the chromosome. However, we recently observed that curing *A. brasilense* Sp7 (ATCC29145) of one of its indigenous plasmids, the 115-MDa plasmid, results in a 30 % decrease of acetylene reduction activity. In addition this cured strain produces a dark brown pigment. Since this particular plasmid is lost at high frequency under stress conditions (up to 25 % at 40°C), it could be of particular relevance for *Azospirillum* inocula in the tropics.

AZOSPIRILLUM - PLANT INTERACTION

Azospirilla dispose of several properties that are thought to promote competitiveness in the rhizosphere. First, they are chemotactically attracted to the rhizosphere by root exudates and they have a highly versatile metabolism that allows them to survive under widely variable conditions. Second, they can adhere to the root cell surface. Third, some *Azospirillum* spp. can penetrate the root surface and colonize intercellular spaces in the root interior.

Motility and chemotaxis. *Azospirillum* spp. are very motile. They possess a single polar flagellum that renders them motile in liquid medium. On solid media they develop in addition several lateral flagella of smaller size and shorter wavelength that are involved in swarming (6). Azospirilla respond chemotactically to a variety of amino acids, sugars and organic acids, irrespective of the ability to metabolize the attractant (13). In monoxenic plant systems, migration distances of several centimetres have been recorded (1). Chemotaxis was shown to be strain-specific and attuned to the composition of the host root exudate (13). We have isolated a number of chemotaxis mutants (Che$^-$) that have a wild-type flagellation pattern (16). These mutants are general chemotaxis mutants since they have lost responses to all amino acid and sugar attractans tested. Two *che* loci map on a second indigenous plasmid of *A. brasilense* Sp7, the 90-MDa plasmid, while a third locus maps on the chromosome. The plasmidic Che$^-$ mutants lack a 110-kDa acidic protein. Synthesis of this protein is also affected in an *exoC* mutant (see below).

Attachment. Adsorption of azospirilla to roots, root hairs or cells in suspension cultures has been reported by several authors and is thought to contribute to a selective enrichment of azospirilla in the rhizosphere. We have obtained mutants that are drastically impaired in attachment to roots of wheat seedlings (see Table 1). These mutants were originally isolated as Calcofluor minus (Cal$^-$) mutants and are therefore affected in extracellular polysaccharide synthesis (EPS or CPS). Since these Cal$^-$ mutants still produce an exopolysaccharide (secreted into the medium) it is believe that these mutants are affected in the synthesis of a surface polysaccharide and that Azospirilla produce both EPS and CPS. Interestingly, these mutants are also affected in floc formation (11). Up to now four different loci involved in the synthesis of this surface polysaccharide have been recognized and they all are located on the chromosome.

Root invasion. Some *Azospirillum* strains penetrate the roots of their host and become established in high numbers in the intercellular spaces between the epidermis and the cortex or in the outer cortical layers. The results with two *A. brasilense* strains isolated from wheat, Sp245 and Sp246, are particularly striking in this respect (2). Strain Sp245, isolated from surface sterilized roots,

Table 1. Characteristics of *Azospirillum brasilense* 7030 Tn*5* mutants.

Strain	Cal phenotype*	Mucoid character	Floc formation in NMF medium	Percent attachment to wheat seedlings
7030	Cal^{+}	+	+	9
7030TN5-1	Cal^{++}	++	+	2
7030TN5-11	Cal^{d}	+	+/-	
7030TN5-12	Cal^{d}	+	+/-	
7030TN5-21	Cal^{d}	+	+/-	0.5
7030TN5-22	Cal^{-}	+	-	0.6
7030TN5-23	Cal^{-}	+	-	0.9

* refers to intensity of fluorescence on Calcofluor-containing medium. Cal^{-}, no fluorescence; Cal^{d}, dimmed fluorescence; Cal^{+}, fluorescence of parental strain *A. brasilense* 7030; Cal^{++}, brighter fluorescence than *A. brasilense* 7030.

is able to penetrate and to establish within wheat roots upon inoculation, whereas strain Sp246, isolated from surface washed roots, is not able to penetrate wheat roots. Whenever strain Sp245 establishes will within inoculated roots, a positive growth response is observed. Experiments in which a reporter gene is fused to the promoters of *Azospirillum* genes that are thought to be important for the plant growth promotion effect are currently carried out in strain Sp245. These experiments should reveal whether *Azospirillum* cells within the plant root are metabolically active.

COMPARATIVE GENETICS OF *AZOSPIRILLUM* AND *RHIZOBIUM*

Because of the lack of a clear phenotype on the plant at the sites of *Azospirillum* interaction, the *Azospirillum brasilense* genome has been probed for the presence of DNA sequences that are recognized to contain information for plant interaction in other microbial systems. This was done either by DNA hybridizations or by functional complementation of mutants. In both cases most information has been obtained with *R. meliloti* Rm1021 DNA or mutants. Elmerich and collaborators (17) have described homology of *A. brasilense* DNA with a *nodPQ* and a *nodG* probe. The nucleotide sequence of the *nodPQ* homolog of *A. brasilense* has been determined (C. Vieille, personal communication) and the deduced amino acid sequence appears to show extensive homology with NodPQ of *R. meliloti*. *nodPQ* is located on the 90-MDa plasmid (17) whereas *nodG* is located on the chromosome. Nothing is known so far on functional complementation of the corresponding *nod* mutants of *R. meliloti*. For *exo* genes on the contrary, identification has been approached by functional complementation of *R. meliloti* mutants. In Table 2 the results of the complementation tests are summarized. Clones that correct *exoB, exoC* and *exoM* mutants of *R. meliloti* contain DNA of the 90-MDa plasmid of *A. brasilense* (9,10). All other correcting loci are located on the chromosome. It should be noticed that cosmid pCal112 corrects the *exoC* mutant only for the Calcofluor phenotype but not for the Fix^{-} phenotype. It was shown before that a mutation in the *exoC* locus in *R. meliloti* and *Agrobacterium tumefaciens* causes a pleiotropic phenotype (8). In addition to

Table 2. Correction of *R. meliloti exo* mutants with *A. brasilense* Sp7 DNA.

R. meliloti exo mutants[a]		*R. meliloti* transconjugants	
genotype	phenotype	recombinant cosmid	phenotype
exoB	Cal^-Fix^-	pCal102	Cal^+Fix^+
exoC	Cal^-Fix^-	pCal112	Cal^+Fix^-
exoG	$Cal^dFix^{+/-}$	pCal301	Cal^+Fix^+
exoK	Cal^dFix^+	pCal201	Cal^+Fix^+
exoM	Cal^-Fix^-	pCal102	Cal^+Fix^+
exoN	Cal^dFix^+	pCal601	Cal^+Fix^+
exoP	Cal^-Fix^-	pCal503	Cal^+Fix^+

[a]*R. meliloti* mutants were obtained from G. Walker (MIT, U.S.A.)

the lack of synthesis of exopolysaccharide, *exoC* mutants fail to produce a periplasmic ß(1-2) glucan. It was recently elucidated that this is due to a deficiency in the activity of the enzyme phosphoglucomutase (15). Proton-NMR analysis of the EPS isolated of the *R. meliloti exoC* mutant containing pCal112 revealed that the EPS produced is quite different from the wild type EPS, particularly in the region of the non-carbohydrate substituents. Moreover this transconjugant is devoid of ß(1-2) glucan, which also explains its Fix^- phenotype. An *A. brasilense* strain with a mutation in this locus still produces EPS. Electrophoretic protein typing of this *A. brasilense* mutant indicates disappearance of the 110-kDa protein, similar to the *che* mutants (see above). However, the *exo* mutant shows a wild-type chemotaxis phenotype. Recently we isolated a second *exoC* correcting locus from an *A. brasilense* cosmid library (pCal134). This cosmid clone fully complements the *R. meliloti exoC* mutant (Fix^+, ß(1-2) glucan synthesis, wild-type EPS). This second *exoC* locus is clearly distinct from the first *exoC* locus, since it is located on the chromosome and shows, in contrast to the first one, structural homology in DNA hybridizations with the *exoC* gene of *R. meliloti*.

CONCLUSIONS

By comparing *Azospirillum* with other diazotrophic bacteria it can be seen that quite a number of genes might be in common, not only those for nitrogen fixation as expected, but also genes involved in plant interaction. This approach will lead to the construction of *A. brasilense* mutants in a site-directed way, that then can be tested in a number of assays like chemotaxis, plant root attachment and even plant root invasion. One of the indigenous plasmids of *A. brasilense*, the 90-MDa plasmid, is of special interest for this approach since this plasmid is highly conserved among *A. brasilense* strains, and secondly because of the presence of *nod* and *exo* homologs. Therefore we decided as a first step to construct a physical map of this plasmid , as shown in Figure 1.

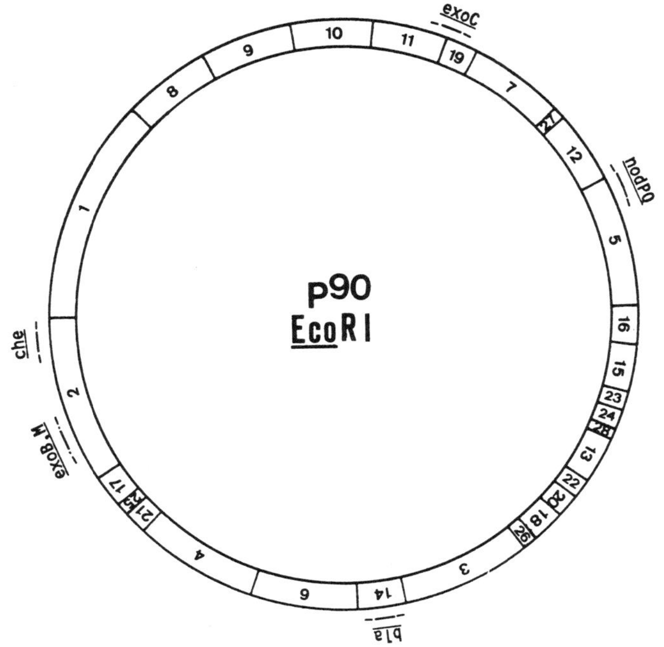

Figure 1. Physical map of the 90-MDa plasmid of *A. brasilense* Sp7. The plasmid is 144-kb in size. *bla* is a locus that confers ampicillin resistance to *E. coli* and shows homology with the beta-lactamase-gene of Tn*3*.

ACKNOWLEDGEMENTS

The authors wish to thank C. Elmerich for communication of results prior publication. This work was supported by research funds from the 'Onderzoeksraad K.U. Leuven' and the 'N.F.W.O.'.

REFERENCES

1. Bashan, Y. & Levanony, H. (1987) J. Gen. Microbiol. **133**, 3473-3480.

2. Boddey, R.M., Baldani, V.L.D., Baldani, J.I. & Döbereiner, J. (1986) Plant and Soil **95**, 109-121.

3. de Zamaroczy, M., Delorme, F. & Elmerich, C. (1989) Mol. Gen. Genet. **220**, 88-94.

4. Fani, R., Allotta, G., Bazzicalupo, M., Ricci, F., Schipani, C. & Polsinelli, M. (1989) Mol. Gen. Genet. **220**, 81-87.

5. Galimand, M., Perroud, B., Delorme, F., Paquelin, A., Vieille, C., Bozouklian, H. & Elmerich, C. (1989) J. Gen. Microbiol. **135**, 1047-1059.

6. Hall, P.G. & Krieg, N.R. (1983) Can. J. Microbiol. **29**, 1592-1594.

7. Kapulnik, Y., Feldman, M., Okon, Y. & Henis, Y. (1985) Soil Biol. Biochem. **17**, 509-515.

8. Leigh, J.A. & Lee, C.C. (1988) J. Bacteriol. **170**, 3327-3332.

9. Michiels, K., De Troch, P., Onyeocha, I., Van Gool, A., Elmerich, C. & Vanderleyden, J. (1989) Plasmid **21**, 142-146.

10. Michiels, K., Vanderleyden, J., Van Gool, A. & Signer, A.R. (1988) J. Bacteriol. **170**, 5401-5404.

11. Michiels, K., Verreth, C. & Vanderleyden, J. (1990) J. Appl. Bacteriol., in press.

12. Pedrosa, F.O. & Yates, M.G., (1984) FEMS Microbiol. Lett. **23**, 95-101.

13. Reinhold, B., Hurek, T. & Fendrik, I. (1985) J. Bacteriol. **162**, 190-195.

14. Singh, M., Tripathi, A.K. & Klingmüller, W. (1989) Mol. Gen. Genet. **219**, 235-240.

15. Uttaro, A.D., Cangelosi, G.A., Geremia, R.A., Nester, E.W. & Ugalde, R.A. (1990) J. Bacteriol. **172**, 1640-1646.

16. van Rhijn, P., Vanstockem, M., Vanderleyden, J. & De Mot, R. (1990) Appl. Environm. Microbiol. **56**, 990-996.

17. Vieille, C., Onyeocha, I., Galimand, M. & Elmerich, C. (1989) in *Nitrogen Fixation with Non-legumes*, eds. Skinner, F.A., Boddey, R.M. & Fendrik, I. (Kluwer Academic Publishers, The Netherlands), pp. 165-172.

18. Von Bulow, J.F.W. & Döbereiner, J. (1975) Proc. Natl. Acad. Sci. USA **72**, 2389-2393.

Gresshoff, Roth, Stacey & Newton (eds.) Nitrogen Fixation: Achievements and Objectives
Chapman and Hall New York • London 1990

Nitrogen fixation in filamentous cyanobacteria

R. Haselkorn, M. Basche, H. Böhme[‡], D. Borthakur*, P. B. Borthakur*, W. J. Buikema, M. E. Mulligan[+], and D. Norris

Department of Molecular Genetics and Cell Biology, The University of Chicago, Chicago, IL 60637 USA

Introduction

Cyanobacteria have the same requirements for nitrogen fixation as other diazotrophs: synthesis of the nitrogenase complex and its cofactors, a source of ATP and low potential electrons, and an environment protected from oxygen. Unique among diazotrophs, cyanobacteria evolve oxygen in the light. Filamentous strains such as Anabaena solve the problem of protecting nitrogenase from oxygen by differentiating specialized cells called heterocysts at regular intervals along the filament. In the heterocyst, the oxygen-evolving reaction of Photosystem II is turned off, CO_2 fixation stops, nitrogenase, cofactors and electron donor proteins are synthesized, and the oxidative pentose pathway is activated. A double-layered envelope of polysaccharide and glycolipid is laid down outside the existing cell wall. New connections between the heterocyst and the adjacent vegetative cells are established; these mediate the transport of carbohydrate and of fixed nitrogen.

In other diazotrophs, various combinations of classical and molecular genetic techniques were used to determine the organization of the genes required for nitrogen fixation (nif). These experiments generally began with the isolation of a few Nif^- mutants and complementation of these mutants to obtain the corresponding wild type DNA fragments. More nif genes were located by walking along the chromosome on either side of the identified genes, either by transposon mutagenesis or DNA sequencing or both. For Klebsiella pneumoniae these methods were sufficient to identify a 24-kb region

*current address: Biotechnology Program, University of Hawaii, 3050 Maile Way, Honolulu, HI 96822

[+]current address: Dept. of Biochemistry, Memorial University, St. John's, Newfoundland A1B 3X9 Canada

[‡]current address: Botanical Institute of Bonn University D-5300 Bonn 1 W. Germany

of the chromosome which, when transferred to E. coli, enabled E. coli to fix nitrogen. Until recently, such methods were not possible to use with cyanobacteria because there was no system for introduction of DNA fragments for complementation or mutagenesis. In the absence of such a system, Anabaena DNA fragments were cloned on the basis of their sequence similarity to known Klebsiella nif genes and DNA sequencing has revealed some information regarding their organization. Such studies additionally showed that large DNA elements (11 kb and 55 kb) interrupt the nif genes in Anabaena vegetative cells and that these elements are excised during heterocyst differentiation (6,7,8). Those results will be reviewed below.

The most significant breakthrough in the study of cyanobacterial nitrogen fixation and heterocyst differentiation is the development of methods for efficient conjugation of DNA from E. coli into Anabaena (17). Using these methods it has been possible to isolate genes by complementation of mutants, to inactivate genes by targeted insertional mutagenesis, to do random mutagenesis with transposons and to insert gene fusions with reporter genes to measure gene expression at the single cell level. These results will also be reviewed below.

Methods

All of the experiments to be described from our laboratory use the strain called Anabaena sp PCC7120. The cells are maintained on medium BG-11 with nitrate or ammonia as N source and induced to differentiate heterocysts by transfer to BG-11 without N source. The various manipulations of DNA have been described extensively. We currently determine DNA sequences by making nested deletions with exonuclease III and nuclease S1 and then using the M13 universal primer in dideoxy synthesis. Mutagenesis involved treatment with DES followed by penicillin selection for cells unable to fix nitrogen (4). The construction of a cosmid shuttle vector for efficient conjugation from E. coli with Anabaena is described elsewhere, as is the procedure for complementation of mutants and the isolation of complementing DNA fragments (W. J. B. & R. H., to be submitted).

Results and Discussion

Organization of the nif genes of Anabaena

The paradigm of nif gene organization is Klebsiella pneumoniae, in which all the known genes uniquely required for nitrogen fixation are clustered within 24 kb (14). In general, genes in the nif cluster that encode proteins with related functions are transcribed together. The nifHDK genes, which encode the components of the nitrogenase complex and the nifEN genes, which encode a structure required for synthesis of the Fe-Mo cofactor, comprise operons. The nifUSV genes are also cotranscribed. The functions of nifU and nifS are not yet known; nifV encodes homocitrate synthase and homocitrate is a part of the Fe-Mo cofactor (10). The nifL and A genes, both participating in regulation of transcription of the other nif genes, are cotranscribed. Finally, nifB and Q, both required for Fe-Mo cofactor synthesis, also comprise an operon (14).

The first studies of Anabaena nif genes indicated that their organization was different. The nifD and K genes were not contiguous and the nifS gene was 5' to nifHD rather than 3' as in Klebsiella (16). Moreover, these nif genes appeared to be interspersed with and surrounded by DNA that was not transcribed under nitrogen fixing conditions and therefore not likely to contain other nif genes. Subsequent work showed that although this description fits the situation in Anabaena vegetative cells, it is not the case in heterocysts. During heterocyst differentiation, two rearrangements occur in this region of the Anabaena chromosome. Both are excisions of DNA elements that interrupt the reading frames of Anabaena nif genes. One excision results in removal of an 11-kb element from the nifD coding region, resulting in the formation of a nifHDK operon very similar to that of Klebsiella (8). The second excision removes a 55-kb element to the 5' side of the nifS gene that interrupts the reading frame of a gene called fdxN (6,7,12). This results in the formation of another operon, nifB-fdxN-nifS-nifU (13).

The rearrangements are summarized in Figure 1, which also shows the Klebsiella nif gene map and the surrounding region of the Anabaena chromosome. The operon furthest to the right in the diagram is rbcLS, encoding the large and small subunits of RuBP carboxylase. These genes are transcribed in vegetative cells but not in heterocysts. The 10 kb between rbcLS and nifB have not been characterized in terms of sequence or transcripts yet.

The nifB, nifS and nifU genes were identified on the basis of sequence similarity to the corresponding genes of Klebsiella, Azotobacter, and several Rhizobium species (13). The nifB gene contains an open reading frame for a 53 kD protein whose sequence includes eight cysteine residues conserved among all nifB gene products sequenced to date. These cysteines are located in three highly conserved regions of the protein and are believed to function in the assembly of the Fe-Mo cofactor. The nifS gene encodes a 47 kD protein similar over its entire length to the corresponding nifS gene products of Klebsiella and Azotobacter. Three cysteines are conserved among the three proteins, one of which (cys 324) is embedded in a fully conserved stretch of 17 amino acids. The function of the nifS gene product is not known because nifS mutations

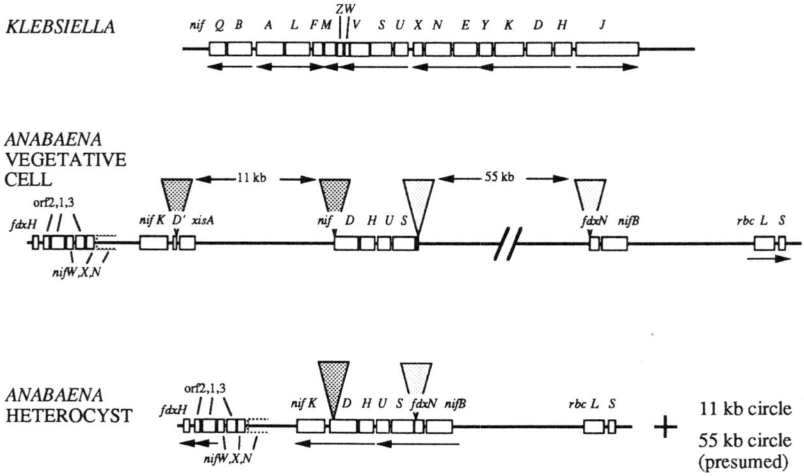

Figure 1. Organization of the nif gene region of Klebsiella, Anabaena vegetative cells, and Anabaena heterocysts after excision of the 11-kb and 55-kb elements. Shaded triangles indicate the location of short direct repeats within which the recombination events leading to excision occur. Horizontal arrows below each map indicate transcription units. Sizes of the transcripts produced by genes between nifW and nifK have not been determined. Sequencing of the nifN-nifK region has not been completed.

in Klebsiella, for the most part, are polar on nifV and so are defective in Fe-Mo cofactor synthesis. The nifU gene of Anabaena is a bit different. The predicted size of the NifU protein is 32 kD and there is considerable sequence similarity with the corresponding proteins of Klebsiella and Azotobacter. However, this similarity is confined to the amino terminal half of the sequence and another short stretch from residues 240 to 272. Within the amino terminal half there are six conserved cysteines, widely spaced, while the 240-272 stretch contains a single conserved cys-X-X-cys. The conserved cysteines suggest that NifU, like NifS, has metal ligands and therefore might be involved in Fe-Mo cofactor synthesis. Neither of these putative proteins, however, has the correct size for the protein (20 kD) found associated with the dinitrogenase apoprotein of Azotobacter (15).

The nifB operon of Anabaena contains an additional open reading frame not seen to date in other diazotrophs. Called fdxN, this gene encodes a 9 kD protein containing two domains, each with appropriately spaced cysteines for binding a 4Fe-4S center (12). Bacterial-type ferredoxins have not been seen in cyanobacteria previously except for the 9 kD Fe-S protein (product of the psaC gene) that forms part of the photosystem I reaction center.

The precise roles of these four genes in Anabaena nitrogen fixation is not known, in part because they have not yet been mutated selectively. Such experiments are now possible using the conjugation system. Transcripts of 4.4, 3.1 and 1.75 kb are seen only under conditions of nitrogenase induction. These transcripts probably start at the same site 283 bp upstream of the nifB translation start. The shortest transcript terminates after nifB, the 3.1-kb RNA probably terminates after nifS and the longest RNA includes nifU. The operon contains an additional unusual feature: tandem repeats of the sequence CCCCAAT following nifB and nifS. Since these are approximately where the 1.75 and 3.1-kb transcripts terminate, it was considered that the tandem repeats are involved in transcription termination. However, a search of a cosmid library of Anabaena DNA fragments revealed many instances of these tandem repeats, one containing 25 copies, so they may not have a specific termination function.

Immediately following the nifB operon is the nifHDK operon. This operon produces three transcripts that correspond to nifH, nifHD and nifHDK in size. The start site for transcription of this operon was mapped by S1 nuclease protection and primer extension (S. J. Robinson, unpublished) and the 5'-flanking sequence of the transcript family was determined. This sequence is not easily related to the characteristic nif gene promoter of Klebsiella or Rhizobium. Both the nifH and the nifB 5'-flanking sequences contain CTA at -12 to -10 with respect to the start site and a few conserved residues in the -70 to -40 region, but there is no obvious "nif" consensus sequence. This situation is not improved by including the fdxH gene transcript (see below).

Following the nifHDK operon are several kb of DNA for which we do not yet have sequence information. Beyond the unsequenced region we find, in the following order, nifN, nifX, orf 3, nifW, orf 1, orf 2 and fdxH. All of these except the last were assigned on the basis of similarity to sequences in Klebsiella, Azotobacter or both. NifN is required for the structure that assembles Fe-Mo cofactor. The nifX gene is in the same relative position in Klebsiella but its function is unknown. Orf 3 corresponds to a sequence (orf 3) in the Azotobacter nif gene region, not found in Klebsiella, of no known function. Orf 2 of Anabaena corresponds to orf 6 of Azotobacter. NifW is found in both Klebsiella and Azotobacter but its function is unknown also. Orf 1 of Anabaena encodes a 30 kD protein not recognized in any other previously sequenced nif gene region. However, sequence similarity to the chlD gene of E. coli was discovered by R. Jones (personal communication) in a detailed search of sequence data bases. The chlD is required for synthesis of molybdopterin, the Mo cofactor of nitrate reductase and other enzymes. Interruption of the orf 1 (chlD) reading frame in Anabaena resulted in very slow growth on N_2, suggesting that the gene product participates in nitrogen fixation (2).

The last Anabaena nif gene shown in the Figure is fdxH, which encodes a 2Fe-2S ferredoxin found exclusively in heterocysts and shown previously to be an efficient donor of electrons to Anabaena nitrogenase in vivo (1). The fdxH gene is represented in two transcripts, 0.59 and 1.85 kb. The smaller transcript contains fdxH alone and initiates between fdxH and orf 2. The larger transcript initiates 42 bp before the ATG of orf 1 and contains orf 1, orf 2 and fdxH. Neither 5'-flanking sequence, fdxH or orf 1, contains information leading to an improved consensus for an Anabaena nif gene promoter.

The rearrangement of the nif genes during Anabaena heterocyst differentiation has been reviewed recently (9). Each of the interrupting elements contains a short directly repeated sequence at its ends, within which site-specific recombination results in excision of the element and fusion of the chromosome. In the case of the 11-kb element, a gene called xisA near the left end of the element encodes a site-specific recombinase that catalyzes its own element's excision (11). The corresponding enzyme for the 55-kb element has not been located yet. Both elements are widely distributed among Anabaena and Nostoc strains from different parts of the world, although it is not at all clear what advantage they confer on their "hosts" in nature.

Not all the genes required for nitrogen fixation in Anabaena are located in the region shown in the Figure. The transposon Tn5 can be introduced into Anabaena by conjugation from E. coli, if it is suitably methylated. The transposon can enter the chromosome at many sites, where it is detected by its ability to confer resistance to neomycin (3). Resistance provides a selectable marker for cloning the interrupted DNA fragment; in this way, a previously unrecognized gene required for nitrogen fixation was cloned. The DNA fragments on which the gene is located are not in the region shown in the Figure.

Heterocyst Differentiation

A new approach to the study of genes required for heterocyst differentiation is based on the conjugation system mentioned briefly in the Introduction. The system developed by Wolk and Elhai has three components: a shuttle vector built from a cryptic plasmid capable of replication in Anabaena fused to part of pBR322, a broad host range plasmid to make the apparatus for conjugal transfer of DNA, and a third plasmid that mobilizes the shuttle vector for conjugal transfer (5,17). The shuttle vector contains cloning sites for inserted fragments of Anabaena DNA. A recent improvement includes genes encoding methylase activities, capable of protecting DNA against the Ava restriction enzymes, in the E. coli strain in which the shuttle vector is maintained. Additionally, a low copy number shuttle vector with a cos site and a multiple cloning site flanked by termination sequences that block transcription into the cloned insert has been constructed (W. J. B., to be published). These principles have been applied in Wolk's laboratory to study the timing and the spatial distribution of transcripts of several genes expressed only in heterocysts (J. Elhai and C. P. Wolk, submitted).

In our laboratory, chemical mutagenesis and three rounds of penicillin selection were used to isolate aerobic Nif^- mutants of Anabaena 7120. Among these mutants were a number that displayed abnormal heterocyst morphology; thickened envelopes, weakened cell attachments, odd shape, or, in one case, no signs of differentiation at all (W. J. B. and R. H., to be submitted). Each of these mutants was complemented en masse with the library of wild type DNA fragments, selecting for ability to fix N_2 and resistance to an antibiotic carried by the cosmid vector. From each complemented colony, plasmid DNA could be re-isolated and used to transform E. coli. Amplification of the complementing plasmid in E. coli provided the starting material for further analysis, i.e. trimming down the insert and repeating the complementation test, in order to define the minimum length of DNA needed for complementation. In this way, individual genes required for heterocyst differentiation have been cloned.

One mutant was analyzed in some detail (W. J. B., to be published). This mutant fails to initiate heterocyst differentiation when transferred to nitrogen-free medium. The wild type Anabaena gene (hetR) that complements the mutation in this strain is expressed at a low level in vegetative cells, although it is not required for vegetative growth. Its transcript increases in abundance within the first few hours of heterocyst induction. The transcript appears to be monocistronic. The gene's translation product is not similar in sequence to any known protein, including sigma factors and DNA-binding transcriptional activator proteins. The most interesting result obtained with this cloned gene to date is the following: when the wild type gene is introduced on a multicopy plasmid into either wild type Anabaena or the original mutant, on nitrogen-free medium the filaments produce clusters of heterocysts, two to five cells in a row, instead of the usual single heterocysts at approximately 10 cell intervals. It is clear that the genetic approach will yield many new insights into the fundamental questions of nitrogen fixation and heterocyst differentiation in cyanobacteria.

References

1. Böhme, H. and Haselkorn, R. (1988). Mol. Gen. Genet. **214**, 278-285.
2. Borthakur, D., Basche, M., Buikema, W. J., Borthakur, P. B. and Haselkorn, R. (1990). Mol. Gen. Genet. **221**, 227-233.
3. Borthakur, D. and Haselkorn, R. (1989) J. Bacteriol. **171**, 5759-5761.

4. Chapman and Meeks, J. C. (1987). J. Gen. Microbiol. **133**, 111-120.
5. Elhai, J. and Wolk, C. P. (1988). Methods Enzymol. **167**, 747-754 (1988).
6. Golden, J. W., Carrasco, C. D., Mulligan, M. E., Schneider, G. J. and Haselkorn, R. (1988) J. Bacteriol. **170**, 5034-5041 (1988).
7. Golden, J. W., Mulligan, M. E. and Haselkorn, R. (1987) Nature **327**, 526-529.
8. Golden, J. W., Robinson, S. J. and Haselkorn, R. (1985) Nature 314, 419-423.
9. Haselkorn, R. (1989) in Mobile DNA, eds. Berg, D. E. & Howe, M. M. (American Society for Microbiology, Washington, D. C.), pp. 735-742.
10. Hoover, T. R., Imperial, J., Ludden, P. W. and Shah, V. K. (1989). Biochemistry **28**, 2768-2771.
11. Lammers, P. J., Golden, J. W. and Haselkorn, R. (1986). Cell **44**, 905-911.
12. Mulligan, M. E., Buikema, W. J. and Haselkorn, R. (1988). J. Bacteriol. **170**, 4406-4410.
13. Mulligan, M. E. and Haselkorn, R. (1989). J. Biol. Chem. **264**, 19200-19207.
14. Orme-Johnson, W. H. (1985). Annu. Rev. Biophys. Chem. **14**, 419-459.
15. Paustian, T. D., Shah, V. K. and Roberts, G. P. (1990). Biochemistry **29**, 3515-3522.
16. Rice, D., Mazur, B. J. and Haselkorn, R. (1982). J. Biol. Chem. **257**, 13157-13163.
17. Wolk, C. P., Vonshak, A., Kehoe, P. and Elhai, J. (1984). Proc. Natl. Acad. Sci. U.S.A. **81**, 1561-1565.

Gresshoff, Roth, Stacey & Newton (eds.) Nitrogen Fixation: Achievements and Objectives
Chapman and Hall New York • London 1990

Hydrogen metabolism and the alternative nitrogenase in the cyanbacterium *Anabaena variabilis*

H. Bothe, G. Haverkamp, T. Kentemich

Botanisches Institut, Universität Köln, Gyrhofstr. 15, D-5-Köln 41, FRG.

The alternative nitrogenase was demonstrated in Azotobacter (1) and also occurs in Anabaena variabilis. This cyanobacterium can be grown in dependence on V in the medium under N_2-fixing conditions(2). The cyanobacterial V-nitrogenase reduces C_2H_2 with lower rates than the Mo-enzyme, forms partly C_2H_6 and produces more H_2 than the Azotobacter V-nitrogenase(3). This communication reports some new data on the expression of nitrogenases in A. variabilis.

The addition of nitrate to N_2-fixing, Mo-grown cultures causes a decrease of Mo-nitrogenase and the expression of nitrate reductase activities. In contrast, V- cultures continue to maximally express V-nitrogenase when supplemented with nitrate. V-grown cells are not able to use nitrate as the N-source. There is no evidence for a V-dependent nitrate reductase.

As shown by experiments with the GS inhibitor MSX, A. variabilis incorporates NH_4^+ by the GS/GOGAT pathway when fixing N_2 by either the Mo- or the V-nitrogenase. MSX causes a significant enhancement of the C_2H_4-formation rates in both cultures.

Tungstate suppresses nitrogenase activities in Mo-cultures as expected. In cultures which have expressed V-nitrogenase W apparently does not reduce C_2H_4-formation (possibly even causes an enhancement) but eliminates C_2H_6-production (Tab. 1). Rhenium suppresses nitrogenase activities in both Mo-and V- grown cultures in a parallel way.

A N_2-fixing culture which had been grown with V was diluted every second day with fresh medium containing neither V or Mo (40% fresh medium, 6o% culture,v/v). After approximately 30 d the culture lost its blue colour indicating N-deficiency. The culture restarted growth after approximately 50 d. Growth experiments (between days 50 and 135) showed a 2-4 fold increase in the protein content within 4 d indicating that the culture had recovered. Therefore a third nitrogenase which utilizes neither Mo nor V possibly exists in A. variabilis. The Mo content in the medium was less than 10^{-8} M.

Mo- and V- nitrogenases of A. variabilis grown in continuous culture catalyze an outburst of H_2 when assayed under argon and in the absence of any other substrate (N_2, C_2H_2) and under strong light intensities. Such H_2-formation, however, proceeds maximally 3-4 h. In this case, the H_2-outburst is larger by Mo-nitrogenase than by V-nitrogenase. Continuous H_2-production can be obtained when the fermenter cultures are assayed under light-limited conditions. In this case, H_2-production is higher by V- than by Mo-nitrogenase. H_2-production can be enhanced by incubating the cells with EDTA which blocks H_2-utilization by the uptake hydrogenases to approximately 50%. The data of this latter paragraph on H_2-metabolism have recently been published (4).

Tab. 1.: The effect of W on N_2-fixing Anabaena variabilis which had expressed either Mo- or V-nitrogenase.

culture condition	formation of C_2H_4	formation of H_2	formation of C_2H_6
A) Mo-culture:			
a) control culture grown with Mo	598	0.5	0
b) medium with 10 µM W	155	≤0.3	0
c) medium with 100 µM W	10	≤0.3	0
B) V-culture:			
a) control culture grown with V	280	5.0	2.8
b) medium with 10 µM W	310	4.5	1.4
c) medium with 100 µM W	340	5.0	0

References:

(1) Bishop,P.E., Jarlenski, D.M.L. & Hetherington, D. R. (1980) Proc. Natl. Acad. Sci. USA 77, 7342-7346

(2) Kentemich, T., Danneberg, G, Hundeshagen, B. & Bothe, H. (1988) FEMS Microbiol. Lett. 51, 19-24

(3) Eady, R. R. et al. (1988) in Nitrogen Fixation: Hundred Years After, eds. Bothe, H., De Bruijn F.J.& Newton, W.E. (Fischer, Stuttgart), pp.81-86

(4) Kentemich T., Haverkamp G. & Bothe H. (1990) Naturwiss. 77, 12-18

Gresshoff, Roth, Stacey & Newton (eds.) Nitrogen Fixation: Achievements and Objectives
Chapman and Hall New York • London 1990

Isolation and characterization of three amino acid aminotransferase genes from *Rhizobium meliloti*

James R. Alfano and Michael L. Kahn

Institute of Biological Chemistry and Department of Microbiology,
Washington State University, Pullman, WA 99164-6340

INTRODUCTION

Nitrogen fixation, the reduction of atmospheric dinitrogen to ammonia, is an energetically-expensive process. Carbon compounds are supplied to the bacteroid by the plant and are the source of reductant and energy needed for nitrogen fixation. In 1985, Kahn *et al.* (2) proposed a model to explain some aspects of this energy transfer. In one implementation, the model utilizes the malate-aspartate (MA) shuttle, a pathway known to operate in mitochondria. In its ideal form, the MA shuttle could transfer reducing power (NADH) into the bacteroids by shuttling organic and amino acids across the peribacteroid and bacteroid membranes. The two major enzymes used in the MA shuttle are malate dehydrogenase and aspartate aminotransferase (AAT). Our strategy to test whether a MA shuttle is operating in the symbiosis is to isolate *R. meliloti* mutants deficient in AAT activity and to determine their nitrogen fixation ability.

RESULTS AND DISCUSSION

Rhizobium meliloti genes encoding amino acid aminotransferase enzymes have been cloned by complementing an *E. coli* auxotroph using a *R. meliloti* gene library. A partial Sau3A digest of *R. meliloti* chromosomal DNA was cloned into the BamH1 site of pUC18. This gene library was then transformed into *E.coli* DL39 (1), a strain that contains mutations in aspartate aminotransferase (*aspC*), aromatic amino acid aminotransferase (*tyrB*), and branched chain amino acid aminotransferase (*ilvE*). Transformants were placed on minimal media in the absence of specific amino acids. Distinct plasmids were isolated that relieved aspartate auxotrophy (pJA10, and to some extent pJA30), leucine auxotrophy (pJA20) and tyrosine and phenylalanine auxotrophy (pJA30). Extracts of DL39 have low aminotransferase activity. However, when oxaloacetate is used as the amino acceptor, DL39(pJA10) has high aspartate aminotransferase (AAT) activity and DL39(pJA30) has high aromatic amino acid aminotransferase activity and some AAT activity. The aminotransferase gene on pJA30 was insertionally inactivated with Tn5 and this mutation was then recombined into *R. meliloti* genome. The resulting transaminase mutant was Fix$^+$.

REFERENCES

1. Berg, C.M., Wang, M., Vartak, N.B. and Liu, L. 1988. Gene 65: 195-202.
2. Kahn, M.L., Kraus, J. and Somerville, J.E. 1985. p. 193-199. *In* H.J. Evans, P.J. Bottomley and W.E. Newton (eds.), "Nitrogen Fixation Research Progess".

Gresshoff, Roth, Stacey & Newton (eds.) Nitrogen Fixation: Achievements and Objectives
Chapman and Hall New York • London 1990

Identification and functional analysis of *fixLJ*-like genes in *Bradyrhizobium japonicum*

D. Anthamatten and H.Hennecke

Mikrobiologisches Institut, Eidgenössische Technische Hochschule, ETH-Zentrum, Schmelzbergstrasse 7, CH-8092 Zürich, Switzerland

By interspecies hybridization using the *Rhizobium meliloti fixLJ* genes (1), we identified and cloned *fixL*- and *fixJ*-like genes of *Bradyrhizobium japonicum*. DNA sequence analysis revealed the presence of two ORFs, *fixL* and *fixJ*, whose closely adjacent arrangement suggests that they are organized in an operon (Fig. 1).

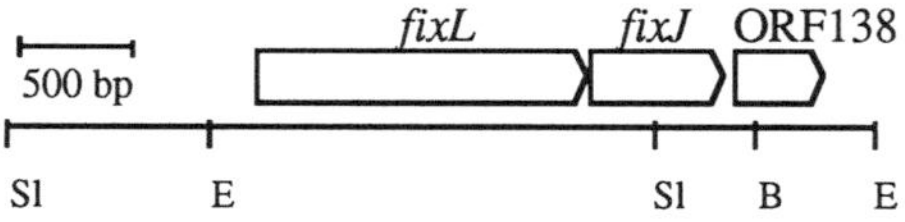

Figure 1: Physical map of the *B. japonicum fixLJ* DNA-region. B: *Bam*HI; E: *Eco*RI; Sl: *Sal*I.

FixL, a protein of 505 amino acids, and FixJ, 205 amino acids long, are highly homologous (~50% identity) to their counterparts in *R. meliloti*. In contrast to FixL of *R. meliloti* no hydrophobic domains are present in the N-terminus of *B. japonicum* FixL. Both genes, *fixL* and *fixJ*, are important for symbiosis between *B. japonicum* and soybean: a nonpolar *fixL* deletion mutant as well as a *fixJ* insertion mutant showed a 90% reduced Fix phenotype. *fixJ* mutant strains are not able to grow anaerobically with nitrate; therefore, FixJ seems to be necessary for switching from aerobic to anaerobic respiration. In *R. meliloti* the *fixLJ* genes are responsible for the activation of *nifA* expression at low oxygen concentrations (1). In contrast to *nifA* regulation in *R. meliloti* the *B. japonicum* regulatory *fixRnifA*-operon is expressed under aerobic conditions, and there is a NifA-dependent 5-fold increase of this expression under microaerobic conditions (2). Using chromosomally integrated *fixR'*- and *nifA'-'lacZ* fusions in *fixL* and *fixJ* mutant backgrounds we showed that the aerobic expression of the *fixRnifA* genes was not dependent on the presence of FixL or FixJ. Furthermore, FixJ (which is supposed to be a DNA-binding protein) does not bind to the *fixR*-upstream DNA. A plasmid-coded *fixJ'-'lacZ* fusion was introduced into a *nifA* mutant strain. We found that the aerobic as well as the anaerobic expression of *fixJ* was independent of NifA, thus excluding the possibility of FixJ being indirectly involved in the NifA-dependent microaerobic activation of *fixRnifA* expression .

Downstream of *fixJ* there is a third ORF (Fig. 1) predicting a protein of 138 amino acids that is homologous (36% identity) to the N-terminal part of FixJ of *B. japonicum*. Disruption of ORF138 led to a Fix^{+}phenotype; therefore, the function of ORF138 is presently unknown.

References:
(1) David et al. (1988), Cell **54:**671-683
(2) Thöny et al. (1989), J. Bacteriol. **171:**4162-4169

Gresshoff, Roth, Stacey & Newton (eds.) Nitrogen Fixation: Achievements and Objectives
Chapman and Hall New York • London 1990

Tn*5*-induced mutants of *Azospirillum brasilense* ATCC 29710 (Cd) altered in cell surface properties

A. Arunakumari and C.A. Neyra*

Biochemistry and Microbiology Dept., Rutgers University, New Brunswick, NJ 08903-0231, USA

Azospirillum brasilense ATCC 29710 (Cd) was mutagenized with Tn*5* using plasmid pRK602 at a frequency of 2.1 x 10^{-7}. Mutants were selected on nutrient agar containing kanamycin (Km) and streptomycin. Analysis of 1000 Km resistant mutants yielded three Tn*5* derivatives impaired in nitrogenase activity (C_2H_2 reduction) in N-free semisolid malate medium. Mutant Cd6A expressed reduced rates of growth and pellicle formation while mutant Cd16A did not grow nor form a pellicle even after prolonged incubation. However, both mutants exhibited similar growth patterns as the wild type in liquid and solid media under aerobic conditions. Cd16A was defective in extracellular polysaccharide production (EPS) as indicated by the failure to yield fluorescence in calcofluor containing agar plates (Cal^-). Flocculation in minimal medium containing fructose and nitrate (MMFN) was significantly reduced by 70% and nitrite accumulated in the medium (140 nmole/ml). Mutant Cd6A compared well with the wild type in terms of calcofluor fluorescence (Cal^+) and floc yield but nitrite did not accumulate in the medium. Mutant Cd6A1, on the other hand was impaired in carotenoid biosynthesis (Car^-) and flocculation in minimal media. However, adherence of mutants Cd16A and Cd6A1 to corn roots in short term *in vitro* assays, or multiplication in corn rhizosphere, in a soil microcosm, were unaltered, indicating that carotenoid production and flocculation ability may not be directly related to corn root colonization under conditions similar to our experiments.

Gresshoff, Roth, Stacey & Newton (eds.) Nitrogen Fixation: Achievements and Objectives
Chapman and Hall New York • London 1990

Genetic analysis of *nifA*- and *nifL*-like genes in *Azotobacter vinelandii*

A. Bali and C. Kennedy

AFRC IPSR Nitrogen Fixation Laboratory, University of Sussex, Brighton BN1 9RQ UK

Mutagenesis and sequencing identified nifA-like regions in A. vinelandii (1,2). The partial sequence (300bp) of a gene upstream from the sequenced nifA indicated an ORF with similarity to nifL of K. pneumoniae (2). Another nifA-like region identified by phenotype and restriction map associated with Tn5 insert (1) is different from the sequenced nifA gene. This region hybridized to a nifA probe but mutants had a more pleiotrophic phenotype than nifA mutants; further restriction and hybridization analysis of the two regions show them to be unambiguously distinct. This second nifA-hybridizing gene is proposed to be ntrD. Plasmids which constitutively produce A. vinelandii NIFA further distinguished the nifA and ntrD regions.

A lambda library clone contained contiguous 12Kb and 4Kb EcoRI fragments hybridizing to nifA and nifB probes, respectively. The 12 Kb fragment carried the N-terminal region of nifA and an upstream nifL-like gene. Sequencing has extended the region of similarity between this gene and the nifL gene of K. penumoniae. A KIXX cartridge was inserted within the nifL-like gene and also within nifA. The effect of these mutations on nitrogenase activity and nifH-LacZ expression was as follows:

		UW136 (wt)	MV378 (nifL.KIXX)	MV383 (nifA.KIXX)
N_2ase Activity	$-NH_4$	48	52	0
	$+NH_4$	0	45	0

The mutation of the nifL-like gene results in relief of ammonium repression of nitrogenase synthesis in A. vinelandii as it does in K. pneumoniae . However, unlike in K. pneumoniae, nifL mutants show persistent nitrogenase synthesis because nifA expression is not controlled by fixed nitrogen. Also, the requirement of molybdenum for full epxression of the nifH operon is partially relieved in the mutants.

1. Santero, E., Toukdarian, A., Humphrey, R., & Kennedy, C. (1988) Mol. Microbio. 2:303-314.
2. Bennett, L., Cannon, F.C. & Dean, D. (1988) Mol. Microbiol. 2: 315-321.

Gresshoff, Roth, Stacey & Newton (eds.) Nitrogen Fixation: Achievements and Objectives
Chapman and Hall New York • London 1990

Construction of broad host range NPT II reporter vectors and their use in analysis of constitutive and inducible promoters in *Bradyrhizobium japonicum* USDA 110

W. Mark Barbour, Maria G. Schell, and Gary Stacey.

Center for Legume Research and Dept. of Microbiology, University of Tennessee, Knoxville, TN 37996-0845, USA.

INTRODUCTION

An important technique for genetic analysis is the use of transcriptional or translational gene fusions to reporter genes. A useful reporter is neomycin phosphotransferase (NPT II), a protein which confers neomycin and kanamycin resistance on its host cells. The presence of NPT II-fusion proteins can be positively selected by resistance to neomycin or kanamycin allowing direct selection of an active fusion from a very large population. The goal of this project was to construct a broad host range vector series which can be maintained in *Bradyrhizobium japonicum* USDA 110 and to use these vectors to isolate constitutive and inducible promoters from *B. japonicum*.

MATERIALS AND METHODS

Plasmid pP375 was partially digested with EcoRI and BamHI and ligated individually to EcoRI/BamHI digests of NPT II reporter vectors pKm109/3, pKm109/9, and pKm109/90 (1,3), yielding vectors pMM100, pMM101, and pMM102. These vectors were used to generate a random clone bank in USDA 110 using electroporation (2) and this bank was screened for clones with constitutive kanamycin resistance. An enrichment method was used to isolate presumptive seed exudate inducible clones.

RESULTS AND DISCUSSION

A broad-host range, NPT II reporter vector series was developed for analysis of inducible and constitutive promoters in the soybean symbiont *B. japonicum*. The vectors pMM100, pMM101, and pMM102 represent the three possible reading frames in a translational fusion and allow the direct isolation of active fusions to NPT II by plating on kanamycin containing medium. Clones were isolated which expressed NPT II in both *B. japonicum* and *E. coli*. This expression was confirmed by ELISA using anti-NPT II antibody and by subcloning the plasmid inserts into a β-galactosidase reporter vector. Work is in progress to isolate inducible promoters from *B.japonicum*.

REFERENCES

1. Beck, E., Ludwig, G., Auerswald, E. A., Reiss,B., and Schaller, H. (1982) *Gene* 19:327-336.
2. Hatterman, D. R. and Stacey, G. (1990) *Appl. Environ. Microbiol.* 56:833-836.
3. Reiss, B., R. Sprengel, and H. Schaller. (1984) *EMBO* 3:3317-3322.

Gresshoff, Roth, Stacey & Newton (eds.) Nitrogen Fixation: Achievements and Objectives
Chapman and Hall New York • London 1990

Identification of Bacteriohopanetetrol in *Frankia*

A.M. Berry[1], R.A. Moreau[3], A.D. Jones[2] and S.F. Osman[3]

[1]Department of Environmental Horticulture, and [2]Facility for Advanced Instrumentation, University of California, Davis, CA 95616, USA; [3]Eastern Regional Research Center, USDA/ARS, Philadelphia, PA 19118, USA.

Cytological evidence indicates that high levels of lipids are associated with nitrogen-fixing tissue in actinorhizal root nodules (1). The purpose of the investigation was to characterize the lipid profile of actinorhizal nodules using a new method to separate lipid classes (2), and then to identify resulting compounds of interest.

Lipid-containing extracts were prepared by conventional methods from root nodules of two hosts, *Alnus rubra* and *Ceanothus griseus* var. *horizontalis*. Extracts were prepared from root tips of the hosts, and from two strains of *Frankia*, HFPArI3 and HFPCcI3. Lipids classes were separated and quantified by a new HPLC method which employs a flame ionization detector (2).

An unusual, major lipid peak was detected in nodule tissue extracts, representing 20-30% of total extracted nodule lipids. The lipid was also present in *Frankia* extracts, but not in host root extracts, or in a plant lipid standard mixture. The unknown lipid was identified by mass spectrometry (with acetylation) to be bacteriohopanetetrol. The molecular ion of the tetraacetate was 714, and major ion fragments had m/z of 191, 369, 493 and 654. In *Frankia*, bacteriohopanetetrol represented 30-50% of total extractable lipids, in ammonium-grown as well as in N_2-grown cells.

Bacteriohopanetetrol ($C_{35}H_{62}O_4$) is a hopanoid, or pentacyclic triterpenoid, with a hydroxylated side chain. Although this is the first report of the occurrence of bacteriohopanetetrol in *Frankia*, hopanoids have been shown to occur in numerous bacterial species (3). This amphiphilic, sterol-like molecule probably functions in stabilizing *Frankia* membranes, and could play a role in maintaining functional nitrogenase activity in the *Frankia* vesicle.

REFERENCES

1. Berry, A.M., Sunell, L.A. & Liu. Q. (1989) in *Physiology, Biochemistry and Genetics of Nongreen Plastids*, eds. Boyer, C., Shannon, J., & Hardison, R. (A.S.P.P., Rockville, MD), 287.
2. Moreau, R.A., Asmann, P.T. & Norman, H.A. (in press) Phytochemistry.
3. Rohmer, M., Bouvier-Navé, P. & Ourisson G. (1984) J. Gen. Microbiol. **130**, 1137-1150.

Gresshoff, Roth, Stacey & Newton (eds.) Nitrogen Fixation: Achievements and Objectives
Chapman and Hall New York • London 1990

Hydrogen-deuterium exchange studies of the two 'iron-only' hydrogenases from *Clostridium pasteurianum*

D. Bonam and L.E. Mortenson

Center for Metalloenzyme Studies, University of Georgia, Athens, GA 30602, USA.

Clostridium pasteurianum, a saccharolytic, anaerobic, N_2-fixing bacterium, expresses two distinct hydrogenases. The bidirectional hydrogenase (H_2ase I) functions to produce H_2 during fermentative growth. The function of the uptake hydrogenase (H_2ase II) is not known. These two "iron-only" hydrogenases contain hydrogen activating Fe-S centers (H-centers), and ferredoxin-like Fe-S centers (F-centers) which shuttle electrons between the H-center and electron carriers. We have compared these two enzymes using the HD exchange technique, which allows the analysis of hydrogenase enzymes in the absence of net electron flux. D_2 in H_2O is converted into a mixture of D_2, HD, and H_2. The rates and ratios of formation of these products provide information about the active-site of the enzyme. Although these enzymes are postulated to contain similar H-centers, they did not show similar HD exchange. H_2ase I catalyzed total exchange at a rate 200-times faster than H_2ase II [790 vs. 3.5 umole (H_2 + HD) produced $min^{-1}mg^{-1}$ at 30°C, pH 8.0]. These rates are correlated with the hydrogen evolution rates [510 vs. 5.9] suggesting that release of H_2 is the rate limiting step in both assays. 1 mM dithionite caused a dramatic decrease in the H_2:HD ratio of H_2ase I between pH 8.0-8.8, without affecting the total exchange rate, suggesting that dithionite can reduce an electron accepting cluster in the enzyme and inhibit double exchange. The H_2:HD ratio for H_2ase II increased from 0.3 to 10, with no change in the total exchange rate, when the enzyme was stored under H_2. This suggests a substrate-induced stable modification of the enzyme (F-center reduction?) which alters the reaction occurring at the H-center, yielding preferential double exchange.

Gresshoff, Roth, Stacey & Newton (eds.) Nitrogen Fixation: Achievements and Objectives
Chapman and Hall New York • London 1990

Ammonia assimilation in *Rhodobacter capsulatus*

Roberto Borghese and Judy D. Wall

Biochemistry Department, University of Missouri. Columbia, Mo 65201 USA

Nitrogen assimilation in prokaryotes has been shown to proceed via the action of glutamate dehydrogenase (GDH) and/or through the coupled action of glutamine synthetase (GS) and glutamate synthase (GOGAT) with the latter operating to assimilate ammonium generated during nitrogen fixation. The phototrophic nitrogen fixing Rhodobacter capsulatus has been shown to possess only the GS/GOGAT assimilatory pathway. This situation implies that, as in other microorganisms, a GOGAT$^-$ mutant should be a glutamate auxotroph. Such a mutant, despite extensive search, has not been isolated in R.capsulatus. We have now isolated a series of mutants (e.g., ADI-1 and ADI-5) from EMS mutagenized cells that grow more slowly with ammonium as nitrogen source under aerobic conditions. Significant differences were seen in the aerobic growth rates between the wild-type B100 strain and the mutants with ammonium or serine but no difference when a number of other amino acids were used. On the other hand photosynthetic growth rates were not different even when ammonium or dinitrogen were used as nitrogen source. The enzymes known to be involved in ammonium assimilation in R.capsulatus, GS, GOGAT and alanine dehydrogenase (ADH), were then tested in crude extracts from cultures grown under aerobic, photosynthetic/anaerobic and nitrogen fixing conditions. GS and ADH activities in the mutants were always close to wild-type levels while GOGAT was undetectable under all conditions tested. The mutants were also characterized for methylammonium uptake which is mediated by a carrier thought to be under an NtrC-type control, since R.capsulatus NtrC mutants show no methylammonium uptake under nif-derepressing conditions.ADI-1 and ADI-5 grown under photosynthetic nif-derepressing conditions, either on N_2 or glutamate, showed methylammonium uptake levels that were between 70% and 90% of those of the wild-type strain B100. Wild-type cells grown aerobically on glutamate (also nitrogen-derepressing conditions) showed uptake values similar to photosynthetic/anaerobic nif-derepressed cells. The mutants grown under the same conditions had no measurable methylammonium uptake. We infer that different factors may be involved in nitrogen regulation under aerobic and anaerobic growth conditions. This new phenotype seems to be the result of at least two mutations; one eliminating GOGAT activity and the second possibly in a regulatory locus. These findings suggest that another, as yet unidentified, ammonium assimilatory pathway may operate preferentially under anaerobic conditions.

Gresshoff, Roth, Stacey & Newton (eds.) Nitrogen Fixation: Achievements and Objectives
Chapman and Hall New York • London 1990

Genetic studies of glutamate metabolism in *Rhizobium meliloti*

J. L. Botsford,* Thomas Lewis, and Rigoberto Gonzalez.

New Mexico State University.
Las Cruces, New Mexico 88003, USA.

We have cloned the gene for glutamate synthase, glt. Using site directed mutagenesis and gene replacement, mutants auxotrophic for glutamate have been isolated. All mutants grow very slowly on minimal medium supplemented with glutamate and are unable to utilize amino acids as a source of assimilated nitrogen (Ntr^-). These mutants, like the wild type, accumulate glutamate when osmotically stressed. All are Fix^+. Suppressor mutants which remain auxotrophic but grow rapidly on minimal medium with glutamate or other amino acids (Ntr^+) can be isolated. Pseudo-revertants, cells no longer auxotrophic for glutamate can be isolated. These appear to have elevated levels of glutamate dehydrogenase. An insertion mutation adjacent to glt and resulting in a Fix^- Glt^+ phenotype has been isolated.

Gresshoff, Roth, Stacey & Newton (eds.) Nitrogen Fixation: Achievements and Objectives
Chapman and Hall New York • London 1990

A *Rhizobium phaseoli* Tn5 mutant pleiotropically affects succinate utilization, GSII activity and N_2 fixation

A. Bravo, L. M. Martínez, Y. Mora, G. Du Pont, J. Calderón* and J. Mora

Centro de Investigación sobre Fijación de Nitrógeno, U. N. A. M., Apdo. Postal 565-A, Cuernavaca, Morelos, México.

A single Tn-5 insertion mutant (AB2) of *R. phaseoli* CFN42 does not grow well on NO_3^- or NH_4^+ in succinate or fumarate; however optimal growth is observed in the presence of glutamine or glucose. This mutant induces the transport of succinate very slowly, the level reaching at most 50% of the parental type. This carbon limitation on succinate results in a very low expression of GSII activity on different nitrogen sources as NH_4^+, glutamate or glutamine; however, on NO_3^- or in the absence of nitrogen, GSII is somehow induced. There is a coordination between the transport of the carbon source and the expression of GSII, since in glucose, where the AB2 mutant grows optimally, it expresses a high GSII activity. This mutant utilizes glucose and glutamine preferentially, rather than succinate as the carbon source. No hybridization was detected with the Tn-5 insertion of AB2 with the following genes: *ntrB*, *ntrC*, *glnB*, and *nifA*. Also, *ntrA* does not complement this mutant, though it is partially complemented by a *dctABD* plasmid. Furthermore, a *dctA-lac* plasmid expresses 7-fold higher ß-galactosidase activity in the AB2 mutant than in the parental type. The pleiotrophic character of the AB2 mutatation is also reflected in a decrease of N_2 fixation. This new mutation coordinately lowers *dctA* expression and increases *dctBD* transcription. Preliminary characterization indicates that the Tn-5 is inserted in the *dctABD* region.

Acknowledgements. We are grateful to Clive Ronson for *dctABD* and *dctA-lac* plasmids, and to Leticia Olvera for technical assistance.

1. Ronson, C.W., Astwood, P.M., Nixon, B.T. & Ausubel, F.M. (1987) Nucleic Acids Res. **15**, 7921-7934.

Gresshoff, Roth, Stacey & Newton (eds.) Nitrogen Fixation: Achievements and Objectives
Chapman and Hall New York • London 1990

Purification of a pyruvate: Oxidoreductase from nitrogen fixing *Rhodospirillum rubrum*

Erica Brostedt and Stefan Nordlund

Department of Biochemistry, Arrhenius Laboratories,
Stockholm University, S-106 91 Stockholm, Sweden

INTRODUCTION

Electron transport to nitrogenase have only been extensively studied in Klebsiella pneumoniae, in which a pyruvate: flavodoxin: oxidoreductase and a flavodoxin transfer electrons from pyruvate to nitrogenase (2,3). In Rhodospirillum rubrum pyruvate will support nitrogenase activity in extracts in the presence of added ferredoxin (1). In this investigation we report the purification of a pyruvate: oxidoreductase from R. rubrum.

MATERIALS AND METHODS

R. rubrum was grown photoheterotrophically with molecular nitrogen as nitrogen source. The activity of pyruvate: oxidoreductase was determined spectrophotometrically with the reduction of methyl viologen (3). Purification was done under anaerobic conditions and included ion-exchange chromatography, gelfiltration and dye-ligand chromatography.

RESULTS

The enzyme obtained had a specific activity of 14 µmol/min mg and the purification was 50 fold compared with the crude extract. The M_r of the native enzyme was estimated to 250 000 by gelfiltration and SDS-PAGE indicated that the enzyme is a dimer. The purified enzyme supported pyruvate dependent nitrogenase activity in the presence of ferredoxin. The results show that electron transfer to nitrogenase from pyruvate in R. rubrum involves an oxidoreductase similar to the one in K. pneumoniae and that ferredoxin probably is the electron donor to dinitrogenase reductase. Further investigations will show if this electron transport is operating under all growth conditions.

REFERENCES

1. Ludden, P.W. & Burris, R.H. (1981) Arch. Microbiol. 130, 155-158.
2. Shah, V.K., Stacey, G. & Brill, W.J. (1983) J. Biol. Chem. 258, 12064-12068
3. Wahl, R.C. & Orme-Johnson, W.H. (1987) J. Biol. Chem. 262, 10489-10496.

Gresshoff, Roth, Stacey & Newton (eds.) Nitrogen Fixation: Achievements and Objectives
Chapman and Hall New York • London 1990

In vitro sporulation of *Frankia*

Stephen H. Burleigh and Jeffrey O. Dawson.

Dept. of Forestry, 1301 West Gregory Dr., Univ. of Illinois, IL, USA.

Sporulation is a common attribute of Frankia strains and has a profound effect on the actinorhizal symbiosis. Sporulation within nodules of actinorhizal plants is associated with decreased nitrogen fixation and spores greatly increase the nodulation potential of the Casuarina cunninghamiana isolate, HFPCcI3. We have studied the effects of ammonium chloride, sodium chloride, potassium phosphate, melibiose, temperature and rotary-shaking treatments on the in vitro sporulation of Frankia strain HFPCcI3. Maximal sporulation (28 million spores per ml of medium) was achieved in nitrogen- and phosphate-free cultures grown at 23-28°C, and rotary-shaken at 100 r.p.m. High temperature (33°C) combined with rotary-shaking greatly inhibited sporulation. Sodium chloride at concentrations ranging from 100-500 mM inhibited the growth and sporulation of HFPCcI3 more than the same molar concentrations of melibiose. This suggests that NaCl inhibition is not attributable solely to osmotic effects. These finding may be useful in obtaining highly infectious inocula for Casuarina species and understanding how environmental effects influence sporulation of Frankia.

Gresshoff, Roth, Stacey & Newton (eds.) Nitrogen Fixation: Achievements and Objectives
Chapman and Hall New York • London 1990

Nitrate reductase activity in plasmid-minus derivatives of *Rhizobium "hedysari"*

S. Casella*, A. Toffanin, N. Rossi

Istituto di Microbiologia Agraria, Università di Pisa
Centro di Studio per la Microbiologia del Suolo, CNR. Pisa, ITALY

Some fast-growing strains of *Rhizobium "hedysari"* can reduce NO_2^- to N_2O after growing at low oxygen concentration. Nitrite reductase activity appeared not to be coupled to energy conservation as shown by the inability of cells to grow anaerobically on nitrite or to link proton translocation to nitrite reduction. The ability to remove nitrite should play a generally important role in maintaining nitrogenase activity in bacteroids.

Plasmid-minus mutant was obtained by thermic shock (4) from *R. "hedysari"* HCNT1 (IMAP 801). Plasmidic patterns were obtained by the Eckhardt procedure (3). N_2O evolution and C_2H_2 reduction were estimated by G.C. as previously described (2). Cytochrome spectra were determined in whole cells (1).

Lack of the two large plasmids in the cured strain HA7 (IMAP 807) resulted in loss of both symbiotic properties and ability to reduce acetylene *in vitro*. However, it showed nitrite reductase activity comparable with that of the parent strain if properly induced under low O_2 concentration, indicating that the enzymes involved on nitrite reduction are chromosomally encoded.
From synthesis of cytochrome aa_3 found in aerobic cells, strain HCNT1 and its plasmid-minus derivative seem to shift to the production of cytochrome *o* when grown under O_2-limited conditions. Under the same growth conditions *R. "hedysari"* CC1335 (IMAP 835), unable to produce N_2O from nitrite, and the "true denitrifier" *Bradyrhizobium japonicum* USDA6 did not synthesize any cytochrome *o*.

REFERENCES.

(1) Casella, S., Shapleigh, J.P., and Payne, W.J. (1986) Arch. Microbiol. **146**, 233-238.
(2) Casella, S., Shapleigh, J.P., Lupi, F., and Payne, W.J. (1988) Arch. Microbiol. **149**, 384-388.
(3) Eckhardt, T. (1978) Plasmid **1**, 584-488.
(4) Zurkowski, W. and Lorkiewicz (1978) Gen. Res. Camb. **32**, 311-314.

ACKNOWLEDGEMENTS.

This work was supported by CNR.

Gresshoff, Roth, Stacey & Newton (eds.) Nitrogen Fixation: Achievements and Objectives
Chapman and Hall New York • London 1990

Morphology and life cycle of the nitrogen-fixing cyanobionts in the seven *Azolla* species

R. Caudales[1], A. D. Antoine[1] and A. C. Vasconcelos.[2],

Dept. Biochemistry and Microbiology[1] and Dept. Biological Sciences.[2] Rutgers-The State University, New Brunswick New Jersey, 08903, USA.

The goal of this research was to establish the true taxonomy of the cyanobacteria found in the cavities of Azolla spp. A comparative study of the morphology, architecture, factors controlling morphological changes, and life cycle of these important nitrogen-fixing cyanobacteria was undertaken.

To study the morphology and life cycle of the cyanobacteria, the apex and the 5th, 10th, 15th, and 20th leaves were dissected from the main stem axes at 2, 4, 6 and 8 week old Azolla plants. The cyanobacteria were observed under phase contrast microscopy.

There exists a dynamic, synchronous morphological differentiation in vegetative cells and heterocysts parallel to the growth of Azolla. In the apex, trichomes are composed entirely of small, vegetative cells (hormogonia). In older leaves, cyanobacteria are enclosed in the cavity and embedded in a mucilaginous material. Trichomes showed a reduction of individual sheath and are straight to curved. Heterocysts are intercalary and terminal. Akinetes, when present, are between vegetative cells or adjacent to one or both sides of the heterocysts. They are single or in chains with variable shapes and sizes depending on the Azolla species. Akinete germination gives rise to hormogonium composed of cells smaller than the parental trichome. Hormogonia are formed adjacent to a single or double heterocyst or between heterocysts. This is the first report of the occurrence of multiple heterocysts and their adjacency to akinetes in the trichomes of the cyanobacteria in Azolla. Plant chromatic changes and light intensity regulate the life cycle and pigment composition of the cyanobacteria. The cyanobionts inside the Azolla cavities receive the light spectrum imposed by the physiological, chromatic changes of the plant. Green light stimulates an increase of heterocyst frequency and acetylene reduction. Yellow light stimulates akinete germination and hormogonial differentiation. In red light, trichomes are short and vegetative cells are reduced in width. They also lack normal pigmentation and acetylene reduction is negligible. Based on microphotographic analysis and the observations presented above, the cyanobionts in Azolla should be classified in the genus Nostoc rather than Anabaena.

Gresshoff, Roth, Stacey & Newton (eds.) Nitrogen Fixation: Achievements and Objectives
Chapman and Hall New York • London 1990

Nitrogen fixation and hydrogen evolution of immobilized *Anabaena azotica* HB686

Cheng Shuangqi, Chen Zhaoping, Mo Ximu, Cao Yuying, Zhang Wenwen and He Xiaoqiong

Department of Biology, South China Normal University, Guangzhou, Peoples Republic of China.

INTRODUCTION. The technique of immobilizing cells has made great progress recently, but there are few reports on the technique being applied to studies of nitrogen fixation. We have used this technique to study nitrogen fixation, H_2 evolution and H_2 uptake by immobilized A. azotica HB868.

METHOD. Utilizing Ca-alginate as a carrier, A. azotica HB686 was embedded to make a particle of 4mm diameter, incubated in nitrogen-free medium, and illuminated with 3000 lux. The acetylene-reducing and hydrogen-evolution activities of immobilized and native cells was estimated under an illumination of 20000 lux with incandescent light in different growth periods.

RESULT AND DISCUSSION. The acetylene-reducing activity of immobilized cells rose steadily to 300 nmol C_2H_4/mg/h by the 10th day and was maintained through the 20th day; in native cells, it reached 200 nmol C_2H4/mg/h by the 10th day and dropped gradually to 100 nmol C_2H_4/mg/h on the 10th day. The amounts of H_2 evolution in immobilized and native cells were activate 50 nmol H_2/mg/h and 25 nmol H_2/mg/h, respectively. The H_2 uptake in immobilized cells was 2.5 times higher than that of native cells, illustrating that energy recycling in immobilized cells was better than in native cells. Tolerance to oxygen for immobilized cells was greater than for native cells. The acetylene-reducing activity of the former decreased 25% while the latter decreased 45% under 10% oxygen for 2 hours. This result provides the basis for industrial ammonia production by nitrogen-fixing microorganisms by packing particles of immobilized cells into columns.

Project supported by National Natural Science Foundation of China.

Gresshoff, Roth, Stacey & Newton (eds.) Nitrogen Fixation: Achievements and Objectives
Chapman and Hall New York • London 1990

The presence and relative amounts of a stress protein in free-living and symbiotic *Bradyrhizobium japonicum*

E. Y. Choi, G. S. Ahn and K. W. Jeon

Department of Zoology, University of Tennessee, Knoxville, TN 37996, USA

Free-living bacteria and symbiotic bacteroids of *Bradyrhizobium japonicum* were found to contain a 65-kDa stress protein which is antigenically related to other known stress proteins such as groEL heat-shock protein of *E. coli*. Thus the 65-kDa protein of *Bradyrhizobium* cells cross-reacted with an anti-groEL antibody prepared against hsp58 of *Tetrahymena* (T. McMullin & R. L. Hallberg, Mol. Cell. Biol. 8:371, 1988) and a monoclonal antibody against a 67-kDa stress protein from obligatory bacterial endosymbionts of the xD strain of *Amoeba proteus*, as determined by immunoblotting. The relative amount of a stress protein contained in symbiotic *Bradyrhizobium* bacteroids from soybean root nodules was about 7-times larger than that of free-living bacteria of the same species, as determined by comparing the amount of ^{125}I-labeled secondary antibody bound on immunoblots. The results suggested that bacterial symbionts were under a stress condition, although the nature of the stress is not known. The synthesis of stress proteins is known to be enhanced in a variety of cells ranging from archaebacteria to human cells when cells are exposed to stress conditions. It is generally believed that the stress proteins protect cells from harmful effects of various stresses. A host cell has been considered to be a hostile environment to intracellular parasites (J. W. Moulder, J. infect. Dis. 130:300, 1974). However, such a view has not been extended to cases where the host and symbionts have a stable and long-lasting relationships, although it has been known that nitrogen-fixing bacteroids are under the host cell's control of oxygen and nutrient availability (T. M. Rolfe & P. M. Gresshoff, Annu. Rev. Plant. Physiol. 39:297, 1988). Our results suggest that symbionts are under a stress condition even in "mutually beneficial" symbiotic relationships and that the stress protein might somehow help stabilize the host-symbiont interactions. [Supported by a grant from the National Science Foundation]

Gresshoff, Roth, Stacey & Newton (eds.) Nitrogen Fixation: Achievements and Objectives
Chapman and Hall New York • London 1990

Genetic analysis of two symbiotic mutants of *Bradyrhizobium japonicum*

Jong-Yoon Chun and Gary Stacey.

Center for Legume Research, Dept. of Microbiology and Graduate Program of Plant Physiology and Genetics, The University of Tennessee, Knoxville, TN 37996-0845 USA.

INTRODUCTION

Previously, two Nod+, Fix- mutants of *Bradyrhizobium japonicum* strain 61A101c were isolated (1). The mutants, ML126 and ML150, have very similar symbiotic phenotypes in that they are defective in release from the infection thread and induce the host cell to fill with empty vesicles (2). The aim of this study was to further characterize these Fix- loci. In addition, this work presented an opportunity to compare the genomic organization between *B. japonicum* 61A101c and USDA110 in the cloned regions.

METHODS

Genomic DNA libraries of mutants ML126 and ML150 were constructed in cosmid pHC79. The cosmid clones corresponding to the mutated regions were isolated by the kanamycin resistance encoded by Tn5. The wild-type regions corresponding to the original mutations were cloned from *B. japonicum* 61A101c and USDA110 and compared. Deletion and insertion mutants were constructed in the cloned DNA by interposon mutagenesis.

RESULTS

Physical mapping of the mutations on cloned DNA from strain ML126 and ML150 indicated that they were approximately 18 kb apart. Strain ML126, but not ML150, was defective in cytochrome oxidase activity as indicated by a negative Nadi test; suggesting a defect in heme biosynthesis or electron transport in this strain. The DNA loci corresponding to the original ML126 and ML150 mutations were cloned from USDA110 and the two regions were found to be at least 80 kb apart. Deletion mutations of the USDA110 DNA locus corresponding to the ML126 homologous region were found to be defective in symbiotic nitrogen fixation. These cloned regions show no hybridization to known nod, fix, or nif loci and therefore identify new fix loci in *B. japonicum*.

REFERENCES

1. So, J.-S., Hodgson, A.D.M., Haugland, R., Leavit, M., Banfalvi, Z., Nieuwkoop, A.J. & Stacey, G. (1987) Mol. Gen. Genet. 207, 15-23.
2. Roth, L.E. & Stacey, G. (1989) Eur. J. Cell Biol. 49, 13-23.

Gresshoff, Roth, Stacey & Newton (eds.) Nitrogen Fixation: Achievements and Objectives
Chapman and Hall New York • London 1990

Isolation and characterization of regulatory sequences of *Azospirillum brasilense*

L. Civardi, M. Delledonne and C. Fogher

Istituto di Genetica, Università Cattolica, Piacenza 29100, Italy.

Bacteria of the genus Azospirillum are diazotrophs associated with the roots of grasses. Considerable progress towards understanding the molecular biology of their association with plants is based on the identification of plant-inducible genes. With this goal, we have isolated and sequenced a set of "constitutive" promoter-containing DNA fragments of Azospirillum brasilense Sp7 (1). Trascriptional fusions to CAT reporter gene of plasmid pKK232-8 (2) allowed us identified in E. coli-constitutive promoters at the frequence of 3.4%. The first 17 sequences, isolated at 20 µg/ml Cm, have been classified in seven MIC classes of resistance from 20 to 280 µg/ml of Cm (Fig. 1).

Table 1. Number of clones, for each MIC classes, and size (bp).

	Concentration of Cm (µg/ml)						
	20	80	100	120	140	200	280
number of clones	2	2	5	1	2	3	2
size in bp	2950	-	500	390	1070	590	467
	213	990	650		-	1100	1300
			700			1600	
			1100				
			2200				

One fragment from each class has been sequenced and analysis of the general feature has, for the moment, revealed only a common TCCTTC sequence. This sequence is present also at position -150 from the start codon of A. b. glnA (3), -232 of nifH (4) and -64 of nodG (5).

1. Civardi L., Fogher C. 1989. XXXIII Conv. S.I.G.A., Alghero, p.126.
2. Brosius J. 1984. Gene, 27:151-160.
3. Bozouklian M., Elmerich C. 1986. Biochimie 68:1181-1187.
4. Fani R. et al. 1989. Mol. Gen. Genet. 220:81-87.
5. Delledonne M. et al. Abstr. this Congress.

Gresshoff, Roth, Stacey & Newton (eds.) Nitrogen Fixation: Achievements and Objectives
Chapman and Hall New York • London 1990

A *Rhizobium leguminosarum* biovar. *viciae* protein homologous to *Escherichia coli* Fnr is involved in regulation of nitrogen fixation

S.Colonna-Romano, A.Schlüter, W.Arnold, A.Pühler and U.B. Priefer
Universität Bielefeld, Lehrstuhl für Genetik, Postfach 8640, 4800 Bielefeld 1 FRG

According to the current model, oxygen regulation of nif and fix genes in Rhizobium meliloti is controlled by a cascade regulatory circuit consisting of the gene products encoded by fixL, fixJ, fixK and nifA.
By interspecific complementation, a gene of R.leguminosarum biovar. viciae strain VF39 was identified which is able to promote microaerobic induction of the nifA-independent R.meliloti fixN gene in both an R.meliloti fixJ and fixK mutant and to complement the Fix$^-$ phenotype of an R.meliloti fixK mutant. Sequence analysis of this open reading frame revealed that the corresponding gene product showed homology to R.meliloti FixK and to Fnr, the transcriptional regulator of anaerobic respiratory genes in E.coli. The R.leguminosarum protein differs in two main aspects from R.meliloti FixK:

- its expression is independent from FixJ in R.meliloti
- its N-terminal domain contains a cysteine motif resembling a metal binding cluster.

A similar domain is also present in E.coli Fnr, where it was shown to be essential for its oxygen dependent activity. We therefore propose to call the R.leguminosarum gene fnrN. However, a Tn5-generated mutant of fnrN showed only reduced nitrogenase activity (40% of wild-type level), although the ability to induce the R.meliloti fixN gene was completely lost. Therefore, additional oxygen regulatory systems must exist in R.leguminosarum biovar. viciae.
The promoter regions of Fnr-regulated E.coli genes contain a highly conserved element which represents an Fnr consensus binding sequence. Analogous motifs were detected in front of fnrN and upstream of the nifA gene in R.leguminosarum biovar. viciae (within the fixC coding region). Constitutive expression of fnrN did not affect the induction of R.leguminosarum nifA-lacZ fusions in E.coli, suggesting that fnrN is not involved in the modulation of oxygen-regulated expression of R.leguminosarum biovar. viciae nifA.

Gresshoff, Roth, Stacey & Newton (eds.) Nitrogen Fixation: Achievements and Objectives
Chapman and Hall New York • London 1990

Uptake of macromolecules by *Frankia*

Benoit Cournoyer and Philippe Normand

Centre de Recherche en Biologie Forestière, Université Laval, Québec (CAN), G1K-7P4. Canada

Frankia are slow-growing nitrogen-fixing actinomycetes which enter in symbiosis with dicotyledenous plants such as members of the Rosaceae, Casuarinaceae and Betulaceae families by forming root nodules. Presently, identification of symbiotic genes is hindered by the lack of a cloning procedure. Frankia spp. genes described in the litterature have been cloned using homologous probes from other bacterial genera (3) thus limiting the comprehension of this symbiosis. The introduction of DNA in Frankia spp. is a necessary step towards a new understanding. Attempts using protoplasts failed, probably because of low regeneration rates (4).
Through electroporation, we have introduced various macromolecules inside Frankia spp. This technique was highly reproducible and simple. The uptake efficiency was similar for strains of all host-specificity groups tested. Electroporation was obtained on intact Frankia cells using a single exponentially decaying pulse. Relative to other Gram positive cells (1, 2, 5), a longer Rc-time and a lower voltage were found to be optimum. About 20% of the initial population survived the electrical shock. After pulse delivery, some pores remained opened at 0°C and showed complete reversibility only at 37°C. Dextran labeled with fluorescein isothiocyanate (70 kDa) was introduced and appeared to be a good marker for uptake detection. This opens possibilities of physiological works using dyes inside cells. Plasmids as large as 26 kb were also introduced but uptake efficiency decreased with size. They were detected for at least 5 days but not observed after two months of incubation. This suggests a replication problem and we are trying to improve the plasmids by reduction of their size and by using an indigeneous cryptic plasmid coming from Frankia strain ArI3. Since it was shown that plasmids were maintained for at least 5 days, transient expression studies can be done.
We have elaborated a way to introduce and follow DNA in Frankia cells. Moreover, we can now study replication and expression. We believe that electroporation will allow us to identify symbiotic genes specific to this interaction.

References

1. Aukrust, T. and Nes, I.F. (1988) FEMS Microbiol. Lett. **52,** 127-132.
2. Mahillon, J., Chungjatupornchai, W., Decock, J., Dierickx, S., Michiels, F., Peferoen, M. and Joos, H. (1989) FEMS Microbiol. Lett. **60,** 205-210.
3. Normand, P., Simonet, P. and Bardin, R. (1988) Mol. Gen. Genet. **213,** 238-246.
4.Normand, P., Simonet, P., Prin, Y. and Moiroud, A. (1987) Physiol. Plantarum **70,** 259-266.
5. Somkuti, G.A. and Steinberg, D.H. (1988) Biochimie **70,** 579-585.

Gresshoff, Roth, Stacey & Newton (eds.) Nitrogen Fixation: Achievements and Objectives
Chapman and Hall New York • London 1990

A single *Bradyrhizobium* locus differentially affects host plant nodule development

Susan Crank, Kate Wilson and Ken Giller,

Wye College, University of London, Wye, Ashford, Kent TN25 5AH. UK.

Introduction: Many *Bradyrhizobium* strains can fix nitrogen with a variety of hosts. Here we study 284, a transposon mutant of wild type *Bradyrhizobium* strain NC92, with three effective hosts of NC92 - groundnut, pigeonpea and siratro (1).

Materials and methods: Plants were grown under nitrogen free conditions and inoculated with NC92, 284 or the transconjugant 284pKW43 which carried a plasmid containing a 3.2kb fragment from NC92. At harvest nodules were either processed for microscopy or their nitrogenase and respiratory activities were measured using a gas flow through system (2).

Results and discussion: After 8 weeks nitrogen fixation by 284 symbioses was reduced relative to wild type to 77% on siratro, 55% on groundnut and 8% on pigeonpea. In all cases nodulation was delayed, also nodule development was impaired and bacteroid density decreased in proportion to the drop in host fixation effectivity. However neither the relative efficiency nor the energy costs of fixation by 284 nodules differed from wild type values implying that those bacteroids present were fixing nitrogen as efficiently as the wild type. Strain 284pKW43 fully complemented the 284 mutation on siratro and pigeonpea but not groundnut, although this was possibly due to secondary effects of the vector on the groundnut symbiosis. As 284 showed a qualitatively similar symbiotic phenotype with each host it seems likely that the 3.2kb wild type DNA fragment in pKW43 is neccessary for effective nodule invasion of all three hosts. This probably involves some aspect of host/bacteroid interaction, such as signalling or metabolic compatibility, which is of particular importance in pigeonpea.

References:

1)Wilson,K.,Anjaiah,V.,Nambiar,P. & Ausubel,F. (1987) J.Bacteriol. **169**, 2177-2186.
2)Witty,J.Minchin,F. & Sheehy,J. (1983) J.Exp.Bot. **34**, 951-963.

Gresshoff, Roth, Stacey & Newton (eds.) Nitrogen Fixation: Achievements and Objectives
Chapman and Hall New York • London 1990

Membrane vesicle-bound nitrate reductase from *Bradyrhizobium japonicum*

M.J. Delgado, J. Olivares and E.J. Bedmar.

Departamento de Microbiología. Estación Experimental del Zaidín, CSIC. Apdo. 419. 18080-Granada, Spain

INTRODUCTION

Although the presence of a membrane-bound nitrate reductase (NR) enzyme in free-living cells and bacteroids of Bradyrhizobium japonicum is well established (1,2), no data are available about the properties of the enzyme. In this work we described a partial purification and characterization of the enzyme from microaerobically nitrate-grown cells and bacteroids of B. japonicum strain L-236.

METHODS

Cells of B. japonicum L-236 were grown microaerobically (1% O_2) in defined liquid medium with KNO_3 (10 mM) as the only nitrogen source. Bacteroids were isolated from nodules of Glycine max L. Merr. Cv. Williams inoculated with L-236. Bacteria and bacteroids were disrupted by sonication. Membranes were collected by ultracentrifugation. Membrane vesicles were obtained by treatment of the membranes with 4% Triton X-100. Aliquots were applied to a Sephacryl S-200 column and eluted with Tris-HCl buffer, pH 7.5, containing 0.1% Triton X-100. Active fractions were pooled together, concentrated by precipitation with ammonium sulfate (80% saturation) and subjected to nondenaturating polyacrylamide gel electrophoresis. Triton-PAGE on gradient (4%-20%) gels was carried out in vertical slabs (3). NR activity in gels was determined as in (4).

RESULTS

Relative molecular mass of the membrane vesicle-bound NR enzyme from microaerobically nitrate-grown cells estimated by gel filtration was of 240000 daltons, whereas it was of 3000000 in PAGE. The NR enzyme from bacteroids had a molecular weight of 80000 after gel filtration. Activity could not be detected in gels. Differences in molecular weight and in sensitivity to several reagents suggest that the two enzymes are different.

REFERENCES

Daniel, R.M. and Gray, J. 1976. J. Gen. Microbiol. 96: 247-251.

Kennedy, I.R., Rigaud, J. and Trinchant, J.C. 1975. Biochem. Biophys. Acta 3970: 24-35.

Laemmli, U.K. 1970. Nature (London) 227: 680-685.

Lund, K. and DeMoss, J.A. 1976. J. Biol. Chem. 252: 2207-2216.

Supported by DGICYT Grant Nº AGR89-564

Gresshoff, Roth, Stacey & Newton (eds.) Nitrogen Fixation: Achievements and Objectives
Chapman and Hall New York • London 1990

Nucleotide sequence of an *Azospirillum nod* homologous region

M. Delledonne, R. Porcari, and C. Fogher

Istituto di Genetica, Università Cattolica, Piacenza 29100, Italy.

A 12 kb DNA segment of the A. brasilense Sp7 genome, that shows homology with an heterologous R. meliloti nod region has been cloned and partially sequenced (1). The construction of the physical map of the cosmid with this fragment and the DNA sequence analysis of some subclones has allowed the positioning and the identification of homology with nodG from R. meliloti. The nucleotide sequence of the putative nodG gene of A. brasilense shares 60% homology with nucleotide sequence of R. meliloti nodG (2). The sequence has a ORF of 765 bp coding for a protein of 255 aa residues and 27.3 kd MW. The predicted aminoacid sequence shows an overall 33% homology with R.m. nodG and two highly conserved regions (residue 69 to 93 and 133 to 190) with, respectively, 86 and 88% similarity, (Fig 1).

Figure 1. Aminoacid sequence of the two highly conserved regions of R. m. nodG and A. b. nodG.

```
69                       KAMVAKIEADLGPVDILVNNAGITRD     nodG  A. b.
                         ::   . ::::  :::::::::::.:
68                       KALGQRAEADLEGVDILVNNAGITKD     nodG  R. m.
133   GRIINISSVNGVKGQAGQTNYSAAKAGVIGFTKALAAELATKGVTVNAIAPGYIGTDM
      :::::..:: :  :. ::::: : ::: :::.: :: :.::. .::: .:::.: . :
132   GRIINVTSVAGAIGNPGQTNYCASKAGMIGFSKSLAQEIATRNITVNCVAPGFIESAM
```

The nucleotide sequence of the A.b. nodG gene was compared against the nucleotide sequence data base. One sequence, for the actIII gene of Streptomyces coelicolor homologous to known oxidoreductases, shows a 63% homology (3). The region upstream of the ATG start codon shows homology to the 16S rRNA of E. coli, and a large inverted repeat of 43 residues is present after the stop codon. The presence of the ribosome binding site, the high G+C percent inside the ORF (65.5), and the G or C preferential presence as third base in codons (82%) give strong indications as coding frame of this DNA sequence (4).

1. Antonelli M. et al. 1988. Abstr. XXXI Conv. S.I.G.A., Como.
2. Debellé F., Sherma S.B. 1986. Nucleic Acid Res. 14: 7453.
3. Hallam S.E. et al. 1988. Gene 74:305-320.
4. Gribskov M. et al. 1983. Nucleic Acid Res. 11:539.

Gresshoff, Roth, Stacey & Newton (eds.) Nitrogen Fixation: Achievements and Objectives
Chapman and Hall New York • London 1990

Cloning and characterization of a gene for production of a symbiotically repressed outer membrane protein antigen of *Rhizobium leguminosarum*

Ruud A. de Maagd, Hayo C.J. Canter Cremers and Ben J.J. Lugtenberg.

Leiden University, Department of Plant Molecular Biology, Botanical Laboratory, Nonnensteeg 3, 2311 VJ Leiden, The Netherlands.

The outer membrane of *R. leguminosarum* undergoes a large number of changes during the differentiation from the free-living form to the bacteroid state (1). We are studying the nature of these changes, their regulation, and their role in normal nodule development. Using polyclonal and monoclonal antibodies, we have defined 4 distinct outer membrane protein antigen groups, present in all studied *R. leguminosarum* strains. The function and expression of one of those groups, group III, is object of this study. This is a group of antigenically related outer membrane proteins with apparent molecular weights from 35 to 45 kDa. The group consists of two major proteins, with Mr's of 35 and 40 kDa, respectively, and their derivatives containing peptidoglycan residues of varying lengths. These proteins form Ca^{2+}-stabilized oligomers, which resist denaturation by SDS at 100°C (2). The quantity of group III antigens is severely diminished in cell envelopes of pea bacteroids of *Rhizobium leguminosarum* biovar *viciae* strain 248 (1). In order to study the function and regulation of expression of this antigen, we have cloned a gene responsible for production of part of this group, i.e. the 35 kDa major protein. Monoclonal antibody 38, recognizing a strain-specific, surface-exposed epitope on the oligomers of group III, was used in a colony blot assay to isolate a clone from a cosmid-bank of strain 248 DNA, expressed in another *Rhizobium* strain. Analysis of the nucleotide sequence of this clone revealed the presence of an open reading frame of 1098 base pairs, encoding a protein of 366 amino acids with a predicted molecular weight of 39 kDa. The N-terminal part of this protein contains a possible signal sequence. The promoter of this gene will be isolated to study its regulation.

1. de Maagd, R.A., de Rijk, R., Mulders, I.H.M. & Lugtenberg, B.J.J. (1989) J.Bacteriol. **171**,1136-1142.
2. de Maagd, R.A., Wientjes, F.B., & Lugtenberg, B.J.J. (1989) J. Bacteriol. **171**,3989-3995.

Gresshoff, Roth, Stacey & Newton (eds.) Nitrogen Fixation: Achievements and Objectives
Chapman and Hall New York • London 1990

Control of cellular nitrogenase expression by the molar malate/ammonium-ratio in *Rhodobacter capsulatus*

M. Dörffler and J. Oelze*
Inst. Biol. II (Mikrobiologie), Universität Freiburg,
Schänzlestr.1, D-7800 Freiburg FRG

Expression of nitrogenase is controlled by the N-status of cells, which depends on the presence of both bound N and an organic carbon source (1, 2). It is not known, however, at which particular C/N-ratio nitrogenase is expressed or repressed, respectively, in Rhodobacter capsulatus. In order to study nitrogenase expression under strictly controlled culture conditions, R. capsulatus was grown in a chemostat on 5mM of NH_4^+ and increasing concentrations of malate in the feed medium. To permit N_2 fixation, the cultures were gassed with N_2. Steady state cultures were subject to analyses of total N as well as of residual NH_4^+. Increasing malate concentrations up to 10 mM resulted in decreasing steady state levels of NH_4^+ in the culture fluid. Above 10 mM of malate, NH_4^+ was completely used up by the organisms, while the concentration of bound N increased with increasing malate up to at least 30 mM. This suggests the occurrence of nitrogenase activity above 10 mM of malate. Subsequently, cellular nitrogenase activity was determined by in situ N_2 fixation and acetylene reduction with cultures adapted to different NH_4^+ and malate concentrations in the feed medium. The results confirmed that nitrogenase activity becomes detectable as soon as malate and NH_4^+ are consumed at a molar ratio $\geq$ 2. Similar results were obtained on the basis of H_2 production by batch cultures. Western immunoblots with nitrogenase iron protein antiserum showed that the activity and the proteins of nitrogenase occurred concomitantly. Calculating malate and NH_4^+ conversion into biomass according to:
$1.43\ C_4O_5H_6 + 0.72\ NH_3 \longrightarrow C_4H_{7.3}N_{0.72}O_2 + 1.71\ H_2O + 1.72\ CO_2$ reveals that the malate/NH_4^+-ratio of two defines the threshold above which the cultures become N-limited.

1. Hallenbeck, P.C. (1987) Crit. Rev. Microbiol. 14:1-48
2. Hillmer and Gest (1977) J. Bacteriol. 129:724-731

Gresshoff, Roth, Stacey & Newton (eds.) Nitrogen Fixation: Achievements and Objectives
Chapman and Hall New York • London 1990

Glutamine cycling and unbalanced growth in *Rhizobium*

S. Encarnación, V. Narváez, E. Martínez, Y. Mora, H. Taboada, A. Bravo, J. Calderón, and J. Mora*
Centro de Investigación sobre Fijación de Nitrógeno, U.N.A.M., Apdo. Postal 565-A, Cuernavaca, Mor., México.

During the infection of the plant by Rhizobium, this bacteria, after proliferating in the nodule, stops growing. It has been found that R. phaseoli and R. meliloti strains after growing optimally a first time on succinate ammonium (M.M.) lose the capacity to grow again on M.M. Although these Rhizobium strains are not auxotrophic for biotin and glutamine, the addition of each compound restores optimal growth on M.M. R. phaseoli CFN42 assimilates ammonium by the GSII-GOGAT pathway when growing on M.M.(1). This cyclic pathway expends more energy than when ammonium is assimilated by the GDH-GS, unidirectional pathway. The inhibition of ammonium assimilation in R. phaseoli CFN42 by the GSII-GOGAT pathway restores partial growth on M.M. However, if the inhibition is on the CFN42 strain harboring the GDH gene (2), full and continuous growth on M.M. is observed. Furthermore, this strain (GDH^+) has a low N_2 fixation (15%) (2). The loss of growth capacity on M.M. is different for each Rhizobium and is related to its ATP content and to its GSII structure and activity. The unbalanced growth of Rhizobium on M.M. can be understood as a process that derives more carbon and energy for growth, and may explain why succinate is a good carbon source for Rhizobium to grow optimally, although temporarily. The effect of glutamine and biotin on Rhizobium growth capacity on M.M. may be the result of a more efficient carbon supply and distribution. Biotin may activate the enzymes that decarboxylate oxaloacetate to phosphoenolpyruvate or to pyruvate. As other microorganisms (4) R. meliloti cycles glutamine. The synthesis of glutamine and its cycling on M.M. derives carbon into amino acids as well as the oxidation of succinate. R. meliloti lacking GSI, GSII and GSIII activity (3) reduces carbon oxidation of succinate and its distribution into amino acids. It is proposed that the energy expenditure in glutamine cycling (4) on M.M. drives an effective carbon flow to support optimal growth, and may be responsible for the proliferative-stop process of Rhizobium. This would explain why the addition of glutamine restores continuous growth of Rhizobium on M.M.

References.

1. Bravo, A. & Mora, J. (1988) J. Bacteriol. 170, 980-984.
2. Bravo, A., Becerril, B. & Mora, J.(1988) J. Bacteriol.170, 985-988.
3. De Bruijn et al. (1989) J. Bacteriol. 171, 1673-1682.
4. Mora, J. (1990) Microbiol. Rev., in press.

Gresshoff, Roth, Stacey & Newton (eds.) Nitrogen Fixation: Achievements and Objectives
Chapman and Hall New York • London 1990

Two functional homologs of the σ^{54} gene (*rpoN*, *ntrA*) in *Bradyrhizobium japonicum*

H.M. Fischer, I. Kullik, T. Bruderer, S. Fritsche, J.Sanjuan, H. Knobel, Ch. Walther and H. Hennecke

Mikrobiologisches Institut, Eidgenössische Technische Hochschule, Schmelzbergstrasse 7, CH-8092 Zürich, Switzerland.

Recognition of -24/-12-type promoters by RNA polymerase requires a specific sigma factor, σ^{54} (RpoN, NtrA). In the nitrogen fixing symbiont, *Bradyrhizobium japonicum,* two separate, strongly conserved *rpoN* genes (78 % identical aa) were identified and sequenced. A comparison of RpoN1 (484 aa) with RpoN2 (537 aa) revealed the presence of an extra stretch of 48 aa in RpoN2 between the short N-terminal and the large C-terminal, highly conserved domains. This region which differs between RpoN1 and RpoN2 corresponds to the domain that is most diverged in all known RpoN proteins. Downstream of *B.japonicum rpoN2* an open reading frame (ORF203) was identified which showed homology to ORFs located downstream of the *rpoN* genes of *Rhizobium meliloti* (4), *Klebsiella pneumoniae* (3), *Azotobacter vinelandii* (3) and *Pseudomonas putida* (2).

Individual mutagenesis of *B.japonicum rpoN1* and *rpoN2* by gene disruption resulted in *B.japonicum* mutants which both were Nod$^+$Fix$^+$ on soybean whereas a *rpoN1/2* double mutant had a Nod$^+$Fix$^-$ phenotype. This suggested that both *rpoN* copies were functional and could replace each other under symbiotic conditions. The number of nodules formed by the double mutant was about 50 % higher as compared to wild type-infected plants; the total nodule dry weight per plant, however, was not affected. The peribacteroid space in plant cells infected by the *rpoN1/2* double mutant was enlarged and the leghemoglobin content in these nodules was reduced to about 25 %. Nodules elicited by the *rpoN1/2* mutant strain did not show necrosis as described for *nifA*$^-$ *B.japonicum* induced nodules (1).

In agreement with the observed Fix$^-$ phenotype of the *rpoN1/2* double mutant no expression of a *nifH'-'lacZ* fusion was observed in this background under anaerobic and symbiotic conditions. Surprisingly, both the aerobic expression and the *nifA*-dependent autoregulation of the proposed -24/-12 *fixRnifA* promoter were unaffected by the *rpoN* mutations (see also Fig.1. in the article by Hennecke et al. in this book). Using chromosomally integrated, translational fusions of *lacZ* to *rpoN1* and *rpoN2* it was found that *rpoN2* was expressed in an apparently constitutive manner under aerobic and anaerobic conditions whereas expression of *rpoN1* was increased about 15-fold upon a shift from aerobic to anaerobic conditions.

1. Fischer, H.M., Alvarez-Morales, A. & Hennecke, H. (1986) EMBO J. **5**, 1165-1173.
2. Inouye, S.,Yamada, M., Nakazawa, A. & Nakazawa, T. (1989) Gene **85**, 145-152.
3. Merrick, M.J. & Coppard, J.R. (1989) Mol. Microbiol. **3**, 1765-1775.
4. Ronson, C.W. Nixon, T.B., Albright, L.M. & Ausubel, F.M. (1987) J. Bacteriol. **169**, 2424-2431.

Gresshoff, Roth, Stacey & Newton (eds.) Nitrogen Fixation: Achievements and Objectives
Chapman and Hall New York • London 1990

The immediate fate of N_2 fixed by *Gloeothece* growing under alternating light and darkness

J.R. Gallon, S.M. Perry & K.J. Flynn

Biochemistry Research Group, School of Biological Sciences, University College of Swansea, Singleton Park, Swansea, SA2 8PP, UK.

Introduction: Under alternating 12h light and 12h darkness, cultures of *Gloeothecee* fix virtually all of their N_2 in the dark. However, most protein synthesis occurs in the light. As a consequence, the immediate fate of newly fixed nitrogen is not clear.
Methods: *Gloeothece* sp. ATCC 27152 was grown as described in [3]. Amino acids were measured by HPLC [2]; cyanophycin [5] and phycobiliproteins [1] by spectrophotometry.
Results & Discussion: Arginine was the only intracellular amino acid to show fluctuations similar to those of N_2 fixation. Its concentration decreased in the light and increased in the dark. The insoluble nitrogen reserve material, cyanophycin [multi-L-arginyl-poly-(L-aspartic acid)], attained its highest concentration coincident with peak N_2 fixation, though fluctuations in cyanophycin were small and some accumulation also occurred in the light, probably at the expense of free arginine. The intracellular concentration of phycobiliproteins, which function both as a nitrogen reserve and as photosynthetic pigments, decreased during the period of N_2 fixation. There were significant changes in extracellular amino acids, with a release of glycine, serine and alanine during N_2 fixation followed by an uptake when fixation had ceased. Assuming a $C_2H_2:N_2$ ratio of 4.2 - 4.8:1 [4], this release constituted 38 - 43% of the nitrogen fixed. The balance could be accounted for by the increases in intracellular arginine and cyanophycin observed during the period of N_2 fixation. It is proposed that some newly fixed nitrogen is released as amino acids into the mucilage sheath that surrounds *Gloeothece*, whilst the rest is converted into insoluble cyanophycin. In neither form would newly fixed nitrogen interfere with metabolism. Consistent with this, cultures of *Gloeothece* appear to be physiologically nitrogen starved throughout the light:dark cycle (intracellular glutamine/glutamate < 0.05), and N_2 fixation under these conditions is regulated by the availability of carbon rather than nitrogen [3].

References: [1] Bennett, A. & Bogorad, L. (1973) J. Cell Biol. **58**, 419-435.
[2] Flynn, K.J. (1988) Chem. Ecol. **3**, 269-293.
[3] Gallon, J.R., Perry, S.M., Rajab, T.M.A., Flayeh, K.A.M., Yunes, J.S. & Chaplin, A.E. (1988) J. Gen. Microbiol. **134**, 3079-3087.
[4] Peterson, R.B. & Burris, R.H. (1976) Anal. Biochem. **116**, 125-132.
[5] Simon, R.D. (1973) J. Bacteriol. **114**, 1213-1216.

Gresshoff, Roth, Stacey & Newton (eds.) Nitrogen Fixation: Achievements and Objectives
Chapman and Hall New York • London 1990

Aspects of *nod*-gene regulation in *Azorhizobium caulinodans* ORS571

D. Geelen, K. Goethals, G. Van den Eede, E. Messens, M. Van Montagu and M. Holsters.

Laboratorium of Genetics, State University Ghent, 9000 Ghent, Belgium.

The stem-nodulating *Sesbania rostrata* symbiont *Azorhizobium caulinodans* ORS571 has been shown to harbour *nodABC* related genes, the expression of which is induced in the presence of host plant exudate or commercially purchased naringenin (Goethals *et al.*, 1989). By screening a pLAFR1 gene library for naringenin-induced expression of the common nod genes in an Agrobacterium background, we have been able to isolate a functional *nodD* gene of strain ORS571 (Goethals *et al.*, 1990). A Tn*5* insertion in the activator locus was homogenotized in the ORS571 genome and resulted in a mutant strain that had lost the capacity to nodulate both roots and stems of *Sesbania rostrata*. Sequencing of the activator locus indicated the presence of a single 942 bp-long open reading frame, ORFD, with significant homology to the *nodD* genes of (brady)rhizobia. Additionally we purified the major inducer from *Sesbania rostrata* seedling exudate and identified it as 4',7-dihydroxyflavanone or liquiritigenin.

References

Goethals, K., M. Gao, K. Tompeke, M. Van Montagu, and M. Holsters. 1989. Common *nodABC* genes in Nod locus *1* of *Azorhizobium caulinodans* : nucleotide sequence and plant-inducible expression. Mol. Gen. Genet. **219**:289-298.

Goethals, K., Guy Van den Eede, M. Van Montagu, and M. Holsters. 1990. Identification of a functional *nodD* gene in *Azorhizobium caulinodans* ORS571. J. Bacteriol. **172**:2658-2666.

Gresshoff, Roth, Stacey & Newton (eds.) Nitrogen Fixation: Achievements and Objectives
Chapman and Hall New York • London 1990

Nod box-related sequences in *Azorhizobium caulinodans* strain ORS571

K. Goethals, M. Gao, D. Geelen, G. Van Den Eede, M. Van Montagu, and M. Holsters

Laboratorium voor Genetica, State University Ghent, B-9000 Gent (Belgium).

Azorhizobium caulinodans strain ORS571 is able to induce nitrogen-fixing nodules on the root and stem of the tropical legume *Sesbania rostrata*. Several *nod* loci involved in this interaction have been isolated and characterized. *Nod* locus *2* encodes functions presumably involved in the production of polysaccharides (1). *Nod* locus *1* carries the common nod genes the expression of which is regulated by a unique *nodD* gene (*nod* locus *3*) in conjunction with plant produced flavonoids or the flavanone naringenin (2, 3). Upstream of the common *nod* locus a sequence (Pa) was detected showing some 50% homology to the *nod* box sequences of (brady)rhizobia. When cloned in the correct orientation in a promoter-probe plasmid a naringenin-induced, *nodD*-dependent *lacZ* expression was evident. Transcript mapping experiments located a major, induced start site at 25 nucleotides downstream of the Pa 3' end. The Pa sequence was functionally equivalent to a typical *nod* box promoter as it became activated in an *Agrobacterium* background in the presence of the ORS571 *nodD* gene as well as the MPIK3030 *nodD* gene. Comparison of the Pa sequence with (brady)-rhizobial *nod* boxes resulted in a new, more stringent consensus.

In this sequence the presence of the inverted repeat ATC-9 bp-GAT (or a variant) in two or more copies suggests that these motifs present the actual recognition sites of consecutive NodD dimers. Comparison of this sequence with a second naringenin-inducible promoter and with other *nod* box-related, but inactive sequences of ORS571 confirmed the importance of the depicted motifs in the NodD/*nod* box interaction.

References

1. Van den Eede *et al.* (1987) Mol. Gen. Genet. 206, 291-299.
2. Goethals *et al.* (1989) Mol. Gen. Genet. 219, 289-298.
3. Goethals *et al.* (1990) J. Bacteriol. 172, 2658-2666.

Gresshoff, Roth, Stacey & Newton (eds.) Nitrogen Fixation: Achievements and Objectives
Chapman and Hall New York • London 1990

Independent regulation of *nifHDK* operon rearrangement and transcription during heterocyst differentiation in the cyanobacterium *Anabaena* sp. strain PCC 7120

J.W. Golden*, L.L. Whorff, and D.R. Wiest

Department of Biology, Texas A&M University, College Station, Texas 77843-3258, USA.

The filamentous cyanobacterium *Anabaena* sp. strain PCC 7120 produces terminally differentiated cells called heterocysts that are highly specialized morphologically and biochemically for nitrogen fixation. The genome of *Anabaena* 7120 undergoes at least two specific developmentally regulated DNA rearrangements during heterocyst differentiation. One rearrangement involves the nitrogen-fixation operon *nifHDK*. During heterocyst differentiation, an 11-kb DNA element is excised from the *nifD* gene by site-specific recombination. The excision results in the formation of the complete *nifD* coding sequence and allows expression of the three genes from a single promoter. The *xisA* gene, located on the 11-kb element, is required for the excision of the element and presumably encodes a site-specific recombinase. Transcription and rearrangement of the *nifHDK* operon both occur late during heterocyst differentiation, about 12-18 hours after induction. We show that these two events are independent of one another. Northern analysis of the *xisA* mutant strain DW12-2.2, which can not excise the *nifD* 11-kb element or fix nitrogen, showed that the *nifH* and *nifD* genes are transcribed on the unrearranged chromosome after heterocysts are formed. Therefore, the activation of the *nifH* promoter in heterocysts is not dependent on rearrangement of the operon. The *nifK* gene was not transcribed in DW12-2.2, indicating that its expression is dependent on the *nifH* promoter and excision of the 11-kb element from the operon. To test the hypothesis that transcription from the *nifH* promoter induces excision of the 11-kb element, we deleted a 1.68-kb fragment containing the *nifH* promoter from the chromosome to produce the mutant strain LW1. LW1 formed heterocysts, but did not fix nitrogen and showed no transcription through *nifD*. Southern analysis of LW1 showed normal rearrangement of the 11-kb element from the *nifHDK* operon, indicating that transcription from the *nifH* promoter is not required for developmentally regulated excision of the 11-kb element.

Gresshoff, Roth, Stacey & Newton (eds.) Nitrogen Fixation: Achievements and Objectives
Chapman and Hall New York • London 1990

Rhizobium meliloti dctD gene product (DCTD) has three functional domains for activating *dctA* gene expression

B. Gu and B.T. Nixon

Department of Molecular and Cell Biology, Pennsylvania State University, University Park, PA 16802 USA.

Rhizobium meliloti dctA, dctB and *dctD* genes have been identified and sequenced. The *dctB* and *dctD* gene products DCTB and DCTD are thought to form a two-component regulatory system which responds to the presence of C_4-dicarboxylates to regulate the expression of a transport protein encoded by *dctA*. Homology with *Rhizobium leguminosarum* DCTD and other σ^{54}-dependent transcriptional regulators suggests that the 460 amino acid DCTD protein possesses three functional domains. We obtained evidence to confirm the existence of these domains by constructing and expressing various deletions of *R. meliloti dctD* in both *E. coli* and *R. meliloti* and measuring the production of ß-galactosidase activity encoded in several reporter genes. Our results showed that the 260 amino acid central domain of DCTD was sufficient to activate *dctA* gene expression. The DCTD amino-terminal domain repressed this activity of the central domain, which was released in *R. meliloti* when the cells were exposed to C_4-dicarboxylates. The C-terminal domain of DCTD was found to be essential for the protein to recognize an activation sequence located upstream of the *dctA* promoter, presumably by encoding a specific DNA-binding function (see Abstract by Ledebur et. al.). In *E. coli,* the DCTD central domain also activated expression of *R. meliloti* and *Bradyrhizobium* sp. [*Parasponiae*] *nifH* genes. In these assays, the *nifH* upstream activation elements were also functional, provided that the DCTD C-terminal domain was intact.

Gresshoff, Roth, Stacey & Newton (eds.) Nitrogen Fixation: Achievements and Objectives
Chapman and Hall New York • London 1990

Iron regulation of the *hemA* genes from *Bradyrhizobium japonicum* and *Rhizobium meliloti*

Mary Lou Guerinot, Erin Connolly and Barbara Anne Morisseau

Department of Biological Sciences, Dartmouth College, Hanover, New Hampshire 03755 USA

INTRODUCTION

The control of rhizobial heme production is of interest as the bacterial partner is believed to synthesize the heme moeity of leghemoglobin, with the plant partner contributing the apo-leghemoglobin, thus forming a truly symbiotic protein. Iron has been implicated as a positive regulator of Δ-aminolevulinic acid synthase (4), the enzyme which catalyzes the first committed and rate-limiting step in the heme biosynthetic pathway. Having previously cloned and sequenced the gene (*hemA*) encoding this enzyme from *B. japonicum* (3), we set out to determine how the iron control was affected.

RESULTS AND DISCUSSION

Using either a *hemA-lacZ* fusion integrated into the *B. japonicum* chromosome (which leaves the endogenous *hemA* gene intact) or a *hemA-lacZ* fusion carried on a broad host range plasmid (pLAFR1), we see approximately three to five fold less ß-galactosidase activity in iron-deficient cells than in iron-sufficient cells. Because we also see iron regulation of the *B. japonicum hemA-lacZ* fusion in *E. coli*, we have conducted experiments to determine whether this regulation is mediated by Fur, a DNA-binding protein which has been shown to regulate a number of genes in response to iron, including iron uptake genes. Indeed, iron regulation of the *hemA-lacZ* fusion is lost in a *fur* mutant. This implicates *fur*, or another gene controlled by *fur*, in the positive regulation of *hemA* transcription. There is a sequence upstream of both the *B. japonicum* (3) and *R. meliloti* (2) *hemA* genes with homology to an "iron box" as described for *E. coli* (1), which is the site at which the Fur regulatory protein binds to DNA. We are currently assaying plasmids carrying deletions of the "iron box" for ß-galactosidase activity in iron-deficient and iron-sufficient cells of both *B. japonicum* and *R. meliloti*. We are also using the *B. japonicum hemA-lacZ* fusion to examine expression *in planta*.

REFERENCES

1. deLorenzo, V., Wee, S., Herrero, M & Neilands, J.B. (1987) J. Bacteriol. **169**, 2624-2630.
2. Leong, S.A.,Williams, P.H. & Ditta, G.S. (1985) Nucleic Acids Res. **13**, 5965-5976.
3. McClung, C.R., Somerville, J.E., Guerinot, M.L. & Chelm, B.K. (1987) Gene **54,** 133-139.
4. Roessler, P.G. & Nadler, K.D. (1982) J. Bacteriol. **149**, 1021-1026.

Gresshoff, Roth, Stacey & Newton (eds.) Nitrogen Fixation: Achievements and Objectives
Chapman and Hall New York • London 1990

Morphology and nitrogenase activity of derepressed *Frankia* in continuous culture

Sharon Harris and Warwick Silvester

Department of Biological Sciences, University of Waikato, New Zealand.

INTRODUCTION. *Frankia* grown in derepressed batch culture shows no stationary phase in which a stable nitrogenase activity can be sustained (1). This makes for difficulties in analysing morphological and physiological changes due to environmental factors as they are always superimposed on a changing background of development. Previous attempts to delay autolysis and maintain activity have proved unsuccessful (1).

METHODS. Cultures of *Frankia* CcI3 and ArI3 were maintained in continuous culture by daily aseptic additions of nutrient and withdrawal of cell mass. Bottles were septum sealed, on a shaker, and could be maintained indefinitely at defined pO_2 by aseptic gas addition. Standard techniques of assay and microscopy were used (2).

RESULTS AND DISCUSSION. Dilutions of 0.125 day^{-1} were ideal for both strains giving stable biomass protein levels of c.40 $\mu g\ ml^{-1}$ maintained for periods as long as 100 days. Nitrogenase activity was sustained and a stable population of vesicles was produced.

Cultures grown at $2kPaO_2$, shocked with $21kPaO_2$ for 30 minutes and returned to 2kPa lost all nitrogenase activity but recovered completely with a short lag in 30h. Cultures grown at $2kPaO_2$, shocked with $21kPaO_2$ and then maintained in $21kPaO_2$ showed a 36h lag before recovery of any nitrogenase, with eventual full recovery after 97h. We believe that nitrogenase is destroyed and recovers in existing vesicles in the former case while in the latter case nitrogenase recovery is dependent on the thickening of existing vesicle envelope or formation of new vesicles. The lipid envelope of vesicles has been shown to thicken dramatically in response to high pO_2 (2). This was tested in continuous culture, where the populations of vesicles were measured for envelope thickness at varying times after a change from $24kPaO_2$ to $21kPaO_2$. Mean apparent wall thickness changed from 0.4 µm to 0.6 µm over a period of 100 hours following the change. This work confirms the response of vesicles to elevated pO_2 and extends the observation to cultures grown in steady state.

1. Murry, M.A., Fontaine, M.S. & Torrey, J.G. (1984) *Plant Soil* **78**, 61-78.
2. Parsons, R., Silvester, W.B., Harris, S., Gruijters, W.T.M. & Bullivant, S. (1987) *Plant Physiol.* **83**, 728-731.

Gresshoff, Roth, Stacey & Newton (eds.) Nitrogen Fixation: Achievements and Objectives
Chapman and Hall New York • London 1990

Efficient high voltage electroporation of *Bradyrhizobium japonicum* USDA 110

Dennis R. Hattermann and Gary Stacey

Center for Legume Research and Department of Microbiology. The University of Tennessee, Knoxville, TN 37996-0845, USA.

INTRODUCTION

Genetic studies of *B. japonicum* are difficult due to the limited methods of genetic transfer to this strain and its high intrinsic antibiotic resistance. Conjugation is the only efficient, widely used technique to introduce DNA into *B. japonicum.* In this report, we describe conditions for the efficient transfer of DNA to *B. japonicum* by electroporation. Furthermore, we describe a heat shock method which can be used to transfer DNA directly from *E. coli* to *B. japonicum*; thus, avoiding difficulties likely due to DNA restriction.

METHODS

The protocol for cell preparation and optimal electroporation conditions for *B. japonicum* has been published (2).

RESULTS AND DISCUSSION

Bradyrhizobium japonicum (USDA 110) was electroporated to a maximum efficiency of 10^7 transformants/µg with a 30 Kb plasmid (pZB32)(1) extracted from the same strain. Highest efficiencies were obtained with field strengths of 12.5 Kv/cm, pulse lengths of 5-8 milliseconds and DNA concentrations of 125 ng/ml (1). Efficient direct cloning in *B. japonicum* with ligated DNA has been observed. (W.M. Barbour et al., this volume). Variable electroporation efficiencies with plasmid DNA prepared from different source strains (*E. coli* as well as *B. japonicum* strains) may be due to a restriction/modification system in *B. japonicum* (1,2). Plasmid DNA isolated from a DAM -, DCM - *E. coli* strain (GM2163) has been shown to transform *B. japonicum* (2). Preliminary results indicate that this procedure may not work in all situations, and the highest efficiencies are not equal to those obtained using DNA extracted from *B. japonicum.* Heating *B. japonicum* (USDA 110) cells to 49°C for 15 min allowed electroporation by 0.1 µg of a 20 Kb plasmid (pP375)(1) extracted from *E. coli* to efficiencies of 4200 transformants/µg DNA. In the absence of heat, less than 4 transformants were observed/electroporation event.

REFERENCES

1. Banfalvi, Z., A. Niewkoop, M. Schell, L. Besl, and G. Stacey. 1988. Mol. Gen. Genet. 214:420-424.
2. Guerinot, M.L., B.A. Morisseau and T. Klapatch. 1990. Mol. Gen. Genet. (in press).
3. Hattermann, D.R. and G. Stacey. 1990. Appl. Environ. Microbiol. 56:833-836.

Gresshoff, Roth, Stacey & Newton (eds.) Nitrogen Fixation: Achievements and Objectives
Chapman and Hall New York • London 1990

Detection of a temperate bacteriophage of Rhizobium leguminosarum bv. trifolii in a root nodule cell

Shiro Higashi, Toshiki Uchiumi, Mikiko Abe

Department of Biology, Faculty of Science, Kagoshima University, Kagoshima 890, Japan.

A temperate bacteriophage (ϕU) was isolated from a wild strain of R. l. bv. trifolii. This phage belongs to the Podoviridae morphologically and has a narrow host range. In lysogenized R. l. bv. trifolii 4S cells with phage ϕU (1), phage DNA was integrated into the chromosome by lysogenization. Two different lysogens were isolated with different symbiotic phenotypes; st. 4S:ϕUn$^-$, which had lost the Sym plasmid (315 kb) and symbiotic ability (Nod$^-$), and st. 4S:ϕUn$^+$, which was not affected symbiotically (Nod$^+$). EcoRI or HindIII digests of total DNA of st. 4S:ϕUn$^-$ and 4S:ϕUn$^+$ were hybridized with a cloned attP fragment or total DNA of phage ϕU as probes. Although the phage oU was integrated into the same chromosomal site in both strains, the prophage in st. 4S:ϕUn$^-$ appeared to incorporate a cryptic DNA fragment that may affect Sym plasmid replication.

Total peptide patterns of the cell homogenate from these two strains were compared by 2D-PAGE. At least one particular protein, 31 kDa, was found only in Nod$^+$ strains (4S:ϕUn$^+$ and 4S).

This peptide might be involved in the infection process of Rhizobium.

The prophage was induced by mitomycin C treatment of both free-living bacteria and bacteroids in root nodule cells at 1-10 ug/ml at 28° and 25°C, respectively, overnight (Figure).

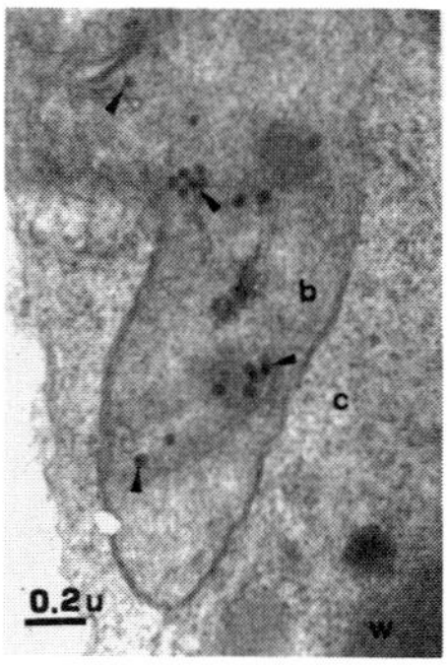

arrowhead=phage particle, b=bacteroid, c=nodule cell, w=nodule cell wall.

It is expected that some of the foreign DNA fused with phage DNA is expressed in nodule cells.

REFERENCE

1. Uchiumi, T., Ono, Y., Abe, M. & Higashi, S. (1989) J. Gen. Microbiol. **135**, 3133-3141.

Gresshoff, Roth, Stacey & Newton (eds.) Nitrogen Fixation: Achievements and Objectives
Chapman and Hall New York • London 1990

Integration host factor binds to *nif* promoter-regulatory regions from a variety of 'purple' bacteria

T. Hoover, E. Santero, S. Porter, and S. Kustu

Departments of Plant Pathology and Cellular and Molecular Biology, University of California, Berkeley, CA, USA 94720.

We recently demonstrated that the integration host factor (IHF) binds to the *Klebsiella pneumoniae nifH* promoter region (2), and that it stimulates activation of transcription by the nitrogen fixation-specific activator NIFA (1). We report here that IHF also binds to the *nifJ*, *nifE*, *nifU*, and *nifB* promoter-regulatory regions from *K. pneumoniae*, but not the *nifF* or *nifL* promoter-regulatory regions. In addition, IHF binds to the *nifH* promoter-regulatory regions of *Azotobacter vinelandii*, *Rhizobium meliloti*, *Bradyrhizobium japonicum*, *Rhodobacter capsulatus*, *Rhodospirillum rubrum*, and *Thiobacillus ferrooxidans*. These organisms represent three of the four subdivisions within the phylum "purple bacteria and their relatives" (3), which is the largest division within the eubacterial kingdom (4) (α subdivision - *K. pneumoniae* and *A. vinelandii* ; β - *T. ferrooxidans* ; and γ - *R. meliloti*, *B. japonicum*, *R. capsulatus*, and *R. rubrum*). In all cases where IHF binds, its binding site is located between an upstream binding site(s) for NIFA and the promoter, a binding site for σ^{54}-holoenzyme. The general occurance of the IHF binding site and its constant relative position within the promoter-regulatory regions of the *nif* genes from a wide variety of bacteria argues strongly for its importance in the regulation of expression of these genes. The length of the region protected by IHF from DNase I or hydroxyl radical cleavage indicates that IHF bends the DNA; this inference is confirmed by direct electron microscopic observation. We hypothesize that the role of the IHF-induced bend is to stabilize a DNA conformation that facilitates productive contacts between NIFA and σ^{54}-holoenzyme. Based on studies at the *K. pneumoniae nifH* promoter, we anticipate that IHF will stimulate NIFA-mediated activation of transcription at many *nif* promoters.

1. Hoover, T.R., Santero, E., Porter, S., and Kustu, S. (1990) Cell, (in press).
2. Santero, E., Hoover, T., Keener, J., and Kustu, S. (1989) Proc. Natl. Acad. Sci. USA **86**, 7346-7350.
3. Stackebrandt, E., Murray, R.G.E., and Truper, H.G. (1988) Int. J. Syst. Bacteriol. **38**, 321-325.
4. Woese, C.R. (1987) Microbiol. Rev. **51**, 221-271.

We thank G. Roberts, D. Dean, R. Haselkorn, D. Rawlings, F. Ausubel and H. Hennecke for providing plasmids. We also thank H. Nash for his gifts of materials and encouragement. T.H. is a Burroughs Wellcome Fund Fellow of the Life Sciences Reseach Foundation. E.S. is a recipient of an EMBO postdoctoral fellowship.

Gresshoff, Roth, Stacey & Newton (eds.) Nitrogen Fixation: Achievements and Objectives
Chapman and Hall New York • London 1990

A strain of *Bradyrhizobium japonicum* with improved symbiotic properties

William J. Hunter

USDA-ARS, Soil-Plant Nutrient Research Unit, P. O. Box E, Fort Collins, CO 80522, USA.

Tryptophan metabolism may play an important role in the symbiotic properties of *Bradyrhizobium*. Mutants that cannot synthesize tryptophan often exhibit poor nodulation (3), whereas bradyrhizobia with enhanced tryptophan catabolism have been obtained that exhibit improved symbiotic properties (2). For this study, cells of *Bradyrhizobium japonicum* USDA I-110 were exposed to 5-methyltryptophan, a toxic tryptophan analog. Bacteria that are resistant to 5-methyltryptophan are often biosynthetic mutants that overproduce tryptophan (1). Resistant clones were isolated and their cultural and symbiotic properties investigated. When grown in the presence of tryptophan, many of these strains accumulated anthranilic acid and indolic compounds, an indication of altered tryptophan metabolism. Enhanced tryptophan catabolism was observed with some strains but not with others. Greenhouse studies showed that at least one strain exhibited improved symbiotic properties. Plants inoculated with this strain formed more nodules and contained more nitrogen than did plants inoculated with wild-type bacteria. This improved strain resembled the tan mutants described by Kaneshiro and Kwolek (2), in that it accumulated indoleacetic, indolelactic and indolepyruvic acids. Unlike the tan mutants it does not show enhanced tryptophan catabolism. These results show that bradyrhizobia which are resistant to 5-methyltryptophan can exhibit improved symbiotic properties.

References:

1. Hoch, S. O., Roth, C. W., Crawford, I. P. & Nester, E. W. (1971) J. Bact. **105**,38-45.

2. Kaneshiro, T. & Kwolek, W. F. (1985) Plant Sci. **42**,141-146.

3. Wells, S. E. & Kuykendall, L. D. (1983) J. Bact. **156**,1356-1358.

Gresshoff, Roth, Stacey & Newton (eds.) Nitrogen Fixation: Achievements and Objectives
Chapman and Hall New York • London 1990

Construction and screening high effective strains of *Bradyrhizobium japonicum* by the method of genetic engineering

Zhou Junchu, Shen Hui, Zhang Zhongming, Hu Fuyun, Chen Huakui, Dou Xintain*, Li Xiaoming* and Li Xinming*

Biotechnology Center of Huazhong Agricultural University, Wuhan, 430070, P.R. China. *Institute of Soil and Fertilizer, Agricultural Academy of Hei Long Jiang Province, Harbin, 150026, P.R. China.

Methods for isolating improved effective Rhizobium commonly involve screening of natural isolates or mutagenized populations. DeJong et al. (1982) transferred the pSym plasmid from *R. leguminosarum* 128C53 to 300 which resulted in an increase in plant dry weight and nitrogen content. Considering the complexity of the symbiotic nitrogen-fixation genes in soybean rhizobia, we designed a new method: 1. construction of a gene library of *Rhizobium fredii* B52 using pLAFR1 as the vector. 3.9 x 10^4 Tc^r clones were obtained with 23% carrying foreign DNA fragments. 2. This library was conjugated *en masse* with the help of pRK2013 into the effective inoculant strain, *B. japonicum* 2-10. 2 x 10^4 transconjugants were obtained on selective media. 3. Plant experiments were performed in pots to compare the transconjugants to the 22-10 parent strain. Four strains (32,43,B-2-6, and B-3-8) were judged superior to 22-10 on the basis of plant growth, nodule weight, and acetylene reduction activity (2-times higher). 4. These four strains were used in field experiments in He Long Jiang province and examined in the V_2, R_1, and R_3 growth periods. The final yields (Kg/ha) for each treatment were 2522.2 (control), 2731.5 (22-10), 2659.5 (B-3-8), 2830.5 (43), 2880 (B-2-6) and 2945.3 (32). Compared to the control, the increased rates were 83% (22-10), 50% (B-3-8), 12% (43), 14.3% (B-2-6) and 16.8% (32). Compared to 22-10, the increases were -2.6% (B-3-8), 3.6% (43), 5.6% (B-2-6), and 7.8% (32). The difference between 22-10 and 32 was statistically significant. 5. The plasmid from strain 32 was transferred to *E. coli* and isolated.

Gresshoff, Roth, Stacey & Newton (eds.) Nitrogen Fixation: Achievements and Objectives
Chapman and Hall New York • London 1990

Genetic evidence for the involvement of cytochrome *d* in respiratory protection

M. J. S. Kelly. R. K. Poole, M. G. Yates and C. Kennedy

Microbial Physiology Research Group, King's College London and AFRC, IPSR Nitrogen Fixation Laboratory, University of Sussex, Brighton, BN1 9RQ, UK.

The genome of *Azotobacter vinelandii* contains DNA sequences homologous to the structural genes (1) for the *Escherichia coli* cytochrome *bd* terminal oxidase complex. Two recombinant clones bearing *cydA*- and *cydB*-like sequences were isolated from an *A. vinelandii* gene library and subcloned into the plasmid vector pACYC184. Physical mapping demonstrated that the *cydA*- and *cydB*-like regions in *A. vinelandii* are contiguous. The *cydAB* and flanking DNA was mutagenized by the insertion of Tn5-B20. Mutations in the *cydB*-hybridizing region resulted in the loss of spectral features associated with cytochromes b_{595} and *d*. A new locus *cydB* encoding cytochromes b_{595} and *d* in *A. vinelandii* is proposed. A second region was found to be involved in the expression of cytochrome *d* in *A. vinelandii*. Mutations in this region resulted in a five-fold increase in cytochrome *d* levels. Both the region involved in cytochrome *d* expression and *cydB* are transcribed in the same direction. Cytochrome *d*-deficient mutants were unable to grow on N-deficient medium when incubated in air, but could fix nitrogen when the environmental O_2 concentration was reduced to 1.5%. It is proposed that the cytochrome *d* branch of the respiratory chain (2) supports the high respiration rates required for the respiratory protection (3) of nitrogenase.

1. Green, G.N., Kranz, J.E., & Gennis, R.B. (1984) Gene **32**, 99-106.

2. Poole, R.K. (1988) in *Bacterial Energy Transduction*, ed. Anthony, C. (Academic Press, London), pp.231-291.

3. Robson, R.L. & Postgate, J. R. (1980) Ann. Rev. Microbiol. **34**, 183-207

Gresshoff, Roth, Stacey & Newton (eds.) Nitrogen Fixation: Achievements and Objectives
Chapman and Hall New York • London 1990

Localization of the reversible hydrogenase in cyanobacteria

T. Kentemich and H.Bothe

Botanisches Institut, Universität zu Köln, Gyrhofstr. 15, D-5000 Köln 41, FRG

The reversible hydrogenase of the cyanobacterium Anacystis nidulans was purified 250fold by classical methods. Activity staining on gels obtained by native PAGE allowed to identify two bands. Antibodies were raised against the electrophoretically homogeneous protein. The molecular weight of the hydrogenase was determined by SDS-PAGE followed by Western-blot analysis with these antibodies. Crude extracts from the unicellular A. nidulans and from heterocysts and vegetative cells of Anabaena variabilis showed precipitation bands of 56 and 17 kD. The reversible hydrogenase appeared to be composed of a larger and a smaller subunit as the NiFeS-hydrogenases from other bacteria. The hydrogenase subunits, particularly the smaller one, were subject to proteolysis. By application of the protein A-immuno-gold labeling technique the reversible hydrogenase was shown to associate with the cytoplasmic membrane in Anacystis. It is concluded that Anacystis has two different hydrogenases: the reversible hydrogenase located at the cytoplasmic membrane and a thylakoid-bound enzyme which catalyzes only the uptake of H_2 (1). Experiments were started to characterize the genes involved in H_2-metabolism of cyanobacteria. For this purpose, DNA-DNA hybridization between DNA isolated from Anacystis or Anabaena and the following hydrogenase DNA-probes from several other bacteria were tested in Southern-blot experiments and in the screening of a gene-bank from A. variabilis: pHU52 (genes for the 'uptake' hydrogenase and for CO_2-fixation from Bradyrhizobium japonicum ; H.J. Evans); pHV150 (gene for the periplasmatic FeS-hydrogenase from Desulfovibrio vulgaris; G. Voordouw); pCMS1 (gene for the 'uptake' hydrogenase from Azotobacter chroococcum ; M.G. Yates); pAC76 (gene for the 'uptake' hydrogenase form Rhodobacter capsulatus; P. Vignais). In contrast to the efforts in other bacteria all attempts to get positive hybridization signals failed so far.

(1) Kentemich, T.; Bahnweg, M.; Mayer, F.; Bothe, H.; Z. Naturforsch. 44c, 384-391 (1989)

Gresshoff, Roth, Stacey & Newton (eds.) Nitrogen Fixation: Achievements and Objectives
Chapman and Hall New York • London 1990

Organization of *nif* genes in *Rhodobacter capsulatus*: Identification of an alternative nitrogenase system

W.Klipp, M.Schmehl, U.Schramm, C.Moreno-Vivian*, S.Hennecke and B.Masepohl

Lehrstuhl für Genetik, Universität Bielefeld, FRG
* Dep. de Bioquimica, Universidad de Cordoba, Spain

By sequence analysis of a DNA fragment of R.capsulatus nif region A hybridizing to Klebsiella pneumoniae nifS, five R.capsulatus genes were identified, which form one transcriptional unit (ORF6-nifUSVW). The deduced amino acid sequences of R.capsulatus nifS, nifV and nifW are homologous to their counterparts from Azotobacter vinelandii and K.pneumoniae over the entire lengths of the proteins. In contrast, R.capsulatus NifU differs from the nifU gene products of A.vinelandii and K.pneumoniae. Only the C-terminal part (73aa) is conserved in NifU of all three organsims whereas the large N-terminal part, which is highly homologous between A.vinelandii and K.pneumoniae NifU, is not present in R.capsulatus NifU. A DNA sequence with striking homology to R.capsulatus nifU was identified in nif region B between the structural genes for nitrogenase (nifHDK) and the regulatory nifR4 gene. Nevertheless, single and double mutants in both copies of nifU exhibit a Nif$^+$ phenotype. In addition to the "common nif genes" four nif associated open reading frames (including ORF6) were identified, which had their counterparts only in A.vinelandii. The map position of ORF6 upstream of nifUSV and ORF3 downstream of nifENX is identical in R.capsulatus and A.vinelandii. However, there is no obvious reason why open reading frames homologous to A.vinelandii ORF2 and ORF8, as well as nifQ are located at different positions in R.capsulatus. R.capsulatus ORF1 and its homologue ORF2 from A.vinelandii code for periodic proteins containing a 24 amino acid motif that is directly repeated seven times. To demonstrate the existence of an alternative nitrogenase system in R.capsulatus, a nifHDK deletion strain was constructed, which is unable to fix nitrogen in the presence of molybdenum. However, slow diazotrophic growth was obseved in molybdenum-free media indicating the existence of at least one alternative nitrogenase system. Since this nitrogenase activity could not be stimulated by vanadium, it is assumed that the alternative nitrogenase in R.capsulatus corresponds to A.vinelandii nitrogenase-3.

Gresshoff, Roth, Stacey & Newton (eds.) Nitrogen Fixation: Achievements and Objectives
Chapman and Hall New York • London 1990

Isolation and characterization of oxygen tolerant *Rhizobium meliloti* *nif*A mutants

R.Krey, L.Sharypova, W.Klipp and A. Pühler

Lehrstuhl für Genetik, Universität Bielefeld, Postfach 8640, D-4800 Bielefeld, FRG

The R.meliloti nifA gene product plays a central role in the regulation of symbiotic nitrogen fixation. In addition to oxygen regulated stimulation of the nifA promoter it was shown that the activation of nif promoters by NifA is oxygen sensitive as well (1,2).
To analyze this oxygen sensitivity, we developed a method for rapid identification of oxygen tolerant nifA mutants. Point mutations in plasmids containing a constitutively expressed R.meliloti nifA gene were generated in vitro by hydroxylamine treatment. Oxygen tolerant mutants were identified by their ability to activate a R.meliloti nifH-lacZ fusion in E.coli under high oxygen concentrations. We have isolated 73 R.meliloti nifA mutants with increased oxygen tolerance. On the basis of specific nifA activity under different oxygen concentrations, these mutants could be devided into several classes. To localize the mutations and to exclude plasmid-born effects, we exchanged different fragments between mutated and wildtype plasmids. In all mutants analyzed so far, we were able to transfer the oxygen-tolerance by replacement of the C-terminal part of NifA including the central domain, the interdomain linker and the DNA binding domain. Two mutants were further characterized by partial DNA sequence analysis. One mutation (Gly → Glu) was found within the interdomain linker, which is characteristic for oxygen sensitive nifA proteins. The other mutation, which was located in the 3'-part of the central domain, did not change the amino acid sequence. However, the mutant triplett is known to have a higher codon usage in E.coli than the wildtype one. Considering the results of the fragment exchange experiments we could not exclude additional mutations in the central domain.

(1) Beynon, J.L., Williams, M.K. & Cannon, F.C. (1988) EMBO J. **7**, 7-14.
(2) Klipp, W., Reiländer, H., Schlüter, A., Krey, R. & Pühler, A. (1989) Mol. Gen. Genet. **216**, 293-302.

Gresshoff, Roth, Stacey & Newton (eds.) Nitrogen Fixation: Achievements and Objectives
Chapman and Hall New York • London 1990

Field isolates belonging to different biovars of *Rhizobium leguminosarum* can share identical chromosomal backgrounds

G. LAGUERRE, S. MAZURIER and N. AMARGER

INRA, Microbiologie des Sols, 21034 DIJON cédex, France.

In order to explore genotypic diversity of a field population of *R. leguminosarum*, four sub-populations, two of biovar *viciae*, one of biovar *trifolii* and one of biovar *phaseoli*, were isolated from root nodules of peas, lentils, red clovers and french beans grown in a single field. One hundred and twenty isolates per sub-population were characterized by plasmid profiles. A sample of one fourth was chosen to be representative of the main plasmid types and classes (grouping plasmid types with at least two common plasmid bands). This sample was further characterized by total DNA restriction digest patterns and hybridization patterns using chromosomal probes (plac12 (1), pCOS309.1 (2) and pRt654 (3))and Sym plasmid probes (pIJ1246 (4), bv. *phaseoli nod*AB genes from M. Flores and pRt587 (3)) originating from strains belonging to the 3 different biovars.

The analysis of the different patterns allowed to define restriction types grouped in restriction classes and chromosomal types. A strong correlation has been etablished between restriction classes, chromosomal types and plasmid classes. Using the pLac12 probe, 3, 5, 8 and 10 distinct chromosomal backgrounds have been etablished for pea, clover, bean and lentil isolates, respectively. One of these chromosomal backgrounds was the most frequent and identical for the 3 biovars (87, 69, 54 and 46% for the pea , clover, lentil and bean isolates, resp.). Three others were also common to different biovars. In addition, from hybridizations with *nod* genes probes and from data of symbiotic assays on plants, it was shown that some isolates from beans were in fact belonging to biovar *viciae*. So, using biovar specific *nod* genes probes, hybridizations appear to be a useful tool to avoid mistaken determinations of biovar only found upon the host range.

The results indicate that rhizobia with identical or close related chromosomes can harbour different host specificity determinants and that a same chromosomal genotype can form the majority of nodules in different species of legumes. The results suggest that the different biovars of *R. leguminosarum* have not evolved separatly but are more probably the results of recent exchange or modifications of plasmid-born genes. They also confirm that biovar cannot be considered as a taxonomic entity.

References :

(1) Young, J.P.W. & Wexler, M. (1988) J. Gen. Microbiol. **134**, 2731-2739.

(2) Cava, J.R., Elias, P.M., Turowski, D.A. & Noel, K.D. (1989) J. Bacteriol. **171**, 8-15.

(3) Schofield, R.R., Gibson, A.H., Dubman, W.F. & Watson, J.M. (1987) Appl. Environ. Microbiol. **53**, 2942-2947.

(4) Downie, J.A., Knight, C.D., Johnston, A.W.B. & Rossen, L. (1985) Mol. Gen. Genet. **198**, 255-262.

Gresshoff, Roth, Stacey & Newton (eds.) Nitrogen Fixation: Achievements and Objectives
Chapman and Hall New York • London 1990

An open reading frame in the *Anabaena* 7120 *nifD* excision element exhibits sequence similarity to cytochromes P-450

Peter J. Lammers*, Sherry McLaughlin, Sophie Papin, Casilda Trujillo-Provencio, and Alexander J. Ryncarz.

Department of Chemistry, New Mexico State University, Las Cruces, N.M. 88003, USA.

Many diazotrophic cyanobacteria undergo chromosomal rearrangements during heterocyst differentiation (3). The *nifD* gene is often split by a large DNA element, the excision of which is necessary for nitrogenase synthesis but not for heterocyst formation (4). Strains that have lost the *nifD* element show no obvious differences in heterocyst development, spacing, or function (2). Thus, any selective advantage associated with the presence of the element must be subtle, or unrelated to nitrogen fixation.

To investigate the genetic potential of the *Anabaena nifD* element we have determined 5000 base pairs of DNA sequence extending from the right border within the *nifD* element toward the distal *xisA* gene. Three open reading frames have been found. ORF4, found in the An207.4 *Hin*dIII fragment encodes a putative polypeptide of approximately 40,000 daltons. Significant homology exists between this sequence and patterns diagnostic of six subfamilies within the cytochrome P-450 monooxygenase superfamily. The homology extends over most of the polypeptide, and is particularly well conserved around an invariant cysteine which acts as a fifth ligand for the heme iron in P-450 proteins.

ORF-specific probes were used to examine DNA sequence conservation in other heterocystous cyanobacteria that contain the *nifD* element. The ORF2-specific probe, spanning the An207.3 and An207.9 *Hin*dIII fragments was the most highly conserved sequence tested with DNA from *Anabaena* (ATCC 29413), *Nostoc* strain MAC, and *Calothrix* (PCC 7601). Signals were readily detectable with ORF1 and ORF4 probes at lower stringency with the single exception of the ORF4 probe and *Calothrix* DNA.

We speculate that the P450-like sequence may confer the ability to degrade exogenous inhibitory molecules or initiate utilization of unusual organic nitrogen sources through N-dealkylation or oxidative deamination reactions. An alternative hypothesis based on work by Appleby (1) with *Bradyrhizobium*, is that the protein may serve as an oxygen carrier to a coupled or uncoupled respiratory oxidase functioning in heterocysts.

1. Appleby, C.A. (1984). Ann. Rev. Plant Physiol. 35, 443-478.
2. Brusca, J.S., C.J. Chastain, and J.W. Golden. (1990). J. Bacteriol. In Press.
3. Golden, J.W., S.J. Robinson, and R.Haselkorn. (1985). Nature 314, 419-423.
4. Golden, J.W. and D.R. Wiest (1988). Science 242, 1421-1423.

Gresshoff, Roth, Stacey & Newton (eds.) Nitrogen Fixation: Achievements and Objectives
Chapman and Hall New York • London 1990

Rhizobium meliloti and *Rhizobium leguminosarum dctD* gene products bind to tandem sites in an activation sequence located upstream of IHF and RNA Polymerase-σ^{54} bound at *dctA* promoters

H. Ledebur, B. Gu, J. Sojda and B.T. Nixon

Dept of Molecular and Cell Biology, Penn State University
University Park, PA 16802, USA.

Rhizobia can import external C_4-dicarboxylates. In the bacteroid state, oxidation of these compounds is thought to fuel nitrogen fixation. In both *Rhizobium leguminosarum* and *Rhizobium meliloti*, transport of C_4-dicarboxylates depends upon the genes *dctA, dctB* and *dctD*, which are thought to respectively encode a transport protein (DCTA) and two regulatory gene products (DCTB and DCTD). In bacteroids, DCTB and DCTD can be replaced by as yet unidentified factors, but in free-living cells only DCTB and DCTD appear to regulate production of DCTA. In a recent comparison of the sequence of DNA between respective *dctA* and *dctB* genes of the two species, we identified blocks of homology which included duplicate regions of partial dyad symmetry upstream of the putative *dctA* consensus promoters.

Primer extension analysis of RNA isolated from free-living cells confirmed that initiation of *dctA* transcription occurred approximately 10 bases downstream of the previously identified σ^{54}-consensus sequences. The construction of chimeric *dctA-nifH* promoters and of *dctA-lacZ* reporter genes demonstrated that full regulation of the initiation of transcription of *R. leguminosarum dctA* required an additional upstream activation sequence. This UAS functioned with either *dctA* or *nifH* σ^{54}-dependent promoters, and could reside at least 180 basepairs upstream of the site of transcription initiation. Both species' DCTD were shown to bind to tandem sites in the upstream activation element. The *E. coli* integration host factor (IHF) and a probable homolog of *R. meliloti* were found to bind adjacent to the *dctA* promoter of *R. leguminosarum*, but not at all, or only weakly, to that of *R. meliloti*.

It seems clear that upstream DCTD binding sites are major determinants of the sensitivity of the *dctA* σ^{54}-promoter for the presence of external C_4-dicarboxylates in both species. However, some sensitivity to activated DCTD appears to reside in the promoter itself. We infer that RNA polymerase-σ^{54} and activated DCTD can directly interact, independent of upstream binding sites. Furthermore, it appears that an IHF homolog facilitates activation by UAS-bound DCTD in *R. leguminosarum*, but perhaps not in *R. meliloti*. These results confirm that σ^{54}-dependent promoters, known to be widely distributed among the purple bacteria, may serve as model systems for understanding at least one mechanism of transcriptional activation by factors that bind at a distance. Further attempts to understand the more subtle interrelationships between σ^{54}-dependent systems like *nif, ntr* and *dct* may reveal how seemingly unrelated pathways are integrated into larger metabolic networks.

Gresshoff, Roth, Stacey & Newton (eds.) Nitrogen Fixation: Achievements and Objectives
Chapman and Hall New York • London 1990

Characterization of an osmotically induced periplasmic glycine betaine-binding protein from *Rhizobium meliloti*

D. Le Rudulier*. K. Gloux. N. Riou and M.C. Poggi.

Lab. Biologie Végétale et Microbiologie. URA CNRS 1114. Université de Nice - Sophia Antipolis, Parc Valrose, 06034 Nice Cedex, France.

In media of high osmolarity, *Rhizobium meliloti* uses glycine betaine (GB) as a powerful osmoprotectant. A high-affinity transport system has been characterized. The apparent K_m for GB was 5.6 and 4.3 μM with low- or high-salt-grown cells, respectively. Maximal velocity was strongly stimulated (x 5) by elevated osmolarity (NaCl 0.3 M). Periplasmic fractions were isolated from the cells by cold osmotic shock. Glycine betaine-binding activity was monitored by three techniques : equilibrium dialysis, filter binding procedure, and detection of [^{14}C] ligand-protein binding by direct non-denaturing polyacrylamide gel electrophoresis followed by autoradiography. The three methods demonstrated the existence of a strong GB-binding activity, but only in periplasmic fractions from cells grown at high osmolarity. The K_D for GB-binding was 2.5 μM and optimum pH was around 7.0. Optimum pH for binding was around 7.0 but approximately 90% of the glycine betaine-binding activity remained at pH 6.0 or 9.0. Both glycine betaine-binding activity and affinity were not significantly modified whether or not the binding assays were done at high osmolarity. Using a non-denaturing PAGE with ^{14}C-GB added to periplasmic fluid, we showed that the complex is sufficiently stable to be detected by autoradiography. Only one radioactive band was observed, indicating the existence of one GB-binding protein is shock fluid from 0.3 M NaCl-grown cells. The osmotically induced GB-binding protein has an apparent molecular weight of 32,000. The specificity of the binding is very strict, only choline and glycine betaine aldehyde were weak competitors. Other betaines or amino acids did not bind to the protein.

Gresshoff, Roth, Stacey & Newton (eds.) Nitrogen Fixation: Achievements and Objectives
Chapman and Hall New York • London 1990

Construction and characterization of mutations in genes coding for the reversible ADP-ribosylation system of dinitrogenase reductase in *Rhodospirillum rubrum*

J.-H. Liang, G. Nielsen, P.W. Ludden, R.H. Burris, and G.P. Roberts*.

Department of Biochemistry, * Department of Bacteriology, University of Wisconsin, Madison, WI 53706, USA

The activity of nitrogenase in *R. rubrum* is regulated by reversible ADP-ribosylation of dinitrogenase reductase in response to exogenous nitrogen sources, or to the light energy. The genes encoding dinitrogenase reductase ADP-ribosyl transferase (*draT*) and dinitrogenase reductase activating glycohydrolase (*draG*) have been cloned and sequenced. In this work, a Km^r cassette, containing transcriptional terminators in both ends, was inserted into *draG*, *draT* and *Orf* at *Eco*RV, *Sac*II and *Hinc*II sites, respectively. Removal of the cassette in *draG* by *Bam*H1 digestion and re-ligation leaves a 24-base pair fragment of the original cassette sequence, encoding an aa-sequence of AsnSerArgGlySerProGlyIle, between asp-82 and ile-83 of the DRAG sequence. Removal of the cassette in *draT* by *Sma*I digestion and re-ligation results in the replacement of the arg-191 of the DRAT sequence with GluPheProGlyIle. Both of these oligonucleotide insertions are in-frame. *R. rubrum* strains carrying mutations in *draG* and *draT* (both polar and non-polar insertions) exhibited the same growth rate and the same pigmentation as that of wild-type cells. When grown under nitrogen-fixing conditions, the $DraT^-$ mutants had C_2H_2 reduction activity comparable to that of wild-type cells. Addition of NH_4Cl to a final concentration of 2 mM, or placing the cells in the dark, did not cause the switch-off of the nitrogenase activity, while it did in the wild-type cells. The dinitrogenase reductase was not modified, as indicated by the absence of ^{32}P in the protein when quickly extracted from these cells. The $DraG^-$ mutants, grown under nitrogen-fixing conditions, had C_2H_2 reduction activity comparable to that of wild-type cells. Addition of NH_4Cl to the culture (final 0.5 mM), or placing it in the dark sharply reduced C_2H_2 reduction activity to less than 10%. This indicates that DRAT is not activated until cells see exogenous fixed nitrogen or receive no light energy. When shifted back to light or exhausted NH_4, $DraG^-$ cells showed a partial recovery (20 to 30%) of C_2H_2 reduction activity. This recovery was slower than that of the wild-type cells and could be resulted from the *de novo* synthesis of dinitrogenase reductase. Neither activity nor protein of DRAG was detectable in the $DraG^-$ cells by *in vitro* assays. No difference was observed between the polar and non-polar insertion mutants. The polar insertion in *Orf* did not give an obvious phenotype except that the C_2H_2 reduction activity decreased at a higher rate in the mutant cells than in the wild-type cells when provided with NH_4 or placed in the dark.

Current address of J.-H. Liang: Dept. of MGCB, The University of Chicago, 920 E 58th St, Chicago, IL 60637, USA.

Gresshoff, Roth, Stacey & Newton (eds.) Nitrogen Fixation: Achievements and Objectives
Chapman and Hall New York • London 1990

Histochemical and immunocytochemical characterization of intercellular matrix during intercellular infection of *Ceanothus* spp. by *Frankia*

Qinqin Liu and Alison M. Berry

Dept. of Environmental Horticulture, Univ. of California, Davis, CA 95616, USA.

Early infection of *Ceanothus* spp. by *Frankia* is characterized by an extensive darkly-staining intercellular matrix in prominent intercellular spaces of the root cortex and gradually decreasing through a transition zone to the nodule cortex. The common occurrence of a similar matrix in both *Frankia* and *Rhizobium* induced intercellular nodulation suggests that it has functional significance in the intercellular infection process. Studies were carried out to provide information on the structure and composition of this intercellular matrix during *Ceanothus* spp. nodule initiation using electron microscopy, histochemistry, and immunocytochemistry.

Histochemical studies showed that high levels of phenolic compounds, proteins and pectic compounds or other polyanions were present in the intercellular matrix. Fluorescence hybridization analysis using both pectate and alginate probes suggested that the matrix consists of methylated and some non-methylated polygalacturonate. Immunogold labeling using a polyclonal anti-polygalacturonic acid antibody (RG-I) was localized in the root mucilage, in the intercellular infection matrix and in the middle lamella of both root and infected nodule cells. Some labeling also occurred in the outermost wall layer of both infected and uninfected root cortical cells. Within the infected nodule cortical cells, the host-derived encapsulation surrounding the *Frankia* endosymbiont and some electron dense material within the endosymbiont were also labeled. The most intensive labeling occurred in mucilage and also in the intercellular matrix of infected root tissue. Thus the host releases largely esterified polygalacturonides, protein and other compounds in association with early *Frankia* colonization of root tissue, which may be involved in cell-to-cell interactions important for successful colonization by *Frankia*.

REFERENCES

1. Lancelle, S.A. & Torrey, J.G. (1984) Protoplasma **123**, 26-37.
2. Miller, M. & Baker, D.D. (1985) Protoplasma **128**, 107-119.
3. Moore, P.J. & Staehelin, L.A. (1988) Planta **174**, 443-445.
4. Vreeland, V. et al. (1989) Planta **177**, 435-446.

ACKNOWLEDGEMENTS

We are grateful for generous gifts of primary antiserum from Ms. M.A. Lynch in Dr. L.A. Staehelin's lab, Univ. of Colorado, Boulder, CO, and fluorescent probes provided by Dr. V. Vreeland, Dept. of Plant Biology, Univ. of California, Berkeley, CA.

Gresshoff, Roth, Stacey & Newton (eds.) Nitrogen Fixation: Achievements and Objectives
Chapman and Hall New York • London 1990

Purification and characterization of a novel glutamine synthetase from *Rhizobium meliloti*

Yuan Liu[1,2], Robert G. Shatters[1,3] and Michael L. Kahn[1,4]

Institute of Biological Chemistry[1] and Departments of Biochemistry[2], Genetics and Cell Biology[3] and Microbiology[4],
Washington State University, Pullman, WA 99164-6340

A *Rhizobium meliloti glnA glnII* mutant that is missing both glutamine synthetase I (GSI) and GSII can grow on minimal medium that contains both ammonium and glutamate (1) and suggests that a third GS may exist in *R. meliloti*. We have purified a novel GS protein from this mutant that we call GSIII. GSIII has considerable glutamine synthetase activity when assayed using a (glutamate and ammonia dependent) biosynthetic assay or a (glutamate and hydroxylamine dependent) semi-biosynthetic assay, but has no detectable (glutamine and hydroxylamine dependent) transferase activity. Semi-biosynthetic activity is optimal in PIPES buffer at pH 6.8 and at 50°C. The K_m for glutamate is 13.3 mM, for NH_4^+ is 33 mM and for hydroxylamine is 5.3 mM. The enzyme is inhibited by ADP and pyrophosphate but not by nitrogen containing metabolites known to cause end-product inhibition of GS activity in other systems. Activity is also inhibited by methionine sulfoximine, a transition state analog that inhibits other GS enzymes, but the K_i was 50 to 100 times higher than required to inhibit rhizobial GSI or GSII. SDS-PAGE analysis of the purified protein shows two bands of 46.5 Kd and 49 Kd. V8 protease digestion of these two bands indicates that they are related proteins. The purified protein has a native MW of 355 Kd indicating the native enzyme is an octamer. Antibodies to GSIII react with a protein of 49 kD in *E. coli* cells that carry pFB6162, a plasmid that carries the *R. meliloti glnT* gene (2) but we can detect GS enzymatic activity only when the cells are grown on certain media.

1. Somerville, J.E., Shatters, R.G. and Kahn, M.L. (1989) J. Bacteriol. 171,5079-5086.
2. de Bruijn, F. J., Rossbach, S., Schneider, M., Ratet, P., Messmer, S., Szeto, W.W., Ausubel, F.M. and Schell, J. (1989) J. Bacteriol. 171,1673-1682.

Gresshoff, Roth, Stacey & Newton (eds.) Nitrogen Fixation: Achievements and Objectives
Chapman and Hall New York • London 1990

Characterization of a new host-inducible genetic locus in *Bradyrhizobium japonicum*

Sohail Luka and Gary Stacey

Center for Legume Research and Department of Microbiology, The University of Tennessee, Knoxville, TN, 37996-0845 USA.

INTRODUCTION

The nod genes of *Bradyrhizobium japonicum* are induced by host produced isoflavones (e.g. genistein and daidzein, 1,2,3). Induction requires a functional nodD gene and the *cis*-acting nod box sequence, the NodD binding site. A Tn*3-lacZ* insertion in a locus linked to the *nodDYABC* genes of *B. japonicum* was shown to be induced 8-fold by genistein or soybean seed extract. This report describes the initial characterization of this new isoflavone-inducible locus.

METHODS

The DNA region localized by the isoflavone inducible Tn*3-lacZ* insertion was cloned into M13 and sequenced. DNA hybridization was done by standard procedures.

RESULTS

A Tn*3-lacZ* fusion inserted in a region linked to the *nodDYABC* genes of *B. japonicum* was found to be inducible 8-fold by genistein or soybean seed extract. The region identified by the Tn3 insertion was sequenced (approx. 2 kb). The sequence revealed that the inducible Tn*3-lacZ* fusion was inserted in an open-reading-frame (ORF3) corresponding to a 40 Kd protein product. The stop codon of ORF3 is followed by an inverted repeat that could be a possible transcription terminator. Upstream to ORF3, two other ORF's were identified, ORF1 and ORF2, each of which is preceeded by an appropriately located Shine-Dalgarno sequence. ORF1 encodes a predicted 6 Kd product, while ORF2 encodes a predicted 14 Kd product. Searches against the DNA sequence database did not reveal any significant homologies to the three ORF's. Computer analysis predicts that ORF2 and ORF3 are likely to be coding sequences. Interstrain and interspecies DNA hybridization using an internal probe from ORF3 showed that it is conserved in all *Bradyrhizobium* species tested, but not in *Rhizobium* species.

References

1. Banfalvi, Z.; Nieuwkoop, A.; Schell, M.; Besl, L. & Stacey, G. (1988) Mol. Gen. Genet. **214**, 420-424.
2. Gottfert, M.; Weber, J. & Hennecke, H. (1988) J. Plant Physiol. **132**, 394-397.
3. Kosslak, R.; Bookland, R.; Barkei, J.; Paaren, H. & Appelbaum, E. (1987) Proc. Natl. Acad. Sci. USA **84**, 7428-7432.

Gresshoff, Roth, Stacey & Newton (eds.) Nitrogen Fixation: Achievements and Objectives
Chapman and Hall New York • London 1990

Excretion of ammonia by *Azospirillum brasilense*

H.B. Machado, S. Funayama, L.U. Rigo and F.O. Pedrosa.

Dept. of Biochemistry, Universidade Federal do Paranã, C. Postal 19046, 81504 Curitiba, PR, Brazil.

INTRODUCTION: *Azospirillum brasilense* usually fail to excrete NH_4^+ during diazotrophic growth. We have isolated ethylenediamine (EDA) resistant mutants of strain FP2 and observed that they fix nitrogen constitutively and excrete NH_4^+ derived from N_2 fixation.

MATERIALS AND METHODS: Enzymes activities were determined as previously described (2). Ammonium ions were determined by the indophenol procedure (1). Diazotrophic growth was carried out in semisolid NFbHP medium under air.

RESULTS: Four spontaneous mutants of *A. brasilense* strain FP2, resistant to EDA (HM053, HM14, HM26, HM210) were found to fix nitrogen in the presence of NH_4^+, and to be prototrophs for Gln, growing well on Arg, Glu or high NH_4^+ (20 mM), but poorly on NO_3^- or low NH_4^+ (2 mM). All excreted NH_4^+ (up to 5 mM) in N-free semisolid medium. Excretion was dependent on N_2 in the gas phase; substitution of N_2 for argon eliminated NH_4^+ excretion. All mutants had wild-type levels of GOGAT and GDH activities. They showed GS synthesis regulated by NH_4^+, similar to the wild-type strain. However, their GS activities showed three regulation patterns, namely: a) low GS, constitutively adenylylated (HM053), b) wild-type GS levels, constitutively adenylylated (HM26 and HM210), and c) low GS, wild-type regulated (HM14). Mutants derepressed on N-free medium showed nitrogenase activity resistant to NH_4^+ switch off. Glutamate-grown cells, on the contrary, showed NH_4^+-dependent switch off. Glutamine failed to switch off nitrogenase activity of the wild type FP2 and its mutants.

CONCLUSIONS: EDA resistant mutants of *A. brasilense* strain FP2 showed constitutively adenylylated GS. This seems to eliminate the effects of NH_4^+ on synthesis and activity of nitrogenase and on NH_4^+ assimilation, leading to NH_4^+ excretion during diazotrophic growth.

1 Chaney, A.L. & Marbach, E.P. (1962) Clin. Chem. **8**, 130.
2 Pedrosa, F.O. & Yates, M.G. (1984) FEMS Microbiol. Lett. **23**, 95.

ACKNOWLEDGMENTS: We thank Brazilian Research Council and United Nations Development Programme for financial support.

Gresshoff, Roth, Stacey & Newton (eds.) Nitrogen Fixation: Achievements and Objectives
Chapman and Hall New York • London 1990

Molecular cloning of the *Rhizobium phaseoli recA* gene and construction of *recA* mutants

J.M. Martínez-Salazar, D. Romero, M.L. Girard and G. Dávila.
Departamento de Genética Molecular, Centro de Investigación sobre Fijación de Nitrógeno, UNAM.; Cuernavaca, Morelos, México.

The *R. phaseoli recA* gene was isolated from a genome bank constructed in lambda 1059 by interspecific complementation of the Fec- phenotype. A region carrying the *recA* gene was obtained by sucloning into pSUP202 and functional complementation of an *Escherichia coli recA* mutant. Approximate localization of the *recA* gene was determined by mutagenesis with MudIIPR13 and heterologous hybridization with an *E. coli* intra-*recA* probe. Complementation studies of *E. coli recA* mutant with the *R. phaseoli recA* gene cloned into pSUP202 restored the resistance to DNA-damaging agents as well as recombination proficiency in Hfr- mediated conjugation and P1-mediated transduction. The interposon Sp^r/Sm^r from pHP45 was inserted into the *recA* gene and introduced in two *R. phaseoli* strains by homogenotization. The *R. phaseoli recA* mutants were more sensitive to DNA-damaging agents and have a 100-fold reduction in the frequency of recombination.

Gresshoff, Roth, Stacey & Newton (eds.) Nitrogen Fixation: Achievements and Objectives
Chapman and Hall New York • London 1990

Ferredoxin-like gene products from *Rhizobium meliloti* and their role in symbiotic nitrogen fixation

B.Masepohl, K.-U.Riedel, M.Schmehl, W.Klipp and A.Pühler

Lehrstuhl für Genetik, Universität Bielefeld,
Postfach 8640, D-4800 Bielefeld, FRG

R.meliloti mega-plasmid II carries several genes necessary for symbiotic nitrogen fixation, which have no counterparts in Klebsiella pneumoniae. Two of these fix genes, fdxN and fixX, encode small ferredoxin-like proteins. A mutant in R.meliloti fdxN exhibits a Fix$^-$ phenotype in contrast to similar mutants in Azotobacter vinelandii and Bradyrhizobium japonicum. The Fix$^-$ phenotype of the R.meliloti fdxN mutant can be complemented by the Rhizobium leguminosarum fdxN gene, whereas no complementation could be observed for nif-associated genes encoding ferredoxins from Azotobacter chroococcum and Rhodobacter capsulatus. In order to overexpress R.meliloti fdxN in E.coli a series of transcriptional and translational fusions were constructed. Analysis of translational fusions revealed, that the ribosomal binding site in front of R.meliloti fdxN is recognized in E.coli. Nevertheless, the R.meliloti fdxN gene product could only be detected in E.coli using the very sensitive T7-promotor/T7-polymerase system. These fusions were tested for their ability to complement an R.meliloti fdxN mutant. Not only the transciptional fusions but also some of the translational fusions restored the FdxN$^-$ phenotype to some extent. For mutational analysis of R.meliloti fixX, a defined non-polar mutation was constructed, which should not effect readthrough into nifA. The corresponding fixX deletion mutant turned out to be Fix$^-$. Since fdxN mutant bacteroids contain the apo-proteins of nitrogenase, the role of FdxN and FixX in symbiotic nitrogen fixation was analyzed in an in vitro system. In contrast to an R.meliloti wildtype cell-free system, fdxN and fixX mutant extracts could not reduce acetylene in the presence of an artificial electron donor, indicating that neither fdxN nor fixX are involved in electron transport to nitrogenase. As a basis for further analysis of R.meliloti FdxN a prediction of the 3-dimensional structure was deduced by molecular modelling.

Gresshoff, Roth, Stacey & Newton (eds.) Nitrogen Fixation: Achievements and Objectives
Chapman and Hall New York • London 1990

Deleterious effects of combined nitrogen on *ex planta* nitrogenase activity, polysaccharide synthesis and lectin-binding ability of cowpea *Rhizobium* sp. JLn(c)

Bina Mody and Vinod Modi

M.S. Univ. of Baroda, Dept. of Microbiology, Baroda, India 390002.

Cowpea Rhizobium sp. JLn(c), a symbiont of the legume Arachis hypogae, showed significant inhibition of ex planta nitrogenase activity, polysaccharide synthesis and lectin binding ability when grown in Ashby's mannitol broth supplemented with various combined nitrogen (CN) salts (KNO_3/NH_4Cl/NH_4NO_3/glutamine). A 20 mM concentration of either NO_3^- or NH_4^+ resulted in complete inhibition of nitrogenase activity. Lectin binding ability as well as exopolysaccharide and capsular polysaccharide synthesis were reduced in the presence of CN salts, both under aerobic and microaerophilic conditions. The tricarboxylate cycle enzymes; viz. isocitrate dehydrogenase and malate dehydrogenase showed higher activities in the presence of CN salts suggesting the channelization of carbon towards CO_2 formation rather than for polysaccharide synthesis. This was further confirmed from the observed increase in the activity of glucose-6-phosphate dehydrogenase and respiration rates of JLn(c) cells in the presence of fixed nitrogen.

Reduction in the synthesis of acidic exopolysaccharide (EPS) in response to CN was accompanied by a decrease in the non-carbohydrate constituents of the exopolymer. The amounts of non-carbohydrate constituents like pyruvate and acetate were significantly lowered in the CN supplemented EPS. These observations complement the results obtained by infrared spectroscopy and chemical analysis. It can be postulated that CN induced alterations in the acidic EPS may be affecting the host-symbiont interaction leading to a decrease in infectivity and nodule formation.

Gresshoff, Roth, Stacey & Newton (eds.) Nitrogen Fixation: Achievements and Objectives
Chapman and Hall New York • London 1990

Regulatory mutants of *Rhizobium leguminosarum* biovar *phaseoli* impaired in nitrogen metabolism

S. Moreno*, R. Meza, M. Chiurazzi**, R. Defez**, E. Patriarca**, M. Iaccarino**, and G. Espín*.
* Unidad de Biología Molecular y Biotecnología Vegetal. Universidad Nacional Autónoma de México. Apdo. Postal 2-246. Cuernavaca, Morelos, México. ** Istituto Internazionale di Genetica e Biofisica, Via Marconi 10, 80125 Napoli, Italia.

In a previous study we described the isolation of two R. l. phaseoli mutants (CFN2012 and GE11) that lost GSII activity. Further studies of these mutants have shown that they are also defective in glnB expression, since the regulatory pII protein, its product, is not detectable. The level of adenylylation of glutamine synthetase I (GSI) is also impaired in these mutants suggesting a role for the PII protein in the adenylylation mechanism. A 12 kb BamHI fragment of the wild type was isolated (plasmid pSM261) and shown to complement in both CFN2012 and GE11, the absence of GSII and PII, and adenylylation of GSI. Sequence analysis of the complementing DNA showed the presence in pSM261 of an operon comprising the ntrB and ntrC regulatory genes. Hybridization of pSM261 with EcoRI digests of the mutants total DNA, indicated that strain CFN2012 carries an ntrC::Tn5 mutation and strain GE11 carries a ntrB::Tn5 mutation. Thus, these results indicate that the above described phenotypes are due to the lack of the ntrC gene product. In contrast to ntrC mutants of other Rhizobeacea, strain CFN2012 grows on nitrate as the sole nitrogen source.

This work was supported by EEC collaborative grant CI1-0410-ME/ CI1-0408-I

Gresshoff, Roth, Stacey & Newton (eds.) Nitrogen Fixation: Achievements and Objectives
Chapman and Hall New York • London 1990

Genetic characterization of symbiotically defective *Bradyrhizobium japonicum* mutants

P. Müller, A. Schultes and D. Werner,

Philipps Universität, FB Biologie/Botanik, Karl-von-Frisch-Str., D-3550 Marburg, F.R.G.

Three *Bradyrhizobium japonicum* strains from different origins have previously been shown to be symbiotically defective: the soybean nodules induced by strains RH-31-Marburg, 61-A-24 or A-9 all have in common an instability of the peribacteroid membrane (PBM). As a consequence the infected plant tissue undergoes a complete lysis (1-4). Mutant RH-31-Marburg was obtained by UV-treatment of strain 61-A-101, whereas mutant 61-A-24 is a naturally occurring ineffective strain. Mutant A-9 is a derivative of 110*spc*4 where an internal fragment of *nifA* is replaced by a kanamycin resistance cassette. The genetic relationsphip between the parental strains was analyzed by pulse-field gelelectrophoresis (PFGE) in combination with DNA-hybridization experiments. Fragments carrying *nodVW, nodCSUIJ* and *nifA* genes of *Bradyrhizobium japonicum* were used as probes.
It became evident that strains 61-A-101 and RH-31-Marburg were identical, indicating that major deletions had not occurred within mutant RH-31-Marburg. In contrast to this result the restriction patterns between 61-A-101, 61-A-24 and 110*spc*4 were clearly different and therefore can be used for strain identification. The hybridization studies revealed that two clusters of symbiotic genes within the genome of *B. japonicum* which so long have been regarded to be unlinked are located on a ca. 100 kb *Spe*I fragment in strains 110*spc*4 and 61-A-24. This method therefore may be suitable for mapping genes within the genome.

References:

1. Werner et al., Planta 147:320-329 (1980).
2. Werner et al., Planta 162:8-16 (1984).
3. Werner et al., Planta 174:263-270 (1988).
4. Fischer et al., EMBO J. 5.6:1165-1173 (1986).

Gresshoff, Roth, Stacey & Newton (eds.) Nitrogen Fixation: Achievements and Objectives
Chapman and Hall New York • London 1990

The ribosomal and nitrogenase genes from *Frankia* amplified using PCR

P. Normand[1*], P. Simonet[2], M.P. Fernandez[2], B. Cournoyer[1] and S. Nazaret[2]

[1]Centre de Recherche en Biologie Forestiere, Universite Laval, Quebec, Canada; [2] Laboratoire de Microbiologie des Sols, U.R.A. CNRS 697, Universite Lyon I, France. *Present address in Lyon.

With the aim of studying Frankia in the field, we first undertook a taxonomic study to define genomic species using DNA/DNA hybridization. Nine genogroups with more than 70% homology were recognized among the 40 strains tested that belong to three host compatibility groups (1). EMBL3 clones for two genomic regions, nifHDK (2) and the rrn region (unpubl. data), in the gene banks of four representative strains (ArI3, HRN18a, EUN1f and ORS020606) were characterized.

The nif region was investigated first and we found that the gene order was nifH, nifD and nifK in all strains tested and that nifH showed some variation between closely related strains. However, since nif genes had been shown to be plasmid-borne in at least one Frankia strain (3), phylogeny based on nifH sequences may only reflect transferable plasmids.

Therefore we decided to study rrn genes which are generally chromosome-borne. Casuarina-compatible strain ORS020606 has two ribosomal gene clusters and both copies have the usual gene order 16S-23S-5S. The two copies could not be differentiated in the coding region, but varied a lot outside of it. The complete nucleotide sequence of one rrn operon was determined and by comparison with the published sequences from other microorganisms, two conserved sequences were chosen as targets for oligonucleotides, framing a divergent region in the second half of the 16S gene. The sequence of this hypervariable region of 280 bp was determined for 30 Frankia strains. A tentative phylogeny of the genus was thus devised using distance matrix analysis which grouped Alnus-compatible with Casuarina-compatible strains, leaving Elaeagnus-compatible strains in a different branch and the Rosaceae Purshia strain PtI1 completely isolated. All strains tested formed a coherent phylum.

1. Fernandez, M.P., Meugnier, H., Grimont, P.A.D. & Bardin, R. (1989) Int. J. Syst. Bacteriol. **39**, 424-429.
2. Normand, P., Simont, P. and Bardin, R. (1988) Mol. Gen. Genet. **213**, 238-246.
3. Simonet, P., Haurat, J., Normand, P., Bardin, R.& Moiroud, A. (1986) Mol. Gen. Genet. **204**, 492-495.

Gresshoff, Roth, Stacey & Newton (eds.) Nitrogen Fixation: Achievements and Objectives
Chapman and Hall New York • London 1990

Identification of downstream activator sequences (DAS) on the *nifHDK* promoter of *Rhizobium meliloti*

F. O'Gara, B. Boesten, Y-P. Wang, K. Birkenhead and A. Dobson.

National Food Biotechnology Centre and Microbiology Department, University College, Cork, Ireland.

Deletion analysis has been carried out on the nifHDK (P1) promoter of Rhizobium meliloti to determine sequences which are involved in the expression of this promoter under both microaerobic and symbiotic conditions. These studies revealed additional sequences (DAS) located downstream from the promoter element (PE) which are involved in regulating expression of P1. It had previously been demonstrated that sequences upstream from -35 on the promoter were essential for the microaerobic expression of P1 (1). We have shown that the DAS (+17 to +61) are also essential for microaerobic expression of P1. However, neither the UAS or DAS alone appear to be essential for the symbiotic expression of P1. Only when both sequences are deleted is a significant reduction in expression of P1 observed during symbiosis with alfalfa plants. Expression studies of the mutated P1 promoters in a nifA mutant background indicate that the DAS require a functional NifA to mediate expression from these promoters. It appears likely that NifA binds to both the DAS and UAS to activate transcription from the nifHDK promoter.

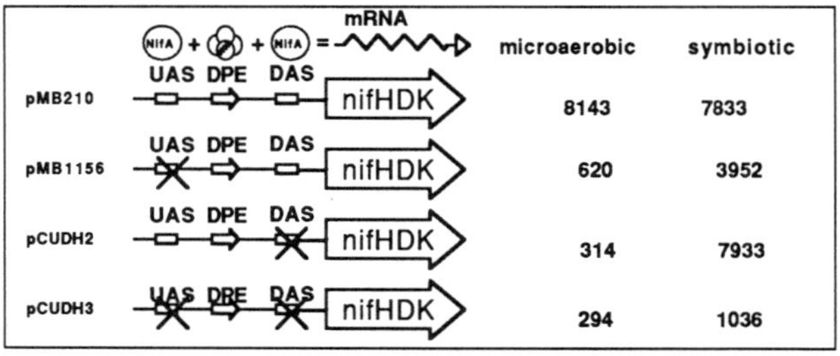

Downstream activator sequences on the P1 promoter.

(1) Better, M., Ditta, G. and Helinski, D.R. (1985). EMBO J. 4: 2419-2424.

Gresshoff, Roth, Stacey & Newton (eds.) Nitrogen Fixation: Achievements and Objectives
Chapman and Hall New York • London 1990

Azospirillum brasilense Sp7: p90 carries essential function

I. Onyeocha* and C. Elmerich

Unite de Physiologie Cellulaire, Dept. des Biotechnologies, Institut Pasteur, 25 rue du Dr. Roux, 75724 Paris, France.

Azospirillum strains contain plasmids. Homology was previously found between DNA fragments containing *Rhizobium meliloti* nodulation genes (*nodPQ*) and the 90 MDa plasmid, p90, of several *Azospirillum* strains. A more detailed characterization of p90 from *A. brasilense* Sp7 was undertaken to establish its physical map and to localize the *nod* homology and other specific regions. In addition to the *nodPQ,* the *R. meliloti exoB* and *exoC* mutation-correcting loci, two loci for ampicillin resistance, and a region necessary for the maintenance of p90 as an independent replicon probably corresponding to the origin of replication were mapped on p90. The physical map of p90 established for *Bam*HI, *Hin*dIII, *Xho*I and *Eco*RI (not shown), and the specific regions mapped are presented in the figure (size scale in kb).

Fragments of p90 hybridized with plasmids of 90 MDa present in most strains of *A. brasilense* and in some of *A. lipoferum,* suggesting the existence of common functions on these plasmids and a plasmid family in *Azospirillum*. Attempts to obtain p90-cured strains were without success, suggesting existence of essential bacterial functions on this plasmid.

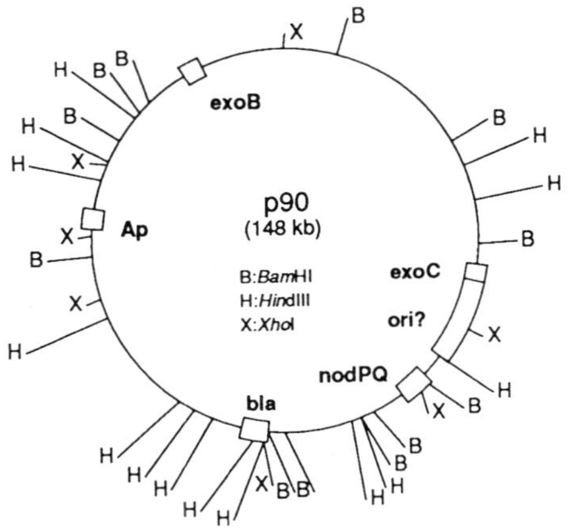

References

Michiels, K., de Troch, P., Onyeocha, I., Van Gool, A., Elmerich, C. and Vanderleyden, J. (1989). *Plasmid* **21,** 142-146.

Onyeocha,I., Vieille, C., Zimmer, W., Baca, B. E., Flores, M., Palacios, R. and Elmerich, C. (1990). *Plasmid* **23,** in press.

Gresshoff, Roth, Stacey & Newton (eds.) Nitrogen Fixation: Achievements and Objectives
Chapman and Hall New York • London 1990

Analysis of positive and negative regulators of *nif* genes in *Rhodobacter capsulatus*

Victor M. Pace, Dawn Foster-Hartnett and Robert G. Kranz

Department of Biology, Washington University, St. Louis, MO 63130 USA.

Most free-living unicellular diazotophs activate transcription of the genes involved in nitrogen fixation only when oxygen and fixed nitrogen become limiting. The fixed nitrogen sensing regulatory circuit of the photosynthetic bacterium *R.capsulatus* is shown to be similar to the *E.coli* circuit. This involves the regulatory genes *glnB, nifR2* (*ntrB*-like) and *nifR1* (*ntrC*-like). The phenotype of both point and transposon insertion mutations in each gene have been used to study the regulatory circuit. *glnB* is in an operon with *glnA*, the gene encoding glutamine synthetase and mutations in *glnB* yield *R.capsulatus* strains that constitutively transcribe *nif* genes with respect to fixed nitrogen. GLNB is shown to be a negative regulator of *nif* genes that transmits the nitrogen status of the cell to NIFR2. *LacZYA* translational fusions have been constructed to the *glnBA* operon and to *nifA* (copy II) and *nifR4*, two positive regulatory genes. It is shown that *nifR1* is required to activate *glnBA* (50%), *nifA*(100%) and *nifR4* (75%) transcription.

Both in vivo and in vitro constructed *nifR4:lacZ* fusions are expressed at very low levels thus complicating the analysis of its regulation. It appears that transcription of both *nifR4* and *nifA* are repressed by oxygen, possibly mediated by the level of DNA supercoiling under anaerobic versus aerobic growth conditions. A detailed analysis of their promoters is in progress. Activation of *nifA* and *nifR4* thus begins the second level of control that is required for oxygen signaling. NIFA, another DNA binding activator protein, and NIFR4, an NTRA-like sigma factor, subsequently activate transcription of all other *nif* genes. NIFR4 is a unique RNA polymerase sigma factor in that it is linked to the *nif*HDK operon, it is controlled by the nitrogen status, and the only effect of either *nifR4* insertion or deletion mutations is the inability to transcribe *nif* genes. A consensus promoter sequence, based on five promoters that have been shown (or presumed) to require NIFR4 in *R.capsulatus* suggest a promoter structure similar to NTRA-activated promoters.

Gresshoff, Roth, Stacey & Newton (eds.) Nitrogen Fixation: Achievements and Objectives
Chapman and Hall New York • London 1990

NAD- and NADP-dependent malic enzymes in *Bradyrhizobium japonicum* USDA110 bacteroids

Rosanne G. Quinnell*[1], Les Copeland[2] and David A. Day[1]

1.Botany Department, The Australian National University, Canberra A.C.T. 2601, Australia.
2. The Department of Agricultural Chemistry, The University of Sydney, Sydney N.S.W.2009, Australia.

It appears that the primary substrates supplied to the bacteroids for fuelling nitrogen fixation in soybean nodules are C_4-dicarboxylates. The co-ordinated activities of malic enzyme (ME) and the TCA cycle enzyme malate dehydrogenase (MDH) provide a possible means of fulfilling the energy and carbon requirements of effective bacteroid metabolism. NADP-ME activity in rhizobia has been reported by McKay *et al* (1) and by Kimura and Tajima (2) but two malic enzymes have been identified in the crude enzyme extract of *Bradyrhizobium japonicum* USDA110 bacteroids; an NAD- and an NADP-dependent form (3). Studies were carried out to obtain more conclusive evidence for the presence of two distinct malic enzymes and to ascertain the possible roles of each ME.

A partially purified bacteroid extract was subjected to gel electrophoresis in non-denaturing 7.5% polyacrylamide gels to physically separate the two malic enzymes and to distinguish between NAD-ME and NAD-dependent MDH. Native gel electrophoresis showed that NADP-ME, NAD-ME and MDH are three distinct enzymes.

An induction study on the free-living *B. japonicum* cells was performed, together with an examination of possible *in vivo* effectors on enzyme activities *in vitro* using the partially purified extract. NAD-ME was induced by growth on succinate and malate but NADP-ME appeared to be constitutive. In addition, each ME exhibited different responses to effectors; the activities of both enzymes were stimulated by NH_4^+ but NAD-ME was also stimulated by K^+.

Activities of both NAD-ME and NADP-ME were found in the crude enzyme extracts of bacteroids isolated from *Parasponia* and siratro nodules suggesting that ME activities may be an essential feature of *Bradyrhizobium* symbioses.

(1)McKay, I.A. Dilworth, M.J. and Glenn, A.R. (1988). C_4-dicarboxylate Metabolism in Free-living and Bacteroid Forms of *Rhizobium leguminosarum* MNF3841. J. Gen. Microbiol. **134**: 1433-1440.
(2)Kimura, I. and Tajima, S. (1989). Presence and Characteriastics of NADP-malic enzyme in Soybean Nodule Bacteroids. Soil Sci. Plant Nutr. **35:** 271-279.
(3)Copeland, L. Quinnell, R.G. and Day, D.A (1989). Malic Enzyme Activity in Bacteroids from Soybean Nodules. J. Gen. Microbiol.**135**: 2005-2011.

Gresshoff, Roth, Stacey & Newton (eds.) Nitrogen Fixation: Achievements and Objectives
Chapman and Hall New York • London 1990

Expression of *Rhizobium galegae* common *nod* clones in various background

L.A. Rasanen, U. Heikkila-Kallio, L. Suominen, S. Kaijalainen, P Lipsanen and K. Lindström

Dept. of Microbiology, University of Helsinki SF-00710 Helsinki, Finland.

INTRODUCTION. The cosmid clone pRg30, carrying common nodulation genes of Rhizobium galegae HAMBI 1174, and pRg33, a subclone of pRg30 containing a 5,7 kb Cla1 insert carrying nodDABC, were conjugated into various backgrounds and used in expression analysis in plant tests.

METHODS. Surface sterilized and germinated seeds of different test plants were transferred onto Jensen-agar slants and inoculated with conjugants. The plants were grown in test tubes in a growth chamber.

RESULTS. pRg30 and pRg33 complemented different Nod^- mutants. Derivatives of R. leguminosarum strain 5039 carrying Tn5 inserts in nodA, nodB and nodC genes complemented with pRg30 and pRg33 showed delayed nodulation in V. sativa. However, root hair deformation and the development of infection threads were similar to the wild type.

The nodulation ability of the nodC mutant strain R. meliloti 1126 was restored almost to wild-type level in M. sativa. Deformation of root hairs was similar to the wild type, but infection threads were difficult to detect.

A. tumefaciens C58C1 into which pRg30 and pRg33 had been introduced, produced root hair deformation on G. orientalis and G. officinalis plants but not on M. sativa. After 6 weeks, there were some (1-7%) tumor-like structures in all test plants inoculated with conjugants.

pRg30 and pRg33 conjugated into R. trifolii strain 60Bs $pSym^-$ did not produce any root hair deformation on M. sativa and on both Galega plants, except 60 Bs pSm^- with pRg30, which deformed the root hairs of both G. officinalis slightly.

Gresshoff, Roth, Stacey & Newton (eds.) Nitrogen Fixation: Achievements and Objectives
Chapman and Hall New York • London 1990

Occurrence of cellulolytic enzymes in *Azoarcus*

B. Reinhold-Hurek, T. Hurek, and M. van Montagu

Laboratorium Genetika, Rijksuniversiteit Gent,
Ledeganckstraat 35, B-9000 Gent, Belgium.

From the root interior of Kallar grass (*Leptochloa fusca* (L.) Kunth), microaerobically N_2-fixing, gram-negative rods have repeatedly been isolated, which are proposed to form a new genus "*Azoarcus*" (see Reinhold et al., this volume). Since they occur in high numbers in the endorhizosphere of field-grown plants (Reinhold et al., 1986), and infect the root cortex of healthy looking Kallar grass (Reinhold and Hurek, 1989), we were interested if plant-polymer degrading enzymes occur in this genus. We cloned and characterized the cellulolytic enzymes of strain BH72.
We screened for activity of three classes of cellulolytic enzymes, ß-glucosidase (glc), cellobiohydrolase (cbh), and endoglucanase (egl). Fluorogenic and chromogenic substrates as well as carboxymethylcellulose were used for cellulase assays.
Among five strains tested, only strain BH72 had well detectable enzyme activities of all three classes, the other strains missing endoglucanase. Screening of a genomic library of BH72 in pUC 19 revealed 4 different clones having simultaneously glc and cbh activity, and five clones showing only endoglucanase activity. Restriction maps gave no evidence for physical linkage of both gene regions. *glc* and *cbh* genes were located on one 2.7 kb *Kpn*I/*Sau*3AI, *egl* on a 1.4 kB *Sst*I/*Kpn*I fragment. In isoelectric focusing gels, glc and cbh activity showed the same mobility (pI 5.4), whereas egl occurred as two isoenzymes (pI 6.0 and 6.4). Enzymes expressed in *E. coli* had the same pI values. Apparently, ß-glucosidase- and cellobiohydrolase-like activity were conferred by one enzyme.
The cellulolytic system of *Azoarcus* had several unusual features: (i) Avicel (crystalline cellulose) is not attacked; (ii) cells do not grow on cellulose, cellobiose or glucose; (iii) activity of endoglucanase or ß-glucosidase is not repressed by glucose, but increased by ethanol as a carbon source; (iv) endoglucanase (and as well ß-glucosidase) is not excreted into the culture supernatant, but cell-surface associated. Whether this complex catabolic enzyme system might contribute indirectly to growth by playing a role in establishemnt inside the host root will be investigated.

Reinhold, B., Hurek, T., Niemann, E.-G.& Fendrik, I. (1986) Appl. Environ. Microbiol.**52**, 520-526.
Reinhold, B. & Hurek, T. (1988) Plant Soil **110**, 259-268.

Gresshoff, Roth, Stacey & Newton (eds.) Nitrogen Fixation: Achievements and Objectives
Chapman and Hall New York • London 1990

Are *Bradyrhizobium lupini* bacteroids able to assimilate sugars?

V.I.Romanov, N.G.Fedulova, V.I.Shramko, I.E.Chermenskaya, B.F.Ivanov, Z.N.Tsaguriya, and W.L.Kretovich

A.N.Bach Institute of Biochemistry, USSR Academy of Sciences, Leninsky pr.33, Moscow, 117071, USSR

Washed cells of the B .lupini N-limited (C/N=120/1) chemostat culture grown on a medium containing molasses assimilated 14-C-sucrose 100 times more actively than the C-limited culture (C/N=6/1). The N-limited culture resembled bacteroids in a number of characteristics.

Isolated B .lupini bacteroids assimilated 14-C-succinate, -glutamate, -sucrose, and -glucose with comparable rates under both aerobic and microaerophilic conditions (in rather long experiments - 1-3 hr incubation). The O_2 concentration significantly influenced both the substrate consumption rate and the substrate carbon distribution among intracellular compounds. The maximum succinate utilization by bacteroids was observed at 5% O_2, while at 0.1% O_2, glucose and sucrose were preferential substrates. At all O_2 concentrations (from 0.1 to 21%) sugar carbon was involved in the amino acid biosynthesis in bacteroids more actively than succinate carbon. The ability of bacteroids to assimilate 14-C-succinate and -sucrose measured by 14-CO_2 evolution depended on the plant developmental stage, curves being different for the two substrates.

The radioactivity ratio of 14-CO_2 derived from 1-14-C- and 6-14-C-glucose in bacteroids incubated under aerobic conditions and at 5% O_2 in the gas mixture is 3:1 and 1:1, respectively. The radiorespirometric pattern obtained at 5% O_2 for glucose is $C1 \geqslant C6 > C3,4 \geqslant C2$ (where C1=18.0%; C6=17.4%,C3,4=15.2%; C2,3=15.0%) and for succinate - $C1,4 > C2,3$ (where C1,4=33.0%; C2,3=17.0%). Bacteroids of "effective" nodules were found to have active key enzymes of the EMP, PP, and ED pathways. The ratio of the enzymatic activities depended to a great extent on the developmental stage, in particular the ED pathway was found only in old nodules. The enzymatic analysis of bacteroids from "ineffective" (Fix^-) nodules showed the prevailence of the ED pathway.

Gresshoff, Roth, Stacey & Newton (eds.) Nitrogen Fixation: Achievements and Objectives
Chapman and Hall New York • London 1990

Amplification and deletion events in the symbiotic plasmid of *Rhizobium phaseoli*

D. Romero, M.L. Girard, S. Brom, J.M. Martínez-Salazar, M. Flores, R. Palacios and G. Dávila

Departamento de Genética Molecular, Centro de Investigación sobre Fijación de Nitrógeno, UNAM.; Apdo. postal 565-A, Cuernavaca, Morelos, México.

We have determined the physical map of the symbiotic plasmid of R. phaseoli CFN42. This plasmid contains two nifHDK operons in direct orientation, separated by 120 Kb. Within this region there is a third nifH gene, located in reversed orientation, as well as some essential nodulation genes. Other reiterated sequences were located in several regions of the physical map. This structure suggests that the symbiotic plasmid must be prone to several kinds of genomic rearrangements. We have designed a system, based on a dose-dependent kanamycin resistance determinant and the sacB gene of Bacillus subtilis; this gene confers sucrose-dependent lethality in gram-negative bacteria. This system allows the positive selection of different types of genomic rearrangements, including amplification and deletion events. The rearrangements more frequently found (10^{-3}) were amplification events, which increase two to three times the copy number of the 120 Kb region. These strains are highly unstable, segregating cells that had returned to the normal state. Higher-order amplifications, increasing eight times the copy number of this region were also found. Deletion events that eliminate the 120 Kb region were also observed, albeit at a lower frequency (10^{-4}). These rearrangements are frequently generated by recombination between the two nifHDK operons. Rearrangements similar to these, as well as loss or alteration of other plasmids present in this strain have been found without any kind of selection. These rearrangements may explain the instability of symbiotic characters frequently observed in inoculant strains, and might be relevant to strain's adaptative potential.

Gresshoff, Roth, Stacey & Newton (eds.) Nitrogen Fixation: Achievements and Objectives
Chapman and Hall New York • London 1990

Isolation of rhizobitoxine mutants of *Bradyrhizobium japonicum*

Xiaoan Ruan and N. Kent Peters

Biotechnology Center and Department ofAgronomy, The Ohio State University, Columbus, OH 43210, USA.

Rhizobitoxine is synthesized by strains of the group II *Bradyrhizobium*. Synthesis of rhizobitoxine by bradyrhizobia causes a transient chlorosis in some cultivars of soybeans (2). Even though the biological function(s) of rhizobitoxine remains unknown, its production has been observed commonly among the *Bradyrhizobium* spp. with a world-wide distribution (3). Rhizobitoxine has been isolated from extracts of root nodules, young leaves of affected plants (7), and from liquid cultures of some chlorosis-inducing strains (6). Its chemical structrue was determined and found to be a structural analog of cystathionine, a precursor of methionine (9). Previous studies have found that rhizobitoxine-producing strains were are able to nodulate soybean rj1 mutants (2),and help protect soybeans from fungal infection (1). Mitchell et al. (5) showed that *Pseudomonas andropogonis*, a chlorosis causing bacterial plant pathogen which infects a wide range of plants, also produces rhizobitoxine. The synthesis of rhizobitoxine by two apparently unrelated plant-associated bacteria suggests that rhizobitoxine plays some role in bacterial infection of plants. To investigate the role of rhizobitoxine in plant-bacterial interactions, we isolated eleven Tn5 generated Rtx^- mutants from *B. japonicum* strain USDA61 using the an enzyme assay. The assay is based on the inhibition of β-cystathionase of *Salmonella*. These mutants were checked and found to be prototrophic. Rtx^- mutants formed normal numbers of nodules and fixed normal levels of nitrogen on *Glycine max* and *G. soja*. Therefore rhizobitoxine synthesis has no obvious role in nodulation of wild -type plants.

1. Chakraborty, Y. U. & Purkayastha, R. P. (1984) Can. J. Microbiol. **30**, 285-289.
2. Devine, T. E. & Weber, D. F. (1977) Euphytica **26**, 527-535.
3. LaFavre, J. S. & Eaglesham, A. R. S. (1986) Plant Soil **92**, 443-452.
4. Kaplan, M. M. & Martin, F. (1966) J. Biol. Chem. **241**, 5781-5789.
5. Mitchell, R. E., Frey, E. J. & Benn, M. K. (1986) **25**, 2711-2715.
6. Owens, L. D. & Wright, D. A. (1965a) Plant Physiol. **40**, 931-933.
7. Owens, L. D. & Wright, D. A. (1965b) Plant Physiol. **40**, 927-930.
8. Owens, L. D., Guggenheim, S. & Hilton, J. L. (1968) Biochem. Biophys. Acta **158**, 219-225.
9. Owens, L. D., Thomson, J. F., Pritcher, R. G. & Williams, T. (1972) J. C. A. Chem. Comm. 714.

Gresshoff, Roth, Stacey & Newton (eds.) Nitrogen Fixation: Achievements and Objectives
Chapman and Hall New York • London 1990

Generation of self-transmissible nodulation plasmids differing in host specificity from the crossing of *Rhizobium meliloti* strains

M.L. Rumayuntseva, N.A. Provorov and B.V. Simarov

All-Union Research Institute for Agricultural Microbiology, Leningrad, 189620, USSR.

By use of suicide vector pSUP5011, 114 independent Tn5-mob recombinants were obtained in strain 135a, an effective R. meliloti strain. This strain is characterized by a broad host range: it forms Fix^+ nodules on cultivated tetraploid alfalfa (Medicago sativa) and wild diploid Arabian alfalfa (M. arabica). After introduction of RP4-4, Tn5-mob mutants were conjugated to strain CXM1-46, a Nod^- R. meliloti mutant defective in the 'common' nod genes. This mutant was originated from strain CXM1, a narrow host range strain, which is able to form Fix^+ nodules only with M. sativa (on M. arabica it forms Fix^- small white swellings). The donor strain was marked by Rif^r and harbored 2 megaplasmids and one medium-size cryptic plasmid (120 MD). Recipient strain was Str^r and had the megaplasmids only. Recombinants forming Fix^+ nodules on both Medicago species were obtained from 2 independent crosses after selection for a Km^r, Str^r phenotype. After passage through nodules of either host species, the ability of recombinants to form Fix^+ nodules on M. arabica was lost, while specificity to M. sativa was retained. By the use of the Eckhardt procedure, the presence of new plasmids of a size nearly 60 MD was detected in these recombinants, while RP4-4 and Tc^r marker were absent. Southern blot hybridization with pKSK5 revealed the presence of the 'common' nod genes in the 60 MD plasmids in R. meliloti recombinants. These plasmids were transferred into GM1-9023 (a plasmidless mutant of Agrobacterium tumefaciens) after selection for Km^r (frequency 10^{-6} per recipient). Obtained recombinants acquired ability to form Fix^-nodules on M. sativa. We suppose that the instability of host range determinants in R. meliloti transconjugants is caused by recombination events which transformed the donor strain megaplasmid into the medium-size self-transmissible form.

Gresshoff, Roth, Stacey & Newton (eds.) Nitrogen Fixation: Achievements and Objectives
Chapman and Hall New York • London 1990

Isolation of *Rhizobium galegae* species-specific DNA probes

A. Saano, S. Kaijalainen and K. Lindström

Dept of Microbiology, University of Helsinki, SF-00710 Helsinki, Finland.

PURPOSE. To apply the DNA-DNA-hybridization for the identification of Rhizobium galegae (1) from soil samples, eight EcoRI fragments from its nod-region were tested for their suitability to serve as species-specific DNA probes.

METHOD. The fragments were separated in gel electrophoresis, isolated, and labeled with dig-dUTP. Every labeled DNA fragment was hybridized with total-DNAs (50 ng-dots on n-cellulose filter) of different rhizobial species and other soil bacteria for 3 hours at 68°C. The hybrids were detected with dig-antibody-AP, NBT and X-phosphate. The best probe fragment was cloned in pBR322.

RESULTS. The 2.4 kb probe (Fig.1) hybridized with the DNAs of R. galegae 59A2; gal1261; 740; gal1; gal3; gal14; galE; G6; G8; G9; and gal302. Slight hybridization was observed with the DNAs of R. loti NZP2213 and Erwinia carotovora subsp. atroseptica 549. No hybridization was detected with the DNAs of R. meliloti MEL2; D44d; Rm1021; Rm1023; Rm1027; Rm1126; R. leguminosarum 60B; HT3; 162P30; 175F9; CB596; NZP5486; R. trifolii 60B51; Agrobacterium tumefaciens AC58; Escherichia coli DSM 2840; Klebsiella pneumoniae; Pseudomonas syringae AD111. The other tested fragments were less species-specific (Fig.2).

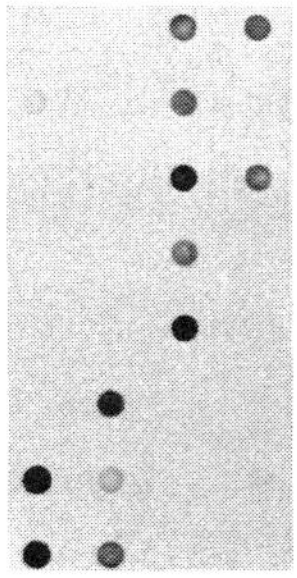

Fig.1.
2.4 kb probe

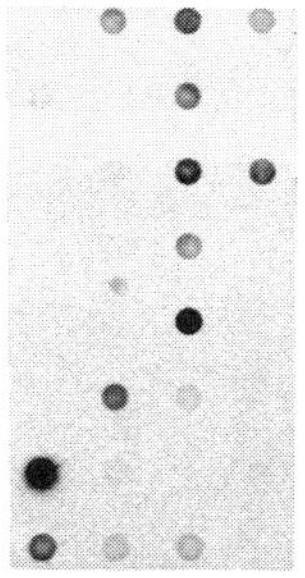

Fig.2.
2.7 kb probe

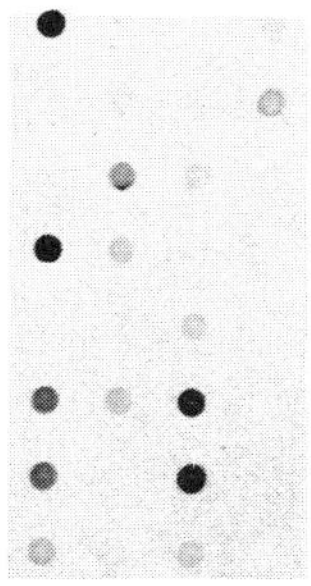

Fig.3.
R. me Rm1126 tot-DNA probe
(blot control)

REFERENCE. Lindström, K.(1989) Int. J. Syst. Bacteriol. 39, 365-367.

Gresshoff, Roth, Stacey & Newton (eds.) Nitrogen Fixation: Achievements and Objectives
Chapman and Hall New York • London 1990

Nitrogenase genes in Frankia HFPCcI3

Amala Reddy and Ann M. Hirsch

Dept. of Biology, University of California, Los Angeles, CA 90024, USA

We are studying symbiotic genes in *Frankia* strain HFPCcI3 which forms a nitogen-fixing association with the roots of the important actinorhizal tree *Casuarina cunninghamiana*. The nitrogenase structural genes (*nifHDK*) are conserved among a number of diverse nitrogen-fixing bacteria, including *Frankia* (1, 2). We have isolated a clone, pAR2, which contains the *nif* genes, from a partial genomic library of HFPCcI3. The *nifHDK* genes are contained on a 5.6 kb *Bam*HI fragment in HFPCcI3, and the restriction map of this fragment is as expected for a *Casuarina* isolate (2). We have localized the different nitrogenase genes to specific restriction fragments in pAR2, and they appear to be contiguous. DNA:DNA hybridization experiments indicate that the *Frankia nifH* and *nifD* genes exhibit approximately 70% and 65% sequence similarity, respectively, to *nifH* and *nifD* probes from *Rhizobium meliloti*, while the *nifK* gene is approximately 65% homologous to a *Klebsiella pneumonia nifK* probe. We have generated subclones of the *Bam*HI *nif* fragment, and are in the process of sequencing them. From the partial sequence information we have obtained to date, the *Casuarina* isolate of *Frankia* appears to have the same level of DNA sequence similarity to other nitrogen-fixing organisms, as was observed in *Alnus* isolates (1, 2). We intend to characterize the *nif* genes in terms of their sequence, as well as their transcriptional organization and control.

1. Normand, P., Simonet, P. & Bardin, R. (1988) Mol. Gen. Genet. **213**, 238-246.
2. Ligon, J.M. & Nakas, J.P. (1988) Nucleic Acids Res. **16**, 11843.

Gresshoff, Roth, Stacey & Newton (eds.) Nitrogen Fixation: Achievements and Objectives
Chapman and Hall New York • London 1990

Analysis of the *glnII* gene from the actinorhizal symbiont *Frankia* CpI1

D. A. Rochefort* and D. R. Benson

Dept. of Molecular and Cell Biology, Univ. of Connecticut, Storrs, CT, 06269-3044, USA.

INTRODUCTION

Frankia is an actinomycete that forms N_2-fixing root nodules on a wide variety of angiosperms that inhabit N-deficient soils. Like members of the *Rhizobiaceae*, *Frankia* has two glutamine synthetases (GSs), GSI and GSII, involved in ammonia assimilation [1]. GSI is produced constitutively under all growth conditions and GSII is produced during nitrogen-limited growth [2].

RESULTS AND DISCUSSION

The cloned GSII gene (*glnII*) from *Bradyrhizobium japonicum* [3] was used to identify similar sequences from *Frankia* sp. strains CpI1, Cc13, and Ai17. A putative *Frankia* CpI1 *glnII* gene was cloned into pUC19 from a size selected *KpnI* library. Once localized by restriction mapping, the gene was sequenced. The sequence has a high degree of similarity to *B. japonicum glnII*, with the high GC content expected from actinomycete DNA. The translated protein was found to be equally similar to both *B. japonicum* GSII and to pea (*Pisum sativum*) GS. Possible substrate-binding sites for ATP and glutamate were observed in the translated sequence, and a possible Ntr-type promoter was identified upstream of the *glnII* gene. The cloned gene was found to be functional, as the *glnII* gene was expressed from the *lac* promoter of pUC19 in an *E. coli glnA* deletion strain. These studies have important implications for the origin of the *glnII* gene and may provide a key to understanding the nature of N_2-fixing root module symbioses.

REFERENCES

1. Edmands, J., Noridge, N. A., Benson, D. R. (1987) Proc. Natl. Acad. Sci. USA. 84, 6126-6130.
2. Tsai, Y.-L., Benson, D. R. (1989) Arch. Microbiol. 152, 382-386.
3. Carlson, T. A. and B. K. Chelm. (1986) Nature (London) 322, 568-570.

Gresshoff, Roth, Stacey & Newton (eds.) Nitrogen Fixation: Achievements and Objectives
Chapman and Hall New York • London 1990

Bradyrhizobium japonicum δ-aminolevulinic acid (ALA) synthase is not essential for the expression of bacteroid heme in soybean root nodules

Indu Sangwan and Mark R. O'Brian

Department of Biochemistry, State University of New York at Buffalo, Buffalo, NY 14214, USA.

The *Bradyrhizobium japonicum* ALA synthase mutant MLG1 grew well in yeast extract-glycerol medium in the absence of exogenous ALA. No heme was discerned spectrophotometrically in extracts of those cells, and thus heme formation is strictly dependent on ALA synthase activity. Conversely, MLG1 bacteroids isolated from soybean root nodules contained significant levels of protoheme and mesoheme relative to that found in bacteroids of the parent strain I110, hence bacterial ALA synthase activity is not essential for heme expression in bacteroids. Activity of the heme synthesis enzyme porphobilinogen (PBG) deaminase was measured because the heme intermediate PBG is a cofactor precursor as well as a substrate for that enzyme in prokaryotes and in eukaryotes. MLG1 cultured cells contained significantly less PBG deaminase activity compared with that of I110, which is expected of a strain deficient in the ability to synthesize PBG *in situ*. However, MLG1 bacteroids possessed wild type levels of PBG deaminase activity, suggesting that PBG is available to the mutant bacteroids. ALA synthase and glutamate-dependent ALA synthesis activities were measured in bacteroid and plant cytosol fractions of nodules incited by strains I110 or MLG1. The plant cytosol fraction from both nodule types was able to synthesize ALA from glutamate, but neither had ALA synthase activity. Strain I110 bacteroids had only ALA synthase activity, and no ALA formation activity was found in MLG1 bacteroids. The data suggest that bacteroid heme can be synthesized from plant-derived ALA, resulting in the rescue of the ALA synthase mutant.

Gresshoff, Roth, Stacey & Newton (eds.) Nitrogen Fixation: Achievements and Objectives
Chapman and Hall New York • London 1990

Involvement of IHF in gene expression from the *nifH* promoter of *Klebsiella pneumoniae*

E. Santero*, T. Hoover and S. Kustu.

Departments of Plant Pathology and Molecular and Cellular Biology, University of California, Berkeley, CA 94720, USA

The regulatory protein NIFA activates transcription of nitrogen fixation operons by σ^{54}-holoenzyme. We recently demonstrated that NIFA activates transcription from the nifH promoter *in vitro*. We also showed that integration host factor (IHF), a small protein that bends DNA in site-specific manner, bound to this promoter region at a site located between the upstream binding site for NIFA and the promoter, a binding site for σ^{54}-holoenzyme.

We report here that IHF greatly stimulates NIFA-mediated activation of nifH transcription *in vitro*. IHF failed to stimulate transcriptional activation by the functionally related activator NTRC, which has no specific binding site in the nifH promoter region. Although the IHF binding site is located just upstream of the nifH promoter, the activator specificity suggested that IHF does not directly stimulate binding of σ^{54}-holoenzyme; this is confirmed by direct DNase I protection assays. Our evidence is consistent with the view that an IHF-induced bend facilitates contacts between NIFA and σ^{54}-holoenzyme, both of which bind to their sites independently of IHF. Our experiments indicate that the use of a "weak" nifH promoter plus IHF confers high fidelity of activation by NIFA together with high efficiency of activation.

For more detailed information, see E. Santero *et al* "Mechanism of initiation from nif promoters: involvement of IHF" in this book.

Gresshoff, Roth, Stacey & Newton (eds.) Nitrogen Fixation: Achievements and Objectives
Chapman and Hall New York • London 1990

DNA sequence and mutational analysis of three operons of *Rhodobacter capsulatus nif* region A containing genes encoding ferredoxin-like proteins

M.Schmehl, A.Meyer zu Vilsendorf, S.Hennecke, B.Masepohl and W.Klipp
Lehrstuhl für Genetik, Universität Bielefeld, Postfach 8640, D-4800 Bielefeld, FRG.

DNA sequence analysis of a 7.2 kb EcoRI fragment located upstream of nifE in R.capsulatus nif region A, revealed the presence of nine open reading frames, which are organized in at least three transcriptional units. Two of these operons, ORF9-ORF10 and ORF14-fdxP-fdxN-ORF15, are transcribed in the same direction as the nifENX operon, whereas the third operon, ORF16-ORF17-ORF18, is transcribed in opposite direction. The predicted amino acid sequence of the fdxP gene product (10 kDal) contains a cysteine motif $C-X_4-C-X_2-C$ typical for plant-type ferredoxins (1). ORF17 (19 kDal) and ORF18 (56 kDal) carry two identical cysteine motifs $C-X_2-C-X_2-C-X_3-C$, which are characteristic for bacterial-type ferredoxins. In contrast, R.capsulatus FdxN (7 kDal) contains two different cysteine motifs $C-X_2-C-X_2-C-X_3-C$ and $C-X_2-C-X_8-C-X_3-C$ (2), like nif-associated ferredoxins from R.meliloti, B.japonicum and A.vinelandii. The ORF16 gene product (20 kDal) carries six transmembrane helices, indicating that some steps in the nitrogen fixation process in R.capsulatus are membrane associated. Typical NtrA-dependent promoters were identified in front of ORF9 (22 kDal), ORF14 (41 kDal), fdxN and ORF16. Furthermore, the nif promoters in front of ORF9 and ORF16 show extended homology to each other and to the promoters in front of R.capsulatus nifE, ORF6 and both copies of nifB. A single NifA-binding site was found between the divergently transcribed ORF14 and ORF16. Mutational analysis revealed, that ORF9, ORF15 (28 kDal), ORF16, ORF17 and ORF18 are absolutely required for nitrogen fixation. Mutants in fdxN show slow diazotrophic growth. Mutants in ORF10 (20 kDal) and ORF14 remained Nif^+. The ferredoxin-like cysteine motifs of four proteins of this region suggest, that these gene products may play a role in electron transport to nitrogenase in R.capsulatus.

(1) Saeki, K., Miyatake, Y., Young, D.A., Marrs, B.L. & Matsubara, H. (1990) Nucl. Acids Res. **18**, 1060.
(2) Schatt, E., Jouanneau, Y. & Vignais, P. (1989) J. Bacteriol. **171**, 6218-6226.

Gresshoff, Roth, Stacey & Newton (eds.) Nitrogen Fixation: Achievements and Objectives
Chapman and Hall New York • London 1990

DNA-sequence of the *hup* structural genes and restriction fragment length polymorphism of the *hup* region from *Rhizobium leguminosarum* bv. *viciae* B10

H. Schmitt, C. Schneider, S. Koll, B. Schellhorn, Ch. Schild, H.V. Tichy* and W. Lotz

Institut fur Mikrobiologie und Biochemie, Universitat Erlangen-Nurnberg, D-8520 Erlangen, FRG; *Institut fur Biologie II, Universitat Freiburg, D-7800 Freiburg, FRG.

Cosmid clone pR1B505 carries the hup region of R. leguminosarum bv. viciae B10 on a 28.2 kb pSym-fragment. We have sequenced the "left" part of this region, which contains two open reading frames coding for the small (SU) and large (LU) subunits of the uptake hydrogenase. ORF1 (length: 1083 bp) and ORF2 (length: 1791 bp) are separated by 23 bp. The deduced amino acid sequence shows significant homology to the amino acid sequences of SU and LU of the uptake hydrogenases from Alcaligenes eutrophus (1), Bradyrhizobium japonicum (2), Escherichia coli (3), Rhodobacter capsulatus (4) and Rhodocyclus gelatinosus (5). Listed is the amino acid sequence homology (in %) between SU/LU of R. leguminosarum bv. viciae B10 with B. japonicum (89/89), R. capsulatus (78/74), A. eutrophus (78/70), R. gelatinosus (76/70) and E. coli (67/63). The high sequence homology among the hup genes indicates co-evolution of these genes.

The restriction fragment length polymorphism (RFLP) of 48 Hup$^+$ R. leguminosarum bv. viciae wild-type strains isolated from two locations has been analyzed using subclones of the cosmid pR1B505 as hybridization probes. The resulting DNA-fingerprints were grouped into 13 patterns (for EcoRI-digested total DNA), which indictated that only the "left" part, carrying the hup structural genes, is conserved on pSym of the R. leguminosarum bv. viciae strains.

1. Kortluke, Ch., Friedrich, B. (1990) pers. communication.
2. Sayavedra-Soto, L.A., Powell, G.K., Evans, H.J. & Morris, R.O. (1988) Proc. Natl. Acad. Sci. USA **85**, 8395-8399.
3. Menon, N.K., Robbins, J., Peck, H.D., Chatelus, C.Y., Choi, E.-S., & Przybyla, A.E. (1990) J. Bacteriol. **172**, 1969-1977.
4. Colbeau, A., Magnin, J.-P., Cauvin, B., Champion, T. & Vignais, P.M. (1990) Mol. Gen. Genet. **220**, 393-399.
5. Uffen, R.L., Colbeau, A., Richaud, P. & Vignais, P.M. (1990) Mol. Gen. Genet. **221**, 49-58.

Gresshoff, Roth, Stacey & Newton (eds.) Nitrogen Fixation: Achievements and Objectives
Chapman and Hall New York • London 1990

Characterization of actinorhizins involved in the development of *Frankia-Alnus* symbiosis

A. Séguin and M. Lalonde

Centre de recherche en biologie forestière. Faculté de foresterie et de géomatique, Université Laval, Ste-Foy, Qué. Canada G1K 7P4

In an effort to investigate the presence and origin of actinorhizae specific proteins (actinorhizins) that are expressed during the development of the actinorhizal nodules, the polypeptide content of *Alnus glutinosa* roots, infected root systems and actinorhizae have been compared.

A. glutinosa host plants were obtained by *in vitro* propagation and transferred into a plastic growth pouch containing a modified nitrogen-free salt solution. The plantlet roots were inoculated, 30 days after the transfer into growth pouch, with a culture of *Frankia alni* subsp. *pommerii* (strain ACN1$^{\underline{AG}}$). Protein extraction from roots and actinorhizae were done as previously described (1). An anti-nitrogenase antibody (kindly provided by Dr. Paul Ludden, University of Wisconsin-Madison) was raised against purified *Rhodospirillum rubrum* nitrogenase (Fe protein) extract. A second anti-nitrogenase antibody (kindly provided by Dr. Barbara Burgess, University of California) was raised against purified MoFe protein of *Azotobacter vinelandii* nitrogenase.

The analysis of the soluble proteins by two dimensional polyacrylamide gel electrophoresis (2D-PAGE), after silver staining, showed many differences between non-infected roots and actinorhizae. Analysis of soluble proteins by SDS-PAGE showed an increased accumulation in actinorhizae, 14 days after innoculation, of three polypeptides of molecular weight 36 kD, 50 kD and 61 kD. Immunoblot analysis with the anti-nitrogenase (Fe protein) antibody revealed a protein at 36 kD. The second anti-nitrogenase (MoFe protein) antibody reacted with a protein of 61 kD. These results suggested to us that the 36 kD and 61 kD proteins, specific to actinorhiza extracts, are constituents of the dinitrogenase reductase of *Frankia*. The nature of the 50 kD protein remains unclear but the presence of this protein in an extract from *Frankia* in pure culture indicates bacterial origin. Using an antibody raised against a culture filtrate from *Frankia* containing pectolytic activity, we have found a 53 kD polypeptide in actinorhiza extracts by immunoblotting. The fact that the antibody reacts slightly in non-infected roots does not allow us to confirm a bacterial origin of this protein. This result is in agreement with previous results on pectolytic activity that have been found in protein extracts from actinorhizae and culture filtrates of *Frankia* (2). In all these findings demonstrate that protein synthesis is altered in response to the morphological and physiological changes resulting from the establishment of actinorhizae.

1. Damerval, C., de Vienne, D., Zivy, M. and Thiellement, H. (1986) Electrophoresis **7**, 52-54.
2. Séguin, A. and Lalonde, M. (1989) Plant and Soil **118**, 221-229.

Gresshoff, Roth, Stacey & Newton (eds.) Nitrogen Fixation: Achievements and Objectives
Chapman and Hall New York • London 1990

Expression of heat induced proteins in wild type and plasmid-cured derivatives of *Rhizobium leguminosarum* bv *trifolii*

Dipankar Sen, Jose Ivo Baldani, and R.W. Weaver

Department of Soil and Crop Sciences
Texas A & M University, College Station, TX 77843, U.S.A.

Analysis of plasmid profiles indicated that inoculant clover rhizobia did not persist well in Texas soils probably due to environmental stresses like heat and drought. Reseeding clovers are nodulated by native rhizobia which are often not very effective in nitrogen fixation. We are investigating whether plasmids play a role in providing resistance to heat and drought stress. We looked for specific proteins expressed by specific plasmids under ambient and elevated temperatures.

An isolate, W14-2, obtained from root nodule of arrowleaf clover grown on a Texas soil was identified to have a greater tolerance to elevated soil temperatures compared to commercial inoculant strains. W14-2 has four plasmids of 150, 176, 261, and 464 MDalton (a,b,c, and d respectively). d is the pSym. W14-2 and four of its derivatives, cured of each of the individual plasmids (-a,-b,-c, and -d) were used in these experiments.

Four sets of the five genotypes were grown in yeast-mannitol broth at 25°C. One set was grown at this temperature for 24 hrs. and the others for 20 hrs. followed by 4 hrs. at 40°C. While shifting to the higher temperature 50μg of rifampicin/ml was added to one set of cultures and 100μg of chloramphenicol/ml was added to another. The cells were harvested and lysed by ten cycles of rapid freezing and thawing in liquid nitrogen and a 50°C water bath. The cell lysates were analysed for their total soluble proteins by SDS-PAGE using 7-17% gradient gels.

A 27kD protein was not expressed in the absence of the smallest(a) or the largest(d) plasmid at 25°C or 40°C. Additionally absence of plasmid d did not express or had a reduced expression of a 17kD protein which was quite well expressed at 40°C. At 40°C two proteins of mol. wts. close to 55kD had much enhanced expression in the wild type and the plasmid-cured derivatives except in the b-cured derivative. These proteins were the most prominent expressions of the heat shock treatment but were not expressed in the heat-treated cultures in the presence of either rifampicin or chloramphenicol indicating that the regulation of these heat shock proteins are at the transcriptional level.

Gresshoff, Roth, Stacey & Newton (eds.) Nitrogen Fixation: Achievements and Objectives
Chapman and Hall New York • London 1990

Localization of the Tn*5* mutations enhancing symbiotic efficiency on *Rhizobium meliloti* megaplasmids

L.A. Sharypova, J.G. Fomina-Eshchenko and B.V. Simarov

All-Union Research Institute for Agricultural Microbiology, Leningrad, 189620, USSR.

Alfalfa root nodule bacteria (R. meliloti) have two large plasmids with similar electrophoretic mobility; M1, controlling nodulation (Nod) and M2, controlling EPS synthesis (Exo). We have isolated 19 Tn5 induced mutants of R. meliloti with enhanced symbiotic efficiency and ability to increase plant mass (Eff^h). Southern blot analysis indicated that in 8 out of 19 cases, the Tn5 insertion had occurred in one of the megaplasmids. Cointegration of the pSUP102::Tn5 vector into the megaplasmid of these 8 strains allowed the transfer of these plasmids to avirulent R. meliloti ZB121 or plasmid-cured Agrobacterium tumefaciens UBAPF2. The frequency of megaplasmid transfer was 10^{-9} to 10^{-8} per recipient, respectively. Mutant ZB121 (Exo^-, Nod^-) carries a cryptic 140 MD plasmid, the M2 plasmid, and a deleted M1 plasmid. The transfer of the Tn5-continuing megaplasmids was easy to detect against this background. Transfer plasmids from three of the Eff^h mutants to strain ZB121 resulted in the loss of the deleted M1 plasmid (likely due to incompatibility) and resulted in a Exo^-, but Nod^+ phenotype. A. tumefaciens transconjugants with the same plasmid also gave a Nod^+ phenotype. We concluded that, in these three cases, the original Tn5 insertion occurred in the M1 plasmid. Transfer of the plasmids from the other 5 Eff^h strains to ZB121 resulted in an Exo^-, Nod^- phenotype. A. tumefaciens transconjugants with the same plasmids were Nod^-. We concluded that the original Tn5 insertions in these strains occurred in the M2 plasmid. The location of the Tn5 insertions of the megaplasmids was confirmed by deletion analysis of the Eff^h mutants. Loss of the kanamycin resistance marker from the M1 insertions resulted in a Nod^- phenotype in 25% of the clones, while similar deletions of the M2 insertions did not affect nodulation. These results indicate that both megaplasmids of R. meliloti are involved in the control of symbiotic efficiency.

Gresshoff, Roth, Stacey & Newton (eds.) Nitrogen Fixation: Achievements and Objectives
Chapman and Hall New York • London 1990

Isolation and genetic characterization of a *Rhizobium meliloti* T37 mutant with delayed nodulation and decreased competitiveness.

B.V. Simarov, O.P. Onishchuk, L.A. Sharypova and O.N. Chesnokova

All-Union Research Institute for Agricultural Microbiology, Leningrad, 189620, USSR.

An R. meliloti T37 mutant with delayed nodulation (Noi^+) was isolated by Tn5-mutagenesis of strain CXM1-105. (Suicide vector pSUP2021 was used; frequency of Km^r recombinants was 10^{-6} - 10^{-7}). In sterile experiments on alfalfa seedings (Medicago varia cv. Zaikevitcha) grown on nitrogen-free agar medium, T37 induced nodules 4-6 days later than the parental strain. This mutant retained the high symbiotic efficiency of CXM1-105, but it's competitiveness was markedly reduced ($Comp^-$). In mixed inoculation (1:10 ratio) with a Nod^+ Fix^- strain (CXM1-48), the parental strain CXM1-105 did form effective symbiosis, while mutant T37 did not. In pot experiments in nonsterile soil, strain CXM1-105 formed 50% of the nodules while T37 formed only 7% of the nodules.

In order to study the relationship between delayed nodulation and decreased competitiveness, the Km^r marker was transduced using phage ϕM12 from T37 to CXM1-105. All of the 15 transductants analyzed in plant experiments inherited the $Noi^{\pm}$ and $Comp^-$ phenotypes. These results suggest that the nodulation rate is one of the factors determining competitiveness. In order to localize the Tn5 insertions in the genome of T37, the Km^r marker was transferred by the use of RP4-4 into R. meliloti ZB121, an avirulent mutant with a large deletion in the symbiotic megaplasmid, or Agrobacterium tumefaciens UBAPF2, a plasmid-cured strain. All recombinants obtained from ZB121 were able to induce nitrogen-fixing nodules on alfalfa. Transconjugants of UBAPF2 acquired a Nod^+ Fix^- phenotype. These results demonstrated that the Tn5 insertion in T37 occurred in the Sym-plasmid. Eckhardt gel electrophoresis of plasmids of recombinants obtained from both ZB121 and UBAPF2 confirmed the transfer of the Sym-plasmid. The $Noi^{\pm}$ mutation in T37 was not complemented by the R-prime plasmid pGR1, carrying the non-nif region of R. meliloti. We conclude that the transposon of strain T37 labels a new locus on the Sym-plasmid which controls both nodulation and competitiveness.

Gresshoff, Roth, Stacey & Newton (eds.) Nitrogen Fixation: Achievements and Objectives
Chapman and Hall New York • London 1990

Studies on the metabolic regulation of nitrogenase activity in *Rhodospirillum rubrum*

Abdelhamid Soliman and Stefan Nordlund

Department of Biochemistry, Arrhenius Laboratories, Stockholm University, S-106 91 Stockholm, Sweden

INTRODUCTION

Nitrogenase activity in the photosynthetic bacterium Rhodospirillum rubrum is controlled on the metabolic level, manifested as the "switch-off" effect. This effect is due to inactivation of dinitrogenase reductase by ADP-ribosylation catalyzed by DRAT, with NAD^+ as donor of ADP-ribose (1). The metabolic signal initiating the inactivation upon the addition of "switch-off" effectors has not been identified yet.

MATERIALS AND METHODS

Rsp. rubrum was grown photoheterotrophically with molecular nitrogen as nitrogen source in 1 l bottles. Nitrogenase activity was measured as acetylene reduction. Glutamine synthetase activity was determined by the transferase reaction in cell extracts produced from cells collected on filters.

RESULTS & DISCUSSION

In this investigation the in vivo effect of NAD(P) and analogs of this nucleotide on nitrogenase activity in Rsp. rubrum was studied. It was shown that NAD^+ caused "switch-off" and that the duration and extent of the effect was dependent on the concentrations of NAD^+ used. "Switch-off" was also obtained in N-starved cultures upon addition of NAD^+. Analogs of NAD modified in the nicotinamide portion and NAD(P)H did not lead to "switch-off", whereas $NADP^+$ and analogs changed in the adenine moiety did. This is in agreement with studies of purified DRAT reported by Ludden and coworkers, except that $NADP^+$ did not function in vitro (1). Our results indicate that the NAD^+ pool is of importance in the regulation of the reactions leading to inactivation of dinitrogenase reductase. The concentration of NAD^+, a substrate of DRAT, would increase upon the addition of e.g. ammonium ions due to the reactions catalyzed by glutamine synthetase and glutamate synthase producing $NADP^+$ and generating NAD^+ through the reaction of transhydrogenase. This suggestion is supported by results showing that "switch-off" by NAD^+ is less pronounced at high light intensities.

1. Ludden, P.W. & Roberts, G.P. (1989) Curr. Top. Cell Regul. 30, 23-56

Gresshoff, Roth, Stacey & Newton (eds.) Nitrogen Fixation: Achievements and Objectives
Chapman and Hall New York • London 1990

Regulation of *nifA* expression in *Azorhizobium caulinodans* ORS571

J. Stigter*, K. Pawlowski, U. Klosse, P. Ratet & F.J. de Bruijn*

Max-Planck-Institut für Züchtungsforschung, 5000 Köln 30, F.R.G.
*Present address : MSU-DOE Plant Research Laboratory, Michigan State University, East Lansing, Michigan 48824, U.S.A.

The *A.caulinodans* ORS571 nitrogen-fixation regulatory gene *nifA* is controlled by the cellular nitrogen and oxygen status (3). Of interest is which factors are involved in signal transduction leading to expression of *nifA*. At least three gene products, NtrC and NtrYX are involved in the activation of *nifA* (2).
The nucleotide sequence of the *ntrY/ntrX* region was determined. The deduced NtrY amino acid sequence showed homology to the sensor proteins of two-component regulatory pairs (1). The regulator, NtrX, had strong homology to NtrC proteins.
The promoter region of *nifA* contains three DNA sequence motifs highly homologous to elements in oxygen and nitrogen regulated promoters: an FNR-box, a NifA-binding motif (UAS) and an *ntr* motif, found in the -24 to -12 region of NtrA dependent promoters. The role of these *cis*-acting elements was examined by site-directed mutagenesis (using PCR). Conserved bases of different boxes were changed and their effects measured with a translational *nifA-lacZ* fusion, integrated in a non-essential region of the chromosome. Mutations in the FNR-box and in the *ntr* consensus reduced *nifA* expression by 99%, mutations in the UAS reduced *nifA* expression by 91%.

References:

1. Albright, L.M., Huala, E. & Ausubel, F.M. (1989) Annu. Rev. Genet. 23: 311-336.
2. de Bruijn, F.J., Pawlowski, K., Ratet, P., Hilgert, U., Wong, C.H., Schneider, M., Meyer z. A., H. & Schell, J. (1988) in *Nitrogen Fixation: Hundred Years After*, eds. Bothe, H., de Bruijn, F.J. & Newton, W.E. (Gustav Fischer, Stuttgart-New York), pp 351-355.
3. Ratet, P., Pawlowski, K., Schell, J. & de Bruijn, F.J. (1989) Mol. Microbiol. 3: 825-838.

Gresshoff, Roth, Stacey & Newton (eds.) Nitrogen Fixation: Achievements and Objectives
Chapman and Hall New York • London 1990

Characterization of the aerobically purified hydrogenase of *Azotobacter vinelandii*

J. H. Sun and D. J. Arp

Laboratory for Nitrogen Fixation Research and Department of Botany and Plant Pathology, Oregon State University, Corvallis, OR 97331 USA.

The hydrogenase of *Azotobacter vinelandii* has previously been purified under anaerobic conditions. In this work, we have purified the enzyme in the presence of air. The properties of aerobically purified enzyme are compared with those of the anaerobically purified hydrogenase.

Hydrogenase was purified to near homogeneity with the same procedure as that used in the anaerobic purification, but the yield was only about 10% of that obtained in the anaerobic purification. This aerobically purified enzyme is partially active in the $^3H_2$ exchange, H_2 evolution and H_2 uptake (Methylene blue, Benzyl viologen, PMS, DCPIP and $Fe(CN)_3$ as electron acceptor) assays. To express full activity, the enzyme required activation by incubation with H_2, dithionite, or dithiothreitol. Once activated, aerobically purified enzyme is similar to anaerobically purified hydrogenase with respect to its catalytic properties (e.g Km for Methylene blue, response to pH changes and the effects of O_2).

The aerobically purified *A. vinelandii* hydrogenase showed two significant absorption peaks at 450 and 345 nm in difference spectra of the isolated vs. activated enzyme, indicating Fe-S centers were oxidized in the "as isolated" state. Unlike O_2 inactivated enzyme, the purified enzyme in the presence of O_2 did not display an absorption peak at 315nm. However, when the enzyme was activated by H_2 and then inactivated by incubation with O_2, the 315nm peak appeared. Upon activation of the aerobically isolated hydrogenase with H_2, the increase in activity was correlated with a decrease in absorption peaks at 435 and 345 nm. The activation of the hydrogenase by H_2 was not inhibited by C_2H_2. However, treatment with cyanide did lead to a time-dependent loss of the ability of H_2 to activate hydrogenase.

Activation of the aerobically purified hydrogenase from *A. vinelandii* appears to involve both the reduction of the metal centers and conformational changes. The lack of oxidation of a chromophore at 315 nm in the aerobically purified hydrogenase seems to be key factor which determines the ability of this enzyme to be reductivly activated.

Gresshoff, Roth, Stacey & Newton (eds.) Nitrogen Fixation: Achievements and Objectives
Chapman and Hall New York • London 1990

Cloning of the host specific gene region of *Rhizobium galegae*

L. Suominen, A. Saano and K. Lindström

University of Helsinki, Dept of Microbiology, SF-00710 Helsinki, Finland.

The 25 bp nod-box sequence was used as a probe in a screening experiment where putative clones carrying unknown nodulation genes were isolated from the cosmid clone bank of Rhizobium galegae HAMBI 1174 (ATCC43677). Three overlapping clones containing sequences with homology to the nod-box were assayed for the presence of various nodulation genes from R. meliloti and R. leguminosarum.

A restriction map of this gene region covering 40 kb of the cloned DNA was constructed. The hybridization analysis was performed in a low stringency and the genes used in this experiment are listed in table 1.

Table 1. A list of probes used in hybridization analysis of the R. galegae nodulation region.

Probe	size kb	plasmid	origin
nifHDK	10.6	pRmRT2	R.meliloti
nodD1	0.3	pRmSL42	R.meliloti
nodD2	9.4		R.meliloti
nodD3	11.0		R.meliloti
hsnCBAD	6.8	pEK10	R.meliloti
nodE	5.3	pRL1JI	R.leguminosarum
nodL	1.0	pRL1JI	R.leguminosarum
nodM	1.0	pRL1JI	R.leguminosarum

The total DNA of Rhizobium galegae contains four EcoR1 fragments carrying the nod-box sequence. One of them is in the 9.2 kb common nod fragment cloned in plasmid pRg30. Others are found in EcoR1 fragments sized 4.6, 3.9 and 3.3 kb. All these, a second nodD gene and the sequences sharing homology with R. meliloti host genes are found in one cosmid clone, pRg10. This indicates that pRg10 is the plasmid carrying the host specific gene region of Rhizobium galegae.

Gresshoff, Roth, Stacey & Newton (eds.) Nitrogen Fixation: Achievements and Objectives
Chapman and Hall New York • London 1990

Purification and characteristics of NADP-malic enzyme in *Bradyrhizobium japonicum*

Shigeyuki Tajima, Noriyuki Tamamura, Makoto Tsukamoto and Isao Kimura

Dept.of Bioresource Science, Kagawa Univ., Miki-cho, Kita-gun,
Kagawa, 761-07 JAPAN

Since soybean nodule bacteroids' respiration depends on the supply of C_4-dicarboxylate from the host plants, the anaplerotic pathway to supply acetyl-CoA to TCA cycle from the organic acids plays a key role for the bacteroid energy metabolism and nitrogen fixation (1,2).

In the soybean nodules, high NADP- and NAD-malic enzyme were detected especially in cortex tissue. Though the enzyme activities in the bacteroids were less than those in the plant tissues, the malic enzyme was dominant in the anaplerotic enzymes detected.

The NADP-malic enzyme was purified to the homogeniety from the culture of Bradyrhizobium japonicum A1017 by ammonia sulfate precipitation, DEAE-cellulose chromatography, preparative IEF, 2'5'-ADP sepharose 4B and NAD agarose affinity chromatographies. The purified enzyme preparation converted 98% of malate in the reaction system to pyruvate. Gel filtration chromatography and SDS-PAGE of the purified enzyme showed the enzyme as a dimer of subunit of MW 59,000. The activity was eluted as one peak during the purification, though denatured enzyme tended to be precipitated or eluted as mixtures of dimer and tetramer.

The optimun pH, the Km values against L-malate and NADP was 7.4, 61 and 4.9, respectively. Among the organic acids as metabolic intermediates, OAA, 2-oxoglutarate, succinate, PEP, glyoxylate and malonate inhibited more than 50% of the activity in 1 mM. Both ADP and ATP inhibited 75% of the enzyme activity in 0.4 mM. The enzyme required Mn^{2+} for the activity. Mg^{2+}, Co^{2+} and Zn^{2+} also stimulated the activity in less extent.

1) Kimura, I. and Tajima, S. (1989) Soil Sci. Plant Nutr., 35,271-279.
2) Copeland. L. et al. (1989) J. gen. Microbiol., 135,2005-2011.

Gresshoff, Roth, Stacey & Newton (eds.) Nitrogen Fixation: Achievements and Objectives
Chapman and Hall New York • London 1990

Protoplast fusion in *Rhizobium* sp. TAL *ure his* and *Rhizobium fredii* USDA192 trp phe

P. Thipayathasana*, S. Sidhipraneed*, S. Sukwattanasinit* and N. Boonkerd**

*Dept. of Biochem., Faculty of Science, Chulalongkorn Univ., **Dept. of Agriculture, Ministry of Agriculture and Cooperatives, Thailand.

INTRODUCTION

A modification of a protoplast fusion technique was performed in two rhizobial species, namely R. fredii USDA 192 Trp Phe and Rhizobium sp TAL 141 ure his. The physiological and biochemical properties of the fusants and their wild-type were reported in this Proceedings.

MATERIALS AND METHODS

S. Sukwattanasinit (1990), "Formation and Fusion of Protoplasts in Rhizobium spp.", Master Thesis of Department of Biochemistry, Graduate School, Chulalongkorn University.

RESULTS

Properties of fusants and the wild type were summarized in Table 1.

Table 1. Properties of fusants and the parental strains.

Properties	Parental strains		Fusant strains			
	TALL141	USDA192	F11	F14	F15	F20
Growth in MMG	-	-	2+	4+	2+	-
YM + 8% sucrose	4+	-	3+	4+	2+	-
YM + 0.3 M NaCl	4+	-	3+	4+	2+	-
YM at 42°C	4+	-	3+	4+	2+	-
Resist to Ab.	Km^r, Sm^r	Km^s, Sm^s	Km^r, Sm^r	Km^s, Sm^s	Km^r, Sm^r	Km^s, Sm^s
Cell surface cross reaction	TAL141	USDA192	TAL141	None	USDA192	USDA192
Soybean nodulation	poor	poor	none	poor	medium	good

MMG = minimum medium having mannitol as C-source, glutamate as N-source

CONCLUSION

Protoplast fusion should be the future of rhizobial molecular genetics in terms of both basic knowledge and as an application to agriculture.

Gresshoff, Roth, Stacey & Newton (eds.) Nitrogen Fixation: Achievements and Objectives
Chapman and Hall New York • London 1990

The relationship between pO_2 and the C_2H_2-induced decline in actinorhizal nodules

J.D. Tjepkema and C.R. Schwintzer,

Department of Botany and Plant Pathology, University of Maine, Orono, ME 04469, U.S.A.

Our objective was to determine the factors that affect the magnitude of the C_2H_2-induced decline in nodules of Myrica gale, an actinorhical plant. We studied water-cultured plants, using methods described previously (1).

The magnitude of the decline was markedly affected by pO_2. In assays at 5 kPa O_2 the decline was very small, at 20 kPa it was large, and at 30 kPa it was very large. When the pO_2 was decreased from 20 kPa to 16 kPa during the assay, the final recovered rate of C_2H_2 reduction (at 60 to 120 min) was reduced as expected, but the rate during the decline was increased (Fig. 1). These results suggest that introduction of C_2H_2 leads to decreased O_2 uptake at the site of nitrogenase which in turn causes a rise in pO_2 that inhibits nitrogenase. A second possibility is that nitrogenase turns over more rapidly in the presence of C_2H_2, leading to a depletion of reductant. A decrease in external pO_2 might relieve competition between nitrogenase and oxidative phosphorylation for reductant.

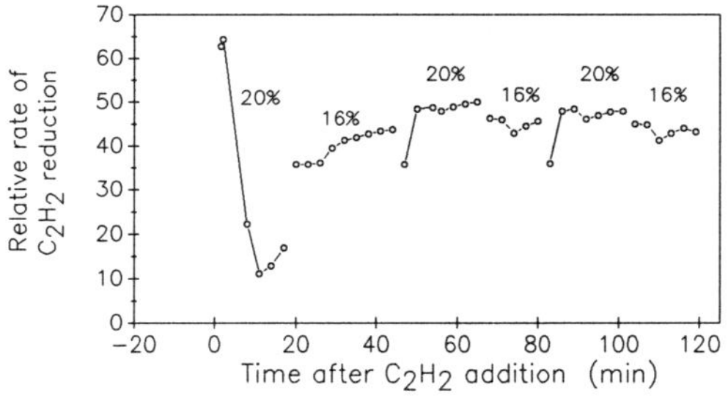

Fig. 1. Effect of a shift in pO_2 on the rate of C_2H_2 reduction during the C_2H_2-induced decline.

1. Tjepkema, J.D., Schwintzer, C.R., & Monz, C.A. (1988) Plant Physiology 88, 581-583.

Gresshoff, Roth, Stacey & Newton (eds.) Nitrogen Fixation: Achievements and Objectives
Chapman and Hall New York • London 1990

Plasmids in *Plectonema boryanum* and a mobilisable vector system for the cyanobacterium

Rakesh Tuli*, Ajay K. Vachhani and R.K. Iyer

Molecular Biology and Agriculture Division, Bhabha Atomic Research Centre, Trombay, Bombay 400 085.

Nitrogen fixing cyanobacteria are not amenable to conventional methods of gene transfer. Mobilisable plasmid vectors constructed by utilising the Nostoc replicon (Wolk et al, 1984) are suitable for a narrow range of heterocystous cyanobacteria closely related to Nostoc. In order to develop a mobilisable vector system for the nonheterocystous cyanobacterium Plectonema boryanum UTEX 596, we have characterised its extrachromosomal DNA.

P. boryanum UTEX 596 contains at least two plasmids of about 1.5 and 14 kb size. The 1.5 kb plasmid was cloned at ClaI site of pBR322. Mapping the sites for restriction enzymes established its similarity to the small plasmid reported from Plectonema sp. PCC6306 (Lambert et al., 1984). The 1.5 kb plasmid is present as 3 to 4 copies per genome in the cyanobacterium grown under dinitrogen fixing or non-fixing conditions. A small proportion of the plasmid is present as multimers. The small plasmid shows no homology to the higher molecular weight plasmid bands or to chromosomal DNA. Attempts to transform P. boryanum for ampicillin resistance using the recombinant plasmid or to mobilise the plasmid by triparental mating were unsuccessful.

A plasmid containing Tn5 and mob functions of RP4 cloned in pBR325 could be mobilised from E. coli to P. boryanum. Transfer and expression of resistance to kanamycin, ampicillin and chloramphenicol was observed. The mobilisable vector plasmid was unstable in the cyanobacterium. However, stability improved after its passage through the cyanobacterium. The stabilised plasmid showed rearrangement of DNA.

Wolk, C.P., Vonshak, A., Kehoe, P. & Elhai, J. (1984) Proc, Natl. Acad. Sci. USA 81, 1561-1565.

Lambert, G.R., Scott, J.G., Carr, N.G. (1984) FEMS Microb. Lett. 21, 225-231.

Gresshoff, Roth, Stacey & Newton (eds.) Nitrogen Fixation: Achievements and Objectives
Chapman and Hall New York • London 1990

A mutant of *Bradyrhizobium* (*Parasponia*) sp. ANU289 affected in nitrogen regulation

Michael K. Udvardi[1], Diane L. Lister, and David A. Day
Botany Department, Australian National University, Canberra A.C.T. 2601, Australia.
[1]Present address the Institute of Biological Chemistry, Washington State University, Pullman WA 99164-6340, USA.

Regulation of a variety of nitrogen assimilation pathways in enteric bacteria is coordinated by a central nitrogen regulation (Ntr) system (3). An analogous system of Ntr control has recently been found in the *Rhizobiaceae*. Both *ntrA* (encoding a unique sigma factor) and *ntrC* (encoding a transcriptional activator) are necessary for expression of several genes involved in nitrogen assimilation in rhizobia (4,5). The number of genes known to be under such control remains meagre, however. Here we describe the isolation and characterization of a Tn-5 induced mutant of *Bradyrhizobium (Parasponia)* sp. ANU289 that is affected in nitrogen regulation.

RESULTS AND DISCUSSION

The mutant was identified by its poor growth on minimal media containing 100µM NH_4Cl as sole nitrogen source. The mutant was unable to induce expression of :

(i) the ammonium transporter (Amt; determined as in 2)
(ii) nitrate reductase (NR; determined as in 1)
(iii) glutamine synthetase II (GSII; determined as in 1)

The mutant retained the ability to fix nitrogen both *in vitro* and in siratro (*Macroptilium atropurpureum*) root nodules. However, growth of siratro inoculated with the mutant was slower than that of control plants inoculated with the wild type. Since the efficiency of nodule initiation by the mutant was not impaired, slower plant growth probably resulted from retarded nodule development.

The results suggest that at least 3 genes those encoding Amt, NR and GSII are under the control of an Ntr system in *B. (Parasponia)*. The phenotype of the mutant is consistent with a defect in the transcriptional activator, NtrC.

REFERENCES

1. Howitt S.M. & Gresshoff, P.M. (1985). Journal of General Microbiology **131**, 1433-1440.
2. Howitt, S.M., Udvardi, M.K., Day, D.A. & Gresshoff, P.M. (1986). Journal of General Microbiology **132**, 257-261.
3. Magasanik, B. (1982). Annual Reviews in Genetics **16**, 135-168.
4. Martin, G.B., Chapman, K.A. & Chelm, B.K. (1988). Journal of Bacteriology **170**, 5452-5459.
5. Ronson, C.W., Nixon, B.T., Albright, L.M. & Ausubel, F.M. (1987). Journal of Bacteriology **169**, 2424-2431.

Gresshoff, Roth, Stacey & Newton (eds.) Nitrogen Fixation: Achievements and Objectives
Chapman and Hall New York • London 1990

Induction of cyst-formation by entrapment of *Azospirillum* spp. in Ca^{2+}-alginate beads

J.Ueckert[1*], M.Stephan[2], J.Doebereiner[2], E.-G.Niemann[1], I.Fendrik[1]

[1]Inst. of Biophysics, University of Hannover, 3 Hannover 21, Herrenhaeuser Str.2, F.R.G.
[2]EMBRAPA-CNPBS, Km 47, Seropedica, RJ 23851, Brazil.

INTRODUCTION

Flocculation and cyst formation has been observed for Azospirillum spp. in the late exponential and during the stationary growth phase (1). The role of bacterial c-forms for infectivity and associations with Kallar grass (2) or sugar cane (3) remains unclear. A new approach to induce cyst formation was made by entrapment of Azospirillum spp. in Ca^{2+}-alginate beads. By this method it is possible to obtain c-forms in pure culture for further research.

MATERIALS AND METHODS

Cells of different Azospirillum brasilense strains were mixed with sterile 2-5 % (w/v) sodium alginate sol. (Protan GmbH, 2 Norderstedt, FRG) and dropwise added to a gently stirred 2% (w/v) $CaCl_2$-solution. Bacterial bead inoculation was 0.01-1 % (w/v). Beads of 1 mm diameter were formed by ionotropic gelation with the multivalent cation calcium. The gel network has a range of pore sizes from 50 to 500 Å, allowing gases, nutrients and metabolic products to diffuse in and out the beads. Cells or cysts can easily be released by dissolving the beads in 5% (w/v) tripolyphosphate.

RESULTS AND DISCUSSION

Azospirilum brasilense strains Sp7, Sp245, and Sp246 showed immediate encystation after immobilization in the gel, depending mainly on the alginate concentration. No encystation occured at 0.5%; best effiency was at 4% alginate concentration and 1% bacterial inoculation. The immobilized nonmotile cells comprise sphere-like and vibroid c-forms, which completely develop into the sphere shape at the end of the encystation process, with sizes from 5.8 to 10.1 μm. The results demonstrate the use of alginate beads to induce and investigate Azospirillum c-forms.

REFERENCES

1. Sadasivan, L. & Neyra, C.A. (1985) J. Bacteriol. **163**, 716-23.
2. Reinhold, B. et al. (1987) Appl. Env. Microbiol. **53**, 889-91.
3. Berg, R.H. et al. (1979) Protoplasma **101**, 143-63.

Gresshoff, Roth, Stacey & Newton (eds.) Nitrogen Fixation: Achievements and Objectives
Chapman and Hall New York • London 1990

Random *lac* transcription fusion mutagenesis in *Bradyrhizobium japonicum*

PETER van BERKUM and MICHAEL J. SADOWSKY

USDA-ARS, Nitrogen Fixation & Soybean Genetics Lab., Beltsville, MD 20705, USA, and Dept. of Microbiology & Soil Science, University of Minnesota, St. Paul MN 55108, USA.

INTRODUCTION. Fusion of nitrogen fixation and nodulation genes to E. coli lacZ genes in vitro has resulted in an understanding of the regulation of these genes in B. japonicum (2,4). Host-inducible genes without apparent homology with the common nodulation region of R. meliloti were indentified in R. fredii by random lac-transcription fusion mutagenesis (3). We describe the development of methods for generating in vivo random lac-transcription fusion mutants in B. japonicum.

RESULTS AND DISCUSSION. A suicide vector with mini-Mu phage MudI 1681 (1) was constructed (pLNJ100) for mutagenesis of B. japonicum. Success for mutagenesis depended upon the recipient strain used. Southern hybridization analysis indicated that MudI 1681 integrated randomly into the genome. Vector survival with frequencies of 30% required transconjugants to be tested for vector loss. Isolation of cytochrome oxidase mutants indicated that the lac fusions generated were useful to investigate operon regulation. Mutants with insertions in operons responding to changes in osmotic pressure were isolated by following beta-galactosidase activity. These operon(s) were shown to respond rapidly to low concentrations of salt. Our methods enable the identification of genes through mutagenesis and the study of their regulation.

REFERENCES.

1. Castilho, B. A., Olfson, P., & Casadaban, M. J. (1984) J. Bacteriol. 158, 488-495.
2. Kosslak, R. M., Bookland, R., Barkei, J., Paaren, H. E., & Applebaum, E. R. (1987) Proc. Natl. Acad. Sci. USA 84, 7428-7432.
3. Sadowsky, M. J., Olson, E. R., Foster, V. E., Kosslak, R. M., & Verma, D. P. S. (1988) J. Bacteriol. 170, 171-178.
4. Yun, A. C., Noti, J. D., & Szalay, A. A. (1986) J. Bacteriol. 167, 784-791.

Gresshoff, Roth, Stacey & Newton (eds.) Nitrogen Fixation: Achievements and Objectives
Chapman and Hall New York • London 1990

The *nodA* gene is physically and transcriptionally separated from the *nodBC* genes in *Rhizobium leguminosarum* bv. *phaseoli* CE-3 strain

Vázquez M., A. Dávalos, A. de las Peñas, J. Miranda, O. Santana, F. Sánchez and C. Quinto
Plant Molecular Biology and Biotechnology Unit, CIFN, UNAM
Apdo. Postal 565-A, Cuernavaca, Mor., MEXICO.

The nodulation region of the *R. leguminosarum* bv. *phaseoli* symbiotic plasmid p42d has been isolated in cosmid pSM991. We have defined two essential regions for nodulation in a deletion-derivative of pSM991, pSM991.25 (Fig.1). The nucleotide sequence of both regions show that region I has the common *nodBC* genes preceded by two ORF's, ORF1 and ORF2. ORF1 is not similar to any other *nod* gene, meanwhile ORF2 shows 52% of similarity to *nodA* gene. At 30 bases upstream of the translational initiation codon of ORF1, there is a *nod* box. Thus it is probable that ORF1, ORF2, *nodB* and *nodC* could be transcribed as an operon although the distance between *nodB* and *nodC* is longer (105 bases) than in other *Rhizobia*, because there are no promoters nor terminators in this intergenic region.
Region II has the *nodA* gene that is separated from *nodBC* by 20kb. Upstream of the 5' end of *nodA* there is a *nod* box that is not identical to the *nod* box found in the region II and in the 3' end of the gene there is a sequence that shows some similarity to the *nod* box consensus sequence. Downstream the *nodA* gene there are 400 nt that show 50% of homology to the *nodB* gene. A mutation in *nodA* homogenotized in the wild type CE-3 background confers a Nod$^-$ phenotype to this strain, showing that this *nodA* is unique and essential for nodulation in this strain. In conclusion, functional common *nodABC* genes are separated in *R. leguminosarum* bv. *phaseoli* CE-3 strain and this particular arrangement could derive from a rearrangement event in the previous organization of this genes in strain CE-3.

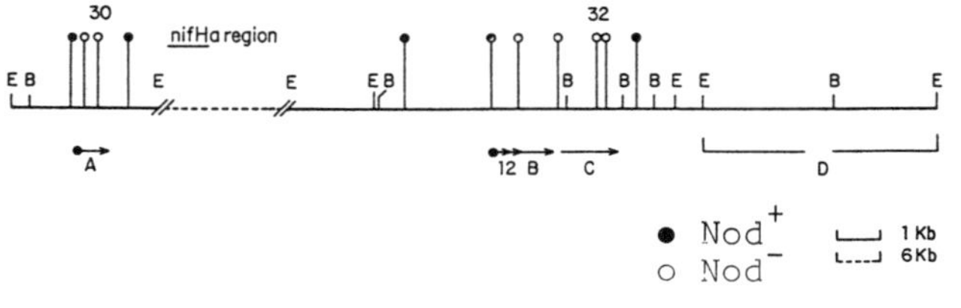

This research was partially supported by grants from EEC and the US NAS/NRC by means of a grant from the USAID.

Gresshoff, Roth, Stacey & Newton (eds.) Nitrogen Fixation: Achievements and Objectives
Chapman and Hall New York • London 1990

Nodulation genes in *Azospirillum*

C. Vieille* and C. Elmerich

Unité de Physiologie cellulaire, Dépt. des Biotechnologies
Institut Pasteur, Paris 75724 Cedex 15, France.

Hybridization was detected with *A. brasilense* Sp7 total DNA and a 8.5 kb *Eco*RI fragment, which contains *nodABC* (nod probe) and a 6.8 kb *Eco*RI fragment, which contains *nodHFEG* and part of *nodP* (hsn probe) from *R. meliloti* 41. Two recombinant plasmids, pAB502 and pAB503, containing *A. brasilense* Sp7 cloned DNA fragments hybridizing with the *hsn* probe were constructed.

Most of *Azospirillum* strains contain a plasmid of about 90 MDa (p90). The pAB502 insert, a 10 kb *Eco*RI fragment, hybridized with p90. The pAB503 insert, a 4kb *Sal*I fragment, hybridized with chromosomal DNA.

Sequence analysis of a 3.6 kb *Eco*RI fragment of the pAB502 insert revealed the existence of ORFs homologous to *R. meliloti nodPQ*. The two genes overlapped on one bp, suggesting that they are cotranscribed. A *nodP-lacZ* translational fusion was constructed in the broad host range plasmid pGD926. No ß-galactosidase activity was detected in *E. coli* nor in *R. meliloti*, but the fusion was functional in *Azospirillum* and constitutively expressed. Analysis of the sequence upstream from *nodPQ* revealed limited *ntr* consensus, both in *Azospirillum* and in *R. meliloti*. However, no ß-galactosidase activity was detected in *K. pneumoniae* harboring the fusion, even in the presence of a constitutive *nifA*. The *nodQ* product shares homology with a family of initiation and elongation factors. In particular, the GTP-binding consensus is well conserved.

Several *nodP* and *nodQ* mutants were constructed by insertion and/or deletion. No significant difference with the wild type Sp7 was observed in the following tests: growth rate in minimum medium, nitrogen fixation, indole-3-acetic acid production, effect on growth of wheat seedlings.

The nucleotide sequence of the 1.4 kb *Pst*I-*Sma*I fragment of the pAB503 insert revealed 57% homology with *R. meliloti nodG*. The two *nodG* products share homology with the *Klebsiella aerogenes* ribitol dehydrogenase and with the *Bradyrhizobium japonicum fixR* product. A *nodG-lacZ* translational fusion was constructed in pGD926. A high level of expression was detected whatever the growth conditions used. The function of *nodG* and *nodPQ* in *Azospirillum is* being studied.

References:

Fogher *et al.* (1985). FEMS Microbiol. Lett. 30:245-249.
Onyeocha *et al.*, Plasmid, in press.

Gresshoff, Roth, Stacey & Newton (eds.) Nitrogen Fixation: Achievements and Objectives
Chapman and Hall New York • London 1990

Involvement of regulatory genes *nifA*, *vnfA*, *nfrX*, *ntrD* and structural genes *nifUSVM* in expression and activity of the three nitrogenases in *Azotobacter vinelandii*

Jean Walmsley, Paul Woodley, Anil Bali, Dennis Dean* and Christina Kennedy

AFRC IPSR Nitrogen Fixation Laboratory, University of Sussex, Brighton UK. *Dept. of Anaerobic Microbiology, Virginia Tech, Blacksburg, VA 24061 USA.

Involvement of nifUSVM gene products in aerobic and microaerobic nitrogen fixation.

We have determined that nifU, nifS and nifV are required for activity of all three nitrogenases in addition to nifM (1) and nifB (2). Since Mo nitrogenase structural (nifHDK) and co-factor (nifEN) genes are present in the same cluster as nifUSVM, this finding suggests that MO nitrogenase was the initial enzyme from which the other two evolved. Duplication and divergence of nifHDK led to vnfHDK and anfHDK while nifUSVM genes retained the ability to function with the other nitrogenases. nifU and nifS mutants grew in 2% O_2 but not in air. This may indicate a role for the nifU and nifS gene products in stabilizing all three nitrogenases in the presence of oxygen.

Roles for nifA, vnfA, anfA, nfrX, ntrD and rpoN in expression of structural genes for three nitrogenases.

Gene fusions (lacZ) in the structural gene operons (nifH, vnfH, vnfD, and anfH) were introduced into regulatory mutants (see accompanying poster and 3). rpoN is required for expression of all four operons; ntrC is required for none of them. nfrX and ntrD are needed to express only nifHDK and anfHDGK operons. ntrD activates nfrX expression.

The activators NIFA, VNFA, and ANFA are specific for expression of the corresponding structural gene operons. VNFA is required for expression of vnfE, vnfH and vnfD promoters. The VNFA protein activates these promoters in E. coli.

1. Kennedy, C., Gamal, R., Humphrey, R., Ramos, J., Brigle, K. and Dean, D. (1986). Mol. Gen. Genet. 205:318-325.
2. Joerger, R.D. and Bishop, P.E. (1988). J. Bacteriol. 170:1475-87.
3. Joerger, R.D., Jacobson, M.R., Bishop, P.E. (1989). J. Bacteriol. 171:3258-67.

Gresshoff, Roth, Stacey & Newton (eds.) Nitrogen Fixation: Achievements and Objectives
Chapman and Hall New York • London 1990

A divergent *nod* box sequence is essential for *nodD1* induction in *Bradyrhizobium japonicum*

Shui-ping Wang and Gary Stacey

Center for Legume Research and Department of Microbiology, University of Tennessee, Knoxville, TN 37996, USA.

INTRODUCTION

Induction of the *nodYABC* transcript in *B. japonicum* requires the *nodD1* gene product and host-produced isoflavones (e.g., genistein and daidzein). Surprisingly, the *nodD1* gene of *B. japonicum* is autoregulatory and inducible. A *nod*-box promotor sequence, the NodD binding site, precedes the *nodYABC* transcript, and a second divergent *nod*-box sequence precedes the *nodD1* transcript. The function of this divergent *nod*-box is unknown.

MATERIALS AND METHODS

The *B. japonicum* and *E. coli* strains and microbiological techniques used were as described previously (1). SDS lysis of the cells and hot phenol extraction were used in mRNA preparation. The method of Henikoff (2) was used for constructing deletions.

RESULTS AND DISCUSSION

The transcriptional start sites of the *nodD1* and *nodYABC* transcripts were determined by primer extension. The synthetic oligonucleotides complementary to the coding regions of *nodD1* and *nodY* were end-labeled with -^{32}P-ATP and T4 polynucleotide kinase in the presence of the four dNTPs, and extended on mRNA template isolated from *B. japonicum* cells which were induced 10 hrs with 2 µM genistein. The transcriptional start site for *nodYABC* is located 28 bp downstream of the *nod*-box. The transcriptional start site for *nodD1* is located 44 bp downstream of its presumptive *nod*-box sequence. The function of the divergent *nod*-box sequence preceding the *nodD1* gene was studied by making a series of deletions of the *nodD1* promotor. The different deletions were fused to a promotorless *lacZ* gene. Analysis of the inducible activity of these deletion-*lacZ* fusions clearly indicated that the divergent *nod* box sequence of *nodD1* was essential for induction by isoflavones and soybean seed extract. Upon analyzing the known *nod*-box sequences, we found a possible repetitive sequence structure existing in all *nod*-box sequences consisting of four 9 bp repeats, arranged as two doublets, and exhibiting significant homology to the OxyR binding repeat sequence(3).

REFERENCES

1. Banfalvi, Z., Nieuwkoop, A., Schell, M., Besl, L. and Stacey, G. (1988) Mol. Gen. Genet. **214**: 420-424.
2. Henikoff, S. (1984) Gene **28**: 351-359.
3. Bolker, M. and Kahmann, R. (1989) The EMBO J. **8**: 2403-2410

Gresshoff, Roth, Stacey & Newton (eds.) Nitrogen Fixation: Achievements and Objectives
Chapman and Hall New York • London 1990

Molecular cloning, sequence analysis and mutagenesis of the *fixA* gene from *Azotobacter vinelandii*

René Wientjens, Walter van Dongen and Huub Haaker

Dept. of Biochemistry, Agricultural University, Dreyenlaan 3, NL-6703 HA, WAGENINGEN, The Netherlands.

The 2.0 kb *Hin*dIII-*Sal*I fragment of plasmid pRleH58, containing *fixW, fixA* and the N-terminal sequence of *fixB* of *Rhizobium leguminosarum* (1) was found to hybridyze to a 7.2 kb *Eco*RI fragment of *A.vinelandii* ATCC478. This fragment was cloned into the *Eco*RI site of pUC9 and a restriction map was constructed. A 2.2 kb *Sst*I-*Stu*I fragment of this plasmid was subcloned in pUC18. The nucleotide sequence of part of this insert was determined from both strands. The sequence contained an open reading frame of 846 basepairs, (281 codons including the AUG start codon and the UGA stop codon). This region had a high homology with the total or partial sequence of *fixA* determined for *Rhizobium meliloti* (2), *Bradyrhizobium japonicum* (3), *Rhizobium leguminosarum* (1) and *Azorhizobium caulinodans* ORS571 (4). A consensus sequence for a *nifA* binding site and a Upstream Activator Sequence (UAS) were found 300 and 400 bp upstream of the start codon, respectively. A Ribosome Binding Site (RBS) was located at pos. -13 to -7 from the start codon of *fixA*.
No terminator sequence was found downstream of the stop codon of the *fixA* gene, but downstream of the stop codon a sequence was found having a homology with the N-terminal sequence of the *R.meliloti fixB* gene (3) together with a RBS. So it seems likely that *fixA* and *fixB* are cotranscribed as a single mRNA transcript under the control of the *nifA* gene product.

The *fixA* gene was interrupted by insertion of the Tn5 kanamycin resistance gene in the central part of the coding region and the mutated gene was transferred to wild type *A.vinelandii* by transformation. Double cross-overs were selected and checked on the DNA level; the genotype was shown to be completely *fixA*$^-$. However, no differences between wild type and mutant were found when cells were grown either on Burk Nitrogen free medium or Burk medium supplemented with : ammoniumacetate, various concentrations of sucrose, succinate, aspartate or lactate.

The observations indicate that the *fixA* gene is not essential for nitrogen fixation in *A.vinelandii* under the conditions tested. This is in contrast with the results found for *Rhizobia*.

References:

1. Hontelez, J.G.J., Klein Lankhorst, R., Katinakis, P.,Van den Bos, R.C. and van Kammen, A. 1990. Mol. Gen. Genet. in press.
2. Earl, C.D., C.W. Ronson and F.M. Ausubel. 1987. J. Bacteriol. **169**:1127-1136.
3. Gubler,M and H. Hennecke. 1986. FEBS **200**:186-192.
4. Kaminski,P.A., F. Norel, N. Desnoues, A. Kush, G. Salzano and C. Elmerich. 1988. Mol. Gen. Genet. **214**:496-502.

Gresshoff, Roth, Stacey & Newton (eds.) Nitrogen Fixation: Achievements and Objectives
Chapman and Hall New York • London 1990

The N- and C-terminal portions of NifV are encoded by two different genes in *Clostridium pasteurianum*

Shu-Zhen Wang*, Dennis R. Dean, Jiann-Shin Chen, and John L. Johnson.

Department of Anaerobic Microbiology, Virginia Polytechnic Institute ans State University, Blacksburg, VA 24061, USA.

While identifying potential nif (nitrogen fixation) genes residing on a DNA fragment located downstream from the nifN-B gene of Clostridium pasteurianum, we observed that the deduced amino acid sequences from two open reading frames (ORFs) had sequence similarity to NifV of Azotobacter vinelandii and Klebsiella pneumoniae. The similarities, however, were to different portions of NifV. Conserved regions were found between the C-terminal 195 residues of the first ORF and the C-terminal portion of NifV, and between the entire sequence of the second ORF (269 residues) and the N-terminal portion of NifV. The first ORF is designated nifVω and the second ORF is designated nifVα. The correspondence of NifVω and NifVα to different portions of NifV of A. vinelandii is illustrated below. The deduced amino acid sequences from nifVω and nifVα were also found to have sequence similarities to that from leuA, the structural gene for α-isopropylmalate synthase in Salmonella typhimurium. Plasmids containing nifVω and nifVα, singly and in combination, were used to transform a nifV deletion mutant of A. vinelandii (strain DJ388). The NifV$^+$ phenotype of A. vinelandii was obtained only when both nifVω and nifVα were introduced into the mutant. The presence of nifVω and nifVα of C. pasteurianum in the rescued strain was verified by Southern analysis of its DNA with the wild type strain as a control. The results suggest that NifV contains two or more essential domains which can involve either one or two polypeptides.

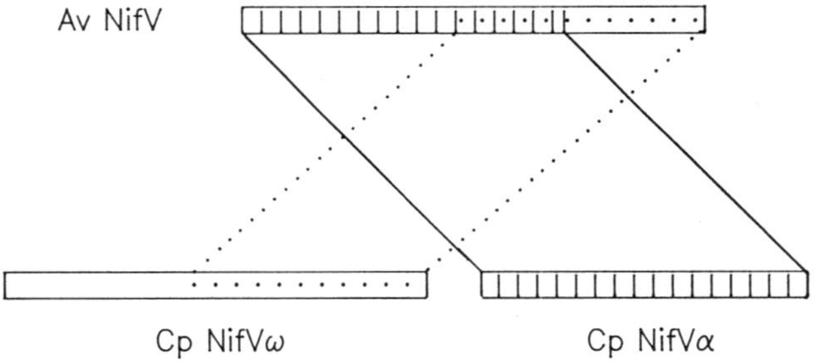

Schematic illustration of corresponding regions of NifVω and NifVα of C. pasteurianum to NifV of A. vinelandii.

Gresshoff, Roth, Stacey & Newton (eds.) Nitrogen Fixation: Achievements and Objectives
Chapman and Hall New York • London 1990

Selection of Nif⁻ mutants in chemostat cultures of *Rhodobacter capsulatus*

John C. Willison and Jérôme Pierrard

Biochimie Microbienne / LBIO (CNRS URA 11300) - Centre d'Etudes Nucléaires de Grenoble - 85 X - 38041 Grenoble cedex - France.

When *Rhodobacter capsulatus* strain B10 is grown photosynthetically in N-limited continuous culture, spontaneous Nif^- mutants accumulate, and eventually outgrow the wild-type strain (1). These spontaneous Nif^- mutants contain lesions in the *nifR4* regulatory gene, which is homologous to the *ntrA* gene of enteric bacteria. Experiments in mixed chemostat cultures, using defined *nifR4* and *nifH* mutants, have shown that selection is related to the loss of nitrogenase synthesis and activity and is not due to an additional regulatory effect of the *nifR4* mutations (2). In order to identify the factors responsible for this selection, we have compared in detail the behaviour of the wild-type and a Nif^- strain during growth in N-limited chemostat cultures. At a dilution rate (D) of 0.06 h^{-1}, cultures of the mutant strain, AP29, contained about 30% more biomass and twice as many colony forming units as did cultures of the wild-type strain, B10. However, microscopic observation indicated that cultures of both strains contained similar numbers of cells, with few aggregates or chains. In B10, the cellular content of nitrogenase was estimated to represent 13% of the total cell protein. Loss of nitrogenase synthesis should therefore result in a significant "NH_4^+-sparing" effect, with a concomitant increase in cell density. However, a more accurate quantification of cell numbers will be necessary to establish this.

At a dilution rate of 0.21 h^{-1}, the rate of selection of Nif^- mutants in cultures of B10 was observed to be more than six times lower than at D = 0.06 h^{-1}. In contrast, the cellular nitrogenase content and the rate of nitrogenase-mediated H_2 production by the cultures, were similar at the two dilution rates. This suggests that the selection of Nif^- mutants may be related in some way to ATP synthesis, since the specific rate of ATP production will be higher at D = 0.21 h^{-1} than at D = 0.06 h^{-1} (to support the increased growth rate), whereas the rate of ATP-hydrolysis linked to the production of H_2 is apparently the same (assuming a constant stoichiometry of ATP/H_2). At both dilution rates, the concentration of free NH_4^+ was below the limits of detection (< 1 μM), showing that growth was limited by the supply of N source. We propose, therefore, that under these conditions, the rate of ATP synthesis is limited by growth rate, so that loss of nitrogenase synthesis and activity results in an "ATP-sparing" effect. This, in turn, might lead to an increase in cell density and/or cell viability and possibly to an increase in the rate of NH_4^+ transport at sub-micromolar concentrations.

(1) Allibert, P., Odom, J.M., Wall, J.D. & Vignais, P.M. (1984) FEMS Microbiol. Lett. *23*, 221-226.
(2) Willison, J.C. & Lac, B. (1988) FEMS Microbiol. Lett. *56*, 61-66.

Gresshoff, Roth, Stacey & Newton (eds.) Nitrogen Fixation: Achievements and Objectives
Chapman and Hall New York • London 1990

Use of ß-glucuronidase (GUS) as a transgenic marker in the genetic manipulation and study of *Bradyrhizobium strains*

Kate Wilson

Wye College, University of London, Wye, Ashford, Kent TN25 5AH, UK.

I am developing the use of ß-glucuronidase (GUS, encoded by the *E. coli gusA* gene) as a transgenic marker for the genetic manipulation and study of bradyrhizobia and other microorganisms. Advantages of GUS are the absence of background activity in higher plants and most soil bacteria, and the versatility of the enzyme: GUS catalyses the hydrolysis of diverse glucuronide conjugates, including the substrate "X-Gluc" resulting in this case in formation of a blue precipitate.

For genetic manipulation I am constructing vectors that are "tagged" with GUS so that their presence - or absence - can be detected unambiguously by formation of blue -or white - colonies on agar plates containing X-Gluc. This approach has many advantages over reliance on vector-encoded drug resistance markers. It will overcome the problem of spontaneous drug-resistant mutants being confused with true transconjugants; and it will also make screening for loss of vector sequences far easier, faster and more reliable as hundreds of colonies can be screened on X-Gluc plates for a single white derivative, thus avoiding the step of restreaking individual colonies on different media to identify those that have lost vector- encoded drug resistance markers.

To make these vectors, I first constructed a cassette that will express GUS constitutively in diverse bacterial species. The promoter and 11 N-terminal codons of the transposon Tn5 kanamycin resistance gene, which has been shown to be active in many different bacterial genera, was fused in frame to *gusA*, and the *E. coli trpA* transcriptional terminator was placed downstream. This construction shows strong GUS activity in *E. coli*. This cassette was inserted into the *PstI* site of the suicide vector pSUP202 to create the GUS-marked suicide vector pKW105. Like pSUP202, pKW105 has a unique *EcoRI* site available for insertion of genomic fragments to be used in site-directed mutagenesis. Its use in the introduction of Tn5-containing fragments into the genome of *Bradyrhizobium* strain NC92 is currently being tested. Other vectors that use GUS in similar ways will be developed.

I am also developing GUS as a marker for competition studies. As a first step I constructed an *R. meliloti* strain carrying the entire *E. coli gus* operon, which includes *gusB*, a gene encoding a glucuronide-specific permease. In this construction, GUS is expressed in *Rhizobium* from its own *E. coli* promoter. Alfalfa nodules induced by this strain can be readily identified as blue nodules following incubation of whole roots in buffer containing X-Gluc. This strategy will greatly simplify assays for rhizobial competition and also enables the position on the roots of nodules induced by a GUS-marked strain to be readily recorded.

These strategies based on the use of GUS as a transgenic marker could be used with many other bacterial species in addition to rhizobia and bradyrhizobia.

Gresshoff, Roth, Stacey & Newton (eds.) Nitrogen Fixation: Achievements and Objectives
Chapman and Hall New York • London 1990

Glutamine synthetase from *Rhodospirillum rubrum* is adenylylated and ADP-ribosylated *in vitro*

Diana L. Woehle and Paul W. Ludden

Department of Biochemistry, University of Wisconsin-Madison, Madison, WI 53706 USA.

When the supernatant fraction of R. rubrum cells is incubated with either [α-^{32}P]NAD or [α-^{32}P]ATP and the proteins separated by two-dimensional gel electrophoresis, ^{32}P-labelled glutamine synthetase (GS) of different mobilities is seen. ^{32}P-labelled GS purified from cells grown in low phosphate glutamate/malate medium with 1 mCi $H_3{}^{32}PO_4$ appears on a 2-D gel as three spots: one unlabelled, presumably unmodified GS, and two labelled spots, the more abundant spot migrating at the position of [α-^{32}P]ATP-labelled GS and the smaller spot at the position of [α-^{32}P]NAD-labelled GS. A ^{32}P-labelled peptide purified from a subtilisin digest of in vivo ^{32}P-labelled GS has the amino acid sequence L-(D)-Y-L-P-P-E-E-L-M, resembling the sequence around the adenylylated tyrosine of GS from Escherichia coli. The tyrosine of the R. rubrum peptide is modified; the identification of the aspartate is uncertain due to interference from ammonium. CNBr cleavage of GS labelled in vitro with [α-^{32}P]NAD gave a ^{32}P-labelled peptide of about 6kD. The sequence of the first 36 amino acids of this peptide is similar to the N-terminal sequence of E. coli GS. A shorter peptide will be needed to identify the ADP-ribosylated amino acid. E. coli GS is also ADP-ribosylated by R. rubrum supernatant. Both R. rubrum and E. coli ADPR-GS are stable to neutral 0.5 M hydroxylamine treatment, with a half-life greater than 19 hours. This result rules out glutamate, cysteine, and arginine as the sites of ADP-ribosylation.

R. rubrum GS purified from cells grown in limited ammonium medium migrates as a single band on SDS-PAGE, while GS from glutamate/malate-grown cells migrates as two bands. However, the limited ammonium GS appears on a 2-D gel as two spots. It has not yet been shown whether this is due to ADP-ribosylation. The specific activity of GS from limited ammonium-grown cells is about three times higher than GS from glutamate/malate-grown cells.

ADP-ribosylation of R. rubrum GS is inhibited by 1 mM ADP, 1 mM $MnCl_2$, and 1 mM EDTA, and not affected or slightly enhanced by 1 mM $MgCl_2$. Adenylylation of R. rubrum GS is strongly enhanced by 1 mM $MnCl_2$.

Gresshoff, Roth, Stacey & Newton (eds.) Nitrogen Fixation: Achievements and Objectives
Chapman and Hall New York • London 1990

Identification of *nod* and *nif* genes on pRaZ15 from *Rhizobium astragali* gene library

Zhang Zhongming, Zhou Junchu, Li Fudi and Chen Huakui

Biotechnology Center of Huazhong Agricultural University, Wuhan, 430070, P.R. China.

Astragalus sinicus L. is extensively grown as a green manure in rice fields, especially in the southern area of China. Prof. Chen Wenxing proved the Rhizobium sp. (Astragalus) is a distinct species of fast-growing rhizobia and proposed the name, R. huakuii. Previous work demonstrated that R. huakuii strains contain 1 to 3 plasmids. Wild-type strain 7653R contains 2 plasmids, while a Nod$^-$ mutant (7653R+1) obtained by heat treatment, was shown to have lost one large plasmid.

Total DNA partially digested by EcoRI was prepared from R. astragali 7653R and 20-30 kb DNA fragments were collected by sucrose gradient centrifugation. Cosmid pLAFR1 was digested with EcoRI, ligated with the 20-30 kb DNA fragments, and packaged into lambda bacteriophage. Bacteriophage were used to infect E. coli C600 and a gene library of 7653R was constructed. 7.4 x 10^4 Tc^r clones were obtained with 30% containing a recombinant plasmid. The size of the cloned DNA fragments varied from 18.8-29.8kb. A positive clone named pRa370 was identified from 2000 clones by DNA hybridization with a 8.7kb nod fragment from pRmSL26.

The 7653R gene library was also conjugated to the Nod$^-$, pSym-cured 7653+1 strain. Trans-conjugants were obtained, inoculated onto A. sinicus L. and were found to restore nodulation and nitrogenase activity. Plasmid pRaZ15 was isolated from the bacteria in the nodules, restricted using SalI or EcoR1 and hybridized to a nifHDK probe from K. pneumoniae. Transconjugants were also obtained by conjugation of pRaZ15 to R. leguminosarum T8K3 or bv. phaseolus 3622-15 and were found to nodulate A. sinicus L. The results indicate that nod, hsn, and nif genes are encoded on plasmid pRaZ15.

Gresshoff, Roth, Stacey & Newton (eds.) Nitrogen Fixation: Achievements and Objectives
Chapman and Hall New York • London 1990

R. meliloti nodD3 region contains two functional sequences

J.B. Zhu, G.Q. Yu and S.C. Shen*

Laboratory of Molecular Genetics, Shanghai Institute of Plant Physiology, Academia Sinica, 300 Fonglin Road, Shanghai, China 200032.

In Rhizobium meliloti, three copies of the nodD gene, i.e., nodD1, nodD2 and nodD3, have been detected. nodD1 and nodD2 genes activate the structural nod genes in conjunction with the inducer flavone and nodD3 activates the nod genes effectively in the absence of inducer. Here we report that two functional promoters exist in the cloned nodD3 region and that the sequences under the different promoters behaved differently in activating the nodABC genes.

Plasmid pGY22 contains a 2.0 Kb BgIII-BgIII fragment, which covers the coding region of nodD3, and plasmid pGY19 caries a 1.6 Kb ClaI-BgIII fragment cleaved from the BgIII fragment of pGY22. When these two plasmids were introduced into R. meliloti JM57, which bears the chromosomal nodC-lacZ fusion, pGY19 was found to activate nodC-lacZ only in the presence of luteolin or seed exudate, while pGY22 exhibited high inducibility of nodC in the presence or absence of the inducer.

A sequence of ca.610 bp and another of 340 bp removed from the 5' portion of the nodD3 coding region were cloned into a promoterless lacZ plasmid, a derivative of pGD499, to create nodD3-1'::lacZ and nodD3-2'::lacZ fusion plasmids. Activities of nodD3-1'::lacZ and nodD3-2'::lacZ fusions were demonstrated. The whole nodD3 region has been sequenced and the transcriptional initiation site has been determined by primer extension. Two transcriptional initiation sites were located within a distance of 277 nucleotides. A series of deletion clones with different sized deletions from the 5' end were recloned and their activities were determined. For full activity of the promoter, the 400 bp region upstream of first transcriptional initiation site seemed important. Deletion from -400 to -300 caused a 50% activity loss, while deletion to -200 resulted in a further activity loss. We applied the T7 RNA polyermase/promoter coupled system in E. coli harboring various subclones of nodD3 coding region. Three proteins of the sizes 35Kd, 37Kd and 41Kd were found to be produced from the two coding sequences of the nodD3 region.

Gresshoff, Roth, Stacey & Newton (eds.) Nitrogen Fixation: Achievements and Objectives
Chapman and Hall New York • London 1990

Genetic variation among edaphic species of the N_2-fixing cyanobacterium *Nostoc*

W. J. Zimmerman and D. E. Culley

Department of Agronomy and Soils, Washington State University, Pullman, WA 99164-6420, USA.

Genetic similarity within Nostoc was evaluated, including comparisons among four isolates originally cultured from facultative symbioses with higher plants (Gunnera and Encephalartos). These cyanobacteria and one of two additional free-living strains were either identified or reputed to be N. punctiforme, a species often cited as a cyanobiont. No consistent correlation was found to either species or symbiotic history from DNA hybridizations to genes coding for phycocyanin (cpcAB)[1], allophycocyanin (apcAB)[1], gas vesicle protein (gvpA1)[2], and dinitrogenase reductase (nifH)[3]. One gene (gvpC)[4] was not found, and gvpA1 was a single-copy gene in all strains. Gas vesicle gene probes were concluded to be potentially useful for characterizing Nostoc or at least a distinct subgroup of that genus. DNA incubations with 22 restriction endonucleases indicated a high degree of and similarity of methylated DNA to that of other heterocystous cyanobacteria. No distinct DNA methylation profile for Nostoc was recorded. Different dcm-like methylases in the Gunnera isolates were inferred from cpcAB hybridizations to DNA digested with the EcoRII or BstNI isoschizomers. However, the presence of an overlapping methylase may also have modified the DNA. The genetic variation of the Nostoc isolates was judged to reflect their different soil origins.

[1]Johnson, T.R. et al. 1988. J. Bacteriol. 170: 1858

[2]Tandeau de Marsac, N. et al. 1985. Nuc. Acids Res. 13: 7223

[3]Mevarech, M. et al. 1980. Proc. Nat. Acad. Sci. USA 77: 6476

[4]Damerval, T. et al. 1987. Gene 54: 83

Gresshoff, Roth, Stacey & Newton (eds.) Nitrogen Fixation: Achievements and Objectives
Chapman and Hall New York • London 1990

Mutants of rhizobia with altered responses of symbiotic N_2 - fixation to combined nitrogen

K. Zlotnikov and M. Kadyrova

Institute of Biochemistry and Physiology of Microorganisms, USSR Academy of Sciences, 142292 Pushchino, Moscow Region, USSR.

The inhibitory effect of combined nitrogen on the nitrogenase activity of legume root nodules is well known (Streeter, J. [1988], CRC Crit. Rev. Plant Sci. 7:1-24). It is believed that the influence of nitrogen on symbiotic N_2 fixation is mediated exclusively by the host plant. This work is an attempt to elucidate the possible active role of the rhizobial microsymbiont in mediating the response of symbiotic N_2 fixation to soil nitrogen.

Bacterial strains and media: B. japonicum 110, R. leguminosarum bv. phaseoli 693. ZN (zero nitrogen): (g/l) K_2HPO_4 - 0.8; D-gluconic acid - 1.0; DL-malic acid - 2.0; $MgSO_4.7H_2O$ - 0.06; $CaCl_2$ - 0.25 trace elements; vitamins; pH 6.0. IPG: ZN+L-glutamic acid - 1.0; L-glutamine - 0.1 g/l. IPG - 0.5: IPG + yeast extract - 0.5 g/l. Spontaneous mutants of B. japonicum 100 (designated as NRJ) and R. leguminosarum bv. phaseoli 693 (designed as NRP), resistant to ammonium or nitrate, were selected on IPG - 0.5 agar medium. The ammonium and nitrate concentrations used were for B. japonicum 110, 3.5-4.0 and 4.0-4.5 g/l and for R. leguminosarum bv. phaseoli 693, 8.0 and 7.0 g/l.

Mutants NRJ-18, NRJ-19, and NRP-1 to NRP-7 were selected with nitrate ; mutants NRJ-13 to NRJ-16, and NRP-8 to NRP-11 were selected with ammonium. However, all the isolated mutants exhibited resistance to both nitrate and ammonium. Nearly all the mutants grew worse than wild-type bacteria on media IPG, IPG - 0.5 with organic sources of nitrogen and on medium ZN with inorganic nitrogen, ammonium, nitrate or nitrite. In comparison with the parental strains, the isolated mutants as a rule showed slower symbiotic N_2 fixation; mutants NRJ-13 and NRP-1 had a Fix^- phenotype. In experiments involving long-term watering of plants with medium containing NH_4NO_3, none of the mutants exhibited any resistance of symbiotic N_2 fixation to NH_4NO_3. Many of them actually showed an increased sensitivity, which in our opinion, proves an active role of rhizobia in mediating the response of symbiotic N_2 fixation to combined nitrogen.

SECTION F:

NITROGEN FIXATION IN AGRICULTURE AND INDUSTRY

Invited papers
Workshop summary
Poster summaries

Gresshoff, Roth, Stacey & Newton (eds.) Nitrogen Fixation: Achievements and Objectives
Chapman and Hall New York • London 1990

Introduction to nitrogen fixation in agriculture and industry: Contribution of BNF to sustainability of agriculture

Ben Bohlool

NifTAL Project, University of Hawaii, Paia, HI 96779. USA

In Praise of BNF: Agricultural sustainability relies heavily on practices that are ecologically sound, while at the same time economically attractive. Biological nitrogen fixation (BNF) technologies fulfil both of these requirements. N-fixing plants offer an attractive means of reducing costly and energy-intensive inputs into managed systems while increasing return from the land. In addition, while biologically-fixed N is immobilized into plant tissue and released slowly, fertilizer N is lost rapidly due to leaching, run-off and volatilization causing water and atmospheric pollution.

BNF in Agriculture and Industry: The practice of using legumes to restore and improve soil quality dates back to pre-biblical times. It's commercial value was realized shortly after the discovery of the root nodule bacteria more than 100 years ago. Patent applications for commercial production were filed before the turn of the century. Today, nitrogen-fixing crops (grain legumes/pulses) place third behind cereals and root-crops in global food production from the land. Their importance is likely to increase as nations move more seriously towards sustainable agriculture and forestry systems. Industrial inoculant production is almost exclusively for legume crops, and mostly for grain legumes. Many of the inoculant producers, however, do market products for selected pasture legumes. A few also have inoculants for tree legumes. Recently, an inoculant company, MEDCOM, was established in Port-au-Prince, Haiti, that specializes in rhizobia for tree legumes and addresses the needs of numerous reforestation projects in the Caribbean region. There is a worldwide demand for inoculants for actinorhizal plants but, to my knowledge, none of the major producers has undertaken production of this product. Few small companies market inoculants consisting of cyanobacteria and free-living and associative bacteria, but these are small operations by comparison and do not survive market pressures for long.

Uses of BNF Systems: Among symbiotic BNF systems, nodulated legumes have dominated both the basic and applied fields of study. However, symbiotic non-legumes (actinorhizal and cyanobacterial) and other N-fixing systems have been shown to have agronomic value in specific circumstances. BNF systems are multipurpose and can be used as primary sources of food, feed, fuel and fiber, or, secondarily, to enrich the soil, preserve moisture and prevent soil erosion. Some of the more common uses of BNF systems are:

1. Primary Food: Pulses lead the way among legumes grown throughout the world. Common bean, pea, chickpea, lentil, limabean, mungbean, peanut and favabean are the most commonly grown, but their distribution is influenced heavily by geography and cultural preferences. Some of the tuber-forming legumes, notably winged bean and Mexican yambean, are quite popular in some cultures for their edible tubers and, in the cased of winged bean, leaves and pods. There are many other legumes which are edible (for listing, see *Tropical Legumes: Resources for the Future*, NAS, 1979).
2. Processed Products: Soybean and peanut lead the way in their use for production of oil and processed foods (tofu, tempe, soy sauce, peanut butter, etc.). Others (e.g., chickpea and favabean) are marketed as flour. Carob and tamarind are also legumes, but neither is known to nodulate and fix nitrogen.
3. Feed: Alfalfa is perhaps the most commonly grown forage legume in the world, used directly in pastures and/or for hay and silage. Soybean is also used commonly for feed. Others adapted to the temperate regions are species of clovers and vetches. In the tropics, *Centrosema* and *Stylosanthes* spp. as well as many shrubs (e.g., *Desmodium* and *Desmanthus*) and trees (e.g., *Leucaena*, *Acacia*, *Albizia* and *Sesbania* species) are used as animal fodder/feed, as well as for cover crops and green manures.
4. Fuel and Wood: Many woody species are used as fuelwood and several have been recommended for reforestation of degraded lands. Some of the most promising in the tropics are the actinorhizal Casuarina and the nodulated legumes *Acacia* spp., *Albizia* spp., *Leucaena leucocephala*, *Gliricidia sepium*, and *Sesbania* spp. Some species make excellent commercial quality lumber. In temperate areas, the black locust, *Robinia pseudoacacia*, grows very fast and fixes large quantities of N.
5. Other: Legumes have a variety of other uses in agriculture and industry. They are used for extraction of fiber, gum and medicinal drugs and for windbreaks, hedgerows, trellis and shade.

BNF Practices for Sustainable Agroecosystems: A variety of uses can be made of nitrogen fixing legumes and non-legumes to contribute to the regenerative processes that, by necessity, must operate in a sustainable system. Some of the more important applications are listed below:

1. Multiple/Rotation/Inter-cropping: N-fixing plants are used, either simultaneously or sequentially, with another crop, usually cereals, vegetables or root crops.
2. Agroforestry/Alley-Cropping: Emphasis here is on the use of N-fixing trees to enrich and stabilize the soil and provide fuelwood, mulch, forage and shade, when needed.
3. Cover Crop/Green Manure: Where N-fixing plants (usually herbaceous) are grown and either incorporated directly to benefit a subsequent crop or cut and transported to be used as a mulch.

Economics of BNF vs. N Fertilizers: In a large number of field trials with four species of legumes at five diverse sites (J. Thies, Ph.D. Thesis, Univ. of Hawaii, 1990), we have found that, overall, N fertilizer (total of up to 1000 kg/ha of N, applied as urea on a weekly basis) increased yields significantly over uninoculated 52% of the time, and over inoculated more than 25% of the time. The table below shows a summary of the combined results, as well as an analysis of economic return for soybean, assuming a market value of $250/ton of soybean and a May, 1990, price of about $335/ton of urea:

	Yield	$US/ha		
Treatment	% of N-treatment	cost*	yield	return
Uninoculated	38	0	380	380
Inoculated	72	7	720	713
N-treatment	100	740	1000	260

*Only for N or inoculant inputs; all other costs being equal.

Priorities for Future Research: Agronomic and commercial success in the use of N-fixing systems is dependent exclusively on how well they perform in a particular agroclimatic environment. A quantitative understanding of the ecological factors that control the fate and performance of symbiotic organisms is essential to the success of practices using BNF systems. Ecological models must be devised that can match a specific N-fixing plant to the conditions of the environment it is intended for. The gap between inoculated and N-fertilized treatments (full growth potential) in the above section represents the window of opportunity for research to improve the performance of symbiotic systems. This may be accomplished either by management practices that enhance symbiotic performance or by genetic engineering to improve the symbionts.

Gresshoff, Roth, Stacey & Newton (eds.) Nitrogen Fixation: Achievements and Objectives
Chapman and Hall New York • London 1990

Improvement of inoculant efficiency by strain improvement and formulation manipulations

Alan S. Paau, Mari Lynn Bennett, Cara J. Kurtenbach and Lori L. Graham

Agracetus, 8520 University Green, Middleton, WI 53562, U.S.A.

Introduction

The use of commercial Rhizobium inoculants in legume culture has long been a common agronomic practice. With years of legume cultivation, the populations of indigenous Rhizobium strains build up and the value obtained by seed- or soil-inoculation with commercial inoculants begins to decline. Most commercial Rhizobium inoculants are now used only on new legume ground or as an insurance. Significant yield increase is rarely obtained. The reasons why commercial inoculants generally lack field efficacy are many.

Firstly, it is generally believed that the Rhizobium strains commonly used in commercial inoculants are outcompeted (with respect to relative nodule occupancy) by the indigenous strains. The absence or minimal presence of the inoculant strains in the nodules accounts for the low contribution of the inoculants to the overall yield. The low nodule occupancy may be attributed principally to the inherent properties of the Rhizobium strains, and/or to the methods by which the strains are delivered to the field. Secondly, commercial inoculant strains selected by laboratory/greenhouse assays are not necessarily more effective in symbiotic nitrogen fixation than the indigenous strains. It remains possible that either or both of these limitations can be overcome.

Strain improvement by genetic manipulations

Maier and Brill (20) were the first to demonstrate that the symbiotic nitrogen fixation activity of a Rhizobium strain can be enhanced by simple genetic manipulations. Since then, genetic manipulation of Rhizobium strains to acquire superior characteristics are common. Methods including the selection of spontaneous derivatives, chemical and radiation mutagenesis, transposon mutagenesis, and recombinant DNA technology are used to obtain mutants which have improved nitrogen fixation activity (both symbiotically or asymbiotically), nodulation, competitiveness, and efficiency of energy usage. Table 1 outlines selected examples of strain improvement based on genetic manipulations. Although genetically improved strains are now widely available, few of them have been tested extensively in the field and developed into commercial products.

The original work of Maier and Brill (20) also illustrated the complexity in developing improved commercial inoculants. Their mutants, although highly efficient in symbiotic nitrogen fixation in laboratory and greenhouse conditions, were subsequently proven not to be competitive in field situations and did not warrant commercial development. Despite the advances in Rhizobium research, the determinants for competitiveness in the field are still largely unknown.

Table 1. Selected examples of Rhizobium strain improvement by genetic manipulations.

Improved traits	Methods	Mechanisms	Reference
Symbiotic fixation	chemical	unknown	20
	chemical	unknown	24
	transposon	unknown	16
	recombinant DNA	*nif*A & *dct* genes	6
	chemical	cytochrome	31
Asymbiotic fixation	recombinant DNA	*dct* genes	3
Nodulation	transposon	unknown	16
	transposon	lectin-binding	38
	culture enrichment	tryptophan metabolism	18
	culture enrichment	succinate sensitivity	36
	transposon	*nod*ABC inducibility in fixed N	13
	transposon	occupancy	16
	Recombinant DNA	trifolitoxin	35
	chemical	rapid nodulation	9
	host selection	host range	5
	recombinant DNA	*nod*D, host range	21
	recombinant DNA	host range, *nod* D & H	14
Hydrogenase	chemical	increase Hup activity	22

Recognizing these limitations, my laboratory preselected competitive indigenous isolates from the fields in the southern U.S. Mutants with enhanced symbiotic fixation activity were then isolated from these competitive wildtypes following chemical mutagenesis as described by Maier & Brill (20). These mutants have been field tested extensively in the last several years. Inoculants prepared with these mutants in a vermiculite carrier have improved field efficacy even in fields with high indigenous Rhizobium populations. Table 2 illustrates some of the field results.

Since the mutants were derived from wildtypes initially isolated from Louisiana, it is anticipated that the inoculants containing the mutants would perform well in tests in the same geographical region.

Table 2. Yield enhancement of soybean after inoculation with several seed-applied Rhizobium inoculants in field tests in Louisiana, 1987.

Inoculants	Ave. Yield Increase (bushel/acre)[1]	
Agracetus mutants	2.4	(P<0.01)
commercial peat inoculant-1	1.2	(P=0.27)
commercial peat inoculant- 2	-.5	(P=0.40)
commercial peat inoculant- 3	0.5	(P=0.39)

[1]Average of 6 tests. Data from Reference (4). Commercial peat inoculants 1 & 2 are produced in the U.S. with non-sterile peat. Commercial peat inoculant 3 is an imported product from England produced with sterile peat. All yield results were compared to an uninoculated control which was fully nodulated by the indigenous Rhizobium populations. P= combined P values of the individual tests.

The same inoculant was tested the next year in North Carolina in 40 separate tests in two different counties. In side-by-side comparisons to a popular commercial granular inoculant, the inoculants containing the mutants significantly (P<0.01) increased yield of 1.1 bushels/acre higher than the commercial granular inoculant (data obtained from reference 12).

In 1989, the same inoculants containing the mutants were field tested in Arkansas. On the average of 6 field tests, the mutants increased yield significantly (P<0.01) by 1.9 bushels per acre more than a commercially available inoculant (unpublished, C.A. Stutte & T.H. Clark). Mutants selected for enhanced symbiotic nitrogen fixation activity, hence, can be used to prepare inoculants with improved field efficacy.

With the advances in recombinant DNA technology, Rhizobium strains with altered gene regulation related to the *nif* and *dct* functions and improved competitiveness have been constructed (personal communication, D.J. Glass, BioTechnica International). These strains showed significant yield increases under greenhouse conditions and are now being field tested as inoculants for soybean and alfalfa [see the paper by C. Ronson for more details].

We can now confidently conclude that Rhizobium strains can be improved through genetic means to obtain better traits related to the symbiotic nitrogen fixation process. However, it has already been demonstrated that the improvement of any one single trait may not be sufficient to significantly improve field efficacy of the inoculant. The simultaneous improvement of several traits may indeed be essential.

Formulation research

Rhizobium strains are delivered commercially as formulated inoculants. An inoculant formulation generally consists of a carrier medium and other additives which provide for an acceptable shelf life, easy application, and the enhancement of the Rhizobium performance. The most commonly used carrier to date in commercial inoculants is peat.

Most early formulation research in Rhizobium focused on the carrier material. Roughley (28) recently reviewed the subject and concluded that 'research into the suitability of different carriers has generally ended with the production of an inoculant with a specified population of rhizobia. Research which ends at this point fails to recognize that the achievement of large populations of rhizobia in an inoculant is not an end in itself but rather a means of introducing rhizobia into the soil to nodulate the host.' It is now increasingly recognized that the fermentation and the formulation processes, the carrier material, and the final application methods are all interrelated. For example, it had been estimated that when peat inoculants, containing high populations of rhizobia, were used as a seed hopper inoculant for legumes, up to 65% of the seeds were not treated. Hence, besides using superior Rhizobium strains and achieving a high population of rhizobia in the carrier, the formulated inoculants must be easy to apply uniformly, and provide a microenvironment in which the rhizobia are stabilized, physiologically ready to respond to the host, safe from predators, and remain viable on exposure to the seed and field conditions. Table 3 illustrates some of the recent advances in formulation research.

Uniform inoculation, and microbiologically pure products probably hold the most promise to improve the field efficacy of Rhizobium inoculants. Hopper box-applied inoculants often suffer from misuse by the growers either because of time constraints or carelessness. Incidence of non-uniform inoculation and/or undesirable exposure of the rhizobia to high concentrations of toxic agricultural chemicals compromises the inoculant performance in the field.

New formulations that can minimize such misuse and ensure that the seeds or plants are uniformly inoculated by providing the user either precoated seeds or an inoculant that is easy to apply will help to improve the overall performance of the Rhizobium inoculants. The use of new carriers, stabilizing agents for the rhizobia, and innovations in packaging concept may contribute to this end.

Products that are not microbiologically pure often contain deleterious organisms to the seeds and the plants. The use of sterile carriers and new production processes that allow fermentation of Rhizobium directly in the end-use container will help eliminate the problem.

My laboratory favors the use of vermiculite as a carrier not only because its lamellar structure allows excellent coating of most seeds without any sticker adhesives, but also because of its inorganic nature. It can be sterilized by steam or gamma-irradiation without undergoing undesirable or unexpected chemical and physical changes, and can be used to support fermentation of many microorganisms, including Rhizobium, directly in the point-of-sale container. Inoculants produced by this method contain rhizobia which are physiologically more active than some inoculants which contain rhizobia in a dormant state. The physiologically active rhizobia should be more responsive and ready to infect their host.

Table 3. Selected examples of formulation research to improve Rhizobium inoculant performance.

	Ingredients or Methods	Improvement and New Features	References
Carrier	water	long shelf life and low cost	32
	filter mud	availability,low cost, sterilization	33
	oil/freeze dry	long shelf life,good seed adherence	19
	mineral soil	long shelf life	7
	vermiculite	inorganic, sterilization ease, low predator stimulation	15
	clay/$CaSO_4$	free-flowing granules, viability	10
	polysaccharide	bacteria stabilization	8
	polyacrylamide	entrapment, slow-release	11
	cellulose gel	seed priming and fluid drill	2
Additives	aminoethoxyvinyl-glycine	ethylene inhibitor to enhance nodulation	26
	copolymers	polyvinylpyrrolidone copolymers	37
	nod genes inducers	improve 'readiness' of inoculant strains to nodulate	27
Organisms	VAM	promote plant growth/N fixation	30
	Erwinia	promote nodulation	17
	Agrobacterium	promote root mass & nodulation	34
Production	direct fermentation in sterile carrier	microbiological purity, low cost fermentation	15
	use of sterile peat	microbiological purity	1
	drying in a porous carrier	friable dry powder with improved extender, shelf life	23
Packaging	compartments	separations of molybdenum, adhesives, and fungicides	commercial products
	measuring device	built-in measuring feature to accurately dispense inoculant	commercial products

One of the unexpected benefits of using vermiculite as a carrier is that it, being inorganic, does not stimulate predator populations as most inoculants based on organic carriers do. This, in turn, helps promote the nodule occupancy of the inoculant strains (29).

Table 4. Direct fermentation of Rhizobium japonicum in vermiculite supplemented with a nutrient source and MoS_2.

Strain	MoS_2 (mg/g)	Initial bacteria no. per gram	Bacteria no./gram after 14 (days)	56
K567	0	10^4	1.7×10^9	1.5×10^9
	49	10^4	1.2×10^9	1.2×10^9
S258	0	10^4	1.7×10^9	1.8×10^9
	49	10^4	1.2×10^9	1.1×10^9
T363	0	10^4	1.7×10^9	1.2×10^9
	49	10^4	1.3×10^9	1.9×10^9

Inoculants were fermented and stored at 23 ^{O} C. The vermiculite carrier was gamma-irradiated at 2.5 MRad in a polyethylene container and supplemented with a yeast extract-mannitol broth to a final moisture of 60%.

We can also successfully ferment Rhizobium strains in vermiculite in the presence of MoS_2. This molybdenum source is less toxic to most Rhizobium strains (Table 4), is less soluble than sodium molybdate, and is utilized by the legumes to promote nitrogen fixation in fields where molybdenum is limiting.

Table 5. Competition between Rhizobium strains from inoculants produced by direct fermentation in vermiculite with or without a soybean meal supplement.

	Strain fermented in the presence of soybean meal	% Nodule Occupancy K567	IA2875
Experiment I			
Control	none	55.4	44.6
Treated	K567	61.3ns	38.7
Experiment 2			
Control	none	68.7	31.3
Treated	K567	90.1**	9.9
Experiment 3			
Control	none	68.2	31.8
Treated	K567	87.8*	9.2
Average			
Control	none	64.1	35.9
Treated	K567	79.7***	20.3

Strains were distinguished by differential resistance to antibiotics. Nodule occupancy was expressed as percent of total viable colonies reisolated from all the nodules from 10 plants each of 4 replicates. Centennial soybean was used. Soybean meal was prepared with a Wiley mill equipped with a 40 mesh screen. Vermiculite to soybean meal was 4 to 1 on dry weight basis. ns=not significant;*P<0.05; **P<0.01; ***combined P value=0.003.

Flavone derivatives, root exudates, seed washings and seed extracts have been reported to induce the expression of *nod* genes in many Rhizobium species (27). We have also demonstrated that root exudate of soybean seedling helps to promote nodulation by inoculant strains in greenhouse experiments (25). When superior mutant Rhizobium strains were pretreated with root exudate, inoculated on soybean seeds, and planted in fields where the mutant strains normally were not competitive, the inoculation significantly increased soybean yield. My laboratory has further demonstrated that Rhizobium japonicum strains can be directly fermented in vermiculite supplemented with soybean meal or an aqueous extract of soybean meal. Rhizobia in these inoculants are more 'ready' to nodulate and are more competitive (higher nodule occupancy) than the identical strains which are fermented solely in vermiculite (Table 5).

These results suggest that by changing the fermentation medium in the inoculant, the physiological 'readiness' of the rhizobia to nodulate can be altered. This has an obvious effect on the competitiveness of the inoculant strains.

Other additives, including the use of ethylene synthesis inhibitor, tackless polymer stickers, and other synergistic microorganisms, have been reported to promote nodulation or allow the inoculant strains to better occupy the nodules. These new formulations, when combined with the use of superior Rhizobium strains obtained by genetic manipulations, promise to provide commercial inoculants with improved field efficacy in the future.

References

1. Araujo, S.C. (1987) Brazil Patent Application No. 86U4811.
2. Berg, R.K. Jr., Jawson, M.D., Franzluebbers, A.J. & Kubik, K.K. (1989) Soil Sci. Scoc. Am. J. 53, 1712-1717.
3. Birkenhead, K., Manian, S.S. & O'Gara, F. (1988) J. Bacteriol. 170, 184-189.
4. Breitenbeck, G.A., Boquet, D.J., Griffin, J.L., Hutchinson, R.L. & Millhollon, E.P. (1988) Louisiana Agriculture 32, 12-13.
5. Buendia-Claveria, A.M., Cubo-Sanchez, T., Romero, F., Perez-Silva, J. & Ruiz-Sainz, J.E. (1988) in Nitrogen Fixation: Hundred Years After Gustav Fischer, eds. Bothe, H., de Bruijn, F.J. & Newton, W.E. (VCH Publishers, Inc., New York), pp. 781.
6. Cannon, C., Beynon, J., Hankinson, T., Kwiatkowski, R., Legocki, R.P., Ratcliffe, H., Ronson, C., Szeto, W., & Williams, M. (1988) in Nitrogen Fixation: Hundred Years After Gustav Fischer, eds. Bothe, H., de Bruijn, F.J. & Newton, W.E. (VCH Publishers, Inc., New York), pp. 735-740.
7. Chao, W.L. & Alexander, M. (1984) Appl. Environ. Microbiol. 47, 94-97.
8. Charley, R.C. (1989) Eur. Patent Application No. 346545.
9. Cho, M.J., Yang, M.S., Yun, H.D., Choe, Z.R., Choe, Y.L. & Kang, K.Y. (1985) Korean J. Appl. Microbiol. Bioeng. 13, 79-86.
10. Day, J.M. & Williams, P.M. (1988) Eur. Patent Application No. 286351.
11. Dommergues, Y.R., Diem, H.G. & Divies, C. (1979) Appl. Environ. Microbiol. 37, 779-781.
12. Dunphy, E.J. (1988) in 1988 Soybean On-Farm Test Report (The North Carolina Agricultural Extension Service, Raleigh, NC), pp.40-43.

13. Dusha, I., Bakos, A., Kondorosi, A., de Bruijn, F.J., & Schell, J. (1989) Mol. Gen. Genet. 219, 89-96.
14. Faucher, C., Camut, S., Denarie, J. & Truchet, G. (1989) Mol. Plant-Microbe Interact. 2, 291-300.
15. Graham-Weiss, L., Bennett, M.L. & Paau, A.S. (1987) Appl. Environ. Microbiol. 53, 2138-2140.
16. Hagedorn, C. (1988) Virg. J. Sci. 39, 28-33.
17. Handelsman, J. & Brill, W.J. (1985) Appl. Environ. Microbiol. 49, 818-821.
18. Kaneshiro, T. & Kwolek, W.F. (1985) Plant Sci. 42, 141-146.
19. Kremer, R.J. & Peterson, H.L. (1983) Agron. J. 75, 139-143.
20. Maier, R. & Brill, W.J. (1978) Science 201, 448-450.
21. McIver, J., Djordjevic, M.A., Weinman, J.J., Bender, G.L. & Rolfe, B.G. (1989) Mol. Plant-Microbe Interact. 2, 97-106.
22. Merberg,D. & Meier, R.J. (1983)Science 220, 1064-1065.
23. Paau, A. (1989) U.S. Patent No. 4875921.
24. Paau, A.S. (1989) Appl. Environ. Microbiol. 55, 862-865.
25. Paau, A., Bennett, M.L., Graham-Weiss, L., Bennett, T. & Park, C. (1987) in XI th North American *Rhizobium* Conference (Agriculture Canada, Sainte-Foy, Quebec, Canada G1v 2J3), pp. 82.
26. Peters, N.K. & Crist-Estes, D.K. (1989) Plant Physiol. 91, 690-693.
27. Rolfe, B.G., Redmond, J.W., Bateley, M. & Djordjevic, M.A.L. (1987) Eur. Patent Application No. 245931.
28. Roughley, R.J. (1988) in Nitrogen Fixation by Legumes in Mediterranean Agriculture (Martinus Nijhoff Publishers), pp. 259-268.
29. Savoy, M.M. & Breitenbeck, G.A. (1988) in Agronomy Abstracts (American Society of Agronomy), pp. 224.
30. Schoeneberger, M.M., Volk, R.J. & Davey, C.B. (1989) Soil Sci. Soc. Am. J. 53, 1429-1434.
31. Soberon, M., Williams, H.D., Poole, R.K. & Escamilla, E. (1989) J. Bacteriol. 171, 465-472.
32. Somasegaran, P. & Halliday, J. (1982) Appl. Environ. Microbiol. 44, 330-333.
33. Stein, M., de Mallorca, M. & Williams, P.M. (1980) Acta Cient. Venez. 31, 374-375.
34. Strobel, G.A., Galvak, A.H. & Haynes, J.M. (1984) U.S. Patent No. 4425150.
35. Triplett, E.W. (1990) Appl. Environ. Microbiol. 56, 98-103.
36. Urban, J.E. (1988) U.S. Patent No. 4720461.
37. Williams, P.M. & Day, J.M. (1989) U.S. Patent No. 4849005.
38. Zablotowicz, R.M., Upchurch, R.G. & Ligon, J.M. (1989) U.S. Patent No. 4863866.

Gresshoff, Roth, Stacey & Newton (eds.) Nitrogen Fixation: Achievements and Objectives
Chapman and Hall New York • London 1990

Agronomic and economic benefits of nitrogen contributed by legumes in live-mulch and alley cropping systems

K. Mulongoy and I.O. Akobundu

International Institute of Tropical Agriculture (IITA), Ibadan, Nigeria

Introduction

The challenge facing research on increasing food production in tropical Africa is to develop low input systems that are not only sustainable but also economically viable. At the International Institute of Tropical Agriculture, substantial resources are devoted to research on alley cropping and live-mulch systems. These are among the potential alternatives to the bush fallow system that would sustain food production with low purchased inputs in the humid and subhumid tropics.

Alley cropping is an agroforestry system in which arable crops are planted in alleys formed by hedgerows of trees. Leguminous woody species reportedly can fix between 134 and 274 kg N ha^{-1} per year under field conditions. This represents an average of 45% of their total N (4, 9, 13, 18). Nitrogen-fixing trees thus have a great potential to supply N and reduce the N requirement of the companion crops.

Live-mulch is a crop production system in which food crops are planted directly in a low-growing cover crop, with minimum disturbance (1). A number of leguminous cover crops, including *Calopogonium mucunoides, Centrosema pubescens, Mucuna pruriens* var *utilis, Psophocarpus palustris* and *Pueraria phaseoloides*, have been tested for their potential in the live-mulch system. Annual dry-matter yields of these legumes range between 1.5 and 7.5 t ha^{-1}, with N yields ranging from 30 to 300 kg ha^{-1} per year. Their soil restorative and protective value is well documented (2).

This paper reviews our findings on the N contribution of legume components in alley cropping and live-mulch systems. Because N-rich leaves of N_2-fixing legumes decompose rapidly in tropical environments and because their N is thus prone to losses through volatilization, denitrification, and leaching, we also discuss management practices that would increase N use efficiency in these systems. Adoption of alley cropping and live-mulch systems depends greatly on their realized economic benefits. We thus present some economic considerations on the profitability of alley cropping and live-mulch systems, relative to the use of inorganic N and to the traditional bush fallow system.

Live mulching

Agronomic aspects

In the live-mulch system of crop production, food crops are established simultaneously with the cover crops. In succeeding growing seasons, crop seeds are planted in narrow strips made in the living mulch either with a cutlass or a contact herbicide. This facilitates crop seedling emergence. Herbaceous legumes with climbing habit require a growth retardant to prevent their smothering the crop. Recent trials include an indigenous cover crop, *Pseudovigna argenteus,* which is nonclimbing.

Experiments conducted in West Africa indicated that competition for nutrients takes place between intercrops in newly established live-mulch plots of *P. palustris* and *C. pubescens* (10, 11). In these experiments, maize N was at least 14% less than in bare plots and maize grain yield 41% less. Similarly, in live mulches of *Pseudovigna argenteus,* maize yields decreased when the cover crop was established simultaneously with maize or a month later (8).

Positive net N contributions and increases in maize yield (up to about 6-fold) occurred after a lag of two years from the time of cover crop establishment (1, 10). This was confirmed as maize responded significantly to 60 kg N ha^{-1} in bare plots but not in the live mulch treatments (1). In long-term plots, *P. palustris* was the best live mulch, with N contributions averaging 31 kg ha^{-1} (11). The beneficial effect of the live mulch resulted, at least in part, from increases in microbial biomass and in organic matter derived from the cover crop in the form of litter and earthworm casts. These casts represented a supplementary N reserve of about 20 kg ha^{-1} and provide better soil physicochemical properties than the surrounding soil (12).

After four years of continuous cropping, the live mulches of *P. palustris* and *C. pubescens* sustained maize yield at about 2.2 t ha^{-1} and 2.1 t ha^{-1}, respectively, without fertilizer N, but maize grown by itself in no-till produced less grain yield with applications of 120 kg fertilizer N ha^{-1} (Table 1). The real impact of this system on soil conservation and sustainability is shown by maize response to residual effects of the live mulch systems. Without fertilizer N, maize yield averaged less than 1.5 t ha^{-1} while yield in the live mulch plots without fertilizer N exceeded 3.0 t ha^{-1} (Table 2).

Table 1. Maize yield (t ha^{-1}) response to live mulch and levels of fertilizer N in the fourth year of cropping.

Land management	Fertilizer N (kg ha^{-1}) 0	30	60	90	120	Mean
Conventional tillage	1.9	2.8	2.4	2.0	2.6	2.3
No-tillage	0.9	2.1	2.1	2.1	1.9	1.8
Centrosema pubescens	2.1	2.2	2.5	2.7	2.5	2.4
Psophocarpus palustris	2.2	2.5	3.1	2.7	2.7	2.6
Mean	1.8	2.4	2.5	2.4	2.4	
LSD (0.05)			0.4			0.7
LSD (0.05)	N levels for same land management					NS
LSD (0.05)	N levels for different land management					NS

NS = not significant

Economics of the live-mulch system

Costs involved in establishing the cover crops include seeds, labor for planting and additional weedings required to ensure adequate establishment, growth retardants used to control the climbing tendency of some cover crops, and herbicides for opening up strips where seeds will be sown. Smallholder farmers, constituting the majority of farmers in tropical Africa, may not be willing to invest scarce cash in herbicides for killing the live mulch. For these farmers, an

Table 2. Residual effect of N fertilizer and cover crops killed with herbicide on maize yield (t ha^{-1})

	Fertilizer N (kg ha^{-1})					
Land management	0	30	60	90	120	Mean
Conventional tillage	1.7	1.8	2.0	2.0	1.7	1.9
No-till	1.2	1.3	1.1	1.2	1.8	1.3
Centrosema pubescens	4.6	4.4	4.2	3.2	4.1	4.1
Psophocarpus palustris	2.7	2.9	3.4	3.9	2.7	3.1
Mean	2.5	2.6	2.7	2.6	2.6	
LSD (0.05)			NS			0.64
LSD (0.05)	N levels for same land management					0.75
LSD (0.05)	N levels for different land management					0.93

alternative method of using a cutlass to open strips in the mulch is being investigated. Labor for cutting strips is negligible compared to labor currently used by farmers to clear a bush regrowth. Establishment of cover crops is very expensive. This cost could be offset by the saving in the N_2 fixed by the legumes over time, and also by other benefits, such as soil conservation.

Drawbacks of the use of live-mulch system include difficulty or incompatibility with using crops such as sweet potato or rice, if these are components of the farming system, and absence of firewood and staking materials obtainable in a bush fallow system. Other disadvantages also include management of the thick mulch layer from such legume cover crops as *Pseudovigna argenteus* which has been reported to reduce maize seed germination and survival of seedlings (8). This will lower the yield potential of the companion crops. While crop yield in the live mulch system may not be better than in sole cropping in the year of cover crop establishment, the advantage of the live mulch becomes evident during subsequent cropping. Compared to no cover at all, the live-mulch system is not profitable in the first two years when the establishing cover crop reduces growth and yield of the associated crop. When soils are poor and degraded, crop performance is generally so low that the investment in legume cover crops will certainly prove beneficial.

Live-mulch is a promising alternative to the bush fallow system characterized in some highly populated areas by one cropping year alternating with 3 years under natural bush. Economic analysis of live-mulch systems has not been done against the bush fallow system. However, it can be speculated that live mulch has the potential to be more profitable than the bush fallow system as it allows a continual cropping of cash crops, while conserving and even improving soil fertility.

Some other benefits in the live-mulch system can be expressed as saving in N and weed control. These result in reduced production costs per unit area and their benefits increase with time.

These benefits, together with improved soil physical properties and reduced soil erosion, are translated into sustained high yields. For correct economic analysis, soil savings due to the cover crops must be valued. Discounting the value of these benefits to the present has been used to avoid overrating them (3).

Alley cropping
Agronomic aspects

In the alley cropping system, hedgerow trees are pruned periodically to minimize shading of, and competition with, the companion crops. The prunings are used as mulch and source of nutrients. Prunings of various hedgerow trees have been shown to increase maize grain yields to the same level as application of fertilizer N up to 90-100 kg N ha^{-1} (5, 7). Nitrogen contribution calculated as the difference between maize N contents in plots with and without prunings, is, however, low, in the range of 40 kg N ha^{-1} per season (5, 9, 14). This represents less than 30% of the N applied as prunings. This low efficiency of the pruning N use is probably due to lack of synchronization between maize N demand and pruning N supply, and possibly to N losses through volatilization and leaching. Figure 1 summarizes the findings reported by Mulongoy and van der Meersch (1988). Additional data on N volatilization were obtained indirectly from pot experiments. Data on N recycled by *Leucaena* were estimated assuming that *Leucaena* derived 50% of its N from the soil. Preliminary data from field experiments indicate that an average of 15% of N applied can be lost through leaching.

Factors that should be considered to improve N-use efficiency in alley cropping include time of application of prunings, methods of pruning placement, and quantity and quality of organic inputs. Application of prunings close to maize planting time increases pruning N-use efficiency and gives the highest maize yields (Figure 2). Incorporation of 10 t prunings ha^{-1} of *Leucaena leucocephala* or *Gliricidia sepium* gave 65 and 25% more grain yield, respectively, than surface application of the prunings (6). It is also possible to influence pruning N-use efficiency by changing the amount of prunings applied. Low C:N ratio materials like leaves of *Leucaena*, which are rich in N, decompose faster than leaves of non-leguminous plants (20, 21). By combining *Leucaena* leaves with *Cassia* leaves, the half life of *Leucaena* pruning N increased by 60% (Mulongoy, unpublished data). Delayed decomposition will likely reduce N losses and improve the residual effects of the prunings.

Despite the low N-use efficiency in alley cropping, Kang and Duguma (1985) concluded from a long-term experiment that over a 6-year period, maize yield could be sustained around 2.0 t ha^{-1} in alley cropping, but maize yield declined to 0.5 t ha^{-1} in the control plot without application of prunings. Thus, inclusion of leguminous trees in cropping systems can be a major strategy for sustainable agriculture.

Root turnover as it is affected by pruning activities, and its contribution to the nutrient cycling in alley cropping is worth investigating. Sanginga *et al.* (1988) found that roots of *Leucaena leucocephala* accounted for about 30% of the total N contribution of this woody species in a sequential cropping system. It should also be noted that factors other than N can account for the success of alley cropping. Trees have a great potential to reduce the downward movement of nutrients by intercepting leached components. The roots also absorb, and return to the soil surface, minerals that are not available to the shallow roots of annual crops. The vegetative cover provided by trees also reduces wind and water erosion and supplies organic matter to replace that lost by microbial action.

Economics of alley cropping

The linear programming model has been used to compare the profitability of upland rice production with inorganic N fertilizer or in an alley cropping system having hedgerows of *Leucaena leucocephala* (16). In this analysis, pruning N-use efficiency was fixed at 33% of that of N fertilizer. Under the conditions of smallholder production in the model, it was consistently more profitable to grow rice with N from leucaena hedgerows than from the mineral source. The break-even price of N fertilizer was estimated at ≤ 3% of present price. In other words, at least 97% of the price of N fertilizer would need to be subsidized for it to become competitive with leucaena-based rice production.

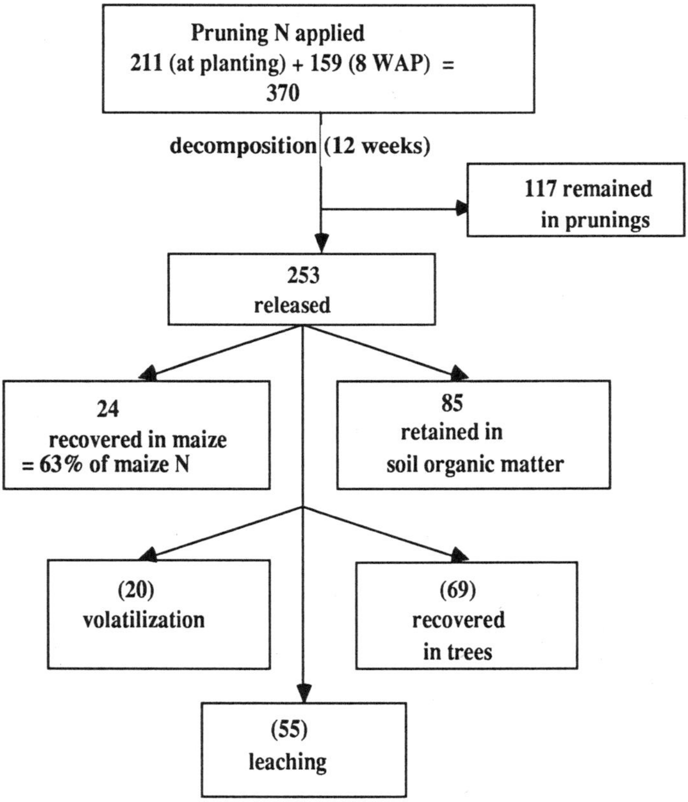

Figure 1. Fate of N from prunings of *Leucaena leucocephala* (adapted from Mulongoy and van der Meersch, 1988). All data are expressed in kg/ha. Values in parentheses were estimated by indirect methods.

It is generally accepted that labor intensity can make the adoption of alley cropping difficult where labor is scarce, or increase the production cost particularly if labour is hired at peak labor times. Alley cropping has been shown to be more profitable than the non-agroforestry system partly through the reduced need for N fertilizer, in spite of its high labor requirement (15). This was further confirmed when alley cropping was compared to a fallow model consisting of 3 years maize cropping followed by fallow (19). In the alley cropping model, prunings were applied at the rate of 3 t ha^{-1} per year, equivalent to 105 kg N ha^{-1}. This low quantity of tree foliage was selected based on difficulties in obtaining good tree stands under village conditions. It also reflects farmers' hesitancy to plant trees densely to obtain high tree populations.

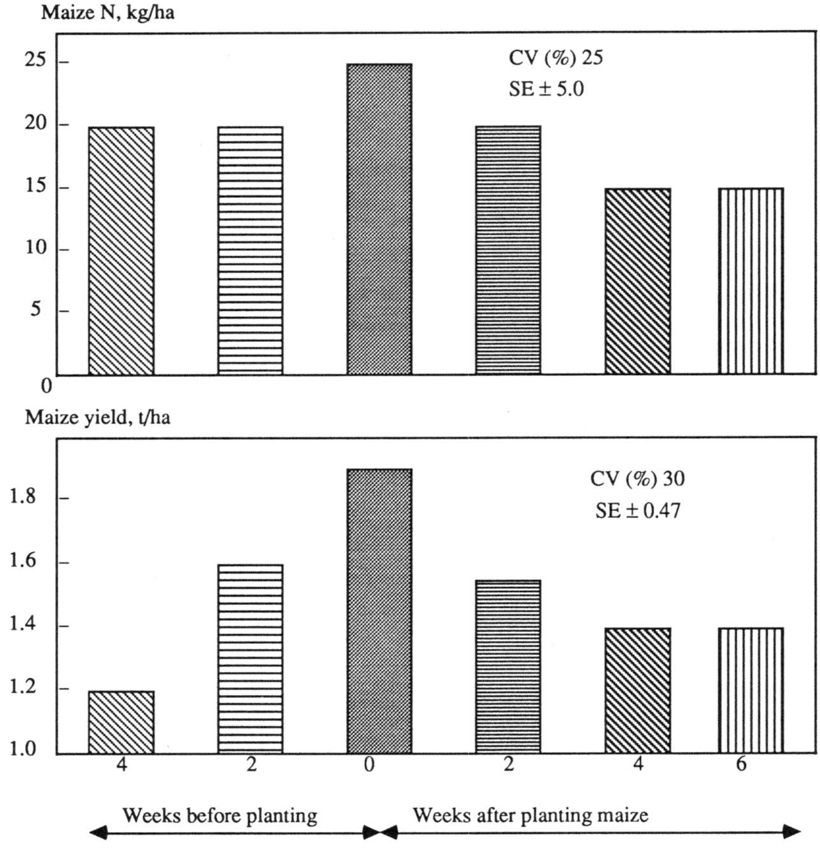

Figure 2. Effect of time of pruning application on maize yield and N content (average for cassia, leucaena and gliricidia).

Marginal physical product (MPP) of foliage applied to the soil surface was 5 kg of maize per kg of mulch N, and its marginal value product (MPV) was US $ 0.63. Buried foliage had a MPP of 10 kg of maize and a MVP of US $ 1.25. Therefore, 20 kg of mulch N was needed to maintain maize yields at the base level of 1 t ha^{-1}. In spite of the higher labor cost of alley cropping for tree pruning from the second year, alley cropping had a present net value 14 to 59% greater than for the fallow model. Pruning incorporation substantially increased profits from alley cropping so that even when labor increased by 50% in alley cropping and decreased by 100% in fallow, alley cropping was still 65% more profitable. Most of these data however, have to be validated in farmers' fields.

Conclusions

Alley cropping and live-mulch systems are both very promising technologies for the tropics. Their success is based on biological processes that use renewable resources to increase and sustain food production, with little or no purchased inputs. The importance of N for adequate plant growth and yields, and the low level of this element in cultivated soils in the tropics are well known. Leguminous cover crops and trees can fix high amounts of atmospheric N_2 and recycle nutrients. They can contribute these nutrients, particularly N, to the companion crops in alley cropping and live-mulch systems, with resulting crop yield increases. Both cropping systems are economically profitable relative to the use of inorganic N fertilizer, an input that is not only expensive but also can acidify soils and make them infertile.

Net N contributions and significant food crop yield increases occur in the live mulch systems only after the cover crop has established for some cropping seasons. There is, therefore, a need to shorten this lag in benefits if the live mulch system is to be adopted by farmers. Competition between the food crop and the establishing cover crop should be minimized. Also, a clear understanding of the pathways of N transfer from the live mulch to the food crop is needed to devise management practices that would increase the amount of N contributed by the legume.

In alley cropping, hedgerow trees can yield prunings containing more than 300kg N ha^{-1}. Depending on the extent of biological N_2 fixation in the trees used, this N yield represents an important gain in the system. The efficiency of N use by the associated crops, however, is usually low, and supplementary N fertilizer is still necessary for high maize yields, though at rates lower than are normally required without alley cropping. Various research approaches are being investigated to improve N-use efficiency in alley cropping, reduce the need for N fertilizer, and extend in time the N benefits. Research is also devoted to describe the interactions between hedgerow trees and the food crops, and to assess nutrient contributions from the root systems regulated by frequent prunings of the trees.

If N contribution is a key function of hedgerow trees, the nutrient cycling ability of trees, their role in replenishing soil organic matter, and the mulching effect of tree prunings and litter should not be overlooked. Finally, on-farm studies show that improved establishment and early growth of hedgerow trees will promote a rapid adoption of the system.

Acknowledgements. The authors thank the Belgian Government for a grant to conduct part of the experiments reported, and M. Raj, R. Carsky, E. Jegede and R. Cabus for technical assistance.

References

1. Akobundu, I.O. (1980) *Proc. British Crop Prot. Conf. - Weeds,* pp. 377-382.
2. Akobundu, I.O. & Okigbo, B.N. (1984) Field Crops Res. **8**, 177-186.
3. Allison, J.R. & Ott, S.L. (1987) in *The Role of Legumes in Conservation Tillage Systems,* ed. Power, J.F. (Soil Conservation Society of America, Athens), pp. 145-150.
4. Herridge, D.F. & Bergersen, F.J. (1988) in *Advances in Nitrogen Cycling in Agricultural Ecosystems,* ed. Wilson, J.R. (C.A.B. I., Aberystwyth), pp. 46-65.
5. Kang, B.T. (1988) in *Advances in Nitrogen Cycling in Agricultural Ecosystems,* ed. Wilson, J.R. (C.A.B. I., Aberystwyth), pp. 338-348.
6. Kang, B.T. & Duguma, B. (1985) in *Nitrogen Management in Farming Systems in Humid and Subhumid Tropics,* ed. Kang, B.T. & Van der Heide, J. (Institute for Soil Fertility, IB, Haren), pp. 269-284.
7. Kang, B.T., Reynolds, L., Atta Krah, A.N. (1990) Adv. Agron. **43**, 315-359.
8. Lawson, T. (1989) in *Resource and Crop Management Program. Annual Report 1987*. (International Institute of Tropical Agriculture, Ibadan, Nigeria), pp. 51-53.
9. Mulongoy, K. (1986) Biol. Agric. & Hortic. **3**, 341-346.
10. Mulongoy, K. (1986) Soil Biol. Biochem. **18**, 395-398.
11. Mulongoy, K. & Akobundu, I.O. (1985) in *Nitrogen Management in Farming Systems in Humid and Subhumid Tropics,* eds. Kang, B.T. & Van der Heide, J. (Institute for Soil Fertility, IB, Haren), pp. 285-290.
12. Mulongoy, K. & Bedoret, A. (1989) Soil Biol. Biochem. **21**, 197-203
13. Mulongoy, K. Owoaje, B.T., Ibewiro, E.B. & Okogun, J.A. (1988) in *Nitrogen Fixation: Hundred Years After*, eds. Bothe, H., de Bruijn, F.J. & Newton, W.E. (Gustav Fischer Verlag, Stuttgart), p. 824
14. Mulongoy, K. & van der Meersch, M.K. (1988) Biol. Fert. Soils **6**, 282-285.
15. Ngambeki, D.S. (1985) Agricultural Systems **17**, 243-258.
16. Raintree, J.B. & Turay, F. (1980) IITA Research Briefs **1**, 5-7.
17. Sanginga, N., Mulongoy, K. & Ayanaba, A. (1988) Plant Soil **112**, 137-141.
18. Sanginga, N., Mulongoy, K. & Ayanaba, A. (1989) Plant Soil **117**, 269-274.
19. Sumberg, J.E., McIntire, J., Okali, C. & Atta-Krah, A. (1987) ILCA Bulletin, **28**, 2-6.
20. Wilson, G.F., Kang, B.T. & Mulongoy, K. (1986) Biol. Agric. & Hort. **3**, 251-267.
21. Yamoah, C.F., Agboola, A.A. & Mulongoy, K. (1986) Agroforestry Systems **4**, 239-246.

Gresshoff, Roth, Stacey & Newton (eds.) Nitrogen Fixation: Achievements and Objectives
Chapman and Hall New York • London 1990

Stem-nodule symbiosis and its unusual properties

J. K. Ladha, R. P. Pareek, R. So and M. Becker
The International Rice Research Institute, P. O. Box 933, Manila, Philippines.

Introduction

Although the phenomenon of stem nodulation was first reported in 1928 in *Aeschynomene* spp. by Hagerup (8), actual interest arose only after the recent discovery of *Sesbania rostrata* (4). *S. rostrata* is a remarkably fast-growing and high N_2-fixing legume with a great potential as a source of N for rice (11). This interest has led to the identification of yet another stem-nodulating species, *Aeschynomene afraspera*, with comparable or higher potential than *S. rostrata* (3). Stem nodulation, a phenomenon that evolved in nature as a result of adaptation to flooding, is characterized by several unique features : 1) increasing number of infection sites with plant growth, 2) cyclic nature of symbiosis, 3) tolerance for combined N, 4) presence of photosynthetic cells around the nodules, 5) dinitrogen-dependent growth and N_2 fixation by stem-nodule rhizobia, and 6) photosynthetic nature of stem-nodule rhizobia. Due to these unusual features, there has been extensive research on stem nodulation recently. This paper reviews the most recent findings on stem nodulation.

Diversity of stem nodulation and stem-nodulating legumes

Our understanding of stem nodulation has become complicated as more and more research findings are published. Stem nodules may be restricted to the lower portion of the stem or they may spread all over the stem. There are 18 species of *Aeschynomene*, 4 of *Sesbania,* and 1 of *Neptunia* reported to bear stem nodules. Moreover, some cultivars of *Arachis hypogeae* (13), *Vicia faba* (7), and tree species *Cassia tora* L. and *Parkinsonia aculeata* L. (17) were also reported to form nodules, mostly on the lower portion of the stem. Recently, we observed nodule formation in the lower stem of intermittently submerged plants of *S. javanica*, *S. sesban*, and *S. speciosa.* Further exploration of legumes, especially the genera *Aeschynomene* and *Sesbania*, may add more stem-nodulating species to the list.

The question remains whether the system or pattern of nodulation shown by *S. rostrata, Ae. afraspera,* and *Ae. nilotica,* which form nodules clearly on the aerial portion of the stem, should be regarded at par with that of species such as *S. speciosa, Ae. elaphroxylon,* and others, which form nodules only on the submerged portion of the stem. Based on the type of stem nodules, legumes can be divided into two major groups : I) Legumes with visible root primordia on the stem, which can be further subdivided into two subgroups -- a) profuse stem nodulation all over the stem (includes *S. rostrata, Ae. afraspera*, and *Ae. nilotica*), and b) scarce nodulation on the aerial stem but good nodulation on the submerged stem (includes *Ae. aspera, ciliata, denticulata, evenia, indica, pratensis, rudis, scabra, sensitiva, tambacoundensis*, and *schimperi*); and II) legumes with invisible nodule primordia on the stem with nodulation occurring on the submerged part of the stem when dormancy of root primordia is broken on flooding (includes *Ae. crassicaulis, Ae. pfundii, Ae. elaphroxylon, S. sesban, S. punctata, S. speciosa, S. javanica, Neptunia oleracea, Ar. hypogeae,Vicia faba* , *Cassia tora* and *Parkinsonia aculeata).*

Another point to consider is the working definition of stem nodules. *S. rostrata*, when waterlogged, produces many adventitious roots from the submerged stem and these roots and the submerged stem bear profuse highly active nodules. Anatomically the submerged stem that produces the adventitious roots more closely resemble root than stem. A comparative situation exists in non stem-nodulating *S. speciosa, S. sesban,* and other species. Therefore, the "root nodules" should include the nodules on the submerged stem and the "stem nodules" are those which form on the aerial portion of the stem. Strictly speaking, the nodule primordia on the aerial stem are also anatomically similar to the roots, but it is the location which differentiates root from stem nodules. Based on this definition, legumes with invisible root primordia on the stem and nodules on the submerged stem (which were classified in the second group) should not be considered as stem-nodulating legumes.

Induction of stem nodules in non stem-nodulating *S.speciosa*
We were able to induce effective, green nodules on the aerial stem of young plants of *S. speciosa* grown in pots. The lower aerial portion of the stem (3-5 cm above the soil) of a 30-day-old plant was wrapped with cotton wool, which was kept continuously moist for several days.

After about 3-5 days, the specific *Rhizobium* strain was inoculated onto the cotton wool. About 7-10 days after inoculation, stem nodules formed from the root primordia, which otherwise would have formed adventitious roots if the stem portion was kept continuously flooded. The number of stem nodules increased and formed an effective N_2-fixing mass of nodules on the aerial stem. What appears important is the formation of well-developed root primordia breaking through the epidermis. It is at this stage that a specific *Rhizobium* should infect the root primordia and floodwater recede; otherwise adventitious root would develop. And probably once the root formation is initiated, nodules do not form on the stem but they appear on newly formed roots. Though this situation may not be specific to *S. speciosa* and *S. sesban,* and may be found in other aquatic *Sesbania* species, we neither observed nor were able to induce stem nodules in *S. cannabina* (Pareek and Ladha, unpublished).

N_2 fixation potential of stem-nodulating legumes

Although there are several stem-nodulating species, only *S. rostrata*, *Ae. afraspera*, and *Ae. nilotica (Ae. nilotica* is, similar to *Ae.afraspera* in habit and habitat), have potential as green manure (GM) for rice. These species are also fast-growing. *Ae. indica*, although used as GM to some extent in China, is slow growing. The fast-growing species belong to stem nodulation group/subgroup I/a. Stem nodulation group/subgroup I/b, although composed of several species, only has *Ae. indica* with some potential use as GM. It appears that the ability to grow fast is associated with the species' ability for profuse stem nodulation.

Rinaudo et al.(15) in Senegal (West Africa) and Ladha et al.(10) in the Philippines (Southeast Asia) estimated N_2 fixation of about 50-day-old *S. rostrata* by N (Kjeldahl) balance in *S. rostrata*/rice sequence to be 190 and 267 kg N/ha, respectively. Recently, Pareek et al.(14) estimated N_2 fixation of *S. rostrata* by ^{15}N dilution using ^{15}N-stabilized soil. Results indicate that *S. rostrata* is possibly the fastest N_2-fixing plant known, fixing 70-90% of the total plant N during 45-55 days (see Table). The calculated rate of N_2 fixation during 55-65 days in wet season is 17 kg/ha per day (almost 100% N from BNF).

Table. Nitrogen fixed (kg/ha) by *S. rostrata* *

Days after seeding	Dry season	Wet season
25	7 (76)	11 (53)
45	100 (88)	140 (71)
55	240 (91)	286 (81)
65	324 (94)	458 (86)

*N_2 fixation of individual plants multiplied by plant density of 500,000/ha. Figures in parentheses indicate % N_2 fixation of total plant N. Source: Pareek et al (14).

Ae. afraspera, which has been identified recently as a potential GM species has two additional advantages over *S. rostrata* : 1) being less photoperiod-sensitive, it produces higher biomass and N year round, and 2) it branches profusely and produces less woody stems, which make incorporation easy (3). Becker et al.(3) compared N accumulation by *S. rostrata* and *Ae. afraspera* at different ages and seasons. Based on the regression equation analysis of the data, iso-N accumulation graphs for 100 kg N/ha were calculated. Results show that to obtain 100 kg N/ha, *Ae. afraspera* will have to be grown for a constant period of 45 to 50 days throughout the year. On the other hand, *S. rostrata* exhibited much variability, requiring only 40 days during the long-day season (May-Jun) and 50-60 days during the short-day season (Dec-Jan) (see Figure) to accumulate 100 kg N/ha.

Cyclic nature of symbiosis

S. rostrata-Azorhizobium caulinodans symbiosis is highly species-specific. *Az. caulinodans* neither forms effective root nodules in other *Sesbania* species nor does it produce stem and root nodules in any of the stem-nodulating *Aeschynomene* species. One reason that *S. rostrata* has limited promiscuity is its restricted occurrence in a few countries in West Africa. It was only recently that this species was introduced in Asia and other parts of the world. However, it has been found to spontaneously nodulate without inoculation in several Asian countries including the Philippines.

Ladha et al.(11) found a high occurrence of azorhizobia in seeds of *S. rostrata* that could not be eliminated by treatments normally used for seed sterilization or scarification. Sometimes a pair of stem nodules would be observed on the first node of plants grown in a leonard jar

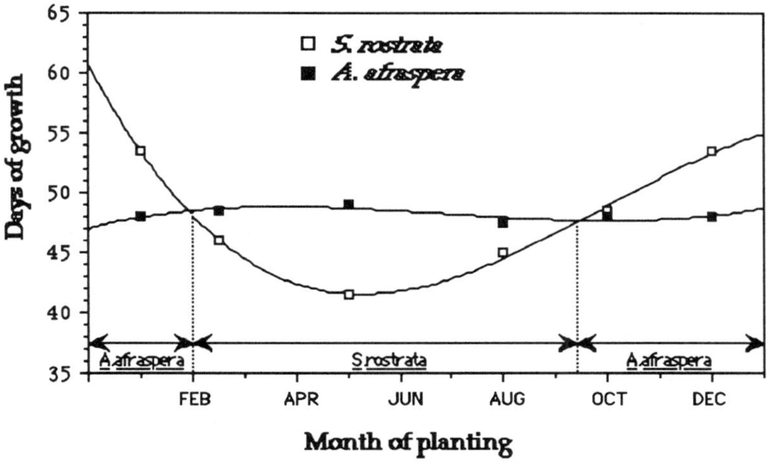

Figure : Growth duration requirement (iso-N accumulation graph) of *S. rostrata* and *Ae. afraspera* for accumulation of 100 kg N/ha. Source: Becker et al (3).

system where the seeds used were sterilized by mercuric chloride. Ladha et al,(11) hypothesized that the azorhizobia in the seeds could originate from senesceing stem nodules at the time of flowering and seed formation. Because nodules inhabit that part of the stem where flowering and seed formation occur, this area may become contaminated with airborne rhizobia and cause rhizobia to be trapped in the seeds. In a subsequent study, a large population of azorhizobia was found on the leaves and flowers of *S. rostrata* (1). This supports the above hypothesis. It can be further hypothesized that the mode of initial entry of azorhizobia into the soil is through the seeds where azorhizobia establish saprophytically (9). Azorhizobia may reach the infection sites on the stem from the soil through rain-splash and insect activity. The nodules thus formed on the stem may release azorhizobia on senescence and survive epiphytically (1). This cyclic nature of symbiosis is a unique feature of *S. rostrata*; that is, the bacterial symbiont is maintained epiphytically on the host plant, is carried on into the flower, and transmitted from one host generation to the next enclosed in the seed. It has been reported that in legumes, microbial infection occurs under the seedcoat and even the most drastic methods of sterilization can not eliminate contamination (2). Additionally, the microorganisms can enter seeds through the hilum scar, pilar groove, and micropyle, all of which are difficult to cleanse by chemical agents (2). However, it is not known if azorhizobia do enter the seeds, and if it does, what mechanism is involved.

Free-living growth and N_2 fixation by stem-nodule rhizobia in N-free medium

Normally, rhizobia are obligate diazotrophs, i.e., fix N_2 only with the host. But a limited number of rhizobia (mostly belonging to the genus *Bradyrhizobium*) show *ex planta* nitrogenase activity. These strains, however, require combined N for growth and N_2 fixation. *Az. caulinodans* ORS 571 (5) and *Rhizobium trifolii* 0403 (16) are exceptions to this finding. Strain ORS 571 differs from 0403 in its N_2 fixation ability at higher O_2 concentration (3%) and temperature (37 ^oC). Recently, in our laboratory, we found rhizobia strains of *Ae. afraspera* (strain IRG2), *Ae. sensitiva* (strain IRG109), *Ae. pratensis* (strain IRG108), and *Ae. indica* (strain BTAi1) all with the ability to grow and fix N under free-living condition with no combined N (So and Ladha, unpublished).

The significance of freeliving growth and N_2 fixation by stem-nodule rhizobia is still not clear. Ludwig (12) has proposed a possible role for N_2 fixation under free-living condition in a biphasic nature of N_2 fixation. Accordingly, in the proliferative phase during early nodulation, N_2 fixation is coupled with rhizobial ammonium assimilation, thus allowing the growth of the invading rhizobia; in the subsequent altruistic phase during late nodulation, N_2 fixation is decoupled from rhizobial ammonium assimilation, thus allowing ammonium export from bacteroids to the host plant. The N_2-dependent growth and N_2 fixation of rhizobia may be required to provide N during the proliferative phase because of inaccessibility to soil N. Recently, Ladha et al.(9) demonstrated a role of azorhizobia in rice-associative N_2 fixation. They found that when *S. rostrata* is grown as a pre-rice GM crop and incorporated before rice planting, the rice rhizosphere harbored a large population of azorhizobia and supported higher nitrogenase activity than that of fallow-rice sequence. Sterile rice seedlings inoculated with *Az. caulinodans* also showed nitrogenase activity comparable to seedlings inoculated with *Azospirillum lipoferum* 34H, a rice root isolate. Rhizobia with no ability for free-living, N_2-dependent growth and N_2 fixation from several legumes did not support nitrogenase activity.

Photosynthetic nature of stem-nodule rhizobia

Another novel feature of stem-nodulating microsymbiont is its ability to form bacteriochlorophyll (BChl) and photosynthetic reaction centers resembling those of purple photosynthetic bacteria. Rhizobia of *Ae. indica* have been shown to be photosynthetic under free-living and symbiotic conditions (Darrell Fleischman, personal communication). We recently found that this feature is characteristic to most stem-nodulating species of *Aeschynomene*. Rhizobia of stem nodule from *Ae. afraspera*, *Ae. denticulata*, *Ae. evenia*, *Ae. indica*, *Ae. nilotica*, *Ae. pratensis*, and *Ae. sensitiva* produce Bchl both under symbiotic and free-living conditions (Ladha and So, unpublished). Unlike photosynthetic bacteria, rhizobia of stem nodules produce pigment only under aerobic/light conditions and failed to grow anaerobically. On the other hand, *S. rostrata-Az. caulinodans* stem-nodule symbiosis lack this property. The rhizobia of stem-nodulating species of *Aeschynomene* can not be placed in the group of photosynthetic bacteria because of their aerobic nature of photosynthesis and symbiotic ability. Perhaps a separate taxonomic placement is warranted.

Since the cortical cells around the stem nodules are also capable of photosynthesis, it was speculated that nodules are "self energy-generating" systems. Data supporting this hypothesis were obtained by Eaglesham et al.(6). They found that defoliation reduced root nodule nitrogenase activity but had no effect on the activity of stem nodules. The recent finding of the photosynthetic ability of bacteroids further supports the hypothesis. The stimulation of N_2 fixation by stem nodules of *Ae. indica* under infrared light has been reported (Darrell Fleischman, personal communication). Nitrogenase activity continued much longer in illuminated stem nodules than those incubated in the dark.

The ability of the stem-nodule bacteroid to use light in a wavelength that is not used by the plant could be an advantage, diminishing competition for energy for C and N reduction. However, whether N_2 fixation is an autonomous system obtaining all its ATP requirements by photophosphorylation within the bacteroid or whether it obtains energy from chloroplast photosynthesis in stem nodules is not known. Research is also needed to understand the interactions between N_2 fixation and dual photosynthesis and how the system manages O_2 inhibition.

Conclusion

Stem nodulation is a fascinating system and has attracted the attention of workers in both applied and basic sciences. Due to submergence tolerance, rapid growth, and high N_2 fixation potentials, *S. rostrata* and *Ae. afraspera* are uniquely suitable for GM production during the pre-rice period when short-term flooding can occur. *S. rostrata* is possibly the fastest N_2-fixing plant known, on the order of 100-285 kg N/ha in 45-55 days. The finding that *Az. caulinodans* occurs on leaves, flowers, and seeds of the host plant (*S. rostrata*) may suggest its cyclic nature and thus explain spontaneous nodulation in the field. In some *Aeschynomene* species the stem-nodule bacteroids fix both N and C; in addition, cortical cells around the bacteroids carry out O_2-generating photosynthesis. This provides a unique opportunity, and a challenge, to study bacterial photosynthesis, chloroplast phostosynthesis, and N_2 fixation at close proximity.

References

1. Adebayo, A., Watanabe, I. & Ladha, J.K. (1989) Appl. Environ. Microbiol. **55**, 2407-2409.
2. Ash, C.G. & Allen, O.N. (1948) Soil Sci. Soc. Proceeding **13**, 279-283.
3. Becker, M., Ladha, J.K. & Ottow, J.G.C. (1990) Soil Biol. Biochem. (in press).
4. Dreyfus, B.L. & Dommergues, Y.R. (1981) FEMS Microbiol. Lett. **10**, 313-317.
5. Dreyfus, B.L., Elmerich, C. & Dommergues, Y.R. (1983) Appl. Environ. Microbiol. **45**, 711-713.
6. Eaglesham, A.R.J. (1984) Abstract of the Fourth Sttenbock Symposium.
7. Fyson, A. Sprent, J.I. (1980) J. Exp. Bot. **31**, 1101-1106.
8. Hagerup, O. (1928) Dan. Bot. Ark. **15**, 1-9.
9. Ladha, J.K., Garcia, M., Miyan, S., Padre, A.T & Watanabe, I. (1989) Appl. Environ. Microbiol. **55**, 454-460.
10. Ladha, J.K., Miyan, S. & Garcia, M. (1989) Biol. Fert. Soils **7**, 191-197.
11. Ladha, J,.K., Watanabe, I & Saono, S. (1988) in Green manure in rice farming. The International Rice Research Institute, Manila, Philippines, pp. 165-183.
12. Ludwig, R.A. (1988) Abstract of an International Symposium on *Sesbania rostrata*, Dakar, Senegal.
13. Nambiar, P.T.C., Dart, P.J., Rao, B.S., & Rao, V.R. (1982) Exp. Agric. **18**, 203-207.
14. Pareek, R.P., Ladha, J.K. & Watanabe I. (1990) Biol. Fert. Soils (in press).
15. Rinaudo, G., Dreyfus, B.L. & Dommergues, Y.R. (1983) Soil Biol. Biochem. **15**, 111-113.
16. Urban, J.E., Davia, L.C. & Brown, S.J. (1986) Appl. Environ. Microbiol. **52**, 1606-1607.
17. Yatazawa, M., Hambali, G.G., Wairiadianta, H., Susilo, H & Uchino, F. (1987) BIOTROP Spec. Publ. **31**, pp. 191-205.

Gresshoff, Roth, Stacey & Newton (eds.) Nitrogen Fixation: Achievements and Objectives
Chapman and Hall New York • London 1990

Workshop summary: What are the applications of our research to ecology and agriculture worldwide?

Dietrich Werner

Philipps-University, Department of Biology, Karl.v. Frischstr., D-355 Marburg, FRG

Applications of research in the biological sciences go far beyond direct economic utilization. From the many subjects of this conference, three will be especially emphasized: (i) the consideration of ecological aspects and their utilization in agriculture in general; (ii) faster utilization of molecular genetic and cell phytological methods for plant breeding; and (iii) transferring our experience with agricultural plants to trees and agroforestry.

With respect to the first topic, a large number of examples were presented in the poster contributions (see Sections D and F. The inclusion of symbiotic and non-symbiotic diazotrophic microorganisms in the plant-soil nutrient system has contributed to a better understanding of root physiology, N-metabolism in plants and plant-microbe communication. It has especially stimulated research with other symbioses, such as VA-mycorrhiza (2) and ectomycorrhiza. However, communication between researchers on the Rhizobium/legumes and those on mycorrhiza is still limited. Two new journals (2,8) dedicated to symbiosis and cell research should improve the transfer of methods and results.

The increased contribution of biological nitrogen fixation to the nitrogen budget in agriculture (4) is still mainly due to conventional techniques; namely, increased productivity of legume crops per hectare, increased production areas, e.g., of soybeans in Brazil and Argentina, and crop rotation. Compared to a corn monoculture with NPK fertilization, a corn/soybean rotation increased corn production from 4.05 $t.ha^{-1}$ to 4.69 $t.ha^{-1}$ (5). The increased study of biological nitrogen fixation as compared to nitrate, ammonia, urea or slurry fertilization has also stimulated methods for evaluation of N balances and N cycles by ^{15}N methods. NH_3 released from agricultural soils in Western Europe, after heavy N-fertilization, varies between 10 and 70 kg $N.ha^{-1}$. This has already contributed to forest declines (7) and to replacement of sensitive vegetation, such as heath, by grasses (9).

Plant breeding and plant biotechnology is a second important field of applied biology. This area can profit from the methods of molecular genetics and cell physiology, which were developed by studying *nif*, fix and nod genes and gene products in the microsymbiont and nutrients and nodule development in the host plants. Overnodulating and super-nodulating soybean cultivars (3) and other legumes have brought, in addition to an appreciation of their own properties, a better understanding of shoot-root interactions in plants.

Work on early nodulins (1) has contributed to a molecular understanding of plant organ morphogenesis, which is important for breeding strategies, in which the site of sinks (seeds, fruits, roots) determines the production value. Breeding for resistance against phytopathogenic fungi and bacteria has a high priority for all major crop plants. At several levels of the plant-microbe interaction, there are similarities (and some decisive differences) between symbiotic and parasitic systems. Specific mutations in the microsymbiont (10) or a different genetic background in the host plant (6) can change a stable symbiosis into a phytopathogenic interaction with phytoalexin accumulation and a hypersensitive reaction in plant cells.

The transfer of experience in the fields of N_2 fixation and plant-microbe interaction from agricultural plants to trees is a third example of applications. Fast-growing and nitrogen-fixing trees (FGNFT) are of special interest (11), producing wood material as an energy source and leaves as fodder. Mycorrhiza research is of basic interest for both nitrogen-fixing and non-fixing trees. If we realize that the area in all continents together covered with forests is larger than the area under agricultural cultivation, then the small number of contributions on trees in these Proceedings, indicates the potential for further application with these long-living plants.

References

1. Fransen, H., et al., these Proceedings.
2. Galun, M., ed. (1990) Symbiosis, Vol. 8 (Balaban Pub., Tel Aviv).
3. Gresshoff, P.M. (1990) in Advances in New Crops, ed. Janick, J. (Timber Press), pp. 113-126.
4. Mengel, H. (1990) in Symposium on Nitrogen Metabolism in Plants, (Oxford University Press, Oxford), in press.
5. Odell, R., Helsted, S. & Walker, W. (1984) Soil Sci. **137**, 160-171.
6. Parniske, M., Zimmermann, C., Creegan, P. & Werner, D. (1990) Botanica Acta **103**, 143-148.
7. Roclofs, J.G.M., Botma, A.W. & vanDijk, H.F.G. (1988) in Air Pollution and Ecosystems, ed. Mathy, P. (Boer Radel Publ., Dordrecht, The Netherlands), pp. 876-880.
8. Schenk, H.O., ed. (1990) Endocytobiosis and Cell Research, Vol. 7 (University Press, Tubingen).
9. Steubing, L. & Buchwald, K. (1989) Natur Landschaft **64**, 100-105.
10. Werner, D., Mellor, R.B., Hahn, M.G. & Grisebach, H. (1985) Z. Naturforsch. **40c**, 179-181.
11. Werner, D. & Muller, P., eds. (1990) Fast-growing and nitrogen-fixing trees (G. Fischer, Stuttgart & New York).
12. Werner, D. (1991) Plant and Microbial Symbioses (Chapman & Hall, London).

Gresshoff, Roth, Stacey & Newton (eds.) Nitrogen Fixation: Achievements and Objectives
Chapman and Hall New York • London 1990

Production and application of legume-inoculants in Egypt

S. M. Abdel-Wahab and M. E. Hassan

Soils and Water Research Institute, ARC, Giza Egypt

Legumes constitute a major portion of food and feed in Egypt. The area cultivated to legumes in Egypt is about 3.5 million acres (60% of the total cultivated area). Research carried out in Egypt showed that the amount of nitrogen fixed varied from 19 to 138 kg N/ha. In 1989/1990 season 350,000 acres were inoculated with rhizobia, which includes 100,000 acres of the cultivated area to soybean.

Legume inoculants have been produced in Egypt since the 1950's, under the commercial name "Okadin". The production technology encounters several factors such as isolation, purification and selection of the most efficient local strains of rhizobia. Mass culturing of rhizobia is achieved using 20 L fermentors. Different carrier material, i.e. nile silt, jiffy-7 discs and imported peat, were tested and the peat was found to be the most proper carrier. The inoculant is packed in polyethylene bags containing 200 g each which is sufficient to inoculate the seeds of one acre. Quality control tests on the inoculants are carried out for each batch before distribution.

Recently, several studies were carried out to improve the quality and efficiency of the legume inoculants including construction of rhizobial strains resistant to pesticides, dissication, salinity and replacement of the imported peat carrier with local carrier materials.

Gresshoff, Roth, Stacey & Newton (eds.) Nitrogen Fixation: Achievements and Objectives
Chapman and Hall New York • London 1990

Nitrogen fixation by *Phaseolus vulgaris* L. in response to inoculation

B. Ahmed and R.H. Phelps

Scientific Research Council, Kingston, Jamaica, and

Caroni Research Station, Carapichaima, Trinidad

Response of bean(**Phaseolus vulgaris** L.) to inoculation has been variable, though positive responses have been recorded (Pacovsky et. al. 1984). This study evaluates the effect of 11 locally isolated **Rhizobium phaseoli** strains on plant growth and nitrogen fixation of two CIAT bean cultivars.

Seeds of bean cultivars CIAT 07/79 and CIAT 17/79, surface sterilized and rinsed with sterile water, were planted 3 seeds per pot, containing 2 kg of either Piarco fine sand (pH 4.2) or River Estate Loam (pH 5.6). Inoculation was done with $1x10^8$ cells/ml of each of 11 strains. Two replicates, arranged in a randomized complete block design, were harvested at 42 days after planting and shoot dry weight, nodule dry weight. Top N and nitrogenase activity were determined.

Inoculation with differing strains resulted in significantly ($p < 0.05$) higher nodule weight, Total N and nitrogenase activity compared to uninoculated control (Table 1). This indicates the significance of inoculating beans with effective rhizobial strains to increase the utilization of symbiotic nitrogen fixation. It was calculated that local isolates can fix around 50 kg N/ha after 6 weeks growth by bean plants.

Table 1. Mean values of shoot wt., nodule wt., Total N and nitrogenase activity of bean as affected by inoculation.

Treatments	Shoot wt. g pot $^{-1}$	Nodule wt. mg pot $^{-1}$	Top N %	Nitrogenase activity u mole C_2H_4/pl/h
Uninoculated	1.48	59.06	1.96	1.78
50 Kg N/ha	2.84	41.33	3.20	3.14
Inoculated (Range of 11 strains)	1.43-2.26	58.01-194.79	2.07- 3.08	9.58-21.58
LSD (0.05)	0.39	20.25	0.14	0.94

In searching for compatible, competitive and productive **Rhizobium** strains for beans, local isolates must be evaluated first.

Pacovsky, R.S., Bayne, H.G. and Bethlenfalvay, G.J. 1984 Crop Sci. 24, 101-105.

Gresshoff, Roth, Stacey & Newton (eds.) Nitrogen Fixation: Achievements and Objectives
Chapman and Hall New York • London 1990

Evidence of *Rhizobium* cross-inoculation among fast-growing tree legumes used in agroforestry

D.S. Amara

Department of Soil Science, Njala University College, University of Sierra Leone, Sierra Leone.

Pure *Rhizobium* isolates obtained from one annual and six tree legumes were used to inoculate seedlings of their host of isolation and other seedlings referred to here as their "heterologous host". The seedlings were sown in sterile sand medium and assessed for nodulation after 30 days.

Cross-inoculation was observed among the following legumes: *Albizia lebbek, Cajanus cajan,* and *Vigna unguiculata* (group 1); *Cajanus cajan, Enterolobium cyclocarpum,* and *Vigna unguiculata* (group 2); *Gliricidia sepium* and *Leucaena leucocephala* (group 3). The legumes in group 1 were nodulated by slow-growing strains while those in groups 2 and 3 were nodulated by both slow and fast-growing strains.

Rhizobia from Vigna, Cajanus, and Albizia exhibited promiscuous and effective nodulation in the majority of the trees. For plants in which nodulation failures occurred, there was good and effective nodulation in reciprocal legume-*Rhizobium* associations. This nodulation pattern is suggestive of different recognition mechanisms between the legumes and the rhizobia.

The identification of the above cross-inoculation groups has provided an initial basis for the characterization of the tree legume rhizobia. It has also provided a basis for the production of single or multistrain inoculants for use by farmers in soils where nodulation problems occur.

Gresshoff, Roth, Stacey & Newton (eds.) Nitrogen Fixation: Achievements and Objectives
Chapman and Hall New York • London 1990

Improved grain protein content in Argentine crops of *Phaseolus vulgaris* through inoculation

J. C. Pacheco Basurco[1], A. Perticari[1], A. Gauna[1], R. N. Diéguez[1], A. Pellegrini[2], P. N. Figueroa[2], J. L. Boiardi[3], E. Arrarás[3], A. Lodeiro[4] and G. Favelukes[4].

[1]Inst. Microbiología INTA, Castelar; [2]EEA INTA, Salta; [3]CINDEFI, FCE, UNLP, La Plata; [4]QBI, FCE, UNLP, La Plata, Argentina.

Phaseolus vulgaris beans (originated from Central and South America) provide low cost, essential protein nutrients to large populations throughout Latin America. Both N-fixation and grain yield in bean crops in this region are generally very poor. Bean cultures in NW Argentina (a center of origin of the primitive *P. vulgaris* var. *aborigineus*) are spontaneously but poorly nodulated, and their moderate grain yields do not generally show a positive response to rhizobial inoculation.

We have studied in two consecutive cropping seasons (late summers in 1988 and 1989) the effects of inoculation with four selected strains of *R. phaseoli* on the symbiotic behavior and productivity of three varieties, in field trials at two locations in the bean cropping area in NW Argentina. Natural populations of infective *R. phaseoli* ranged from $<10^2$ to $>10^4$ per g soil. In comparison to non-inoculated control plots, the response to inoculation was highly variable in grain yield, with 6 increases, 3 decreases, and 31 cases of no significant difference. Total protein yield showed a better response, with 33 increases out of 40 - 7 being statistically significant. The most important response was the increase in the protein content (PC) of grain, which in all 40 averages of subtreatments was larger (with high significance, $p<0.01$) than the respective controls. Thus, 99% of all tests (160 observations) showed a net increase in PC over the respective noninoculated plots, even in those cases of a decrease in grain yield. The average increase in PC throughout all trials was 11.1% (S.D. = 8.1%) of the PC in the respective controls. The basis for this effect is not known; at present we are studying whether it correlates with the nitrogen harvest index and the nodule occupancy by inoculated strains. The generalized improvement of the protein content of beans and probably of their nutritional value, would warrant by itself the systematic practice of inoculation in bean cropping - which in turn might be stimulated by introducing grain protein content as a quality criterion in bean commerce.

-Supported by Organización de Estados Americanos (OEA).

Gresshoff, Roth, Stacey & Newton (eds.) Nitrogen Fixation: Achievements and Objectives
Chapman and Hall New York • London 1990

Rhizobium inoculation studies on faba beans, soybeans and haricot beans in Ethiopia

Desta Beyene

Institute of Agricultural Research (IAR).
P.O. Box 2003, Addis Abeba, Ethiopia

In recent years, food legumes are receiving increased attention in the national development program for sustainable agriculture in Ethiopia. They are considered important due to their symbiotic association with rhizobia and consequently their role in reducing the dependence on imported nitrogen fertilizer. Because of the economic significance of nitrogen fixation in grain legumes, increased research emphasis has been placed on the use of *Rhizobium* inoculants. Field and greenhouse experiments have been carried out since 1980 to study the effects of imported *Rhizobium* inoculants on the yields of haricot beans, faba beans and soybeans. There was a general tendency for crop response to inoculation to be influenced by location variables. In the central region of the country (Menagesha and Ada districts), inoculant application significantly increased the grain yields of haricot bean and faba beans. Haricot bean yield was improved with inoculation by 36.5% over the control (1.6 t/ha). The amount of N fixed by haricot beans was also increased from 5 to 18 kg/ha with inoculation. Grain yield of the control treatment (2.4 t/ha) for faba beans was increased by 7 to 20% with inoculation. The magnitude of response became even greater when the inoculant was applied in combination with N and P fertilizers. In the southwestern region of the country (Awassa and Jima districts), the control yield of 1.5 t/ha for soybeans was improved by 15.0 to 22.0% with inoculation. Evidence of positive response to inoculation in different parts of the country prompted interest into the isolation and testing of indigenous *Rhizobium* strains for use as inoculant sources. Systematic collection of nodules from host legumes in the major production regions of the country was then initiated in 1986. Currently, 55 isolates of *Rhizobium leguminosarum*, from *Vicia faba*, 42 isolates of *R. leguminosarum* from *Cicer arietinum*, 23 isolates of *R. leguminosarum* from *Lens culinaris*, 20 isolates of *R. phaseoli* from *Phaseolus vulgaris* and 18 isolates of *R. japonicum* from *Glycine max* have been identified. A preliminary study to evaluate *Rhizobium* strains from faba beans showed that of the eleven indigenous strains tested, seven were found to outyield the exotic strains. More research work is required to characterize the indigenous strains so as to identify the most effective and efficient ones for use in inoculant production.

Gresshoff, Roth, Stacey & Newton (eds.) Nitrogen Fixation: Achievements and Objectives
Chapman and Hall New York • London 1990

Quantification of biological nitrogen fixation associated with sugar cane

Robert M. Boddey, Segundo Urquiaga and Johanna Dobereiner

EMBRAPA-CNPBS, Km 47, Seropédica, 23851, Rio de Janeiro, Brazil

The total nitrogen balance 15N isotope dilution techniques were applied together to estimate the contribution of plant associated biological nitrogen fixation to 7 Brazilian commercial sugar cane varieties and cultivars of Saccharum barberi and S. spontaneum. The plants were grown in soil (0.108% total N) mixed with 15N-labelled organic matter in a concrete tank (20x6x0.6m) in order to restrict root growth to the same soil depth. The grass Brachiaria arrecta was utilized as a non-N_2-fixing control (ex B. radicans, Boddey & Victoria, 1986). At 3 successive annual harvests the commercial varieties CB 45-3 and SP 70-1143 maintained high cane yields (equivalent to more than 200 t ha^{-1}) with no N fertilizer addition, and these two varieties, and the S. spontaneum cultivar Krakatau, accumulated more than 250 kg ha^{-1} of nitrogen at each annual harvest. The soil analyses showed that not more than 220 kg N ha^{-1} was removed from the soil over the three year period, indicating that these cane varieties obtained 180 to 270 kg N ha^{-1} from BNF. The interpretation of the 15N data was complicated by carry-over of nitrogen of higher label from one harvest to the next in the stem bases and roots which were not harvested, but the weighted mean 15N enrichments of the plant material showed that up to 70% of plant nitrogen was obtained from BNF. In ideal conditions of water, and macro and micro-nutrient supply (except nitrogen), it seems that N fertilizer can be dispensed with while still maintaining high yields of certain cane varieties. This possibility holds out the perspective of further increasing the positive energy balance already associated with Brazilian ethanol biofuel program. This program provides the 4 million motor vehicles in this country with ethanol (95% pure), which in contrast to fossil fuels, causes no net release of carbon dioxide into the atmosphere.

Reference

Boddey R.M. & Victoria R.L. (1968) Plant Soil 90,265-292.

Gresshoff, Roth, Stacey & Newton (eds.) Nitrogen Fixation: Achievements and Objectives
Chapman and Hall New York • London 1990

Relationship between nodule development and nitrogenase activity during soybean pod setting

S. Bona*, R. Voltan and G. Mosca

Institute of Agronomy, University of Padova, Via Gradenigo 6, 35131 Padova, Italy

Nitrogenase activity during pod setting of soybean variety "Hodgson 78" was tested in a growth chamber experiment to evaluate its relationship with nodule number, nodule weight and specific nitrogenase activity. Seeds were normally inoculated using strain (INRA-G49) and plants were grown in a N-free nutrient solution. A water stress incidentally occurred at the R4 stage . Total nitrogenase activity was affected by the development stage; it reached its maximum at R4 and it rapidly decreased in the last part of the cycle. The number of nodules did not change significantly during the reproductive phase while the total nodule dry weight increased (from 0.75 to 1.25 g) because of the increase in the average dry weight per nodule.

A multiple regression analysis of total nitrogenase activity per plant versus total nodule dry weight, nodule average dry weight, number of nodules and specific nitrogenase activity was carried out. The results of this analysis showed that the main factors affecting total nitrogenase activity were the specific activity and the nodule mass ($P \leq 0.001$). Number of nodules although significant in the multiple regression model ($P \leq 0.05$) did not have an important role in determining the total nitrogenase activity because of the low value of its partial regression coefficient (B=0.2). The coefficient relative to the nodule average dry weight was significantly different only in the sample taken after the water stress and was negatively related to the total nitrogenase activity ($P \leq 0.05$). This showed that, under water stress conditions, the soybean plant that possesses a large number of small nodules is able to guarantee a good rate of nitrogen fixation. Owing to the strict relationship that exists among the specific and the total nitrogenase activity the weighing of the nodules cannot be a means for quantifying the fixation rate.

Gresshoff, Roth, Stacey & Newton (eds.) Nitrogen Fixation: Achievements and Objectives
Chapman and Hall New York • London 1990

Legume nodules are efficient scavengers of molybdenum for N_2-fixation

S.J. Brodrick, F. Amijee and K.E. Giller

Department of Biochemistry and Biological Sciences, Wye College, University of London, Wye, Ashford, Kent TN25 5AH, UK

Introduction: Molybdenum is thought to be of intermediate mobility in the phloem and this may limit N_2-fixation by restricting the supply of Mo to the nodules of legumes. The seed can be an important source of Mo and may make significant contribution to the nutrition of the host-Rhizobium symbiosis. The aim of this work was to observe how molybdenum was distributed in plants grown from seed of large and small molybdenum content with and without an external supply of Mo, and the effects of these on N_2-fixation.

Materials and Methods: Shoot, root and nodule dry weight, and Mo content by nitric acid digestion (1) and graphite furnace atomic absorption spectrophotometry were measured at sequential harvests over the first 48 days of growth on Phaseolus vulgaris cv BAT 1297 receiving -N nutrient solution with or without Mo. All plants were inoculated with Rhizobium strain CIAT 899. Acetylene reduction activity (ARA) was measured on detached roots and nodules using a flow-through gas system (2).

Results: Plants grown from seed of large molybdenum content (1.5-2.1 μg $seed^{-1}$) used the seed reserve as the major source of Mo for the initiation, exponential increase and maximum rate phases of N_2-fixation irrespective of external molybdenum supply. There was no difference in ARA between +Mo and -Mo treatments, showing that the large seed content was adequate to establish and maintain N_2-fixation. Plants grown from seed of small molybdenum content (0.09-0.4 ug $seed^{-1}$) without an external source of Mo had smaller shoot and root dry weight at 38 days after planting (DAP), smaller Mo content at all harvests, and smaller rates of ARA at 20DAP. Molybdenum in the shoot was translocated to the roots and nodules in plants grown from both large and small molybdenum content seed without Mo in the feeding solution. Molybdenum applied to the unifoliolate leaf was readily absorbed and translocated to shoots, roots and nodules. The most probable pathway for this movement is the phloem.

1 Williams E V (1978). J Food Technology 13, 367-384
2 Minchin F R et al. (1982). J Exp Bot 34, 641-649

Gresshoff, Roth, Stacey & Newton (eds.) Nitrogen Fixation: Achievements and Objectives
Chapman and Hall New York • London 1990

Effect of green manures on rice crop at Tarhara during 1989-90

Shankar L. Chaudhary

Agriculture Research Center, Tarhara Sunsari, Nepal.

Nepal is predominantly an agricultural country where mineral fertilizer is very scarce and costly due to the lack of a mineral fertilizer production facility. A cheaper alternative is to harness locally available green manures, such as *Sesbania aculeata, Sesbania rostrata, Azolla pinnata*, and *Vigna radiata*. These green manures are fast growing and fix nitrogen to the level of 40-100 KgN/ha within 60-70 days. These green manure species easily decompose and can increase rice grain yield 25-40% above the control. These species also contain high nitrogen levels (3.5-5.0%) in their leaves, are easy to propagate, and resist adverse environmental conditions. Therefore, green manure species represent one of the best alternatives to chemical fertilizer in rice farming systems in the tropics.

Gresshoff, Roth, Stacey & Newton (eds.) Nitrogen Fixation: Achievements and Objectives
Chapman and Hall New York • London 1990

Screening of wheat genotypes for associative nitrogen fixation

M.I. Chumakov

Institute of Biochemistry and Physiology of Plants and Microorganisms, USSR Academy of Sciences, pr. Enthuziastov 13, Saratov, 410015, USSR.

The N_2-fixation supportive trait (nis) of 78 *Triticum* L. cultivars with different ploidy levels was investigated. Soil bacteria population isolated from dark-brown soils was used for inoculation. Nitrogen-fixation was determined by acetylene reduction method by using "spermosphere" model (Thomas-Bauzon et al., 1982). Nis-activity of tetraploid wheats (13 varieties) was 38% higher than that of diploid wheats (6 varieties). Nis-activity level of hexaploid wheat group (10 varieties) was 59% higher than that of diploid wheat groups (6 varieties). The highest activity among 9 tested wheat species was detected in *T. aestivum*. Nis-activity of *T. timonovum* (2n=56) was 38% higher than that of diploid wheats (with significance of 99.88%).

The old selection varieties (Rusak, Poltavka) with hexaploid chromosome type possess the maximum nitrogenase activity. Nis-activity variation coefficients were significantly higher within cultivars than within species. Nis-activity differences between winter and spring wheat cultivars (at P 0.05) were not significant. Correlation between N_2-fixation supportive trait and N-accumulation by various wheat cultivars was not found.

Nis-activity screening of 21 monosomic lines of wheat Saratovskaya 46 (ABD genome) showed the trait inheritance to have polygenic character. The following order A D B can be made up according to the degree by which different genomes affect N_2-fixation activity. The most significant reduction of nis-activity was registered in the 6D line. Individual plants, incapable of supporting N_2 fixation by the soil bacteria population ("0-plants"), were noted within the same wheat cultivar.

Thomas-Bauzon, B., Weinhard, P., Villencourt, P., Balandreau, J. (1982). Can. J. Microbiol., 28: 922-928.

Gresshoff, Roth, Stacey & Newton (eds.) Nitrogen Fixation: Achievements and Objectives
Chapman and Hall New York • London 1990

Nodulation and nitrogen-fixation improvement in soybean over-inoculated with *Bradyrhizobium japonicum* at the three-node stage

G. Ciafardini, L. Busi, G. Marinelli and L.M.R. Arena

Soil Microbiology Laboratory, Istituto Sperimentale per le Colture Industriali,Via di Corticella n.133, 40129-Bologna, Italy.

INTRODUCTION

Symbiotic N_2 fixation in soybean occurs in root nodules produced following root colonization by *Bradyrhizobium japonicum*. In clay soils with low populations of *B. japonicum* seed inoculation of soybean often results in a small number of nodules distributed mainly on the primary root, due to poor soil mobility of the inoculated simbiont (1). In previous work, we increased nodulation of seed-inoculated soybean by cover inoculating at the V_3 stage with *B. japonicum* borne in the irrigation water (2). This study examined the application of cover inoculation on soybean grown as a secondary crop.

MATERIALS AND METHODS

The experiment entailed a randomized-block design with four repetitions, employing the Hodgson soybean cultivar grown as a secondary crop after barley in clay loam soil (Haploxeroll aridic) free of *B. japonicum*. Inoculation treatments were i) seed inoculation with peat inoculum; ii) cover inoculation with liquid inoculum at the V_3 stage; iii) seed+cover inoculation; iv) uninoculated control.

RESULTS

Cover inoculation of soybean previously seed-inoculated with peat inoculum showed a 2.4-fold increase in the number of nodules per plant compared to seed inoculation alone and a 1.7-fold increase compared to cover inoculation alone. This increased nodulation produced by the combination of seed+cover inoculation significantly increased the ureide content as well as the total nitrogen in stems and leaves of the Hodgson soybean cultivar.

REFERENCES

1. Ciafardini et al.(1984) L'Informatore Agrario, 19:63-65.
2. Ciafardini et al.(1987) Agronomy Journal, 79:645-648.

Gresshoff, Roth, Stacey & Newton (eds.) Nitrogen Fixation: Achievements and Objectives
Chapman and Hall New York • London 1990

Inoculation by *Azospirillum brasilense* and *Rhizobium ciceri* of chickpea (*Cicer arietinum*)

Maddalena Del Gallo and Paola Fabbri

Dept.of Agrobiotechnology, ENEA-Casaccia, 00060 Rome, Italy.

According to data reported in the literature, inoculation with Azospirillum spp. has a positive effect on the Rhizobium-legume symbiosis (1).

Two Italian cultivars of chickpea, "Calia" (spring cv.) and "Califfo" (winter cv.), were inoculated with R. ciceri USDA 3379 (from Mexico)and Azospirillum brasilense Cd in pots containing a mixuture of sand and perlite (2:1 vol). Controls were carried out with plants either uninoculated or inoculated with a single strain. Plants were watered weekly with a Jensen N-free solution and analyzed after 30 day growth. In experiment 2, Azospirillum was also inoculated on "Califfo" cv. either 66 hrs before or together with Rhizobium and the plants sampled after 22 and 52 days growth.

The results of experiment 1 showed a statistically significant benefit of Azospirillum on both varieties inoculated with Rhizobium. The highest increase was in the number of nodules/ plant (about 34% in "Califfo" and 23% in "Calia") and in nodule dry weight (70% and 53%, respectively) in comparison to plants inoculated with Rhizobium alone. Root and shoot dry weight increased 26% and 46% in "Califfo" and 27% and 21% in "Calia"; no effect was observed on acetylene-reduction activity (ARA). The mM of C_2H_4 reduced per plant was the same with Rhizobium alone or with both Rhizobium and Azospirillum. The winter cultivar "Califfo" showed a better response. In experiment 2, Azospirillum enhanced all parameters when inoculated before Rhizobium. However, after 52 days, the ARA, although higher on a per plant basis, is lower for nitrogenase activity per nodule and per g of nodule.

In conclusion, Azospirillum stimulated early plant development. On root hair proliferation,it may increase Rhizobium attack sites and, consequently, the number of nodules.

1. Yahlom et al. (1987) Can. J. Microbiol. **33**, 510-514.

Gresshoff, Roth, Stacey & Newton (eds.) Nitrogen Fixation: Achievements and Objectives
Chapman and Hall New York • London 1990

Effect of *Azospirillum* inoculation on the amino acid content in the roots and shoots of wheat, barley, peas and lupin

Mehreshan T. El Mokadem and Anssaf M. Badawi

Botany Department, Women's College and Faculty of Education, Ain Shams University, Heliopolic, Cairo, Egypt

This experiment was performed to study effects of Azospirillum brasilense 245 inoculation on the protein-bound amino acid concentration in wheat, barley, peas and lupin at 30, 60 and 90 days after planting. Cereals were inoculated with Azospirillum, legumes were inoculated with either Rhizobium or with Azospirillum or with a mixture of both inoculants; uninoculated plants were used as control. Amino acids were extracted and measured in both roots and shoots of plants. Determinations were compared with uninoculated control plants of similar growth stages.

Results indicate that single inoculation with both diazotrophic cultures had favourable influence on the concentration of the bound amino acids present in roots and shoots of peas and lupin compared to uninoculated control during plant growth. The effect of mixed inoculants was much more pronounced than single inoculation of either diazotrophs. Inoculation did not significantly affect the quality of the different amino acids in roots and shoots. Aspartic and glutamic acids were present in considerably higher amounts than other amino acids in peas and lupin tissues.

The presence of Azospirillum brasilense has shown to have positive influence on the protein-bound amino acid content of inoculated wheat and barley; the highest increase in quantity was observed at 30 days old in roots and shoots of both plants. There was a marked change in concentration increase in each amino acid as a result of Azospirillum inoculation throughout the experimental period.

Gresshoff, Roth, Stacey & Newton (eds.) Nitrogen Fixation: Achievements and Objectives
Chapman and Hall New York • London 1990

Nitrogen contribution of *Azolla* and *Sesbania rostrata* to rice

N. ESIOBU[1], K. MULONGOY[2], and C. VAN HOVE[3]

[1]University of Lagos, Akoka, Nigeria, [2]International Institute of Tropical Agriculture, Ibadan, Nigeria, and [3]Université Catholique de Louvain-la-Neuve, Belgium.

Azolla spp. and the stem-nodulated *Sesbania rostrata* thrive in aquatic and swamp environments. Diazotrophy has been unequivocally demonstrated in both cases. They can thus constitute an excellent source of N for lowland rice. Using the ^{15}N dilution technique, we compared their N-fertilizer value and investigated management practices that would enhance their efficiency as green manure in rice production.

For labelling the green manures with ^{15}N, *Azolla* was grown for 14 days in 2/5 Hoagland supplemented with 100 ppm (10% ^{15}N a.e.) ammonium sulphate N, and *S. rostrata* was grown for 30 - 60 days in a N-poor soil supplemented with 10% a.e. ^{15}N ammonium sulphate. The experiment was carried out in the greenhouse using a split design with 3 replications, 4 N sources (*Azolla caroliniana, A. pinnata, Sesbania rostrata* and ammonium sulphate) as main treatments, and 2 incorporation times (at transplanting, and at panicle stage) as subtreatments. A no-N control was included for comparison.

Both species of *Azolla, Sesbania rostrata,* and ammonium sulphate increased rice grain yield over control by 41, 82 and 58% respectively. The highest rice yields were obtained with *S. rostrata*. In fact, differences in yields were related to amounts of N applied ($R = 0.87$ significant at $P = 0.002$). The two *Azolla* species had similar effects on rice growth and yield. Proportion of added fertilizer N assimilated by rice into grains averaged 44% for *Azolla*, 47% for *S. rostrata* and 37% for ammonium sulphate. Nitrogen was more effeciently utilized and recovered in grains when applied at panicle (53% in average) than at transplanting (averaging 31%). The fertilizer value of the green manures was equivalent to that of ammonium sulphate.

Gresshoff, Roth, Stacey & Newton (eds.) Nitrogen Fixation: Achievements and Objectives
Chapman and Hall New York • London 1990

Dinitrogen fixation and apparent nitrogen transfer from legume to grass

G. H. Heichel* and K. I. Henjum

*Dep. of Agronomy, Univ. of Illinois, 1102 S. Goodwin Ave., Urbana, IL 61801 (formerly USDA-ARS, St. Paul, MN), and Dep. of Agronomy, Univ. of Minnesota, St. Paul, MN 55108, USA.

The extent of dinitrogen (N_2) fixation in legume-grass mixtures grown in the north central U.S.A. and the importance of legume to grass nitrogen (N) transfer are not well understood. The objective of this 4-yr field experiment was to determine seasonal N_2 fixation and N transfer by stable isotope methods in swards of alfalfa (*Medicago sativa* L.), birdsfoot trefoil (*Lotus corniculatus* L.), red clover (*Trifolium pratense* L.) and ladino white clover (*Trifolium repens* L.) with reed canarygrass (*Phalaris arundinacea* L.). Alfalfa in mixture with grass fixed the most N_2 per season (82 to 254 kg N/ha) and ladino white clover the least (1 to 20 kg N/ha). For all mixtures except birdsfoot trefoil-grass, N_2 fixation was greatest during the second year of the stands. The proportion of legume N from symbiotic fixation increased progressively from Year 1 through 4 and averaged over 93% for all legumes in Year 4. Legume to grass N transfer was greatest in Year 2 of the stands and averaged about 36% of grass N across all mixtures. On a land area basis, N transfer in Year 2 ranged from 29 to 53 kg N/ha and was greatest for the ladino white clover-grass mixture. The decline in N transfer in Years 3 and 4 was associated with a loss in sward legume content, reduced whole plant dry matter accumulation in legume and grass, and large losses in root plus crown dry matter accumulation by grass in Year 4. The results indicate that forage legumes in mixture with grass are virtually self-sufficient for N and can concurrently transfer appreciable N to the companion grass.

Gresshoff, Roth, Stacey & Newton (eds.) Nitrogen Fixation: Achievements and Objectives
Chapman and Hall New York • London 1990

Reliability and limitations of repeated acetylene reduction measurements

JOHN IMSANDE

Departments of Genetics and Agronomy, Iowa State University, Ames, Iowa 50011, U.S.A.

To gain additional insight into the potential usefulness and/or limitations of the acetylene reduction (AR) assay in estimating N_2 fixation rates, well nodulated soybean [Glycine max. (L.) Merr.] plants were grown hydroponically. At beginning pod fill, fertilizer N was withdrawn from the nutrient solution, and the growing plants were subjected to the nondestructive AR assay twice weekly for five weeks. When the experimental plants were shifted to nitrogen-free medium, control plants were examined for both fresh weight and total nitrogen content. A linear relationship ($r = 0.95$) was found between these two parameters. Because the fresh weight of each experimental plant was known, its N content at beginning pod fill could be calculated.
At physiological maturity plants were dried, weighed, and subjected to Kjeldahl digestion. Nitrogen acquired during pod fill was measured by Kjeldahl analysis, and this value was compared to nitrogen accumulation predicted by nitrogenase activity obtained from the AR assay.

Assuming a theoretical electron allocation coefficient of 0.75, the average AR assay underestimated nitrogen acquisition by approximately 57%. To determine whether the AR assay procedure itself, which included repeated exposure to the plant hormone ethylene, might alter plant growth rate and/or rate of N acquisition, N acquisition was measured by Kjeldahl digestion in the absence of the AR assay. The AR assay altered neither plant growth nor N acquisition.

Nitrogen determination by Kjeldahl digestion showed that plants reducing 0 to 5 μmol of acetylene h^{-1} $plant^{-1}$ underestimated N acquisition by approximately 80% whereas plants reducing 20 to 25 μmol acetylene h^{-1} $plant^{-1}$ underestimated N accumulation by approximately 40%. When the rate of acetylene reduction was approximately 50 μmol h^{-1} $plant^{-1}$, the AR assay predicted approximately 80% as much N fixed as was found by Kjeldahl digestion. Thus, for the 168 plants examined, the reliability of AR assay increased dramatically with increasing rates of measured AR. Indeed, the correlation between accumulated AR activity measurements and acquired nitrogen was approximately 0.70 with $P<0.0001$.

Gresshoff, Roth, Stacey & Newton (eds.) Nitrogen Fixation: Achievements and Objectives
Chapman and Hall New York • London 1990

Influences of benlate on the growth and nitrogen fixation in three species of *Azolla*

S. Kannaiyan

Department of Agricultural Microbiology, Tamil Nadu Agricultural University, Coimbatore-641 003, Tamil Nadu, India.

Azolla is an aquatic water fern that fixes atmospheric nitrogen in symbiotic association with a cyanobacterium, Anabaena azollae. The agronomic potential of Azolla is quite significant particularly for rice when applied as a biofertilizer for increasing rice yields. Azolla is widely used as a dual crop with rice and it can contribute 40-60 kg N/ha per rice crop. Azolla fronds are commonly infected by a soil-borne pathogen, Rhizoctonia solani. Benlate, a systemic fungicide, was used for treating three species of Azolla with a view to assess its effect on growth, chlorophyll synthesis and nitrogen-fixing potential besides checking the R. solani infection.

Benlate at 2.5 ppm induced the growth of A. microphylla-BR-GI, A. caroliniana-WT-V and Azolla sp.-ST-SI. The growth-rate induction in Azolla sp-ST-SI due to the Benlate treatment was significant. The culture of Azolla sp.-ST-SI recorded the maximum nitrogenase activity followed by A. microphylla-BR-GI due to Benlate treatment. Among the three species of Azolla tested, A. caroliniana-WTV and Azolla sp.-ST-SI have significantly increased total chlorophyll content due to Benlate treatment. Benlate treatment reduced the nitrogen content in A. microphylla-BR-Gi and Azolla sp.-ST-SI, while the carbon content was little affected.

Gresshoff, Roth, Stacey & Newton (eds.) Nitrogen Fixation: Achievements and Objectives
Chapman and Hall New York • London 1990

Utilization of biofertilizers compensates for yield reduction of vaccinated tomato plants

S.I. Kotb[1], F.M. Hashem[1], M.S. Montasser[2], & J.M. Kaper[3]

[1]Dept. of Agronomy, University of Maryland, College Park, MD 20742 USA
[2]Dept. of Plant Pathology, University of El Menia, Egypt.
[3]Microbiology and Plant Pathology Laboratory, BARC-W, USDA, MD 20705 USA.

The use of satellite viral RNA [cucumber mosaic virus (CMV) associated RNA 5 (CARNA 5)] (1) as a biological control agent against severe CMV strains on tomato (Lycopersicon esculentum cv. UC82B) plants was found to cause yield reduction of about 15-20% of preinoculated or "vaccinated" plants.

To overcome this problem, a greenhouse experiment was conducted using a mixture culture of nitrogen fixing bacteria of azospirillum (Azospirillum brasiliensis N040, Azospirillum brasiliensis SP7 and Azospirillum lipoferum MRB16) and one strain of blue-green algae [Anabaena oryzae (Fritsch)]. Treatments were control (unvaccinated) and vaccinated plants in combination with algae and/or azospirillum. Results showed that inoculation with azospirillum and/or algae gave significant differences in number and fresh weight of tomato fruits and plants' dry weight when compared to the control or vaccinated plants. Growth-stimulating compounds and/or nitrogen-fixation are the reasons for yield increase (2). Polyacrylamide gel electrophoresis analysis of total nucleic acid extractions revealed that biofertilization did not affect the accumulation of viral satellite RNA required for plant protection.

1. White, J.L. & Kaper, J.M. (1989) J. of Virological Methods, 23, 83-94.
2. Rodgers, G.A., Bergman, B., Henriksson, E. & Udris, H. (1979) Plant and Soil, 52, 99-107.

Gresshoff, Roth, Stacey & Newton (eds.) Nitrogen Fixation: Achievements and Objectives
Chapman and Hall New York • London 1990

'Incomplete symbiosis' of Azolla

C. Lin, C.C. Liu and I. Wantanabe*

Azolla Res. Centre, Fuijan Academy of Agric. Sci., Fuzhou, 350013, Fuijan, P.R. China, *Soil Microbiol. Dept., Intern. Rice Research Institute, P.O. Box 933, Manila, Philippines.

Azolla is a small aquatic fern with a symbiotic association with the N_2-fixing cyanobacterium, Anabaena azollae Strasburger. The host and symbionts grow synchronously. This paper describes an incomplete symbiosis with Azolla due to infection by free-living cyanobacteria.

Nostoc spp. and Anabaena azollae, isolated by Newton and Bai, were cultured in BG-11 and broken into single or oligo-cells by sonication. At a concentration of 4-5 x 10^4 cells/ml, they were inoculated onto decapitated megaspores of Azolla microphylla Kaulfuss during the greening stage of the germinating megasporocarp. Sporelings were cultivated in IRRI's culture solution containing 75 mg/l ammonium nitrate.

Anabaena-free Azolla plants, derived from decapitated megasporocarps, were infected with either Nostoc spp. or Anabaena azollae. Some had algal filaments inside the leaf cavity of the 1st through the 6th leaf. Most algal cells disappeared as sporelings matured, however, a few infected Azolla plants survived on N-containing culture solution. In these, some Nostoc or Anabaena filaments entered the leaf cavities, but most stayed on the surface and formed aggregates connecting one leaf to another. Some Anabaena cells settled at the apical portion in synchronization with the host's growth. Azolla thus infected failed to grow in N-free culture solution. Acetylene-reduction activity (ARA) showed that these infected Azolla plants fix N_2 but only about 20% that of normal symbiotic Azolla plants.

Immunological studies showed that Anabaena filaments from leaf cavities of these Azolla plants reacted positively to FA prepared from Newton's or Bai's isolates, but negatively against FA from the original symbiotic Anabaena azollae. This seems to indicate that, even though Anabaena cells were able to enter leaf cavities, the symbiotic relationship is not fully reestablished, a condition we call "Incomplete symbiosis".

Gresshoff, Roth, Stacey & Newton (eds.) Nitrogen Fixation: Achievements and Objectives
Chapman and Hall New York • London 1990

Proteins of lupin seeds binding molybdenum, tungsten and radionuclides from the Chernobyl fall-out

N.P. L'vov, K.L. Kalakoutsky and A.I. Zabolotny

A.N. Bach Institute of Biochemistry, USSR Academy of Science, Moscow II707I, Leninsky pr.33, USSR and Institute of Experimental Botany, Byelorussian Academy of Science, Minsk 220733, Skoriny str.27, USSR.

INTRODUCTION. Legumes are known to accumulate significant amounts of Mo (1). The biochemical mechanism of Mo-accumulation (Mo-storage proteins) has been studied to some extent in microorganisms (2,3) but not in legumes, although by the end of the onthogenesis, they accumulate up to 80% of the total plant Mo in seeds (4).

METHODS. We have developed a highly sensitive procedure for detecting Mo(W)-binding proteins by PAGE using an *in vitro* labelling procedure for lupin seed extract with the radioactive isotope 185-W (see ([5] for details).

RESULTS AND DISCUSSION. Two proteins (protein I and protein II) and a low-molecular-weight "yellow" fraction were detected as binding 185-W *in vitro*. We have shown that 185-W and Mo, when introduced into seeds *in vivo* (addition via a nutrient solution), are also bound to the proteins and the "yellow" fraction. Moreover, the two proteins and the fraction also bind the Chernobyl fall-out radionuclides, using seeds obtained from plants grown within a 30-km zone of the Chernobyl nuclear power station. Protein I, when tightly binding microelements, undergoes proteolytic degradation to give a product which preserves the ability to hold microelements. Protein II is characterized by highly labile binding with 185-W. These proteins and the low-molecular-weight fraction are not specific for Mo or W, but are also able to bind other microelements as shown for the Chernobyl fall-out radionuclides.

REFERENCES

1. L'vov, N.P. (1989) *Molybdenum in Nitrogen Assimilation of Plants and Microorganisms*, (Russ.) Nauka, Moscow.
2. Pienkos, P.T., Brill, W.J. (1981) J. Bact. **145**, 743-751.
3. Elliott, B.K., et al. (1979) Annual Rev. Biochem. **48**, 387-418.
4. L'vov, N.P. (1988) in *Nitrogen Fixation: 100 Years After*, eds. H. Bothe, et al. (Gustav Fischer, Stuttgart & New York), p. 659.
5. Kalakousky, K.L., et al. (1990) FEBS Lett., in press.

Gresshoff, Roth, Stacey & Newton (eds.) Nitrogen Fixation: Achievements and Objectives
Chapman and Hall New York • London 1990

Persistent effects on biological nitrogen-fixing microorganisms by the herbicide 2,4-D

Anna Mårtensson

Department of Soil Sciences, Division of Plant Nutrition, Swedish University of Agricultural Sciences, P. O. Box 7014, S-750 07 Uppsala, Sweden.

Introduction

Caution must be observed regarding the possible occurrence of hazardous effects on soil microbes owing to the application of pesticides. In order to investigate persistent effects on biological nitrogen fixing micro organisms by pesticides the following study was conducted.

Experimental background

In 1976, an accidiental aerial application of approx. 40 kg 2,4-D was made in Potmäk, Sweden. The affected area is situated in an old spruce stand (30·60 m). Three years later, 50% of the 2,4-D remained in the 0-5 cm soil layer (100%=day of discovery). An adaptation of the soil microbes had taken place and the capacity for 2,4-D decomposition had improved. Soil respiration was reduced by 60-90% cf. control. Thirteen years later, the free-living diazotrophs bacteria was studied by glucose additions to soil samples and determining the acetylene reduction activity (ARA). Establishment and ARA of cyanobacteria were studied by incubating soil samples under permanent light and measuring ARA at regular intervals. Plant-dilution tests for investigating red clover rhizobia were also carried out.

Results and conclusions

Free-living diazotrophs were adversely affected. No ARA was detectable within 17 m of the deposition centre and within 72 m ARA was significantly reduced. Cyanobacteria established themselves except within 10 m of deposition. ARA was retarded at plots close to deposition, but this was later overcome. No legume bacteria were found. It is suggested that diazotrophs, particularly free-living ones, are useful indicator organisms when investigating the effects of various environmental pollutants. The presence of measureable effects on soil biological parameters by accidental additions of pesticides warrants the utmost caution, considering the long-term persistence of the observed adverse effects.

Gresshoff, Roth, Stacey & Newton (eds.) Nitrogen Fixation: Achievements and Objectives
Chapman and Hall New York • London 1990

Late N-fertilization effects on nodulation, N_2-fixation and yield in determinate and indeterminate soybeans (*Glycine max* L. Merr.)

F. Miceli, M. Martin and G. Mosca.

Istituto produzione vegetale, Udine University, Piazzale Kolbe, 33100 Udine (Italy).

INTRODUCTION

The eco-physiology of the *B. japonicum* - soybean symbiosis is better known in Maturity Group I indeterminate (ID) soybeans if compared with that involving determinate (DT) ones. Field information on nodule persistance and N_2-fixation indices can help in explaining the yield performance of new DT varieties.

MATERIALS AND METHODS

A field experiment with Nod^+ soybean varieties and a Nod^- line, all seed-inoculated with commercial 'Rhizosarb' (Lipha), was performed in Udine on a typic xerofluvent soil in two years. A standard ID variety (Hodgson78) and a DT line (INRA 172/16), both of Maturity Group I, were tested in 1987, while in 1989 a second DT genotype (Sioux) was added. The Nod^- line (L65) was always used as non-fixing, reference crop for calculations of the % Ndff by the difference method (Ruschel et al., 1979; Vasilas and Ham, 1984). Two mineral N treatments were imposed: none (N0) and 150 Kg N ha^{-1} applied at R4 growth stage (N150). The objectives were: i) to estimate the nitrogen amount obtained from cumulated N_2-fixation of the crops; ii) to compare nodulation and C_2H_2 reduction activities (ARA) persisting on late growth stages in a small number of ID and DT soybean genotypes.

RESULTS

In 1987, about 60% and 43% of the biomass total N at harvest came from N_2-fixation, respectively in N0 and N150 treatments. In the latter, about 32 % of plant N came from soil and 25 % from fertilizer. In 1989 the % Ndff was similar to that of 1987 in N0 plants, while in N150 ones it was lower, probably because of the higher soil N status. Total nodule mass per plant on R7 was clearly depressed by N150, but at a lesser extent in DT varieties. In 1987, INRA seemed to perform a higher late ARA with N150, but in 1989 this treatment lowered the specific ARA in all the plant-bacterium combinations. A highly significant multiple regression (R^2= .93) of Ndff on nodule mass per plant and specific ARA at R7 was found. Seed yield and shoot biomass were influenced by the soybean variety, while late N fertilization did not affect yield.

1. Ruschel, A.P., Vose, P.B., Victoria, R.L. and Salati, E. (1979). Plant Soil 53:513-525.
2. Vasilas, B.L. and Ham, G.E. (1984). Agron.J. 76:759-764.

Gresshoff, Roth, Stacey & Newton (eds.) Nitrogen Fixation: Achievements and Objectives
Chapman and Hall New York • London 1990

Nitrogen fixation by leguminous trees: Fixed nitrogen availability in mixed-crop systems and its effects on soil nitrogen capital

D. O. Nyamai

Agroforestry Research Department, Kenya Forestry Research Institute, P. O. Box 20412, Nairobi, Kenya.

The amount of N fixed by *Leucaena*, *Calliandra*, *Gliricidia* and *Prosopis* and its effects on the yield and growth of Millet (*Eleusine coracana* Lam.) under different management systems was determined by the nitrogen accumulation technique (the 'difference method'). Significant differences occurred in the amount of N-fixed among the tree species studied. *Leucaena lecocephala* fixed the greatest amount, followed by *Calliandra callothyrsus*, *Prosopis juliflora* and *Gliricidia sepium* in this order. All the species fixed more N as a pure crop than as mixed crops.

The yield and growth of millet differed significantly depending on the treatment. Millet mixed with trees, particularly *Leucaena*, produced more grain, while pure millet gave significantly lower grain yield compared with any other mixed crops. Similarly, *Leucaena*, in a mixed crop system, in which prunings were applied/retained, resulted in 10% higher foliar Millet N than mixed crops where pruning was carried out but the prunings were removed.

The results also showed that the total soil N was greatly affected by the various treatments. Treatments where pruning and subsequent return of manure was initiated had significantly higher soil N levels than those which were pruned but the prunings removed as well as unmanured pure millet crop.

The better performance of millet in manured treatments was principally caused by the relatively higher soil and millet foliar N content. These relationships were further confirmed by the significant linear relationships and positive correlations between millet grain yield with soil and foliar N contents [$Y=^{-}2.3+23X$ ($R^2=0.64$) and $Y= -4.0+4.36X$ ($R^2=0.5$)] respectively.

Gresshoff, Roth, Stacey & Newton (eds.) Nitrogen Fixation: Achievements and Objectives
Chapman and Hall New York • London 1990

The kinetics of nitrogen fixation and microbial growth in soils

N.S. Panikov and I.K. Kravchenko

Institute of Microbiology, USSR Academy of Sciences, Moscow, 117811 USSR.

Fresh samples of arable gray forest and alluvial meadow soils were incubated with glucose and cellulose under controlled laboratory conditions (25°C, water content 60% from WHC). During incubation, the following parameters were measured: CO_2-evolution rate by GC, residual glucose with glucooxidase, residual cellulose by anthrone method, nitrogenase activity (NA) by acetylene reduction, the content of NO_3^- and NH_4^+ and microbial biomass by physiological methods.

NA was observed only transiently when NO_3^- and NH_4^+ content dropped below 5 ug N/g of soil and respiration activity reached the maximum values. In the glucose-amended soil CO_2 evolution delayed the glucose consumption. In cellulose-amended soil maximum NA and respiration took place when no more than 50% of added substrate were decomposed (15-20 days of incubation). Afterwards the cellulose decay proceeded slowly without concomitant nitrogen fixation. Air-drying and N-amendment stimulated cellulose decomposition by 2-5 times.

To describe experimental curves several hypothetical mechanisms in the form of mathematical models were tested. Adequate agreement was achieved when the following mechanisms were taken into account: (i) glucose is catabolized via fermentation (85%) and oxidative consumption (15%); (ii) oxidation of fermentation products is inhibited by glucose (Crabtree-effect); (iii) potential NA is dependent on the physiological state of cells and controlled by NH_3 and NH_4 content of the medium; (iv) the stoichiometric ratio between energy consumed and N_2-fixed is not constant and is related to O_2 partial pressure; (v) some fractions of N and fermentation products are available for immediate microbial uptake because of steric effects (localization of sterile or anaerobic microsites of soil); (vi) cellulose decomposition is carried out by microbial community which consists of hydrolytics (produce cellulases), diazotrophs and olygotrophs (scavenge monomeric compounds escaping from decomposition zone); and (vii) hydrolytic microorganisms undergo the substrate-accelerated death caused by N-starvation under conditions of high monomer concentrations and low O_2 partial pressure.

Gresshoff, Roth, Stacey & Newton (eds.) Nitrogen Fixation: Achievements and Objectives
Chapman and Hall New York • London 1990

Non-sulfur purple bacteria as associative diazotrophs

T.V. Red'kina

Institute of Microbiology, USSR Academy of Sciences, Moscow, USSR.

Phototrophic non-sulfur purple bacteria were shown to be associative nitrogen fixers in the rice rhizosphere. A binary association that includes two new species of nitrogen-fixing non-sulfur bacteria (cultures 1 and 11) was isolated. Both species of bacteria can grow and fix nitrogen under heterotrophic conditions in the dark. Nitrogen fixation under phototrophic conditions goes most intensively and is accompanied by hydrogen evolution. The combination of phototrophs utilizes a wide spectrum of carbon sources, while each culture taken separately assimilates a more limited range of carbon sources. The maximum productivity of nitrogen fixation was found on malate and succinate.

High nitrogenase activity in the presence of high doses of mineral nitrogen was observed for both cultures. There were some differences between cultures in utilization of nitrogen sources in the light and in the dark. The growth activating effect of these bacteria on plants was shown as well as their ability to produce antibiotic substances which inhibit the growth of phytopathogenic fungi.

Gresshoff, Roth, Stacey & Newton (eds.) Nitrogen Fixation: Achievements and Objectives
Chapman and Hall New York • London 1990

Infection of sugar cane by *Acetobacter diazotrophicus* is aided by VA mycorrhyzae

V.M. Reis, M.A. de Paula, S. Urquiaga and J. Dobereiner

EMBRAPA-CNPBS, Km 47, Seropédica, 23851, Rio de Janeiro, Brazil.

In recent N balance and 15N isotope dilution experiments certain sugar cane genotypes were shown to be able to incorporate more than 200 kg $N.h^{-1}yr^{-1}$ derived from biological fixation in three consecutive cuts (Boddey et al 1990). Acetobacter diazotrophicus was isolated from roots, stems and leaves of sugar cane (Cavalcante & Dobereiner 1988, Gillis et al. 1989) but not from soil. In order to confirm the endophytic habit of this diazotroph, experiments were conducted with micropropagated sterile cane plants. Roots of such plants inoculated in vitro were infected by A. diazotrophicus and the organism could be reisolated after 30 days from the aerial parts in high numbers (10^5 $cells.g^{-1}$ fresh wt.). Inoculation of such plants with spores of the VAM fungus Glomus clarum containing A. diazotrophicus resulted in 20 times higher numbers of the bacteria in tops. Micropropagated cane plants inoculated at transplanting into non-sterile soil with the bacteria, or with G. clarum spores containing it, also resulted in root infection and translocation to the tops. A. diazotrophicus inoculation enhanced % VAM colonization of roots and specially the formation of spores within roots.

References

Boddey R.M., Urquiaga S. and Dobereiner J. (1990) See the poster at this congress.
Cavalcante V.A. and Dobereiner J. (1988) Plant and Soil 108:23-31
Gillis N., Kerters K., Hoste B., Janssens D., Kroppnstedt R.M. Stephan M.P., Teixeira K.R.S., Dobereiner J. and Deley J. (1989) Int. J. Syst. Bacteriol., 39(3):361-364

Gresshoff, Roth, Stacey & Newton (eds.) Nitrogen Fixation: Achievements and Objectives
Chapman and Hall New York • London 1990

Root growth on *Vicia faba* in relation to its nodulation by *Rhizobium leguminosarum*

Peter Roskothen

Institute of Plant Agronomy and Plant Breeding, University of Gottingen, D-3400, Gottingen, Germany FRG.

17 European Vicia faba varieties were inoculated with 5 Rhizobium leguminosarum strains. The strains were characterized in symbiosis in an earlier experiment (unpublished) with other faba bean genotypes.

Table 1. Characterization of wild-type Rhizobium strains.

Strain	Characteristics of symbiosis with all (a,b,d) or different plant genotypes (c) tested earlier
a) Hoh Erb2	nod^-
b) 128C13 Hup^+	nod^+fix^-
c) 3853 Hup^+	$nod^-/nod^-fix^+/nod^+fix^+$
d) 14(1)	nod^+fix^+

Five replications of each combination were grown under greenhouse conditions using a root sterile plant culture method. Plants were harvested after 12 weeks. Dry matter of shoot and root as well as number and weight of nodules were recorded.

Table 2. Absolute and relative production of shoot (root=1) in inoculated Rhizobium strains.

Rhizobium strain	n	mean* (relation)	mean* [dry matter root (in g)]
Hoh Erb2	133	1.1	1.233
128C13 Hup^+	134	1.1	1.27
3853 Hup^+ (only nod^+fix^+ combinations)	56	1.7	0.895
14(1)	139	2.3	1.035

*Differences significant at p=0.01, except for Hoh Erb2 & 128C13 Hup^+.

Absolute as well as relative production of the roots (Table 2) of plants inoculated with ineffective strains (nod^-, nod^+fix^- were increased compared to those inoculated with effective strains (nod^+fix^+). Therefore, it is possible that, with plant nodulation by strains of high symbiotic performance, the root architecture is so changed that under water stress conditions, the growth of faba beans may be limited by the root system because it was established under optimum conditions during early plant growth and development.

Gresshoff, Roth, Stacey & Newton (eds.) Nitrogen Fixation: Achievements and Objectives
Chapman and Hall New York • London 1990

Selection of suitable carriers for rhizobia incubating under different temperatures

S.H. Salem and S.A.Z. Mahmoud

Dept. of Agric. Botany, Faculty of Agricutlure, Zagazig Univ. and Microbiology Dept., Faculty of Agriculture, Ain Shams University, Egypt.

Introduction:

A vast majority of the scientific community consider peat base inoculants for rhizobia to be superior to the others. Due to lack of peat in our country Egypt, it was necessary to find suitable material from our local resources to be used as a carrier for rhizobia. The selected carrier should fulfill several requirements: It must be cheap in price, light in weight, having a good adsorption capacity and preserve the number of rhizobia for a long time without significant decrease in number.

Material and Methods:

Nine carriers were tested in laboratory under sterile conditions for the suitability of Rhizobium leguminosarum. The used carriers were silt clay loam soil, sandy soil, saw dust, talcum powder, charcoal and compost either alone or mixture of them. Known weight of each carrier was put in 250 ml. bottles after mixing it thoroughly with gelatin and sterilized in the autoclave. Each bottle was inoculated with the Rhizobium suspension and incubated, thereafter, during 4 months either on room temperature or in the refrigerator. Serial counts of rhizobia were determined after 15, 30, 45, 90 and 120 days.

Results and Discussion:

Results show that rhizobia grew well and increased in their numbers in most carriers up to 30-45 days. The numbers decreased, thereafter, but maintained considerable numbers of viable rhizobia at the end (3-4 months). The number of rhizobia decreased rapidly when saw dust and talcum powder were used as carriers, either alone or mixture with clay soil. The growth of rhizobia showed better results with longer periods of viability (3 months) when grown on mixture of silt clay loam soil + charcoal + compost (5:3:2) and incubated on room temperature. Therefore, the latter could be used as a carrier for rhizobia to the farmer use within 3 months from the production.

References:

Saleh, A.S.E. (1981). Development of new methods for preparing legume inoculant. M. Sc. Thesis, Fact. of Agric., Ain Shams Univ., Cairo.

Salem, S. H. (1969). Microbiological studies on root nodule bacteria. Ph.D. Thesis, Fac. of Agric., Ain Shams Univ., Cairo.

Gresshoff, Roth, Stacey & Newton (eds.) Nitrogen Fixation: Achievements and Objectives
Chapman and Hall New York • London 1990

Nickel is essential for active hydrogenase in free-living *Frankia* isolated from *Casuarina*

A. Sellstedt* and G.D. Smith

Dept. of Biochemistry, Faculty of Science, Australian National University, GPO, Box 4, Canberra, ACT 2601, Australia, *Permanent address: Dept. of Plant Physiology, Univ. of Umea, S-901 87 Umea, Sweden.

Five free-living Brankia strains isolated from Casuarina were investigated for occurrence of hydrogenase activity. Nitrogenase activity (acetylene reduction) and hydrogen evolution were also evaluated.

Acetylene reduction was detected in the Frankia strains. None of them had any hydrogenase activity when grown on nickel-depleted medium, and they all released hydrogen in air.

After addition of 0.68 uM nickel to the medium, the Frankia strains were shown to possess an active hydrogenase, which resulted in hydrogen uptake and loss of hydrogen evolution.

The results show that the absence of hydrogenase activity in free-living Frankia isolated from Casuarina is caused by nickel deficiency. Frankia living in symbiosis with Casuarina show hydrogenase activity and therefore the results also indicate that hydrogenase activity is to some extent regulated by the host plant and/or that the host is supplying the microorganism with nickel.

Gresshoff, Roth, Stacey & Newton (eds.) Nitrogen Fixation: Achievements and Objectives
Chapman and Hall New York • London 1990

Selection of mungbean (*Vigna radiatea*(L.) Wilczek) for high nitrogen fixation

R. Senaratne and N.D.L. Liyanage

Dept. of Agronomy, Faculty of Agriculture, University of Ruhuna, Mapalana, Kamburupitiya, Sri Lanka.

Mungbean is an important food legume in Sri Lanka which is mainly grown by resource-poor farmers on N-poor soils. Fertilization of this crop is almost non-existent. Consequently, its productivity has remained very low (i.e., 0.75 mt/ha as opposed to 2.25 mt/ha under well-managed conditions). When mungbean establishes effective symbiosis with rhizobia, it could derive over 100 kg N/ha from the atmosphere (Myers and wood 1987). Hitherto no systematic effort has been made in Sri Lanka to exploit the potential of biological nitrogen fixation (BNF) in improving mungbean yields when N limited.

The study involved screening of over 100 genotypes to ascertain the magnitude and nature of genotypic variability in several traits associated with BNF of mungbean (i.e., nodule number, nodule mass, plant vigour and seed yield) under three cultural regimes, namely, without rhizobial inoculation, with rhizobial inoculation and N fertilization at the rate of 100 kg/ha. versus a local control.

A great variability in the above parameters was observed under each regime, and genotypes superior to the local check were identified. The ranking of genotypes in terms of nodulation and seed yield depended upon the cultural regimes adopted suggesting an interaction between genotype and cultural regime (Table 01).

Table 01. Genotypic variability of some BNF traits under varying cultural regimes

Cultural Regime	No. of nodules/ plant	Fresh nodule wt. mg/plant	Seed yield mt/ha	Seed yield of Type77
uninoculated	0-47	0-423.7	0.1933-2.306	0.8714
inoculated	0-40*	0-719.0	0.3033-3.257	0.8231
N fertilized	ND	ND*	0.5467-2.600	1.0939

* - not determined

Seed inoculation in general produced greater nodule mass and seed yield compared to the control. It appears that there is considerable scope in improving productivity of mung bean.

Myers, R.J.K. and Wood, I.M. (1987) in Food legume improvement for Asian Farming Systems, ACIAR Proc. No. 18, Canberra, pp 46-52.

Gresshoff, Roth, Stacey & Newton (eds.) Nitrogen Fixation: Achievements and Objectives
Chapman and Hall New York • London 1990

Stimulation of plant growth by inoculation with nitrogenase-depressed mutants of phototrophic bacteria

S. Shestakov, L. Mikheeva, V. Frolova, N. Belavina, T. Mitronova

Department of Genetics, Moscow State University, Moscow 119899, USSR.

Mutant strains of phototrophic bacteria derepressed for nitrogen fixation and blocked in NH_4^+ assimilation can potentially serve as suppliers of nitrogen fertilizer (1). We have investigated the effect of inoculation with nitrogenase-derepressed (ND) mutants of purple bacteria and cyanobacteria on the growth of some non-legume plants in microvegetative experiments.

A number of *Rhodobacter sphaeroides* mutants have been obtained which are characterized by derepression for nitrogenase in the presence of ammonium ions and by excretion of fixed N_2 in the form of NH_4^+ into the medium (3). Seedlings of rape, rice and tomato inoculated with these mutant cells exporting NH_4^+ showed (after 6 days of growth) a 20-30% increase of dry weight over the control plants incubated in NH_4^+ -free medium or with the wild type strain. Inoculation with ND-mutants stimulates significantly the growth of rape in 3-week microvegetative experiments. Plants inoculated with ND-mutants showed a greater dry weight and extension of shoots than when inoculated with wild type cells. No detectable NH_4^+ was found in the plant growth medium supplemented by ND-mutants. Reproduction of purple bacterium cells occurs mainly in the rhizosphere area. Formation of red cell congestion on some part of the roots indicated a possibility of colonization of rape plants by purple bacteria.

Cyanobacterium *Anabaena azollae* mutants (ED-12, ED-25) characterized by derepressed nitrogenase, a low level of glutamine synthetase biosynthetic activity and a high rate of NH_4^+ excretion have been revealed among strains resistant to ethylenediamine (NH_4^+ analogue) (2). Inoculation of cress seedlings with these mutants stimulated the growth of plant hypocotyl and the 1st leaf in N-limiting conditions. There was a 60-70% increase of total dry weight and extension of rice shoots during 7 days of growth in the presence of mutant cells actively excreting NH_4^+; wild type cells do not produce such an effect. Ammonium ions released from the cells of mutants ED-12 and ED-25 were completely assimilated by plants.

1. Latorre, C., Lee, J.H., Spiller, H. & Shanmugan, K.T. (1986) Biotechnol. Lets. **8**, 507-512.

2. Mikheeva, L. Priamchuk, S., Gogotov, I. & Shestakov, S. (1990) Microbiolgia **59**, 35-39.

3. Shestakov, S., Zinchenko, V. *et al.* (1988) in *Nitrogen Fixation; Hundred Years After*, eds. Bothe, H., de Bruijn, F. J., Newton, W. E. (Gustav Fisher, Stuttgart/ New York), pp. 163-169.

Gresshoff, Roth, Stacey & Newton (eds.) Nitrogen Fixation: Achievements and Objectives
Chapman and Hall New York • London 1990

Increases of nitrogen fixation and phosphorus in rice plant inoculated with *Azospirillum amazonense* Kpl

S.D. Song, T.G. Park and K.B. Lee

Dept. of Biology, Kyungpook National University, Taegu, Korea.

Azospirillum spp., gram negative, vibroid, and microaerobic N_2-fixing soil bacteria are known as effective biofertilizers. They colonize and symbiotically associate with the roots of several economically important cereal crops and forage grasses and contribute to the increase of yields by providing biologically fixed nitrogen and by improving root development, the rate of water and mineral uptake.

We inoculated young rice seedlings with a culture of the acid tolerant diazotroph, A. amazonense Kpl isolated with a culture by the semisolid method from the rhizosphere of the grasses in Korea. During the growing period the plant parts were analyzed for inoculation effects on growth and compared for dry matter productivity and changes in nitrogen and phosphorus contents between bacterial association and combined nitrogen applications.

The plants treated with A. amazonense Kpl in the rhizosphere showed increases in height, leaf area, and dry weight by 15.4%, 65.4% and 33.7% respectively on the 56th day after sowing in comparison with untreated plants. The net assimilation rates for the bacterial inoculation treatment, combined nitrogen treatment, or untreated plants were 0.32, 0.47, and 0.56 mg DW cm^{-2} day^{-1}, respectively.

The maximum rate of acetylene reduction activity showed 230 nM C_2H_4 $plant^{-1}$ $hour^{-1}$ in the inoculated rhizosphere. The exogenous supply of combined nitrogen inhibited denitrogenase activity in the rhizosphere but enhanced optimum plant growth and productivity. The symbiosis of the associative nitrogen fixing bacteria in the roots resulted in an increase of the total nitrogen content of the rice plant by 38.8%. The total phosphorus content in each organ of the plant showed two- to three-fold higher levels than the control plant. The associative bacteria accelerated the growth of the underground parts more than those of the upper grown parts of the plant and caused significant decreases in T/R ratios.

1. Lee, K.B., Park, T.G., and Song, S.D. (1989) Effects of Azospirillum amazonense Y1 on the growth corn. Kor. J. Bot. **32(3)**, 137-144.

Gresshoff, Roth, Stacey & Newton (eds.) Nitrogen Fixation: Achievements and Objectives
Chapman and Hall New York • London 1990

Nitrogen fixation and cycling of N in a pure *Leucaena leucocephala* stand

C. van Kessel*[1] and J.P. Roskoski[2], [1]University of Saskatchewan, Saskatoon, Canada S7N 0W0, Canada, & [2]National Science Foundation, Washington DC, USA.

Leucaena leucocephala (Lam) de Wit is a fast growing tropical leguminous tree which is able to derive most of its N from N_2 fixation; it is not known if an established leucaena stand will continue fixing N_2. The objectives were: (1) to estimate N_2-fixation by leucaena seedlings, and (2) to determine whether established leucaena continued to fix N. A pure stand of *L. leucocephala* was established in 1986 on Maui, HI. Seedlings were planted 1.5 by 1.5 m apart. The % N derived from N_2 fixation was estimated using the $\delta^{15}N$ method. Deep and shallow rooted annual grasses near sampling trees were used as reference plants. One year after planting, the average $\delta^{15}N$ value of the oldest and new leaves was 3.8 ± 0.4 and 3.9±0.5, respectively. The average $\delta^{15}N$ value for the reference plants was 7.3±0.4, which indicated that 43.0±4.9% of the N was derived from N_2 fixation (Fig. 1). Two years after planting, average $\delta^{15}N$ values for older and new leaves was 3.7±0.2 and 3.0±0.2, respectively. Branches showed an average $\delta^{15}N$ value of 0.6±0.2. Average $\delta^{15}N$ values for reference plants (14 species) was 3.4±0.2. Two years after planting, total above ground N accumulation in leucaena was estimated at 271±33 kg/ha, of which 201±25 kg N/ha (74%) was derived from N_2 fixation. Four years after planting, the $\delta^{15}N$ values for leaf, pod and seed were not significantly different and averaged 1.7±0.1. Stems had a significantly lower $\delta^{15}N$ value of -0.9±0.1. The average $\delta^{15}N$ value of the reference plants (6 species) was 0.7±0.3. It appears that N in two and four-year-old leucaena stands and grasses was derived from recycled N and four years after planting 89 % of the N in the grasses was derived from fallen abscised leucaena plant components. Further decreases in the $\delta^{15}N$ value of the reference plant and leucaena leaves four years after planting indicate that N_2 fixation had not ceased.

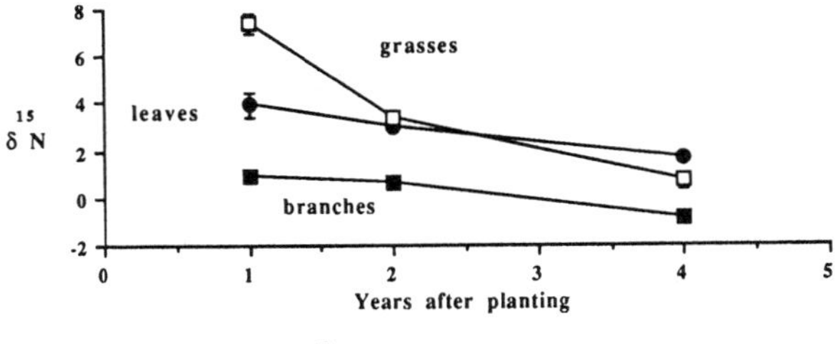

Fig. 1 Changes in the natural ^{15}N abundance of leaves, branches and grasses in a pure *L. leucocephala* stand.

Gresshoff, Roth, Stacey & Newton (eds.) Nitrogen Fixation: Achievements and Objectives
Chapman and Hall New York • London 1990

Field evaluation of nitrogen fixation potential and agronomic performance of three hypernodulating mutants of soybean

Shenchuan Wu and James E. Harper

Dept. of Agronomy and USDA/ARS, Plant Physiology and Genetics Research, University of Illinois, 1102 S. Goodwin Avenue, Urbana, Illinois 61801, USA.

Soil N (primarily NO_3^-) is a major deterrent to increased N_2 fixation in soybean (*Glycine max* [L.] Merr.] in the midwest US area. Three hypernodulating mutants (NOD1-3, NOD2-4, and NOD3-7) and a nonnodulating mutant (NN5), selected from mutagenized Williams, were compared with the Williams parent and a nonnodulating mutant (NN5) in a two-year field study. Urea fertilizer treatments of 0 and 200 kg N ha^{-1} were imposed to test whether nodule formation and N_2 fixation of these hypernodulating soybean mutants would respond similarly to added N. Nodule activity was determined by C_2H_2 reduction activity of nodulated roots and relative ureide abundance in vacuum-extracted xylem exudate. In the absence of N fertilizer, two year data based on the C_2H_2 reduction assay indicated that all mutants fixed more N_2 than did Williams, and that NOD1-3 was capable of fixing more N_2 than the other two mutants. In contrast, no difference was observed among hypernodulating mutants and Williams with either N treatment based on the relative ureide abundance technique in 1988. Both the C_2H_2 reduction assay and the relative ureide abundance assay (comparing NN5 with other lines in 1989) confirmed that nodulation of Williams was completely inhibited by N fertilizer through the initial sampling date, while the three hypernodulating mutants showed detectable N_2 fixation activity. The relative ureide abundance assay indicated that both NOD1-3 and NOD2-4 lines were more tolerant of added N fertilizer than was Williams in 1989. In the absence of N fertilizer and at later stages of plant development, N input from N_2 fixation was similar for Williams and the three hypernodulation mutants. Our results indicated that when using relative ureide abundance to assess N_2 fixation, it is desirable to include a nonnodulating soybean line as a control. Seed yields from the hypernodulating mutants were significantly less than that from Williams in both years (10 to 30% less). There was no difference among soybean lines in seed protein and oil concentrations in 1988. NOD1-3 actually had lower seed protein concentration in 1989. Therefore, total N harvested from seeds was less for the hypernodulating lines than for Williams. The results indicate that the hypernodulating mutants are currently inferior to Williams from an agronomic standpoint, although backcrossing is expected to improve their performance.

SECTION G:

PLANT FUNCTION

Invited papers
Workshop summary
Poster summaries

Gresshoff, Roth, Stacey & Newton (eds.) Nitrogen Fixation: Achievements and Objectives
Chapman and Hall New York • London 1990

Protein synthesis and protein phosphorylation in *Bradyrhizobium japonicum* bacteroids

D.B. Karr and D.W. Emerich

Department of Biochemistry, University of Missouri, Columbia, MO 65211 USA

Introduction

Biological nitrogen fixation requires a large portion of an organisms energy metabolism. Free-living diazotrophs fixing atmospheric nitrogen have longer doubling times than cultures supplied fixed nitrogen compounds. During symbiotic nitrogen fixation, the microsymbiont does not grow significantly during the most active phase of dinitrogen reduction. The non-growing condition of the microsymbionts could provide the necessary energy for nitrogen fixation. As much as 80% of the metabolic energy of a cell may be used for protein synthesis (6). The switching of metabolic energy from protein synthesis to dinitrogen reduction could supply the energy needs of nitrogen fixation without requiring additional energy sources or invoking supplemental metabolic pathways. This concept of switching energy from protein synthesis to dinitrogen fixation was first proposed by Perry Wilson in his studies on metabolism in non-growing cultures (12).

Protein synthesis has been reported in bacteroids from indeterminate nodules (2,9,11) but not from bacteroids of determinate nodules. In an indeterminate symbiosis, the microsymbiont population consists of a mixture of mature, nitrogen-fixing bacteroids, undifferentiated bacteria and transforming bacteria (7). In *Rhizobium leguminosarum* bacteroids, incorporation of ^{35}S-sulfate from the plant nutrient solution into protein was greatest in young nodules and declined with age (2,11). The period of rapid incorporation corresponds to the age when the bacteria are rapidly transforming into mature bacteroids and the required symbiotic proteins are being synthesized. The nitrogenase component proteins were rapidly labeled with ^{35}S-sulfate but synthesis of the two component proteins was not strictly coordinated (2). Shaw and Sutton (9) demonstrated that isolated lupin bacteroids were capable of incorporating ^{35}S-methionine into protein. The labelling patterns were dependent on the stage of symbiotic development, were dependent on the strain of *Rhizobium*, but independent on the type of host plant.

In a determinant symbiosis, such as that between *Bradyrhizobium japonicum* and soybean, the majority of the microsymbiont population consists of a uniform developmental stage (3). This feature of determinant symbioses permits a clearer picture of events during symbiotic development as a single developmental species can be examined.

The capacity of *B. japonicum* bacteroids to incorporate ^{35}S-methionine or ^{3}H-leucine into protein throughout symbiotic development was investigated. The results with either isotope gave the same results, so only the results with ^{35}S-methionine will be presented here.

Results and Discussion

^{35}S-Methionine Accumulation

The accumulation of ^{35}S-methionine by *B. japonicum* bacteroids was linear for at least 30 minutes (Figure 1). The kinetics of the accumulation of labeled methionine consisted of a non-linear, saturable process and a linear, unsaturable process similar to that described by Reibach and Streeter (8) for dicarboxylate uptake. By analogy, the saturable process is an active transport and the unsaturable process is diffusion. Both processes maintain their respective rates during the development of nitrogen fixation activity but the saturable process is reduced to a greater extent during senescence than the non-saturable process. The decline in the non-saturable process approximates the loss of viability as nodules age (3). The $S_{0.5}$ value of ^{35}S-methionine for the saturable process remained unchanged, at ~ 250 pM, throughout nodule development. Azide, chloramphenicol and acridine, but not cycloheximide inhibited ^{35}S-methionine

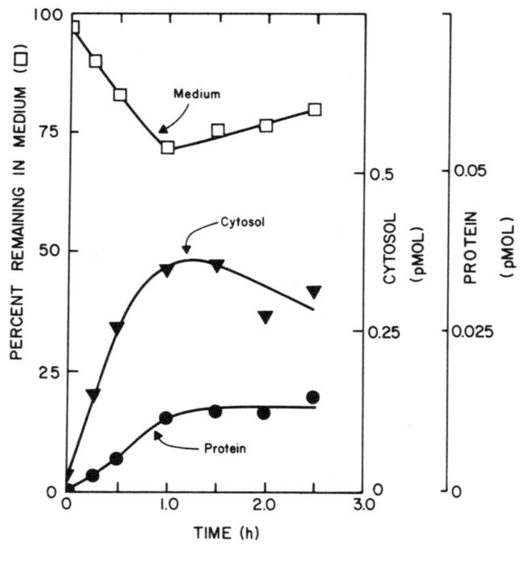

Figure1. Partitioning of exogenous ^{35}S-methionine by *B.japonicum* bacteroids isolated from 18 day old nodules. The initial concentration of methionine was 3.0 nM and the assays contained 10^9 bacteroids. The percent methionine remaining in the medium (□) is indicated by the left axis, amount in the cytosol (▽) is indicated by the near right axis the amount in protein (0) is indicated by the far right axis.

accumulation (Table 1). Bacteriods isolated from nodules which had been frozen at -20°C or isolated bacteroids stored at 4°C for 48 hours could accumulate only 10% or less of the ^{35}S-methionine that was accumulated by bacteroids isolated from fresh nodules and assayed immediately. ^{35}S-methionine accumulation requires metabolically active bacteroids.

Table 1 Inhibition of Methionine Accumulation and Protein Synthesis by *B. japonicum* Bacteriods

	% Inhibition	
Treatment	Methionine Accumulation	Protein Synthesis
Azide (10mM)	94	93
Chloramphenical (750 μg/ml)	85	90
Acridine (60 μg/ml)	84	99
4°C	95	93

Protein Synthesis

Only 1 to 2% of the ^{35}S-methionine accumulated by *B. japonicum* bacteriods was recovered in the soluble protein fraction. The bulk of the radioactivity accumulated was partitioned approximately equally between the cytosolic and membrane fractions of the cell. The radioactivity of the protein fraction was pronase-sensitive and was diminished in intensity when chased with unlabeled methionine. Incorporation of ^{35}S-methionine into protein was inhibited by azide, chloramphenicol and acridine (Table 1). Protein synthesis was greatly reduced in bacteroids obtained from frozen nodules or in bacteroids stored at 4°C for 48 hours. The rate of incorporation of ^{35}S-methionine into protein was more dependent on the age of the nodules from which the bacteroids were isolated than was methionine accumulation (Table 2). Protein synthesis declined ~75% between 14 and 39 days compared to ^{35}S-methionine accumulation which declined by ~50 % over the same period. The calculated V_{max} of protein synthesis declines rapidly during the development of acetylene reduction activity but the protein content of the bacteriods does not decline until senescence begins (Figure 2).

The autoradiographic patterns of ^{35}S-methionine labeled proteins were dependent on the procedure by which the proteins were extracted from the bacteroids (Figure 3). When extracted by the same procedure, few differences were seen in the autoradiographic patterns of ^{35}S-methionine labeled proteins of bacteriods isolated from nodules from 14 to 39 day old plants. The observable differences were in the 33 KDa and 60 KDa regions. The intensity appeared to decrease relative to the other labeled proteins in older nodules at the 33 KDa region, whereas it appeared to increase in the 60 KDa region. Additions of amino acids, malate and potassium to the reaction medium did not affect incorporation of ^{35}S-methionine. Overnight incubation in a complete medium increased incorporation in bacteriods isolated from older nodules but not from those of younger nodules. A starvation period prior to assay increased incorporation of ^{35}S-methionine into proteins of bacteriods obtained from nodules of any age.

Table 2 Kinetic Parameters for Protein Synthesis by *B. japonicum* Bacteroids

Nodule Age (days)	$S_{0.5}$ (μM)	V max (pmol/h/10^9cells)
14	0.55	2.2
19	0.35	0.8
39	0.24	0.6

Figure 2. The calculated V_{max} for protein bacteroid synthesis, bacteroid protein content and acetylene reduction activity during soybean nodule development. Calculated V_{max} (0); protein content (■); and acetylene reduction activity (Δ).

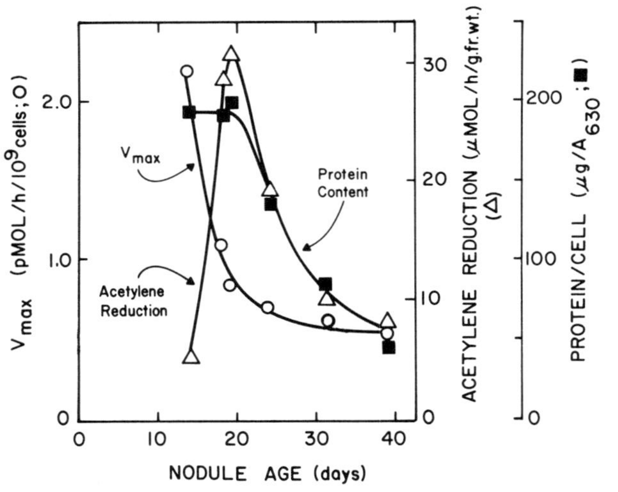

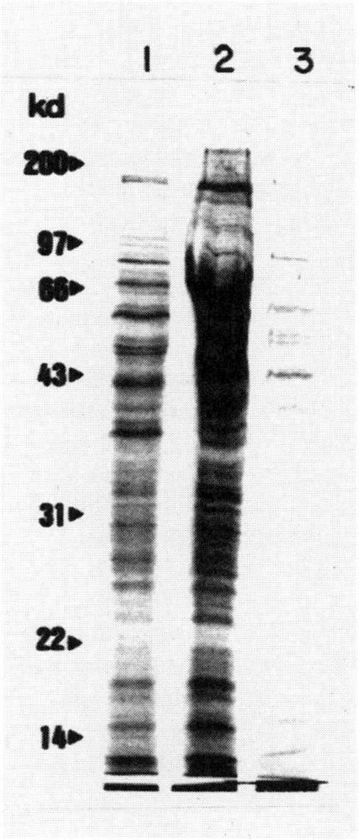

Figure 3. Autoradiographic comparison of three methods for extracting ^{35}S-methionine labeled proteins of *B. japonicum* bacteroids. Lane 1, French press ruptured cells; Lane 2, acetone precipitation of cell lysate; Lane 3, phenol extraction of cell lysate.

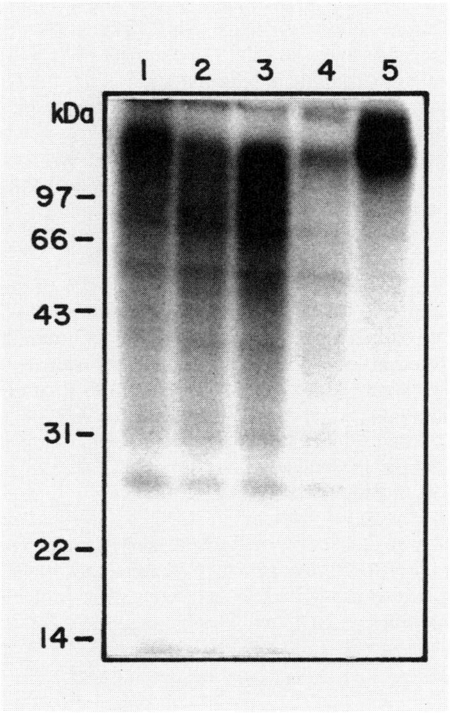

Figure 4. Autoradiogram of *in vitro* phosphorylated *B. japonicum* bacteroid proteins. Assays contained 0.1 mM γ-^{32}P-ATP, 2 mM $MgCl_2$, 0.4 mM EDTA for 0 min (Lane 1), for 2.5 min (Lane 2) and 5 min (Lane 3). Lane 4 was the same as lane 3 but with 20 mM EDTA. Lane 5 was the same as 3 but followed by a 5 min chase with unlabeled 2 mM ATP.

Protein Phosphorylation

Protein phosphorylation in procaryotes regulates signal transduction and carbon metabolism, two processes fundamental to symbiotic nitrogen fixation. Krishnan and Pueppke (5) demonstrated that the plant cytosol of soybean root nodules contained Ca^{2+}-dependent and Ca^{2+}-independent protein kinases that phosphorylate several endogenous proteins. Suzuki and Verma (10) reported three types of protein kinases in soybean nodule nuclei which were distinguished by their dependence on calcium and/or calmodulin. Bassarab and Werner (1) reported that the peribacteroid membrane of soybean root nodules possessed a Ca^{2+}-and phospholipid-dependent protein kinase and demonstrated the phosphorylation of at least three different membrane-associated proteins. We have shown that protein phosphorylation occurs in *B. japonicum* bacteroids (4).

Similar to methionine labeling, only ~2% of the radiolabeled phosphate accumulated by bacteroids was incorporated into protein (4). The amount of ^{32}P-phosphate incorporated into protein was at least 2.5 times greater in bacteroids isolated from younger nodules than those from older nodules. The amount of ^{32}P incorporated into bacteriod protein *in vitro* increased with length of exposure to γ-^{32}P-ATP and was decreased when chased with unlabeled ATP or when excess EDTA was added to the reaction (Figure 4). The amount of incorporation could be increased up to 9-fold by a 24 hour preincubation in yeast extract and malate. A period of starvation prior to labeling did not affect incorporation.

The position of ^{35}S-methionine labeled proteins were compared to the position of ^{32}P-phosphate labeled proteins to determine whether any newly synthesized proteins were also phosphorylated. Newly synthesized proteins, as well as pre-existing proteins, were phosphorylated. Two ^{32}P-labeled peptides between 22 and 32 KDa in both *in vivo* and *in vitro* assays appeared in positions similar to ^{35}S-labeled proteins. The region between 43 and 97 KDa appeared to contain several ^{32}P-labeled and ^{35}S-labeled proteins.

Summary

Isolated bacteriods of *B. japonicum* retain their ability to accumulate methionine and synthesize proteins. *B. japonicum* bacteroids are capable of accumulating phosphate and incorporating it into proteins indicating the presence of several protein kinases. Both bacteroid protein synthesis and protein phosphorylation decline with increasing nodule age.

Acknowledgements

We thank Judy Wall and Doug Randall for helpful suggestions and Jackie Kimsey for typing this manuscript. Support for this research was provided by USDA Competitive Grants Program, Missouri Soybean Merchandising Council and the University of Missouri Weldon Springs Competitive Grants Program.

References

1. Bassarab, S. & Werner, D. (1987) J. Plant Physiol. **130**, 233-241.
2. Bisseling, T., van den Bos, R.C., Weststrate, M.W., Hakkaart, M.J.J. & van Kammen, A. (1979) Biochim. Biophys. Acta **562**, 515-526.
3. Karr, D.B. & Emerich, D.W. (1988) Plant Physiol. **86**, 693-699.
4. Karr, D.B. & Emerich, D.W. (1989) J. Bacteriol. **171**, 3420-3426.
5. Krishnan, H. & Pueppke, S.G. (1989) Symbiosis **7**, 127-138.
6. Lehninger, A. (1965) Bioenergetics (W.A. Benjamin, Inc., New York).
7. McRae, D.G., Miller, R.W. & Berndt, W.B. (1989) Symbiosis **7**, 67-80.
8. Reibach, P.H. & Streeter, J.G. (1984) J. Bacteriol. **159**, 47- 52.
9. Shaw, B.D. & Sutton, W.D. (1979) Biochim. Biophys. Acta **563**, 216-226.
10. Suzuki, H. & Verma, D.P.S. (1989) The Plant Cell **1**, 373-379.
11. van den Bos, R.C., Bisseling, T. & van Kammen (1978) J. Gen Microbiol. **109**, 131-139.
12. Wilson, P.W. (1938) J. Bacteriol. **35**, 601-623.

Gresshoff, Roth, Stacey & Newton (eds.) Nitrogen Fixation: Achievements and Objectives
Chapman and Hall New York • London 1990

Some nodulin and Nod proteins show similarity to specific animal proteins

Niels N. Sandal and Kjeld A. Marcker

Laboratory of Gene Expression, University of Aarhus, Department of Molecular Biology and Plant Physiology, Gustav Wieds Vej 10, DK-8000 Aarhus C, Denmark.

Nodulins and Nod proteins

During the development of a legume root nodule several genes encoding nodule-specific plant proteins (nodulins) are specifically activated (18). The nodulins can be divided into early and late nodulins. Early nodulins are primarily involved in development and progression of the infection (30,35), while the late nodulins are involved in nodule function. The functions of some of the late nodulins like leghemoglobin (1), uricase (4,24), nodule-specific glutamine synthase (3,7) and sucrose synthase (21,33) are known. However the function of the majority of the nodulins are at present unknown. The bacterial symbiotic partner encodes several proteins (Nod proteins) that are important for the formation of a functional nodule and for host specificity, but the biochemical functions of the Nod proteins are not known. Leghemoglobin is related to animal globins, and soybean uricase shows similarity to animal uricases (14,22). An indication of the function of some of the nodulin and Nod proteins can be found by searching the databases for similar proteins. In this paper we demonstrate similarity between nodulin 26 and several specific proteins. In addition we report a surprising similarity between the *Rhizobium* NodC protein and the *Xenopus* DG42 protein.

Nodulin 26.

Nodulin 26 is an integral part of the peribacteroid membrane (pbm) (8). We have determined the sequence of a partial cDNA clone encoding nodulin 26 (28). This sequence differs from the nodulin 26 cDNA sequence published by Fortin et al. (1987) (8) by a few changes which extends the reading frame by 66 amino acids. Surprisingly, we found by computer analysis that the amino acid sequence of nodulin 26 is similar to the amino acid sequence of the major intrinsic protein of the bovine eye lens fiber membrane (MIP) (10,28) (fig. 1). The similarity between the two proteins is 35% and if conservative changes are taken into account the similarity is 48%.

The hydropathic profiles of the two proteins are remarkably similar, each showing 6 hydrophobic regions large enough to span a membrane (fig. 2). Based on the hydropathic profiles and proteolytic degradation studies, a model for the organisation of MIP in the membrane was proposed (10). This model is supported by circular dichroism studies (13). The localization of antigenic sites also supports the model. An essentially similar model can be envisaged for the organization of nodulin 26 in the peribacteroid membrane (fig. 3). In this model the two most conserved regions are located on either side of the membrane. There is similarity between these two regions (S/T G A/G X y N P A X S/T y A/G (y=hydrophobic residue, X=any residue)). These regions also contain two conserved histidines and a conserved arginine respectively, which probably are of importance for MIP and nodulin 26 function.

MIP is phosphorylated *in vivo* on serines located 20-40 residues from the carboxyl terminus (16). In

```
NOD 26  M A D Y S A G T E S Q E V V V N V T K N T S E T I Q R S D S L V S V   34
MIP                                                 M W E L R S A    7
FACIL                                                 M S Q T S    5
GYLA                                                  M S S S D    5

NOD 26  P F L Q K L V A E A V G T Y F L I F A G C - A S L V V N E N Y Y N M   68
MIP     S F W R A I C A E F F A S L F Y V F F G L G A S L R W A P - - - G P   38
FACIL.  T L K G Q C I A E F L G T G L L I F F G V G C V A A L K V A G A S F   39
GYLA    I F - - - - I G E T I G T A L L I L L G G G V C A AVTLK A S K A R N   37

NOD 26  I T F P G I A I V W G L V L T V L V Y T V G H I S G G H F N P A V T  101
MIP     L H V L Q V A L A F G L A L A T L V Q A V G H I S G A H V N P A V T   72
FACIL.  G Q W E - I S V I W G L G V A S A I Y L T A G V S G A H L N P A V T   72
GYLA    A G W L A I A F G W G F A V M T A A Y I S G P L S G A H L N P A V T   71

NOD 26  I A F A S T R R F P L I Q V P A Y V V A Q L L G S I - - - - - L A S  130
MIP     F A F L V G S Q M S L L R A I C Y M V A Q L L G A V A G A A V L Y S  106
FACIL.  I A L W L F A C F D K W K V I P F I V S Q V A G A F     98

NOD 26  G T L R L L F M G N H D Q F S G T V P N G T N L Q A F V F E F I M T  164
MIP     V T P P A V - R G N L A L N T L H - P G V S V G Q A T I V E I F L T  138
FACIL.                        141       P H I N F V Q A F A V E M V I T  151

NOD 26  F F L M F V I C G V A T - D N R A V G E L A - - G I A I G S T L L L  195
MIP     - - L Q F V L C I F A T Y D E R R N G R L G S V A L A V G F S L T L  170
FACIL.  A I L M - G L I L A L T V D G N G V P R G P L A P L L I G L L I A V  156

NOD 26  N V I I G G P V T G A S M N P A R S L G P A F V H G E Y E G I W I Y  229
MIP     G H L F G M Y Y T G A G M N P A R S F A P A I L T R N F T N H W V Y  204
FACIL.          198     T G F A M N P A R D F G P   210

NOD 26  L L A P V V G A I A G A W V Y N I V R Y T D K P L S E I T K S A S F  263
MIP     W V G P V I G A G L G S L L Y D F L L F P - R L K S V S E R - L S I  236
FACIL.  L F G P I V G A I V G A F A Y     251

NOD 26  L K G R A A S K *                                                    271
MIP     L K G S R P S E S N G Q P E V T G E P V E L K T Q A L *            263
```

Fig. 1. Alignment of soybean nodulin 26 (28), bovine MIP (10), *E. coli* glycerol facilitator (23) and *Streptomyces coelicolor* GylA (31).

the corresponding region of nodulin 26 there are several serines which could serve as phosphorylation sites. MIP is found in animal eye lenses and is used as an eye lens differentiation marker (37). Several lines of evidence suggested that MIP is a gap junction protein (5,9,20,29), but recent experiments have not supported this (11,32,38).

A root-specific cDNA clone with similarity to MIP and nodulin 26 has been isolated from tobacco. *In situ* hybridization of the cDNA probe revealed that the gene is highly expressed at meristematic regions of root tips (36).

Baker and Saier (1990) (2) observed similarity between the *E. coli* glycerol facilitator (23), nodulin 26 and MIP. The glycerol facilitator is the only known membrane channel in the *E. coli* cytoplasmic membrane. The glycerol facilitator allows the passage through the membrane of polyols (e.g. glycerol), urea, glycine, DL-glyceraldehyde and xylitol (12). Baker and Saier (1990) (2) found

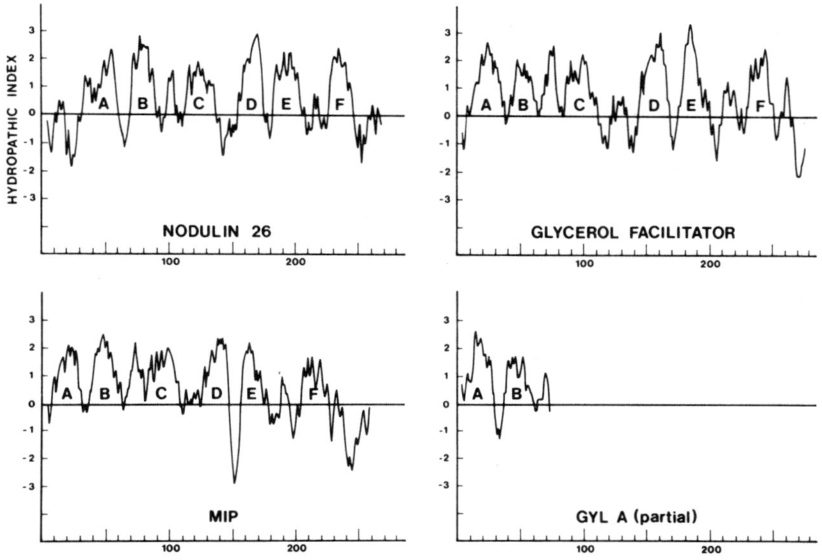

Fig. 2. Hydropathic profiles of A: nodulin 26, B: MIP, C: glycerol facilitator and D: GylA. The hydrophobicity values given by Kyte and Doolittle (17) were used. A window of nine residues was used. The letters A to F indicate the predominantly hydrophobic regions that are large enough to span a membrane.

that the similarity between the glycerol facilitator, nodulin 26 and MIP was greatest in the two N-terminal regions that are highly conserved between nodulin 26 and MIP, but we have also found great similarity in the last highly conserved region (fig. 1).

In addition we have found similarity between nodulin 26, MIP and the partial N-terminal sequence of the GylA protein from *Streptomyces coelicolor* (fig. 1). The *GylA* gene is part of the glycerol operon in this species and it was tentatively proposed as a glycerol kinase gene (31). However we suggest based on the observed similarity that the GylA protein from *Streptomyces coelicolor* also functions as a glycerol facilitator protein.

An alignment of the four related proteins is shown in fig. 1. The glycerol facilitator and the GylA protein are only shown in the regions where some similarity can be demonstrated. In addition to the conserved amino acids in the alignment, conservative changes are found at many positions. The very similar hydrophobic profiles of the four proteins are shown in fig. 2. Nodulin 26, MIP and the glycerol facilitator each have six hydrophobic domains large enough to span a membrane. The partial GylA protein sequence only allows the detection of two hydrophobic domains, but we predict that this protein also will contain four additional hydrophobic domains.

During the preparation of this manuscript similarity betweeen the *Drosophila* neurogenic gene *big brain (bib)* and the group of related proteins described here was published (25). The *bib* protein is much larger than the other proteins. The N-terminal half of the *bib* protein shows the similarity whereas the C-terminal half is not related to any known protein sequence. The regions that are highly

conserved among the related proteins are also highly conserved in the *bib* protein. In addition the hydropathic profile of the N-terminal half of *bib* is very similar to the related proteins.
The similarity between the glycerol facilitator and the related proteins described here indicates that these proteins constitute membrane channels. The characteristics of these proteins would then determine which types of molecules are allowed to pass through the membrane channels.
Nodulin 26 is located in the peribacteroid membrane which separates the bacteroid space from the plant cytoplasm within the plant cell. Nodulin 26 is therefore probably involved in the transport of some metabolites across this membrane.

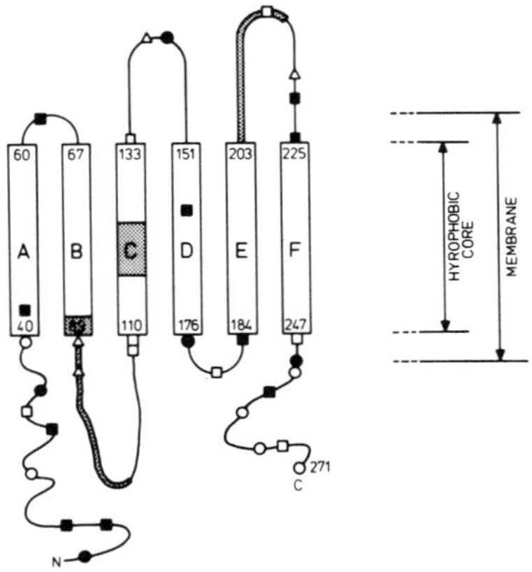

Fig. 3. Diagrammatic representation of nodulin 26 based on the deduced amino acid sequence. The six putative transmembrane helices, labeled A to F, are shown as boxes. Numbers refer to amino acid number in the peptide chain. Transcore regions of the protein are considered to have an α-helical configuration. All charged residues are indicated. Arg:□ Lys: o, His: Δ, Glu:■ Asp:•. Highly conserved regions are shown as dotted regions. The proposed model for nodulin 26 in the pbm is based on the model for organisation of MIP in the bovine lens membrane. In analogy to the MIP-model the N- and C-terminus are expected to be located on the plant cytoplasmic side of the pbm.

Similarity between the gene products of *Rhizobium nodC* and *Xenopus* DG42

The *Rhizobium meliloti* NodC protein is a membrane protein, which is necessary for the production of a sulphated, acetylated and acylated glucosamine tetrasaccharide that mediates root hair curling activity of *Rhizobium meliloti* extracts (19). A hydrophobic region close to the C-terminus of NodC functions as a membrane-anchor domain. Proteolysis experiments suggested that the C terminus is an intracellular domain, while a region of about 295 amino acids containing a cysteine-rich cluster is extracellular (15). Using the computer programs of the University of Wisconsin Genetics Group (UWGCG) (6) we found a surprising similarity between the DG42 protein from *Xenopus* (26) and the NodC proteins of *Rhizobium meliloti* (34) and *Rhizobium leguminosarum* (27). See fig. 4 for an alignment of the regions of similarity. It is noteworthy that this similarity (37%) is present in the regions which are most highly conserved (app. 90%) between the NodC proteins of *Rhizobium meliloti* and *Rhizobium leguminosarum.*

DG42 mRNA is present in the *Xenopus* embryo from just after midblastula, peaks by the end of late gastrula, and decays by the end of neurulation . Antibodies showed that the level of the 70 kDa DG42 protein follows the course of appearence and decay of the mRNA with a delay of several hours (26) Immunochemical staining localized the DG42 protein at the periphery of cells (26) and the similarity to the NodC protein suggests that it is membrane bound as both proteins have highly hydrophobic regions in the C terminal part. Right after the three C terminal hydrophobic regions there is a short stretch of similarity (LPSKYFALLTL versus LPLKAYALCTL). This region is expected to be intracellular in the NodC protein.

The biochemical functions of the NodC and DG42 proteins are not known, although a receptor function has been suggested for the NodC protein (15). The structural similarity between the two types of proteins of widely different origins implies a partial functional relationship and further suggests that the two proteins are members of a ubiquitous gene family which serves a very important function in the cell.

```
DG42 109  E N P E Y L I K C L E S C K Y V K Y P K D K L K I I L V I D G   139
NODC  60  E D P G I L S A C L A S I A D Q D Y P G E - L R V Y V V D D G    89

DG42 217  G K R E V M Y T A F Q A I G - T S V D Y V Q V C D S D T K L D   246
NODC 119  G K R K - - - A Q I A A I G Q S S G D L V L N V D S D S T I -   145

DG42 247  E L A T V E M V K V L E S - - - N D M Y G A V G G D V R I L N   274
NODC 146  - - - A F D V V S K L A S K M R D P E V G A V V G Q L T A S N   173

DG42 275  P Y D S F I S F M S S L R Y W M A F N V E R A C Q S Y F D C V   305
NODC 174  S G D T W L T K L I D M E Y W L A C N E E R A A Q S R F G A V   204

DG42 306  S C I S G P L G M Y R N N I L Q V F L E A W Y R Q K F L G T Y   336
NODC 205  M C C C G P C A M Y R R S A L A S L L D Q Y E T Q L F R G K P   235

DG42 337  C T L G D D R H L T N R V L S M G Y R T K Y T H K S R A F S E   367
NODC 236  S D F G E D R H L T I L M L K A G F R T E Y V P D A I V A T V   266

DG42 368  T P S L Y L R W L N Q Q T R W T K S Y F R E   389
NODC 267  V P D T L K P Y L R Q Q L R W A R S T F R D   288

DG42            476  L P S K Y F A L L T L   486
NODC            377  L P L K A Y A L C T L   387
```

Fig. 4. Alignment of regions of similarity between the *Xenopus* DG42 (26) and *Rhizobium meliloti* NodC (34) proteins

Acknowledgement.
This research was supported by the Danish Biotechnology Programme.

1. Appleby, C.A. 1984 Ann. Rev. Plant Physiol. 35:443-478
2. Baker, M.E. & Saier, M.H. 1990 Cell 60:185-186

3. Bennett, M.J., Lightfoot, D.A. & Cullimore, J.V. 1989 Plant Mol. Biol. 12:553-566
4. Bergmann H., Preddie E. & Verma D.P.S. 1983 EMBO Jour. 2:2333-2339
5. Bok, D., Dockstader, J. & Horwitz, J. 1982 J. Cell Biol. 92:213-220
6. Deveraux, J., Haeberli, P. & Smithies, O. 1984 Nucl. Acids Res. 12:387-395
7. Forde, B.G., Day, H.M., Turton, J.F., Wen-jun, S., Cullimore, J.V. & Oliver, J.E. 1989 The Plant Cell 1:391-401
8. Fortin, M.G, Morrison, N.A. & Verma, D.P.S. 1987 Nucl. Acids Res. 15:813-825
9. Gooden, M., Rintoul, D., Takehana, M. & Takemoto, L. 1985 Biochem. Biophys. Res. Comm. 128:993-999
10. Gorin, M.B., Yancey, S.B., Cline, J., Revel, J.P. & Horwitz, J. 1984 Cell 39:49-59
11. Gruijters, W.T.M. 1989 Jour. Cell Sci. 93:509-513
12. Heller, K.B., Lin, E.C.C. & Wilson, H. 1980 Jour. Bact. 144:274-278
13. Horwitz, J. & Bok, D. 1987 Biochemistry 26:8092-8098
14. Ito, M., Suzuki, M. & Yasuyuki, T. 1988 Eur. J. Biochem. 173:459-463
15. John, M., Schmidt, J., Wieneke, U., Krüssmann, H.-D. & Schell, J. 1988 EMBO Jour. 7:583-588
16. Johnson, K.R., Lampe, P.D., Hur, K.C., Louis, C.F. & Johnson, R.G. 1986 Jour. Cell Biol. 102:1334-1343
17. Kyte, J. & Doolittle, R.J. 1982 Jour. Mol. Biol. 157:105-132
18. Legocki, R.P. & Verma, D.P.S. 1980 Cell 20:153-163
19. Lerouge, P., Roche, P., Faucher, C., Maillet, F., Truchet, G., Promé, J.C. & Denarie, J. Nature 344:781-784
20. Menko, A. S. & Boettiger, D. 1988 Mol. Cell. Biol. 8:1414-1420
21. Morell, M. & Copeland, L. 1985 Plant Phys. 78:149-154
22. Motojima, K., Kanaya, S. & Goto, S. 1988 Jour. Biol. Chem. 263:16677-16681
23. Muramatsu, S. & Mizuno, T. 1989 Nucl. Acids Res. 17:4378
24. Nguyen, T., Zelechowska, M., Foster, V., Bergmann, H. & Verma, D.P.S. 1985 Proc. Natl. Acad. Sci. USA 82:5040-5044
25. Rao, Y., Jan, L.Y. & Jan, Y.N. 1990 Nature 345:163-167
26. Rosa, F., Sargent, T.D., Rebbert, M., Michaels, M.L., Jamrich, M., Grunz, H., Jonas, E., Winkler, J.A. & Dawid, I.B. 1988 Dev. Biol. 129:114-123
27. Rossen, L., Johnston, A.W.B. & Downie, J.A. 1984 Nucl. Acids Res. 12:9497-9508
28. Sandal, N.N. & Marcker, K.A. 1988 Nucl. Acids Res. 16:9347
29. Sas, D.F., Sas, M.J., Johnson, K.R., Menko, A. S. & Johnson, R.G. 1985 Jour. Cell Biol. 100:216-225
30. Scheres, B., Horvath, B., Van de Wiel, C., Zalensky, A., Horvath, B., Spaink, H., Van Eck, H., Zwartkruis, F., Wolters, A.-M., Gloudemans, T., Van Kammen, A. & Bisseling, T. 1990 Cell 60:281-294
31. Smith, C.P. & Chater, K.F. 1988 J. Mol. Biol. 204:569-580
32. Swenson, K.I., Jordan, J.R., Beyer, E.C. & Paul, D.L. 1989 Cell 57:145-155
33. Thummler, F. & Verma, D.P.S. 1987 Jour. Biol. Chem. 262:14730-14736
34. Török, I, Kondorosi, E., Stepkowski, T., Pósfai, J. & Kondorosi, A. 1984 Nucl. Acids Res. 12:9509-9524
35. Van de Wiel, C., Scheres, B., Franssen, H., Van Lierop, M.-J., Van Lammeren, A., Van Kammen, A. & Bisseling, T. 1990 EMBO Jour. 9:1-7
36. Yamamoto, Y.T., Acedo, G.N. & Conkling, M.A. 1989 Jour. Cell. Biochem. Suppl. 13D:313
37. Yancey, S.B., Koh, K., Chung, J. & Revel, J.P. 1988 Jour. Cell Biol. 106:705-714
38. Zampighi, G.A., Hall, J.E., Ehring, G.R. & Simon, S.A. 1989 Jour. Cell Biol. 108:2255-2275

Gresshoff, Roth, Stacey & Newton (eds.) Nitrogen Fixation: Achievements and Objectives
Chapman and Hall New York • London 1990

Alfalfa root nodule aspartate aminotransferase (AAT): Biochemical importance and genetic control

C.P. Vance, M.W. Farnham, N. Degenhart, R.J. Larson, S.S. Miller, D.K. Barnes and J.S. Gantt.

USDA-ARS Plant Science Research Unit, Departments of Agronomy and Plant Genetics, and Plant Biology, University of Minnesota, St. Paul, MN 55108, USA.

INTRODUCTION

Aspartate aminotransferase (AAT:EC 2.6.1.1) plays an important role in nitrogen (N) and carbon (C) metabolism in all plants and is particularly important in the assimilation of fixed-N derived from the legume-Rhizobium symbiosis (24). The enzyme catalyzes the reversible reaction Glutamate (Glu) + Oxaloacetate (OAA) ---▷ Aspartate (Asp) + α-Ketoglutarate (α-KG) (6). AAT plays an essential role in: 1) the assimilation of fixed N and C into asparagine (Asn) in legume root nodules (24); 2) the transfer of fixed C from mesophyll cells to bundle sheath cells in C_4 plants (28); and 3) the flow of reducing equivalents to chloroplasts, mitochondria and perhaps bacteroids through a malate-aspartate shuttle (1,28). AAT has been isolated from both legume and nonlegume species and has been demonstrated to be a dimeric enzyme (5,7,16,19,22). Cytosolic, mitochondrial, and plastid forms of the plant enzyme have been documented and suggested to be controlled by different genetic loci. Independent segregation of soluble, mitochondrial and glyoxysomal isozymes in maize and wheat support the idea that these forms of AAT are controlled by separate genes (3,23). Animal, cytosolic and mitrochondrial forms of AAT have been identified and the genes encoding these proteins have been isolated (9,17). Plant genes encoding AAT have not been isolated.

Effective alfalfa, pea, and birdsfoot trefoil assimilate recently fixed N_2 primarily into aspartate and asparagine (13,21,25,27). By contrast, ineffective nodules fix little, if any, N_2 and produce little, if any, aspartate and asparagine (13,21). Effective alfalfa and pea nodules rapidly convert exogenously applied [^{14}C]CO_2 into [^{14}C] Asn and [^{14}C] TCA cycle acids while ineffective nodules do not. Furthermore, treatment of alfalfa nodules with aminooxyacetate (AOA), an inhibitor of AAT activity, showed that the flow of C between the organic acid and amino acid pools was regulated by AAT activity (25). Such data has been interpreted to suggest that multiple forms of AAT regulate C and N pools in effective root nodules.

Alfalfa root nodule AAT is comprised of two major forms, AAT-1 and AAT-2 (7). AAT-2 is associated with nodule development and effective N_2 fixation, while AAT-1 appears to predominate in roots (4,7). Similarly, two analogous forms have also been identified in lupine (19), pea (1), and soybean (22) nodules. In pea, lupine, alfalfa, and soybean nodules 90% or more of the AAT activity is of host origin (1,7,19,22). Because of the importance of AAT in nodules and since it occurs as multiple forms, we initiated a program to biochemically and genetically characterize the enzyme. Our approach has been to isolate and purify AAT-1 and AAT-2, produce antibodies to the enzyme proteins, evaluate expression of AAT-polypeptides, select plants for increased nodule AAT activity, and isolate AAT cDNAs through antibody screening of λgtll libraries.

METHODS

Alfalfa AAT-1, from root suspension cells, and AAT-2 from effective root nodules were purified to near homogeneity using conventional protein purification procedures with a final step of FPLC (4,7). High titre polyclonal antibodies to native and denatured forms of AAT-1 and AAT-2 were produced in NZW rabbits. Nondenaturing PAGE was performed using the standard Tris-glycine system of Ornstein (18). AAT activity and enzyme protein after native-PAGE were identified as previously described (7). Electrophoresis of proteins by SDS-PAGE was performed as described by Maizel (11). Proteins were transferred to nitrocellulose and probed with antibodies as described by Miller et al. (14). Rocket immunoelectrophoresis as described by Laurell (10) was used to quantitate AAT-2 enzyme protein. Methods for immunoprecipitation of AAT activity were similar to those described by Miller et al. (14) for phosphoenolpyruvate carboxylase. Alfalfa nodule polyA$^+$ RNA (10 μg) was used to construct a cDNA expression library in λgtll according to established protocols (12). The library was screened with AAT-2 antibodies using goat-antirabbit IgGs conjugated to horseradish peroxidase as the secondary antibody.

RESULTS AND DISCUSSION

Native-PAGE of alfalfa effective nodule and root suspension cell soluble proteins showed two major regions when stained for AAT activity (Fig. 1). The slower migrating band was designated as AAT-1, while the faster migrating bands were designated as AAT-2. When equal amounts of total AAT activity from crude extracts of nodules and root suspensions were loaded on the gels, nodules stained more intensely for AAT-2 than for AAT-1. By comparison, root suspension cell extracts stained more intensely for AAT-1 than for AAT-2. Electrophoresis and activity staining of purified preparation of AAT-1 and AAT-2 showed that the two major forms of AAT could be effectively separated from one another.

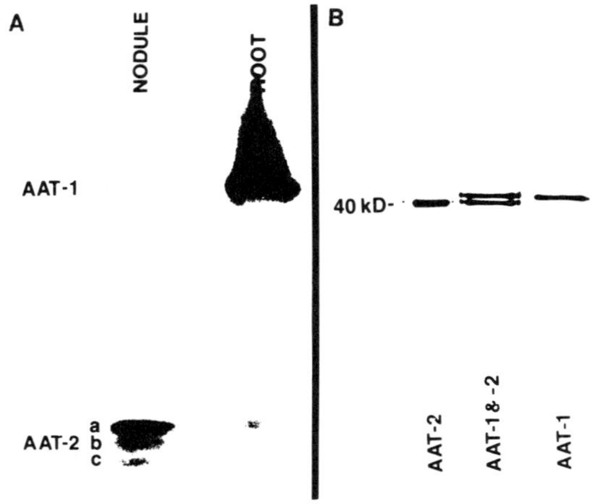

Figure 1. Native- and SDS-PAGE of AAT-1 and AAT-2 from alfalfa nodules and root suspension cell cultures. A) Isozyme staining for AAT activity from cell free preparations of effective nodules and root suspension cells. B) Purified preparations of AAT-1 and AAT-2 separated by SDS-PAGE. Native lanes contain the same amount of AAT activity while SDS lanes contain the same amount of AAT protein. This figure is representative of three separate experiments.

A relatively simple five step procedure involving hydroxylapatite, $(NH_4)_2SO_4$, fractionation, gel filtration, and FPLC gave 369- and 447-fold purification of AAT-1 and AAT-2, respectively (Table 1). Exclusion chromatography and SDS-PAGE showed that AAT-1 had a native MW of 90 kD and a subunit MW of 42 kD. By comparison the native and subunit MWs of AAT-2 were 80 and 40 kDs, respectively. Purified AAT-2 consistently was resolved into three bands AAT-2a, 2b, and 2c. Griffith and Vance (7) purified all three forms of AAT-2 and showed they had identical MWs. Thus, alfalfa AATs are similar to other AATs in MW and the holoenzymes are dimeric being comprised of two similar MW polypeptides.

Table 1. Purification of AAT-1 From Alfalfa Root Suspension Cells and AAT-2 From Effective Root Nodules.

Enzyme Form	Specific Activity[a]	-fold Purification[a]	Yield[a]	Native MW[a]	Subunit MW[a]
	units·mg protein^{-1}		%	--------kD---------	
AAT-1	181	369	16	90±4	42±2
AAT-2	169	447	30	80±2	40±2

[a]Data are representative of four separate determinations.

Western blots and immunoprecipitations of in vitro AAT activity showed that AAT-1 and AAT-2 are strikingly different. Western blot analysis of equal amounts of AAT-1 and AAT-2 after SDS-PAGE showed that AAT-1 antiserum (Fig. 2A) readily recognizes the 42 kD subunit of AAT-1, but not the subunit form of AAT-2. Likewise, AAT-2 antiserum (Fig. 2B) readily recognizes the 40 kD subunit of AAT-2 but does not recognize the subunit form of AAT-1. Immunoprecipitation of in vitro AAT activity further verified the dissimilarity of the two major forms of AAT. Farnham et al. (4) demonstrated that AAT-1 antibodies immunoprecipitated only AAT-1 activity, while AAT-2 antibodies immunoprecipitated only AAT-2 activity.

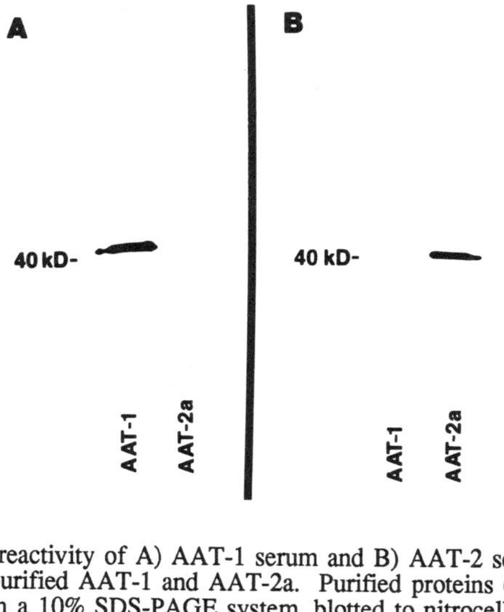

Figure 2. Cross-reactivity of A) AAT-1 serum and B) AAT-2 serum with subunit polypeptides of purified AAT-1 and AAT-2a. Purified proteins (1 μg each lane) were electrophoresed in a 10% SDS-PAGE system, blotted to nitrocellulose and probed with A) AAT-1 serum or B) AAT-2 serum. Blots are representative of four experiments.

These data conclusively demonstrate that alfalfa AATs are physically and immunologically distinct. Immunological distinction between AAT-1 and AAT-2 indicates that the amino acids comprising their various antigenic epitopes are strikingly different and suggests that the isozymes are encoded by separate distinct genes. Similar immunological distinction between animal mitochondrial (mAAT) and cytosolic (cAAT) AATs have been noted previously (15). In recent years antibodies to animal AATs have been used to isolate mAAT and cAAT cDNAs and characterize the proteins expressed by these AATs (9,17).

Immunological distinction has been noted for other plant enzymes including glutamine synthetase (GS) and NADH- and Fd-glutamate synthase (GOGAT) (8,26). However, such distinction is seldom as complete as that observed for AAT-1 and AAT-2. Numazawa et al. (16) have purified an AAT from leaves of the C_4 species Panicum maximum and produced polyclonal antibodies to the enzyme protein. Their Panicum AAT appears to be similar to alfalfa AAT-1, in that it has a native and subunit MW of 100 and 42 kD, respectively, and appeared not to recognize a second form of AAT in Panicum leaves.

Since AAT-1 and -2 antibodies are so highly specific they can be used to quantitate the relative and absolute contribution of each form to total AAT activity in various organs and tissues.

Our laboratory has previously shown that total AAT activity in alfalfa increases dramatically during effective nodule development and this increased activity is associated with enhanced AAT-2 activity staining on native-PAGE (7). We, therefore, used rocket immunoelectrophoresis to quantitate the contribution of AAT-2 enzyme protein to this increase (Table 2). Total nodule AAT activity increased 7-fold between 7- to 17-days after planting, coincident with N_2 fixation (7). The concentration of AAT-2 enzyme protein showed an almost identical increase. The data verified our earlier suggestion that expression of AAT-2 is closely linked to alfalfa nodule development while AAT-1 is root associated.

Table 2. Total AAT Activity and Concentration of AAT-2 Enzyme Protein During Development of Effective Alfalfa Nodules.

Days After Planting	Tissue	Acetylene Reduction	Total AAT Activity[a]	AAT-2 Concentration[a-b]
		-----µmole·min·g FW^{-1}------		µg·g FW^{-1}
5	roots	0.0	0.5	23
7	nodules plus small root segments	0.0	1.2	45
10	nodules	13.5	4.7	172
17	nodules	17.0	8.2	291
38	nodules	17.0	8.6	373

[a]Activity values are the average of three separate extractions while concentration values are the average of two separate extractions.
[b]Rocket immunoelectrophoresis was used to quantitate AAT-2 enzyme protein.

Immunoprecipitation with AAT-2 antibodies allowed us to determine the relative contribution of AAT-2 to total AAT activity in organs of alfalfa, soybean, wheat, and maize (Table 3). In alfalfa and pea nodules AAT-2 comprised 70% or more of the total AAT activity, while in soybean and common bean AAT-2 comprised only 25-30% of the total AAT activity. These data indicate that in nodules of soybean and common bean, two ureide forming species, AAT-2 contributes much less to total AAT activity than it does in nodules of pea and alfalfa, two amide forming species. This suggests that biochemical and genetic regulation of AAT in nodules of ureide types may be quite different than that in amide types.

In all species tested AAT-2 comprised more of the total leaf AAT activity than of the total root activity (Table 3). This was particularly true for alfalfa, soybeans, and wheat rather than for maize. This provides additional support for AAT-1 being the predominant form of AAT associated with root development. The data also indicates that since leaves as well as nodules express AAT-2 the increased AAT-2 activity in nodules probably does not reflect a nodule specific induction phenomenon. This is further supported by the fact that all three forms of AAT-2 detected in alfalfa nodules can also be detected in alfalfa leaves (4). Furthermore, on western blots probed with AAT-2 antibodies leaves and nodules of all species tested had strongly reacting polypeptides of near 40 kD while roots were deficient in this band.

Table 3. Contribution of AAT-2 to total AAT Activity of Nodules from Alfalfa, Soybean, Pea, and Common Bean and From Roots and Leaves of Alfalfa, Soybean, Wheat and Maize.

Species	Tissue	Total AAT Activity[a] Preimmune Control[b]	Total AAT Activity[a] After Precipitation	Contribution of AAT-2
		----nmoles product·min^{-1}----		%
Alfalfa	nodules	31.0	9.0	71
Soybean	nodules	33.0	24.8	25
Pisum	nodules	33.0	9.5	71
Common Bean	nodules	29.0	18.8	35
Alfalfa	roots	26.8	18.3	32
Alfalfa	leaves	25.2	6.0	76
Soybean	roots	24.8	23.1	7
Soybean	leaves	22.6	8.4	63
Wheat	roots	32.3	31.1	4
Wheat	leaves	22.5	8.2	64
Maize	roots	17.6	9.4	47
Maize	leaves	23.1	9.7	58

[a]Activities are the average of two separate determinations.
[b]Preimmune controls were standardized to near similar activities.

For several years we have been interested in selecting alfalfa plants for traits associated with N_2 fixation and assimilation (2). We recently completed bidirectional selection for root nodule AAT activity in two diverse germplasm sources (Table 4). Divergent selection resulted in subpopulations that were significantly different in total AAT activity. Immunoprecipitations showed that high AAT subpopulations had a greater contribution of AAT-2 to total activity than did low subpopulations. Such data provides additional support for the importance of AAT-2 to root nodule AAT activity in alfalfa. In these glasshouse experiments, nodule mass compensated for changes in nodule AAT activity, therefore, selection in either direction appeared not to alter plant biomass. The molecular and biochemical bases for selection remain to be established.

Table 4. Selection for Alfalfa Total Root Nodule AAT Activity and Contribution of AAT-2 to Selection.

Germplasm Source	Subpopulation	Total AAT Activity	Contribution of AAT-2
		nmol·min·g FW^{-1}	%
Saranac	Hi	10,488	78.5
	Base	8,690	74.6
	Lo	8,665	73.5
MnPl-10 x MnNC-7	Hi	11,696	80.8
	Base	10,817	79.9
	Lo	9,927	75.2
	LSD	1,646	5.4

In efforts to isolate an AAT-2 cDNA clone we prepared a cDNA library in λgtll from poly A^+ RNA isolated from 17 to 20 day old effective nodules. Using AAT-2 antibodies we immunoscreened 1.4 x 10^5 plaques and isolated four clones which express AAT-2 crossreactive polypeptides. The largest cDNA clone is approximately 1.8 kb, a sufficient size to encode the 40 kD polypeptide of AAT-2. A comparison of total root and nodule RNA on a Northern blot probed with this 1.8 kb cDNA showed that hybridizing mRNA is much more abundant in nodules than in roots. Sequence analysis of one 500 bp alfalfa nodule AAT-2 cDNA showed 40% homology with chicken mitochondrial AAT. Reynolds et al. (20) presented preliminary data on the isolation of an AAT clone from lupine, however, we are not aware of a report characterizing this clone. Lastly, we are making a cDNA expression library to root suspension cell poly A^+ RNA and it will be screened with AAT-1 antibodies.

AAT plays a key role in the flow of N and C in alfalfa root nodules. The enzyme occurs as two distinct forms (AAT-1 and AAT-2) with each probably being encoded by two separate genes. Our antibodies to AAT-1 and AAT-2 offer exceptional opportunities to evaluate AAT regulation at the biochemical and molecular level.

Acknowledgements: The authors acknowledge the support of USDA-ARS (C.P.V., M.W.F., D.K.B.), USDA-CRGO Grant 87-CRCR-1-2588 (C.P.V.), and NSF:DCB Grant 8905006 (C.P.V., R.J.L. S.G.) for research discussed in this manuscript.

1. Appels, M.A. (1989) Ph.D. Thesis (Agricultural University, Wageningen, The Netherlands), pp. 53-78.
2. Barnes, D.K., Heichel, G.H., Vance, C.P. & Ellis, W.R. (1984) Plant Soil **82**, 303-314.
3. Diaz, F. & Jouve, N. (1986) Euphytica **35**, 129-135.
4. Farnham, M.W., Miller, S.S., Griffith, S.M. & Vance, C.P. (1990) Plant Physiol. **93**, in press.
5. Forrest, J.C. & Wightman, F. (1972) Can. J. Biochem. **50**, 813-829.
6. Givan, C.V. (1980) in The Biochemistry of Plants, Vol. 5, eds. Stumpf, P.K. & Conn E.E. (Academic Press, New York), pp. 329-357.
7. Griffith, S.M. & Vance, C.P. (1989) Plant Physiol. **90**, 1622-1629.
8. Hirel, B., McNally, S.F., Gadal, P., Sumar, N. & Stewart, G.R. (1984) Eur. J. Biochem. **138**, 63-66.
9. Joh, T., Nomiyama, H., Maeda, S., Shimada, K. & Morino, Y. (1985) Proc. Natl. Acad. Sci. USA **82**, 6065-6069.

10. Laurell, C.B. (1966) Anal. Biochem. **15**, 45-52.
11. Maizel, M.V. (1971) in Methods in Virology, Vol. 5, eds. Maramorosch, K. & Koprowski, H. (Academic Press, New York), pp. 179-224.
12. Maniatis, T., Fritsch, E.F. & Sambrook, J. (1982) Molecular Cloning: A Laboratory Manual (Cold Spring Harbor Laboratory, Cold Spring Harbor, NY)
13. Maxwell, C.A., Vance, C.P., Heichel, G.H. & Stade, S. (1984) Crop Sci. **24**, 257-264.
14. Miller, S.S., Boylan, K.L.M. & Vance, C.P. (1987) Plant Physiol. **84**, 501-508.
15. Morino, Y., Kagamiyama, H. & Wada, H. (1964) J. Biol. Chem. **239**, PC943-944.
16. Numazawa, T., Yamada, S., Hase, T. & Suqiyama, T. (1989) Arch. Biochem. Biophys. **270**, 313-319.
17. Obaru, K., Nomiyama, H., Shimada, K., Nagshima, F., & Morino, Y. (1986) J. Biol. Chem. **261**, 16976-16983.
18. Ornstein, L. (1964) Ann. N.Y. Acad. Sci. **121**, 404-427.
19. Reynolds, P.H.S., Boland, M.J. & Farnden, K.F. (1981) Arch. Biochem. Biophys. **209**, 524-533.
20. Reynolds, P.H., Jones, W.T., Grant, M.R., Rodbeu, H.A. & Smith, L.A. (1988) in Nitrogen Fixation: Hundred Years After, eds. Bothe, H., deBruijn,F.J. & Newton, W.E. (Gustav Fischer, Stuttgart), pp. 574.
21. Rosendahl, L. & Vance, C.P. (1990) Plant Physiol. **92**, in press.
22. Ryan, E., Bodley, F. Fottrell, P.F. (1972) Phytochemistry **11**, 957-963.
23. Scandalios, J.G., Sorenson, J.C. & Ott, L.A. (1975) Biochem. Gen. **13**, 759-768.
24. Schubert, K.R. (1986) Annu. Rev. Plant Physiol. **90**, 1622-1629.
25. Snapp, S. & Vance, C.P. (1986) Plant Physiol. **82**, 390-395.
26. Suzuki, A. & Gadal, P. (1984) Physiol. Veget. **22**, 471-486.
27. Ta, T.C., Faris, M.A., & Macdowall, F.D.H. (1986) Plant Physiol. **80**, 1002-1005.
28. Wallsgrove, R.M., Keys, A.J., Lea, P. & Miflin, B.J. (1985) Plant Cell Environ. **6**, 301-309.

Gresshoff, Roth, Stacey & Newton (eds.) Nitrogen Fixation: Achievements and Objectives
Chapman and Hall New York • London 1990

Regulation of host gene expression during nodule development in soybeans

C. Sengupta-Gopalan, E. Estabrook, H. Gambliel, W. Nirunsuksiri and H. Richter
Dept. of Agron. & Hort./Mol. Biol. Prog., New Mexico State Univ., Las Cruces, NM 88003, USA.

Introduction

The processes leading to the development of the nitrogen fixing root nodule in legumes is highly complex and involves many discrete events. In this symbiotic association between the leguminous plant and bacteria in the genus Rhizobium/Bradyrhizobium, the symbionts are in constant communication, both emitting and responding to a variety of signals that influence the progression of this association. Root nodule development is accompanied by the expression of several host and rhizobial genes. These genes are expressed in a very regulated manner at very distinct developmental stages, suggesting involvement of multiple but distinct induction signals.

The first known signals that affect changes in gene expression during symbiosis are phenylpropanoid derived compounds produced by the plant that stimulate the transcription of the nod genes (8). Furthermore, these compounds also function as host defense molecules when plants encounter biotic or abiotic stress. Two key enzymes in the synthesis of flavanoids and isoflavanoids, are phenylalanine ammonia lyase (PAL) and chalcone synthase (CHS). The first part of this chapter will deal with the expression of PAL and CHS genes during nodule initiation and development in soybeans leading to effective or ineffective symbiosis.

The second part of this chapter will focus on the regulatory aspect of soybean genes encoding for proteins of known biochemical function and known cellular location with the aim of emphasizing the multiplicity of regulatory signals. Some of the highly expressed nodulin genes in soybean can be placed in two distinct families based on extensive stretches of sequence homology in their coding region. We will also focus on the structure, organization and regulation of members of these gene families.

Results and Discussion

Expression of PAL and CHS genes during nodule development: Since the legume:Rhizobium interaction resembles a plant:pathogen interaction so closely and since the products derived from the phenypropanoid pathway, act both as symbiotic signals and as defense molecules, it is appropriate to ask the question -- how does the host differentiate between a symbiont and a pathogen? Towards addressing this problem, we have focused on genes encoding for PAL and CHS. PAL and CHS are encoded by multigene families. Since we are interested in genes that are expressed in nodules, cDNA clones for the two enzymes were isolated from a soybean nodule cDNA library. The cDNA clones were sequenced and the sequence compared with the published sequence for other PAL and CHS genes. Based on sequence comparison to all the bean PAL genes (3), the 3' untranslated region was found to be gene specific, and as such this region of the nodule PAL cDNA clone was used as a gene specific probe. For the CHS cDNA clone, we still have not identified a gene specific region. We have in our analysis for expression of PAL and CHS genes used a conserved domain of PAL, the 3' gene specific region of PAL and a region of the CHS gene as probes.

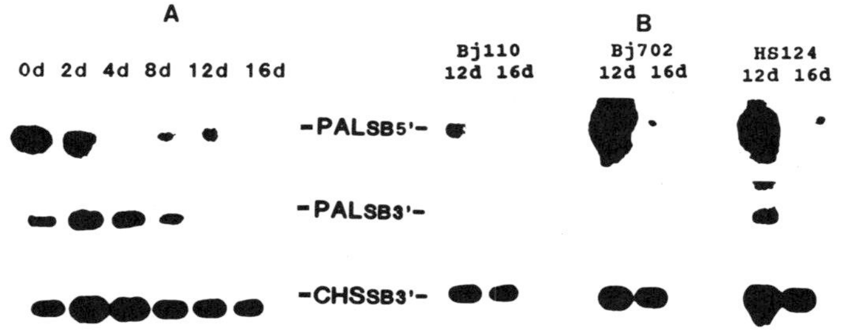

Fig. 1. Expression of PAL and CHS genes during nodule development leading to effective or ineffective symbiosis. A. Poly(A)-RNA from roots of a supernodulating mutant line of soybeans at different times following infection with a WT strain of B.japonicum, was subjected to Northern blot hybridization using the 3 different probes as indicated; B. Poly(A)-RNA from roots of the same line infected with the WT strain, Bj 702 and HS124, 12d and 16d following infection was subjected to Northern blot hybridization using the probes as indicated.

To study the expression of CHS and PAL genes during the different stages of nodule development, we analyzed the root system of a super-nodulating mutant strain of soybean (2) at different times following infection with an effective strain of B. japonicum, for the corresponding transcript level. As seen in Fig. 1A, with both the CHS probe and the nodule PAL gene specific probe, a transient but significant increase in the level of the corresponding transcript was seen during the early days following infection (day 2 and 4). The common PAL probe, however, did not show this transient increase in the transcript level. That this transient increase in expression of CHS and the specific PAL gene is due to interaction with the compatible bacteria was confirmed by demonstrating that infecting soybean plants with R. meliloti, showed no change in expression of CHS and PAL genes. Analysis of soybean roots infected with mutant strains of B. japonicum, Bj 702 (defective in structural nitorgenase) and HS 124 (defective in bacterial release) showed a transient but dramatic increase in the expression of PAL genes (Fig. 1B). The increase in expression of PAL genes, however, cannot be attributed to the specific nodule PAL gene. Expression of CHS at the time points tested, showed just a minor increase. Taken together these results indicate that there are at least two PAL genes that have a role in the development of an effective nodule. It also appears that the CHS genes are regulated differently

transcript is drastically reduced while the level of N-35 transcript in the same nodules is reduced to a much lesser extent. In the absence of gene specific probes for GS and Lb, the different forms of Lb and GS proteins were identified by immunoblotting following 2D gel electrophoresis of the soluble fraction of the different nodules. As seen in Fig. 2B, the Lb and GS profile was identical for the effective nodules and nodules made by bacteria defective in functional nitrogenase suggesting that the induction of the Lb genes and GS_n genes is independent of N_2 fixation. However, not all the Lb gene products and the GS_n proteins were made in nodules formed by the strain Bj2101. These results suggest that the GS_n genes and the late Lb genes are not expressed until after the stage of bacterial transformation, a stage by which N-35 and many other late nodulins including the early Lb genes have been induced. However, interestingly enough, analysis of the isozyme pattern for GS, shows that in the absence of nitrogenase activity, the GS_n polypeptides do not assemble into an active holoenzyme suggesting that NH_3, somehow, influences the activation of the GS_n holoenzyme. It is interesting to note that the expression of the N-35 gene in the uninfected cells is strongly influenced by events occurring in the infected cells.

Characterization of 2 nodulin gene families in soybean:

i. By comparing the nucleotide sequence of several nodulin cDNA clones, a nodulin gene family has been identified. These genes are related to one another by domains of conserved sequences in their coding region. Based on the RNA hybridization pattern, the genomic DNA hybridization pattern and the profile of the hybrid-select translation products corresponding to the different members of this gene family, we would predict that there are at least seven different functional members and two to three truncated or pseudogenes (Table 1). Based on quantitative RNA dot blot hybridization

Table 1: Characterization of a small nodulin gene family

Nodulins	Protein KD	RNA Kb	EcoRI Genomic fragments Kb
N-44	44*	1.6	7.0
N-44 (4)	44		7.0
N-29	29	1.1	
N-23 (4)	(24.3)		
N-28	28	1.2	5.0 2.7
N-26b (4)	(25.5)		
N-27	27*	1.3	5.2 4.3 2.7
N-27 (4)	27*		5.8
N-22 (9)	(22.7)	1.2	5.0
N-26,25,23	26 25 23	1.0 0.8	10 7 5 4.6 4.2
N-20 (9)	(20)	1.0 0.8	12 7 5 4.5 4.2

The underlined nodulins were characterized in our laboratory;
* Immunoreactive with Ab to the soluble-fraction of nodules;
Values in ( ) is MW based on predicted amino acid sequence.

from the PAL genes during nodule development. Furthermore, by analyzing the root tips and the root hair zone, we have demonstrated that the transient increase in the expression of CHS is localized to the root hair zone and that the factor responsible for this induction is secreted by the free living B. japonicum into the culture medium. Studies are in progress to determine the nature of signals that trigger PAL and CHS genes during nodule development.

Genes encoding for leghemoglobin, glutamine synthetase and nodule specific uricase: To elucidate the multiplicity of signals that induce nodulin genes, we have chosen to study genes encoding for leghemoglobin (Lb), glutamine synthetase (GS) and nodule-specific uricase (N-35). The globin component of Lb is encoded by the host and the heme component by the bacteria. It is localized in the infected cells and functions in buffering the O_2 concentration so as to enable the O_2 labile nitrogenase to function. In soybeans, Lb is encoded by four functional genes (11) and some of the gene products undergo post translational modification. In soybeans, two sets of GS genes are expressed in the nodule, a set that is expressed both in the roots and nodules (GS_r) and a set of 4 genes that are expressed in the nodules and not in the roots (GS_n) (10). GS is the first plant encoded enzyme that functions in ammonia assimilation. N-35 is the nodule specific form of uricase which catalyzes the conversion of uric acid to allantoin and is localized in the peroxisomes of the uninfected cells (1).

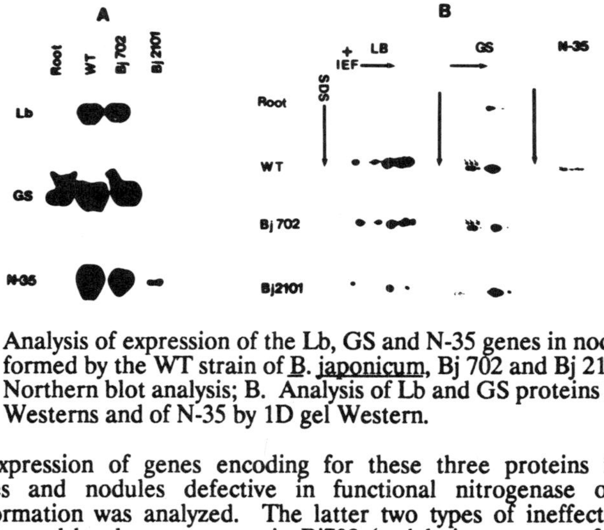

Fig. 2. Analysis of expression of the Lb, GS and N-35 genes in nodules formed by the WT strain of B. japonicum, Bj 702 and Bj 2101. A. Northern blot analysis; B. Analysis of Lb and GS proteins by 2D gel Westerns and of N-35 by 1D gel Western.

The expression of genes encoding for these three proteins in effective nodules and nodules defective in functional nitrogenase or bacterial transformation was analyzed. The latter two types of ineffective nodules were formed by the mutant strain Bj702 (a deletion mutant of the nif KD gene) and Bj2101 (a deletion mutant of the nif A gene), respectively. As seen in Fig. 2A, analysis of the steady state level of RNA in nodules 14d after infection, shows that the expression of Lb and GS genes was not affected by defect in nitrogenase activity while the level of N-35 transcript was decreased slightly in the same nodules. Nodules defective in the step immediately following bacterial release (formed by Bj2101), the level of Lb and GS

studies, we find at least a 4 fold higher level of accumulated RNA for these genes than for the leghaemoglobin genes suggesting the presence of a high level of these gene products in soybean nodules. All members of this gene family that have been sequenced so far, show conservation in the 5' coding region which translates into a hydrophobic amino terminal end, typical of a signal peptide. However, as indicated in table 1, some of the protein members of this gene family have antigenic determinants in the soluble fraction of the nodule while others have it in the pbm fraction. A conserved domain of this gene family used as a probe, showed hybridization to cowpea nodule RNA but not to alfalfa nodule RNA. Whether the results imply that this class of proteins is made only in the nodules of ureide transporting legumes, is not clear at this stage.
Analysis of genomic clones containing the gene for one of the members of this gene family, N-20, showed that the gene is flanked by a related gene, N-20'. As seen in Fig. 3, the flanking gene is a truncated version of the N-20 gene containing the promoter region and 5' region of the first exon. A stretch of 458bp 3' to the N-20 gene follows the N-20' gene maintaining a sequence identity of >99%. Hybridization of soybean genomic DNA with this repeat unit indicates that the repeat unit is associated with DNA fragments containing N-20 related coding sequences. These results would suggest that probably the N-20 related genes, which would include 3 functional genes (based on HST results) and the N-20' gene are all clustered and have resulted from gene duplication. The near absolute identity in the sequence of this 3' repeat units would suggest that the region has a regulatory role and is under functional constraints.

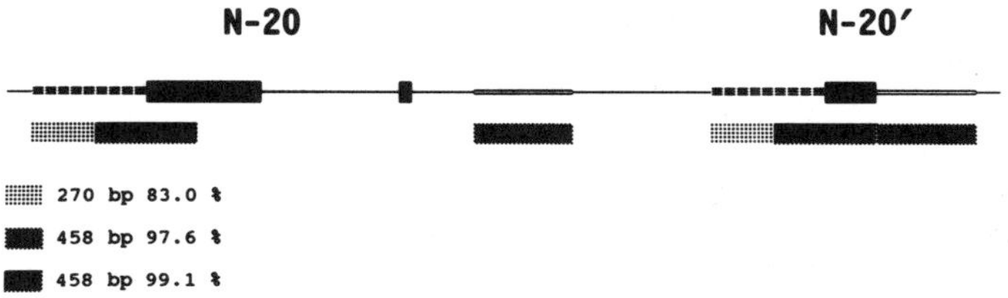

Fig. 3. Comparison of the nucleotide sequence of N-20 and N-20'. The exons are represented as solid boxes, the hatched region is the promoter region that shares sequence homology and double line indicates the 3' repeat region. The percent homology is shown in the figure.

ii. DNA sequence analysis of a cDNA clone which encodes for nodulin-16, shows high degree of localized sequence similarity with the coding sequence of a soybean pbm nodulin, N-24 (7). The comparison of the N-16 and N-24 genes is shown in Fig. 4. Sequence similarity between the two genes is limited to the entire 90bp long coding region in the first exon and the first 54 bp in the second exon of the N-16 gene which is repeated as the 2nd, 3rd, 4th and part of the 5th exon of the N-24 gene. However, the 2nd, 3rd, and 4th exons of the N-24 gene are flanked by intron sequences that also share high degree of sequence homology, making the repeat unit 180 bp long. The only intron present in the N-16 gene shares no sequence similarity with the introns of the N-24 gene. Since this 180 bp repeat unit in the N-24 gene resembles

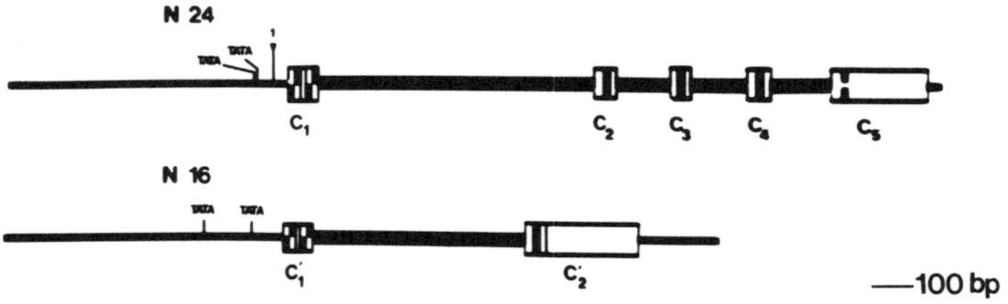

Fig. 4. Diagrammatic representation of the N-24 and N-16 genes. The coding region in the 1st exon (c_1/c'_1) and the remaining exons (c_2/c'_2, c'_3, c'_4 and c'_5) are represented as boxed-in regions and the introns are represented by thick lines while the thinner lines represent the 5' and 3' flanking regions. Sequence similarity in the first exons is represented while the sequence similarity between the second exon of N-16 gene and the 2nd, 3rd and 4th exons of N-24 genes is represented as . The low level of sequence similarity with the 5th exon which does not translate into a significant aa sequence similarity is shown as

an insertion element, it is postulated that the N-24 gene is generated via duplication of this inserted element (7). One of the hypothetical ancestral gene for N-24 would be one with only one of the 180 bp repeat unit. However, the N-16 gene, even though it contains a 54 bp stretch which is similar to the repeat exons of N-24, does not fit the description of this ancestral gene, since it lacks the flanking intron sequences that forms part of the 180 bp repeat unit in the N-24 gene.

Both N-16 and N-24 genes have a similar hydrophobic amino terminal end, which in the case of N-24 is cotranslationally cleaved (7), suggesting that N-16 and N-24 proteins have to pass through a vesicular route. Hybrid select translation products corresponding to N-24 and N-16, can be immunoprecipitated with antibody to the soluble fraction of nodules. Nodulin-24 is immunoreactive also with antibody to the pbm fraction of nodules (7). Taken together, these results suggest that N-24 and N-16 are probably associated with pbm surface. It is interesting to note that the two genes, even though they appear to be unrelated phylogenetically, have maintained a high degree of similarity in the 90 bp of coding sequence contained in the first exon which encodes for the hydrophobic amino terminal domain. This is comparable to the other nodulin gene family represented by N-44, N-20 etc. where also, one of the conserved domains is the region encoding for the hydrophobic amino terminal end.

Analysis of soybean genome for N-16 and N-24 related genes showed the presence of multiple copies of the genes. In fact, analysis of a genomic clone containing the gene for N-16 shows the presence of another related sequence in the 13Kb soybean DNA insert, suggesting that some of these closely related genes are linked. That some of these genomic sequences infact encode for proteins is evident from the fact that two dimensional gel electrophoresis of the hybrid-select translation products with N-16 or N-24 coding sequences, shows multiple spots in each case.

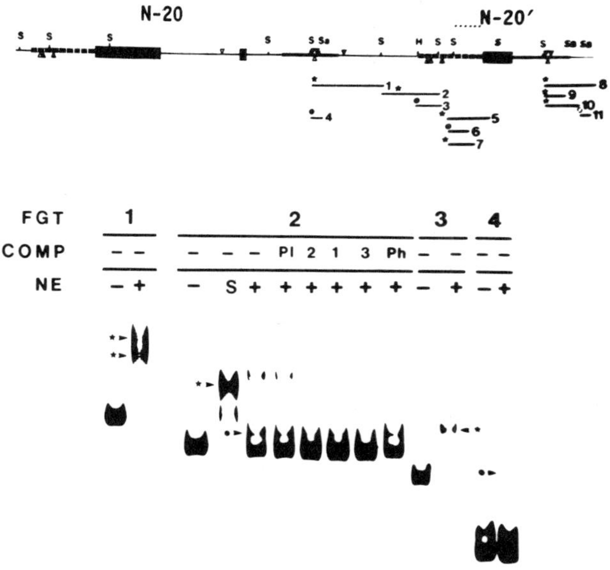

Fig. 5. Delimitation of binding sites on the N-20/N-20' gene cluster. A. Restriction map indicating the sites for enzymes used to generate DNA probes. Abbreviations: S, Ssp1; Sa, Sau 1; H, Hind III; Hh, Ha1; Sc, Sac 1; D, Dra 1; X, Xba1. The DNA fragments are numbered binding sites 1 and 2 are represented as ▽ and △, respectively. The symbols ★, •, ○ present HMC, LMC and no binding, respectively. B. Gel retardation analysis of representative DNA fragments (FGT). The presence of competitor DNA (COMP) and nuclear extract (NE) is indicated. Abbreviations: Pl, plasmid; Ph, β-phaseolin promoter region; S, seed nuclear extract. DNA-protein complexes are indicated by ➤.

Promoter analysis of the N-20/N-20' and the N-16 genes: In order to understand how nodulin gene expression is regulated, we have undertaken to analyze the 'cis' -acting and 'trans' -activating factors that regulate representative members of the two gene families. Conservation of some of the regulatory sequences would imply involvement of a universal regulatory mechanism underlying organ specific expression of nodulin genes while regulatory sequence unique to the two genes would suggest that they function as the receptor for a specific signal. As a first approach, we have initiated DNA binding studies with the different regions of the N-20/N-20' gene cluster and N-16 gene cluster. As shown in Fig. 5, we have identified several DNA binding regions in the N-20/N-20' gene cluster, one of which is in the 458bp region just 3' of the N-20' gene. Two types of DNA-protein complexes can be identified, a low molecular weight complex (LMC) which appears less specific than the other high molecular weight complex (HMC) which is specific. Whether, some of this binding can be attributed to HMG proteins as indicated by Jacobson et al. (5) or whether they represent DNA binding to nuclear scaffold is not known at this stage. However, using nuclear extract from bean seeds with fragment 2, a protein-DNA complex is formed which has a different migration from the DNA-protein complex formed with the nuclear extract from nodules. Furthermore, in a competition experiment, the DNA binding domain of the β-phaseolin gene did not compete with fragment 2 for the same nuclear factor. Taken together, the results suggest that the binding is specific for nodule-nuclear factors. In most cases, the DNA fragments that show specific binding contain potential DNA binding sites as reported for the Lbc_3 gene of soybean (6) except fragment 5 which is also the conserved region for the different members of this gene family. Based on the binding properties of the different fragments, we predict that the different regions of the DNA act cooperatively to interact with the nuclear proteins. Analysis of the N-16 gene cluster for protein binding domains, reveal some DNA binding regions. Work is in progress to determine if there is any commonality in DNA binding domains between the two gene clusters. Furthermore, to understand the functional role of the different binding domains, several promoter deletion constructs are being made for insertion into the alfalfa genome.

References

1. Bergmann, H., Preddie, E. and Verma, D.P.S. (1983) EMBO J. 2, 2333-2339.
2. Carroll, B.J., McNeil, D.L. and Gresshoff. P.M. (1985) Proc. Natl. Acad. Sci. 82, 4162-4166.
3. Cramer, C.L., Edwards, K., Dron, M., Liang, X., Dildine, F.L., Bolwell, G.P., Dickson, R.A., Lamb, C.J., Schuch, W. (1989) Plant Mol. Biol. 12, 367-383.
4. Jacobs, F.A., Zhang, M., Fortin, M.G. and Verma, D.P.S. (1987) Nucl. Acids Res. 15, 1271-1280.
5. Jacobsen, K., Laursen, N.B., Jensen, E.O., Marcker, K., Poulsen, C. and Marcker, K.A. (1990) Plant Cell 2, 85-94.
6. Jensen, E.O., Marcker, K.A., Schell, J. and DeBruijn, F.J. (1988) EMBO J. 7, 1265-1271.
7. Katinakis, P. and Verma. D.P.S. (1985) Proc. Natl. Acad. Sci. 82, 4157-4161.
8. Long, S.R. (1989) Cell 56, 203-214.
9. Sandal, N.N., Bojsen, K. and Marcker, K.A. (1987) Nucl. Acids Res. 15, 1507-1519.
10. Sengupta-Gopalan, C. and Pitas, J.W. (1986) Plant Mol. Biol. 7, 189-199.
11. Wiborg, O., Hyldig-Nielson, J.J., Jensen, E.O., Paludan, K. and Marcker, KA. (1982) Nucl. Acid Res. 10, 3487-3494.

Nodulins and nodule development

H. J. Franssen, B. Scheres, C. van de Wiel, B. Horvath, M. Moerman, W.C. Yang, F. Govers and T. Bisseling

Department of Molecular Biology, Agricultural University, Dreijenlaan 3, 6703 HA Wageningen, The Netherlands

Introduction

Nodule formation on the roots of leguminous plants has been well analyzed morphologically. Based on these analyses, root nodule formation has been divided into 3 distinct steps; (i) the pre-infection stage, (ii) infection and nodule formation, and (iii) nodule function and maintenance. During the pre-infection stage (brady)rhizobia interact with growing root hairs and induce root hair curling. During the second stage, the bacteria invade the root hair cell and the root cortex through the infection thread. Meanwhile, but independently from the infection process, cortical cells start dividing at several places and at these sites, the nodule primordia are formed. Infection threads grow towards these primordia and upon contact rhizobia are released from the tips of the infection threads into the cytoplasm of the plant cells. During the third stage, the bacteria differentiate into pleomorphic bacteroids that fix atmosphere nitrogen.

There are two main categories of legume nodules, determinate and indeterminate. In general, temperate legumes such as *Pisum*, *Vicia*, *Trifolium* and *Medicago* develop indeterminate nodules, while determinate nodules occur on the roots of tropical legumes such as *Glycine*, *Phaseolus* and *Vigna*. Indeterminate and determinate nodules are composed of similar tissues. The central tissue contains infected and uninfected cells, and is surrounded by several other tissues that are also formed from the nodule meristem. The outermost tissue is the nodule cortex, formerly named the outer cortex. The nodule cortex is separated from the nodule parenchyma, previously named the inner cortex, by an endodermis. Vascular bundles are located in the nodule parenchyma. Indeterminate nodules are characterized by a persistent apical meristem. Consequently, the indeterminate nodule can be divided into distinct zones which differ in developmental stage. The most distal zone is the apical meristem, which is bounded proximally by a zone containing enlarging cells that become infected by *Rhizobium*. This is the invasion zone. In the early symbiotic zone, host cells differentiate into infected and uninfected cells, while infected cells fully packed with bacteroids together with uninfected cells are found in the late symbiotic zone. In this late symbiotic zone, nitrogen fixation and ammonia assimilation occur. The most proximal zone in a mature indeterminate nodule is the senescent region where both plant cells and bacteroids degenerate.

In contrast to indeterminate nodules, determinate nodules do not have a persistent meristem. After release of (brady)rhizobium from the infection thread, the infected cells continue to divide until about one week after the onset of nitrogen fixation. When mitotic activity has ceased, increase in nodule size is caused by cell expansion. As a consequence, the developmental phases in a determinate nodule are separately temporally rather than spatially.

The formation of both determinate and indeterminate nodules is accompanied by nodule-specific expression of genes, the so-called nodulin genes. Nodulin genes are expressed exclusively during the development of the symbiosis and consequently are neither expressed in uninfected roots nor in other parts of the host plant.

Nodulin genes are differentially expressed during nodule development. The majority of the nodulin genes are first expressed around the onset of nitrogen fixation. These genes are named late nodulin genes. Nodulin genes that are expressed at earlier stages of development are named early nodulin (ENOD) genes. Early nodulin genes are expressed when the plant becomes infected and the nodule structure is being formed.

We have focused on the expression of early nodulin genes during development of both a determinant nodule (on soybean) and an indeterminate nodule (on pea). Five early nodulin cDNA clones (i.e., pPsENOD2, pPsENOD3, pPsENOD5, pPsENOD12, and pPsENOD14) have been isolated from a cDNA library prepared from polyA$^+$ RNA isolated from pea nodules (6,7,8). Four early nodulin cDNA clones (i.e. pGmENOD2, pGmENOD13, pGmENOD40, and pGmENOD55) have been isolated from a cDNA library prepared from polyA$^+$ RNA from soybean nodules (1.2). In the following sections, we will discuss our data obtained from studies on the expression of nodulin genes represented by the isolated cDNA clones during nodule development, the characteristics of the encoded nodulins and their possible functions and how these cDNA clones can be used as tools to identify the signals involved in the bacterium-plant interactions.

Early nodulin gene expression during root nodule development

In situ hybridization using the isolated pea nodulin cDNA clones as probes demonstrated that pea early nodulin genes are expressed at different stages during root nodule development and in different cell types. Cells, which prepare for or are involved in , infection thread formation and cells which start to divide in the inner cortex to form the nodule primordium, express the ENOD12 gene. The cells of the nodule parenchyma express the ENOD2 gene as soon as this cell layer is formed by the activity of the nodule meristem or primordium (8) ENOD5 gene expression is switched on in cortex cells containing infection thread tips and reaches its highest level in the invasion zone consisting of elongating infected cells containing proliferating bacteria. ENOD3 and ENOD14 gene expression starts in infected cells of the early symbiotic zone, when ENOD5 transcripts are already present in these cells, but before late nodulin transcripts are detectable. This illustrates that ENOD12, ENOD5, ENOD3 and ENOD14, and the late nodulin transcripts mark successive developmental stages of the infected cell type in the central tissue of the mature nodule (7). Both the soybean nodulin genes GmENOD2 and GmENOD13 are expressed in the nodule parenchyma, while the GmENOD40 gene is expressed in the pericycle of the nodule vascular bundle. This strongly suggests that, in addition to the central tissue, other nodule tissues will have a specific function in the symbiotic nitrogen fixation process. Hence, different steps in nodule formation and several nodule tissues are characterized by the expression of specific nodulin genes. These results substantially extend the cytological criteria used thus far to describe nodule development. Our search for (early) nodulins has not been extensive. Nevertheless, we were able to isolate nodulin sequences representing genes that are expressed in 3 different nodule tissues. Therefore it is likely that, beyond these three tissues, the development of

Early nodulin function

Both soybean and pea ENOD2 early nodulin as well as the GmENOD13 early nodulin, marking the nodule parenchyma, and the PsENOD12 early nodulin, present in cells preparing for the infection process, are (hydroxy)-proline-rich proteins. These proteins are built up of pentapeptide repeat units, consisting of blocks of three amino acids flanked by two prolines, and are therefore most likely cell wall proteins. The occurrence of several specific cell wall proteins in nodule tissue with different functions points to an important role of the plant cell wall in tissue differentiation. Developmental regulation of genes encoding hydroxy-proline-rich cell wall proteins has recently also been demonstrated in soybean (4), in agreement with the suggestion that (hydroxy)-proline-rich proteins can play a general role in differentiation. Precise functions of the early nodulins are unknown, although the ENOD2 early nodulin might give the cell wall properties that enable it to function as an oxygen barrier. The function of the PsENOD5 protein is unknown, but the primary structure derived from the cDNA sequence allows some speculation as to the nature of the protein. The amino acid sequence of the proline-rich region in the ENOD5 polypeptide chain has some analogy to arabinogalactans. On the other hand, hydrophobic regions in the amino acid sequence point to the possibility of PsENOD5 being a membrane protein. The most striking feature of the PsENOD3 and PsENOD14 early nodulins are the cysteine clusters which indicate that these early nodulins might be capable of binding a metal ion. There is not sufficient amino acid sequence homology in the sequence between and around the cysteine clusters to allow ENOD3 and ENOD14 to be put in a particular class of proteins with cysteine clusters; like, for example, the group of zinc-finger proteins, which are in several cases known to be DNA binding proteins. We hypothesize that ENOD3 and ENOD14 might function in the transport of metal ions toward the bacteroids and the proteins are expected to occur in the peribacteroid space since, after cleavage of the signal peptide, the small ENOD3 and ENOD14 proteins are hydrophylic and presumably easily soluble. Based on the amino acid sequence derived from the nucleotide sequence of the soybean cDNA clone pGmENOD40, the encoded nodulin most likely represents a soluble protein since a signal peptide is lacking. The nodulin is probably involved in intracellular transport in the cells of the pericycle, since the main function of the pericycle is transport of nutrients from and to the vascular bundle (3). In the GmENOD55 encoded proteins, prolines are confined to an internal domain of 32 amino acids in which proline and serine occur alternating. Based on this amino acid sequence, the encoded protein may reside in the cell wall.

Early nodulins as tools to identify signals involved in bacteria-plant interaction

The multistep nature of root nodule formation indicates that at several stages specific signals from *Rhizobium* are involved in inducing the next step(s) in nodule development. The elucidation of the signal transduction from the bacterium to the plant is becoming topical. At present, only a *Rhizobium meliloti* signal molecule (5) inducing root hair deformation has been purified and characterized. This deformation factor can elicit root hair deformation in the absence of bacteria and so a bioassay could be used to purify this compound. However, it is very unlikely that bioassays can be developed to purify factors triggering the infection process or different steps of nodule development. Therefore, the availability of cloned nodulin genes that are markers for specific steps in nodule development will be essential to purify bacterial factors that induce these steps.

References

1. Franssen, H.J., Nap, J.P., Gloudemans, T., Stiekema, W., van Dam, H., Govers, F., Louwerse, J., van Kammen, A. & Bisseling, T. (1987) Proc. Natl. Acad. Sci. USA **84**, 4495-4499.
2. Franssen, H.J., Scheres, B., van de Wiel, C. & Bisseling, T. (1988) in *Molecular Genetics of Plant-Microbe Interactions*, eds. Palacios, R. & Verma, D.P.S. (APS Press, St. Paul) pp. 321-326.
3. Gunning, B.E.S., Pate, J.S., Minchin, F.R. & Marks, J. (1974) Symp. Soc. Exp. Biol. **28**, 87-126.
4. Hong, J.C., Nagao, R.T. & Key, J.L. (1989) Plant Cell **1**, 937-943.
5. Lerouge, P., Roche, P., Faucher, C., Maillet, F., Truchet, G., Prime, J-C. & Dénarié, J. (1990) Nature **344**, 781-784.
6. Scheres, B.,van de Wiel, C., Zalensky, A., Horvath, B., Spaink, H., van Eck, H., Zwartkruis, F., Wolters, A.M., Gloudemans, T., van Kammen, A. & Bisseling, T. (1990) Cell **60**, 281-294.
7. Scheres, B., van Engelen, F., van der Knaap, E., van de Wiel, C., van Kammen, A. & Bisseling, T. (1990) The Plant Cell in press.
8. van de Wiel, C., Scheres, B., Franssen, H., van Lierop, M.J., van Lammeren, A., van Kammen, A. & Bisseling, T. (1990) EMBO J. **9,** 1-7.

Gresshoff, Roth, Stacey & Newton (eds.) Nitrogen Fixation: Achievements and Objectives
Chapman and Hall New York • London 1990

Regulation of expression of the GLN-τ gene of *Phaseolus vulgaris* L.

Julie V. Cullimore, J. Mark Cock, Kevin J. F. Farnden* and Malcolm J. Bennett

Department of Biological Sciences, University of Warwick, Coventry, CV4 7AL, England.
*Department of Biochemistry, University of Otago, Dunedin, New Zealand.

In *Phaseolus* vulgaris glutamine synthetase (GS) is encoded by four expressed, nuclear genes (1), three of which (*gln-α*, *gln-β* and *gln-τ*) encode cytosolic GS polypeptides (α, ß and τ) whereas the fourth gene (*gln-δ*) encodes a precursor to the plastid-located GS polypeptide (δ). During nodulation the activity of GS increases about 8-fold in specific activity (2), presumably in order to assimilate the ammonium derived from dinitrogen fixation. Studies on isoenzymes of GS have shown that this increase is largely due to the production of τ-containing isoenzymes (2). The regulation of activity of the *gln-τ* gene therefore appears to be important for the assimilation of dinitrogen-fixed ammonium.

Previous work on the *gln-τ* gene had detected high abundances of its mRNA in nodules of *P. vulgaris* but not in roots and leaves, leading to the suggestion that this gene shows nodule-specific expression (3). We have used the extra sensitivity of an RNase protection technique to measure the abundance of the *gln-τ* mRNA in different organs of the plant and during nodule development. The results have shown that the *gln-τ* gene is expressed in roots but the abundance of the mRNA is between 150- and 750-fold lower than in nodules (4). The mRNA was not detected in leaves but surprisingly was present in stems and petioles (at an abundance about 50-fold lower than in nodules) and in germinating cotyledons (about 10-fold lower than nodules). Therefore this gene does not encode a nodulin but does however show a very strong enhancement of expression during nodulation.

We examined the factors that are involved in this nodule-enhancement of expression. In soybean it has been suggested that the increase in GS gene expression in nodules is due to the production of ammonium from dinitrogen fixation (5). In *P. vulgaris* this seemed unlikely as the *gln-τ* mRNA had been detected in nodules before the onset of dinitrogen fixation (6). We grew nodules under an atmosphere of argon/oxygen and found that the abundance of the *gln-τ* mRNA increased during nodulation, in a manner initially similar to the increase in nitrogen-fixing nodules (7). This increase also occurred in nodules formed with two Fix$^-$ *Rhizobium* mutants, which formed nodules with a normal morphology, including the presence of intracellular bacteroids. However in both types of non-fixing nodules the abundance of the *gln-τ* mRNA was about 2- to 4-fold lower than in fixing nodules and declined as the nodules aged. These results suggest that the ammonium from dinitrogen fixation is not the primary factor involved in the enhancement of expression of *gln-τ* during nodulation. However dinitrogen fixation is clearly required to increase and maintain the abundance of this mRNA in nodules.

In order to investigate the factors involved in the initial expression of this gene, plants were infected with two *Rhizobium* mutants which produce small white nodules, lacking intracellular bacteroids. In both cases the abundance of the *gln-τ* mRNA failed to increase over that found in roots (7). Therefore some factor produced only in more developed nodules, perhaps associated with intracellular *Rhizobia* or the differentiation of the bacteroids, is required to initiate the enhancement of expression of this gene.

Finally in experiments in which we added ammonium exogenously to nodules we found that, following the addition, the abundance of the *gln-τ* mRNA declined at a rate that was apparent by one-day and was much more pronounced by three-days. This occurred in both normal, fixing nodules and in the *Rhizobium*-determined and the argon/oxygen determined, non-fixing nodules. However in all these cases the ammonium appeared to cause the senescence of the nodule and therefore it is difficult to distinguish between whether the decrease in the *gln-τ* mRNA is due to an indirect consequence of nodule senescence or a more direct effect of ammonium. Moreover addition of nitrate caused an even more pronounced decrease in the *gln-τ* mRNA whereas the abundance of the *gln-ß* mRNA was largely unaffected by this treatment.

1. Forde, B.G. & Cullimore, J.V. (1989) in *Oxford Surveys of Plant Molecular & Cell Biology*, ed. Miflin, B. J. (Oxford University Press, Oxford), pp. 247-296.
2. Bennett, M. J. & Cullimore, J. V. (1989) *Planta* **179**, 433-440.
3. Gebhardt, C., Oliver, J. E., Forde, B. G., Saarelainen, R. & Miflin, B. J. (1986) *EMBO J.* **5**, 1429-1435.
4. Bennett, M. J., Lightfoot, D. A. & Cullimore, J. V. (1989) *Plant. Mol. Biol.* **12**, 553-565.
5. Hirel, B., Bouet, C., King, B., Layzell, D., Jacobs, F. & Verma, D. P. S. (1987) *EMBO J.* **6**, 1167-1171.
6. Padilla, J. E., Campos, F., Conde, V., Lara, M. & Sanchez, F. (1987) *Plant Mol. Biol.* **9**, 65-74.
7. Cock, J. M., Mould, R. M., Bennett, M. J. & Cullimore, J. V. (1990) *Plant Mol. Biol.* in press.

Gresshoff, Roth, Stacey & Newton (eds.) Nitrogen Fixation: Achievements and Objectives
Chapman and Hall New York • London 1990

The glutamine synthetases of *Rhizobium leguminosarum* and their regulatory genes

R. Defez, M. Chiurazzi, M. Manco, E. Patriarca, A. Lamberti, A. Riccio, C. Lopes, S. Colonna-Romano, S. Moreno, R. Meza, G. Espin and M. Iaccarino

International Institute of Genetics and Biophysics, via Marconi 10, 80125 Napoli, Italy and Centro de Fijacion de Nitrogeno, UNAM, Ap.do Postal 565-A, Cuernavaca, Mor, Mexico

The regulation of *Rhizobium* genes involved in nitrogen metabolism is important for the bacterial developmental process leading to the symbiotic interaction with leguminous plants. We are engaged in a study of the different glutamine synthetases of *R.leguminosarum* and their regulation, and describe our current results here.

We previously reported the cloning and sequencing analysis of the *R.leguminosarum* genomic region containing the *gln*A and *glnB* genes (1, 2) coding for GSI and P_{II}, respectively. We now find the presence of a *glnB-glnA* and of a *glnA* mRNA, neither of which changes more than three-fold in concentration when bacteria are grown on different nitrogen sources. Primer extension analysis shows unique transcriptional initiation sites upstream of *glnB* and *glnA*. The *glnB* promoter is of the *rpoN*(*ntrA*) type and requires, in *Klebsiella pneumoniae*, an active *ntrC* gene, but it is not preceded by NtrC binding sites. The *glnA* promoter does not contain a typical consensus sequence for previously described promoters and is, in *K.pneumoniae*, *ntrC*-independent. We report elsewhere (see the abstract of Moreno *et al.*) that in a *ntrC* mutant of a different biovar of *R.leguminosarum glnB* is not expressed and adenylylation of GSI is reduced. A *glnA* insertion mutant of *R.leguminosarum* was shown to be devoid of GSI enzymatic activity and of cross-reacting material, using an antiserum previously described (1). The growth rate of this mutant grown on NH_4Cl, glutamate, glutamine or nitrate is similar to that of the wild type, showing that GSI is not essential to the metabolism of *R.leguminosarum* grown under these conditions.

Using a plasmid overproducing a hybrid protein containing the *R.leguminosarum glnB* gene product we obtained a pure preparation of it and raised an antiserum which gives a single band in an immunoblot after non-denaturing PAGE of a crude extract of *R.leguminosarum* grown in NH_4Cl as a nitrogen source. When *R.leguminosarum* is grown in a poor nitrogen source (glutamate, nitrate etc.) the immunoblot shows two bands (see Fig. 1) and a 3-hour shift from glutamate to NH_4Cl causes the disappearance of one of them. We believe that the presence of the two bands may be related to different states of aggregation of the monomer and/or to its uridylylation.

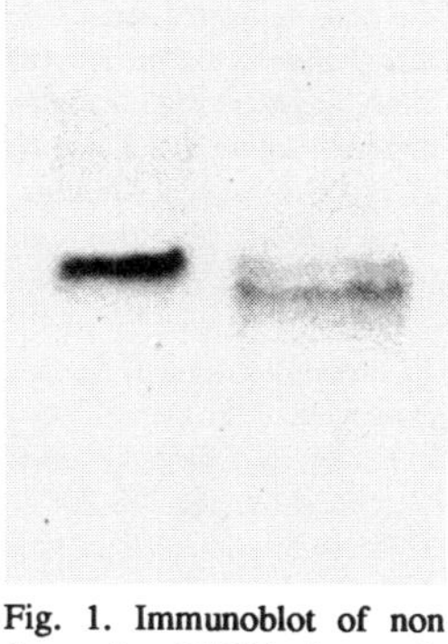

Fig. 1. Immunoblot of non denaturing PAGE of a crude extract of *R.leguminosarum* grown in NH_4Cl (left lane, 2 mg) or in glutamate (right lane, 0.5 mg) as a nitrogen source.

We have cloned and sequenced the *glnII* gene, coding for GSII, and find a high degree of homology with the *glnII* genes of *R.meliloti* and *B.japonicum*. Primer extension analysis shows a unique transcription initiation site, preceded by a *rpoN*(*ntrA*) promoter sequence and by three NtrC binding sites. The intracellular concentration of *glnII* mRNA (as assayed in a primer extension experiment) and of GSII (as assayed by enzymatic activity or cross-reaction with a specific antiserum) decreases to undetectable values in bacteria growing in NH_4Cl. The same results are obtained with a *glnA* mutant.

GSII may be post-translationally modified. A rapid removal of intracellular activity takes place when a glutamate growing culture of

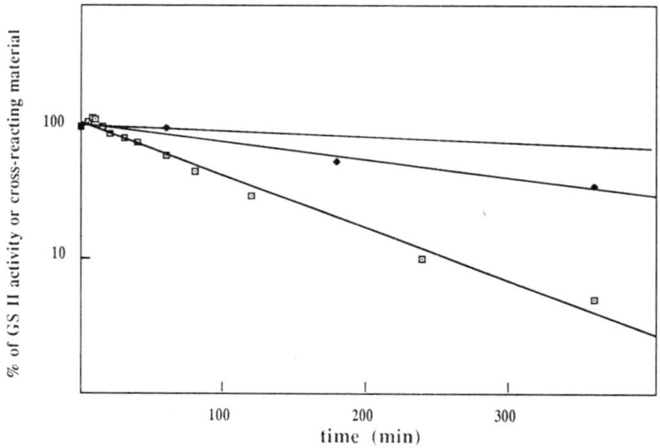

Fig. 2. NH_4Cl was added at zero time to a culture of *R.leguminosarum* grown in glutamate as a nitrogen source and samples taken at different intervals were assayed for GSII activity (□) or cross- reacting material (■). The upper line is the theoretical rate of disappearance of GSII assuming that NH_4Cl causes complete repression of the *glnII* gene and that there is no degradation of GSII.

R.leguminosarum is treated with NH_4Cl (3). We find (Fig. 2) that the rate of decrease in enzymatic activity ($t_{1/2}$: 90 min) is not paralleled by a concomitant decrease in cross-reacting material, as measured either by ELISA or by scanning an immunoblot after SDS-PAGE ($t_{1/2}$: 240 min). The enzyme did not show molecular weight alterations, as measured either under denaturing or non denaturing conditions, by SDS-PAGE or by gel filtration. No charge modifications of the GSII monomer were observed by immunoblot after two-dimensional PAGE.

In *R.meliloti* a locus, called *glnT*, which complements the glutamine auxotrophy of *E.coli glnA* mutants has been described (4), but no glutamine synthetase activity could be demonstrated in cell extracts of the complemented strains. Using *glnT* DNA of *R.meliloti* as a hybridization probe we identified a *R.leguminosarum* locus (*glnT*) expressing glutamine synthetase activity in *K.pneumoniae*. A 11.0 kb DNA fragment of *R.leguminosarum* was cloned to give plasmid pMW5a, which shows interspecific complementation of *K.pneumoniae glnA* mutants, and expressed a novel activity, which we call GSIII. The cloned sequence did not show cross-hybridization to *glnA* or *glnII*. GSIII activity, expressed in a *K.pneumoniae glnA* strain, shows a ratio of biosynthetic to transferase activity 10^3-fold higher than that observed for GSI or GSII. GSIII is similar in molecular weight and heat stability to GSI.

1) Colonna-Romano, S., Riccio, A., Guida, M, Defez, R., Lamberti, A., Iaccarino, M., Arnold, W., Priefer, U. & Puhler, A. (1987) *Nucleic Acids Research* **15**: 1951-1963

2) Holtel, A., Colonna-Romano, S., Guida, M., Riccio, A., Merrick, M.J. & Iaccarino, M. (1989) *FEMS Microbiol lett* **58**: 203-208

3) Rossi, M., Defez, R., Chiurazzi, M., Lamberti, A., Fuggi, A. & Iaccarino, M. (1989) *J Gen Microbiol* **135**: 629-637

4) de Bruijn, F.J., Rossbach, S., Schneider, M., Radet, P., Messmer, S., Szeto, W.W., Ausubel, F.M. & Schell, J. (1989) *J Bacteriol* **171**: 1673-1682

Acknowledgments. This work was partially supported by grant "Tecnologie avanzate in agricoltura" of MAF, DPGA and by a EEC Collaborative grant CI1-0408-I/CI-0410-ME.

Gresshoff, Roth, Stacey & Newton (eds.) Nitrogen Fixation: Achievements and Objectives
Chapman and Hall New York • London 1990

Workshop summary: How can we couple the genetics and physiology of nodule function?

Peter M. Gresshoff[1] and Champa Sengupta-Gopalan[2]

[1]Plant Molecular Genetics and Center for Legume Research, The University of Tennessee, Knoxville ,TN 37901-1071, USA. and [2]Plant Genetic Engineering Laboratory, New Mexico State University, Las Cruces, NM 88003, USA.

The basic challenge of science is to ascertain the causal relationships between structure and function. How does a molecule of certain structure function? How does the observed function reflect the restrictions of the various molecular structural alternatives capable of fulfilling that function?

In biological science this relationship is difficult to analyze because of the plasticity of many biological molecules and structures, and because of the complicating factor introduced by biological selection and evolution. In other words, whereas in an abiotic system one finds direct causality from structure to function, e.g. a certain crystal will have a defined ability to refract light, the biological situation is more complex, because function can influence the structure through the well-known cascade involving the genetic system, the selection of improved variants in a population, and the final establishment of these in the gene pool.

The analysis of the genetic determinants controlling nodulation and nitrogen fixation is faced with the same challenge. The last decade has seen the application of genetic and molecular biological approaches to the analysis of the microbe itself, and to the interaction between microbe and plant.

The genetics of nodulation are well-reviewed in this volume. It becomes clear that a large body of information deals with the physiological description of mechanisms, e.g. oxygen barrier regulation, flavone induction of *nod* genes, competition in soils, ability to denitrify, etc. Opposed to that is the data pool dealing with pure molecular information, usually in the form of nucleotide sequences. Often the function of the encoded proteins is not known. Searches of the computer data banks do suggest homologies to known sequences so that association of function is possible. The model coupling the nodulation genes of *Rhizobium meliloti* and their presumed function to the nodulation factor work by the Dénarié group, presented by Alan Downie at this meeting is an excellent example of this.

Many nodulin encoding genes are defined by their nucleotide sequence and tissue specificity only. The challenge at hand seems to be the linking of these two data bases, so that molecular genetic knowledge has genetical and physiological significance.

What are the approaches available to couple the analysis of the phenotype to the genotype?

The first that comes to mind is the general area of molecular physiology. This is based on the physiological, biochemical description of processes at the molecular level. Excellent examples of this approach is the careful study involving the localization of uricase or sucrose synthase in specific nodule cell types. Similar work using *in situ* hybridization is reported here at the Congress by the Bisseling group and by Triplett. The research by Verma as well as Weaver and Roberts on the function of nodulin 26, showing that the protein is possibly involved in symbiosome-membrane transport functions, and that it is the substrate for calcium-dependent phosphorylation are important steps in the search for a function for a protein molecule, whose structure (as amino acid sequence at least) had been known for some time.

Significant promise appears to come from the application of reverse genetics. Normal genetics is based on the recognition of a phenotypic difference, which then to various degrees of analysis is explained at the molecular level. Reverse genetics generates the known molecular difference and then tests for the phenotypic change.

Promising advances have occurred with the use of antisense gene expression. In this procedure the non-coding strand of a molecularly defined gene is expressed in a transgenic organism, so that the normal gene expression is lowered by the formation of double-stranded RNA intermediates. This results in a changed phenotype, which can be correlated to the gene under investigation. Recent research on the genes of the phenyl propanoid biosynthetic pathway in *Petunia* , antibiotic resistance genes in tobacco, and photosynthetically important genes suggest that this approach may be of value for nodulation and nitrogen fixation genes of either plant or bacterium.

A more speculative, yet feasible approach is to use ribozyme 'gene shears'. This approach takes advantage of the autocatalytic cleavage activity of specific RNA sequences. If these are incorporated in a gene sequence, this catalytic activity can be targeted to a transcript of interest. Loss of transcript level, as with the antisense approach, will lead to a loss of phenotype.

The third approach, still unrealized, is the elegant way of gene replacement. Existing transformation procedures for plants rely on gene addition, not gene replacement (i.e. homologous recombination). Thus the added gene must be dominant or its molecular expression must be discernable from the expression of the 'resident' gene.

All the above approaches base their success on the availability of genetic transformation in plants (or legumes in specific). This is possible for white clover, alfalfa, soybean, and Lotus. However, success rates are not as high as for the well-described tobacco system, making progress slow and restricted to few research groups. This volume (see work by J. Kijne, F. deBruijn and K. Marcker) describes some of the success stories. The above approaches require knowledge of the gene in question; in other words, they attempt to link the known molecule to an yet unknown function.

What about the reverse direction? There are many functions for which we do not know the gene product(s). Many of these phenotypes deal with molecular processes that are reliant on very low level expression of genes. A gene may only express itself for a few hours in a few cells making perhaps 100 protein molecules of short half-life. How can we detect such molecular biological changes? 'Classical' molecular approaches like *in vitro* translation of polyA$^+$ RNA followed comparison of the translation products by gel electrophoresis, differential screens of cDNA populations, cloning into expression vectors and detection of tissue-specific products, subtraction hybridization, or specific antiserum based enrichment of tissue-specific gene products have been very valuable of the last 10 years, but are restricted to those biological situations where differences are clearly pronounced, or where the protein in question (or its mRNA) are abundant. It is tempting to suggest that most biologically important molecules, i.e those that regulate processes rather than being regulated, fall into this difficult category.

One approach to couple the phenotypic knowledge of such processes to the molecular mechanism is the approach, which follows function for most of its steps. Examples of the success of this type research come from human genetics, where the mapping of genes involved in complex mental and physiological disorders has been possible because of the combination of restriction fragment length polymorphism (RFLP) mapping, isolation of overlapping genomic clones, DNA sequencing, and data-base searches.

The same avenue to link phenotype to genotype is possible in the analysis of plant genes involved in nitrogen fixation. The initial step is the establishment of a RFLP map. This is achieved by determining the cosegregation (linkage) of polymorphic probe homologies in families or F2 segregants from known parental stock. Such maps are rapidly being build up for a variety of plants. In soybean we have access to an excellent map generated by the research group at Ames (Iowa) under the direction of Randy Shoemaker. This map at present is calculated to cover about 90% of the soybean genome of 1.8-1.9 x 10^9 base pairs at an average resolution of 10 centimorgans (cM). The polymorphisms were detected between cultivated soybean (*Glycine max*) and *Glycine soja*. It was possible to cross *Glycine soja* with the supernodulation and nonnodulation mutants isolated in *Glycine max* (see Landau-Ellis et al., this volume) and to determine mendelian segregation of the wild-type and mutant phenotypes in the F2 generation. Preliminary results show that this strategy for mapping symbiotically important genes is feasible. Landau-Ellis and Gresshoff (unpublished data) have detected linkage between the supernodulation controlling *nts* locus and a polymorphic probe on linkage group I of the Ames RFLP map. Research is in progress to order the loci by reprobing the same blots with other nearby probes from linkage group I. Similar research successes were presented at this meeting by T. LaRue working with pea (*Pisum sativum*). Here many symbiotically important genes have been mapped using mutants and RFLP probes. Most significantly, several genes were clustered in one region of chromosome 1.

The second step in the detection of linked genes is the isolation of overlapping DNA fragments. In other words the genetic linkage needs to become physical linkage. Traditional DNA isolation and cloning procedures result in DNA fragments of about 15 to 20 kb in length. Such clones allow 'chromosome walking'. In doing this one looks for an overlap in DNA sequence homology from the one end of a known fragment and another one. This second fragment will then be used for a renewed cycle of probing. The procedure is laborious and time-consuming, and may run into difficulty if repeated DNA sequences are encountered. A new modification of this scheme involves chromosome jumping, in which very large fragments of DNA are isolated after restriction with rare cutting endonucleases (like *Sfi*1, *Not*1) and separation by pulse field gel electrophoresis. Research along these lines is presented by the poster of Funke et al. (G55, this volume), who separated large soybean DNA fragments (about 450 kb). Such fragments allow the traversing of large DNA regions. Since the soybean genome has a size of about 1.9x10^9 bp and since the present map spans about 1870 cM, one can extrapolate to an average physical distance per centimorgan of about 1 megabase (Mb). The reality of the conversion factor will depend on the region of DNA, as recombination frequencies can be increased or decreased by hot spots of recombinations, centromeres, and heterochromatin. Since the soybean supernodulation locus is about 10cM distant from the diagnostic probe, we require further mapping to other RFLP markers before attempting a chromosome jump.

Even more speculative at present is the verification that a gene of interest has been isolated following a linkage and chromosome jump/walk approach. Clearly sequencing DNA regions and assignment of open reading frames may be of value, but this represents a laborious long-shot without the promise of an absolute answer. More feasible may be the transformation of homozygous recessive plants with the putative wild-type gene sequences, so that the function of the wild-type allele can be selected for. This strategy of gene identification by complementation reflects the work done with bacterial genes, where transposon inactivation yields the mutant, which then is complemented by a cosmid library for wild-type function. For this reason the transformation frequency in higher plants needs to be increased, permitting more routine and large scale approaches with multiple clones. Research in our laboratory in Knoxville is presently focussed on the application of microprojectile-mediated transformation of soybean. This approach is hoped to complement the *Agrobacterium tumefaciens* and *A. rhizogenes* transformations reported by others, while giving the opportunity to insert larger DNA fragments.

To support some of the molecular approaches discussed above, a few papers were selected in the discussion session. The next part of this paper will briefly summarize these discussions.

The group working on glutamine synthetase (GS) in frenchbeans have linked the functional data base to isolate several GS genes. J. Cullimore presented data on the different GS genes , which includes the two genes that are expressed in nodules, gln-ß and gln-γ. The major activity in the nodule is attributed to gln-γ, which is expressed in a nodule-specific manner. However, the expression of gln-γ genes did not appear to be regulated by ammonia. Furthermore, chimaeric constructions between gln-γ upstream sequences with the *gus* reporter gene, when assayed in transgenic *Lotus* plants, showed nodule-specific expression independent of the onset of expression from the nitrogenase genes (Shen et al, this volume).

Work from D. P. S. Verma's laboratory refutes the presence of nodule-specific GS genes in soybean and furthermore claims that ammonia mediates induction of GS genes in soybean and transgenic *Lotus* roots. However, work from our laboratory in New Mexico has shown the presence of a nodule-specific isozyme, which contributes to the major part of the nodule GS activity in soybeans and which is made up of at least four nodule-specific polypeptides. Furthermore, we have also shown that the induction of the GS_n gene is independent of nitrogen fixation, as evident from the fact that these genes are expressed normally in Fix^- nodules formed by *Bradyrhizobium japonicum* strains deficient in the structural nitrogenase genes.

The tale of nodulin 26 started merely as a cDNA sequence a few years back and today we have evidence to believe that it functions as a channel or transporter on the peribacteroid membrane in soybean nodules (Miao et al, this volume). The amino acid sequence of N-26 shows significant homology to the *E. coli* glycerol facilitator which is the only known pore-type protein in the *E. coli* membrane and which functions in the transport of small molecules across the membrane. Recently, a N-26 homologue was discovered in the membrane of seed protein bodies in many legumes (M. Chrispeels, pers. comm.), suggesting that this protein has a much broader role in the plant. Weaver et al (this volume) have gone a step further in the characterization by providing evidence suggesting that N-26 may be regulated by calcium-dependent phosphorylation.

A nodule is made up of more than one cell type and localizing the site of synthesis of a certain nodulin transcript may suggest a function for the particular gene product. *In situ* hybridization of RNA has been very successfully used by various laboratories, including those of A. Hirsch, K. Dunn and T. Bisseling (see this volume) to localize transcripts in nodule tissue. H. Franssen presented data on the localization of ENOD2 transcript in the inner cortex of soybean nodules. Based on the critical role that the inner cortex might play as an oxygen barrier, it is suggested that ENOD2 as a cell-wall protein might contribute to towards forming this barrier.

Based on purely biochemical assays of the different cellular and subcellular fractions of nodules, Schubert and Boland have in the past proposed a pathway for ureide biogenesis in nodules. Thus immunolocalization of a protein at the subcellular level may aid in the determination of function of that protein. Furthermore, localization of an enzyme of known function may help to confirm or refute the proposed pathway for ammonia assimilation in nodules. This technology becomes even more powerful if monoclonal antibodies to specific forms of a protein are used.

E. Triplett presented data on the localization of xanthine dehydrogenase (XDH) in cowpea nodules. The protein was localized in the cytoplasm of both infected and uninfected cells, with higher levels in the uninfected cells, suggesting that purine biosynthesis is taking place in both cell types. It is not a surprising finding considering the other functions that purines have besides being involved in ureide biosynthesis. Higher levels of XDH in the uninfected cells might suggest that a considerable portion of purine biosynthesis in uninfected cells is channeled towards ureide production. This finding also illustrates that the biosynthetic activities important in legume nodules may arise by functional duplication of existing metabolic reactions.

One of the approaches to determine function for a protein is to relate its absence/presence in a tissue with the physiological state of the tissue at that time. A. Legocki reported data on a very interesting protein in lupin roots (R18), which is present in high levels in roots and other tissues, but not in nodules. Its disappearance from the developing nodule coincides with the appearance of the leghemoglobin protein. Partial amino acid sequence analysis did not show homology to any known protein in the protein data bank, but its pattern of expression suggests that it could be a phytoferritin or an iron transport protein functioning to transfer iron to the hemoglobin protein.

Very pertinent to nodule function is the aspect of regulation of expression of nodulin genes. Laboratories of K. Marcker and F. de Bruijn have done very elegant work to analyze the regulatory elements of leghemoglobin genes from soybean (*lbc3*) and Sesbania (*srLb*) in order to determine what controls nodule-specific expression. F. de Bruijn presented data on the expression of *gus* genes driven by different promoter elements of the SrLb genes in *Lotus* and could analyze regions that had different regulatory functions. Most interestingly, the Lb promoter showed considerable level of GUS expression in tobacco roots suggesting that the activating control of leghemoglobin is found not only in legume nodules. Another very interesting finding reported by F. de Bruijn was that an 11 KD *Rhizobium* -derived protein showed specific binding to an unique GC-rich element in the SbLb 5' region. This finding could be very significant in unraveling a new mechanism of communication between the two symbiotic partners.

Gresshoff, Roth, Stacey & Newton (eds.) Nitrogen Fixation: Achievements and Objectives
Chapman and Hall New York • London 1990

Early nodulation signals in wild-type and symbiotic mutants of soybean

B. J. Bassam*, T. D. Sutherland, L. J. Schuller, and P. M. Gresshoff

Plant Molecular Genetics, Institute of Agriculture and Center of Legume Research, The University of Tennessee, Knoxville, TN 37901-1071 USA.

Introduction A bioassay measuring expression of the *B. japonicum nodYABC* operon was characterized then used to assay *nod* gene inducers from different tissues and root exudates of wild-type, autoregulatory (supernodulating nts382) and non-nodulating mutants (nod49, nod139) of soybean cv. Bragg[2]. This was done to determine if alterations in levels of plant signal compounds contribute to the phenotype of the soybean mutants or if plant signal levels vary in response to inoculation.

The Signal Bioassay Expression of the *nod(Y)ABC* operon was measured by assaying ß-galactosidase activity in strain ZB977[1]. Before using induced ß-galactosidase activity as a bioassay for plant signals, parameters affecting the assay were examined in detail: inducibility of the operon was maximal between pH 5.3 and 6.5.at about 10 μM isoflavone measured after an 8 hr sample incubation time; titration curves were made from all plant extracts to ensure the amount of extract used in comparing samples was in the range most sensitive for the bioassay; native ß-galactosidase activity and toxicity of plant samples were negligible. Because *B. japonicum* cells exhibit a critical "window of inducibility", only cell cultures with an OD_{600} absorption between 0.09 and 0.11 were used in bioassays. To extend the window of inducibility of strain ZB977 cells, we buffered the growth medium with MOPS (5.8 mg/ml). The validity of the bioassay was confirmed using HPLC; peak size of the flavonoid inducer daidzein in extract fractions correlated with *nod* gene inducibility.

Signal Levels Analyses of extracts from seeds, cotyledons, roots, stems and leaves showed signal levels characteristic for each tissue type but no significant variation between each plant genotype. Extracts from the first 2 cm of roots of 8-11 d old cv. Bragg seedlings compared at 0, 24, 36 and 72 hours after inoculation with strain USDA110 showed identical *nod* gene inducing ability. The condition of plant material was important to the accuracy of these experiments; plants suffered extensive browning if over-watered resulting in extracts with artificially high inducing ability. Exudates from each plant genotype showed no significant differences in signal levels.

Conclusion Our results show that although plant signals to the *nod* genes in the soybean-*Bradyrhizobium* symbiosis are involved in the control of early infection events, they have little or no involvement with autoregulation and thus nodulation frequency in cv. Bragg. Furthermore, autoregulatory and nonnodulating soybean mutants do not have significantly different levels of bacterial *nod* gene signalling molecules compared to wild-type plants. It appears unlikely that alterations in levels of plant signals are responsible for the mutant symbiotic plant phenotypes examined.

1 Banfalvi, Z. *et al.*, (1988) *Mol. Gen. Genet.* 214: 420-424.
3 Sutherland T.D. (1990) *Mol. Plant-Microbe Interact.* 3: 122-128.

Gresshoff, Roth, Stacey & Newton (eds.) Nitrogen Fixation: Achievements and Objectives
Chapman and Hall New York • London 1990

Two-dimensional PAGE of proteins in the *Alnus-Frankia* symbiosis

Helen Belefant and Beth Mullin

Dept. of Botany, University of Tennessee
Knoxville, TN 37996

Introduction: Protein characterization of Frankia by 1D-PAGE has been useful in separating 43 Frankia strains into 6 groups (2). Proteins in the rhizosphere from root exudates may be active in the induction of the Alnus-Frankia symbiosis (7).

Material and Methods: Frankia strain WgAvcI1 was grown in either DPMN (containing NH_4Cl) or DPM (- nitrogen) (1). Sterile alder (Alnus glutinosa) trees were grown in test tubes containing a complete (+ nitrates) or -N nutrient solution.

Frankia extraction - method of Gardes and Lalonde (1987). **Alder root and nodule extracts** - A] TCA extraction (4) B] Extraction using protease inhibitors. Roots and nodules were dry crushed in LN_2, added to 2D-MH (6) or a modified buffer containing E-N-caproic acid, benzamidine, and PMSF. C] Phenol extraction (5). Tissues were dry crushed in LN_2. **Exudates** - The nutrient solution left after removal of the trees or bacteria was used directly as the protein source.

Conclusions: Phenol extraction was the best overall procedure for protein extraction from both roots and nodules of alder. Frankia proteins, produced or excreted, varied with nitrogen presence during growth. Unique Frankia proteins were predominately low MW while the unique excreted proteins were of middle MW. Alder roots had few proteins unique to each growth condition (+N, -N, or -N,+Frankia). As the -N,+Fr proteins closely resembled -N proteins, it appears that nodule specific proteins and new nitrogen metabolism proteins were not synthesized. It is possible that the enhanced low MW proteins of the N,+Fr roots are involved in nitrogen metabolism.

1 Baker, D. & O'Keefe, D. (1984) Plant and Soil **78**,23-28
2 Benson, D.R. & Hanna, D. (1983) Can. J. Bot. **61**,2919-2923
3 Gardes, M. & Lalonde, M. (1987) Physiol. Plant. **70**,237-244
4 Granier, F. (1988) Electrophoresis **9**,712-718
5 Hurkman, W.J. & Tanaka, C.K. (1986) Plant Physiol. **81**,802-806
6 Mayer, J.E., Hahne, G., Palme, K., & Schell, J. (1987) Plant Cell Rep. **6**,77-81
7 Vergnaud, I., Chaboud, A., & Rougier, M. (1987) Physiol. Plant. **70**,319-326

Gresshoff, Roth, Stacey & Newton (eds.) Nitrogen Fixation: Achievements and Objectives
Chapman and Hall New York • London 1990

The "vesicle pore": a structure unique to stalked, septate symbiotic vesicles in actinorhizal plants

R. Howard Berg and Lisa A. Nichol

Department of Biology, Memphis State University, Memphis, TN, 38152, USA.

We are interested in the role of Frankia cell wall surface laminae in symbiotic vesicle (SV) development. These laminae are constitutive components of the Frankia cell wall, and are particularly numerous around the SV and subtending stalk, where they are overlayed, to form a thickened, multilamellate layer (MLL) (1). Ultrastructural study of MML development is hampered by the unreactivity of the lamina with most, if not all, cytochemical stains. We present evidence that laminae are not extracted during tissue preparation, and can be imaged using electron density of adjacent tissue components. The "void space" shrinkage artifact interferes with this imaging, but we have been able to discover that the MLL is remarkably thick around the SV stalk, rivaling or even surpassing the MLL thickness around the SV *per se*. The latter case occurs only in stalked, septate SVs, as formed in the alders or members of the Elaeagnaceae. In nonseptate SVs, such as those formed in Ceanothus spp. or members of the Rosaceae, the MLL of SV and stalk are of similar thickness. The abnormally thick MLL of septate SV stalks compresses the stalk cell cytoplasm into a narrow channel we have termed the "vesicle pore", which can be observed in published micrographs, including freeze substituted Frankia vesicles (2). Nonseptate SVs do not have the vesicle pore, resembling immature, nonseptate SVs of symbionts that eventually form vesicle pores. The thickened MLL of stalks with vesicle pores is appressed to the septum joining stalk cell with vegetative hypha. In combination with the vesicle pore, this effectively isolates the SV from both the external environment and, to a large extent, vegetative cells of the symbiont.

(1) Newcomb, W. & Wood, S.M. (1987) Intl. Rev. Cytol. **109**, 1-88.
(2) Lancelle, S.A., Torrey, J.G., Hepler, P.K., & Callaham, D.A. (1985) Protoplasma **127**, 64-72.

Gresshoff, Roth, Stacey & Newton (eds.) Nitrogen Fixation: Achievements and Objectives
Chapman and Hall New York • London 1990

Supernodulation mutants in Elgin 87 soybean

R. I. Buzzell[1], B. R. Buttery[1] and G. Ablett[2]

[1]Agriculture Canada Research Station, Harrow, Ontario and [2]Ridgetown College of Agricultural Technology, Ridgetown, Ontario, Canada

Elgin 87 soybean (*Glycine max* (L.) Merr.) seeds were treated with ethylmethane sulfonate at 0.04M after soaking for 12 hr in water. Field grown M_1 plants were harvested individually. Ten M_2 seeds of each of 999 plants were planted in vermiculite/perlite culture with 14mM combined N. Five M_2 families with supernodulation mutants were identified.

Each of the mutants was crossed with Elgin 87, and mutant line E391 was crossed to the other four mutants.

The supernodulation mutants were pure-breeding for the supernodulation trait.

The F_2 from crossing E300, E420 and E592 to Elgin 87 gave a good fit to a 3:1 ratio expected for the segregation of a dominant allele for normal nodulation and a recessive allele for supernodulation. However, in the cross of E391 to Elgin 87 one F_2 gave a good fit to a 3:1 ratio and two other F_2's gave a good fit to a 15:1 ratio which would be expected with segregation of two recessive alleles controlling supernodulation. The cross of E714 to Elgin 87 did not fit a 3:1 ratio but were closer to a 3:1 than to a 15:1 ratio.

F_3 tests of four F_2 families gave a good fit to the expected 1:2:1 ratio of normal nodulating: segregating: supernodulating for those that had a 3:1 ratio, and a good fit to an expected ratio of 7:8:1 for the F_1 that had a 15:1 F_2 ratio.

Tests indicated that the allele(s) for supernodulation in E391 are allelic with those of E300, E420, and E592.

The results in general indicate the presence of a single recessive allele controlling supernodulation as found by Carroll *et al.* (1988). However, there were two F_2 segregations one of which has been confirmed in the F_3 that indicated the presence of a second gene. That is, two recessive alleles are needed for supernodulation to be expressed. Since crosses with E391 did not consistently result in 3:1 or 15:1 segregations either it or Elgin 87 must vary genetically.

Carroll, B. J., Gresshoff, P. M. and Delves, A. C. (1988) Theor. Appl. Genet. 76:54-58.

Gresshoff, Roth, Stacey & Newton (eds.) Nitrogen Fixation: Achievements and Objectives
Chapman and Hall New York • London 1990

Alternative mechanisms control nodulation in legumes

Gustavo Caetano-Anollés and Peter M. Gresshoff

Plant Molecular Genetics (OHLD), Institute of Agriculture and Center of Legume Research, The University of Tennessee, Knoxville, Tennessee 37901-1071, USA.

The establishment and function of a nodule is regulated by the host plant as well as by the environment. The formation of first nodules on primary roots inhibits subsequent nodule formation by a process termed autoregulation or feedback suppression of nodulation. Autoregulatory responses are systemic, and require of the shoot to be active. In soybean, nodule initiation is restricted to a region of the root that coincides with the zone of elongation and appears to be relatively constant throughout root development. However, a large number of infections become arrested or aborted early in organogenesis. This suggests that maturation of nodule primordia into emergent nodules is the target of host feedback control in soybean. In this study, we show that excision of first nodules releases the autoregulatory suppression of nodulation in soybean and alfalfa. However, the location where newly emergent nodules were formed indicates the existence of at least two alternative mechanisms that control nodulation in legumes. One favours multiplicity of infections and the selective arrest of infection development, the other controls nodule initiation.

Emergent nodules suppress nodule initiation in alfalfa: Following inoculation, almost all infections associated with cortical cell division events that are initiated in 'Vernal' alfalfa develop into mature nodules. Only 9-15% of total cell division foci failed to become normal emerging nodules 14 or 22 d after inoculation, and were located around the initially susceptible region of the root. Removal of fully developed nodules from the primary root resulted in nodules emerging on lateral roots almost exclusively around the location of the lateral root tips at the time of nodule removal. Apparently, this region regained its susceptibility to infection within the first hours after excision of primary nodules, but suppression of nodulation was restored a day after excision probably by the development of new infection foci. The number of nodules formed was similar to the number of nodules that was originally excised, but increased if excision was delayed. A second removal of newly formed lateral nodules caused nodules to emerge again in the susceptible regions of primary and secondary laterals.

Emergent nodules suppress nodule development in soybean: Excision of fully developed nodules resulted in new nodules emerging on primary and lateral roots of soybean cv. Bragg and the supernodulating derivative nts382 in those same regions from where nodules were initially removed. Nodule occupancy studies following inoculation with antibiotic-resistant bacteria at the time of excision indicate that these nodules arose from infections that were already initiated before the time of nodule removal. Our results suggest that nodulation in both genotypes is controlled by the arrest of successfully initiated infections. Apparently, the supernodulating mutant was still able to regulate nodule number despite its diminished ability to arrest cell division in the cortex and pericycle during infection development. This suggests that other mechanisms that are not tied to the supernodulating phenotype control soybean nodulation.

Gresshoff, Roth, Stacey & Newton (eds.) Nitrogen Fixation: Achievements and Objectives
Chapman and Hall New York • London 1990

Spontaneous nodules induce autoregulation of nodule formation in alfalfa

Gustavo Caetano-Anollés, Priyavadan A. Joshi and Peter M. Gresshoff

Plant Molecular Genetics (OHLD), Institute of Agriculture and Center for Legume Research, The University of Tennessee, Knoxville, TN 37901-1071, USA.

The formation of nitrogen-fixing nodules in alfalfa is strictly controlled by a systemic feedback regulatory mechanism that suppresses nodule initiation in younger parts of the root system. Bacteria-free organized nodular structures induced by exopolysaccharide-deficient *Rhizobium meliloti* were able to elicit and be target of this systemic response, indicating that extensive bacterial infection and proliferation were not required for elicitation of feedback and that the target of the response was the *Rhizobium*-induced dedifferentiation of the host root cells (1). To examine whether feedback regulation is not induced by diffusible substances from the rhizobia but rather results indirectly from the formation of a nodule primordium, we studied the ability of nodules formed in the absence of *Rhizobium* to control *Rhizobium*-induced nodulation in alfalfa. Root-derived structures that fulfill most of the histological and molecular criteria defining an indeterminate nodule arise spontaneously in the absence of *Rhizobium* and combined nitrogen in less than 3% of pouch-grown 'Vernal' alfalfa plants. The spontaneous nodules resulted from the dedifferentiation of root cells from the inner cortex much as normal nodules do (see Joshi *et al.*, these proceedings), but lacked intra-or intercellular bacteria and did not fix atmospheric nitrogen. Nodules apparently initiated early during root development but their emergence was considerably delayed. As a consequence, nodulation was restricted to the primary roots and nodules emerged between the second and third week after germination with a comparable rate.

Spontaneous nodules elicit autoregulatory responses: Very few if any nodules formed in the absence of *Rhizobium* appeared in younger regions of the primary root. This provides the first evidence that these nodules are able to induce and be the target of feedback responses in the roots. *R. meliloti* inoculation of nonnodulated plants 19 d after seed imbibition resulted in nodules appearing exclusively in lateral roots clustered around the location of the root tips at the time of inoculation. The presence of spontaneous nodules on the primary root suppressed nodulation in lateral roots by 56-80%. The excision of spontaneous nodules at the time of inoculation eliminated the suppressive response, confirming that the spontaneous nodules were inducing feedback regulation and not that the spontaneously nodulated plants were intrinsically defective in nodulation. Our results suggest that *Rhizobium* is not absolutely required for nodule organogenesis and elicitation of nodulation control.

Feedback suppression of spontaneous nodulation: We were unable to find the typical white multilobed spontaneous nodule forming in plants that were inoculated with wild-type *R. meliloti* early during plant development. After extended growth periods, inoculated plants formed only pink indeterminate nodules. We presume that spontaneous nodulation is suppressed by the *Rhizobium*-induced nodules appearing shortly after inoculation.

1. Caetano-Anollés, G., Lagares, A. & Bauer, W.D. (1990) *Plant Physiol.* **91**, 368-374

Gresshoff, Roth, Stacey & Newton (eds.) Nitrogen Fixation: Achievements and Objectives
Chapman and Hall New York • London 1990

Glyceraldehyde-3-P dehydrogenase from the plant cytosol of soybean nodules

Les Copeland and Adrian Zammit

Department of Agricultural Chemistry, University of Sydney, NSW, Australia 2006

NAD-dependent glyceraldehyde-3-P dehydrogenase (EC 1.2.1.12; GAP DH) catalyzes the oxidative phosphorylation of D-glyceraldehyde-3-P in the glycolytic pathway. GAP DH from animals and bacteria has been studied extensively but information on the enzyme from plant sources is limited. We have purified to homogeneity GAP DH from the plant cytosol of soybean (*Glycine max* L. Merr. cv Williams) nodules and studied physical, kinetic and immunological properties of the enzyme. These studies are part of a continuing investigation on the metabolism of carbohydrates in association with symbiotic nitrogen fixation in root nodules of legumes.

The plant cytosolic GAP DH of soybean nodules was very labile, especially above pH 8. Decay of activity was greatly reduced when the enzyme was kept in the presence of NAD^+ at pH 7.4 and NAD^+ and Pi at pH 8.5. The enzyme had a complex kinetic mechanism, a feature of which was strong competitive inhibition by NADH when either NAD^+ or Pi were the varied substrates. Comparison of the respective Km and Ki values suggests that the ratio of $NADH/NAD^+$ in the plant cytosol of soybean nodules will be important in determining the rate of the reaction catalyzed by GAP DH *in vivo.*

Polyclonal antibodies raised against the plant cytosolic GAP DH of soybean nodules were shown to be monospecific by Western blot analysis. Immunotitration of enzyme activity, Ouchterlony double diffusion tests and Western blots showed that the antibodies also cross-reacted with GAP DH in extracts from roots, leaves and stems of soybeans. Thus, the increased specific activity of GAP DH in the plant cytosol of soybean nodules compared to roots (1) may be due to enhanced synthesis of an enzyme form normally present in other parts of the plant, rather than the expression of a nodule specific form of the enzyme. GAP DH in the bacteroids was immunologically distinct from the enzyme in the plant fraction of the nodules.

1. Copeland, L., Vella, J. and Hong, Z.Q. (1989) Phytochemistry **28**, 57-61.

Gresshoff, Roth, Stacey & Newton (eds.) Nitrogen Fixation: Achievements and Objectives
Chapman and Hall New York • London 1990

Functional changes in cowpea nodules grown in different pO_2

Felix D. Dakora and Craig A. Atkins

Botany Department, University of Western Australia, Nedlands WA 6009 Australia.

Five day old cowpea seedlings preinfected with Bradyrhizobium strain CB756 were grown with their roots in 1 to 80% (v/v) O_2 (in N_2). Structural and biochemical analyses were performed on 28-day plants.

Lenticel development on nodules varied with pO_2, being most enhanced in plants from sub-ambient pO_2 and least in those at supra-ambient concentrations. The lenticel content decreased when plants were transferred from sub-ambient pO_2 to air. Similarly, nodules cultured in high pO_2 (40-80%) showed an increase in lenticel formation on transfer to air while those changed to supra-ambient pO_2 from air showed a decreased formation of lenticels.

Nodules from these plants exhibited a relatively constant nitrogenase activity from 2.5 to 60% O_2 while those from 1 and 80% pO_2 appeared inhibited (2). Sustained symbiosis in support of plant growth in both sub- and supra-ambient pO_2 involved a change in gaseous permeability of nodule tissues and a range of morphological and anatomical adjustments (1,3,4). Nodules cultured in sub-ambient pO_2 were smaller and showed increased intercellular air space of both cortex and central, infected tissue. The size of bacteroid-containing tissue and the frequency of infected cells or bacteroids was also significantly reduced.

Nodules formed in supra-ambient pO_2 were bigger, showed reduced intercellular air space and enlarged central infected tissue. The frequency of infected cells was greater than that of nodules from sub-ambient pO_2. These changes would have enhanced O_2 utilization as a further means to protect nitrogenase from O_2 excess. The amount of Lb per infected cell in nodules at 1% O_2 was enhanced 2.5 times compared to air-grown nodules. The net result of these modifications was an increase in the efficiency of N_2 fixation by infected cells and bacteroids in nodules from pO_2 as low as 1 or 2.5% compared to air or supra-ambient levels.

1. Atkins, C.A., Dakora, F.D. & Storer, P.J. (1990) Planta (in press).
2. Dakora, F.D. and Atkins, C.A. (1990a) Plant Physiol. (in press).
3. Dakora, F.D. & C.A. Atkins (1990b) Plant Physiol. (in press).
4. Dakora, F.D. and Atkins, C.A. (1990c) Planta (in press).

Gresshoff, Roth, Stacey & Newton (eds.) Nitrogen Fixation: Achievements and Objectives
Chapman and Hall New York • London 1990

Monodehydroascorbate reductase from soybean root nodules: Purification and effects of elevated oxygen concentrations

David A. Dalton, Lorene Langeberg, and Chris Post

Biology Department, Reed College, Portland, OR 97202 USA.

INTRODUCTION

Legume root nodules may be damaged by activated forms of oxygen such as H_2O_2.[4] H_2O_2 is scavenged by ascorbate peroxidase and the ascorbate is regenerated by monodehydroascorbate reductase (MR)[1-3]. The purpose of this study was to purify and characterize MR and to determine if external oxygen concentrations affect activity in nodules.

MATERIALS AND METHODS

Purification procedures involved $(NH_4)_2SO_4$ ppt., DEAE Sephacel chromatography, gel filtration chromatography, affinity (reactive blue 2-sepharose 6L-6B) chromatography and native PAGE. The effects of external oxygen concentration were examined by growing plants for 16 days in chambers flooded with nutrient solution and continuously aerated with gas mixtures containing 5, 21, 50, or 80 kPa of O_2.

RESULTS AND DISCUSSION

MR was purified to near homogeneity and found to be a single subunit.

Properties of MR

Molecular weight	43,000
FAD content	0.71 mol FAD / mol protein
pH optimum	8.3 to 8.7
K_m : NADH	5.6 µM
NADPH	150 µM
monodehydroascorbate	7 µM
pI	5.1

Dehydroascorbate was ineffective as an electron acceptor, however, DCPIP and KFe$_6$CN functioned well. 50 µM *p*-chloromercuribenzoate inhibited activity by 90%, indicating a thiol requirement. Activities of MR and related enzymes of the ascorbate-glutathione cycle were not elevated in nodules exposed to supra-ambient O_2 concentrations.

References:

1. Dalton, D.A., Russell, S.A., Hanus, F.J., Pascoe, G.A. and Evans, H.J. (1986) PNAS, USA 83,3811-3815.
2. Dalton, D.A. (1990) in *Peroxidases in Chemistry and Biology*, Vol II, eds., Everse, J., Grisham, M.B. and Everse, K. (CRC , Boca Raton), [In press].
3. Hossain, M. and Asada, K. (1985) J. Biol. Chem. 260,12920-26.
4. Puppo, A. and Halliwell, B. (1988) Planta 173,405-410.

Gresshoff, Roth, Stacey & Newton (eds.) Nitrogen Fixation: Achievements and Objectives
Chapman and Hall New York • London 1990

Effect of nitrate on peroxisome ultrastructure and catalase activity in lupin nodules (*Lupinus albus* L. cv. Multolupa)

C. de Lorenzo, M.M. Lucas, A. Vivo and M.R. de Felipe

Instituto de Edafologia y Biologia Vegetal, C.S.I.C., Madrid, Spain.

This paper reports on an ultrastructural study of peroxisomes in infected nodule cells of lupin plants (*Lupinus albus* L. cv. Multolupa), grown with and without nitrate. Cytochemical localization of catalase and peroxidase was performed using the diaminobenzidine (DAB) technique in these tissues. Two aims were pursued: to corroborate previous results relating catalase activity and nitrogen fixation (1,2), and to observe the features of peroxisomes in lupin nodules.

Surface-sterilized seeds of *L. albus* were planted in Leonard jars and inoculated with *Bradyrhizobium* sp (*Lupinus*) strain ISLU 16. Nitrate-treated plants were supplied with Hoagland's solution containing 20 mM KNO_3, for five and ten days before nodule harvesting. Nitrogenase activity was determined using the acetylene reduction assay. Localization of catalase and peroxidase was performed as described in (3).

The infected cells presented a mean of seven peroxisomes, usually located on the cell periphery. Their shapes were round or egg-shaped, with maximum and minimum diameter means of 0.35 and 0.18 um. A strong catalase reaction was obtained in peroxisomes. Bacteroids showed small osmiophilic precipitates that could be due to catalase activity. Nitrate feeding exerted a negative effect on nitrogenase activity (31.6% decreased after five days of treatment). A reduction in catalase localization in peroxisomes, together with the existence of degenerative forms of these, were also observed under nitrate conditions.

1. Francis, A.J. & Alexander, M. (1972) Can. J. Microbiol. **18**, 861-864.
2. Becana, M., Aparicio-Tejo, P., & Sanchez-Diaz, M. (1988) Physiol. Plant. **38**, 65-69.
3. de Felipe, M.R., Lucas, M.M., & Pozuelo, J.M. (1988) J. Plant Physiol. **132**, 67-73.

Gresshoff, Roth, Stacey & Newton (eds.) Nitrogen Fixation: Achievements and Objectives
Chapman and Hall New York • London 1990

Functional expression of pea lectin on root hair tips of transgenic white clover hairy roots

C.L. Diaz, P.J.J. Hooykaas, B.J.J. Lugtenberg and J.W. Kijne

Dept. Plant Molecular Biology and Center for Phytotechnology RUL/TNO, Leiden University, Nonnensteeg 3, 2311 VJ Leiden, The Netherlands

INTRODUCTION AND METHODS

Introduction of the gene encoding pea lectin (psl) into white clover (hairy) roots extended the range of infection of clover roots to Rhizobium leguminosarum biovar viciae (R.l. viciae), the pea symbiont (1). To test the presence of pea lectin, Psl, on the surface of transgenic clover roots, a light microscopic localization method was developed, based on the specific binding of the carbohydrate moiety of monoclonal rat IgE to Psl. The Psl-rat IgE complex was visualized by signal amplification, comprising the use of biotinylated antibodies, a streptavidin-gold probe and silver enhancement.

RESULTS AND DISCUSSION

In white clover hairy roots transformed with the psl gene as well as in pea roots, Psl appeared to be localized on the surface of elongated epidermal cells and on the tips of emerging and growing root hairs. Hairy roots without the psl gene and wild type white clover roots did not show any label. These results indicate that (i) psl is expressed in transgenic white clover hairy roots, (ii) Psl is present at the target cells for rhizobial infection, similar to the situation with pea roots, and (iii) Psl is expressed in a form allowing binding of a specific saccharide.

REFERENCE

1. Diaz, C.L., Melchers, L.S., Hooykaas, P.J.J., Lugtenberg, B.J.J. & Kijne, J.W. (1989) Nature **338**, 579-581.

Gresshoff, Roth, Stacey & Newton (eds.) Nitrogen Fixation: Achievements and Objectives
Chapman and Hall New York • London 1990

Nitrate inhibits nodulation of supernodulating soybean mutants

D.L. Eskew, J. Kapuya, and S.K.A. Danso.

Joint FAO/IAEA Division, A-1400 Vienna, Austria.

Nitrate inhibits nodulation and nitrogen fixation in soybean (*Glycine max* (L.) Merr.). Thus growing a soybean crop often removes more N from the soil than is added by N_2 fixation(3). Carroll *et al.* (1) isolated several supernodulating mutants of soybean which were tolerant to 5 mM NO_3 in sand-vermiculite culture. In these experiments nodulation and nitrogen fixation of three of these mutants and the parent cultivar Bragg were determined in response to increasing levels of NO_3 when grown to physiological maturity in soil.

METHODS: Plants were grown in pots with 3 kg of soil. Additions of 7.5, 150, 300, or 450 mg N as nitrate (equivalent to 5, 100, 200, and 300 kg N/ha) were made at the time of planting. Nitrogen fixation was determined by the ^{15}N isotope dilution technique.

RESULTS AND DISCUSSION: Nodule dry weight decreased progressively as the NO_3-N level was increased. Nodulation was almost completely inhibited for all genotypes at the 450 mg-N level. Nodulation of mutant nts1007, however, was considerably more nitrate tolerant, as nodule dry weight was inhibited by only 27% at the 300 mg-N level as opposed to 76% for nts382 and 71% for nts1116. Estimation of N_2 fixation by the ^{15}N isotope dilution method indicated that nts1007 fixed 29% more N_2 than Bragg at the 7.5 mg-N level, and 63% more at the 150 mg-N level.

Since nodulation of all genotypes was inhibited by the highest level of NO_3, this suggests that the *nts* mutations are not mutations in the nitrate response mechanism *per se*. The greater nitrate tolerance of nts1007, however, is particularly intriguing as genetic complementation analysis (2) has shown that *nts*382 and *nts*1007 are presumably mutations in the same gene. Field tests of nts1007 would be needed to determine if the greater N_2 fixation has a significant agronomic advantage.

REFERENCES

1. Carroll, B.J. *et al.* (1985). Proc. Natl. Acad. Sci. USA 82:4162-4166.
2. Delves, A.C. *et al.* (1988). J. Genet. 67:1-8.
3. Eskew, D.L. *et al.* (1989). Crop Sci. 29:1491-1496.

Gresshoff, Roth, Stacey & Newton (eds.) Nitrogen Fixation: Achievements and Objectives
Chapman and Hall New York • London 1990

Expression of genes encoding for phenylalanine ammonia lyase and chalcone synthetase and their possible role during nodule development in soybean

E. Estabrook and Sengupta-Gopalan,

Mol. Biol. Prog./Dept. of Agron. & Hort., New Mexico State Univ., Las Cruces, NM 88003, USA.

The first known signals that affect changes in gene expression during symbiosis are the phenylpropanoid-derived compounds produced by the plant that stimulate the transcription of the nod genes. These compounds also function as host defense molecules when plants encounter stress. Two key enzymes in the synthesis of flavanoids/isoflavanoids are phenylalanine ammonia lyase (PAL) and chalcone synthetase (CHS). This study will focus on the expression of genes for CHS and PAL genes during nodule initation and nodule development in soybeans. cDNA clones for PAL and CHS have been isolated from a soybean nodule cDNA library and their sequence compared with all the published sequences to define conserved and unique regions. Analysis of expression of CHS and PAL nodules at different developmental stages leading to effective symbiosis, does not show any significant changes in the level of transcripts except at the very early stages following infection. However, there is a dramatic but transient increase in expression of these genes coincident with the timing of the onset of N_2-fixation when tracing nodule development incited by ineffective strains of *B. japonicum*. The roots of a nodulating mutant of soybean when infected with effective strains of *B. japonicum* shows a 5 to 10 fold increase in the expression of PAL but not CHS, suggesting that the two sets of genes are regulated differently. Expression of these genes has also been studied in the tripartite association between soybean/*B. japonicum*/*P. syringae* pv 'tabaci'. This association results in enhanced nodulation. Experiments are in progress to distinguish between a host defense response and signaling for nod gene induction.

EE is the recepient of graduate fellowship from NSF

Gresshoff, Roth, Stacey & Newton (eds.) Nitrogen Fixation: Achievements and Objectives
Chapman and Hall New York • London 1990

First steps towards the analysis of the soybean genome using pulsed field gel electrophoresis (PFGE)

Roel P. Funke, Luis Sayavedra-Soto, and Peter M Gresshoff

Plant Molecular Genetics, Institute of Agriculture and Center of Legume Research, The University of Tennessee, Knoxville, TN 37901-1071 USA.

Introduction. PFGE is used widely in the analysis of the genomes of mammals and microorganisms, and recently a procedure for its application to a higher plant (tomato) was published (1,2). This requires the preparation of unsheared chromosomal DNA, digestion with rare-cutting restriction enzymes and, following electrophoresis, transfer of the large DNA fragments to Nylon membranes for probing with clones from a genomic library. The technique is being developed in this laboratory as part of a reverse genetics strategy (3) to locate and characterize the genes responsible for the super- and nonnodulating phenotypes of previously isolated soybean mutants (4).

Methods. To obtain intact chromosomal DNA of soybean, protoplasts were prepared from young trifoliate leaves and mixed with an equal volume of LMP agarose (INCERT, FMC) before being pipetted into a prechilled mold. Plug-embedded cells were lysed with 0.5M EDTA, 1% sarkosyl and 1 mg/ml proteinase K and the plugs were equilibrated at length with TE buffer prior to restriction digestion using 10 units enzyme/mg DNA. Plugs were loaded directly onto the gel and running conditions were typically a 60s pulse for 15 h followed by 90s for 9 h, 200V in 0.5xTBE (CHEF apparatus, BioRad). DNA was UV-nicked after staining with ethidium bromide and vacuum blotted onto nylon membranes. Probes (courtesy of Dr. Shoemaker, Ames, IA) were labelled to high specific activity using the random priming method.

Results. Photographs of a gel and an autoradiograph show that this procedure can be used with soybean to produce unsheared DNA embedded in agarose blocks (as demonstrated by hybridization of probe to position of the well in the undigested lane). Restrictability of the embedded DNA was proven by hybridization of the same probe to DNA fragments in the size range 100 to 400 kb produced by the enzymes *Sma*I and *Sfi*I. We plan to check the utility of the technique in long range mapping of defined regions of the soybean genome, and to investigate the correlation between physical distance along the DNA and distances on the genetic map of soybean.

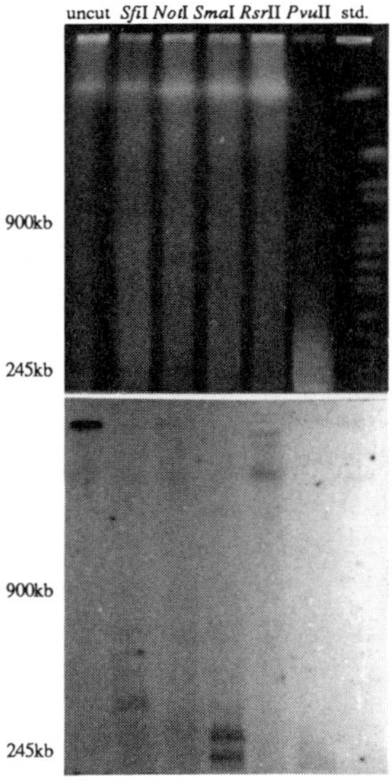

PFG of soybean DNA restricted with various enzymes and autoradiogram of transferred DNA hybridized with pK390.

1. Carle, G.F., and Olson, M.V. (1985) Proc. Nat. Acad. Sci. USA **82**: 3756-3760
2. Ganal, M.W. , Young, N.D., and Tanksley, S.D. (1989) Mol. Gen. Genet. **215**: 395-400
3. Smith, C.L.,*et al*.. (1987) Meth. Enz. **151:** 461-489
4. Carroll, B.J., McNeil, D.L., and Gresshoff, P.M. (1985a) Proc. Nat. Acad. Sci. **82**: 4162-4166

Gresshoff, Roth, Stacey & Newton (eds.) Nitrogen Fixation: Achievements and Objectives
Chapman and Hall New York • London 1990

Distribution of *Bradyrhizobium*-induced cell divisions in soybean

N. Gerahty*, G. Caetano-Anollés, P. A. Joshi and P. M. Gresshoff

Plant Molecular Genetics, Institute of Agriculture and Center of Legume Research, The University of Tennessee, Knoxville, TN 37901-1071 USA.

The anatomical changes related to the early developmental stages of soybean nodules are easily observed[1] and provide a method to distinguish between control of nodulation at initiation or later stages. Induced cell division foci from the initial nodulation windows of main roots from *Glycine max* (L.) Merr. cv Bragg were scored in serial sections four times over a period from 1 to 21 days post-inoculation to obtain a kinetic profile of nodule stages (fig.1b). Foci were also scored in three nodulation windows (fig.1c) , 24 hours apart, 72 hours after each window had opened, to obtain a profile of stages in autoregulated windows.

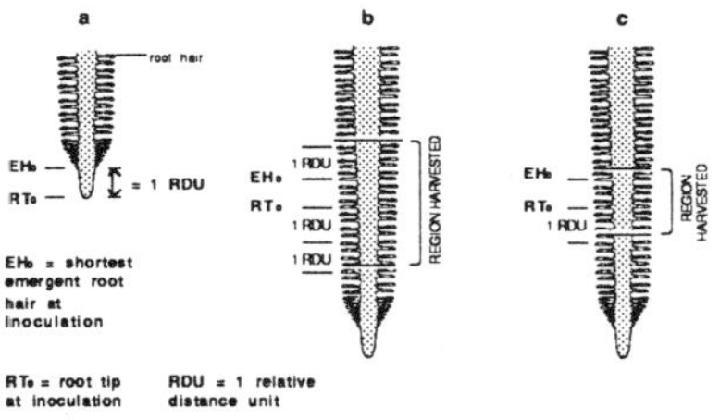

figure 1

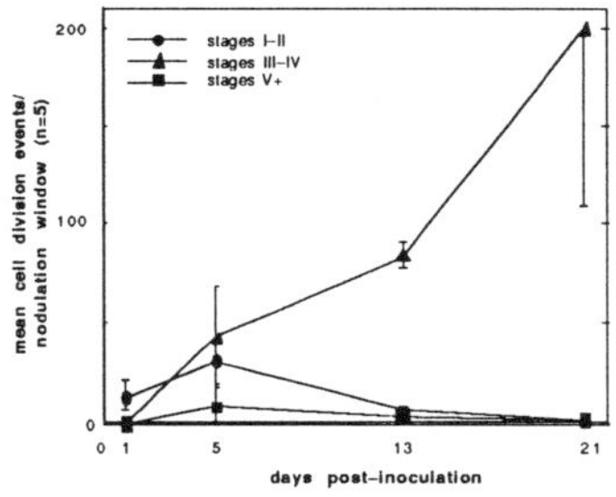

figure 2

The kinetics profile (see figure 2) shows a constant frequency of stages III-IV after nodule initiation. This accumulation suggests a developmental blockage at about stage V and also suggests autoregulation affects nodule initiation. This is supported by the result of the profile of autoregulated windows which shows stages I-II are not constantly initiated in newly opened susceptible zones while the development of stages III-IV is inhibited.

Since most cell division stages in fully autoregulated roots were at stage III-IV and the number of cell division events in wild-type and an autoregulation-minus mutant is similar[2], autoregulation in soybean blocks nodule development and possibly nodule initiation. The developmental blockage of nodule emergence which results in the accumulation of stages III-IV appears to coincide with meristem formation.

1 Calvert *et al.* (1984) Can J. Bot. **62**, 2375-2384.
2 Mathews *et al.* (1989) Protoplasma **150**, 40-47.

Gresshoff, Roth, Stacey & Newton (eds.) Nitrogen Fixation: Achievements and Objectives
Chapman and Hall New York • London 1990

Effect of nitrate on malate dehydrogenase and aspartate aminotransferase activity and on several metabolites in lupin root nodules (*Lupinus albus* L. cv Multolupa)

M.P. Golvano* and P. Lang

Instituto de Edafología y Biología Vegetal,C.S.I.C.Madrid, Spain.

INTRODUCTION

A main function of MDH and AST is to participate in the malate/aspartate shuttle, which can operate between plant cytoplasm and bacteroids (1,2). In this work we present data on the effect of nitrate on the specific activity of MDH and AST in cytosol, bacteroids and mitochondria and on the level of several metabolites.

MATERIAL AND METHODS

Lupin plants, well nodulated with Bradyrhizobium sp. (Lupinus) ISLU 16 were supplied with nitrate (20 mM) for 10 days. Separation of the nodule cellular fractions were carried out in sucrose gradients. MDH and AST activities were analyzed spectrophotometrically.

RESULTS

The bacteroids showed higher MDH specific activity than cytosol, 3-fold in the direction of oxaloacetate reduction and 5-fold in the direction of malate oxidation.The enzyme AST operates in the cytosol mainly in the direction of aspartate utilization and favouring aspartate formation in the bacteroids. In the mitochondria, the predominant reactions were reduction of oxaloacetate and aspartate formation. Supply of nitrate inhibited N_2 fixation and produced a trasient increase of the bactéroids MDH and AST specific activity after 2 d of treatment. Oxaloacetate and cetoglutarate accumulated in bacteroids after 1 day of treatment. Aspartate decreased in bacteroids and cytosol. If, as has been reported (3), nitrate restrict O_2 diffusion to bacteroids, the trasient increases in MDH and AST activities might be a mechanism to oxidize the internal NADH which can not be oxidized by electron transport chain.

1. Streeter, J.G. & Salminen, S.O. (1988) in Nitrogen Fixation Hundred Years After, eds. Bothe,H., de Bruijn, F.J. & Newton, W.E. (Gustav Fisher, Stuttgart).
2. Appels, M.A. (1989) The symbiois between Rhizobium leguminosarum and Pisum sativum: regulation of the nitrogenase activity. Doctoral Thesis. Wageningen.
3. Minchin, F.R., Minguez, M.I., Sheehy, J.E., Witty, J.E. & Skot, L. (1986) J. Exp. Bot. **37**, 1103-1113.

Gresshoff, Roth, Stacey & Newton (eds.) Nitrogen Fixation: Achievements and Objectives
Chapman and Hall New York • London 1990

Stress-induced changes in nodule enzyme activities and levels of leghaemoglobin and Lb mRNA

A.J. Gordon and H.J. Ougham

Cell Biology Department, Institute for Grassland and Environmental Research, Welsh Plant Breeding Station, Aberystwyth, Dyfed, Wales, UK.

Changes in key nodule proteins following defoliation of white clover plants were assessed by measurements of enzyme activities and by use of antibodies to selected nodule proteins.
Removal of all leaves and petioles caused major declines in protein and leghaemoglobin levels and in the activities (per g fresh weight) of invertase, sucrose synthase, UDP glucose pyrophosphorylase, aspartate amino transferase, glutamine synthetase, PEP carboxylase and malate dehydrogenase. However only PEP carboxylase and glutamine synthetase declined to a significantly greater extent than the general decline in soluble protein levels. Nodule alcohol dehydrogenase (per mg protein) increased above control levels following defoliation. All enzyme activities returned to control levels as new leaves grew. In contrast, continuous defoliation resulted in almost complete loss of all enzyme activities within 16-18 days.
Antibodies to GS and Lb confirmed that the declines in levels of these proteins were due to disappearance of antigen. Similarly, the use of antibodies to nitrogenase components I and II indicated that bacteroid proteins declined following defoliation and subsequently recovered in the same way as host plant encoded proteins.
In parallel studies with dark-stressed soybeans an Lba cDNA* probe was used to monitor Lb mRNA levels in total RNA extracted from nodules. Lb mRNA declined by 58, 94 and 100% after 1,2 and 3 days respectively of continuous darkness. The disappearance of Lb mRNA was in marked contrast to a decline in Lb protein of only c.20% over the same time scale. After returning the 3 day dark plants to normal light/dark cycles, Lb mRNA was again detectable after only 1 and 3 days (7.3 and 63% respectively of control levels).
We conclude from these studies that severe stress brought about major declines in a spectrum of nodule enzyme activities and leghaemoglobin. In addition, in soybean, dark stress caused the even more rapid disappearance of mRNA coding for leghaemoglobin. The observed declines in protein and enzyme content during stress may be mediated at the transcriptional and/or post-transcriptional level, and these possibilities are currently being investigated.

* Provided by J. Stougaard, Aarhus University, Denmark.

Gresshoff, Roth, Stacey & Newton (eds.) Nitrogen Fixation: Achievements and Objectives
Chapman and Hall New York • London 1990

Mutation for nodulation which also affect the endomycorrhizal symbiosis in pea (*Pisum sativum* L.)

J.P. GUILLEMIN, G. DUC, A. TROUVELOT, V. GIANINAZZI-PEARSON and S. GIANINAZZI
S.A.P., INRA, BV 1540, 21034 Dijon Cédex, France.

Two independant mutagenesis programs with EMS aimed at the selection of plant mutants for the symbiosis of pea with *Rhizobium* were conducted in Dijon (France) (1)and Roskilde (Denmark) (2). The mutated pea lines were cv-Frisson and cv-Finale.
Independant mutants with deficient *Rhizobium* symbiosis have been selected : 45 with no nodules (nod-) and 21 with inefficient nodules (nod^+fix^-). These mutants were evaluated for their ability to establish vesicular-arbuscular (VA) endomycorrhizal associations in roots. All the mutants as well as the control lines were inoculated with a mixed inoculum of *Glomus intraradices* and *Glomus mosseae*.
The control lines express a typical endomycorrhizal association named (myc^+). This is characterised by appressoria formation followed by epidermal cell penetration and intracellular arbuscule development in the cortical parenchyma layer. Some of the studied mutants have shown a resistance to complete establishment of this symbiosis which is blocked at an early stage, just after appressorium formation. In this mutant phenotype, named (myc^-), penetration does not progress further than one or a few plant cells in depth. All 21 mutants with a (nod^+fix^-) root phenotype expressed a (myc^+) character. Out of 45 (nod^-) mutants, 24 had a (myc^+) phenotype, and 21 are characterized as (myc−). This proportion was equivalent in each mutagenesis program (7 myc^-/15 nod^- in (1) and 14 myc^-/31 nod^- in (2)).
Analyses of the F1 from the diallel crosses between (nod^-,myc^-) mutants has revealed 5 groups of complementation named A, C, P, Q and R with strict coincidence between complementation groups of individually (nod^-) and (myc^-) characters. This result suggests at least 5 loci involved in the control of the (nod^-,myc^-) character.
Complete genetic analysis has been performed for A and C complementation groups. Results support in each complementation group the hypothesis of a single recessive gene determining the (nod^-, myc^-) phenotype. The genetic analysis of other complementation groups is presently being carried out.
These results strongly suggest common genes involved in two plant symbiotic processes. Their function and eventually their application in pathogenic relationship remain to be studied.
Acknowledgement: many thanks are due to Dr K.C. Engvild (RISO, Denmark) who kindly provided the mutants in cv-Finale.
References :(1)G. Duc *et al.*,1989. Plant Science,60,207-213.
(2)K.C. Engwild *et al.*,1987. Theor.Appl.Genet.,74,711-713.

Gresshoff, Roth, Stacey & Newton (eds.) Nitrogen Fixation: Achievements and Objectives
Chapman and Hall New York • London 1990

Photosynthate partitioning and regulation of soybean (*Glycine max* L. Merr.) nodule development

J.I. Hacin, B.B. Bohlool* and P.W. Singleton.

NifTAL Project, University of Hawaii, Paia, HI 96779. USA.

One of the distinct physiological phenomena, by which the host plant controls the number of nodules formed, is a process called autoregulation. This phenomenon has attracted considerable research attention over the past 50 years, but the basic mechanism(s) of autoregulation remain as yet unknown. From the vast number of studies, two concepts of regulatory substance(s) have emerged: I) production of an inhibitor by early nodules and root meristems suppressing the development of late infections (2); and II) a shoot derived systemic inhibitor, the production of which is induced by early infections and its effect(s) exerted on late infections (3). However, other findings (1) suggest a possible regulatory role of photosynthates and their selective partitioning in regulation of nodulation. Our present study examines this role in more detail. Photosynthate (PS) partitioning to developing soybean roots and nodules was evaluated in a split-root growth system consisting of paired growth pouches. Two inoculation treatments with *B. japonicum* (USDA 110) were imposed: 1) E/U -one side inoculated early, the other remained uninoculated; 2) E/D - one side inoculated early, the other side inoculated with a four day delay. Plants were labeled with $^{14}CO_2$ and harvested at 24 h intervals, from inoculation until N_2 - fixation (acetylene reduction). After staining with Eriochrome Black, roots and developing nodules were separated into eight categories (4 root categories and 4 symbiotic categories), based on structure and developmental stage. Selective partitioning of PS to these categories was characterized by autoradiography and scintillation counting. Individual nodule developmental stages occurred at the same time from inoculation on early and delayed side. Similar numbers of initial (cortical cell divisions) and transitional (nodule primordia + emerging nodules) stages on both sides but lower number of nodules on the delayed side suggest that nodule development can be slowed or arrested at any stage following cortical cell divisions. At any developmental stage, photosynthate sink-strength (specific radioactivity) of symbiotic categories could be up to three times that of root categories. Root tips among root categories and nodules among symbiotic categories were the most intense sinks. Sink-strength of symbiotic categories increased with their development causing a decrease in specific radioactivity of root tips. Increased PS partitioning to developing nodules eventually resulted in decreased root growth on both sides of the split root system, implying that nodule development takes place at the expense of root growth. Selective partitioning of current PS to early inoculated side was clearly demonstrated in both treatments. The beginning of selective PS partitioning to early inoculated root coincided with the onset of nodule development on that side, indicating that developing nodules create a strong sink for current PS. Progressive PS partitioning to the early side most likely plays a major role in autoregulation by depriving late infections of current PS.

(1) Kosslak, R.M. and Bohlool, B.B. (1984) Plant Physiol. 75,125-130.
(2) Nutman, P.S. (1952) Annals. of Botany 1,79-101.
(3) Rolfe, B.G. and Gresshoff, P.M. (1988) Ann. Rev. Plant. Physiol. Plant Mol. Biol. 39,297-319.

Gresshoff, Roth, Stacey & Newton (eds.) Nitrogen Fixation: Achievements and Objectives
Chapman and Hall New York • London 1990

Immunolocalization of nitrogenase in *Frankia* of *Alnus icana* root nodules

Kerstin Huss-Danell[1] and Birgitta Bergman[2]

[1]Dept of Plant Physiology, Univ. of Umeå, S-90187 Umeå, Sweden, and [2]Dept of Physiological Botany, Univ. of Uppsala, Box 540, S-75121 Uppsala, Sweden

In *Alnus* root nodules *Frankia* occurs as hyphae and symbiotic vesicles, and in some cases also as spores. There are several lines of evidence for the localization of nitrogenase activity in the vesicles of *Frankia* in nodules of *Alnus*, *Myrica* and *Elaeagnaceae*, as well as in vesicles of free-living *Frankia* (Huss-Danell 1990). However, nitrogenase activity in symbiotic (*Casuarina, Allocasuarina*) as well as free-living *Frankia* without vesicles does occur (Huss-Danell 1990). It cannot be excluded, therefore, that nitrogenase occurs also in hyphae of *Frankia* in *Alnus* root nodules having vesicles.

Alnus incana (L.) Moench in symbiosis with a local source of *Frankia* was studied. Antisera against Fe-protein (*Rhodospirillum rubrum*) and MoFe-protein (*Azotobacter vinelandii*) were used. Immunogold label density was quantified in all cell types. Nitrogenase was localized only in vesicles; not in hyphae and not in spores of *Frankia*. However, very young vesicles without septa were not labelled. Young vesicles with a few septa had significant but still a low label density. Mature vesicles, highly compartmentalized due to transversing septa, contained more nitrogenase than all other cell types. Old degenerating vesicles always contained some nitrogenase. It thus appears that vesicles of a certain stage of development are needed prior to onset of nitrogenase synthesis.

Reference

Huss-Danell, K. (1990) in *The Biology of* Frankia *and Actinorhizal plants*, eds. Schwintzer, C.R. & Tjepkema, J.D. (Academic Press, New York), pp. 129-156.

Gresshoff, Roth, Stacey & Newton (eds.) Nitrogen Fixation: Achievements and Objectives
Chapman and Hall New York • London 1990

Intercellular location of glycoprotein in soyabean nodules: Effect of altered rhizosphere oxygen concentration

E.K. James, [*] N.K. Brewin, [+] F.R. Minchin and J.I. Sprent

Department of Biological Sciences, University of Dundee, Dundee, DD1 4HN, Scotland. *John Innes Institute, Norwich, NR4 7HU, England. +Welsh Plant Breeding Station, Aberystwyth, SY23 3EB, Wales

INTRODUCTION

Soyabean nodules were grown in atmospheres with 10% or 40% oxygen and compared structurally and physiologically with nodules grown in air. Immunogold labelling was used to locate glycoprotein in intercellular spaces of nodules.

MATERIALS AND METHODS

Nodulated roots of soyabean cv. Clarke were exposed to altered pO_2 for 28 days. Nodule slices were fixed in 2.5% glutaraldehyde and embedded in LR White resin. Semi-thin sections were incubated in the monoclonal antibody MAC236, followed by incubation in 5nm goat anti-rat gold. The gold particles were then silver enhanced for viewing under the light microscope. Thin sections for electron microscopy were collected on parlodion coated nickel grids and treated as for semi-thins but with 10nm gold particles and no silver enhancement.

RESULTS AND DISCUSSION

The micrographs clearly show that glycoprotein occlusions of intercellular spaces are widespread in the inner cortex of soyabean nodules and may block a significant number of spaces that would otherwise be utilised as air passageways to the infected zone. Thus, this glycoprotein appears to be a structural component of the oxygen diffusion barrier. Further, levels of the glycoprotein are altered in soyabean nodules subjected to altered pO_2 (higher levels of glycoprotein at above and lower levels at below ambient). This presumably acts as a means of altering the resistance of the cortex to oxygen diffusion over the long term.

Gresshoff, Roth, Stacey & Newton (eds.) Nitrogen Fixation: Achievements and Objectives
Chapman and Hall New York • London 1990

Ultrastructural development of spontaneous nodules in alfalfa

Priyavadan A. Joshi, Gustavo Caetano-Anollés, Effin T. Graham and Peter M. Gresshoff

Plant Molecular Genetics and Ornamental Horticulture and Landscape Design, Institute of Agriculture and Center for Legume Research, The University of Tennessee, Knoxville, TN 37901-1071, USA

Alfalfa (*Medicago sativa* L.) forms nodules in the absence of *Rhizobium* and fixed nitrogen (1). The ontogeny of spontaneous nodules in alfalfa cv. Vernal was studied in serial longitudinal plastic sections of roots at different stages of growth using light and electron microscopy. These nodules, on uninoculated roots, were white multilobed structures, histologically composed of nodule meristems, cortex, endodermis and a central zone with vascular strands (Fig. 1). Early development of nodulation was initiated by anticlinal followed by periclinal divisions of dedifferentiated cells in the inner cortex of the root. These cells formed the nodular meristem from which the nodule developed. Each cell division in the meristem formed one large and one small cell which persisted in the central zone. The large cells were characterized by cytoplasm packed with starch and a large central vacuole. Starch was present only occasionally in the small cells. Cells in the meristem and central zone contained fragmented rough endoplasmic reticulum, numerous free ribosomes, and few organelles. There were no infection tubes or bacteria in spontaneous nodules at any stage of development. The large cells became infected with bacteria in normal nodules of inoculated plants whereas they accumulated starch in spontaneous nodules. Alfalfa plants apparently have an endogenous mechanism for inducing early cell divisions which initiate nodular meristems and ultimately develop bacteria-free nodules. Our results suggest that bacteria and therefore their *nod* genes are not necessarily involved in the ontogeny and morphogenesis of spontaneous and normal nodules in alfalfa.

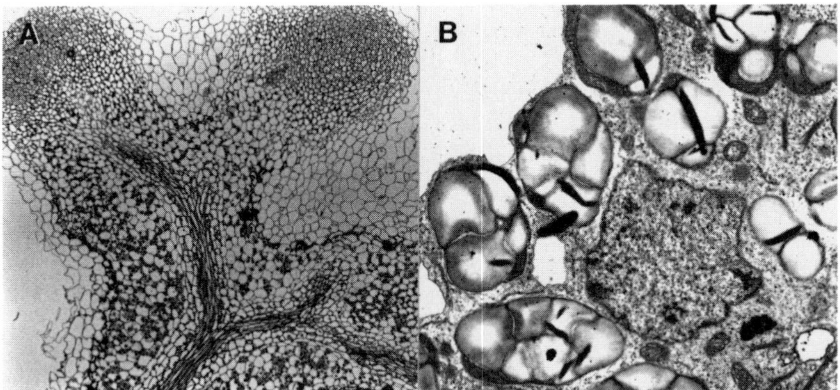

Fig.1. Multilobed spontaneous nodule showing meristems, cortex, central zone and multivascular strands (A). x 100. The central zone shows many starch grains and no bacteroids (B). x 7,000.

1. Truchet, G., Barker, D.G., Camut, S., de Billy, F., Vasse, J., Huguet, T. (1989) *Mol. Gen. Genet.* **219**, 65-68

Gresshoff, Roth, Stacey & Newton (eds.) Nitrogen Fixation: Achievements and Objectives
Chapman and Hall New York • London 1990

Presence of an insoluble nitrate reductase in legume root nodules

A. Kaur, C. Arrese-Igor, A.J. Gordon and F.R. Minchin

Department of Environmental Biology, Welsh Plant Breeding Station, Aberystwyth, U.K.

Introduction:
Measurements of plant-derived nitrate reductase activity (NRA) in legume nodules are normally performed on the supernatant following nodule grinding and centrifugation, based on the assumption that this is a soluble enzyme. This poster presents data for an insoluble component of nodule NRA which can be detected in the pellet following centrifugation at 8,000g.
Results:
A constitutive NADH dependent NRA was detected in the supernatant of soybean (Clarke/RCR 3407) nodule extracts, which increased with time of exposure of the plants to 10 mM NO_3. The same assay procedure also detected a constitutive NRA in the 8,000g pellet, which also increased with NO_3 feeding. This pellet activity represented from 32 to 60% of total nodule NRA. Bacteroids from soybean nodules also had a constitutive NRA of particulate nature. This was lower than the cytosolic activity, but also increased with NO_3 feeding. However, the bacteroidal NRA required methyl viologen as an electron donor and no activity was detected with NADH. Therefore the pellet activity cannot be due to bacteroid contamination.
The insoluble NRA of nodules was not affected by sonification or repeated washing of the pellet, or disruption of nodule cells using a ground glass homogenizer. This treatment broke between 90 and 95% of the nodule cells, therefore, the pellet activity cannot be explained by contamination with unbroken cells.
Following mechanical separation of soybean nodules into cortical and infected regions most NRA was in the infected region, with very low pellet activities in the cortex. Homogenates of roots contained no constitutive NRA, but following 4 days exposure to 10 mM NO_3 soluble NRA was 66.3 nmol min^{-1} g^{-1}FW. Activity in the root pellet was 3.5, and represented only 5% of total activity, which could be explained by observed levels of unbroken cells in the pellet. Therefore, the assay procedures do not appear to be producing spurious data.
Conclusion:
There appears to be an insoluble nitrate reductase in the infected region of soybean nodules. If these reports are confirmed it would suggest that previous measurements of nodule NRA have been underestimated by up to 50-60%.

Gresshoff, Roth, Stacey & Newton (eds.) Nitrogen Fixation: Achievements and Objectives
Chapman and Hall New York • London 1990

α-glucosidase and other enzyme activities in the peribacteroid space of soybean root nodules

A. Kinnback* and D. Werner.

Philipps-University of Marburg, Dept. of Biology, 3550 Marburg, FRG.

Introduction

Bacteroids of *Bradyrhizobium japonicum* in root nodules from *Glycine max* are surrounded by a host derived membrane, the peribacteroidmembrane (PBM) (1). The peribacteroid space (PBS) is a lytic compartment between the outer membrane of the bacteroid and the PBM (2). Signal exchange, nutrient exchange from and to the bacteroids have to pass these compartments.

Methods

PBS was isolated by sucrose gradient centrifugation and the proteins were precipitated with ammoniumsulphate. α-Glucosidase was partially purified by ammonium sulphate precipitation and affinity chromatography.

Results

The PBS contains a discrete set of proteins as revealed after one-dimensional SDS-PAGE. Marker enzyme activities proved no cross contamination with bacteroid cytoplasm and host cytoplasm. Several hydrolytic enzymes could be detected in the PBS. Specific activity of α-glucosidase and of α-trehalase was 27-fold and 10-fold respectively higher compared to host cytoplasm. Very low activity could be detected in bacteroid cytoplasm and the enzymes were not present in root cytoplasm.

α-Glucosidase hydrolyzes maltose, maltotriose and isomaltose. Soluble starch, α-phenylglucoside and trehalose were not hydrolyzed. The enzyme has a pI of 7.2 and a pH-optima of 4.5.

Enzymes involved in the organic acid-metabolism, malatdehydrogenase and glutamate-oxalacetat-transaminase, were highly enhanced in the PBS.

Discussion

The PBS contains hydrolytic enzyme activities that are probably involved between the invading bacteria and the host cell preventing the symbiotic interaction becoming a parasitic one. The enzymes that are involved in organic acid metabolism of the symbiosis are demonstrate the significance of the PBS for the regulation of nutrient exchange between the host plant and the bacteroid.

References

(1) Mellor ,R.B., and Werner, D. (1987) Symbiosis, **3**, 75-100
(2) Mellor, R.B. (1989) J. Exp. Botany, **40**, 831-839.

Gresshoff, Roth, Stacey & Newton (eds.) Nitrogen Fixation: Achievements and Objectives
Chapman and Hall New York • London 1990

The accumulation of lectin in peanut nodules in relation to the symbiotic effectiveness of *Bradyrhizobium* sp. strains

B. Kishinevsky, Chaya Nemas, Y. Friedman and G. Meromi

Dept. of Legume Inoculation, Agricultural Research Organization, The Volcani Center, Bet Dagan, 50 250, Israel.

Peanut (Arachis hypogaea L.) nodules have recently been found to contain a lectin which appears to be galactose-specific (GNL) (1). The function of GNL is not yet known, although it may play a role in the development and symbiotic function of peanut nodules. The purpose of this study was to determine whether the synthesis of GNL in nodules is affected by the nitrogen-fixing activity of Bradyrhizobium strains employed for inoculation of peanut plants.

Three peanut varieties (var. hypogaea, var. vulgaris, var. fastigiata) were inoculated in all possible combinations with three highly effective, three mediocre and one noneffective Bradyrhizobium sp. strains. Nodules of 53 day old plants were used for quantitative detection of GNL (1). The bacteroids were isolated as described elsewhere (2).

The determination of plant yield and N content in the plant tops, as well as the acetylene-ethylene reduction assay for N_2 fixation, showed a variability in the symbiotic effectiveness of the rhizobial strains tested. The GNL levels in fresh tissue mass of nodules formed by effective strains (40-42 µg/g) on all three varieties tested were significantly higher than those in the nodules formed by mediocre (2.5-8.6 µg/g) and noneffective (1.4 µg/g) strains, pointing up the relationship between symbiotic function and lectin expression in this specialized structure. Likewise, the nodules formed by Tn5 mutants contained only 37% of the GNL detected in the more effective wild-type strain nodules. Within individual nodules the bacteroid zone was found to contain 315 GNL µg/g tissue mass as compared with 0.6 µg GNL/g tissue in the cortex. Detectable amounts of GNL were found to be present in membrane-enclosed bacteroids isolated from effective nodules. For most of the strains tested, the number of bacteroids was much lower in noneffective nodules than in effective ones. Whether the bacteroids per se are responsible for lectin activity in the nodules is still an open question and should be investigated in future experiments.

1. Kishinevsky, B. D., Law, I.J. and Strijdom, B.W. (1988) Planta, 176, 10-18.
2. Brewin, N.J., Robertson, J.G., Wood, E.A., Wells, B., Larkins, A.P., Galfre, G. and Butcher, G.W.(1985). EMBO Journal, 605-611.

Gresshoff, Roth, Stacey & Newton (eds.) Nitrogen Fixation: Achievements and Objectives
Chapman and Hall New York • London 1990

The isolated Nms-25 (nodulin-25) promoter of alfalfa ensures nodule-specific expression in transformed *Medicago*

György B. Kiss, Irute Meskiéne, Janina Zabelina, and Anna Gutkina.

Institute of Genetics, Biological Research Center, Szeged
P. O. Box 521, H-6701, Hungary

The promoter and the signal peptide region of the Nms-25 (nodulin-25) gene of alfalfa (1, 2) was placed in front of the structural gene of GUS (beta-glucuronidase, ref.3) and cloned into a plant cloning vector pGA471 (4). Different tissues from the kanamycin resistant individual transformants of Medicago varia A2 (5) was tested for the copy of the transformed DNA and for GUS activity. Using GUS and neomycin phosphotransferase specific probes, single, double and triple insertions could be detected. The position of the hybridizing bands were identical for both probes. The majority of the transformants showed nodule specific expression of GUS. Different level of GUS activity was detected in nodule, root, stem and leave tissue of some transformants which can be explained by "position effect" of the inserted DNA. Some out of frame constructions gave GUS activity. In control experiments CaMv-GUS constructions were used (3). In these cases the transformed calli hardly regenerated. Few regenerated plants showed, however, high GUS activity in all tissue tested.

We concluded from these experiments that the isolated 5' untranscribed sequence of the Nms-25 gene is an active promoter and ensures nodule specific expression of downstream promoters.

1. Kiss et al. Plant Mol. Biol. (1990 in press).
2. Végh et al. Plant Mol. Biol. (1990, submitted).
3. Jefferson et al. EMBO J. 6: 3901-3907 (1987)
4. An et al. EMBO J. 4: 277-284 (1985).
5. Deák et al. Plant Cell Rep. 5: 97-100 (1986)

Gresshoff, Roth, Stacey & Newton (eds.) Nitrogen Fixation: Achievements and Objectives
Chapman and Hall New York • London 1990

Novel nodulin cDNA clones isolated from soybean root nodules by a subtractive hybridization technique

Hiroshi Kouchi and Shingo Hata

National Institute of Agrobiological Resources, Tsukuba, Ibaraki, 305 Japan.

Subtractive hybridization is a method to isolate cDNA clones of very rare messages. We applied this method to isolate nodule-specific (nodulin) genes from soybean root nodules of different developmental stages. ^{32}P-labelled cDNA prepared from poly(A+) RNAs of mature (21-day after Bradyrhizobium inoculation) and young (8-day) nodules were subtracted with large excess of poly(A+) RNA of uninfected roots and used to screen the cDNA libraries of the nodules. By this procedure, we isolated more than 20 different nodule-specific cDNAs as well as most of the nodulin clones previously reported.

Isolated nodulin genes were tentatively classified into three groups by developmental Northern blots. A) Late nodulins (1) expressed concomitantly with Lb genes but not or very slightly detected in the nodules before the onset of N_2 fixation. B) Early nodulins (1,2) expressed in the root segments of infected zone 4-days after the Bradyrhizobium inoculation, when only root hair curling and cortical cell division were present. C) Intermediate nodulins, which were not detected in 4-day infected root segments, but abundantly expressed just after emergence of nodule structure (slightly before the onset of N_2 fixation).

Three novel early nodulin cDNAs were isolated. Two (clones #3 and #37) of them encoded Pro(Hyp)-rich proteins reported as eNod2 family (3) but these were weakly expressed in the uninfected roots. One (#18) had a quite different sequence from eNod2 family and was weakly expressed in the stems but not in the leaves and uninfected roots. Four novel nodulin cDNA clones belonged to intermediate nodulins. One (#315) of them had a Pro(Hyp)-Ser alternate sequence reported previously as eNod55 (4). However, the amino acid composition derived from the DNA sequence of #315 clone was different from eNod55. Further characterization of isolated clones is currently underway.

1. Verma, D.P.S. et al. (1988) in "Molecular Genetics of Plant-Microbe Inernations", pp. 315. APS Press.
2. Gloudemans, T. and Bisseling, T. (1989) Plant Science 65:1-14.
3. Franssen, H.J. et al. (1987) Proc. Natl. Acad. Sci. USA 84: 4495-4499.
4. Franssen, H.J. et al. (1988) In "Molecular Genetics of Plant-Microbe Interactions", pp. 321. APS Press.

Gresshoff, Roth, Stacey & Newton (eds.) Nitrogen Fixation: Achievements and Objectives
Chapman and Hall New York • London 1990

RFLP analysis of symbiotic mutants of soybean

D. Landau-Ellis, S. A. Angermüller, L. J. Schuller, L. Sayavedra-Soto, and P. M. Gresshoff
Plant Molecular Genetics, Institute of Agriculture and Center of Legume Research, The University of Tennessee, Knoxville, TN 37901-1071 USA.

The objective of this study is to determine linkage of soybean (*Glycine max* (L.) Merr.) nodulation phenotype to RFLP markers. Four soybean nodulation mutants were included: *nts*382 and *nts*1007 are supernodulators (1), *nod*49 and *nod*139 are nonnodulators (2). Each of the four mutants were originally isolated from M2 and M3 populations of ethyl methanesulfonate mutagenized seeds of a *G. max* cultivar Bragg. This common origin resulted in a high level of homology among Bragg and the four mutants. A more diverse population was needed in order to detect RFLP linkage (4).

The genus *Glycine* subgenus *Soja* consists of two species, the cultivated soybean (*Glycine max* (L.) Merr.) and the wild annual soybean (*Glycine soja* Sieb. and Zucc.). In order to obtain a more diverse population, crosses were made between Bragg, the four mutants, and a *G. soja* plant introduction (PI 468.397; courtesy of Dr. Gary Stacey, University of Tennessee). Linkage analysis is being conducted on the segregating F2 populations resulting from these crosses.

Iowa State University supplied 120 plasmid clones (courtesy of Drs. Paul Keim and Randy Shoemaker) which detect polymorphisms between their experimental line A81-356022 (*G. max*) and PI 468.916 (*G. soja*). These probes make up a large portion of a linkage map which covers 30 linkage groups, 168 loci, and 1800 centimorgans. Given the genomic size of soybean as 1.8×10^9 bp (3), this represents about 10 Mb DNA per locus.

Some of these probes did not detect polymorphisms between the Bragg genotypes and PI 468.397. At this time we have seen 17 out of 30 discernable RFLPs between *G. max* and *G. soja*; these are located on 13 of the 30 linkage groups. None of these RFLP patterns have shown a difference among Bragg, *nts*1007, *nts*382, *nod*139, and *nod*49, emphasizing the common origin of these symbiotic mutants.

F1 plants, from each cross, were confirmed by plant morphology, nodulation phenotype, flower color, and seed size and color. Heterozygous RFLP patterns were also detected for F1 plants when using the probe pK19 and restriction endonucleases *Eco*RI and *Hin*dIII.

1. Carroll, B. J., McNeil, D. L., and Gresshoff, P. M. 1985. Proc. Natl. Acad. Sci. USA. 82:4162-4166.
2. Carroll, B. J., McNeil, D. L., and Gresshoff, P. M. 1986. Plant Sci. 47:109-114.
3. Goldberg, R. B., 1978. Biochem. Gen. 16:45-68.
4. Keim, P., Shoemaker, R. C., and Palmer, R. G. 1989. Theor. Appl. Genet. 77:786-792.

Gresshoff, Roth, Stacey & Newton (eds.) Nitrogen Fixation: Achievements and Objectives
Chapman and Hall New York • London 1990

Affinity chromatography used to study nodule-specific proteins from soybean (*Glycine max*)

Knud Larsen and Bjarne Jochimsen

Department of Molecular Biology and Plant Physiology, University of Aarhus, DK 8000 Aarhus, Denmark.

Nodule-specific proteins (nodulins) were first identified in soybean using in vitro translation of nodule mRNA followed by immunoprecipitation of translation products with nodule-specific antibodies and later also by cDNA cloning followed by in vitro translation.
Nodule-specific proteins are assumed to be involved in the development, structure and metabolism of the nodule, but the precise function of nodulins has only been established for a few: some having enzymatic activity (eg., glutamine synthetase, uricase and sucrose synthase) or being oxygen carriers (leghemoglobins).

In order to identify nodulins present in too low amounts to be detected by the above mentioned techniques, we used affinity chromatography to concentrate nodule and root extracts in order to detect nodulins occuring at around 0.1% of total protein.
Nodule and root extracts were loaded onto Blue Sepharose columns and eluted with 10 mM NAD, followed by 1 M NaCl. One polypeptide eluting with NAD of molecular mass 40 kDa seems to be nodule-specific & might be a subunit of a dehydrogenase or a kinase.

Similarly, extracts were eluted from a Concanavalin A Sepharose column by the addition of 0.5 M mannose. Polypeptides with molecular mass of 30, 79 and 94 kDa were identified as nodule-specific and represent glycoproteins. However, the 30 kDa polypeptide might be Con A comming off the column due to proteolytic cleavage.
Specific elution from more specialized affinity materials might help to identify the function of such nodulins.

Two affinity materials, constucted by coupling of 5-amino-imidazole-4-carboxamide riboside (AICAR) and L-arginine to activated CH Sepharose, were tested and prooved excellent in purifying uricase (nodulin-35). Uricase was retarded on both materials and repeated passage on the affinity columns resulted in fractions containing only two polypeptide bands (33 and 31 kDa) both representing uricase activity as demonstrated by isoelectric focusing. The rationale why these two materials have affinity towards uricase is unclear to us and the result might indicate that the use of affinity materials for identifying nodulin function is not a simple matter.

Gresshoff, Roth, Stacey & Newton (eds.) Nitrogen Fixation: Achievements and Objectives
Chapman and Hall New York • London 1990

Comparison of mannose- and galactose-binding lectins from nodules and other tissues of peanut

I.J. Law[1], T. Haylett[2], A.J. Mort[3] and B.W. Strijdom[1].

[1]Plant Protection Research Institute, P/Bag X134, Pretoria 0001, South Africa. [2]Processing and Chemical Manufacturing Technology Div,. CSIR, Pretoria, South Africa; [3]Dept. of Biochemistry, Oklahoma State University, Stillwater, USA

Peanut nodules contain a mannose-binding lectin (ML) and a galactose-binding lectin (GL) [1] whose relevance to the peanut-Bradyrhizobium symbiosis is under investigation. Lectin related to nodule ML was purified from all vegetative tissues of 6-week-old nodulated plants but only nodules contained sufficient GL for comparison with the related peanut seed GL.

The isolectin profile of nodule ML closely resembled that of ML from other tissues and nodule ML was antigenically identical to ML from all tissues except cotyledon ML. The amino acid composition of nodule and cotyledon ML was similar and both were glycoproteins containing mannose, fucose, galactose, xylose and N-acetyl-glucosamine although their respective sugar ratios and glycan contents differed. Nodule and cotyledon ML had similar sugar specificities but differed in their reaction with certain glycoproteins. Nodule and leaf ML were identical in this respect.

Despite similar sugar specificities seed and nodule GL differed in their reaction with certain glycoproteins. Nodule GL contained a single major isolectin that differed from the six isolectins of seed GL. Nodule GL contained the same sugars that were identified in peanut ML supporting earlier evidence that it is a glycoprotein; seed GL is non-glycosylated. Despite similar amino acid composition, only four of the first 10 N-terminal amino acids of nodule GL corresponded in position to those of seed GL. This indicates that the two lectins are products of different genes.

The nodule lectins do not appear to function as recognition determinants in the peanut symbiosis as neither reacted with cells or bacteroids of peanut bradyrhizobia. While synthesis of nodule GL may be specific to this tissue the role of the nodule lectins is presently unknown.

[1] Law, I.J., Haylett, T. & Strijdom, B.W. (1988) Planta **176**, 19-27.

Gresshoff, Roth, Stacey & Newton (eds.) Nitrogen Fixation: Achievements and Objectives
Chapman and Hall New York • London 1990

Malate transport and metabolism in infected cells of soybean nodules

Youzhong Li, Li-Jun Ou Yang, Rosanne G. Quinnell, Michael K. Udvardi and David A. Day
Botany Department, Australian National University, Canberra A.C.T. 2601, Australia.

Nitrogen fixation in soybeans is supported by sucrose translocated to the nodule via the phloem. Recent studies of enzyme localization (1, 2) suggest that this sucrose is metabolized in the uninfected cortical cells to organic acids which are then transported to the infected zone and oxidized to support nitrogenase and ammonia assimilation. Malate is likely to be one of these organic acids and we have investigated its uptake by isolated infected cells, and the subcellular compartments within those cells, from soybean nodules.

Isolated infected cells (2) are able to accumulate malate from the external medium; this uptake shows saturation kinetics and is inhibited by succinate and a range of other organic acids, indicating the presence of a carrier on the plasmalemma. Sucrose, glucose and fructose, on the other hand, are not taken up at significant rates by the isolated cells. Inhibition by the mitochondrial inhibitor antimycin A and the uncoupler CCCP suggests that malate uptake by the cells is driven by oxidative phosphorylation.

Malate oxidation by isolated nodule mitochondria is rapid and tightly coupled to oxidative phosphorylation. These mitochondria lack malic enzyme and malate oxidation is catalysed by malate dehydrogenase (MDH); the oxaloacetate (OAA) formed is transported to the external medium via the specific mitochondrial OAA transporter. MDH is also present in large quantities in the cell cytosol. Thus malate oxidation by infected-cell MDH will yield ATP, OAA and NADH to support ammonia assimilation.

Malate is rapidly taken up by isolated symbiosomes via two distinct dicarboxylate carriers, one on the peribacteroid membrane and another on the bacteroid membrane. Bacteroids from soybean contain two malic enzymes, one NAD and the other NADP-linked, which convert malate to pyruvate. Acetyl-CoA from pyruvate can then either enter the tricarboxylic acid cycle or be used in poly-β-hydroxybutyrate production, while the NAD(P)H generated can be used to support nitrogenase.

It is clear that infected cells of soybean nodules possess the transport systems and enzymes necessary to utilize malate as sole carbon substrate, providing the carbon skeletons and energy required for both nitrogen fixation in the bacteroid and ammonia assimilation in the plant cytoplasm.

(1) Copeland, L., Vella, J. & Hong, Z. (1989) Phytochem. **28**, 57-61.
(2) Kouchi, H., Fukai, K., Katagiri, H., Minamisawa, K. & Tajima, S. (1988) Physiol. Plant. **73**, 327-334.

Gresshoff, Roth, Stacey & Newton (eds.) Nitrogen Fixation: Achievements and Objectives
Chapman and Hall New York • London 1990

Developmental and biochemical regulation of constitutive nitrate reductase activity in leaves of nodulating soybean

Zhi Z. Li and Peter M. Gresshoff

Plant Molecular Genetics, Institute of Agriculture and Center of Legume Research, The University of Tennessee, Knoxville, TN 37901-1071 USA.

Leaf nitrate reductases (NR) of soybean (*Glycine max* (L.) Merr) have been well-described (Jolly et al., 1976; Srivastava, 1981; Nelson et al., 1984; Streit et al., 1986; Carroll and Gresshoff, 1986), but have not been examined sufficiently in plants of different symbiotic status. Not enough is known about the effect of root nodules and nitrogen fixation on iNRA and cNRA (c_1NR and c_2NR, Streit et al., 1986) in leaves. Here we show that a product of biological nitrogen fixation is capable of directly affecting cNRA in soybean leaves (Li and Gresshoff, 1990).

Seedlings of cultiver Bragg were divided into four experimental groups (inoculated, uninoculated, uninoculated + nitrate, and uninoculated + urea) and monitored for the developmental profile of *in vivo* and *in vitro* total cNRA and iNRA in leaves. The youngest leaves had the highest cNRA which dropped off as a newer leaf developed. Really young leaves had no measurable activity. Each leaf had its distinct activity period of cNRA. This pattern was different in urea-grown and symbiotically-grown plants (inoculated with *Bradyrhizobium japonicum* strain USDA110), where the latter had no detectable cNRA in older leaves. Urea-grown plants maintained considerable activity in such older leaves. When symbiotically-grown plants had their nodules removed, cNRA reappeared in older leaves within one day of removal, nearly reaching activity levels of young leaves at 3 days after removal . Allantoic acid (at 400 μM) completely inhibited *in vitro* cNRA. Allantoin, in contrast, had no effect. Inhibition was by a competitive process. Allantoic acid also had slight stimulating effect of *in vitro* iNRA (up about 25% at 400 μM). Allantoic acid may be a signal molecule for nodule to alter the physiological nature of leaves (Carroll et al., 1985; Gresshoff and Delves, 1986). It is possible that cNRA of soybean leaves possesses another biological function, perhaps related to ureide metabolism.

Carroll, B.J., McNell, D.L., & Gresshoff, P.M. (1985) Proc. Natl. Acad. Sci. USA.82, 4162-6.
Gresshoff, P.M. & Delves, A.C. (1986) in Plant Gene Research III. A genetical approch to plant biochemistry., eds. Blonstein, A.D. & King, P.J. (Wien Springer Verlag), pp. 159-206.
Jolly, S.O., Campbell, W. & Tolbert, N. E. (1976) Archieves Biochemistry Biophysics, 174, 431-9.
Li, Z. Z. & Gresshoff, P. M. (1990) J. Exp. Bot. (in press)
Nelson, R. S., Streit, L. & Harper, J. E. (1984) Physiologica Plantarum, 61, 384-90.
Srivastava, M. S. (1981) Phytochemistry, 19, 725-33.
Streit, L. & Harper, J.E. (1986) Plant Physiology, 81, 593-6.

Gresshoff, Roth, Stacey & Newton (eds.) Nitrogen Fixation: Achievements and Objectives
Chapman and Hall New York • London 1990

Nodulin-26: Topology and function in root nodules

G.-H. Miao, C. P. Joshi and D. P. S. Verma

Department of Molecular Genetics and Biotechnology Center, The Ohio State University, Columbus, Ohio 43210 U.S.A.

The amino acid sequence of nodulin-26 (N-26), a peribacteroid membrane (PBM) protein from soybean root nodule, was deduced from the analysis of a full-length cDNA clone. Structure analysis of this clone identified a 271 amino acid open reading frame and the encoded protein has six potential transmembrane domains. Like many transmembrane proteins, this protein lacks a transient amino terminus signal sequence; instead, two internal putative signal sequences were identified. In order to determine the topology of N-26, a truncated sequence of the coding region corresponding to amino terminal 91 amino acids (N-26T) was cloned behind a T3 promoter in Bluescript plasmid and transcribed *in vitro*. The corresponding polypeptides were synthesized in rabbit reticulocyte as well as in wheat germ cell-free systems and their processing and insertion into dog pancreatic microsomal membranes were assayed. These results suggested that N-26 is co-translationally inserted into the microsomal membranes. Since the N-26T is unable to insert into microsomes, the insertion of this nodulin in PBM may involve an internal signal sequence near the carboxyl terminus. The translocated protein is further modified by glycosylation as membrane-inserted N-26 binds to Con-A. Chemical cleavage mapping at cystein residues of trypsinized and untreated N-26 showed that both amino (N) and carboxyl (C) termini of the *in vitro* synthesized N-26 are sensitive to trypsin digestion, indicating that both N- and C-termini of this protein face outside of the PBM vesicle (i.e., facing the cytoplasm) and the glycosydic residue seems to occur on the side of bacteroids. The carboxy end of N-26 contains three phosphorylation sites which may intereact with the acidic lipids present in this menbrane.

Amino acid sequence comparision of N-26 with bovine eye lens intrinsic membrane protein (MIP-26), a turgor-regulated gene from pea (7a), a *Drosophila* neurogenic gene, big brain (*bib*), and *Escherichia coli* glycerol facilitator (*glp*F) shows significant homologies among this group of proteins. Six unique transmembrane spaning regions have been identified from the hydropathy profile analysis of all these proteins. Secondary structure analysis of these proteins also shows extensive similarity. The N-26 and glpF have a similar isoelectric point while MIP-26 and JM7a are similar. It has been shown that the glpF is the only pore-type ion channel in the cytoplasmic membrane of *E.coli*. Based on our secondary structure model for N-26 on PBM and its comparison to MIP-26, 7a, *bib* and *glp*F, it is very likely that N-26 may function as a channel in soybean root nodules. Since the expression of this gene occurs prior to and independent of nitrogen fixation, N-26 may participate in translocating specific metabolites such as succinate to support bacteroids. Such a passive translocation of dicarboxylic acid through PBM would allow equilibration of the peribacteroid fluid with the cytoplasm and active uptake by bacteroids.

Gresshoff, Roth, Stacey & Newton (eds.) Nitrogen Fixation: Achievements and Objectives
Chapman and Hall New York • London 1990

Soybean nodule glutamine synthetase genes are regulated by ammonia in roots of transgenic *Lotus corniculatus* but not in transgenic tobacco

G.-H. Miao[1], M.C. Marsolier[2], B. Hirel[2] and D.P.S. Verma[1]

[1]Department of Molecular Genetics and the Biotechnology Center, The Ohio State University, Columbus, Ohio 43210 U.S.A.; [2]Laboratoire du Metabolisme - INRA, Route de Saint-Cyr, 78000 Versailles Cedex, France

We had earlier demonstrated (1) that glutamine synthetase (GS) gene expression in soybean root nodules may be regulated by the availability of ammonia, since the root tissue treated with ammonia for a short time transcribes high levels of GS mRNA. Isolation of a GS gene corresponding to this transcript (pGS15) followed by its insertion into *Lotus corniculatus* as well as tobacco along with a reporter gene (GUS) showed that the expression of this GS gene occurs in a root specific manner in transgenic legume and non-legume plants. Strong GUS activity was also found in root nodules of transgenic *Lotus*. Histochemical localization of GUS activity showed that this gene is primarily expressed in the infected cells of nodules with less activity in the central region of the nodules. That ammonia may be diffused in the infected zone is evident by the low activity of GUS in the uninfected cells of this zone but there is no activity in the cortical region out side the endodermis. The low level of activity in the central region of the infected zone may be due to the lack of oxygen, since this tissue is highly anaerobic. This is demonstrated by the expression of alcohol dehydrogenase - GUS gene fusion experiments (2). In roots, the GUS activity is mainly located in the root meristems in both tobacco and *Lotus*. The GUS activity driven by GS promoter also marks the incipient root primordia prior to any visible morphological demarcation on root surface. Treatment with ammonium sulfate induces this gene uniformly in the root tissue. Another GS gene (pGS21) is also ammonia-inducible and specifcally expressed in the infected cells. Neither of these GS genes are inducible by ammonia when expressed in transgenic tobacco roots.

These data demonstrate that soybean cytosolic root and nodule GS gene retains its tissue specificity in both legume and non-leguminous plants. However, its ammonia inducibility is only observed in a legume background. These results suggest that soybean cytosolic GS15 gene promoter carries at least two unique cis-acting regulatory elements: root-specific and ammonia inducible. The root-specific cis element can be recognized by the *trans*-acting factor(s) present in the roots of both legume and non-legume heterologous plants. However, the *cis*-element conferring ammonia inducibility apparently only function in the legume background and not in a non-legume, as the specific *trans*-acting factor(s) may be lacking in the non-legume plants. Since all root tips including lateral roots preferentially and constitutively express this enzyme , this suggests a high demand for reduced nitrogen in the cells entering into active metabolism and cell division.

1. Hirel,B.,Bouet,C.,King,B.,Layzell,D.,Jacobs,F.& Verma,D.P.S. (1987) EMBO J. 5:1167-1171
2. Hu,C.A.& Verma.D.P.S. (1990) Plant Physiol. 93,83

Gresshoff, Roth, Stacey & Newton (eds.) Nitrogen Fixation: Achievements and Objectives
Chapman and Hall New York • London 1990

Cellular localization of xanthine dehydrogenase (XDH) in root nodules of cowpea (*Vigna unguiculata* (L.) Walp.)

E. H. Newcomb, D.B. Datta* and E.W. Triplett

Departments of Botany and Agronomy, University of Wisconsin, Madison, WI 53706 USA.

Cowpea, like soybean, is a ureide transporter. In ureide production via purine synthesis, early steps take place in the infected cells, and later steps in the uninfected. As plastids are abundant in both infected and uninfected cells, localizing XDH will help identify the intermediate in ureide formation that moves from infected to uninfected cells. If XDH is entirely in the infected cells, this would rule out a role for the plastids of the uninfected cells in purine synthesis for ureide production and would point to uric acid as the intermediate. But if XDH is partly or wholly in uninfected cells, then the intermediate could be xanthine or a precursor, and purine synthesis as a component of purine production in the uninfected cells remains a possibiliy. Recent attempts to localize root nodule XDH to cell type, and even to confirm its localization in the cytoplasm, are contradictory. Using polyclonal antibodies against cowpea XDH in an immunogold labelling procedure, we have obtained labelling in the ground cytoplasm of both cell types, with greater labelling in the uninfected cells. Controls with pre-immune serum showed no labelling. An artifactual origin for the labelling is ruled out by the virtual absence of label in the bacteroids, peribacteroid spaces, cell organelles, cell walls and intercellular spaces. We can conclude that while some purine biosynthesis takes place in the infected cells and uric acid is an intermediate that moves into the uninfected cells, the possibility remains open that some considerable purine biosynthesis is taking place also in the plastids of the uninfected cells.

Gresshoff, Roth, Stacey & Newton (eds.) Nitrogen Fixation: Achievements and Objectives
Chapman and Hall New York • London 1990

Molecular analysis of chickpea (*Cicer arietinum* L.) nodulation mutants

V. S. Paramadevanahalli and T. M. Davis

University of New Hampshire, Plant Biology Department, Durham, NH 03824, USA.

We are using heterologous nodulin cDNA clones as probes to look for restriction fragment length polymorphisms (RFLPs) in radiation-induced chickpea nodulation mutants PM233 (non-nodulating), PM665 and PM679 (temperature-sensitive nodulation), PM405 and PM638 (white, non-fixing nodules), and PM796 (poorly fixing, prematurely senescent nodules) in comparison to their wild-type parent, ICC640 (1). The mutants are isogenic to ICC640 differing only with respect to single nodulation gene mutations. Non-nodulating mutant ICC435M, a spontaneous mutant of wild type ICC435(3) has also been studied.

Southern blots of soybean, alfalfa, wild type and mutant chickpea genomic DNA were probed with alfalfa cDNA clones A2ENOD2, leghemoglobin (Lb), and N-32/34 (9A) (2) and pea asparagine synthetase (AS) (4). A2ENOD2 hybridized to chickpea, alfalfa and soybean genomic Southerns, with Eco R1 and ECO R1 + ECO RV digests each giving a single 5Kb band and ECO RV and Xba 1 each giving a single 11KB band for chickpea DNA. The AS probe gave two-band hybridization patterns in Eco R1-cut chickpea DNA and no detectible hybridization to alfalfa DNA. No A2ENOD2 or AS homologous RFLPs were detected in chickpea mutants. Failure to detect such an RFLP suggests, that the A2ENOD2 and AS homologous chickpea genes are unaltered in these mutants.

Clones 9A and Lb did not hybridize to chickpea and alfalfa genomic Southerns, perhaps due to experimental conditions, or inherent characteristics of the genomic target sequences. Gene expression in appropriate chickpea nodulation mutants will be evaluated using the A2ENOD2, AS and other heterologous probes that hybridize to chickpea DNA.

1) Davis,T.M., Foster, K.W.& Phillips, D.A (1985) Crop Sci. 25, 345-348.
2) Dunn, K., Dickstein, R., Feinbaum, R., Burnett, R.K., Peterman, T.K., Thoidis, G., Goodman, H.M. & Ausubel, F.M. (1988) Molecular Plant-Microbe Interactions, 1, 66-74.
3) Rupela, O.P. & Sudarshana, M.R. (1986) Chickpea NewsLetter, No. 15, 13-14.
4) Tsai, F. & Coruzzi, G.M. (1990) EMBO J. 9, 323-332.

Gresshoff, Roth, Stacey & Newton (eds.) Nitrogen Fixation: Achievements and Objectives
Chapman and Hall New York • London 1990

Two dimensional polyacrylamide gel analysis of total leaf protein from wild type and supernodulating mutant of soybean

R. Prabhu*, L. A. Sayavedra-Soto and P.M. Gresshoff

Plant Molecular Genetics, Institute of Agriculture and Center of Legume Research, The University of Tennessee, Knoxville, TN 37901-1071, USA.

The study of root nodule symbiosis between *Bradyrhizobium* and legumes like soybean has important potential towards understanding the regulation of nitrogen fixation in the context of both, basic and applied research. Grafting studies by Delves et al (1) showed that all supernodulating mutants of soybean (Glycine max [L.] Merr. cv. Bragg) controlled their nodule phenotype through the action of the shoot thus leading to the search for any protein differences in leaves between wild type and the supernodulating mutant of soybean. In this abstract, we present a two dimensional gel analysis of total leaf protein from leaves of supernodulating and wild-type soybean after inoculation with *Bradyrhizobium japonicum* USDA110.

The two dimensional gel electrophoresis was carried out according to (3). Wild type Bragg and the supernodulating mutant *nts*382, 96 hours after inoculation with *Bradyrhizobium japonicum* USDA110 showed a similar pattern and no polypeptide differences between mutant and wild-type soybean could be detected. The total number of spots that can be resolved is above 200. The efficiency of the extraction procedure is about 80% when normalized per ml of extraction buffer added. For soybean leaves the method of Flengsrud and Kobro (2) gives better reproducibility than the method of Mayer et al (4). The 2 dimensional protocol as described by (8) was followed very closely with a few minor modifications. In case of total protein from soybean leaves, factors that improved resolution on a 2-D gel were use of a different crosslinker, diacrylpiperizine instead of bis, thorough dialysis of the sample before isoelectric focussing, and use of CHAPS instead of Triton X-100 in the sample preparation. In case of mini gels freezing the tube gels before the second dimension produced a complete lack of resolution (data not shown).

The lack of detectable differences between total *in vivo* proteins of wild-type Bragg and *nts*382 can be interpreted in several ways; no major protein change is associated with autoregulation, a result that is consistent with the findings of Sayavedra-Soto et al (this Congress). Another possibility is that the minor proteins that are associated with autoregulation are below the sensitivity limits of the extraction and silver staining procedures.

1. Delves, A. C., Higgins, A. Gresshoff, P. M. (1987) J. Plant Physiol. 128:473-478
2. Flengsrud, R. and Kobro, G. (1989) Anal. Biochem. 177, 33 - 366.
3. Hochstrasser, D. F., Harrington, M. G., Hochstrasser, A., Miller, M. J., Merril, C. R. (1988) Anal. Biochem. 173, 424 - 435
4. Mayer, J. E., Hahne, G., Palme, K. Schell, J. (1987) Plant Cell Reports 6:77-81

Gresshoff, Roth, Stacey & Newton (eds.) Nitrogen Fixation: Achievements and Objectives
Chapman and Hall New York • London 1990

Inter- and intracultivar variation of legumes for the activity of symbiosis with *Rhizobium meliloti*

N.A. Provorov and B.V. Simarov

All-Union Research Institute for Agricultural Microbiology, Leningrad, 189620, USSR.

The nitrogen-fixing (acetylene-reduction) activity and symbiotic efficiency (percent increase of plant mass) in 99 cultivars of alfalfa (60 - Medicago sativa, 27 - M. varia, 12 - M. falcata), 8 of sweet clover (4 - Melilotus albus, 4 - M. officinalis) and 29 wild and cultivated genotypes of fenugreek (Trigonella foenum graecum) inoculated with 5 different R. meliloti strains were examined. N_2-fixing activity was characterized by a lower level of intercultivar variation than symbiotic efficiency (in alfalfa coefficients of variation for these properties were 22% and 48%, in sweet clover - 24% and 103%, respectively). In alfalfa, the distribution of the cultivar mean values for symbiotic efficiency had right-sided asymmetry (As = - 0.72), while their distribution for N_2-fixing activity had left-side asymmetry (As = + 0.55). In 6 cultivars of M. varia, the distribution for the N_2-fixing activity of individual plants also showed left-sided asymmetry while, for plant mass, a decrease of asymmetry coefficients after inoculation was observed. Two-factor analysis of variance reveals that in the "alfalfa - R. meliloti" symbiosis the major source of plant mass variation is intercultivar variation for symbiotic activity. The data obtained allow one to suppose that the symbiotic efficiency is more strictly controlled by the host genotype than the N_2-fixing activity. In fenugreek genotypes of Indian subspecies are more susceptible for R. meliloti inoculation than genotypes of Mediterranean subspecies (78% and 20% of nodulated plants and 61% and 0% of nitrogen-fixing plants, respectively). Differences between R. meliloti strains for efficiency of symbiosis with highly heterogenic natural fenugreek population are expressed to a lower extent than with homogenic cultivar population (in the first case the change of plant mass after inoculation varied from -27% to +45% only, while in the second case from -27% to +124%). We suppose that the decrease of genetic variation of legume cultivars for their symbiotic properties will open additional possibilities for the selection of symbiotically effective rhizobial strains.

Gresshoff, Roth, Stacey & Newton (eds.) Nitrogen Fixation: Achievements and Objectives
Chapman and Hall New York • London 1990

Characterization of a nodulin gene family in soybean

H. E. Richter and C. Sengupta-Gopalan

Molecular Biology Program, Dept. of Agronomy and Horticulture, New Mexico State University, Las Cruces, NM 88003, USA.

Root nodule development is accompanied by the expression of several sets of plant genes encoding for nodule specific proteins called nodulins. Several of the more highly expressed nodulins in soybean can be categorized into a gene family based on the presence of extensive stretches of sequence similarity within the coding and the flanking regions. The conservation within the coding sequences is mainly under functional constraints, whereas the constraint to conserve the 5' and 3' region of a gene, may be for regulatory reasons. Duplication and subsequent diversion, including deletion, mutations, and conversion are the basis for the origin of gene families. Characterization of cDNA and genomic clones belonging to this gene family includes: i. Nucleotide sequencing and alignments of highly conserved domains; ii. Southern analysis of the soybean genome; iii. SDS PAGE profile of the translation products of nodule RNA, hybrid-selected with gene specific and gene family common DNA fragments. Based on these results we predict a nodulin gene family of at least seven functional genes. Within this family we predict a subfamily of at least four members of which two genes are found to be linked, separated by only one kb. Their sequence similarity can be subdivided into three domains: i. Promoter -482 to -212 (83.0 %); ii. Promoter/coding -211 to +248 (99.1 %); iii. 3' region of 458 bp (97.6 %).

This work was supported by USDA competitive grants and Ag. Expt. Station at NMSU.

Gresshoff, Roth, Stacey & Newton (eds.) Nitrogen Fixation: Achievements and Objectives
Chapman and Hall New York • London 1990

Autoregulation of nodulation in soybean is not associated with major polypeptide changes in leaves

Luis A. Sayavedra-Soto, Sieglinde A. Angermüller and Peter M. Gresshoff

Plant Molecular Genetics (OHLD), Institute of Agriculture and Center for Legume Research, University of Tennessee, Knoxville, TN 37901, USA.

Previous studies have demonstrated the involvement of the plant's shoot in the establishment and maintenance of autoregulation of the nodulation system in legumes (1). Total leaf RNA and poly A^{+}RNA of soybean (*Glycine max* (L.) Merrill) cultivar Bragg and its supernodulating mutant *nts*382 (2) were *in vitro* translated using a wheat germ translation system and [^{35}S]-methionine. Labeled polypeptides were compared by 2-dimensional gel electrophoresis. Whether plants were inoculated with *Bradyrhizobium japonicum* strain USDA110 (nodules appearing after 7-10 days) or grown uninoculated, no "on-off" differences in polypeptide pattern were detected. Differences were limited to changes in relative intensity of some peptides due to inoculation. The translation products of the supernodulator mutant *nts*382 (which lacks the factors controlling nodulation) were similar to those of Bragg during the 25 day experimental period. Developmental differences were clearly resolvable. Total protein analysis by 2-dimensional gel electrophoresis confirmed this result. Translatable mRNA populations, as well as accumulated leaf proteins, are similar for the supernodulation mutant and the wild type in both the inoculated and the uninoculated state. Thus no association of autoregulation of nodulation with a major change in polypeptides was detected.

Special care was taken to eliminate experimental variation by basing all reported data on multiple translations as well as independent isolations. In conclusion: a) Translation of the RNA populations and the resolution of the polypeptide patterns were reproducible in our hands in 6 replications of the experiment, b) The physiological differences for the establishment of nodulation control, which may involve either regulation at the biochemical level (enzyme activation, substrate availability, etc.) or transcriptional/synthesis changes, are too subtle to be detected by the *in vitro* (and total) protein separation procedures, c) At least 5 peptides changing in relative intensity, due to inoculation, were detected and d) We also recognize that artifactual differences due to growth conditions can be easily confused with true responses to the treatments, and care needs to be taken when evaluating polypeptide differences of *in vitro* translated products.

(1) Gresshoff P.M. and Delves A.C. (1986). Plant Gene Res. 3:159-206.
(2) Carroll B.J., McNeil D.L. and Gresshoff P.M. (1985). Plant Physiol. 78:34-40.

Gresshoff, Roth, Stacey & Newton (eds.) Nitrogen Fixation: Achievements and Objectives
Chapman and Hall New York • London 1990

Effect of indirect nitrate supply on nodulation of pea roots

W. Schmidt and P. Martin

Institut für Pflanzenernährung, Universität Hohenheim
Postfach 700562, D-7000 Stuttgart 70, F.R.G.

Higher nitrate concentration in the rooting medium reduces the nodule number of legumes. It is still an open question, however, whether this is due to local inhibitory effects of substrate nitrate on early infection events or whether it is caused more indirectly by an internal regulating system in the host plant, induced by uptake of high amounts of nitrogen. To examine the possible role of an internal regulating system, split-rooted seedlings of *Pisum sativum* were cultivated in a constant environment cabinet. After 14 d of growth one root portion was supplied with a 25 mM nitrate solution for 2 d (expt. I) or 4 d (expt. II). Subsequently the nitrate treated root half was cut off and the other root part was inoculated with *Rhizobium leguminosarum* (DSM 30132). Nodules were counted 6 d after inoculation.

High nitrate supply to one part of the root system reduced nodulation at the other part significantly when inoculation was performed after a 4-day (expt. II) nitrate supply. No effect was evident after 2 days. This reduction in nodule number at the inoculated root part following a 4-day treatment with 25 mM nitrate on the other root part appeared not to be a result of transport of nitrogen from one root portion to the other since there was no or only a very small increase in nitrogen compounds in the inoculated root parts. Nitrate treatment lowered the level of water soluble carbohydrates in both root parts and the shoot. The degree of this reduction was similar after the 2-day and the 4-day nitrate treatment. Nodulation, however, at the other root part was influenced only after 4 days. Therefore it is suggested that neither the N level nor the level of water soluble carbohydrates at the inoculated (not nitrate-treated) root part are responsible for the observed decrease in nodule number. The effect of nitrate-treatment prior to inoculation may be due to an information carrier, possibly of hormonal nature, activating an autoregulation system for nodulation.

Gresshoff, Roth, Stacey & Newton (eds.) Nitrogen Fixation: Achievements and Objectives
Chapman and Hall New York • London 1990

Metabolite regulation of partially purified soybean nodule phospho-enol-pyruvate carboxylase

Kathryn A. Schuller, David H. Turpin and William C. Plaxton

Biology Department, Queen's University, Kingston, Ontario, CANADA

Phoshoenolpyruvate carboxylase (PEPC) may play a role in the following processes in legume nodules: (a) provision of carbon skeletons for NH_4^+ assimilation; (b) synthesis of dicarboxylic acids used as respiratory substrates by the bacteroids; and (c) synthesis of organic acids to maintain charge balance and neutral pH. Most of the evidence for these functions comes from physiological studies of $^{14}CO_2$ fixation by detached nodules or nodulated roots. Comparatively little is known about the biochemistry of legume nodule PEPC. In the present study we have partially purified and characterized PEPC from soybean (*Glycine max* L. Merr.) nodules. The response of enzyme activity to substrate (PEP) concentration and to various effectors was influenced by assay pH and glycerol addition to the assay. At pH 7 in the absence of glycerol, the K_m(PEP) was two-fold greater than at pH 7 in the presence of glycerol or at pH 8. Glucose-6-phosphate, fructose-6-phosphate, glucose-1-phosphate and dihydroxyacetone phosphate activated PEPC at pH 7 in the absence of glycerol but had no effect under the other assay conditions. Malate, aspartate, glutamate, citrate and 2-oxoglutarate were potent inhibitors at pH 7 in the absence of glycerol but their effectiveness was decreased by raising the pH to 8 and/or by adding glycerol. In contrast, 3-phosphoglycerate was a less effective inhibitor at pH 7 in the absence of glycerol than under the other assay conditions. Pi was an activator at pH 7 in the absence of glycerol but an inhibitor under the other assay conditions. Metabolite regulation of PEPC is likely to play an important role in relation to the proposed functions of this enzyme in legume nodule metabolism.

Gresshoff, Roth, Stacey & Newton (eds.) Nitrogen Fixation: Achievements and Objectives
Chapman and Hall New York • London 1990

Studies of glutamine synthetase gene expression in nodules of french bean (*Phaseolus vulgaris* L.)

W.-J. Shen, J. Freeman, M. Pineda, M.S. Williamson and B.G. Forde

Department of Biochemistry and Physiology, AFRC Institute of Arable Crops Research, Rothamsted Experimental Station, Harpenden, Herts AL5 2JQ, UK.

In French bean, almost all the glutamine synthetase (GS) activity in nodules is specified by just two of the five plant GS genes. One gene, *gln-β*, is expressed at high levels in both roots and nodules, while the second, *gln-γ*, is strongly induced during nodule development. We are investigating the genetic and physiological factors regulating these genes. Analysis of chimaeric gene fusions in transgenic *Lotus corniculatus* plants has shown for both genes that the sequences necessary for developmentally regulated expression are located within the 2 kb region upstream of the coding sequences (2). A 5' deletion analysis of *gln-γ* in transgenic *Lotus* nodules has indicated the presence of several positive regulatory elements located between -354 and -2000.

To study the role of the microsymbiont in regulating *gln-γ* expression, transgenic plants were nodulated with mutant strains of *Rhizobium loti* (kindly provided by Prof. Barry Scott). One Fix$^-$ strain (PN237), which produces small nodule-like swellings, was unable to activate expression of a *gln-γ::gus* chimaeric gene, while a second Fix$^-$ strain (PN1010), which elicits normal-sized white nodules, did induce expression, although at a lower level than the wild type strain. The latter result is in agreement with other evidence that N-fixation *per se* is not required for induction of *gln-γ*, although it may be necessary for maximal levels of expression.

In vitro analysis of DNA-protein interactions has shown that a nuclear factor from French bean nodules (PNF-1) binds to several A/T-rich sequences between -860 and -154 in the *gln-γ* promoter; a related but distinct factor (PRF-1) was detected in uninfected roots. Both factors also have a high affinity for a protein-binding site from a soybean leghaemoglobin gene and appear to be closely related to the soybean nodule factor, NAT2 (3). A correlation between the location of positive regulatory elements and PNF-1 binding sites suggests that this factor may have a role in regulating expression of *gln-γ* and perhaps other late nodulin genes.

(1) Forde, B.G. & Cullimore, J.V. (1989) Oxf. Surveys Plant Molec. Cell Biol. **6**, 247-296.
(2) Forde, B.G., Day, H.M., Turton, J.F., Shen, W.-J., Cullimore, J.V. & Oliver, J.E. (1989) Plant Cell **1**, 391-401.
(3) Jacobsen, K., Laursen, N.B., Jensen, E.Ø., Marcker, A., Poulsen, C. & Marcker, K.A. (1990). Plant Cell **2**, 85-94.

Gresshoff, Roth, Stacey & Newton (eds.) Nitrogen Fixation: Achievements and Objectives
Chapman and Hall New York • London 1990

Plasmodesmata frequency selectively drops as infected soybean-nodule cells mature

Ernest O. Sutton, L. Evans Roth, and Gary Stacey

Center for Legume Research, University of Tennessee, M313 Walters Building, Knoxville, TN 37996-0810, USA.

Plasmodesmata are small confluences of cytoplasm penetrating cell walls and providing pathways for intercellular movement of various nutrients. Selker (*Protoplasma* 147: 178-190, 1988) showed that soybean nodule cells are organized both to facilitate the transport of fixed nitrogen in the plant and partition the nitrogenous compounds from infected cells to facilitate nitrogen fixation. Therefore, the frequencies of plasmodesmata were determined at various stages of development in both wild type and mutant induced nodules and in several cultivars and growth conditions. In the earliest stages of wild type-induced nodule infection, plasmodesmata occur at similar frequencies between contiguous infected-infected, infected-uninfected, and uninfected-uninfected cells. As infections progress, two changes occur. First, plasmodesmata per unit surface area between infected cells decreases to less than 5% of that between uninfected cells. Second, plasmodesmata on uninfected cells tend to cluster. The frequency of plasmodesmata in nodules infected with a nif- mutant strain of *Bradyrhizobium* was also studied. Results similar to those of the wild type-infection were obtained, a decrease to about 15%. These results yield three conclusions. First, since adjacent infected cells lack plasmodesmata and do not have confluent cytoplasms, they do not have plasmodesmata-related transfer; conversely, infected cells have confluent cytoplasms with adjacent uninfected cells. As Selker suggested, nitrogen is transported through infected cells into uninfected cells where more complex nitrogenous compounds are synthesized for distribution throughout the plant. Second, mutant-induced nodules show a progressive loss of plasmodesmata, similar to wild-type nodules. Thus, we have evidence that nitrogen fixation does not cause the frequency of plasmodesmata to be reduced between adjacent infected cells. Finally, the ability of plant cells to both eliminate and produce plasmodesmata is indicated, a concept that extends our understanding of the importance and plasticity of plasmodesmata.

Gresshoff, Roth, Stacey & Newton (eds.) Nitrogen Fixation: Achievements and Objectives
Chapman and Hall New York • London 1990

Functional analysis of the *Sesbania rostrata* leghemoglobin *glb3* gene 5' upstream in transgenic *Lotus corniculatus* and *Nicotiana tabacum* plants

L. Szabados, P. Ratet, B. Grunenberg and F. J. de Bruijn*

Max-Planck-Institut für Züchtungsforschung, Carl-von-Linne-Weg 10, 5000 Köln 30, F.R.G., *Present address: MSU-DOE Plant Research Laboratory, Michigan State University, E. Lansing, Mich. 488245, USA.

Expression of the Sesbania rostrata leghemoglobin glb3 gene was analysed in transgenic Lotus corniculatus and Nicotiana tabacum plants, harboring chimeric glb3-gus (ß-glucuronidase) gene fusions. A 1.9 kb fragment of the glb3 5' upstream region was found to direct a high level of nodule-specific GUS activity in L. corniculatus , restricted to the Rhizobium infected cells of the nodules (1,2). The same fragment directed a low level of GUS activity in N. tabacum, primarily restricted to the roots and to phloem cells of the stem and petiole vascular system (2). A deletion analysis revealed that the region between coordinates -429 and -48 (relative to the ATG) was sufficient for nodule specific expression. Replacement of the -161 to -48 region (containing the glb3 CAAT and TATA boxes) with heterologous truncated promoters (Cauliflower Mosaic Virus 35S, Δ-p35S; nopaline synthase, Δ-pnos) resulted in a loss of nodule-specificity and reduction of GUS activity in L. corniculatus, but a significant increase in N. tabacum, primarily in the roots. The same fragment (-161 to -48) could not direct nodule specific expression when fused to a heterologous enhancer in cis. The -1914 to -162 or -431 to -162 regions could not confer nodule specific expression on truncated promoters in L. corniculatus, in spite of the fact that DNA sequences homologous to the Organ Specific Element (OSE) of the Glycine max gene are entirely contained on these fragments (3). By fusing further upstream fragments to the Δ-p35S and Δ-pnos promoters, two positive regulatory regions were delimited between coordinates -1601 and -670, as well as -429 and -162 (2).

References:
1. de Bruijn, F.J., Felix, G., Grunenberg, B., Hoffmann, H.J., Metz, B., Ratet, P., Simons-Schreier, A., Szabados, L., Welters, P., and Schell, J. (1989) Plant Mol. Biol. 13: 319-325.
2. Szabados, L., Ratet, P., Grunenberg, B. and de Bruijn, F.J. (1990) The Plant Cell, submitted.
3. Stougaard, J., Sandal, N.N., Gron, A., Kühle, A., and Marcker, K.A. (1987) EMBO J. 6: 3565-3569.

Gresshoff, Roth, Stacey & Newton (eds.) Nitrogen Fixation: Achievements and Objectives
Chapman and Hall New York • London 1990

Arachis hypogaea L., root nodules: Ultrastructure reveals the presence of uninfected cells and enlarged peroxisomes

Shiv R. Tandon

Department of Biology, University of Wisconsin-Platteville, Platteville, WI 53818 USA.

INTRODUCTION

The Ultrastructural and Physiological studies of uninfected cells in legume root nodules have received a major boost since 1981 when it was hypothesized that these cells play a key role in ureide biogenesis in soybean root nodules. Until now these uninfected cells in peanut nodules have not been reported.

RESULTS

Mature peanut root nodules infected with Rhizobium spp. var. "P" processed for electron microscopy reveal a multilayered cortex without a scleroid layer and a compact central bacteroidal tissue. We have observed rows of uninfected cells traversing across this infected tissue and they are not interspersed among infected cells as commonly observed in legume root nodules. The infected and uninfected cells also appear to be of similar size (2N), a condition markedly different from other amide- or ureide- transporting legumes. The ratio of these uninfected to infected cells has not been determined. However, the number of these cells appear to be low. The peroxisomes in cortical and infected cells are few and remain small, but in uninfected cells they appear to increase in number and slightly enlarge during nodule development. A diaminobenzidine test reveals catalase activity in these enlarged peroxisomes. Tubular endoplasmic reticulum, which is absent in infected and cortical cells, also appears in uninfected cells.

DISCUSSION

Arachis hypogaea L. is a unique tropical legume in the sense that it transports most of the nitrogen as an amide 4-methyleneglutamine. To the contrary, a high level of asparagine, allantoin and allantoic acid in the xylem sap of nodulated peanut plants has been observed (1). The uninfected cells in the soybean and cowpea root nodules have been reported to carry out the final steps in ureide biogenesis (1). Therefore, based on this information and the fact that uninfected cells with enlarged peroxisomes and tubular ER do exist in peanut nodules, it is reasonable to suggest that these cells may have a role in allantoin and allantoic acid biogenesis in older peanut root nodules.

1. Schubert, K.R. (1986) Annu. Rev. Plant Physiol. 37, 539-574.

Gresshoff, Roth, Stacey & Newton (eds.) Nitrogen Fixation: Achievements and Objectives
Chapman and Hall New York • London 1990

Stabilization of glutamine synthetase protein by the toxin, tabtoxinine-ß-lactam

S.J. Temple[1], T.J. Knight[2], P.J. Unkefer[2], and C. Sengupta-Gopalan[1]

[1]Plant Genetic Engineering Labs/Department of Agronomy and Horticulture, New Mexico State University, Las Cruces, NM 88003. [2]Los Alamos National Labs, Los Alamos, NM 87545, USA.

Pseudomomonas syringae pv. tabaci, a commonly recognized leaf pathogen of tobacco and soybean where it is responsible for the occurrence of wildfire disease, can also infest the rhizosphere of many plants including oats. Normal oats do not survive this infestation as a consequence of the rapid inactivation of all their leaf and root glutamine synthetase (GS) isoforms by tabtoxinine- β - lactam (TBL), a novel amino acid toxin released by the "tabaci". Inactivation of GS by TBL is irreversible, it requires ATP, and is active site directed. It results in one molecule of TBL binding per subunit of completely inactivated GS.

We report here a significant increase in all GS protein isoforms (based on immunological determination) within a few hours following the addition of the pathogen to the soil around the growing oat plants. This increase in total GS protein is accompanied by a rapid drop in GS activity (< 15% control activity after 24 hours for both leaf and root GS) and concurrent increase in NH_3 levels. Essentially identical results for GS enzyme inactivation and protein accumulation were obtained with the commonly studied GS inhibitor, methionine sulphoximine. Quantitative northern blot hybridization of poly(A)-RNA from both oat leaf and root tissues for infested and uninfested plants, suggests that the increase in GS protein resulting from pathogen infestation is not due to increased transcription of GS genes. Furthermore, the results of short term in vivo labeling experiments using (^{35}S) methionine and cysteine, rules out the possibility of translational control as the mechanism for TBL mediating the increase in GS protein.

Incubation of GS with and without TBL bound, with extracts of oat leaf soluble protein showed that binding of TBL results in an inactive GS which is resistant to protease degradation. These results suggest that the increase in GS protein in "tabaci' treated oat plants is due to the stabilizing effect of TBL on GS with respect to proteolytic degradation. Maurizi and Ginsburg ((1982) J. Biol. Chem. 257: 7246-51) reported stabilization of the quaternary enzyme structure (ie. inhibition of dissociation of subunits) following its inactivation by another active site directed irreversible inhibitor of GS, methionine sulfoximine. Our results suggest that TBL stabilizes the plant GS holoenzyme by blocking subunit dissocition; which further suggests that the initial step regulating plant GS proteolysis is subunit dissociation.

This work was supported by the Southwest Consortium and USDA competitive grants.

Gresshoff, Roth, Stacey & Newton (eds.) Nitrogen Fixation: Achievements and Objectives
Chapman and Hall New York • London 1990

Nodulins in the course of effective and ineffective pea-*Rhizobium* symbiosis

I. Tikhonovich*, M. Fedorova, O. Kulikova, N. Kruchinina, A. Borisov, Yu. Volichenco, A. Gutkina and A. Filatov

All Union Research Institute for Agricultural Microbiology, Podbelsky Chaussee 3, Leningrad-Pushkin 6, 189620 USSR.

Polyclonal antibodies, raised against cytoplasmic proteins of pea nodules and purified on the CNBr sepharose coupled with sterile root proteins, were used to test nodulins. About 10 polypeptides (excluding leghemoglobin) could be identified by Western blotting. Most of them belong to the late nodulins as judged by the time course of accumulation during development. The gel filtration data showed that several of the immunoreactive polypeptide were subunits of high molecular weight proteins. Antibodies were used to follow the accumulation of nodulins during symbiosis development in two Nod Fix systems; one was based on an ineffective Rhizobium strain, the other on a pea mutant line obtained by chemical mutagenesis. A majority of the nodulins were repressed in both systems, while the quantity of a specific 55kD protein increased significantly in nodules derived by ineffective rhizobia.

In highly effective pea lines obtained by selection for increased N_2 fixation, we observed minor changes in the nodulin pattern and coordinated enhancement of expression of several late nodulin genes.

Gresshoff, Roth, Stacey & Newton (eds.) Nitrogen Fixation: Achievements and Objectives
Chapman and Hall New York • London 1990

Nucleotide sequence of *nifD*, the structural gene coding for the α subunit of the MoFe protein of the nitrogenase complex from the actinomycete *Frankia*

Paul Twigg,[1] Chung-Sun An[2] and Beth C. Mullin[1]

[1]Department of Botany, The Center for Legume Research and the Plant Physiology and Genetics Program, The University of Tennessee, Knoxville, TN, USA, and [2]Department of Botany, Seoul National University, Seoul, Korea.

Frankia is an actinomycete capable of participating in nitrogen-fixing symbioses with a wide range of woody angiosperms. It is the only actinomycete which is known to be capable of nitrogen fixation. Early studies using a heterologous probe from Klebsiella demonstrated that the nif genes of Frankia shared some sequence homology with other nitrogen-fixing organisms. In order to determine the extent of sequence homology and possible phylogeny of nif sequences in this actinomycete, we have determined the nucleotide sequence and derived amino acid sequence of nifD from Frankia strain FaCl.

A 2.7kb Sal 1 fragment from pFAH-P7.2 (Fig. 1), a subclone of FaCl cosmid clone pFAH1 was cloned into M13mp18 in both orientations. Deletion clones were generated from both orientations using the 3' to 5' processive exonuclease activity of T4 DNA polymerase of the Cyclone I Biosystem (IBI).

Deletion clones were ordered by direct electrophoresis of recombinant M13 DNA on agarose gels. Template was prepared from selected phage supernatants and clones were sequenced by the Sanger method using Sequenase (USB). Sequences of individual clones were merged by hand or with the aid of Microgenie software (Beckman).

Approximately 2.1kb of pFAH-P2.7 has been sequenced at this time. The sequence of nifD is complete except for what we expect will be approximately 10 bases at the 3' terminus. Upstream from nifD the 3' 81 nucleotides of nifH have been determined and downstream from nifD regions of nifK have been sequenced. The FaCl nifD gene has 70% sequence similarity with Parasponia rhizobium, 69% with R. japonicum, 67% with K. pneumoniae and 50% with Clostridium. Derived amino acid sequence identities are 62-65% for Parasponia rhizobium, 62-63% for Anabaena and 43% for Clostridium.

Gresshoff, Roth, Stacey & Newton (eds.) Nitrogen Fixation: Achievements and Objectives
Chapman and Hall New York • London 1990

Extraction and cDNA cloning of mRNA from roots and nodules of European Black Alder (*Alnus glutinose* L. Gaetra.)

Paul G. Twigg and Beth C. Mullin

Department of Botany and The Center for Legume Research, The University of Tennessee, Knoxville, TN 37996-1100, USA.

Total RNA was extracted from Alnus glutinosa nodule and root tissue using a modification of the protocol developed by Dellaporta et al. (1983) for the extraction of DNA from plant leaves. This method was adopted after the efficacy of more conventional methods of RNA extraction [i.e., guandine thiocyante based on Chirgwin et al. (1979) and direct phenol extractions] were found to vary with the physiological status of the plant. Following RNA extraction, poly A mRNA was isolated from the total RNA of both root and nodule samples by oligo d(T) cellulose chromatography. These poly A fractions were used to produce nodule and root libraries in specific cDNA libraries by employing the subtractive hybridization method of Duguid et al. (1988). The resulting nodule-specific clones will be further studied in order to determine their function and regulation during symbiosis.

References

Chirgwin, J.M., A.E. Przybyla, R.J. MacDonald, and W.J. Rutter. 1979. Isolation of biologically active ribonucleic acid from sources enriched in ribonuclease. Biochemistry, Vol. 18, No. 24, pp. 5294-5299.

Dellaporta, S.L., J. Wood, and J.B. Hicks. 1983. A plant DNA minipreparation: Version II., Pl. Mol. Biol. Rep., Vol. 1, No. 4, pp. 19-21.

Duguid, J.R., R. G. Rohwer, and B. Seed. 1988. Isolation of cDNAs of scrapie-modulated RNAs by substractive hybridization of a cDNA library. PNAS USA, Vol. 85, pp. 5738-5742.

Gresshoff, Roth, Stacey & Newton (eds.) Nitrogen Fixation: Achievements and Objectives
Chapman and Hall New York • London 1990

ENOD2 localization in effective, nitrogen-fixing and ineffective, bacteria-free alfalfa nodules

C. van de Wiel[1], J.H. Norris[2], B. Bochenek[3], R. Dickstein[4], T. Bisseling[1], and A.M. Hirsch[3].
[1]Dept. of Molecular Biology, Agricultural University, Wageningen, The Netherlands; [2]Dept. of Botany, Univ. of Rhode Island, Kingston, R.I. 02881 USA; [3]Dept. of Biology, University of California, Los Angeles, CA 90024 USA; and [4]Dept. of Molec. Biology, Mass. General Hospital and Dept. of Genetics, Harvard Medical School, Boston, MA 02114 USA.

Alfalfa (*Medicago sativa* L.) plants form bacteria-free nodules in response to *N*-(1-naphthyl)phthalamic acid (NPA) or to 2,3,5-triiodobenzoic acid (TIBA), compounds which function as auxin transport inhibitors (ATIs) (1). These nodules resemble the bacteria-free nodules induced by *Rhizobium meliloti exo* mutants and *Agrobacterium tumefaciens* transconjugants carrying cloned *R. meliloti* symbiotic genes. We checked nodulin gene expression in nodules induced by two *A. tumefaciens* transconjugants, one harboring the *R. meliloti nod/nif* megaplasmid and the other carrying both the *nod/nif* megaplasmid and the megaplasmid carrying *exo* genes, and found that both Nms30 and ENOD2 are expressed. Thus, all of the "empty" nodules examined so far contain transcripts for the nodulins Nms30 and MsENOD2; transcripts for late nodulins like leghemoglobin are not detected. *In situ* hybridization studies demonstrated that ENOD2 transcripts are localized to the nodule parenchyma cells of normal alfalfa root nodules, a pattern in agreement with the recent observations of pea nodules (2). The ENOD2 gene is also expressed in a tissue-specific manner in nodules elicited by the ATIs, NPA and TIBA. In "empty" nodules elicited by *R. meliloti exoB*, *exoA* and *exoF* mutants and *A. tumefaciens* transconjugants, ENOD2 transcripts are also detected in nodule parenchyma cells, but usually are localized to the base of the nodule around the vascular bundles instead of along the periphery. On the basis of the pattern of ENOD2 gene expression, we conclude that the ATIs in their action very likely mimic some factor(s) responsible for nodule development.

(1) Hirsch, A.M., Bhuvaneswari, T.V., Torrey, J.G. & Bisseling, T. (1990) Proc. Natl. Acad. Sci. USA **86**, 1244-1248.

(2) van de Wiel, C., Scheres, B., Franssen, H., van Lierop, J.J., van Lammeren. A., Van Kammen, A. and Bisseling. T. (1990) EMBO J **9**, 1-7.

Gresshoff, Roth, Stacey & Newton (eds.) Nitrogen Fixation: Achievements and Objectives
Chapman and Hall New York • London 1990

Phosphorylation of nodulin-26 by a calcium-dependent protein kinase

C. David Weaver[1], Gary Stacey[2], and Daniel M. Roberts[1]

Dept. Biochem.[1] and Microbiol.[2], University of Tennessee, Knoxville, TN 37996.

Nodulin-26 is a symbiosome membrane (SM) protein that is developmentally expressed in soybean nodules (Fortin et al., 1985). In order to test whether nodulin-26 is a target of calcium-dependent phosphorylation, we have synthesized a peptide, CK-15, with the carboxyl terminal fourteen amino acid residues of nodulin-26. Our findings show that CK-15 is phosphorylated by a calcium-dependent, calmodulin/phosphatidylserine/diacylglycerol-independent protein kinase from soybean root nodules. The CK-15 kinase activity was detected in soluble and particulate fractions including the SM and is widely distributed throughout the plant. Anti-CK-15 antibodies recognize a protein from the SM which migrates with an apparent molecular weight of 26,000 on SDS-PAGE. This protein is phosphorylated in a calcium-dependent manner. Consistent with the proposal that nodulin -26 is a nodule-specific protein, anti-CK-15 crossreactivity is observed with the SM fraction and not with samples from other tissues. Overall, our data suggest that nodulin-26 may be a target of calcium-dependent regulation.

Fortin M.G., Zelechowska, M. Verma, D.P.S. (1985) EMBO J. 4, 3041-3046.

Gresshoff, Roth, Stacey & Newton (eds.) Nitrogen Fixation: Achievements and Objectives
Chapman and Hall New York • London 1990

Specific interactions of a rhizobial protein (*trans*-acting factor) with distinct DNA sequences in the promoter region of a *Sesbania rostrata* leghemoglobin gene

P. Welters, G. Felix and F.J. de Bruijn*

Max-Planck-Institut für Züchtungsforschung, Carl-von-Linne-Weg 10, 5000 Köln 30, F.R.G., *Present address: MSU-DOE Plant Research Laboratory, Michigan State University, E. Lansing, Mich. 48824, USA.

We have previously reported the cloning and DNA sequencing of two Sesbania rostrata leghemoglobin genes (Srglb2, 3; 3,4). The first 400 bp of the Srglb2 and 3 promoters share a high degree of homology with the correspondent soybean (Glycine max) lbc3 gene, including a distinct AT motif (BS2*) found to interact specifically with DNA-binding proteins from nodules, roots and leaves (2,3). The BS2* AT motif in the glb3 5´upstream region repeated several fold, and the corresponding glb3 restriction fragments interact with the same nuclear factors from nodules and leaves (1). In addition to the AT binding site motifs, a unique GC-rich element was found, which interacts specifically with a DNA binding protein not of plant origin, but derived from the S. rostrata symbiont Azorhizobium caulinodans (Ac). This bacterial protein was found to be present in other rhizobia as well and to be extremelly heat resistant. The Ac binding protein (BBP) was purified from bacterial cultures by heat treatment, heparine-sepharose and affinity chromatographic methods and found to be ~11Kd.
Its binding site was delimited by footprinting and a corresponding oligonucleotide was used for genetic selection schemes in E. coli in order to clone the corresponding gene. If this protein turns out to play a significant role in lb gene induction, an interesting novel aspect of endosymbiont-mediated plant gene activation will have been uncovered.

References:

1. de Bruijn, F.J., Felix, G., Grunenberg, B., Hoffmann, H.J., Metz, B., Ratet, P., Simons-Schreier, A., Szabados, L., Welters, P., and Schell, J. (1989) Plant Mol. Biol. 13: 319-325.
2. Jensen, E.O., Marcker, K.A., Schell, J., and de Bruijn, F.J. (1988) EMBO J. 7, 1265-1271.
3. Metz, B.A., Welters, P., Hoffmann, H.J., Jensen, E.O., Schell, J., and de Bruijn, F.J. (1988) Mol. Gen. Genet. 214, 181-191.
4. Welters, P., Metz, B.A., Schell, J., and de Bruijn, F.J. (1989) Nucl. Acids. Res. 17, 1253.

SECTION H:

NITROGEN FIXATION AND EVOLUTION

Invited papers
Poster summaries

Gresshoff, Roth, Stacey & Newton (eds.) Nitrogen Fixation: Achievements and Objectives
Chapman and Hall New York • London 1990

Evolving a *Rhizobium* for non-legume nodulation

B. G. Rolfe and G. L. Bender

Plant Microbe Interaction Group, RSBS, Australian National University, Canberra ACT 2601, Australia (fax: 61 6 2490754)

Naturally occurring rhizobia in the soil form nodules on legumes and not on non-legumes with the exception of *Parasponia*, a member of the Ulmaceae family (1). Induction of legume nodules by rhizobia involves a multi step interaction, following an exchange of signals between the *Rhizobium* strain and its host (2,3,4). *Parasponia* nodulation resembles the nitrogen-fixing root nodulation formed between other non-legumes and actinomycetes of the genus *Frankia*. In non-legumes, nodule formation requires the initiation of cell division within the root cortex followed by the induction of a lateral root below the site of cell division. The final structure is a lateral root swollen on either side of the central vascular tissue by cortical cells filled with bacteria actively engaged in nitrogen fixation. An important finding was that some strains of rhizobia can nodulate both legumes and *Parasponia* suggesting that the widely differing nodulation processes are largely plant determined and require the correct trigger(s) from rhizobia or *Frankia*. The crucial early steps of *Parasponia* infection involved erosion of the epidermal cell primary cell wall, initiation of cell division in the outer cortex of the root, the formation of a mass of dividing cells called a prenodule which protrude through the epidermis and become infected via *Rhizobium* induced infection threads (5). The *Rhizobium* nodulation genes, *nodABC*, were not essential for the induction of this cell division in *Parasponia* but were needed for the formation of prenodules. However, the initiation of *Parasponia* infection required the *Rhizobium* regulatory gene, *nodD*, and host specific nodulation (*hsn*) genes (6). Thus, nodulation of *Parasponia* requires the coordinated expression of *nodABC* and *Parasponia hsn* genes under the regulation of the *nodD* product. The genetic analysis of the *Parasponia* symbiosis enabled us to construct a *Rhizobium* strain (ANU536) which

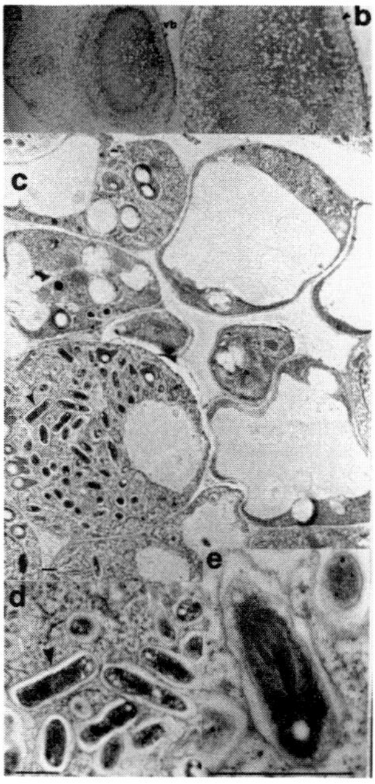

could interact with *Parasponia* roots causing erosion of epidermal cell walls and trigger cortical cell division. Earlier studies showed that specific *Rhizobium* strains could cause root hair curling on rice (7) and that the type and copy number of *nodD* was important. Recently, *nodD* alleles have been cloned which are capable of interacting with exudate produced by the roots of *Parasponia* and other non-legumes including rice wheat, barley,sunflower and cotton (8,9). We have since identified that vanillin and isovanillin are the most active inducers in wheat root exudates which can interact with specific *Rhizobium nodD* products (10). Our bacterial construction contained a *Rhizobium nodD* allele which could interact with rice exudates. When strain ANU536 was inoculated onto rice seedlings of variety Calrose without any other additions, it occasionally caused small spherical nodules to form. When these rice nodule-like structures were sectioned and examined under light and electron microscopy, the most interesting characteristics were the presence of vascular bundles and membrane encapsulated bacteria (see Figure). Whether an infection thread was formed or an invagination took place at the cell membrane, which leads to encapsulated rhizobia, is unknown. At present, this nodulation is infrequent and unpredicable which adds to the difficulties of the analysis. Furthermore, no evidence of nitrogen fixation has ever been found. Recently, Al-Mallah *et al.* (11) have induced "nodules" on rice by pretreating seedlings with a mixture of cellulase and pectinase before adding rhizobia mixed with PEG. The nodulation frequency was low and the bacteria present in these nodules were not packaged and were often associated with dead and dying plant cells. This *Rhizobium* infection in enzyme-PEG-treated rice roots may be analogous to the early steps of *Parasponia* invasion (5,11). Thus, future advances in our understanding of *Parasponia* infection could enable the construction of new engineered rhizobia. These strains should be more refined in their initial interactions with various non-legumes and thus, invade plant tissues more successfully without extensive death of plant cells.

References

(1) Rolfe, B.G. and Gresshoff, P.M. (1988) Ann. Rev. Plt. Phys. Plt. Mol. Biol. **39**, 297-319.
(2) Long, S. (1989) Cell. **56**, 203-214.
(3) Redmond *et al.* (1986) Nature **323**, 632-635.
(4) Lerouge *et al.* (1990) Nature **344**, 781-784.
(5) Bender, G.L. and Rolfe, B.G. (1985) Plant Sci. **38**, 135-140.
(6) Bender, G.L. *et al.* (1987) Mol. Gen. Genet. **210**, 299-306.
(7) Plazinski, J. *et al.* (1985) J. Bacteriol. **163**, 812-815.
(8) Bender, G.L. *et al.* (1988) Molec. Plant-Microbe Inter. **1**, 259-266.
(9) McIver *et al.* (1989) Molec. Plant-Microbe Inter. **2**, 97-106.
(10) Le Strange *et al.* (1990) Molec. Plant-Microbe Inter. (in press).
(11) Al-Mallah *et al.* (1989) J. Exp. Bot. **40**, 473-478.

Figure 1. Small spherical nodule-like structures formed on rice seedlings of variety Calrose inoculated with ANU536; (a) and (b) LM sections showing a region of cells infected with bacteria and the presence of vascular bundles; (c), (d) and (e) EM section of plant cells in the infected zone containing membrane encapsulted bacteria.

Gresshoff, Roth, Stacey & Newton (eds.) Nitrogen Fixation: Achievements and Objectives
Chapman and Hall New York • London 1990

Hypotheses for the evolution of actinorhizal symbioses

Beth C. Mullin, Susan M. Swensen and Paula Goetting-Minesky

Department of Botany, Plant Physiology and Genetics Life Sciences Graduate Program and the Center for Legume Research, The University of Tennessee, Knoxville, TN, 37996, USA.

Actinorhizal symbioses include symbiotic associations between actinomycetes of the genus Frankia and host plants belonging to 24 genera distributed among eight higher plant families. Although all host plant species are woody dicots, there is no close phylogenetic affinity among most host plant groups. Nor is there agreement among actinorhizal biologists as to whether host plants are likely to have evolved from a common nodulated ancestor or whether symbioses have more likely arisen independently more than once resulting in the current diversity of host plant species. Knowledge of the phylogenetic relationships among actinorhizal plants as well as comparative analyses of nodule structure and function should contribute to the formulation of working hypotheses for the evolution of actinorhizal symbioses.

Phylogenetic relationships among host plants. At first glance it might appear feasible to propose a common nodulated ancestral plant among the Magnoliidae[3] (Fig. 1) that gave rise to plants which later formed the subclasses Hamamelidae, Rosidae and Dilleniidae. If that were the case, then one might expect to find nodulated genera among the primitive members of these subclasses. However, this is not the case as seen in Figure 2, a highly modified tree based largely on relationships among the angiosperms according to Cronquist.[3] The Trochodendrales and the Hamamelidales, which are considered to be ancestral to the Myricales, Fagales and Casuarinales, do not have any extant nodulated species. Within the Rosales the family Rosaceae, which contains five nodulated genera (Cowania, Dryas, Purshia, Chamaebatia and Cercocarpus), is considered to be an advanced family rather than a primitive one and the two tribes which contain these genera are considered not to be primitive tribes within the family. Furthermore, there is no indication that the extant nodulated members of the Rosaceae are more closely related to the Rhamnaceae and the Elaeagnaceae than these families are to other non-nodulated families within the Rosalaes.

It is possible to argue that there are no extant nodulated members of the more primitive orders because they evolved rapidly into the orders which we now recognize as containing actinorhizal plants. However, it appears that most actinorhizal species have not evolved

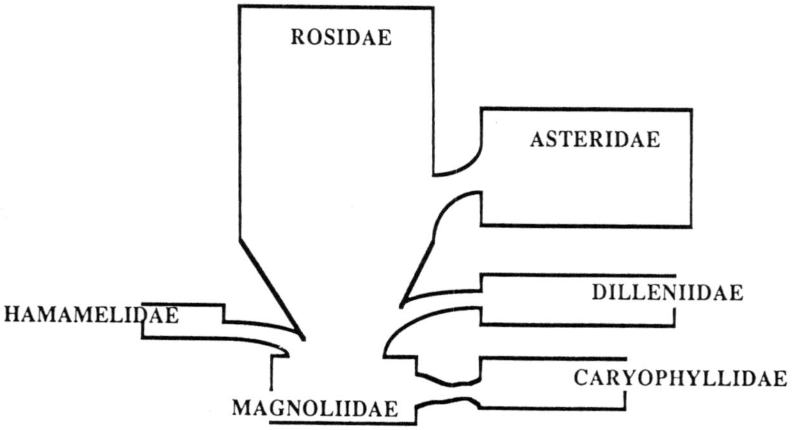

Figure 1

rapidly. Actinorhizal families contain relatively few genera and most actinorhizal genera contain relatively few species. For example, Myricaceae contains 3 genera (40 species) Betulaceae contains 2 genera (54 species), and Cauarinaceae contains 4 genera (70 species) compared to Fagaceae which contains 8 genera and 400 species. The comparison is even more striking in the Rosaceae where actinorhizal tribes Dryadeae and Cercocarpeae contain 8 and 3 genera respectively, while neighboring tribes Potentilleae, Poterieae and Roseae contain 16 genera (347 species), 14 genera (299 species) and 25 genera (1346 species), respectively. Paleontological evidence also indicates that Myrica and Alnus have changed little since the Upper Cretaceous (74 million years before present 'BP') indicating a very slow rate of change. Datisca and Coriaria are the single actinorhizal genera found in their respective subclasses Dilleniidae and Magnoliidae. Furthermore, they are placed in orders not considered to be primitive orders within the subclasses. It must be said, however, that neither is firmly placed in their respective orders and may in fact have phylogenetic affinities elsewhere.

One method of analysis which has been found to be useful to systematic and evolutionary biologists is that of searching for the most parsimonious solution to a set of events. For example, if we assume that nodulation occurred once in an ancestral plant, the number of loss-of-nodulation events to achieve the current distribution of extant actinorhizal plants could be calculated. Likewise, if we assume that nodulation occurred as a result of

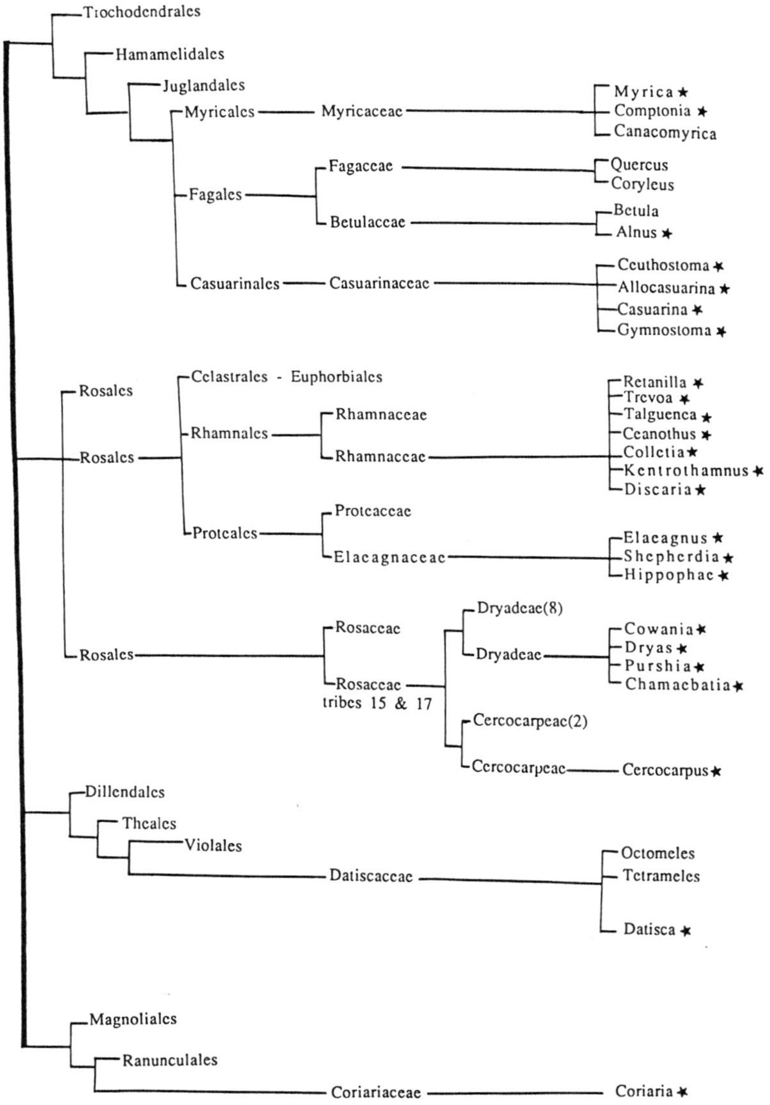
Trochodendrales
Hamamelidales
Juglandales
Myricales
Myricaceae
Myrica ★
Comptonia ★
Canacomyrica
Fagales
Fagaceae
Quercus
Coryleus
Betulaceae
Betula
Alnus ★
Casuarinales
Casuarinaceae
Ceuthostoma ★
Allocasuarina ★
Casuarina ★
Gymnostoma ★
Rosales
Celastrales - Euphorbiales
Rhamnales
Rhamnaceae
Retanilla ★
Trevoa ★
Talguenea ★
Ceanothus ★
Colletia ★
Kentrothamnus ★
Discaria ★
Proteales
Proteaceae
Elaeagnaceae
Elaeagnus ★
Shepherdia ★
Hippophae ★
Rosaceae
Rosaceae tribes 15 & 17
Dryadeae(8)
Dryadeae
Cowania ★
Dryas ★
Purshia ★
Chamaebatia ★
Cercocarpeae(2)
Cercocarpeae
Cercocarpus ★
Dillendales
Theales
Violales
Datiscaceae
Octomeles
Tetrameles
Datisca ★
Magnoliales
Ranunculales
Coriariaceae
Coriaria ★

Figure 2

several independent events, we could count each of these events plus any loss-of-nodulation events. The path which involved the least number of events would be the most parsimonious path and could be considered to be the most likely to have actually occurred.

Figure 2 is a superficial phylogeny for actinorhizal plants (*) based on the assumption that all extant nodulated species were derived from a single nodulated ancestor. The tree has been forced into the most parsimonious tree for loss of symbiosis while still maintaining the relationships among the orders as proposed by Cronquist.[3] In this tree for example, all of the orders of the Rosales which do not contain nodulating genera are shown to branch from the orders which do contain nodulating genera at a single point. Within the Rosaceae, tribes Dryadeae and Cercocarpeae are shown to branch from the 22 non-nodulating tribes at a single point. Based upon this tree, 20 "events" would have had to occur if a single nodulated ancestor had given rise to extant actinorhizal species whereas 10 "events" would have had to occur if nodulation occurred independently in each. Legitimate phylogenies of the tribes of the Rosaceae indicate a very complex branching pattern which would add at least 8 loss-of-nodulation events to that portion of the tree.[4] Furthermore, there is no evidence to support the early branching of nodulated and non-nodulated Rosales as is shown in Figure 2. At this point we can say that given our current level of understanding of the phylogenetic relationships of actinorhizal and non-actinorhizal species, the principles of parsimony do not support the existence of a common nodulated ancestor.

Comparative development and physiology of actinorhizal nodules. If all actinorhizal plants were derived from a single nodulation event it would be reasonable to expect that common mechanisms would exist for nodule initiation, development and regulation of physiological processes. Both root hair infection and intercellular penetration are known to occur in actinorhizal species. It is well documented that members of the Myricaceae, Betulaceae and Casuarinaceae are infected via root hairs and that members of the Elaeagnaceae are infected via intercellular penetration.[2] More recent evidence suggests that members of the Rosaceae and Rhamnaceae are also infected via intercellular penetration.[2] Although all actinorhizal nodules are modified lateral roots, those infections initiated by intercellular penetration do not form the prenodule structures common to nodules initiated by root hair infection.[6] In most actinorhizal nodules, infected and non-infected host cells are interspersed within nodule tissue. In *Coriaria* and *Datisca*, however, no uninfected cells are found within the infection zone and the arrangement of *Frankia* filaments and vesicles differs markedly from all other actinorhizal nodules.

The oxygen environment in the region of infected nodule cells varies greatly among host plant species.[9] There appear to be no host-mediated barriers to the diffusion of oxygen in *Alnus* and infected cells are in direct contact with ambient levels of oxygen. In *Myrica* few air spaces are found adjacent to infected nodule cells and nodule ventilation decreases with increasing levels of oxygen. Cell walls of infected Casuarina cells change in composition and there is physiological evidence that they provide a strong barrier to the diffusion to oxygen. Hemoglobin is found in the highest concentrations in the nodules of these latter two genera. In Coriaria nodule cells are surrounded by a dense suberized periderm, the thickness of which varies depending upon the oxygen concentration at which the nodules are grown. In addition, there appears to be a second mechanism which can respond rapidly to changes in oxygen. Thus the oxygen regulation strategies of actinorhizal plants are as diverse as the plants themselves and provide no evidence for common ancestry.

The status of Frankia during early angiosperm evolution. Ribosomal RNA phylogenies indicate that Frankia existed as a group long before the evolution of the angiosperms.[4] Although it has not been possible to monitor the growth and physiology of Frankia in the soil, all evidence points to the likelihood that frankiae not only survive but may flourish in soil in the absence of host plants.[10] Soils on which host plants are not known ever to have grown are able to serve as an inoculum, and it has been observed that soil under non-host plants is sometimes a better inoculum than soil under host plants. Frankia is able to fix nitrogen asymbiotically at atmospheric concentrations due to the elaboration of remarkable vesicles that provide a diffusion barrier to oxygen. Frankia thus could survive on nitrogen-poor soils as long as a carbon source is available. It is likely that Frankia was a common soil bacteria in the Jurassic and Lower Cretaceous periods but was limited to moist soils with some organic matter. The establishment of symbiosis with selected angiosperms may have provided an additional niche for Frankia and a way for these actinomycetes to survive and multiply in dry soils devoid of organic matter.

Early angiosperm evolution. The first appearance in the fossil record of non-magnoliid dicots as evidenced by the presence of tricolpate pollen is in the Lower Cretaceous at the border of the Barremian and Aptian stages (113 myr BP) in Northern Gondwanaland.[7,11,12] During the Aptian stage there was an extensive diversification of tricolpate pollen and during the Albian stage, the appearance of extant families within the magnoliids, hamamelids, rosidids, caryophyllids and dilleniids was evident. It was at this time that the primary adaptive radiation of the angiosperms occurred. By the mid-Cenomanian, Myricaceae was evident in southern Laurasia and by the end of this stage (92 myr BP) Betulaceae had also appeared. Of particular interest to this study is the observation that the Normapolles pollen characteristic of Myricaceae and Betulaceae is always associated with sandy porous soils typical to costal regions or stream banks where erosion or deposition has drastically altered the soil type. This is a major habitat of these species today.

The breakup of Gondwanaland, which commenced in the Early Cretaceous and led to the formation of the continents of South America and Africa, produced new coastal regions that may have provided conditions conducive to the establishment of symbioses between Frankia and one or several groups of Normapolles species, leading to the current symbioses with members of the Myricaceae, Betulaceae and Casuarinaceae. The Betulaceae spread quickly through Laurasia, first along costal areas and then across the present continents of Europe and North America, almost always appearing in the fossil record associated with land upheavals, erosion or stream banks. The Myricaceae spread throughout Laurasia and Gondwanaland; Casuarinaceae, which appears infrequently in the fossil record in South America and Africa during the Paleocene (66 myr BP), reached Australia via Antarctica before the latter's southward migration and drastic climate change. Reaching Australia by the middle Eocene (49 myr BP), the Casuarinaceae rapidly radiated forming most of the species which are recognized today. The fossil record of the Casurinaceae clearly establishes this family as an early successional family. Its relative abundance increased after glaciation events and then decreased as non-nitrogen-fixing species were able to move into now stabilized soils.

Other actinorhizal families do not appear in the fossil record until much later although certainly their progenitors were identifiable as long ago as the Albian stage. Rhamnaceae and Elaeagnaceae do not appear until the Upper Oligocene (25.2 myr BP) and Rosaceae not until the Pliocene (5.2 myr BP). Most Rosaceae are north temperate species and all of the actinorhizal plants within this family are thought to have originated in North America. These three actinorhizal families contain species nodulated by intercellular penetration and it is our belief that this occurred as a separate set of events from those that led to the symbioses with plants from the hamamelid group.

Host specificity of Frankia isolates. Frankia isolates have been made from only half of the known actinorhizal plant species. Most isolated frankiae may be grouped into one of four host specificity groups (HSG) as defined by Baker.[1] HSG 1 strains are those which nodulate Alnus, Comptonia and Myrica. HSG 2 strains are those which nodulate Casuarina, Gymnostoma, some Allocasuarina and Myrica. HSG 3 strains are those which nodulate Elaeagnus, Hippophae, Shepherdia and Myrica and HSG 4 have the same host specificity as HSG 3 except they cannot nodulate Myrica. As of yet no Frankia strains have been isolated which are capable of nodulating the additional actinorhizal genera. It is interesting to note that the actinorhizal family Myricaceae, which appeared first in the fossil record and has the widest geographical distribution of all actinorhizal species, can be nodulated by most Frankia strains regardless of their additional specificity to nodulate other actinorhizal genera. Perhaps the genus Myrica has served and continues to serve as a repository for symbiotic frankiae.

Summary. Current phylogenies based on morphological characters support a polyphyletic origin for actinorhizal symbioses. It is possible, however, that molecular phylogenies will provide greater insight into relationships among actinorhizal genera and lead us to alternate hypotheses. _Datisca_ and _Coriaria_ whose current tenuous placements are distant from one another may upon molecular analysis show a closer relationship. Molecular phylogenies should also help clarify the relationship between the Elaeagnaceae and the Rhamnaceae. In the construction of molecular phylogenies it will be of little use to confirm the position of a particular order or family without confirming the placement of an individual actinorhizal genus within that family.

Our present knowledge about the physiology of actinorhizal nodules also supports a polyphyletic origin for actinorhizal symbioses. Knowledge about host plant genes involved in symbiosis will provide information about the relative complexity of the symbioses and that in turn will influence our thinking about the likelihood of single or multiple origins of symbiosis.

References

1. Baker, D. (1987) Physiologia Plantarum 70 245-248.

2. Berry, A. & Sunell, L.A. (1990) in The Biology of Frankia and Actinorhizal Plants, eds. Schwintzer, C.R. & Tjepkema, J.D. (Academic Press, New York), pp 61-81.

3. Cronquist, A. (1988) The Evolution and Classification of Flowering Plants (The New York Botanical Garden, Bronx, NY).

4. Hahn, D., Lechevalier, M.P., Fisher, A. & Stackebrandt, E. (1989) System. Appl. Microbiol. 11 236-242.

5. Kalkman, C. (1988) Bot. J. Linnean Soc. 98 37-59.

6. Miller, I.M. & Baker, D.D. (1986) Protoplasma 131, 82-91.

7. Muller, J. (1984) Ann. Missouri Bot. Gard. 71, 419-443.

8. Parsons, R., Silvester, W.B., Harris, S., Gruijters, W.T.M. and Bullivant, S. (1987) Plant Physiol. 83, 728-731.

9. Silvester, W.B., Harris, S.L. & Tjepkema, J.D. (1990) in The Biology of Frankia and Actinorhizal Plants, eds. Schwintzer, C.R. and Tjepkema, J. D. (Academic Press. New York) pp 157-176.

10. Smolander, A. (1990) Plant and Soil 121, 1-10.

11. Walker, J.W. & Walker, A.G (1984) Ann. Missouri Bot. Gard. 71 464-521.

12. Zavada, M.S. (1984) Ann. Missouri Bot. Gard. 71, 444-463.

Gresshoff, Roth, Stacey & Newton (eds.) Nitrogen Fixation: Achievements and Objectives
Chapman and Hall New York • London 1990

Genetics of alternative nitrogen fixation systems in *Azotobacter vinelandii*

Paul E. Bishop, Sheela I. MacDougal,
Elizabeth D. Wolfinger, and Casendra L. Shermer

Department of Microbiology and USDA/ARS,
North Carolina State University, Raleigh, NC 27695-7615
U.S.A.

INTRODUCTION

The diazotroph, *Azotobacter vinelandii*, harbors three genetically distinct nitrogenases(2,13,16). These consist of the extensively studied conventional molybdenum-containing nitrogenase (nitrogenase-1) and two alternative nitrogenase complexes (nitrogenases-2 and -3) that do not contain molybdenum.

Nitrogenase-2 is a vanadium-containing enzyme complex which is expressed in cells cultured under diazotrophic conditions in medium lacking Mo but containing vanadium (V)(7,8,18). This enzyme complex consists of two components: dinitrogenase reductase-2, a dimer of two identical subunits, and dinitrogenase-2, which is now thought to be a hexamer (M_r of about 240,000) of two dissimilar pairs of large subunits (α and β) and a pair of small subunits (δ)(20). Dinitrogenase reductase-2 has an M_r of about 62,000 and contains four Fe atoms and four acid-labile sulfide groups per dimer(5,8). Dinitrogenase-2 contains two V atoms, 23 Fe atoms and 20 acid-labile sulfide groups per molecule(4). A cofactor (FeVaco) analogous to FeMoco has been extracted from dinitrogenase-2 using N-methylformamide(21).

Nitrogenase-3 does not appear to contain either Mo or V and is synthesized in cells cultured under Mo- and V-deficient conditions(2). This enzyme is composed of two components, dinitrogenase reductase-3 and dinitrogenase-3. Dinitrogenase reductase-3 is a dimer (M_r of approximately 65,000) of two identical subunits while dinitrogenase-3 is a tetramer (M_r of about 216,000) composed of two dissimilar pairs of subunits (α and β). However, dinitrogenase-3 may actually be a hexamer since the structural gene operon for nitrogenase-3 contains an open reading frame (ORF) whose predicted product is similar to the δ subunit for dinitrogenase-2(13). Dinitrogenase reductase-3 contains four Fe atoms and four acid-labile sulfide groups per dimer. Dinitrogenase-3 contains approximately 24 Fe atoms and 18 acid-labile sulfide groups per molecule; however, it lacks significant amounts of Mo or V.

GENES SHARED BY THE THREE NITROGENASE SYSTEMS

Two *nif* genes are known to be shared by the three nitrogenase systems. One of these is *nifB* since NifB⁻ mutants (strains CA30 and UW45) are unable to grow under Mo-sufficient, Mo-deficient plus V, and Mo-deficient diazotrophic conditions(10,12). In the wild type a *nifB*-hybridizing transcript (4 kb in size) is also observed under all three diazotrophic conditions(10). Since NifB is known to be required for FeMoco synthesis, this gene product undoubtedly plays an analogous role in the formation of FeVaco [the V-containing cofactor of dinitrogenase-2(21)]. With respect to dinitrogenase-3, NifB must be involved in some function other than one that relates to Mo or V. The other *nif* gene shared by the three nitrogenase systems is *nifM*. NifM is required for maturation of dinitrogenase reductase-1 and a NifM⁻ mutant (strain MV21) was unable to grow diazotrophically either in the presence or absence of Mo or V(15). *nifV* may also be essential for N_2 fixation by the three nitrogenase systems since growth of strain DJ71 [a NifV⁻ mutant (9)] was slow or non-existent under all N_2-fixing conditions tested (unpublished results).

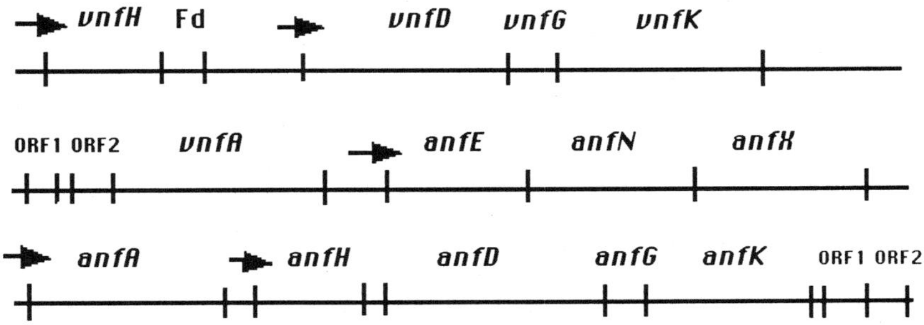

Figure 1. Organization of some genes involved in the two alternative nitrogenase systems of *A. vinelandii*. Arrows indicate promoters and direction of transcription.

GENES ENCODING NITROGENASE-2

The structural genes encoding dinitrogenase-2 and dinitrogenase reductase-2 have been cloned, sequenced and mutagenized for both *A. chroococcum* (19,20) and *A. vinelandii* (14,17). These genes have been designated as *vnf* (*v*anadium *n*itrogen *f*ixation) (20). In contrast to the single operon (*nifHDK*) encoding the subunits for nitrogenase-1, the genes encoding nitrogenase-2 are split into two operons (Fig. 1). *vnfH* encodes the dinitrogenase reductase-2 subunits and is part of a two gene operon. The ORF that is 3' to *vnfH* encodes a ferredoxinlike protein which has not been ascribed a function. *vnfH* is preceded by a potential promoter sequence that would be predicted to interact with core RNA polymerase containing the sigma 54 factor (*ntrA, rpoN,* or *glnF* gene product). The *vnfDGK* operon, located 1.0 kbp (*A. vinelandii*) or 2.5 kbp (*A. chroococcum*) downstream from the *vnfH-Fd* operon, encodes the subunits for dinitrogenase-2. *vnfD* encodes the α-subunit and *vnfK* encodes the β-subunit. In *A. vinelandii*, the 1.0-kbp region between the *vnfH-Fd* and *vnfDGK* operons does not appear to contain any identifiable ORFs(14), while the 2.5 kbp region in *A. chroococcum* contains an ORF of unknown function (R.L. Robson, personal communication).

The third subunit, δ, (M_r of 13,337) is encoded by *vnfG* (20). This gene does not have a counterpart in the nitrogenase-1 system; however, it does share some sequence similarity to *anfG*, an ORF located between *anfD* and *anfK* (Fig. 1). We have used site-directed mutagenesis to change the Cys-17 codon (TGC) in *vnfG* to a stop codon (TGA). This substitution results in a Vnf$^-$ phenotype for *A. vinelandii* (unpublished results). It is unlikely that the stop codon exerts a polar effect on translation of *vnfK* mRNA since it does not appear that *vnfG* and *vnfK* are translationally coupled (14). Thus it is likely that the δ subunit is required for active dinitrogenase-2.

The removal of a 1.4-kbp *Bgl*II fragment, which spans all of *vnfG* and the 3' and 5' ends of *vnfD* and *vnfK*, from the genomes of both *A. chroococcum* (20) and *A. vinelandii* (16) results in deletion strains that lack dinitrogenase-2. When this deletion was transferred to *A. chroococcum* strain MCD1155 (carrying a deletion of the structural genes for nitrogenase-1), the resulting double deletion strain was unable to grow under any N_2-fixing condition. This result indicates that *A. chroococcum* does not contain a third nitrogenase(20). A similar double deletion strain of *A. vinelandii* (strain RP206) grew in N-free media lacking Mo(16). This finding corroborates previously published results describing the isolation and partial characterization of a second alternative nitrogenase(2).

GENES ENCODING NITROGENASE-3

The structural genes encoding nitrogenase-3 in *A. vinelandii* have been cloned, sequenced and mutagenized(13). These genes have been designated *anf* (*a*lternative *n*itrogen *f*ixation) and they are organized in a single operon, *anfHDGK, ORF1, ORF2* which is preceded by a potential promoter sequence that would interact with RNA polymerase containing sigma 54 (Fig. 1). The subunits of dinitrogenase reductase-3 are encoded by *anfH* while the α and β subunits of dinitrogenase-3 are encoded by *anfD* and *anfK*, respectively. *anfG* probably encodes a third subunit (δ) for dinitrogenase-3. The predicted product of *anfG* contains a Cys residue (Cys-36) which corresponds to Cys-17 in the *vnfG* gene product. Replacement of the Cys-36 codon (TGC) with an Ala codon (GCG) results in a mutant strain with an Anf^+ phenotype, however, under Mo-deficient N-free conditions, growth of this strain is somewhat slower than that of the wild type. In contrast, substitution of this codon with a stop codon (TGA) results in an Anf^- phenotype (unpublished results). The Anf^- phenotype may, however, be due to polarity on *anfK* expression since *anfG* and *anfK* are likely to be translationally coupled (13). These observations lead to the preliminary conclusion that AnfG, although not essential for growth of *A. vinelandii* in Mo-deficient N-free medium, is nevertheless required for optimum growth under these conditions.

The predicted protein products of the two ORFs that are located 3' to *anfK* (Fig. 1) do not show overall similarity to any known *nif* gene products. However, the predicted ORF1 product contains some sequence identity to the NH_2-terminal part of dinitrogenase reductase and another region exhibits identity to presumed heme-binding domains of P-450 cytochromes. The predicted product of ORF2 does not seem to show any interesting similarity with other amino acid sequences in the Bionet data base(13).

Deletions plus insertions placed in several regions of the *anfHDGK* operon resulted in Anf$^-$ mutants that were unable to grow in N-free, Mo-deficient medium. However, growth in media containing Mo or V was normal(13). The absence of nitrogenase-3 proteins in these Anf$^-$ mutants was also confirmed by two-dimensional gel electrophoresis(13).

NON-STRUCTURAL GENES

Recently two *nifA*-like ORFs were identified in DNA cloned from *A. vinelandii*. One of these ORFs was recognized in the DNA sequence flanking the Tn*5* insertion carried by a Vnf$^-$ mutant (strain CA46) that is unable to express nitrogenase-2 when derepressed in N-free medium containing V(11,12). Since this mutant synthesizes both nitrogenases-1 and -3 under Mo-deficient N-free conditions in the presence or absence of V, the *nifA*-like ORF was designated *vnfA*(11). The other *nifA*-like ORF was located approximately 700 bp upstream from *anfH* (Fig. 1). A mutant (strain CA66), carrying a deletion plus insertion in this *nifA*-like ORF, synthesized only nitrogenase-2 proteins after derepression in Mo-deficient media with or without V. Thus this ORF was designated *anfA*(11). The highest degree of similarity between the predicted products of *nifA, vnfA* and *anfA* is in the C-terminal half of the proteins where the potential RNA polymerase-sigma 54 interaction sites and ATP-binding domains are located. A potential DNA-binding domain(6) that includes a helix-turn-helix motif is present in predicted products of both *vnfA* and *anfA*(11). The DNA-binding sites for VnfA and AnfA are, in all likelihood, different than the consensus UAS [TGT-N_{10}-ACA(1)] for *nif* operons since a characteristic feature of alternative system operons is the apparent lack of this UAS.

A *vnfENX* operon is located immediately downstream from *vnfA* (Fig. 1)(21). Preliminary results with a strain containing Tn*5-lacZ* inserted towards the 3' end of *vnfN* indicate that *vnfN* and/or *vnfX* is required for diazotrophic growth in Mo-deficient media with or without V. However, in the presence of Mo this strain shows wild-type growth (unpublished results). These results imply that the *vnfENX* operon is required for functional nitrogenases-2 and -3 but not nitrogenase-1.

Our work is supported by the USDA Competitive Grants Program.

REFERENCES

1. Buck, M., Miller, S., Drummond, M., and Dixon, R. (1986) Nature (London) **320**, 374-378.
2. Chisnell, J.R., Premakumar, R., and Bishop, P.E. (1988) J. Bacteriol. **170**, 27-33.
3. Eady, R.R. (1986) eds. Broughton, W.J., Puhler, A., (Clarendon Press, Oxford) **4**, pp. 1-49.
4. Eady, R.R., Robson, R.L., Richardson, T.H., Miller, R.W., and Hawkins, M. (1987) Biochem. J. **244**, 197-207.
5. Eady, R.R., Richardson, T.H., Miller, R.W., Hawkins, M., and Lowe, D.J. (1988) Biochem. J. **256**, 189-196.
6. Fischer, M.-H., Bruderer, T., and Hennecke, H. (1988) Nucleic Acids Res. **16**, 2207-2224.
7. Hales, B.J., Case, E.E., Morningstar, J.E., Dzeda, M.F., and Mauterer, L.A. (1986) Biochemistry **25**, 7251-7255.
8. Hales, B.J., Langosch, D.J., and Case, E.E. (1986) J. Biol. Chem. **261**, 15301-15306.
9 Jacobson, M.R., Brigle, K.E., Bennett, L.T., Setterquist, R.A., Wilson, M.S., Cash, V.L., Beynon, J., Newton, W.E. and Dean, D.R. (1989) J. Bacteriol. **171**, 1017-1027.
10. Joerger, R.D. and Bishop, P.E. (1988) J. Bacteriol. **170**, 1475-1487.
11. Joerger, R.D., Jacobson, M.R., and Bishop, P.E. (1989) J. Bacteriol. **171**, 3258-3267.
12. Joerger, R.D., Premakumar, R., and Bishop, P.E. (1986) J. Bacteriol. **168**, 673-682.
13. Joerger, R.D., Jacobson, M.R., Premakumar, R., Wolfinger, E.D., and Bishop, P.E. (1989) J. Bacteriol. **171**, 1075-1086.
14. Joerger, R.D., Loveless, T.M., Pau, R.N., Mitchenall, L.A., Simon, B.H., and Bishop, P.E. (1990) Bacteriol. 172, 3400-3408.
15. Kennedy, C., Gamal, R., Humphrey, R., Ramos, J., Brigle, K., and Dean, D. (1986) Mol. Gen. Genet. **205**, 318-325.
16. Pau, R.N., Mitchenall, L.A., and Robson, R.L. (1989) J. Bacteriol. **171**, 124-129.
17. Raina, R., Reddy, M.A., Ghosal, D., and Das, H.K. (1988) Mol. Gen. Genet. **214**, 121-127.
18. Robson, R.L., Eady, R.R., Richardson, T.H., Miller, R.W., Hawkins, M., and Postgate, J.R. (1986) Nature (London) **322**, 388-390.

19. Robson, R.L., Woodley, P.R., and Jones, R. (1986) EMBO J. **5**, 1159-1163.
20. Robson, R.L., Woodley, P.R., Pau, R.N., and Eady, R.R., (1989) EMBO J. **8**, 1217-1224.
21. Smith, B.E., Eady, R.R., Lowe, D.J. and Gormal, C. (1988) Biochem. J. **250**, 299-302.
22. Wolfinger, E.D., Pau, R.N., Bishop, P.E. (1989) Ann. Mtg. Amer. Soc. Microbiol. **185** (Abstract H-97).

Gresshoff, Roth, Stacey & Newton (eds.) Nitrogen Fixation: Achievements and Objectives
Chapman and Hall New York • London 1990

Nitrogen fixation in *Methanosarcina barkeri* strain 227

S. H. Zinder and A. L. Lobo

Department of Microbiology, Stocking Hall, Cornell University, Ithaca, NY 14853, USA

Introduction

The ability of methanogenic archaebacteria to fix N_2 was independently reported by four laboratories in 1984-85. Murray and Zinder (9) and Belay, Sparling and Daniels (2) published concurrent papers describing diazotrophy in *Methanosarcina barkeri* strain 227 and *Methanococcus thermolithotrophicus* respectively, followed soon afterward by the report by Bomar, Knoll and Widdell (3) on diazotrophy in *Methanosarcina barkeri* strain Fusaro. Sibold et al. (16) found that eubacterial *nif* probes, especially *nifH* probes, could hybridize with restriction-enzyme digested methanogen DNA. Since then, diazotrophy has been described in several other methanogens including members of all three major methanogenic orders (1), indicating that this ability is widespread in methanogens. Diazotrophy has not been reported in the other branches of the archaebacteria, the halobacteria and sulfur-dependent thermophiles (20).

The finding of diazotrophy in methanogenic archaebacteria raises many questions about the evolutionary relationship between methanogen nitrogenases and those in eubacteria. An important contribution to answering these questions came from Sibold and colleagues who sequenced open reading frames (ORF's) homologous to eubacterial *nifH* genes from several methanogens (17, 18, 19). More recently, Normand and Bousquet (11) summarized in a dendrogram the sequence similarities of several *nifH* genes, including those from methanogens. These results show that methanogen *nifH* sequences cluster outside eubacterial *nifH* sequences, and also show considerable divergence from each other. The one exception is the *nifH1* sequence from *Methanococcus thermolithotrophicus* which shows a distinct similarity (ca 70% at the amino acid level) with *anfH* from *Azotobacter vinelandii* (5) which codes for component 2 of the non-Mo non-V nitrogenase, and with *nifH3* from *Clostridium pasteurianum*. Adjacent to the *nifH1* in *M. thermolithotrophicus* are ORF's homologous to *nifD* and *nifK*, which are not co-transcribed with *nifH1*. A second ORF, called *nifH2*, has not been found to be expressed in *M. thermolithotrophicus* and shows considerable sequence divergence from *nifH1*. The great divergence of methanogen nitrogenases suggests that nitrogenase evolved prior to the splitting of the eubacterial and archaebacterial lines, and the sequence of the *nifH1* from *M. thermolithotrophicus* suggests that lateral transfer of genes occurred, perhaps between methanogens and clostridia.

We have been studying nitrogen fixation by *Methanosarcina barkeri* strain 227. This presentation will briefly summarize previous results and describe new data.

The discovery of N_2 fixation in *M. barkeri* strain 227.

Graduate student Patti Murray was studying accumulation of a glycogen-like polysaccharide by *Methanosarcina thermophila* strain TM-1 in response to NH_4^+ limitation. Since cells only stored around 1% of their dry weight as this polysaccharide (10), we tested another culture, *M. barkeri* strain 227 to determine whether it might store more polysaccharide. The culture continued to grow in ammonium-limited medium. Subsequent tests showed that replacing the N_2 in the gas phase with an inert gas, such as He or Ar, led to considerably decreased growth and methane production unless NH_4^+ was also added. Cells were shown to incorporate $^{15}N_2$, and that this incorporation was abolished in the presence of NH_4^+ in the growth medium, definitive proof of N_2-fixation The culture was carefully checked for contamination to eliminate the possibility of crossfeeding of fixed N by a eubacterial diazotroph. Since these studies, several mesophilic *Methanosarcina* cultures were shown to grow diazotrophically (3, 14, Lobo and Zinder, unpublished results), while thermophilic cultures tested by us have uniformly been negative. *M. barkeri* is an interesting methanogen in which to study nitrogen fixation since it is the most metabolically versatile of the methanogens, being able to grow on several substrates, including H_2-CO_2, methanol, methylamines, and acetate. It is cosmopolitan in its distribution, and a diazotrophic culture of *M. barkeri* was recently cultured from a rice paddy (14).

Physiology of N_2 fixation by *M. barkeri* strain 227.

Diazotrophy is an energetically costly process in eubacteria, typically consuming 16 or more moles of ATP per mole of N_2 fixed (13). We found that growth rates and yields were roughly halved in strain 227 when methanol was the methanogenic substrate (7), while the the effect was even more pronounced when the substrate was acetate, which provides less energy and reducing power. Therefore, the energetic toll of diazotrophy was similar to that in eubacteria.

We also examined the effects of trace metals on diazotrophic growth. We found that molybdate stimulated diazotrophic growth, while tungstate inhibited it (7). This suggests that the nitrogenase in strain 227 is a molybdoenzyme. We could not detect any stimulation by vanadate or vanadyl sulfate, while Scherer (15) demonstrated definite stimulation of diazotrophic growth by either molybdate or vanadium trichloride in both *M. barkeri* strains 227 and Fusaro.

Similar to other researchers (2, 3, 8), we found that C_2H_2 was inhibitory to methanogenesis by cultures of strain 227 and must be used at partial pressures (0.003 atm) considerably lower than the 0.1 atm typically used in the C_2H_2 reduction assay. The rates of C_2H_2 reduction at this lowered partial pressure were extremely low in the case of strain 227 (7) , about 1 nmol ethylene produced h^{-1} $[mg\ protein]^{-1}$, and were similar to rates found for other methanogens (2, 3, 8), but were 100-fold lower than rates that would be needed to support growth, assuming that C_2H_2 reduction and N_2-reduction occur at similar rates per electron, as they do in eubacteria.

Properties of nitrogenase in *M. barkeri* strain 227.

Higher rates of C_2H_2 reduction would be expected from crude extracts of methanogens since the enzyme would be supplied with exogenous ATP and reducing power (from dithionite) and therefore C_2H_2 inhibition of methanogenesis would no longer be a factor. However, rates of ethylene production by crude extracts in initial studies (7), about 5 nmol h^{-1} [mg protein]$^{-1}$, were only marginally higher. More recently, we have achieved higher rates, 20-50 nmol h^{-1} [mg protein]$^{-1}$, when we used cell-free extracts with higher protein concentrations (Lobo and Zinder, submitted), and achieved rates of about 130 nmol h^{-1} [mg protein]$^{-1}$ from a extract from a cell pellet which was never frozen and thawed. Higher protein concentrations also lead to increased activity and stability of the nitrogenase. C_2H_2-reducing activity in crude extracts required ATP, and was strongly inhibited by ADP, similar to eubacterial nitrogenases (7).

Initial attempts at purification were hampered by the low growth yields of the culture and the instability of the enzyme. Another problem was that nitrogenase is one of the less abundant iron sulfur proteins in methanogen cells. The typical situation in eubacterial diazotrophs is that nitrogenase is one of the most abundant iron-sulfur proteins. Because of this, the usual method for purification of nitrogenase components is to visually follow the brownish iron-sulfur proteins on an anaerobically operated chromatography column, and to collect those fractions in a gassed container (4). This method would have been impossible in the case of strain 227, in which over 20 peaks with absorbance at 405 nm elute from a Mono Q anion exchange column.

We have recently developed a three step procedure which affords a 20 to 40-fold purification of the nitrogenase from strain 227 with roughly 30% recovery of activity (Lobo and Zinder, submitted). The first step is precipitation with protamine, which results in ca. two-fold purification. This is followed by column chromatography using a Pharmacia-LKB Fast Protein Liquid Chromatograph (FPLC) operated inside an anaerobic glove-box.

Initial experiments with this apparatus demonstrated that the enzyme had two components. Activity was detected in the two fractions corresponding to molecular weights near 120 kdal and 240 kdal. In later studies, the redissolved protamine precipitate was chromatographed on a Mono Q anion exchange column, and appropriate fractions were tested against each other. Fractions with activity towards each other eluted at 0.42 and 0.46 M NaCl. These fractions were then chromatographed on the Superose 12 gel filtration column.

SDS gel electrophoresis of the purified components showed that they were more than 50% pure (Lobo and Zinder, submitted). For the component 1 analogue, there were two predominant bands with molecular weights near 57 and 62 kdal, consistent with the typical $\alpha_2\beta_2$ subunit structure of component 1 enzymes. The component 2 analogue showed a single major band near 31 kdal. This band reacted with antibodies to component 2 from *Rhodospirillum rubrum* (anti Rr-2), kindly provided by P. Ludden. This molecular weight is considerably different from the 26 kdal band we detected (7) in Coomassie blue-stained SDS gels from crude extracts of N_2-grown cells but which was absent from NH_4^+-grown cells.

This 26 kdal band, which is always darker than the 31 kdal band in N_2-grown cells, did not co-purify with nitrogenase activity, and did not react with the anti-Rr2, and its identity is not known. If the molecular weight of 120 kdal for the component 2 analogue, as determined by Superose 12 chromatography, is correct, this would make it a tetramer of the 31 kdal subunit, which would be in contrast to all known eubacterial component 2 proteins, which are dimers.

The un-optimized specific activity of the purified components was 500-1000 nmol h^{-1} [mg protein]$^{-1}$, considerably lower than values of 60-150,000 nmol h^{-1} [mg protein]$^{-1}$ typically obtained for purified Mo nitrogenases (4). The nitrogenase components reduced N_2 to NH_4^+ at a rate 1.7-fold higher per electron than they reduced C_2H_2 to ethylene. No evidence for significant production of ethane from C_2H_2 was obtained.

Switch-off of nitrogenase activity.

Several eubacteria, especially purple photosynthetic bacteria, switch-off C_2H_2 reducing activity in response to added NH_4^+, as well as to some other preferred nitrogen sources and in response to other environmental perturbations (6). This reaction has been studied best in *Rhodospirillum rubrum* in which it has been shown that switch-off is mediated by reversible ADP-ribosylation of one subunit of the component 2 (12).

We reported that adding 5 mM NH_4^+ resulted in a rapid switch off of C_2H_2-reducing activity in whole cells of strain 227 (7). This switch-off was not affected by the addition of the glutamine synthetase inhibitor methionine sulfoximine. We have since found that as little as 10 μM NH_4^+ causes transient switch-off of C_2H_2 reducing activity in strain 227 (Lobo and Zinder, submitted). We had previously reported that 5 mM glutamine caused switch-off, but now attribute this effect to small amounts of contaminating NH_4^+.

We have examined switched-off cells for modification of the component 2 analogue in Western blots using anti Rr-2. In *R. rubrum* covalent modification leads to the appearance in SDS gels of a second band with slightly retarded migration due to the addition of the ADP ribose to one subunit. We examined Western blots from switched-off cells and found no splitting of the component 2 band (Lobo and Zinder, submitted). Also, extracts from cells harvested an hour after NH_4^+ addition had identical C_2H_2 reducing activity as those from cells harvested before NH_4^+ addition. These results argue against ADP-ribosylation, but it is possible that the modification was reversed during harvesting and sample preparation, or that a different modification occurred not detectable by SDS gel electrophoresis.

Discussion.

Table 1 compares the properties of the Mo-nitrogenase from *Azotobacter vinelandii*, the V-nitrogenase from *Azotobacter chroocuccum*, the non Mo-non V nitrogenase of *A.vinelandii* and the properties we have determined for the nitrogenase from strain 227. The nitrogenase from strain 227 is clearly binary with subunit structure resembling all other described nitrogenases with the possible exception that the component 2 analogue could be a tetramer rather than a dimer. More extensive studies should be performed to ascertain its molecular weight. While no metal analysis has been done on purified component 1 from strain 227, stimulation of diazotrophic growth by Mo suggests that it is a molybdoenzyme.

Table 1. Properties of various nitrogenases.

Property	Mo	V	non-Mo,V	Strain 227
Subunit composition (kilodaltons)	58,54,34	55,50,31.5	58,50,32.5	62,57,31
Metal content	Fe, Mo	Fe, V	Fe	Fe?, Mo?
Average C_2H_2 reducing activity of components (nmol min^{-1} [mg protein]$^{-1}$)	1,100	430	33	13
Relative rate of N_2 vs. C_2H_2 reduction[a]	0.9	0.8	1.9	1.7
Ethane produced	–	+	+	–
Reference	4	16	4	Lobo and Zinder, submitted

[a] per electron, assuming that NH_4^+ production requires 3 electrons and C_2H_4 production requires 2 electrons.

The kinetic properties of the nitrogenase from strain 227 resemble those of the non-Mo non-V nitrogenase of *A. vinelandii*. The specific activity of both enzymes is considerably lower than either the Mo or V nitrogenases. Since the preparations of strain 227 nitrogenase were only about 50% pure and because the activity of the components were not optimized against each other, it is likely that the specific activity of the nitrogenase components are higher than the value presented, but it is also likely that they are still much closer to values of the non-Mo non-V nitrogenase than those for the Mo and V nitrogenases. Both the non-Mo non-V and 227 nitrogenases show a kinetic preference for N_2 over C_2H_2. This may be related to their low overall activity, or could be due to intrinsic differences in the way they bind substrate. In contrast to the two alternative nitrogenases, no ethane production was detected from *M. barkeri* nitrogenase, perhaps due to the likely presence of Mo in the enzyme indicated by growth studies (7).

The resemblance between the nitrogenase from strain 227 and the non-Mo non-V nitrogenase of *A. vinelandii* has some interesting evolutionary implications, depending on whether its sequence groups with *anfH* as the *M. thermolithotrophicus nifH1* does, or whether it groups with the other methanogens. If it groups with *anfH*, its poor specific activity and preference for N_2 over C_2H_2 is explained by its evolutionary relationship, although the likely presence of Mo in the enzyme complicates this interpretation. It should be noted that despite the sequence similarity between *anfH* and the *M. thermolithotrophicus nifH1* ORF, Mo stimulates diazotrophic growth of this organism indicating that its nitrogenase is also a molybdoenzyme (8). If the *M. barkeri* nitrogenase does not group with *anfH*, it may represent a primordial enzyme that although it contains Mo, has not evolved the high specific activity found in eubacterial Mo nitrogenases.

Diazotrophy in *M. barkeri* strain 227 is energetically costly, and as such is highly regulated, with both synthesis and activity of the enzyme components apparently regulated by the availability of NH_4^+. We have obtained no evidence for covalent modification of the component 2 in strain 227, as has been found in some eubacteria exhibiting NH_4^+ switch-off (21), and future studies are required to determine the mechanism involved.

Acknowledgements

This work was supported by USDA/Hatch funds and by grant N00014-86-K-0876 from the Office of Naval Research.

References

1. Belay, N., Sparling, R., Choi, B., Roberts, M., Roberts, J.E. and Daniels, L. (1988) Biochim. Biophys. Acta. **971**, 233-245.

2. Belay, N., Sparling, R. and Daniels, L. (1984) Nature. **312**, 286-288.

3. Bomar, M., Knoll, K. and Widdel, F. (1985) FEMS Microbiol. Ecol. **31**, 47-55.

4. Eady, R.R. (1980) in *Methods for evaluating biological nitrogen fixation.*, ed. Bergersen, F.J. (John Wiley & Sons Ltd. New York), 216-264.

5. Joerger, R.D., Jacobson, M.R., Premakumar, R., Wolfinger, E.D. and Bishop, P.E. (1989) J. Bacteriol. **171**, 1075-1086.

6. Kanemoto, R.H. and Ludden, P.W. (1984) J. Bacteriol. **158**, 713-720.

7. Lobo, A.L. and Zinder, S.H. (1988) Appl. Environ. Microbiol. **54**, 1656-1661.

8. Magot, M., Possot, O., Souillard, N., Henriquet, M. and Sibold, L. (1986) in *Biology of anaerobic bacteria.*, ed. Dubourgier, H.C., et al. (Elsevier Science Publishers B.V. Amsterdam), 193-199.

9. Murray, P.A. and Zinder, S.H. (1984) Nature (London). **312**, 284-286.

10. Murray, P.A. and Zinder, S.H. (1987) Arch. Microbiol. **147**, 109-116.

11. Normand, P. and Bousquet, J. (1989) J. Mol. Evol. **29**, 436-447.

12. Pope, M.P., Murrell, S.A. and Ludden, P.W. (1985) Proc. Natl. Acad. Sci. USA. **82**, 3173-3177.

13. Postgate, J.R. (1982) *The fundamentals of nitrogen fixation* (Cambridge University Press, London, State).

14. Rajagopal, B.S., Belay, N. and Daniels, L. (1988) FEMS Microbiol. Ecol. **53**, 153-158.

15. Scherer, P. (1989) Arch. Microbiol. **151**, 44-48.

16. Sibold, L., Pariot, D., Bhatnagar, L., Henriquet, M. and Aubert, J.P. (1985) Mol. Gen. Genet. **200**, 40-46.

17. Sibold, L. and Souillard, N. (1988) in *Nitrogen Fixation: Hundred Years After.*, ed. Boethe, H., de Bruijn, F.J. and Newton, W.E. (Gustav Fischer Verlag. Stuttgart), 705-710.

18. Souillard, N. and Sibold, L. (1986) Mol. Gen. Genet. **203**, 21-28.

19. Souillard, N. and Sibold, L. (1989) Molec. Microbiol. **3**, 541-551.

20. Woese, C.R. (1987) Microbiol. Rev. **51**, 221-271.

21. Yoch, D.C., Li, J., Hu, C. and Scholin, C. (1988) Arch. Microbiol. **150**, 1-5.

Gresshoff, Roth, Stacey & Newton (eds.) Nitrogen Fixation: Achievements and Objectives
Chapman and Hall New York • London 1990

The first photosynthetic N_2-fixing *Rhizobium*: Characteristics

Allan R.J. Eaglesham[1], Joan M. Ellis[1], William R. Evans[2], Darrell E. Fleischman[2], Mariangela Hungria[1], and Ralph W.F. Hardy[1]

[1]Boyce Thompson Institute for Plant Research Inc., Ithaca, NY 14853, USA; [2]Battelle-C.F.Kettering Res. Lab., Yellow Springs, OH 45387, USA.

Introduction

Biological N_2 fixation requires large amounts of energy based on either *in vitro* biochemical studies of nitrogenase or *in vivo* studies of root nodulated legume plants. For example, calorimetric experiments with effectively root nodulated soybeans suggest that the aerobic metabolism of the equivalent of about 12 g of carbohydrate to CO_2 and H_2O is required to fix 1 g of N_2 by the rhizobial-legume process (11). This carbohydrate is provided through foliar photosynthesis by the legume macrosymbiont and is used by the rhizobial microsymbiont. A photosynthetic rhizobium that could meet directly its energy needs for biological N_2 fixation could be advantageous since the legume plant might use the saved fixed carbon to produce additional harvestable plant product.

Although photosynthetic N_2-fixing bacteria and blue-green algae are well known, no example of a photosynthetic rhizobium has been described prior to that reported here. Since most rhizobia occur in the relative darkness of soil, either as free-living organisms or symbionts in root nodules, there was little reason for scientists to seek photosynthetic rhizobia. Possible exceptions were the rhizobia that form stem nodules.

Over sixty years ago Hagerup observed nodules on submerged stems of the legume *Aeschynomene aspera*, along the shores of the Niger River. He remarked on the green color of the nodules and the presence of large amounts of starch grains and suggested that a very efficient photosynthesis was occurring (8). Stem nodules have been found and studied on other *Aeschynomene* and on *Sesbania* and *Neptunia* (2, 4, 5, 16, 17, 21) including a report in this 8th International Congress on Nitrogen Fixation (Ladha, JK) on the use of stem-nodulated legumes as green manure. The same rhizobia that produce stem nodules on the above legumes will produce root nodules on them under appropriate cultural conditions.

---

*Present address: Wright State University, Dayton, Ohio 45435, USA.

Investigations from the early 1980's to the present at the Battelle-C.F. Kettering Laboratory and the Boyce Thompson Institute led to the discovery that an *Aeschynomene indica* stem-nodulating rhizobium identified as BTAi 1 appears to be photosynthetic while that of *Azorhizobium caulinodans* strain BTSr 3 that stem-nodulates *Sesbania rostrata* is not photosynthetic.

Rhizobial strain BTAi 1 was isolated at Boyce Thompson Institute from sand used for plant cultures (16). The sand originated in West Virginia near the Potomac and was strip-mined with some soil contamination; the fact that *Aeschynomene virginica* is indigenous to fresh and brackish tidal shores from New Jersey to Virginia (7) may explain the presence of BTAi 1. Stem nodules have not been reported on *A. virginica*, although it seems fitting that a rhizobium that can utilize light energy can also form nodules on illuminated parts of its legume host. It is possible that photosynthetic capabilities confer competitive advantage *ex planta* in aquatic C-limited conditions.

Stem nodules, but not root nodules, formed by BTAi 1 on *Aeschynomene* were found to maintain N_2-fixing activity for several days following defoliation (6) supporting Hagerup's early suggestion of efficient photosynthesis of the stem nodules. In contrast N_2-fixing activity of both stem and root nodules formed by BTSr 3 declined following defoliation. These observations suggested a possible photosynthetic uniqueness of BTAi 1.

Independent and parallel studies at the Battelle-C.F. Kettering Laboratory of endophyte(s) from *A. indica* stem nodules collected in India revealed an elaborate internal membrane system, resembling that of purple photosynthetic bacteria, as well as bacteriochlorophyll and photosynthetic reaction centers. A similar ultrastructure has been seen in *A. indica* stem nodules collected in Mississippi (19). However, the endophyte could not be cultured. The Battelle-C.F. Kettering Laboratory followed up studies on this/these endophyte(s) with experiments on BTAi 1. Evidence from the Battelle-C.F. Kettering Laboratory and Boyce Thompson Institute which support the photosynthetic and N_2-fixing activities of BTAi 1 but not BTSr 3 are summarized in the Results section.

Results

A brief summary of compositional and functional studies of free-living BTAi 1 and symbiotic BTAi 1 in root and stem nodules of *Aeschynomene* will be presented. Because of the uniqueness of a photosynthetic rhizobium, many different types of measurements were made to support this conclusion. Comparisons will be made with the non-photosynthetic BTSr 3 and other photosynthetic bacteria. Complete information will be published elsewhere by Boyce Thompson Institute and Battelle-C.F. Kettering groups.

Bacteriochlorophyll. In appropriate media such as 0.15% malate and 0.05% yeast extract, colonies of free-living BTAi 1 were pink-red when grown in a light-dark regimen in contrast to white when grown in the dark (16). Both pigmented and white colonies were capable of nodulating *Aeschynomene*.

Spectrophotometric examination of the pigmented cells showed an 870 nm absorption band similar to that of the long-wavelength of Bacteriochlorophyll *a*-containing purple photosynthetic bacteria (18). The endophyte of stem nodules exhibited an absorption spectrum similar to BTAi 1 *in vitro*. BTAi 1 is an aerobic organism; it could not be grown anaerobically in continuous light or dark or a light-dark regimen. The amount of bacteriochlorophyll synthesized aerobically by BTAi 1 is similar to that of aerobic photosynthetic bacteria such as *Erythrobacter* sp. OCh 114 (14) and less than 10% of that of anaerobically-grown *Rhodospirillaceae* (12).

Koch Postulates. The following steps were used to assure that BTAi 1 was a single organism and could produce stem nodules and be reisolated from stem nodules. A broth culture of BTAi 1 was serially diluted in yeast-mannitol broth with 0.01% Tween 80 and low dilutions were checked for absence of cell-aggregation by light microscopy. Cell number was determined using the Petroff Hauser and spread plate methods. All resultant colonies from the spread plate of yeast-fructose agar were pink. Six well separated colonies were cultured in yeast-mannitol broths and inoculated onto the submerged parts of surface-sterilized cuttings of *Aeschynomene*. All of the single colony isolates induced stem nodules whereas uninoculated cuttings remained nodule free. Six nodules were selected randomly and the endophyte plated to provide single colonies. All colonies were pink.

Free-Living Growth. Cultures of each of BTAi 1 and BTSr 3 were grown in 16 h/8 h light/dark, or continuous dark for 20 days. There was no beneficial effect of the light/dark regimen for log phase growth of BTAi 1 and BTSr 3 but the light/dark regimen was able to retain viable cell numbers of BTAi 1 from 10 to 20 days after the exogenous carbohydrate had been utilized while viable cell number declined after 10 days in continuous dark. The viable cell number declined after 10 days for both light/dark and continuous dark regimens for BTSr 3. Light was unable to support log-phase growth of BTAi 1 or BTSr 3 in the absence of carbohydrate in the media. Photosynthesis in BTAi 1 is adequate to maintain steady-state cell numbers but not to support log-phase growth. A similar pattern has been observed for *Erythrobacter* (13).

Free-Living Photosynthetic Functions. Uptake of CO_2 by BTAi 1 was stimulated several fold under light for cells grown in a light/dark regimen versus those under dark and grown in continuous dark. Similar light stimulation of CO_2 uptake was shown for *Erythrobacter* whereas BTSr 3 showed no response to light. Southern-blot hybridization results for BTAi 1 performed by Antonius Suwanto and Samuel Kaplan (University of Illinois, Urbana) provided evidence for the ribulose bisphosphate carboxylase (RBC) gene form I as found in plants, cyanobacteria, and some eubacteria, but not for the RBC gene form II as found in purple photosynthetic bacteria. These hybridization results are consistent with photosynthetic ability in BTAi 1 but are not conclusive proof since some rhizobia can fix CO_2 chemiautotrophically (15).

The rates of O_2 uptake by BTAi 1 and *Erythrobacter* were decreased by 50% or more by illumination. Light had no effect on O_2 uptake by BTSr 3. The decreased need for terminal oxidation to provide energy, a consequence of ATP formation via photosynthesis, has been reported with *Erythrobacter* (9). Inhibitors of photosystem II (DCMU and o-phenanthroline) and an uncoupler of photophosphorylation (NH_4Cl) overcame the O_2-sparing effect of light on BTAi 1 and *Erythrobacter* but had no effect on BTSr 3. Those inhibitors and uncouplers also decreased the light stimulation of CO_2 uptake.

Southern blot hybridizations of BTAi 1 by Suwanto and Kaplan using probes from the purple non-sulfur photosynthetic bacteria gave no positives for various structural genes including *pufBALMX* and *pucBA* for light harvesting proteins and reaction centers L and M polypeptides, *puhA* for reaction center H polypeptide and *cycA* for cytochrome c_2. A positive response was found for *pufBALMX* with *Erythrobacter*. Two strong positive hybridizations and a weak one were obtained for *hemA* for genes for aminolevulinic acid synthesis in BTAi 1. The probes were from *Rhodobacter sphaeroides*.

Free-Living N_2 Fixation Functions. Various media used to obtain free-living N_2 fixation by different rhizobia were evaluated with BTAi 1. A modified Dreyfus *et al.* medium (3) in which the C source was changed to 10 $g \cdot L^{-1}$ of glucose and 2.0 $g \cdot L^{-1}$ of agar was added to provide a suitable microaerobic zone, supported free-living N_2 fixation by BTAi 1 as well as BTSr 3. The specific N_2-fixing activity (measured as C_2H_2 reduction) of BTAi 1 was 10-20% of that of BTSr 3 and was more variable. Free-living N_2 fixation was found for BTAi 1 from one day after inoculation to ten days with maximum activity from two to six days.

Several experiments were performed to define the effect of light, if any, on N_2 fixation by BTAi 1. The glucose C-source was reduced to 1 $g \cdot L^{-1}$ for these tests. There was no short-term (one hour) effect of light or dark on N_2 fixation. Fixation occurred in the light coincident with photosynthesis. There was a positive long-term effect of light on N_2 fixation. BTAi 1 grown for two days on 16 h/8 h light/dark regimen had 2-5 times as much N_2-fixing activity as those grown for two days in continuous dark. Several hours of darkness were required to reduce the activity of the light/dark grown cultures to that of the continuous dark grown ones. A stimulatory effect of light on N_2 fixation occurred, but the slow time-response suggested that it is indirect.

The protein content of BTAi 1 was increased about 10% for light/dark-grown over continuous dark-grown cells.

Symbiotic Functions. Characteristics of stem and root nodules of *A. scabra* formed by inoculation with BTAi 1 and of *S. rostrata* formed by inoculation with BTSr 3 were determined. The starch and soluble carbohydrate contents were greater in stem than root nodules for both *A. scabra* and *S. rostrata* with those for *A. scabra* stem nodules the larger. Light increased the CO_2 uptake rate of excised stem, but not root, nodules and the rate of *A. scabra* stem nodules was about three times that of *S. rostrata* stem nodules. Possibly of greatest interest is the specific N_2-fixing activity of nodules expressed as µmol $C_2H_4 \cdot g^{-1}$ nodule$\cdot hr^{-1}$. The specific activities of the root nodules were similar while that of *A. scabra* stem nodules was about four times that of *S. rostrata* stem nodules. Transmission electron micrographs of bacteroids of BTAi 1 in Aeschynomene stem nodules show numerous vesicles believed to be chromatophores.

For several parameters stem nodules of *A. scabra* functioned more efficiently than did those of *S. rostrata*. These data are consistent with the photosynthetic ability of BTAi 1. More direct evidence for the positive effect of light on symbiotic N_2 fixation by BTAi 1 is the doubling of acetylene reduction produced by illumination of stem nodules with light in the spectral region beyond 730 nm.

Discussion

Substantial information has been gathered to support the photosynthetic activity of the rhizobium BTAi 1 (Table 1). In several areas such as the positive but modest bacteriochlorophyll content, aerobic nature, inability to grow under light in the absence of a carbon source, ability for stationary phase growth under light, and light-sparing of O_2 uptake, BTAi 1 is similar to the photosynthetic bacteria, *Erythrobacter*. The effects of inhibitors and uncouplers on O_2 and CO_2 uptake in light further solidify the conclusion of the photosynthetic activity of BTAi 1. In all of the above characteristics BTAi 1 is unlike BTSr 3. However, they both have the somewhat unusual ability to form N_2-fixing stem as well as root nodules on their respective legume hosts. Both BTAi 1 and BTSr 3 fix N_2 asymbiotically under microaerobic conditions although BTSr 3 is more active. The co-occurrence of N_2 fixation and photosynthesis in BTAi 1 is unusual. The source of electrons for photosynthesis in BTAi 1 has not been determined. Many other interesting questions can be asked about BTAi 1.

Although *Rhizobium, Bradyrhizobium, Azorhizobium*, and probably the new genus *Sinorhizobium* (1), are phylogenetically linked to the purple photosynthetic bacteria (4, 10) including *Erythrobacter* (20), these rhizobia have not been found to contain bacteriochlorophyll and/or to be photosynthetic. The relatively low content of bacteriochlorophyll, the inability to grow in the light without exogenous C, and the lack of stimulation of logarithmic growth by illumination in the presence of exogenous C indicate that BTAi 1 is limited in its photosynthetic capability. It may be a primitive form, with features in common with present-day rhizobia, bradyrhizobia, azorhizobia and also the purple photosynthetic bacteria. However, the lack of structural similarity between key photosynthesis genes of BTAi 1 and of *Erythrobacter* and those of *Rhodobacter sphaeroides* may indicate a horizontal rather than vertical route of acquisition of these genes by the aerobic photosynthetic bacteria. The absence of plasmids (> 400 kb) in BTAi 1, based on Suwanto and Kaplan's analysis, would appear to rule out the possibility of an aerobic photosynthetic organism (e.g.

Erythrobacter) having obtained plasmid bearing nodulation genes from *Rhizobium*. Pending full taxonomic description, we propose that BTAi 1 be named *Photorhizobium thompsonianum*.

Nitrogen fixation by the legume-rhizobium symbiosis is important in natural ecosystems and in agriculture. Study of this novel organism and its symbiosis with *Aeschynomene* may shed light, not only on the evolution of the rhizobia,

TABLE 1

Comparison of the Photosynthetic and N_2-Fixing Characteristics of *Photorhizobium thompsonianum* (proposed) with a Stem-Nodulating Bacterium and an Aerobic Photosynthetic Bacterium

Characteristics	*Photorhizobium thompsonianum* (Proposed) Strain BTAi 1	*Azorhizobium caulinodans* Strain BTSr 3	*Erythrobacter sp.* Strain OCh 114
N_2-Fixing Symbiosis			
Root and Stem Nodules	*A. indica* *A. scabra*	*S. rostrata*	N.D.*
Free-Living Bacteria			
Bacteriochlorophyll *a*	+		+
Light/Dark Regimen	+		
Aerobic Conditions	+		
Light-Stimulated CO_2 Uptake	+	—	+
Light-Decreased O_2 Uptake	+	—	+
Activity of Inhibitors of Photosystem 2 and Uncouplers	+	—	
N_2 Fixation	+	+++	—
Light-Stimulated N_2 Fixation	+	N.D.	N.D.
Light Effect on Growth			
Log Phase	None	None	None
Steady State	Maintain	None	Maintain
Symbiotic Functions			
Soluble Carbohydrate and Starch			
Stem Nodules	++++	++	N.D.
Root Nodules	+	+	N.D.
Light-Stimulated CO_2 Uptake			
Stem Nodules	++++	++	N.D.
Root Nodules	—	—	N.D.
Specific N_2-Fixing Activity			
Stem Nodules	++++	++	N.D.
Root Nodules	++	++	N.D.

*N.D. is Not Determined.

but also on the evolution of the infection process and of plant-bacterium interactions in general. Photosynthesis by bacteroids may explain the energy sufficiency of stem nodules on *Aeschynomene* spp. (6). If energy-sufficient nodules could be induced to form on the stems of agriculturally important legumes, e.g., soybean, N_2 would be fixed with less carbon drain on the host plant with the possibility of concomitant yield improvement.

References

1. Chen, W.X., Yan, G.H. & Li, J.L. (1988) Int. J. Syst. Bacteriol. 38, 392-397.
2. Dreyfus, B.L. & Dommergues, Y.R. (1981) FEMS Microbiol. Lett. 10, 13-317.
3. Dreyfus, B.L., Elmerich, C. & Dommergues, Y.R. (1983) Appl. Environ. Microbiol. 45, 711-713.
4. Dreyfus, B.L., Garcia, J.L. & Gillis, M. (1988) Int. J. Syst. Bacteriol. 38, 89-98.
5. Eaglesham, A.R.J. & Szalay, A.A. (1983) Plant Sci. Lett. 29, 265-272.
6. Eardly, B.D. & Eaglesham, A.R.J. (1985) in *Nitrogen Fixation Research Progress*, eds. Evans, H.J., Bottomley, P.J. & Newton, W.E. (Martinus Nijhoff, Boston), p. 324.
7. Fernald, M.L. (1950) *Gray's Manual of Botany* (American Book Company, New York).
8. Hagerup, O. (1928) Dansk Bot. Ark. 15, 1-9.
9. Harashima, K., Nakagawa, M. & Murata, N. (1982) Plant Cell Physiol. 23, 185-193.
10. Hennecke, H., Kaluza, K., Thöny, M., Fuhrmann, M., Ludwig, W. & Stackebrandt, E. (1985) Arch. Microbiol. 142, 342-348.
11. Heytler, P.G., Reddy, G.S. & Hardy, R.W.F. (1985) in 14th Steenbock Symposium on Nitrogen Fixation and CO_2 Metabolism, eds. Ludden, P.W., Burris, J.E. (Elsevier Press, New York), pp. 283-292.
12. Lascelles, J. (1963) in *Bacterial Photosynthesis*, eds. Gest, H., San Pietro, A. & Vernon, L.P. (The Antioch Press, Yellow Springs, OH), pp. 35-52.
13. Shiba, T. (1984) J. Gen. Appl. Microbiol. 30, 239-244.
14. Shiba, T. (1987) Plant Cell Physiol. 28, 1313-1320.
15. Simpson, F.B., Maier, R.J. & Evans, H.J. (1979) Arch. Microbiol. 123, 1-8.
16. Stowers, M.D. & Eaglesham, A.R.J. (1983) J. Gen. Microbiol. 129, 3651-3655.
17. Subba Rao, N.S., Tillak, K.V.B.R. & Singh, C.S. (1980) Plant Soil 56, 491-494.
18. Thornber, J.P., Trosper, T.L. & Strousse, C.E. (1978) in *The Photosynthetic Bacteria*, eds. Clayton, R.K. & Sistrom, W.R. (Plenum Press, New York), pp. 133-160.
19. Vaughn, K.C. & Elmore, C.D. (1985) *Cytobios* 42, 49-62.
20. Woese, C.R. (1987) Microbiol. Rev. 51, 221-271.
21. Yatazawa, M. & Yoshida, S. (1979) Physiol. Plant. 45, 293-295.

Gresshoff, Roth, Stacey & Newton (eds.) Nitrogen Fixation: Achievements and Objectives
Chapman and Hall New York • London 1990

Nodulation of non-legumes by rhizobia

Edward C. Cocking, Mozahim K. Al-Mallah, Erica Benson and Michael R. Davey
Plant Genetic Manipulation Group, Department of Botany, University of Nottingham NG7 2RD, UK

INTRODUCTION

In discussing required inputs from basic studies to applied nitrogen fixation research, Gibson (1) has highlighted three major areas which must be investigated if much of the basic research is to become of practical benefit. These include (i) the development of improved nitrogen fixing strains of rhizobia, cyanobacteria, Azospirillum and Frankia species by molecular biology techniques, coupled with physiological studies relating to the ability of such engineered microorganisms to survive in the soil, (ii) the production, storage, transport and application of inoculants and (iii) the control by the host plant of the development of various symbiotic associations and subsequent nitrogen fixation. His belief was that the objective of nitrogen self-sufficiency for agriculturally important plants was most likely to be achieved through symbiotic associations similar to those occurring in legume or actinorhizal nodules. Gibson also suggests that it is worthwhile to consider the possibility that many plant species already possess the basic genetic constitution to enable them to participate in a number of the reactions which normally occur during symbiotic nitrogen fixation. Only their inability to undertake as few as three or four steps, probably those involved in infection, prevent plants entering into symbiosis with soil microorganisms. More recently, Sprent (2) in a perceptive analysis of the steps essential for the formation of functional legume nodules, has pointed out that there is little reason to suppose that nodulating legumes possess unique genes ready to be switched on when appropriate strains of Rhizobium interact with such plants. While it is well documented that the non-legume Parasponia is nodulated by Rhizobium (3), there appears to be no clear reason, on current evidence, why other non-legumes should not nodulate with rhizobia, given the correct environment and genetic background. The possibility has also been discussed of inducing nitrogen fixing nodules on plants, such as wheat, by treating roots with auxin and inoculation with rhizobia (4).

Experiments performed in our laboratory have addressed this question of the interaction of Rhizobium with non-legumes, with the aim of developing new associations between this soil bacterium and a range of crop plants. Initially, these investigations involved the development of enzymatic techniques which permitted removal of the barrier to Rhizobium-host plant specificity in white clover, followed by the development of nodular structures on the roots of cereals. More recent studies have entailed nodulation of oilseed rape by rhizobia which naturally nodulate the non-legume Parasponia. This paper reviews our progress in developing these Rhizobium/non-legume associations.

ENZYMATIC DEGRADATION OF THE CELL WALLS AT THE TIPS OF ROOT HAIRS

In 1985, it was found possible to achieve rapid enzymatic degradation of the cell wall at the apices of root hairs of seedlings from a wide range of crop species of both legumes and non-legumes, including the cereals (5). Specifically, rapidly elongating roots with well developed root hairs of axenically grown seedlings were immersed in cell wall degrading enzymes dissolved in hormone-free Murashige and Skoog cell culture medium containing 6.0% w/v sucrose. Examination of root hairs exposed to such enzymes by bright field microscopy and by ultra violet illumination following staining of roots with the fluorescent brightener Tinopal, confirmed a reduction in fluorescence at the tips of root hairs indicative of wall degradation in these regions. A mixture of 1.0% w/v CEL cellulase (Worthington, Freehold, New Jersey, USA), 0.5% w/v Novozyme 234 (Novo Industries, Bagsvaerd, Denmark) and 0.1% w/v Pectolyase Y23 (Seishin Pharmaceutical, Tokyo, Japan) resulted in wall degradation within 5 minutes of incubation, with protoplasts being released from the tips of root hairs. Replacing the CEL cellulase with 1% w/v Cellulysin (Calbiochem, La Jolla, California, USA) or Cellulase R10 (Yakult Honsha Co., Nishinomiya, Japan) still resulted in degradation of cell walls, but protoplast release was decreased to about 5% of that obtained with the original mixture.

It was possible, by balancing the osmotic pressure of the enzyme solution with that of the plant cell cytoplasm, to expose the plasma membranes at the tips of root hairs and to control protoplast release while maintaining the functional integrity of enzyme-treated seedlings. Indeed, this ability to expose the plasma membranes at the tips of root hairs provided the stimulus for investigations of the interaction of Rhizobium with enzyme-treated roots of legumes and non-legumes.

ENZYMATIC TREATMENT OF THE ROOT HAIRS OF WHITE CLOVER REMOVES A BARRIER TO RHIZOBIUM-HOST SPECIFICITY

It is well established that Rhizobium loti and Rhizobium strain 101/u induce effective nitrogen fixing nodules on the roots of Lotus species, but not on the roots of Trifolium repens (white clover). Indeed, this specificity was confirmed for many thousands of white clover seedlings inoculated with these rhizobia under axenic conditions. However, the Rhizobium-host plant specificity could be overcome by enzyme treatment of seedling roots prior to inoculation with the bacterium. Thus, the roots of 2 day-old axenic seedlings were treated for 5 minutes with 1.0% w/v Cellulase YC (Seishin Pharmaceutical) and 0.1% w/v Pectolyase Y23 in 6.0% w/v mannitol solution, followed by incubation of the roots of the enzyme treated seedlings in a suspension of Rhizobium in the presence of 25% w/v polyethylene glycol (PEG) of molecular weight 6000 and 50 mM calcium chloride. Rhizobium-PEG treated roots were washed in 6% w/v mannitol solution prior to transfer of the seedlings to the surface of agar-solidified nitrogen-free medium contained in 120 mm square Petri dishes. Within 4 weeks of this procedure, 1.1-1.2% of the enzyme-treated white clover seedlings developed root nodules following exposure to R. loti or Rhizobium 101/u in the presence of PEG (6). In the absence of PEG, the number of seedlings which developed nodules was reduced to about 0.1%. The presence of both enzymes in the mixture was essential for nodulation. Seedlings treated with only Cellulase or only Pectolyase prior to exposure to Rhizobium and PEG failed to nodulate. Seedlings which had not been enzyme treated prior to Rhizobium/PEG treatment also failed to nodulate. Transmission electron microscopy confirmed the ultrastructure of nodules induced by R. loti and Rhizobium 101/u to be similar to that of nodules induced on white clover by the compatible R. trifolii, with bacteria within infection threads and enclosed in membrane-bounded vesicles in the cytoplasm of infected nodule cells. Such nodules induced on white clover by R. loti and Rhizobium 101/u were pink in colour and their

nitrogenase activity, as assayed by acetylene reduction, was comparable to that of pink nodules from white clover seedlings induced by R. trifolii (2.18±0.49 compared to 2.20±0.14 nmole ethylene/nodule/hour). Re-isolation of rhizobia from white clover nodules confirmed the specificity of R. loti and Rhizobium 101/u to nodulate Lotus corniculatus, their natural host, without enzyme treatment of seedling roots.
These studies on the nodulation of white clover seedlings by strains of Rhizobium which are normally incompatible with this forage legume, identified the apical cell wall of root hairs as a plant target site for the genetic manipulation of both legumes and non-legumes. With this in mind, the aim of subsequent experiments was to develop novel nitrogen fixing symbiotic associations. While rhizobial infection of non-legumes to form an effective nodular nitrogen-fixing symbiosis is presently restricted to plants within the dicotyledonous genus Parasponia, much interest centers on the extent to which similar symbiotic associations might be extended to other non-legumes, including cereals such as rice. This led us to investigate whether rice seedlings could be induced to nodulate when seedling roots were enzyme treated and inoculated with rhizobia.

INDUCTION OF NODULAR STRUCTURES ON THE ROOTS OF CEREALS BY RHIZOBIUM

The Cellulase-Pectolyase enzyme solution used to treat the roots of white clover seedlings (but with mannitol increased to 8.0% w/v), was also effective in removing the cell walls at the tips of root hairs of 2 day-old rice seedlings within 10 minutes of incubation. Inoculation of enzyme-treated seedlings of the rice variety Taipei 177 with R. loti, Rhizobium 101/u, R. trifolii, or with mixtures of R. loti and Rhizobium 101/u or R. loti and Bradyrhizobium 32H1 in the presence of PEG and 50 mM calcium chloride, resulted in 7.3% of the seedlings developing nodular structures on their roots when the seedlings were subsequently grown on agar solidified nitrogen-free medium (7). Rice seedlings failed to develop such outgrowths when enzyme treatment of seedling roots was omitted from the procedure. The nodular structures which developed on host plants within 4 weeks of enzyme-Rhizobium-PEG treatment were either spherical, approximately 4.0 mm in diameter, or elongate and about 6.0 mm in length. Some elongate structures were lobed, with approximately three times as many elongate structures being produced compared to spherical ones. Spherical nodular structures were composed of irregularly shaped cells of which about 1% contained bacteria. Bacteria were also aggregated on cell walls bordering air spaces. Rhizobia within the elongate structures were located mainly between the cell layers, although some cells were also infected. Bacteria within both types of structure were often associated with fibrillar material. As in the experiments with white clover, rhizobia re-isolated from nodular structures which developed on rice seedlings retained their specificity to nodulate their normal host plants. Thus, a mixed culture of reisolated bacteria induced nodules on Lotus corniculatus, but failed to nodulate the roots of untreated rice seedlings. Similar studies to those described for rice, but with Triticum aestivum varieties Brock, Rendezvous, Sapphire and Wembley, showed that nodular structures could also be induced on the roots of this cereal following enzyme treatment of seedling roots and incubation with Rhizobium in the presence of PEG (8). R. loti, Rhizobium 101/u, Bradyrhizobium 32H1 and mixtures of these bacteria were also employed in this study, nodulation being influenced by the Rhizobium strain and the plant variety. Acetylene reduction (only 0.01 nmole ethylene/nodule/hour), was associated with the nodular structures, which developed on the roots of rice seedlings. However, this was at the limit of sensitivity of the assay procedure. It was suggested that the inability to detect acceptable rates of acetylene reduction might be related to the low number of host cells that became infected by rhizobia and also to the tendency for degeneration of host cell cytoplasm (7). In discussing these results, it was emphasized that further investigations would be required to determine whether nodular structures effective in nitrogen fixation could be induced on rice and other cereals by strains of rhizobia such as those which naturally nodulate the non-legume Parasponia (3).

NODULATION OF OILSEED RAPE BY RHIZOBIA

Investigations have been carried out to determine whether the dicotyledonous non-legume oilseed rape (Brassica napus), taxonomically more closely related to the dicotyledonous non-legume Parasponia than either rice or wheat, would be more responsive to nodulation by rhizobia following enzyme-PEG treatment of seedling roots (9). Recently, we have also investigated whether rhizobial strains nodulating the non-legume Parasponia would be more efficient in nodulating seedlings of oilseed rape than those of rhizobia which normally only nodulate legumes.

In such studies, the roots of 2 day-old oilseed rape seedlings, grown on nitrogen-free agar solidified medium, were treated with the same enzyme mixture as that used previously for rice (1.0% w/v Cellulase YC and 0.1% w/v Pectolyase Y23), but dissolved in 6.0% w/v mannitol solution. Enzyme-treated seedling roots were inoculated with the bacterial suspension mixed with an equal volume of 25% w/v PEG 6000 containing 50 mM calcium chloride. The legume-specific rhizobia included R. leguminosarum bv trifolii I/RCR, R. loti NZP 2037, Bradyrhizobium 32H1 and R. leguminosarum bv viciae strains Nod^+ 8401pRL1JI (=B556) and Nod^- B401 (=B554). When a mixture of R. loti and Bradyrhizobium 32H1 was used as inoculum, the bacterial suspensions were mixed 5 minutes prior to addition of the PEG solution. R. parasponium RP501 was employed as the bacterial strain which naturally nodulates the non-legume Parasponia. Seedling roots were incubated in the bacteria for 25 minutes, followed by washing in 8.0% w/v mannitol solution before culture of the seedlings on the surface of nitrogen-free medium contained in square Petri dishes. The latter were particularly useful for ease of handling of the inoculated seedlings. In some experiments, the nitrogen-free medium used for seedling maintenance was supplemented with 0.1% w/v sucrose, 5.6 mM mannitol, 25 mM succinate, 25 mM arabinose or β-hydroxybutyrate.

R. leguminosarum and Bradyrhizobium 32H1 induced nodules on the roots of enzyme treated oilseed rape seedlings of the winter varieties Rafal, Ariana and Lictor. R. leguminosarum also induced nodules on the spring variety Librawell, but the latter failed to respond to Bradyrhizobium 32H1. R. loti was ineffective in nodulating the four oilseed varieties tested. However, when R. loti was mixed with Bradyrhizobium 32H1, there was a synergistic interaction between the two bacteria, resulting in the highest nodulation frequency on all four plant varieties tested. The Nod^+ strain of R. leguminosarum also induced nodules on Rafal, but the Nod^- strain failed to elicit this response (9).

Nodules induced on oilseed rape were morphologically similar, when examined by light microscopy and cryo-scanning electron microscopy (Figs 1A, B), to those induced on the roots of the forage legume white clover by the compatible R. trifolii. Transmission electron microscopy confirmed the presence of rhizobia within cells of the nodules, with infected cells interspersed with non-invaded cells. Infected cells contained large numbers of bacteria (Fig 1C), with the latter being

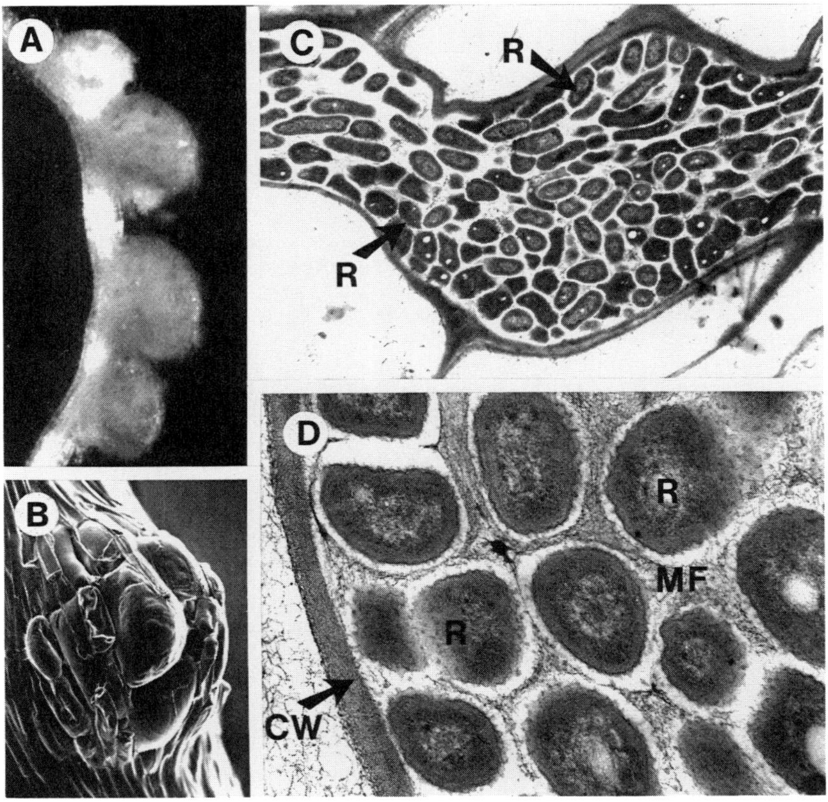

FIGURE 1 A - Nodules on the root of an oilseed rape seedling following enzyme treatment of the root and inoculation with a mixture of Rhizobium loti and Bradyrhizobium 32H1 in the presence of PEG (x10).

B - Cryo-scanning electron micrograph showing the epidermal cells of a nodule developing on the root of an oilseed rape seedling. The seedling root was enzyme treated prior to Rhizobium inoculation as in A (x280).

C - Transmission electron micrograph showing rhizobia (R) in a cell within one of the nodules shown in A (x7,000).

D - Rhizobia (R) enclosed within microfibrillar material (MF) in a nodule cell. CW = cell wall (x44,800).

surrounded by microfibrillar material, frequently forming a dense matrix around the bacterial cells (Fig 1D). Spherical nodules lacked an apical meristem. Nodules which developed on oilseed rape exhibited rates of acetylene reduction of 10 nmole ethylene/g nodule fresh weight/hour; reducing activity associated with rhizobia within nodule cells was confirmed electron microscopically by the reduction of potassium tellurite to insoluble electron opaque deposits. Nodules were not induced on oilseed rape seedlings whose roots had not been enzyme treated prior to inoculation with Rhizobium alone, with Rhizobium in the presence of PEG, or when treated with PEG alone. Rhizobium reisolated from nodules induced by a mixture of R. loti and Bradyrhizobium 32H1 induced nodules on Lotus corniculatus and Vigna sinensis (the natural hosts for R. loti and Bradyrhizobium 32H1 respectively), confirming the ability of re-isolated rhizobia to nodulate their normal host plants.

NODULATION OF OILSEED RAPE BY PARASPONIA-INFECTING RHIZOBIA

Our most recent investigations concern the nodulation of oilseed rape by rhizobia known to nodulate the non-leguminous, tropical tree Parasponia. Initial studies indicate that the nodulation characteristics of oilseed rape inoculated with Rhizobium parasponium RP501 are very different to those following the inoculation of oilseed rape by legume-specific rhizobia. Thus, while nodulation in the latter was only possible after enzyme treatment of seedling roots, nodulation of the roots of oilseed rape seedlings (27% of the seedlings) occurred with R. parasponium RP501 in the absence of enzyme/PEG treatment. Nodules also developed on the roots of seedlings which had been treated with enzyme prior to incubation in the Rhizobium/PEG mixture, but this was at a reduced level (17%). The frequency of nodulation of seedling roots was considerably higher following infection with R. parasponium RP501 than with a mixture of R. loti and Bradyrhizobium 32H1 (about 6% in the case of this mixed inoculum). Several different nodule types were observed on seedling roots infected with R. parasponium. White, brown, spherical and elongate nodules had the same morphological characteristics as those found on roots inoculated with legume-specific rhizobia. However, about 2% of the spherical nodules produced red/pink pigmentation, suggesting the presence of "leg"haemoglobin. Preliminary microscopical studies have shown that oilseed rape nodules arising from treatment of seedling roots with R. parasponium contain both inter- and intra-cellular rhizobia. Rhizobia were also associated with the surface of nodules. These preliminary studies will form the basis for future investigations which will examine, in greater detail, the characteristics of enzyme and non-enzyme mediated nodulation by R. parasponium, evaluate nitrogenase activity and identify the red pigmentation produced by such root nodules.

CONCLUSIONS AND FUTURE PERSPECTIVES

To date, the nodulation of non-legumes by rhizobia has been confined to members of the tropical tree genus Parasponia. The findings presented in this paper suggest that it may be possible to circumvent those barriers, both within legumes and in non-legumes, which normally restrict Rhizobium-host plant specificity. This possibility may have important consequences for both basic and applied aspects of nitrogen fixation. Nodulation in non-legumes may be achieved, in the first instance, by the application of rhizobia in the presence of PEG/Ca^{2+} to enzyme-treated roots of both monocotyledons, such as wheat and rice, and dicotyledons (eg. oilseed rape). In this event, it may be that nitrogen fixing capabilities may not be confined to plants such as oilseed rape, but may, in due course, be extended to agronomically important monocotyledons such as cereals. It also appears possible to induce nodulation in oilseed rape, in the absence of enzyme treatment, when roots are inoculated with R. parasponium RP501. Our early observations with this Rhizobium have been confined to in vitro studies, but it will be essential, in future work, to determine the effects of plant growth conditions on nodulation. The possibility of extending a symbiotic relationship, similar to that which occurs naturally between the non-legume Parasponia and R. parasponium, to other agronomically important species, will, undoubtedly, prove important in applied research.

In the case of more basic studies, the nodulation of non-legumes may provide an important insight as to the evolutionary development of Rhizobium/plant interactions. Sprent and Raven (10) comment that the mechanism of root hair infection typical of temperate legumes is highly advanced and, presumably, more evolved than the primitive modes of Rhizobium infection which occur in Aeschynomene, Parasponia, Sesbania, and Stylosanthes. Rhizobia which infect these plants invade wounds and points of emergence of lateral roots. Hypersensitive disturbances induced by rhizobia, which weaken the epidermis, can also allow bacterial entry. The finding, in the present study, that nodulation in the absence of enzyme treatment is possible with *R. parasponium* RP501, but not with rhizobia which normally enter the host plant through root hairs, may reflect the different routes of entry of these rhizobia. Similarly, the possible presence of haemoglobin in oilseed rape nodules infected with *R. parasponium*, but not in those inoculated with rhizobia whose normal plant host range comprises temperate legumes, is noteworthy, suggesting that the establishment of functional nodules in oilseed rape may be more likely with *R. parasponium*.

At the molecular level, it is also apparent that the specificity of the *Rhizobium*-legume association at certain stages of nodulin gene expression may not be as critical as was first thought. Investigations have demonstrated the expression of the nodule morphogenesis-related gene *ENOD2* in uninoculated plants of *Medicago sativa* (11), while functional haemoglobin genes have been characterised in *Trema tomentosa*, a relative of *Parasponia* which fails to develop root nodules (12). Additionally, root hair curling has been observed in maize and rice following inoculation with transconjugants of *R. trifolii* ANU843 carrying *R. trifolii* root hair curling (*Hac*) genes (13). Future studies may benefit from an understanding of the molecular and physiological nature of the novel *Rhizobium*/non-legume associations described in this paper.

ACKNOWLEDGEMENTS

This work was supported by grants from the Overseas Development Administration, the Rockefeller Foundation and the Agricultural and Food Research Council. *R. leguminosarum* bv *trifolii* I/RCR, *R. loti* NZP2037 and *Bradyrhizobium* 32H1 were obtained from Rothamsted Experimental Station, Harpenden, UK, University of Arhus, Arhus, Denmark and the Nitragin Co., Milwaukee, Wisconsin, USA, respectively. *R. leguminosarum* bv *viciae* Nod^+ and Nod^- were kindly provided by Professor A.W.B. Johnston, University of East Anglia, Norwich, UK.

REFERENCES

1. Gibson, A.H. (1980) in Current perspectives in nitrogen fixation, eds. A.H. Gibson & W.E. Newton, (Elsevier/North Holland Biomedical Press, Amsterdam, New York, Oxford), pp 6-7.
2. Sprent, J.I. (1989) New Phytol. 111, 129-153.
3. Trinick, M.J. & Hadobas, P.A. (1989) in Nitrogen fixation with non-legumes, eds. F.A. Skinner et al. (Kluwer Academic Publishers, Dordrecht, Boston, London), pp 25-33.
4. Tchan, Y.T. & Kennedy, I.R. (1990) Agricultural Sci. 2, 57-59.
5. Cocking, E.C. (1985) Bio/Technology 3, 1104-1106.
6. Al-Mallah, M.K., Davey, M.R. & Cocking, E.C. (1987) Bio/Technology 5, 1319-1321.
7. Al-Mallah, M.K., Davey, M.R. & Cocking, E.C. (1989) J. Exp. Bot. 40, 473-478.
8. Al-Mallah, M.K., Davey, M.R. & Cocking, E.C. (1989) Internat. J. Plant Genetic Manipulat. in Plants 5, 1-7.
9. Al-Mallah, M.K., Davey, M.R. & Cocking, E.C. (1990) J. Exp. Bot. (submitted).
10. Sprent, J.I. & Raven, J.A. (1985) Proc. Royal Soc. Edinburgh 85B, 215-237.
11. Truchet, G., Barker, D.G., Camut, S., de Billy, F., Vasse, J. & Huguet, T. (1989) Mol. Gen. Genet. 219, 65-68.
12. Bogusz, D., Appleby, C.A., Landsmann, J., Dennis, E.S., Trinick, M.J. & Peacock, W.J. (1988) Nature 331, 178-180.
13. Plazinski, J., Innes, R.W. & Rolfe, B.G. (1985) J. Bateriol. 163, 812-815.

Gresshoff, Roth, Stacey & Newton (eds.) Nitrogen Fixation: Achievements and Objectives
Chapman and Hall New York • London 1990

nifH, *D* and *K* are contiguous in *Frankia* strain FaC1

Chung Sun An[1], Paul Twigg[2], James M. Ligon[3], and Beth C. Mullin[2]

[1]Botany Department, Seoul National University, Seoul, Korea, [2]Botany Department and The Center for Legume Research, The Plant Physiology and Genetics Program, The University of Tennessee, Knoxville, TN USA, [3]Biotechnology Research, CIBA-GEIGY, Research Triangle Park, North Carolina USA

Earlier reports concluded that nifH was not contiguous with nifD and K in Frankia strain FaC1 (Ligon and Nakas, 1989). This result was in contrast to other reports that nifH,D, and K were contiguous in at least three other Frankia strains examined. In Frankia strains ArI3, ACON24d and CeD nifH, D, and K have been shown to be contiguous on the basis of restriction mapping and hybridization with gene-specific probes (Normand et al., 1988; Simonet et al. 1988). FaC1 was considered to be an exceptional case because of reports that nifH did not cluster with nifD and K (Ligon and Nakas, 1987). Nucleotide sequence analysis of the putative FaC1 clone revealed that it did not contain legitimate Frankia sequences and pointed to the necessity of looking further in the FaC1 cosmid library for legitimate Frankia nif sequences (Ligon and Nakas, 1988; 1990).

A relatively weak nifH hybridizing clone was selected from the library and a 7.2kb nifH, D and K hybridizing PstI fragment from that clone was subcloned into pBR322. That subclone, designated pFAH-7.2, was mapped using a series of restriction enzymes and the location of nifH, D and K were determined by hybridization with specific fragments from Klebsiella nif sequences.

It was evident from the restriction map of pFAH-7.2 that indeed nifH, D and K are contiguous and are all contained within approximately 5kb of DNA in the pFAH1 cosmid clone. Sequence analysis of the 3'end of nif H clearly identifies it as a Frankia nifH sequence. To date all Frankia strains which have been examined, including FaC1 contain clustered nifHDK genes.

References

Ligon, J. M. and Nakas, J. P.(1987) Isolation and characterization of Frankia sp. strain FaC1 genes involved in nitrogen fixation. Appl.and Env. Micro. 53(10) 2321-2327.

Ligon, J.M. and Nakas, J.P. (1988) Nucleotide sequence of nifK and partial sequence of nifD from Frankia species strain FaC1. Nucleic Acids Research 16 11843.

Ligon, J.M. and Nakas, J.P.(1990) Corrigendum: Ibid. Nucleic Acids Research 18 1097.

Gresshoff, Roth, Stacey & Newton (eds.) Nitrogen Fixation: Achievements and Objectives
Chapman and Hall New York • London 1990

Formation of nodule-like structures on the roots of the non-legumes rice and wheat

G.L. Bender, L. Preston, D. Barnard and B.G. Rolfe*.

Plant Microbe Interaction Group, RSBS, Australian National University, Canberra ACT 2601, Australia (fax: 61 6 2490754)

INTRODUCTION AND RESULTS

Specific *Rhizobium* strains caused root hair curling on rice (1). Research on the genetic basis of *Rhizobium* nodulation of *Parasponia* (2) led to the engineering of a *Rhizobium* strain that occasionally forms nodules on rice. Currently, nodulation is infrequent (about 0.25% of inoculated plants) and unpredictable, but the internal structure of nodules closely resembles that found in legumes and contains membrane encapsulated bacteria. Addition of the synthetic auxin 2,4-D to wheat seedlings resulted in the formation of nodule-like structures, which are probably modified lateral roots. Within a narrow concentration range, 2,4-D induced the formation of structures on wheat roots which resembled nodules formed on legumes. Furthermore, there appears to be a narrow concentration window of 2,4-D addition that induces these structures on wheat roots without apparently interfering with adequate lateral root growth and good shoot growth. Seedlings with 1 x 10^{-6}M auxin form both lateral roots and nodule-like structures and have normal top growth. Plants in the range of 5 x 10^{-6}M 2,4-D and above form nodules, have few to no lateral roots and have yellowing and reduced top growth (see Figure). Different strains of *Rhizobium, Azospirillum, Agrobacterium* and *E. coli* have been added separately with auxin to wheat roots. The various bacteria do not behave in the same manner; some strains added with 2,4-D at 5 x 10^{-6}M caused a marked stunting of the roots and killing of the seedlings. Other strains have a negligible effect or show a slight decrease in the level of added auxin required to induce nodule structures, lateral roots and good shoot growth.

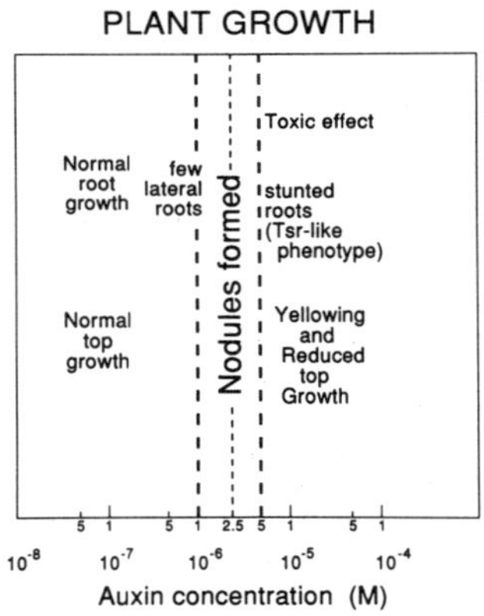

CONCLUSIONS

When nodules do form on rice roots, *Rhizobium* bacteria can enter and be "packaged" within membranes, as found in legume nodules. Although as yet, there is no evidence of infection thread formation, these structures do have vascular bundles. In contrast, nodules on wheat roots do not have substantial levels of bacteria present and the internal structure is not highly organized like legume nodules.To date, no evidence of nitrogen fixation has ever been found with any of these induced nodule-like structures on either rice or wheat.

(1) Plazinski, J. *et al.* (1985) J. Bacteriol. **163**, 812-815.
(2) Bender, G.L. and Rolfe, B.G. (1985) Plant Sci. **38**, 135-140.

Gresshoff, Roth, Stacey & Newton (eds.) Nitrogen Fixation: Achievements and Objectives
Chapman and Hall New York • London 1990

Alternative nitrogenases in *Azotobacter paspali*

E. Fallik, P. G. Hartel and R. L. Robson

Departments of Biochemistry and Agronomy, University of Georgia, Athens, GA 30602 USA.

Azotobacter paspali forms a specific nitrogen fixing association with the subtropical forage grass *Paspalum notatum* (Batatais) (1). We are interested in developing methods which will allow the genetic components of this association to be identified.

We first examined whether *A. paspali* strains have alternative nitrogenases. Genomic DNA from *A. paspali strains* ATCC 23367 and ATCC 23833 hybridized strongly with DNA probes specific for the structural genes for components 1 of molybdenum nitrogenase (*nifDK*), vanadium nitrogenase (*vnfDGK*) (4), and the "third" nitrogenase (*anfDGK*) (2).

P. notatum (Batatais) can be germinated and grown in a gnotobiotic plant growth assembly (3). Wild-type and genetically marked strains of *A. paspali* strain ATCC 23367 proliferated in the rhizosphere and produced a 65% increase in plant dry weight in 10 days. This system provides a means of studying the association and for examining the transfer of fixed N from the bacterium to the plant.

Genetic systems are being developed in order to construct mutants defective in one or more of the nitrogenases systems or in the association with the plant. The Tn5 suicide vector pGS9 (5) was used as a vehicle to induce Tn5 insertion mutants in *A. paspali* ATCC 23367. Kanamycin^r and neomycin^r *A. paspali* exconjugants were obtained at 10^{-7} after conjugation with *Escherichia coli* (pGS9). Tn5 inserts were found to be random in the genome. Several Nif$^-$ strains of *A. paspali* isolated after Tn5 mutagenesis are being characterized and assessed for their ability to proliferate in the rhizosphere of *P. notatum*.

1. Dobereiner, J., Day, J.M., & Dart, P. J. (1972) J. Gen. Microbiol. **71**, 103-116.
2. Joerger, R. D. *et al.* (1989) J. Bacteriol. **171**, 1075-1086.
3. Hartel, P. G. *et al.* (1989) Appl. Environ. Microbiol. **55**, 1291-1294.
4. Robson, R. L. *et al.* (1989) EMBO J. **8**, 1217-1224
5. Selveraj, G. & Iyer, V. N. (1983) J. Bacteriol. **156**, 1291-1300.

Gresshoff, Roth, Stacey & Newton (eds.) Nitrogen Fixation: Achievements and Objectives
Chapman and Hall New York • London 1990

16S ribosomal ribonucleic acid sequence analysis of *Rhizobium* and *Bradyrhizobium* strains

T. Gamo, H. Oyaizu* and K. Mori

National Institute of Agrobiological Resources, Tsukuba, Ibaraki 305, Japan. * Toyama University, Toyama City, Toyama 930, Japan.

The phylogenetic status of members of the genera Rhizobium and Bradyrhizobium was examined by reverse transcriptase sequencing of 16S ribosomal RNA using approximately 150 strains collected from tropical, sub-tropical and temperate climate zones. Partial sequences for the region between 1220 to 1377 nucleotides could clearly separate the rhizobia into two clusters, Rhizobium and Bradyrhizobium by the sequence of specific 10 nucleotides.
Rhizobium could be grouped into three species, viz. R. leguminosarum, R. meliloti(= R. fredii) and R. loti, and 5 new species were identified by substituting some other base sequences. R. spp. 1, 3 and 4 required specifically pyridoxine and thiamine as growth factors.
Bradyrhizobium strains isolated from many leguminous plants showed the same base sequences as B. japonicum except for 7 new species which were identified by substituting some base sequences.
16S rRNA sequence analysis of Rhizobiaceae gave almost same classification as already proposed[1] and new species identified were found specifically in the isolates from the nodules of leguminous plants in tropical and sub-tropical zones. 16S rRNA sequence analysis, therefore, can be used as a classification standard of Rhizobiaceae.

Table 1. Classification of *Rhizobiaceae* by 16S rRNA sequences

Rhizobium	Host plant	*Bradyrhizobium*	Host plant
R. leguminosarum	13 genera	B. japonicum	29 genera
R. meliloti (= R. fredii)	Gycine & Medicago	B. sp. 1	Cassia, Centrosema & Stylosanthes
R. loti	12 genera	B. sp. 2	Pterocarpus & Samanea
R. sp. 1	Mimosa & Sesbania	B. sp. 3	Neonotonia & Derris
R. sp. 2	Entada	B. sp. 4	Indigofera
R. sp. 3	Pterocarpus	B. sp. 5	Aeschynomene
R. sp. 4	Crotalaria & Sesbania	B. sp. 6	Alycicarpus & Centrosema
R. sp. 5	Cicer	B. sp. 7	Pueraria

References

1)Jordan,D.C.(1984) in *Bergey's Manual of Systematic Bacteriol.*, eds. Krieg, N.R. & Holt, J.G.(Williams & Wilkins, Baltimore), Vol. 1:234-244.

Gresshoff, Roth, Stacey & Newton (eds.) Nitrogen Fixation: Achievements and Objectives
Chapman and Hall New York • London 1990

Factors regulating transcription of the structural genes for nitrogenase-2 in *Azotobacter vinelandii*

Susanne Jacobitz and Paul E. Bishop

Dept. of Microbiology and USDA/ARS, N.C. State University, Raleigh, NC 27695-7615 USA.

Azotobacter vinelandii contains three genetically distinct nitrogenases which are expressed under different diazotrophic growth conditions (1, 2, 3, 4). Nitrogenase-1 is expressed in the presence of Mo, whereas nitrogenase-2 is expressed in the presence of V. A second alternative nitrogenase, nitrogenase-3, is expressed when neither Mo nor V are present. The genes encoding nitrogenase-2 are divided between two operons: *vnfH-Fd* and *vnfDGK*. In the present study we used *lacZ*-fusion strains to examine the effect of metals on transcription of these operons under diazotrophic growth conditions. Expression of *vnfH* and *vnfD* occurred in the presence of V and in the absence of both V and Mo (conditions under which nitrogenase-3 is expressed). In addition, transcription (as measured by β-galactosidase accumulation) of both *vnfH* and *vnfD* occurred in the presence of NH_4^+. This is in contrast to the situation with *nifHDK* and *anfHDGK* (the structural genes for nitrogenase-1 and nitrogenase-3, respectively) where NH_4^+ is observed to repress these operons. Furthermore, transcription of *vnfH* and *vnfD* are both VnfA dependent since β-galactosidase does not accumulate in a VnfA$^-$ mutant (CA46). However, CA46 is still able to grow diazotrophically in the presence of V. Transcription of *vnfH* is also AnfA dependent, whereas *vnfD* seems to be AnfA independent. Deletion of *anfA* prevents accumulation of β-galactosidase activity in a strain carrying a *vnfH-lacZ* fusion but not in a strain containing a *vnfD-lacZ* fusion. Surprisingly, the *anfA* deletion strain still grows in the presence of V. Neither *vnfH-lacZ* nor *vnfD-lacZ* fusions in a double mutant strain containing *nifHDK* and *anfA* deletions (CA11.66) accumulated β-galactosidase activity. These results show, that the operons for the structural genes of nitrogenase-2 are independently regulated and expression of *vnfH* requires both VnfA and AnfA whereas *vnfD* expression requires only VnfA.

References

1. Chisnell, J.R., Premakumar, R., and Bishop, P.E. (1988) J. Bacteriol., **170**, 27-33.
2. Joerger, R.D., Jacobson, M.R., Premakumar, R., Wolfinger, E.D., and Bishop, P.E. (1989) J. Bacteriol. **171**, 1075-1086.
3. Pau, R.N., Mitchenall, L.A., and Robson, R.L. (1989) J. Bacteriol. **171**, 124-129.
4. Joerger, R.D., Loveless, T.M., Pau, R.N., Mitchenall, L.A., Simon, B.H., and Bishop, P.E. (1990) J. Bacteriol. **172**, 3400-3408.

Gresshoff, Roth, Stacey & Newton (eds.) Nitrogen Fixation: Achievements and Objectives
Chapman and Hall New York • London 1990

Rice root nodules with acetylene reduction activity

Yuxiang Jing, Guangshan Li, Guohua Jin, Xueqin Shan, Baotian Zhang, Changhui Guan & Jiage Li.

Department of Nitrogen Fixation, Institute of Botany, Chinese Academy of Sciences, Beijing 100044, China.

It has been confirmed that *Parasponia*, a woody member of the elm family, is able to be nodulated by *Rhizobium*. This means the possibility of extending rhizobial symbiosis beyond the *Leguminosae* (3). Recently, it has been also shown that the rice and wheat seedling roots which were treated with cell wall degrading enzymes in the presence of polyethylene glycol, have been nodulated by rhizobia (1-2). Here we provide the results for the nodulation of rice seedling roots without the treatment with the enzymes mentioned above.

The rice seedlings at the 0.5 cm shoot stage were inoculated with rhizobia and nodule primordia appeared on the roots, then developed into small nodules of about 0.5 mm in diameter after 15 days, large nodules 2-3 mm in diameter developed after 35 days. The form of the nodules spherical in shape was similar to that of sesbania and soybean. Transmission electron microscopic observation shows the rhizobia and the presence of peribacteroid membrane-like structures. In the cytoplasm of rhizobia there are electron transparent regions which may represent the accumulation of poly β- hydroxybutyrate.

By gas chromatographic determination the rice root nodules made acetylene to reduced into ethylene with specific activity about 204 nmole/g fresh nodule/h, which corresponds to 1.4% of that of *Sesbania cannabina*. The low level of acetylene reduction in rice nodules is much higher as compared with that of the nodules from non-legumes due to actinomycete-like organisms and of association of nitrogen-fixing bacteria with grasses and cereals.

Western dot blotting experiments qualitatively indicate the weak positive colour reaction both of Mo-Fe protein and hemoglobin with the extract of rhizobia and of rice nodules.

The bacteria isolated from rice nodules have the existence of nitrogenase structure genes (*nifKDH*) and nodulation genes (*nodABC*) by Southern blot with the probe labeled by biotin and ^{32}P-dATP.

These results imply that *Rhizobium* may have the possibility of extending its host range outside the *Leguminosae*, especially the monocot crop plants. From the point of view of recent advances in this research, it should be completely conceivable that there are several ways to make *Rhizobium* and *Bradyrhizobium* infection of non-legumes with resultant nodulation.

REFERENCES

1. **Al-Mallah, M.K., Davey, M.R. & Cocking, E.C.** (1989) J. Exp. Bot. 213, 473-478.
2. Al-Mallah, M.K., Davey, M.R. & Cocking, E.C. (1989) Inter. J. Genet. Manipul. in Plants 5,1-7.
3. **Postgate, J.R.** (1987) J. Appl. Bact. Sym. Suppl. 85S-91S.

Gresshoff, Roth, Stacey & Newton (eds.) Nitrogen Fixation: Achievements and Objectives
Chapman and Hall New York • London 1990

Regulation of the vanadium nitrogenase in *Azotobacter chroococcum*

Lori L. Kerber, Elazar Fallik & Robert L. Robson

Department of Biochemistry, University of Georgia, Athens, Georgia 30602, U.S.A.

Azotobacter chroococcum expresses a vanadium nitrogenase (V-N_2ase) under conditions of ammonium and molybdenum deprivation. Structural genes for this enzyme are arranged in two operons. One contains the vnfDGK genes encoding alpha, delta, and beta subunits of the component 1 ($Ac2^V$). An adjacent operon contains vnfH for the subunit of component 2 ($Ac2^V$) (3). Northern analysis shows that vnfH is co-transcribed with two other ORFs. ORF1 encodes a small ferredoxin-like protein and ORF2 encodes a protein of 515 amino acid residues with no strong resemblence to any other reported protein. Strains bearing chromosomal disruptions in ORF2 are phenotypically indistinguishable from the parent strain. Promoters for both operons only show sequence identity at two positions. Regions -12 to -24 match well with NTRA-dependent promoters. The sequence, 5'-AGCCCNGNNCGACGC-3', is located upstream of the promoter with a spacing corresponding to upstream enhancer sequences identified in NIFA activatable promoters.

We have cloned a region of the genome which contains reiterations of nifE and nifN (1). Strains with a chromosomal disruption in a region upstream of the reiterated nifEN are blocked in transcription of both vnfH and vnfDGK operons. DNA sequence analysis reveals that the mutation lies in a gene resembling transcriptional activators of the nifA/ntrC family and most closely resembles vnfA from A. vinelandii (2). This gene probably encodes the transcriptional activator for the V-N_2ase structural genes and may recognize the upstream sequence conserved in both promoters identified above.

1. Evans D.J. et al. (1988) J. Gen. Microbiol. **134**, 931-942.
2. Joerger R.D., Jacobson M.R. & Bishop P.E. (1989) J. Bacteriol. **171**, 3258-3267.
3. Robson R.L. et al. (1989) EMBO J. **8**, 1217-1224.

Gresshoff, Roth, Stacey & Newton (eds.) Nitrogen Fixation: Achievements and Objectives
Chapman and Hall New York • London 1990

Genetic relatedness and taxonomic considerations of *Rhizobium* strains that nodulate *Phaseolus vulgaris* (L.)

E. Martínez[1], M.A. Pardo[1], F. Martins[2], P. Graham[3], A. Franco[2], R. Palacios[1] and L. Segovia[1].

[1]Centro de Investigación sobre Fijación de Nitrógeno, UNAM, Ap. 565-A, Cuernavaca, México; [2]CNPBS EMBRAPA Km. 47, Río de Janeiro, Brasil; [3]Dept. of Soil Science, University of Minnesota, USA.

Phaseolus vulgaris (L.) (bean) may be nodulated by a wide range of *Rhizobium* strains(2,5) as a consequence, most bacterial collections of *Rhizobium leguminosarum* bv. *phaseoli* harbor heterogeneous groups of bacteria. Two distinct types of strains isolated from bean nodules have been characterized(1,3,4). *Nif* gene reiteration is a genetic marker of type I strains (2), while the remarkable characteristic of type II strains is their ability to nodulate *Leucaena* trees(2). Multilocus enzyme electrophoresis(5,6) of 65 type II strains was performed using eight metabolic enzymes; 0.86 was the genetic distance among type I and type II strains. Type II strains are separated into two groups at 0.79 genetic distance. By the criterion applied in defining species limits, these values would represent distinct evolutionary lineages. By DNA-DNA hybridization types IIA and IIB did not resemble any other *Rhizobium* species.

1. Brom, S., Martínez, E., Dávila, G., & Palacios, R. (1988). Appl. Environ. Microbiol. **54**:1280-1283.
2. Martínez, E., Pardo, M.A., Palacios, R., & Cevallos, M.A. (1985). J. Gen. Microbiol. **131**:1779-1786.
3. Martínez, E., Palacios, R. & Sánchez, F. (1987). J. Bacteriol. **179**:2828-2834.
4. Martínez, E., Flores, M., Brom, S., Romero, D., Dávila, G. & Palacios, R. (1988). Plant and Soil **108**:179-184.
5. Piñero, D., Martínez, E. & Selander, R.K. (1988). Appl. Environ. Microbiol. **54**:2825-2832.
6. Selander, R.K., Caugant, D.A., Ochman, H., Musser, J.M., Gilmour, M.N. & Whittam, T.S. (1986). Appl. Environ. Microbiol. **51**:873-884.

ACKNOWLEDMENTS

The authors acknowledge Ma. de los Angeles Moreno and M. Antonio Rogel for technical help.

Gresshoff, Roth, Stacey & Newton (eds.) Nitrogen Fixation: Achievements and Objectives
Chapman and Hall New York • London 1990

Characteristics of two evolutionary lines of *Bradyrhizobium japonicum*

K. Minamisawa, T. Tsuruta, K. Fukai and S. Onodera

Faculty of Agriculture, Ibaraki University, Ami, Ibaraki, 30003 Japan.

DNA hybridization studies indicated that B. japonicum comprises two evolutionary lines consistent with the status of individual species (1-3). We have investigated various characteristics of two lines of B. japonicum. To distinguish two evolutionary lines, radiolabeled nifDKE of USDA 110 were hybridized to Southern blots of restriction digests of their genomic DNAs. Extracellular polysaccharides (EPSs) and rhizobitoxine (RT) were analyzed by GC and amino acid analyzer, respectively. Indole-3-acetic acid (IAA) in culture supernatant (Tris-YM supplemented with casamino acid) was quantified by GC/MS using deuterium-labeled IAA as an internal standard. Genotype and phenotype of hydrogenase were determined by hybridization with hup structural genes and amperometric technique, respectively.

The sequence divergence in and around nifDKE clearly divided B. japonicum strains into two different groups, designated genotype I and II (GTI and GTII). A strict correlation was observed between the genotype grouping, EPS composition, RT production and IAA production. GTI strains produced no RT, no IAA and an EPS of type A composed of glucose, mannose, galactose, 4-O-methyl galactose and galacturonic acid, whereas GTII strains produced RT, IAA and an EPS of type B composed of rhamnose and 4-O-methyl glucuronic acid. Hydrogen-uptake (Hup) positive strains, possessing the hup structural genes, were confined exclusively to GTI. When various cultivars of soybeans and siratro were inoculated with a soil sample containing GTI and GTII strains, Peking and siratro were nodulated mainly with GTII strains. These results revealed that two evolutionary lines (GTI and GTII) of B. japonicum define a marked division of various characteristics. Thus, this genotype grouping would represent a meaningful criterion for ecological studies on indigenous B. japonicum. Partial sequences of 16S rRNA suggested that the division between GTI and GTII is a recent evolutionary event as compared with the division between Bradyrhizobium and Rhizobium.

REFERENCES

1. Hollis, A. B. et al. (1981) J. Gen. Microbiol. 123:215-222.
2. Minamisawa, K. (1990) Plant Cell Physiol. 31:81-89.
3. Stanley, J. et al. (1985) J. Bacteriol. 163:148-154.

Gresshoff, Roth, Stacey & Newton (eds.) Nitrogen Fixation: Achievements and Objectives
Chapman and Hall New York • London 1990

Inoculation of non-host plants with *Frankia*

Beth C. Mullin[1] and Chung Sun An[2]

[1]Department of Botany, The Center for Legume Research and The Plant Physiology and Genetics Program, The University of Tennessee, Knoxville, TN and [2]Department of Botany, Seoul National University, Seoul, Korea.

The actinomycete Frankia is capable of participating in symbioses with plants from at least seven different phylogenetic orders. Although specificity does exist, host specificity groups frequently include host plant genera from more than one order. Unlike rhizobia which have no intrinsic oxygen protection mechanism, Frankia strains are able to adapt remarkably well to a variety of host plant environments because of their ability to regulate oxygen diffusion into the nitrogen-fixing vesicle. We believe that frankiae should be more amenable to induced symbioses that rhizobia. Following the observations of Al-Mallah et al. (1989) that rice seedling roots treated for a short time with cellulase and pectinase formed nodular structures in response to inoculation with Rhizobium, we have initiated such studies using Frankia as an inoculum.

Two days (occasionally longer) after the emergence of the radical, seedlings of Parasponia, tomato, corn were treated approximately as in Al-Mallah et al. (1989) and were observed for up to three months after treatment. Several week old pine seedlings were treated in a like manner. No nodules were observed on the roots of any plants; however, Frankia appeared to cause excessive early branching of the Parasponia radicle.

Although we were not able to induce nodules to be formed on the non-host plants Parasponia, pine, corn and tomato, by the methods reported here, our enthusiasm to initiate further studies has not been dampened. Relatively few plants of each species were subjected to treatment and no variation in the concentration of Frankia inoculum was tried. Several logical non-host plants were not tested. For example Betula which is very closely related to the host plant Alnus, cannot normally be nodulated by AvcI1. Rice which has been shown to form nodular structures in response to Rhizobium was not tested. Nor were any of the many actinorhizal plants not normally nodulated by AvcI1 tested.

The observation that incubation with Frankia, appeared to cause excessive early branching of Parasponia roots, must be confirmed by additional studies. If it is indeed a real phenomenon, it will be a very interesting observation to pursue.

Gresshoff, Roth, Stacey & Newton (eds.) Nitrogen Fixation: Achievements and Objectives
Chapman and Hall New York • London 1990

Carbon and nitrogen metabolism and the regulation of nitrogen fixation in *Rhizobium meliloti*

B. Noonan, K. Birkenhead, B. Boesten, A. Dobson and F.O'Gara

Microbiology Department and National Food Biotechnology Centre, University College, Cork, Ireland

The transport and metabolism of dicarboxylic acids (DCAs) have been shown to be directly involved with the efficiency of nitrogen fixaction by Rhizobium meliloti. R. meliloti mutants have been isolated which are unable to transport DCAs (1). CM12 is an NTG mutant with a mutation in the dctB gene. Complementation studies indicate that CM12 contains an intact dctD gene. CM12 is fix⁻, whereas CM20, which has a Tn5 insertion in the dctBD region, exhibits a reduced ability to fix nitrogen. Using nif::lacZ fusions, the expression of the symbiotic promoters P1 (nifHDK), P2 (fixABCX) and PnifA (nifA) (2,3,4) was examined. All three promoters were expressed to less than 2% of wild type levels in CM12. However, in CM20, PnifA was expressed to 80% of wild-type levels, while P1 and P2 were expressed to wild-type levels. Complementation of CM12 with cloned dct genes also completely restored nif gene expression and nitrogen-fixation ability.

Under free-living, microaerobic conditions, nifA can be induced to high levels in the wild-type strain CM2 and in the mutants CM12 and CM20. P1 can be induced in CM2 and CM20 but not in CM12. There appears to be a regulatory link between the expression of the P1 promoter and the dct genes (5). A possible explanation is that the inactive DctD present in CM12 could interfere with nifHDK expression analogous to the way it is proposed to interfere with dctA (6).

Given that carbon metabolism may be involved in nif gene regulation, the possibility of a "nitrogen effect" on the regulation of these genes was also investigated. The effect of a variety of nitrogen sources on nif gene expression in wild type R. meliloti under free-living microaerobic conditions was examined. With malate as carbon source, gultamate, glutamine and nitrate were found to support high levels of nifA::lacZ fusion expression, but ammonia was found to have a repressive effect. In the case of nifHDK, however, expression was reduced by nitrate in addition to ammonia. The mechanism by which this regulation occurs is under investigation.

1. Wang et al. (1990) Gene 85: 135-144.
2. Better et al. (1985) EMBO J. 4:2419-2424.
3. Ditta et al. (1987) J. Bacteriol. 169:3217-3223.
4. Kim et al. (1986) Gene 50:141-148.
5. Birkenhead et al. (1990) M.P.M.I. In press.
6. Ronson et al. (1987) Nucl. Acids Res. 15:7921-7934.

Gresshoff, Roth, Stacey & Newton (eds.) Nitrogen Fixation: Achievements and Objectives
Chapman and Hall New York • London 1990

Mutagenesis of the gene coding for a ferredoxin-like protein downstream of the *vnfH* gene in *Azotobacter vinelandii*

RAMESH RAINA and H.K. DAS

Genetic Engineering Unit, Jawaharlal Nehru University, New Delhi-110 067, INDIA.

An open reading frame in the same operon, but downstream of vnfH in Azotobacter vinelandii codes for a protein of the following amino acid sequence: MNAIDGYECTVCGDCEPVCPTGSIVFRDDHYAIEADSCNECTDVGETRCLGVCPV DLCIQPLDD. The arrangement of cysteine residues suggests that this could be a ferredoxin-like protein. Does this gene have any role to play in the vnf (vanadium dependent alternative) pathway of nitrogen fixation? More than 1000 Tn5 mutations in the cloned 6 kb fragment (including flanking regions) were checked, but none could be found in the gene for the ferredoxin like protein. Tn5 could be detected in all other regions, except this 0.4 kb region. Several mutations were detected in a fragment smaller than 0.4 kb, so the small size of the fragment was not the limiting factor. Site-directed mutagenesis was next done and one base each in the 20th and 21st amino acid codon were changed to generate a unique BglII site. An interposon (BamHI fragment) coding for B-lactamase was cloned into the BglII site. This clone was introduced into A. vinelandii CA12 (Δ nifHDK) and a marker exchange (result of two point crossover) was attempted. Colony hybridization and subsequent Southern hybridization using the vector as probe, revealed single point crossovers in all of 500 colonies tested. A kanamycin resistant cartridge was next cloned into the vector and marker exchange attempted afresh. Again 1000 colonies screened revealed single point crossovers in all.

Reference: Raina, R., Reddy, M.A., Ghoshal, D. & Das, H.K. (1988) MGG, **214**, 121-127.

Gresshoff, Roth, Stacey & Newton (eds.) Nitrogen Fixation: Achievements and Objectives
Chapman and Hall New York • London 1990

Evidence for the existence of an alternative nitrogenase in *Rhodobacter capsulatus*

Klaus Schneider and Achim Muller

Lehrstuhl für Anorganische Chemie I, Universität Bielefeld, F.R.G.

Several methods to remove traces of molybdenum from the growth medium of Rhodobacter capsulatus were compared. The most effective and reproducible method appeared to be the treatment of nutrient solutions with activated carbon (filtration through a layer of activated, pulverized charcoal). Whereas the untreated growth medium was contaminated with 1.2 ppb Mo, the final Mo-concentration in the activated carbon-treated medium was under the detection limit (< 0.05 ppb) of the inductively coupled plasma mass spectrometry (ICP-MS).

Using this purified growth medium,we were able to demonstrate that Rhodobacter capsulatus contains, in addition to a conventional Mo-containing nitrogenase, a second, Mo-independent nitrogenase. This enzyme was shown to be present in wild-type cells as well as in cells of a nifHDK deletion mutant (provided by Dr. W. Klipp, Lehrstuhl für Genetik, Universität Bielefeld). The alternative nitrogenase of R. capsulatus is strongly repressed by molybdenum. Mo-concentrations >1 ppb in the growth medium, prevented diazotrophic growth of $nifHDK^-$ cells and the expression of nitrogenase activity. The Mo-independent nitrogenase was maximally derepressed in activated carbon-treated media which contained only minimal amounts of Mo, high concentrations of iron (1 mM Fe-ci-trate) and 5 mM serine as N-source. Under N_2-fixing and optimal Mo-deficient growth conditions, the doubling time of $nifHDK^-$ cells was 9 h. The highest activity achieved with whole cells was 1.2 nmol ethylene/ min• mg of protein. Vanadium neither stimulated nor inhibited growth and activity.

The alternative nitrogenase reduced acetylene to both ethylene and ethane. With whole cells ($nifHDK^-$), the proportion of ethane varied from 2 to 5 % depending on the amount of residual traces of Mo in the medium. The addition of Mo to a growing, nitrogenase-active culture resulted in a slow decrease of total activity but also in a simultaneous increase in the ethane production of up to 40 %.

The partially purified dinitrogenase (component 1) had a pH-optimum (acetylene reduction) of 7.8-8.0, a relatively high affinity to acetylene (K_m=0.055 mM) and was analyzed by ICP-MS to contain approximately 20 Fe atoms and negligible amounts of Mo, V, W and Re. The reduced dinitrogenase appeared to be ESR-silent. The results indicate that the alternative nitrogenase of Rhodobacter capsulatus is not a vanadium-enzyme but an Fe-nitrogenase without a heterometal atom.

Gresshoff, Roth, Stacey & Newton (eds.) Nitrogen Fixation: Achievements and Objectives
Chapman and Hall New York • London 1990

Comparison of the properties of the three nitrogenases from *Azotobacter vinelandii*

Kathleen A. Scorsone, Virginia G. Moore and Brian J. Hales

Department of Chemistry, Louisiana State University, Baton Rouge, LA 70803 USA.

In 1986, the first alternative nitrogenase (V-nitrogenase) was isolated and purified from both *Azotobacter chroococcum.* (1) and *Azotobacter vinelandii* (2) and shown to contain vanadium instead of Mo. More recently, the second alternative nitrogenase (Fe-nitrogenase) from *A. vinelandii* was purified (3) and metal analysis suggests that this enzyme only contains Fe. In order to better understand the three nitrogenases, we have undertaken a comparative analysis of them focusing on their purification, chemical and spectroscopic properties and enzymatic behavior.

The expression of each of the three enzymes in *A. vinelandii* appears to be regulated by the metal composition of the growth medium. We have found that growing strain LS15 of *A. vinelandii* (this strain contains a deletion in the structural genes for Mo-N_2ase) on medium containing rhenium also induces the synthesis of Fe-N_2ase and yielded a protein with component 1 specific activity over twice as great as that previously published (3). Purification of the proteins from the three enzymes showed that the salt concentration necessary for the elution of component 2 was invariant while the concentration for elution of component 1 increased in the order Mo-N_2ase < V-N_2ase < Fe-N_2ase. The EPR spectrum of Fe-N_2ase component 1 showed it to be similar to that of V-N_2ase in that it possessed an axial $S = 1/2$ signal with $g_{av} < 2$ but lacked a detectable $S = 3/2$ signal, normally associated with the nitrogenase cofactor. Attempts to crystallize component 1 from the alternative enzymes, using variations on the procedure employed for the conventional enzyme, proved unsuccessful. Complementation studies showed enzymes to be quite selective, with the strongest complementing pair not always being the natural components. For example, component 2 from V-N_2ase complements all of the component 1 proteins better than does component 1 from Mo-N_2ase, even when the component 1 come from Mo-N_2ase. On the other hand, the component proteins from Fe-N_2ase were not observed to be as selective as previously found by others.

1. Robson, R. L., Eady, R. R., Richardson, T. H., Miller, R. W., Hawkins, M. & Postgate, J. R. (1986) Nature (London) *322*, 388-390.
2. Hales, B. J., Case, E. E., Morningstar, J. E., Dzeda, M. F. & Mauterer, L. A. (1986) Biochemistry *25*, 7251-7255.
3. Chisnell, J. R., Premakumar, R. & Bishop, P. E.(1988) J. Bacteriol. *170*, 27-33.

Gresshoff, Roth, Stacey & Newton (eds.) Nitrogen Fixation: Achievements and Objectives
Chapman and Hall New York • London 1990

The gene encoding dinitrogenase reductase-2 is required for expression of the second alternative nitrogenase from *Azotobacter vinelandii*

Elizabeth D. Wolfinger, Rolf D. Joerger and Paul E. Bishop

USDA-ARS and Dept. of Microbiology, N.C. State University, Raleigh, NC 27695-7615 U.S.A.

Azotobacter vinelandii is capable of expressing three genetically distinct nitrogenases(1,2,3). Nitrogenase-1 is encoded by *nifHDK* and is expressed only in the presence of Mo. Nitrogenase-2 is encoded by *vnfH* and *vnfDGK* and is expressed in the presence of V and in the absence of Mo. Nitrogenase-3 is expressed under Mo-deficient conditions and is encoded by *anfHDGK*. We have observed that whenever nitrogenase-3 is expressed, dinitrogenase reductase-2 is also present. Mutant strains CA80 and CA46 that lack dinitrogenase reductase-2 grow slower than the wild-type strain CA in N-free, Mo- and V- deficient media. Mutant strains that are unable to synthesize either dinitrogenase-1 or dinitrogenase-2 also lack subunits for nitrogenase-3 (as seen by 2-dimensional gel electrophoresis) and thus exhibit an Anf^- phenotype. An *anfH-lacZ* fusion strain [constructed with Tn*5*-B21(4)] was used to monitor transcription of the *anfHDGK* operon in mutant backgrounds lacking either or both *nifH* and *vnfH*. Strains with either *nifH* or *vnfH* genes accumulated β-galactosidase activity (2298-7535 Miller units) while strains lacking both *nifH* and *vnfH* accumulated background levels (176-196 Miller units). Since *nifH* is not normally expressed in the absence of Mo this result indicates that the *vnfH* gene product is required for transcription of the *anfHDGK* operon.

REFERENCES

1. Chisnell, J.R., Premakumar, R., and Bishop, P.E. (1988) J. Bacteriol. **170**, 27-33.
2. Joerger, R.D., Jacobson, M.R., Premakumar, R., Wolfinger, E.D., and Bishop, P.E. (1989) J. Bacteriol. **171**, 1075-1086.
3. Pau, R.N., Mitchenall, L.A., and Robson, R.L. (1989) J. Bacteriol. **171**, 124-129.
4. Simon, R., J. Quandt, and W. Klipp (1989) Gene **80**:161-169.

Gresshoff, Roth, Stacey & Newton (eds.) Nitrogen Fixation: Achievements and Objectives
Chapman and Hall New York • London 1990

Serological and *Nod* genic differences of *Bradyrhizobium japonicum* collected from Thailand, Japan and USA

T. Yokoyama, T. Murakami and H. Imai

Tropical Agriculture Research Centre, 1-2 Oowashi, Tsukuba, Ibaragi, 305 Japan.

A total of 252 isolates of Bradyrhizobium japonicum was collected from Japanese and Thai soils (85 from Japan, 167 from Thailand). Genomic DNAs of these isolates and 11 USDA strains were digested with a restriction enzyme BamHI and hybridized to a 32P-labeled 5.4 Kb DNA probe of the common nod genes (D,A,B,C) obtained from Rhizobium trifolii. The isolates were also tested for their serological relatedness using USDA antisera.

Agglutination tests indicated that 61 Japanese isolates were separable into 11 known cross-reactive groups. Out of the total of 167 Thai isolates, only 87 isoaltes reacted with USDA antisera and fell into 7 known cross-reactive groups and several unknown cross-reactive groups. The serological characteristics of Thai isolates were greatly different from the Japanese isolates and USDA strains.

Hybridizing fragments for each of the isolates examined varied from 1 to 4 by Southern hybridization analysis (SHA). The hybridization patterns from the 61 Japanese isolates and the 11 USDA strains were classified into 6 markedly different groups. The hybridizing pattern was the same within the categorized serogroup except for the Japanese serogroup 110, but greatly different between serogroups. Therefore, the genetic diversity of the nodulation genes of Bradyrhizobium corresponded to the Bradyrhizobium serogroups from the temperate region. Out of 167 Thai isolates, only 2 isolates had the same hybridization pattern as USDA 110. The remaining isolates were categorized into 10 hybridization patterns unique to Thai isolates. These results indicate that there is great diversity in nod gene structure of Thai Bradyrhizobium japonicum strains. The Thai isolates are greatly different from those of Japan or the USA. The remaining 82 isolates were categorized into 8 hybridization patterns which were unique to Thai isolates. These results indicate that there is a great difference in the nod gene structure of Bradyrhizobium japonicum strains from temperate and tropical regions.

Gresshoff, Roth, Stacey & Newton (eds.) Nitrogen Fixation: Achievements and Objectives
Chapman and Hall New York • London 1990

The phylogeny of *nifH*: Gene duplication or lateral transfer?

J. P. W. Young.

John Innes Institute, Colney Lane, Norwich NR4 7UH, UK.

Does the phylogeny of *nifH* reflect the ancestry of the bacteria that now carry it, or must we postulate that it has been transferred laterally between widely diverged branches of the evolutionary tree? There has recently been a wealth of new molecular data that bear on questions of this kind. Phylogenetic trees for most groups of bacteria are emerging from studies of 16S ribosomal RNA (rRNA) sequences, and these probably represent the evolutionary relationships of the basic bacterial genomes. They can be compared with trees based on *nifH* sequences: on the whole, the two trees agree remarkably well .

All 28 published dinitrogenase reductase sequences are recognisably related, but they can be divided into four classes on the basis of amino acid sequence. Class A is the only one that is known to occur in both eubacterial and archaeobacterial kingdoms: it is represented by the product of the *Azotobacter anfH* gene and related genes in *Clostridium* and *Methanococcus*. Class B comprises most known eubacterial enzymes, including the *vnfH* gene product associated with the vanadium nitrogenase of *Azotobacter*. *Clostridium* has a number of closely-related *nifH* genes which are very distinctive (class C). Their only known relative is the *nifH* of *Desulfovibrio*, which resembles class C in the first half of the gene and class B in the remainder (an ancient recombinant between duplicate genes?). Archaeobacterial *nifH* genes are diverse but, with the exception of the class A gene mentioned above, they form a branch (class D) that is separate from the eubacterial genes.

Since some bacteria have more than one *nifH* gene, we know that gene duplication has occurred during their evolution. Indeed, it is possible to account for the present distribution of the different classes of *nifH* genes by postulating that they are all descended lineally from a gene in the common ancestor of all bacteria by means of a number of gene duplication events followed by divergence and by loss in some lineages. Of course, it is also possible that lateral gene transfer has played a key role in *nifH* evolution, but at present there is no clear evidence for this.

1. Young, J. P. W. (1990) in *Biological Nitrogen Fixation*, eds. Stacey, G., Burris, R. H. and Evans, H. J. (Chapman and Hall, New York), in press.

Gresshoff, Roth, Stacey & Newton (eds.) Nitrogen Fixation: Achievements and Objectives
Chapman and Hall New York • London 1990

Response of rice to inoculation with diazotrophs

C.S. Zhang, J.P. Li, M. Lin, Y.D. Wang, H.X. Wang and C.B. You

Institute for Application of Atomic Energy, CAAS., P.O. Box 5109, Beijing 100094, Peoples Republic of China.

Five species of diazotrophs, Alcaligenes faecalis (AF)A15, Klebsiella planticola(Kp)DWUL2, K. oxytoca (Ko)NG13, Enterobacter cloacae(Ec)E26, and Azospirillum brasilense(Ab)Sp245 were selected as inoculants for 8 varieties of dryland and wetland rice (subspecies indica and japonica) and hybrid rice. Different responses were obtained with the different inoculants, the rice-Af and rice-Ab associations giving the highest and lowest N_2-fixing activities, respectively. There was significant variation in the N_2-fixation activities among the rice-bacteria associations, suggesting some specificity in these interactions. The root exudate of rice was examined and found to contain 2.25 mg organic acids, 1.27 mg carbohydrates, and 2.61 mg amino acids per gram root fresh weight. From high to low, the organic acids found were citrate, malate, succinate, and lactate. Among the 15 amino acids found, basic amino acids predominated. The carbohydrates consisted of glucose, fructose, and sucrose. Gibberelin-GA_3 was the only phytohormone detected. The root exudate composition varied depending on the rice variety. The culture media of Af contained 9 amino acids, indoleacetic acid (IAA), GA_3 and malic acid, but no carbohydrates. Rice inoculation with Af stimulated root excretion and modified the root exudate, e.g., IAA was found in root exudates only after inoculation. Measurements indicated that uninoculated rice excreted 2.63% of total photosynthate, while inoculated rice excreted 3.52%. Higher N_2-fixation activity correlated with greater excretion of photosynthate. The N_2-fixation rate by diazotrophs was estimated in field experiments using the ^{15}N dilution method. Rice plants were inoculated with Af, Kp, Ko, Ec, Ko/73A, and Ec/73A the latter two strains containing plasmid pCK3 encoding the K. pneumoniae nifA gene. Over a two-year period, the results indicated that inoculation increased plant nitrogen content by 8%, the nitrogen fixed was approximately 20-30 kg/ha, and the grain yield was equivalent to plants grown with a 1/3 to 1/2 reduction in nitrogen fertilizer application.

SECTION I:

SUMMING UP THE CONGRESS

Gresshoff, Roth, Stacey & Newton (eds.) Nitrogen Fixation: Achievements and Objectives
Chapman and Hall New York • London 1990

Congress summary: the molecular perspective

Robert H. Burris

Department of Biochemistry, University of Wisconsin-Madison, Madison, WI 53706, USA.

It is obvious that W. D. Bauer and I cannot summarize a week of interesting talks and an impressive array of posters in 20 minutes each. It also is doubtful that what we picture as major advances are the same as the ones that struck you.

New Nitrogenases. Bishop et al. reported in 1980 that they had evidence for an alternative nitrogen fixation system in *Azotobacter vinelandii*, and that it was expressed in the absence of Mo. The concept that a nitrogenase could function without Mo was met with pronounced skepticism. Now others have verified that Bishop and colleagues were correct and have joined in the study of these newly described nitrogenases. Not only is there a V nitrogenase but also one lacking both Mo and V; it apparently is dependent on Fe at its active center. Each of these enzymes has 2 components, a dinitrogenase and a dinitrogenase reductase. The V and Fe dinitrogenases contain an additional subunit designated δ. The nucleotide sequence now is known for the structural genes of the *A. vinelandii* modified nitrogenase components. The Sussex group has mounted a major program, and Smith and Eady each reported on purification of the Mo-independent nitrogenases, their homology, their spectra and their kinetics. The purified enzymes also have been examined by EPR, ENDOR and for the role of specific amino acids they carry.

X-Ray Crystallography. We have heard for years about crystallization of proteins from nitrogenases and investigations of their tertiary structures by X-ray crystallography. At this Congress it was reassuring finally to see 3 dimensional images of the proteins that catalyze the nitrogenase reactions. Millie Georgiadis projected pictures of the Fe protein dimer at 3 Å resolution. They revealed the expected 2-fold symmetry with a cleft between the 2 subunits. It was suggested that the ATP probably was bound in this cleft, and that residues 98-115 at the "top" of the molecule might bind the MoFe protein. Bolin and colleagues from Purdue and the University of Georgia found that their X-ray studies at 5 Å resolution on the *Clostridium pasteurianum* MoFe protein required a modification of the usual model for P clusters. They suggested that the 2 FeMoco units are separated by about 70 Å.

FeMoco. There has been much progress on FeMoco since the last Congress, and Shah described the observations he and his associates have made. FeMoco has been biosynthesized and then incorporated *in vitro* into apodinitrogenase to create active holodinitrogenase. It was surprising to find that homocitric acid was incorporated in the *in vitro* synthesis of FeMoco. When citrate, homoisocitrate, isocitrate and other homologs are substituted for homocitrate, the properties of nitrogenase are altered so that its inhibition pattern is changed, and it may be relatively ineffective in reducing N_2 but very effective in reducing alternative substrates. In an evening session, Madden described his synthesis and testing of fluorine-containing analogs of homocitrate as substitutes for homocitrate. Again the responses were altered relative to homocitrate FeMoco, and the fluoro compounds were excellent for NMR investigations. Dean described collaborative research on the genes required for FeMoco biosynthesis and their use in formulating a model for the metalloclusters in *A. vinelandii* and *C. pasteurianum*. They have employed site-directed mutagenesis and biochemical reconstitution in testing the validity of their model.

ATP Hydrolysis and Electron Transfer. Students of muscle physiology have faced a problem similar to that we have faced in establishing whether ATP hydrolysis follows or precedes electron transfer from dinitrogenase reductase to dinitrogenase. They solved their problem calorimetrically. In collaboration with the group, Thorneley made measurements of ATP hydrolysis in imidazole and cacodylate buffers at 6°C. The amount of heat generated in these 2 buffers is quite different. Application of the calorimetric technique indicated that cleavage of MgATP precedes electron transfer from *Klebsiella pneumoniae* dinitrogenase reductase to dinitrogenase. Thus a refractory problem was solved.

Regulation of the Dissolved Oxygen in Nodules. Some years ago, Tjepkema measured the concentration of dissolved oxygen in leguminous root nodules.

Recently there has been a burst of activity in this field; measurements both of O_2 and H_2 have been made with microelectrodes. Layzell, and later Witty and Minchin, reported their measurements and emphasized that legume nodules always are O_2 limited. This imposes a limitation on their ability to generate ATP. There is a substantial barrier to uptake of O_2 by the nodule, and this barrier likewise blocks the exit of H_2 from the nodule. As O_2 is utilized rapidly, the H_2 concentration measured was 1.3-3 times that of O_2 in soybean nodules. There is a precipitous drop in O_2 concentration at the barrier to its movement into the nodule, but once inside the barrier its diffusion is facilitated by leghemoglobin. The nodule adapts quite rapidly when the external pO_2 is increased from 0.2 to 0.4 atm., an adjustment that is accomplished by increasing the resistance of the diffusion barrier. Apparently an increase in cell turgidity can block pores in the nodule and decrease gas diffusion; i.e., osmoregulation is involved. Witty and Minchin suggested that glycoprotein occlusions in the intercellular spaces could form a barrier to gas diffusion.

In his lecture, deBruijn pointed out that the stem nodulating legumes and *Azorhizobium caulinodans* hold a special interest because: a) the association in stem nodules is lodged in a tissue capable of generating O_2 photosynthetically, and b) certain of the symbiotic bacteria also can fix N_2 as free-living organisms. In addition, mutants lacking a second glutamine synthetase or a GOGAT system are available.

Hennecke and colleagues have applied a genetic approach in their investigation of the role of O_2 in the metabolism of soybean bacteroids, and they reported evidence that certain of the genes were expressed only when the dissolved O_2 was low. Their study of the respiration of *Bradyrhizobium japonicum* indicated that the organism had 3 respiratory chains, one of which resembled the mitochondrial pathway. A mutant devoid of the cytochrome bc_1 complex was Fix-, whereas cytochrome c and cytochrome aa_3 mutants were fix+; a bacteroid-specific bc_1 electron transport pathway was suggested.

Control of Nitrogenase. G. P. Roberts reported on the genetics and biochemistry of control of nitrogenases in Rhodospirillum rubrum and Azospirillum lipoferum. In each organism, nitrogenase can be turned off in the presence of ammonium (or darkness for R. rubrum) by covalent modification (ADP ribosylation) of one of the 2 subunits of dinitrogenase reductase. The process is catalyzed by ADP-ribosyl transferase (DRAT) and the system is reactivated by removal of the inactivating group by dinitrogenase reductase activating glycohydrolase (DRAG). Not all N_2-fixers have this control system, e.g., K. pneumoniae lacks it. However, Roberts reported that when draG and draT are transferred to K. pneumoniae they are functional, and one then can tap the fund of genetic information on Kp in studies of DRAT and DRAG. A. lipoferum and Azospirillum brasilense carry functional DRAT and DRAG, but within the same genus, A. amazonense lacks them. However, draG and draT can be transferred from A. lipoferum to A. amazonense and DRAG and DRAT are expressed.

Nitrification and Dinitrification. These topics characteristically draw limited attention among researchers on biological nitrogen fixation, despite their importance in the nitrogen cycle. Hooper reported his studies on Nitrosomonas spp. which are aerobic and obligately autotrophic. At low pO_2 they produce considerable amounts of NO and N_2O in addition to HNO_2. HNO_2 proved to be an obligatory intermediate in the production of N_2O from NH_3. All of the N_2O was produced by a reductive process, and none of it came from the oxidation of N_2; this was established with ^{18}O as a tracer. Nitrosomonas spp. proved to be surprisingly rich in cytochromes.

Zumft has utilized Pseudomonas stutzeri in his investigations of denitrification. Four enzymes are involved in the reduction of nitrate to N_2, namely nitrate reductase, nitrite reductase, nitric oxide reductase and nitrous oxide reductase. Zumft described the properties of these enzymes. Much of the information was derived from the enzymes of his genetically altered P. stutzeri cultures.

Nitrogen Cycling. Are we fixing too much nitrogen? This is a question seldom raised in this group, but it was posed because of its ecological implications. The Haber process consumes large quantities of natural gas, clearly a limited resource. Natural gas serves both as a source of heat and hydrogen in the Haber process. We all would prefer when possible to harness solar energy to drive biological N_2 fixation and thus to conserve our fossil fuels. As Tredje pointed out, the use of excess nitrogenous fertilizers may be followed by its conversion to nitrite and nitrate; when these leach into our water supplies they pose a hazard to infants. More extensive use of biological N_2 fixation and judicious choice of fertilizer levels in a sustainable agriculture can aid in solving problems involving nitrite toxicity.

Related to this theme was the fear that has been voiced that the introduction of plants and microorganisms genetically modified by recombinant DNA methods may pose hazards. Careful examination of many case histories does not substantiate such fears. Field testing of recombinant plants and microorganisms appears to adequately and effectively controlled in the USA by the NIH, USDA, EPA and FDA. These agencies have approached control problems rationally, scientists have acted responsibly, and to date no hazards can be documented.

Ronson of BioTechnica International described their efforts to improve the competitiveness and effectiveness of _Rhizobium meliloti_ and _Bradyrhizobium japonicum_ by genetic alteration of the organisms. They have run field tests in Louisiana and at 2 sites in Wisconsin with permission from EPA and USDA APHIS on the national scene and the Wisconsin Department of Agriculture and Consumer Protection and the comparable agency in Louisiana. The field tests were monitored carefully, and added rhizobia spread less than a foot in the soils. The results to date are promising, and there is no hint that the field trials present a hazard.

Methods. Methods received limited coverage in the Conference, although we all are directly dependent on them for our research. Bergersen reported his applications of the ^{15}N normal abundance technique. In carefully performed field tests with thousands of samples, he has demonstrated that the method can be made to work. But it isn't easy! The technique is demanding, requires extensive replication, necessitates proper use of control plants, and demands skill in mass spectrometry. To date there is no clear evidence that the Dumas method for preparation of N_2 from samples meets the accuracy required by the normal abundance method. D. Kohl, who pioneered the method, was pleased to see that Bergersen's conscientious trials had converted him from a skeptic to a believer.

Harold Evans. The many and diverse contributions of Harold Evans to biological nitrogen fixation were imaginatively portrayed by Dan Arp. Arp helped us visualize Evans' career by walking us through the halls and through the laboratory at Oregon State, following a trail of Evans' papers. It was impressive. There was the early work on nitrate reductase, work on the functional role of a variety of micronutrient elements, and then the plunge into many aspects of N_2 fixation. Arp emphasized the recent work on the role of H_2 in N_2 fixation. It was a great surprise to find that certain strains of

Bradyrhizobium could grow lithoautotrophically with H_2 as the sole source of energy. As H_2 released to the atmosphere represents energy lost from the N_2 fixing system, Evans explored the possibility that legumes might recapture some of this energy through the action of hydrogenase. Careful experiments showed that in fact this could occur; soybeans with *hup*$^+$ rhizobia accumulated demonstrably greater dry weight and total nitrogen than those infected by rhizobia that were identical except for their lack of hydrogenase. Turning back to his interest in micronutrient elements, Evans studied the function of Ni in hydrogenase. The delegates all participated in a rousing tribute to Harold Evans for his many contributions to their favorite field of research.

Gresshoff, Roth, Stacey & Newton (eds.) Nitrogen Fixation: Achievements and Objectives
Chapman and Hall New York • London 1990

Congress summary: the organismal perspective

Wolfgang D. Bauer

The Ohio State University, Columbus, OH 43210, USA

Introduction. This report is intended to summarize several of the major advances presented at the Knoxville meeting and to provide context and implications for readers out of their specialty. The primary focus here will be findings related to establishment of the *Rhizobium* symbiosis. A range of special advances in this area justify such a focus. Readers should consult the accompanying article by Professor Burris for a summary of advances in areas where I have no useful expertise.

Identification of *Rhizobium*-legume signal substances. During the past twenty years, there have been many studies describing the sequence of events which culminate in the formation of functional legume root nodules. In approximate order of their occurrence, these events include chemoattraction of the bacterial symbiont to the host root, host-stimulated expression of *nod*ulation genes in the bacteria, adhesion of bacteria to the root surface, bacterial induction of both root hair curling and localized cell division in the host root cortex, consequent entrapment of bacterial cells between apposed host cell wall surfaces, penetration through localized disruption of the root hair cell wall combined with bacterial multiplication and formation of a tubular ingrowth of the hair cell wall called the infection thread, development of a new meristem, host cell differentiation to form the outer and inner cortex, vascular elements and infected and uninfected cell types of the nodule, release of rhizobia from branch tips of the infection thread into host cytoplasm but separated from it by envelopes of host cell plasmamembrane, differentiation of the rhizobia into bacteroids capable of reducing N_2 to NH_3, and induction of the later nodulin genes in the host required for oxygen protection, carbon and energy supply for bacteroids and NH_3 assimilation/transport.

It is evident that this sequence of events requires that each partner in the symbiosis trigger major changes in the behavior and development of the other partner. But what are the triggering substances? When are they required? What set of responses do they elicit? The Knoxville meeting has made long awaited contributions to these questions by clearly establishing the identity of several major signal compounds and the responses they elicit. Reports at this meeting dealt with five classes of signal substance. Each merits comment.

1. *Rhizobium* inducers of root hair deformation and cortical cell division. First and foremost among the achievements reported at Knoxville was the isolation and identification by Lerouge et al. (1) of NodRm1, a tetrasaccharide of N-acetylglucosamine which is 6-O-sulphated at the reducing sugar and 2-O-acylated with an unsaturated fatty acid at the non-reducing terminal sugar. As described by Jean Dénarié (L15), this compound actively induces both root hair branching and cortical cell division in alfalfa at concentrations as low as 10^{-8} to 10^{-11} M. Synthesis of NodRm1 requires active expression of the *nodABC* genes, although their specific role in this synthesis is unknown. Transfer of the sulfate group to the tetrasaccharide requires *nodH*. Without the sulfate group, the compound (NodRm2) is active on vetch plants, but not alfalfa. With the sulfate group, the compound loses activity on vetch, but gains activity on alfalfa. These observations define at least one important determinant of host range at the biochemical level.

With their most elegant analysis, the French scientists have accomplished what others have attempted for two decades. With the purified compound, they have given us a powerful tool for investigating the molecular mechanisms controlling host plant cell division and cell wall biosynthesis. As described by Alan Downie , the structure of NodRm1 provides us with a molecular framework for understanding the actual function of some of the *nod* genes we have known only through ambiguous nodulation phenotypes and sequence homologies. The French workers have also provided us with direct evidence that a single *Rhizobium* signal substance elicits two very different symbiotic responses in the host: the initiation of a new branch in a root hair cell and the initiation of cell divisions at localized centers within the root cortex. This suggests that secondary messengers may be required to link the receptor for NodRm1 signals to diverse host responses.

The results described by Hollingsworth et al. add an important dimension to our early understanding of the signal compounds which induce root hair branching and cortical cell division. They show that *Rhizobium* culture filtrates can contain several active substances, some which have structures (e.g. N-acetyl glutamate) very different from that of NodRm1, and some which act synergistically with each other.

Such multiplicity of functionally related signals exemplifies one of the central take home lessons of the Knoxville Congress: symbiotic organisms have diverse, complex and often redundant molecular mechanisms for dealing with a given problem. We may not yet appreciate the logic behind such multiplicity, but we can certainly help ourselves by thinking about it when designing and interpreting experiments.

2. Lectins. Lectins have been considered symbiotically relevant molecules since 1974 when Bohlool and Schmidt demonstrated a strong correlation between the ability of various rhizobia to bind soybean lectin and ability to nodulate soybean. However, clear proof of a symbiotic role came only with the very recent demonstration by Diaz et al. (2) that white clover roots acquire the ability to be nodulated by *R. leguminosarum* bv. *viciae* when transformed with the pea lectin gene. Strains of bv. *viciae* normally nodulate pea, but not clover. Thus, transformation of clover with the lectin gene extended the range of *Rhizobium* symbionts which it could recognize and respond to. As described by Jan Kijne , pea lectin is associated with the plasmamembrane of young root hair tips and is developmentally regulated in root epidermal cells, appearing in small patches destined to be the tips of emergent root hairs. Since spot inoculation studies showed that infection initiation requires rhizobia to be in contact with these same pre-emergent root hairs, it appears that pea lectin is present at the right time and place for symbiotic signal transduction. One intriguing possibility is that this lectin serves as the host receptor for the NodRm2 tetrasaccharide signal from *R. leguminosarum* bv. *viciae*.

Complexity and redundancy surface again with regards to lectins. There are two additional twists that deserve mention here. First, earlier studies by Halverson and Stacey showed that soybean roots secrete a lectin which corrects the symbiotic defect of a slow-to-nodulate *Bradyrhizobium japonicum* mutant and significantly enhances the nodulation efficiency of the wild-type parent. Thus, while host lectins can serve as a membrane-associate receptor for rhizobial signals, as indicated for the pea symbiosis, Halverson and Stacey's results indicate that a host lectin can (concurrently?) act as a transmitted signal to elicit symbiotic responses in the bacteria. Another facet of the lectin story comes from the report by Ho et al. , which shows that *B. japonicum* cells also have a lectin. This bacterial lectin has a sugar binding specificity very similar to, yet distinct from that of the host soybean lectin. The *B. japonicum* lectin was shown to play an important role in the attachment of the bacteria to each other and to the host, as bacterial lectins do in E. coli and other species. Thus, it appears that lectins can act as intercellular glues as well as signals and receptors!

3. *Rhizobium* Polysaccharides. The first evidence that *Rhizobium* polysaccharides might be symbiotically relevant signal substances came from work 30 years ago by Ljunggren and Fahraeus. Yet it wasn't until 1987 that Djordjevic et al. (3) offered convincing proof of this possibility by showing that purified *Rhizobium* exopolysaccharide (EPS) and the basic oligosaccharide (OS) repeat unit of the EPS could correct the symbiotic defects of *exo* mutants, enabling them to initiate and sustain infection thread development in a host specific manner.

The report by Battisti et al. provides proof that the EPS from *R. meliloti* also functions as a symbiotic signal. This was a crucial advance because various labs had previously attempted to complement the symbiotic defects of *R. meliloti exo* mutants with added EPS, but had no success, causing concern about the reliability and generality of the Djordjevic results. Battisti et al. clarified the apparent conflict by showing that the active substance is not the intact EPS or the basic repeat unit OS, but rather the tri- and tetra-repeat unit oligomers. The activity of such oligomers raises an earlier report by Fjellheim and Solheim indicating that root exudates from host legumes contained polysaccharide-degrading enzymes that were about 10-fold more active in cleaving the EPS of compatible rhizobia than exudates from non-hosts. Battisti et al. also confirmed Glazebrook and Walker's demonstration that *R. meliloti* has a second, cryptic EPS of very different structure but with comparable symbiotic activity, another example of signal redundancy.

The report by Bhagwat and Keister offers further insight into the signal role of EPS. These authors isolated Tn*5* mutants of *B. japonicum* which were deficient in EPS/CPS synthesis. One of these mutants was found to be capable of normal nodule initiation on soybean, at least at high inoculum dosages, but was about 50-fold less competitive, cell for cell, than its parent in direct competition. Remarkably, the addition of about 1 μg/L of crude CPS from the wild type restored the competitiveness of the *exo* mutant. This strongly suggests a signal role for the EPS/CPS of *B. japonicum,* but perhaps as a determinant of nodulation efficiency rather than as a signal required for nodule formation.

4. Host root flavonoids. In 1986, Peters et al. first identified luteolin as a flavonoid from alfalfa which specifically stimulates *nodD*-mediated expression of *nodABC* genes in *R. meliloti,* analogous to the stimulation of *vir* genes in *A. tumefaciens* by host phenolic compounds. While an extensive literature has developed regarding *nod* gene activation by flavonoid signals from host legumes, this Congress should be remembered as giving us three new views of flavonoid involvement.

First, Ton Van Brussel and colleagues have turned the whole *nod* gene inducer story back on itself. They have shown that *R. leguminosarum* bv. *viciae* makes a biovar-specific extracellular signal compound which induces the synthesis of 8 new *nod-*

inducing flavonoids in its host. Thus flavonoid synthesis in host plants is not simply constitutive, as previously thought. Synthesis of the extracellular factor requires a full range of functional *nod* genes in the bacterium, similar to those required for synthesis of NodRm2. Perhaps the factor is NodRm2! In any case, *Rhizobium*-stimulated synthesis of new flavonoids has two important biological effects: it creates a reciprocal amplification of symbiotic signal strength and also adds a new layer of host specificity.

In addition to providing us with a systematic analysis of *nod*-inducing phenolic compounds in alfalfa , Don Phillips and colleagues have made the important observation that several of the seed flavonoids are able to significantly stimulate the cell division rate of rhizobia. This stimulation occurred at micromolar concentrations of the flavonoids, did not require functional *nodD1/D2* genes, did not involve detectable metabolism of the flavonoids themselves, and was observed at relatively low as well as normal levels of carbon compounds in the media. It is not yet clear whether this effect of certain flavonoids on cell division rate in rhizobia is some direct effect on bacterial physiology and metabolism or some indirect signal effect that enhances cell division independently of nutrient supply.

Dietrich Werner's lab made the observation that the soybean symbiont, *B. japonicum,* is quite sensitive to the pterocarpan phytoalexin of soybean, glyceollin, an antibiotic compound produced in response to pathogen attack or stress. Curiously, *B. japonicum* was found to be fully protected from lethal dosages of glyceollin by prior growth in media containing low concentrations of two of the *nod*-inducing isoflavonoids secreted by soybean roots, daidzein and genistein. The mechanism of this protective activity is unknown, but does not appear to involve bacterial metabolism of the isoflavonoids. The biological effect of protection against the toxic effects of glyceollin might be to enhance the ability of *B. japonicum* to colonize stressed or pathogen-infected roots. Such protection might also help to explain the apparent immunity of rhizobia to host defense responses.

5. *Rhizobium* proteins. One of the most unexpected and exciting discoveries reported at the Eighth Congress was that rhizobia produce proteins that may act directly as trans acting factors to control gene expression in the host plant. Welters et al. found that *Azorhizobium caulinodans,* which forms stem nodules on *Sesbania*, secretes an 11 kd protein which specifically binds to a GC-rich sequence present in the 5' upstream region of the leghemoglobin gene and corresponding regions of other nodulin genes. The isolation and analysis of mutants defective in the synthesis of this protein should clearly establish whether the protein is required for nodulin gene expression and should help determine the mechanisms which regulate its synthesis and export.

Improvement and extension of the *Rhizobium* symbiosis. In addition to these advances at the basic level, several reports describe findings of potentially great practical importance. Three different labs reported that rhizobia capable of nodulating *Sesbania* or the non-legume *Parasponia* were also capable of forming nodule-like structures on rice and rape. Electron micrographs showed some intracellular invasion by bacteria apparently enclosed by membrane envelopes, an important feature of functional nodules in legumes. There was no convincing evidence yet that the rice or rape nodules were capable of efficient N fixation, but this is probably far too much to expect of two organisms just getting to know each other intimately. The discovery that rhizobia can initiate nodulation on both monocots and non-legume dicots qualifies as a genuine breakthrough and is certain to provide a focus of intense research and development during the next few years.

Truchet's finding that some legumes can form spontaneous nodules in the absence of microbes is also relevant to the question of extending the host range of rhizobia to non-leguminous crops. Spontaneous nodule formation implies that the process of nodule development is a pre-programed cascade, requiring just one initiating signal from rhizobia rather than a separate signal for each developmental change in the host. The lovely ultrastructural analysis of spontaneous nodule development by Joshi et al. indicates that essentially all aspects of nodule architecture are dictated by the host without further input from the bacteria. Perhaps it will be considerably easier to transfer the basic gene package for nodule development to non-legumes than we have recently supposed. Interestingly, spontaneous nodules are able to induce the autoregulation of nodulation response and are themselves suppressed in the presence of high nitrate (Caetano-Anollés et al.).

Successful genetic engineering to enhance the efficiency of nitrogen fixation in *R. meliloti* through manipulation of *nifA* gene expression was also reported. Preliminary greenhouse tests gave significant yield increases of alfalfa dry matter after inoculation of plants with the engineered strains compared to wild-type controls, and field trials are underway. This work provides valuable evidence that information about symbiotic functioning obtained through basic research can be rapidly and powerfully exploited for commercial and social benefit.

The work of Beattie and Handelsman may prove equally important in this regard. They showed that it is possible to identify genes for nodulation competitiveness and then transfer them from one *Rhizobium* strain to another, significantly enhancing the competitiveness of the less competitive strain under both laboratory and field conditions. Genetic engineering to enhance nodulation competitiveness in symbiotically superior strains of rhizobia was also recently reported by Eric Triplett (4). Enhanced competitiveness was achieved in this case by the stable interstrain transfer of genes required for production of, and resistance to, trifolitoxin, an antibiotic peptide highly specific for *Rhizobium*. Based on these reports, the prospect of *Rhizobium* inoculants which combine a few cassettes for competitiveness and a few other cassettes for enhanced fixation efficiency does not seem so far from our grasp.

One other discovery related to the establishment of symbiotic associations deserves special mention. Guillemin et al. found that almost half of the pea mutants which fail to form nodules when inoculated with rhizobia also fail to form symbiotic associations with endomycorrhizal fungi. Infection by the fungi was blocked at an early stage after appresorium formation. As described by LaRue, the defects responsible for the development of *Rhizobium* infections in these host mutants remain a mystery. Nonetheless, the identification of host mutants defective in mycorrhizal infection promises to provide us with our first real handle on the most important symbiosis of all.

References

1. Lerouge, P., Roche, P., Faucher, C., Mailet, F., Truchet, G., Promé, J.C. & Dénarié, J. (1990) Nature **314**, 781-784
2. Diaz, C.L., Melchers, L.S., Hooykaas, P.J.J., Lugtenberg, B.J.J. & Kijne, J.W. (1989) Nature **338**, 579-581
3. Djordjevic, S.P., Chen, H., Batley, M., Redmond, J.W. & Rolfe, B.G. (1987) J. Bacteriol. **169**, 53-60
4. Triplett, E.W. (1990) Appl. Environm. Microbiol. **56**, 98-103

Gresshoff, Roth, Stacey & Newton (eds.) Nitrogen Fixation: Achievements and Objectives
Chapman and Hall New York • London 1990

Computer networking and biological nitrogen fixation

Robert Harper, Eng-leong Foo, Thomas Collet.

University of Helsinki, Department of Microbiology, SF-00710 Helsinki, Finland.
UNEP/UNESCO/ICRO Microbiological Resources Center, Department of Bacteriology, Karolinska Institute, S-10401 Stockholm, Sweden.
Department of Chemistry, Massachusetts Institute of Technology, Cambridge, MA 02139, USA.

Introduction: Computers are being used to share information over computer networks. Following the initiative of the MIRCEN in Stockholm a working group has been set up to test the utility of exchanging messages between different laboratories around the world.

Membership Requirements: If you have an account at the computer center of your local University, and can send and receive E-mail on computer networks, then you can participate on BNFNET. To date most of the discussions on BNFNET have been about how to make this service known to a wider user group. The BNFNET (Working Group) drew up a five point plan to promote the awareness of BNFNET at Knoxville. 1) Poster, 2) Flyer, 3) Questionnaire, 4) Workshop, 5) Demo.

The Future. When the "critical mass" of the people involved in BNFNET reaches 40-50 members then the operation will be moved to LISTSERV@FINHUTC, and a new list BNFNET-L@FINHUTC will be established. This will bring with it a number of advantages. Namely, self subscription to the list from your normal E-mail address, monthly notebooks which will keep a record of all the messages that have been entered into BNFNET-L. These notebooks can be queried "interactively" over the network to recover old messages from the database using the LDBASE programme available from LISTSERV. If you have any questions about BNFNET then you can contact the coordinators at the following E-Mail addresses:

Robert Harper : HARPER@FINFUN.BITNET
Eng-leong Foo : eng-leong_Foo_@KOM.komunity.se
Thomas Collet : COLLET@MITWCCF.BITNET
Padma Somasegaran : niftal@uhccux.uhcc.Hawaii.Edu.

Gresshoff, Roth, Stacey & Newton (eds.) Nitrogen Fixation: Achievements and Objectives
Chapman and Hall New York • London 1990

Author Index